# STRUCTURE AND DEVELOPMENT OF THE GREENLAND–SCOTLAND RIDGE

## New Methods and Concepts

## NATO CONFERENCE SERIES

I Ecology
II Systems Science
III Human Factors
IV Marine Sciences
V Air–Sea Interactions
VI Materials Science

## IV MARINE SCIENCES

*Volume 1* Marine Natural Products Chemistry
edited by D. J. Faulkner and W. H. Fenical

*Volume 2* Marine Organisms: Genetics, Ecology, and Evolution
edited by Bruno Battaglia and John A. Beardmore

*Volume 3* Spatial Pattern in Plankton Communities
edited by John H. Steele

*Volume 4* Fjord Oceanography
edited by Howard J. Freeland, David M. Farmer, and Colin D. Levings

*Volume 5* Bottom-interacting Ocean Acoustics
edited by William A. Kuperman and Finn B. Jensen

*Volume 6* Marine Slides and Other Mass Movements
edited by Svend Saxov and J. K. Nieuwenhuis

*Volume 7* The Role of Solar Ultraviolet Radiation in Marine Ecosystems
edited by John Calkins

*Volume 8* Structure and Development of the Greenland–Scotland Ridge
edited by Martin H. P. Bott, Svend Saxov, Manik Talwani, and Jörn Thiede

# STRUCTURE AND DEVELOPMENT OF THE GREENLAND–SCOTLAND RIDGE

## New Methods and Concepts


Edited by

**Martin H. P. Bott**
University of Durham
Durham, United Kingdom

**Svend Saxov**
Aarhus University
Aarhus, Denmark

**Manik Talwani**
Lamont-Doherty Geological Observatory of
Columbia University
Palisades, New York

and

**Jörn Thiede**
University of Oslo
Oslo, Norway



Published in cooperation with NATO Scientific Affairs Division

PLENUM PRESS · NEW YORK AND LONDON

Library of Congress Cataloging in Publication Data

NATO Advanced Research Institute on Structure and Development of the Greenland–Scotland Ridge: New Methods and Concepts (1981: Padua University Student Center)
Structure and development of the Greenland–Scotland Ridge.

(NATO conference series. IV, Marine sciences; v. 8)
"Proceedings of a NATO Advanced Research Institute on Structure and Development of the Greenland–Scotland Ridge: New Methods and Concepts, held May 11–15, 1981, at Padua University Student Center, in Bressanone, Italy"—P.
Includes bibliographical references and index.
1. Geology—Greenland–Iceland Rise—Congresses. 2. Geology—Faeroe–Iceland Ridge—Congresses. I. Bott, Martin Harold Phillips, 1926– . II. North Atlantic Treaty Organization. Division of Scientific Affairs. III. Title. IV. Series.
QE350.22.G73N37 1981 551.46′08′09324 82-9021
ISBN 0-306-41019-2 AACR2

Proceedings of a NATO Advanced Research Institute on Structure and Development of the Greenland–Scotland Ridge: New Methods and Concepts, held May 11–15, 1981, at Padua University Student Center, in Bressanone, Italy


A Division of Plenum Publishing Corporation
233 Spring Street, New York, N.Y. 10013



Printed in the United States of America

PREFACE

A group of geoscientists from a number of NATO countries met under NATO sponsorship in Copenhagen on February 27 and 28, 1978, and formulated a proposal entitled "EVOLUTION OF THE GREENLAND-ICELAND-FAEROE-SCOTLAND RIDGE, A KEY AREA IN MARINE GEOSCIENCE". This part of the North Atlantic Ocean is of particular interest because of its anomalously shallow bathymetry which has profoundly influenced many aspects of the evolution of the North Atlantic. The proposed investigations therefore aim to study the deep crustal structure including relationship of continental and oceanic crust, history of subsidence of the ridge including its past role as a land bridge, age of the oceanic basement along it and its history of formation, and the influence of the ridge on Tertiary and Quaternary depositional palaeoenvironments. In furtherance of this proposal, it is intended to carry out a series of seismic and drilling operations on the Ridge during the coming years. These major marine investigations will be mainly funded from national sources.

An important preliminary stage to the project is the collection and synthesis of available data. NATO has already approved a small budget for this purpose which has enabled a geoscientist to work partly at the Department of Geological Sciences of Durham University, UK, and partly at Lamont-Doherty Geological Observatory of Columbia University, Palisades, USA, for about six months to compile the data. The most important map showing magnetic anomalies and lineations in the area, is included in a pocket at the back of this volume.

Another important step was the proposal to hold a scientific meeting where research personnel who have worked or intend to work on the problems of the Ridge can meet to synthesize present information and highlight the problems. Points included in the programme for this proposed meeting were:

Review of present geophysical knowledge
Review of present state of techniques of investigation
Review of present geological knowledge
Discussion and planning session

# PREFACE

In the fall of 1979 the NATO Science Committee agreed to fund a NATO Advanced Research Institute which then took place at Padua University Student Centre in Bressanone, Italy, during May 11-15, 1981. Besides a series of lectures there were several poster sessions and much time (especially in the evenings) was spent on discussing future work. The closing chapter in this book contains a summary of the conclusions.

The meeting in Bressanone provided a valuable opportunity for presentation of state of the art and for exchanges of information and ideas both from within and between the various disciplines represented. A fruitful discussion on new ideas and questions followed the papers and the poster sessions.

We are particularly grateful to the NATO Science Committee for their generous support of the Research Institute and for the publication of this book.

Martin H.P. Bott
Svend Saxov
Manik Talwani
Jörn Thiede

CONTENTS

## CHAPTER 1: REVIEW OF GEOPHYSICAL KNOWLEDGE

SUPPLEMENT:

# CHAPTER 1

# REVIEW OF GEOPHYSICAL KNOWLEDGE

# DEEP STRUCTURE AND GEODYNAMICS OF THE GREENLAND-SCOTLAND RIDGE: AN INTRODUCTORY REVIEW

Martin H.P. Bott

Department of Geological Sciences
University of Durham
South Road, Durham DH1 3LE, U.K.

The Greenland-Iceland-Faeroe-Scotland ridge is an anomalous shallow bathymetric feature which crosses the North Atlantic from Denmark Strait to the north Scottish shelf (Fig. 1). It forms an aseismic transverse ridge across the ocean, except where it is crossed by the active spreading centre in Iceland. The whole ridge is underlain by unusually thick crust for an oceanic region and this is the main reason for its high elevation relative to the regions to the north and south. Iceland, however, is also underlain by a highly anomalous upper mantle even in comparison with normal ocean ridges. The Greenland-Scotland ridge is a composite feature, but the majority of it has formed at the spreading axis as Greenland and North Europe have separated. The main geophysical problems are to determine the crustal and upper mantle structure beneath the ridge, and to infer the geodynamic processes which led to its origin and development. These are relevant to the broader problems of mantle hot spot activity and the mechanism of continental splitting. The topographical evolution of the ridge, as the underlying oceanic lithosphere has cooled, is also relevant to the problems of early Tertiary land bridges, and the development of the oceanic circulation and sedimentation of the North Atlantic, as discussed in the next section of the book. This introductory review is mainly based on the succeeding papers which describe the present state of geophysical knowledge.

The Greenland-Scotland ridge is divisible into three main sections of contrasting origin as follows: (1) the Icelandic transverse ridge of Tertiary oceanic origin which extends from Greenland to the edge of the Faeroe Block, (2) the Faeroe Block and associated shallow banks to the southwest, probably underlain at depth by continental crust, and (3) the Wyville-Thomson Ridge which seems to be a Palaeocene volcanic feature built up above older oceanic crust. The shallow Davis Strait region, separating the Labrador Sea from Baffin Bay, may be a feature of related origin lying west of Greenland.

The Icelandic transverse ridge is the most prominent and best understood part of the whole ridge system. It comprises Iceland itself, the Iceland-Greenland ridge to the west and the Iceland-Faeroe ridge to the southeast (Fig. 1). It is underlain by anomalously thick oceanic crust. This youngest part of the Greenland-Scotland ridge formed at the spreading axis during the last 55 My as Greenland separated from North Europe but because of the lack of clearly dateable magnetic lineations, its evolutionary history has until recently been obscure in detail. The oceanic region to the south has evolved by symmetrical spreading about the Reykjanes ridge, but the evolution of the region to the north is complicated, with the splitting off of the Jan Mayen ridge micro-continent from Greenland involving westward migration of the spreading axis and contemporaneous spreading occurring on both sides of the microcontinent for a period of time. The evolutionary history of the Icelandic transverse ridge has recently been clarified by recognition of dateable magnetic anomalies from anomaly 22 (or possibly from anomaly 24) onwards on the East Greenland shelf (Larsen, evidence presented at this N.A.R.I.). This suggests that the transverse ridge formed by approximately symmetrical spreading in harmony with the region to the south, rather than by the more complicated type of evolution occurring to the north.

Iceland, crossed by the mid-Atlantic ridge, forms the presently active section of the Icelandic transverse ridge. The study of the deep structure of Iceland and of the processes now occurring at the spreading axis is crucial to our understanding of the origin of the ridge. The Icelandic crust is similar in layering to the oceanic crust beneath the flank of the Reykjanes ridge, except that it is much thicker and more laterally variable. It is being formed at a sub-aerial spreading centre, rather than below sea level. It appears to form by episodic magmatic activity associated with central volcanoes, as recent evidence from the activity of the Krafla volcanic system indicates. The upper mantle beneath Iceland also differs fundamentally from that beneath the flanks of the adjacent Reykjanes ridge, in that the region of anomalously low seismic velocities and density is much more pronounced and extensive than beneath normal oceanic ridges. Seismological, gravity and magnetotelluric evidence suggests that this relates to the widespread occurrence of partial fusion in the mantle beneath most of Iceland. There appears to be a fairly abrupt transition to more typical upper mantle at the junction with the Reykjanes ridge and also at the junction with the Iceland-Faeroe ridge. All this suggest that there is a hot spot caused by mantle

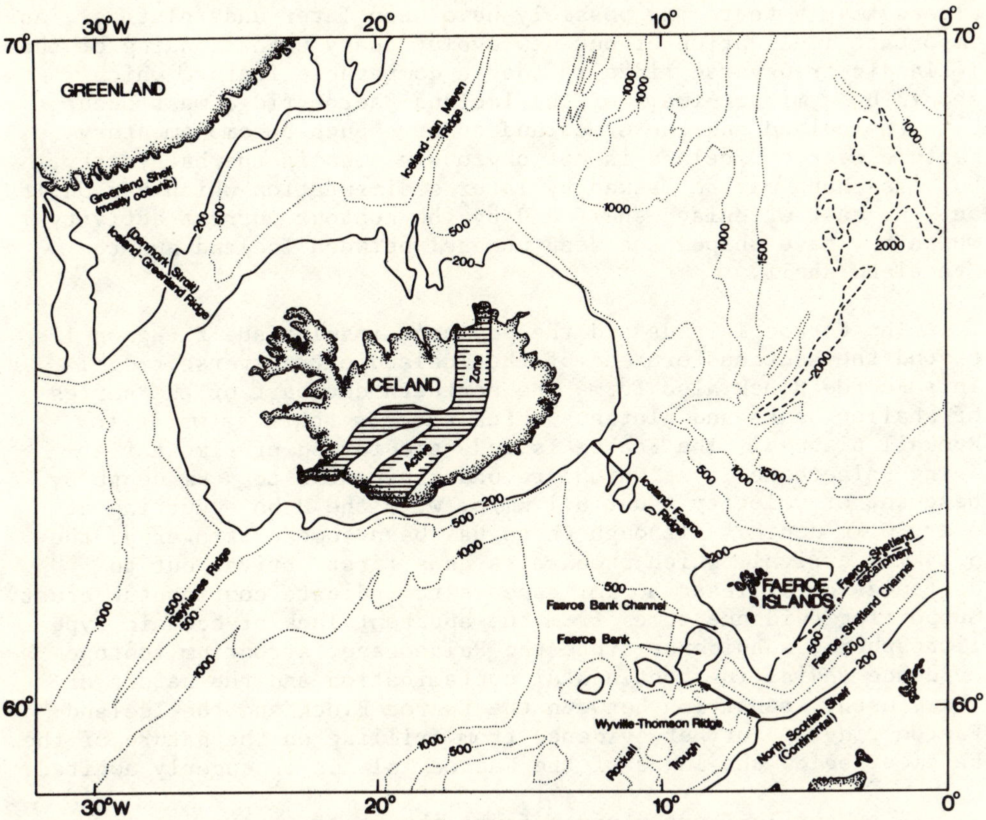


Fig. 1. The morphology of the Greenland-Scotland ridge showing the major features discussed. The depth contours in fathoms are mainly after Grønlie and Talwani (1978).

upwelling at present beneath Iceland, although it is controversial whether or not this is related to the rise of a deep mantle plume.

The older parts of the Icelandic transverse ridge, which have presumably subsided below sea level as the underlying anomalous oceanic lithosphere has cooled, form the Iceland-Greenland ridge crossing the Denmark Strait to the west, and the Iceland-Faeroe ridge to the southeast. Of these two complementary regions, the crustal structure of the Iceland-Faeroe ridge alone has been investigated. It is underlain by Icelandic type crust considerably thicker than beneath Iceland, being more than 30 km thick at the southeastern end and thinning slightly towards Iceland. This may indicate an even more pronounced underlying mantle hot spot during the early stages of formation of the Icelandic transverse ridge;

alternatively there may possibly have been later underplating. An important implication of our new evolutionary understanding of the Icelandic transverse ridge is that a comparable feature which should be a mirror-image of the Iceland-Faeroe ridge must occur between Iceland and the Greenland coast. Such a complementary feature west of Iceland is not obviously seen in the bathymetry, but its shape may be masked by later sedimentation which has built out the east Greenland shelf and by the contour current activity which may have shaped the deep channel between Iceland and the Greenland shelf.

The Faeroe Islands and the adjacent insular shelf region lie beyond the southeastern end of the Icelandic transverse ridge. This Faeroe Block also forms the northernmost part of the series of shallow banks and plateaus which extends southwestwards to Rockall plateau. The Faeroe Islands are formed of almost flat-lying Palaeocene lavas which are underlain at 2 to 4 km depth by basement of velocity about 6.1 km/s, with the Moho occurring at over 30 km depth. Although there has been some controversy, the 6.1 km/s arrival, which propagates as a first arrival out to distances of at least 80 km, appears to indicate continental crust. Supporting evidence comes from the apparent lack of oceanic type lithospheric subsidence since the Palaeocene, strontium isotope evidence indicating continental contamination and the nature of the crustal transition between the Faeroe Block and the Iceland-Faeroe ridge. Further evidence from drilling on the nature of the basement below the lavas of the Faeroe Islands is eagerly awaited.

The Wyville-Thomson ridge forms the easternmost, and least well understood, part of the Greenland-Scotland ridge. It joins the Faeroe Bank, which is separated from the main Faeroe Block by the Faeroe Bank Channel, to the north Scottish shelf. The ridge also forms a barrier between the northeastern end of the Rockall Trough and the southwestern end of the Faeroe-Shetland Channel. The Rockall Trough and the Faeroe-Shetland Channel appear to have been a once continuous oceanic feature, probably underlain by pre-Tertiary ocean floor of Cretaceous or possibly even earlier age. The narrowness of the present Faeroe-Shetland Channel may result from the extensive outpouring of the Palaeocene Faeroe lavas over the west side of the Channel. The contraversial Faeroe-Shetland escarpment appears to represent the eastern feather edge of these lavas, rather than a fundamental crustal boundary as previously supposed. The Wyville-Thomson ridge itself may have formed as a pile of fissure-erupted Palaeocene lavas about 10 km thick formed on top of the earlier oceanic trough, causing depression of the underlying crust to produce a gentle root below the ridge.

The shallow Davis Strait region, which separates the Labrador Sea from Baffin Bay, may possibly form a comparable feature to the Icelandic transverse ridge lying to the west of

Greenland. The region appears to be oceanic in origin but its exact relationship to the Greenland-Scotland ridge remains problematical.

Because of the importance of the subsidence history to the problems of land bridges and North Atlantic oceanic circulation, the principles are here briefly reviewed. The subsidence of the oceanic parts of the Greenland-Scotland ridge is probably substantially caused by the progressive cooling of the underlying lithosphere. Parsons and Sclater (1977) have shown that the depth $d$ in metres of normal ocean floor can be related to its spreading age $t$ in My by the empirical relationships:

$$d(t) = 2500 + 350\, t^{\frac{1}{2}} \quad (t \text{ less than about 50 My})$$

and $d(t) = 6400 - 3200\, exp(-t/62.8)$ ($t$ greater than about 30 My).

If the crust is initially formed partly above sea level, so that a load of increasing water depth is not present to assist the subsidence until the surface is carried below sea level, the initial sub-aerial stage of subsidence is slower by a factor of $(1 - \rho_w/\rho_a)$, where $\rho_w$ is the density of seawater and $\rho_a$ is the density of the underlying asthenosphere. The sub-aerial subsidence at time $t$ after formation is approximately given by $240t^{\frac{1}{2}}$. If the age of formation and the present elevation are known, it is possible to use the above formulae to infer the history of elevation since formation, even when the surface was initially above sea level (Fig. 2). It needs to be borne in mind, however, that the empirical formulae may not apply accurately to Icelandic type lithosphere, and that other effects such as erosion, sedimentation, mantle convection and underplating may modify the elevation. Applying the simple theory indicates that the Iceland-Faeroe Ridge and the Iceland-Greenland ridge probably stood at least 1.5 to 2 km above sea level at time of formation, and that the oldest part of the Iceland-Faeroe ridge probably subsided below sea level about 20 My ago (Vogt, 1972). Application to the Wyville-Thomson ridge is much more uncertain as the age of the underlying oceanic crust is not definitely known and the influence of the Palaeocene volcanism on the thermal state of the lithosphere introduces further uncertainty. On the simplest considerations, however, it seems probable that the Wyville-Thomson ridge stood above sea level for a short time after its Palaeocene formation but that the adjacent parts of Rockall Trough and the Faeroe-Shetland Channel have been continuously marine since formation. The Wyville-Thomson ridge connecting an emergent Faeroe Bank to an emergent North Scottish shelf may have provided a short-lived Palaeocene land bridge connecting Greenland to North Europe, either via the Faeroe Islands (if the Faeroe Bank Channel was emerged) or directly from the Faeroe Bank to Greenland.

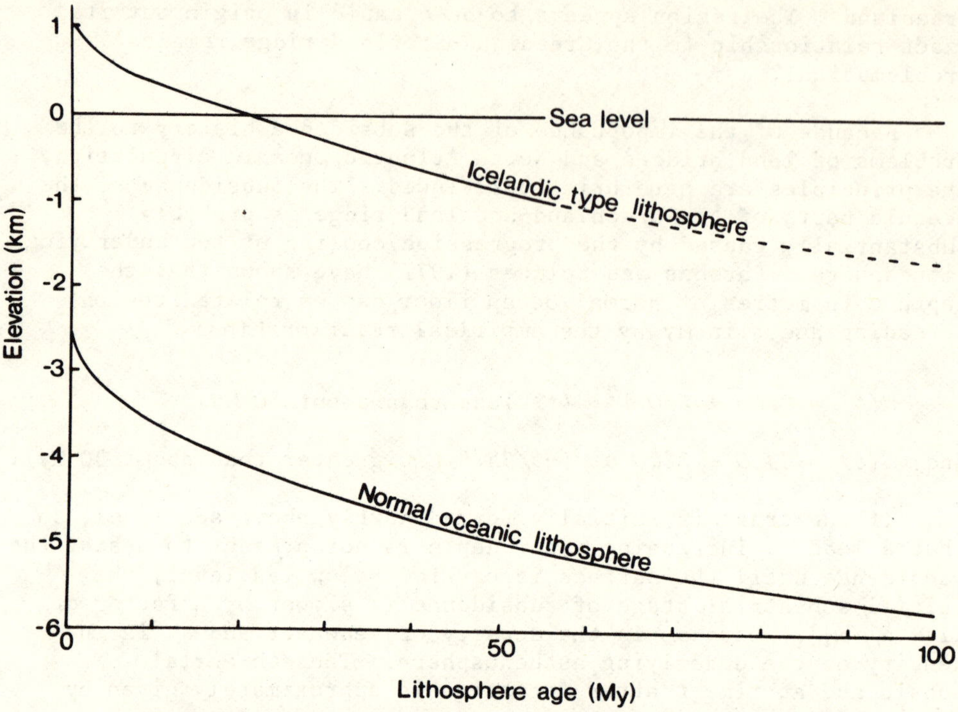


Fig. 2. Empirical curves based on Parsons and Sclater (1977) showing subsidence caused by cooling of the oceanic lithosphere as a function of age, shown for normal oceanic lithosphere and for Icelandic type lithosphere where the initial elevation is about 1100 m above sea level.

The papers in this section show that knowledge of the deep structure and evolution of the Greenland-Scotland ridge has advanced substantially during the last few years but many problems remain. Some of the major outstanding problems are as follows:

(1) The tracing of the early stages in evolution of the anomalous ridge by geochemical and petrological sampling of the Iceland-Faeroe ridge, and associated seismic investigations to gain deeper insight into the structure and the relationship to Iceland.

(2) The search for a complementary feature to the Iceland-Faeroe ridge, between Greenland and Iceland, formed on the west side of the spreading axis.

(3) The proving by drilling, of the basement (believed to be continental) beneath the Faeroe lavas.

(4) The elucidation of the deep structure and origin of the Faeroe-Shetland Channel and the transverse Wyville-Thomson Ridge.

REFERENCES

Grønlie, G. and Talwani, M., 1978, Geophysical Atlas of the Norwegian-Greenland Sea, Vèma Research Series IV, Lamont-Doherty Geological Observatory, Palisades, N.Y.

Parsons, B. and Sclater, J.G., 1977, An analysis of the variation of ocean floor bathymetry and heat flow with age, J. geophys. Res., 82:803.

Vogt, P.R., 1972, The Faeroe-Iceland-Greenland aseismic ridge and the western boundary undercurrent, Nature, Lond., 239:79.

# PLATE TECTONIC EVOLUTION OF THE GREENLAND-SCOTLAND RIDGE AND SURROUNDING REGIONS

Alan G. Nunns

c/o Gulf Research and Development Co.
P.O. Drawer 2038
Pittsburgh, Pennsylvania 15230

## ABSTRACT

A series of plate tectonic reconstructions is presented for the northeast Atlantic and Norwegian-Greenland Sea, showing key magnetic lineations, fracture zones, marginal escarpments and present day bathymetric contours. Under the assumption that the age relationships along the Greenland-Scotland Ridge (GSR) are similar to those across the Reykjanes Ridge, the GSR appears to have evolved in three phases:

(i) From -56? to -44 Ma, there was a major transform zone along the northern flank of the GSR, offsetting the mid-oceanic ridge dextrally to the Norway Basin. A westward shift of axis over the GSR may have taken place before anomaly 22 time. A small amount of spreading between the southern Jan Mayen block and Greenland may have occurred.

(ii) From -44 to -26 Ma, following an anticlockwise reorientation of spreading direction, the Jan Mayen block separated from Greenland. There was a ridge-ridge-transform triple junction over the GSR and complementary fan shaped spreading took place about the Kolbeinsey and Aegir axes. The present Icelandic insular platform began to form during this phase.

(iii) From -26 Ma to the present time, following the extinction of the Aegir axis, the triple junction over the GSR has been replaced by a wide axial rift zone extending to the east of the main line of the Kolbeinsey and Reykjanes Ridges.

## INTRODUCTION

The Greenland-Scotland Ridge (GSR) stretches across the northeastern part of the North Atlantic from Greenland to the northern Scottish shelf (Fig. 1). The main part of the ridge consists of the Greenland-Iceland Ridge (GIR), the Icelandic insular platform, the Iceland-Faeroe Ridge (IFR) and the Faeroe insular platform. South of the Faeroe-Shetland Channel, the Wyville-Thomson Ridge provides a shallow water connection between the Faeroe platform and the Scottish shelf.

It is generally accepted that the GSR has been created, along with the ocean basins to the north and south, by a continuing process of crustal accretion during the separation of Greenland and Eurasia. Dateable "sea floor spreading" magnetic anomalies have not been recognized over much of the GSR. It is necessary to integrate data from a wide area and to make assumptions in order to obtain a model for the evolution of the GSR.

A series of reconstructions for an extended Norwegian-Greenland Sea area is presented in this paper. These are based on the Talwani and Eldholm (1977) tectonic synthesis but include new data and concepts, some of which bear directly on the problem of the evolution of the GSR.

## THE EXTENDED NORWEGIAN-GREENLAND SEA AREA

Fig. 1 shows spreading axes, fracture zones, marginal escarpments and important magnetic anomaly lineations between Greenland and Eurasia (see Table 1 for key). The magnetic anomalies have been depicted as positive bands to convey information about anomaly widths and because it is easier to define the edges of the wide bands (7-12 and 15-18) than it is to pick the locations of the individual peaks. All the lineations and outlines were digitized and plotted automatically, in Figs. 1 to 6. Data sources for the magnetic anomaly lineations are listed in Table 2. The lineations in the vicinity of the GSR are taken from the detailed chart accompanying this volume.

The oldest anomaly present south of the GSR is 24 as it is elsewhere between Greenland and Eurasia. When spreading ceased in the Labrador Sea the direction of spreading between Greenland and Eurasia re-oriented in an anticlockwise fashion (Vogt and Avery, 1974; Talwani and Eldholm, 1977), and closely spaced fractures formed across the Reykjanes Ridge (Vogt and Avery, 1974; Voppel and Rudloff, 1980). These disappeared subsequent to anomaly 7 time.

It has been recently suggested that some of the magnetic lineations about the Reykjanes Ridge continue northward onto the GSR

Table 1. Key to Figures

The figures have been automatically plotted on a polar stereographic projection. Coastline data are from the National Center for Atmospheric Research and the bathymetric contours (Grønlie and Talwani, 1979) are marked in hundreds of fathoms. Sources for magnetic anomaly data are listed in Table 2. Anomaly overlaps are shaded in black in the reconstructions. The ages cited in the captions are from Lowrie and Alvarez (1981).

Bathymetric Features:

JMR = Jan Mayen R., GIR = Greenland-Iceland R., IFR = Iceland-Faeroe R., WTR = Wyville-Thomson R., FSC = Faeroe-Shetland Channel, HB = Hatton Bank, RoB = Rockall Bank, RoT = Rockall Trough.

Active Spreading Axes (Talwani and Eldholm, 1977; Saemundsson, 1979; Vogt et al., 1981 b):

KnR = Knipovitch R., MR = Mohns R., KR = Kolbeinsey R., RR = Reykjanes R.

Extinct Spreading Axis (Talwani and Eldholm, 1977):

AA = Aegir axis.

Fracture Zones (Talwani and Eldholm, 1977; Saemundsson, 1979; Nunns, in press):

HFZ = Hovgaard F.Z., GFZ = Greenland F.Z., SFZ = Senja F.Z., WJMFZ = Western Jan Mayen F.Z., CFZ = Central Jan Ma n F.Z., EJMFZ = Eastern Jan Mayen F.Z., TFZ = Tjornes F.Z.

Marginal Escarpments (Talwani and Eldholm, 1977; Sundvor and Eldholm, 1979; Talwani and Mutter, pers. comm.):

HE = Hornsund Esc., GE = Greenland Esc., VPE = Voring Plateau Esc., FSE = Faeroe-Shetland Esc.

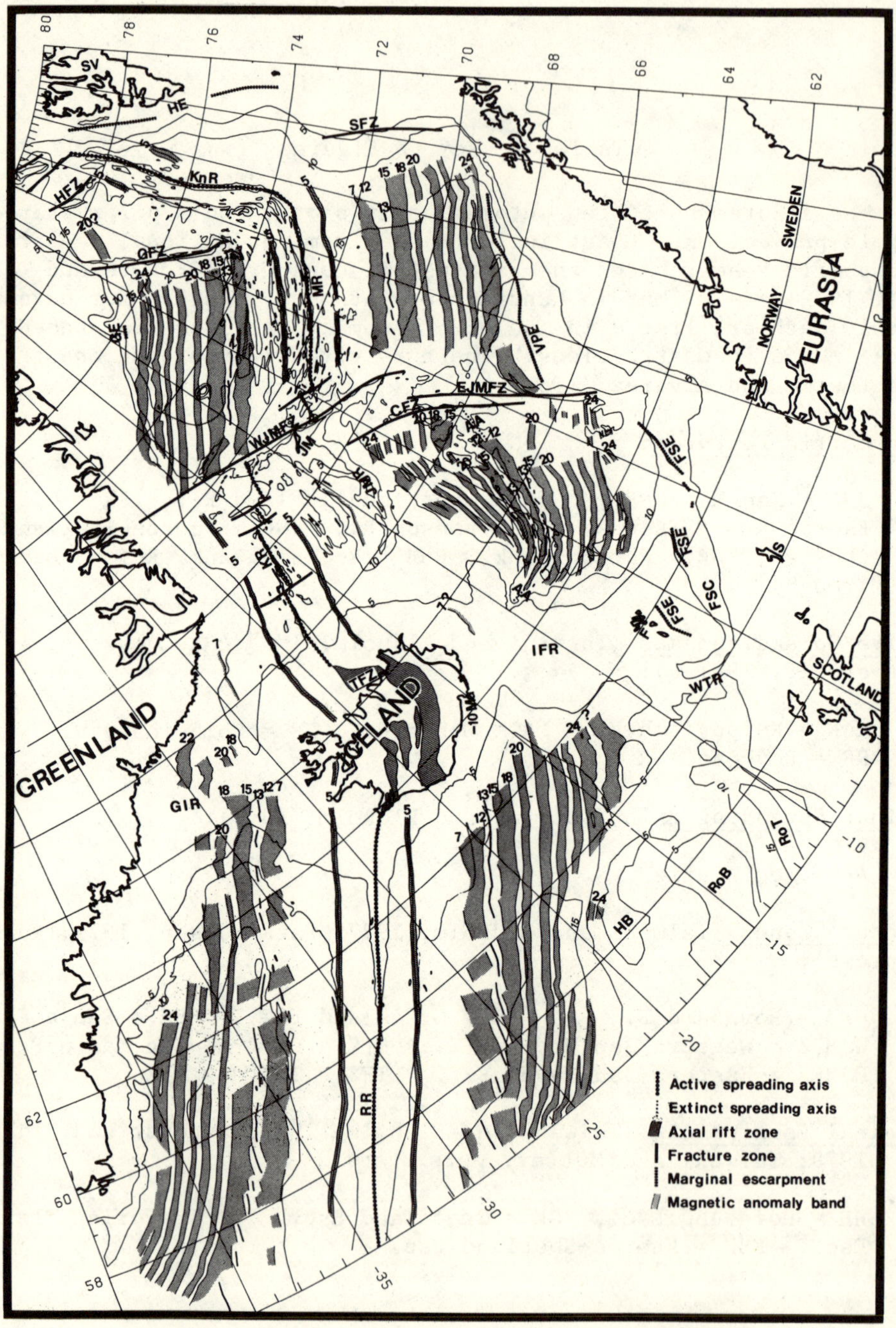


Fig. 1. Present time (see Table 1 for key).

Table 2. Sources of Magnetic Anomaly Data

(a) Data available in digital form

Lamont-Doherty Geo. Obs.: VEMA 23, 27, 28, 29, 30; CONRAD 21 -- Talwani et al. (1971), Talwani and Eldholm (1977), Kristofferson and Talwani (1977), Talwani, Mutter and Eldholm (1981), Talwani and Mutter (pers. comm.)

U.S. Navy: LYNCH 12, 21, 32; KANE 343406, 343522; WILKES 933014; Project MAGNET (Iceland Plateau) -- Johnson et al. (1975), Vogt et al. (1980).

JOIDES: GLOMAR CHALLENGER 38, 49 -- Talwani, Udinstev et al. (1976), Luyendyk, Cann et al. (1979).

Centre National pour L'Exploitation des Oceans: JEAN CHARCOT 20, 21, 37, 38, 51, 55, 56, 57.

Deutsches Hydrographisches Institut: KOMET 28 (E. Reykjanes Basin); METEOR 45 (West Reykjanes Basin) -- Voppel et al. (1979), Voppel (pers. comm.).

Durham University: SHACKLETON 9/77 (Norway Basin) -- Nunns and Peacock (in press).

(b) Data available on charts

U.S. Navy: KANE (East Reykjanes Basin); Project MAGNET (Norwegian-Greenland Sea) -- Vogt and Avery (1974), Vogt et al. (1978), Vogt et al. (1981a,b), Phillips et al. (in prep).

Deutsches Hydrographisches Institut: METEOR 48, 54 (West Reykjanes Basin, Norway Basin) Voppel and Rudloff (1980), Voppel (pers. comm.).

Geological Survey of Greenland: Project EASTMAR (GIR) -- Larsen and Thorning (1980), Larsen (pers. comm.).

Reykjavik University/USDMATC: PROSPEKTA, ISBORG, ALBERT (Iceland insular margin) -- Kristjansson (1976 a, b).

Norwegian Geological Survey: Vøring Plateau aeromag. -- Talwani, Mutter and Eldholm (1981).

without significant offsets. Detailed aeromagnetic surveys (Vogt et al., 1980; Larsen and Thorning, 1980) have revealed the presence of linear anomalies over the GIR. The anomaly identifications shown in Fig. 1 are similar to those of Voppel et al. (1979), Vogt et al. (1980) and Larsen (in prep.).

It is less clear whether there are identifiable linear anomalies over the IFR. Fleischer et al. (1974) presented a magnetic contour map of the ridge which was based on closely spaced marine survey lines running perpendicular to the ridge. The contoured field is highly irregular but some transverse positive and negative bands can be distinguished.

In light of their detailed survey south of the IFR, Voppel et al. (1979) suggested that anomaly 21 continues in a rather disturbed fashion over the ridge; similar continuations of anomalies 20 and 18 are possible (Nunns, in press).

It is assumed that these anomaly identifications are substantially correct and that the GSR has had a development similar to the Reykjanes Ridge, rather than to the area to the north where a major ridge jump occurred. A corollary of this assumption (Voppel et al., 1979) is that a major transform zone existed along the north flank of the GSR during the early stages of its development. If the GSR shared the development of the area to the north a transform zone would have existed along its south flank. Bathymetric data (Bott et al., 1971; Fleischer et al., 1974) show that the northern flank of the IFR is more scarp-like than the southern flank, supporting the former hypothesis.

North of the GSR, spreading about the now extinct Aegir axis in the Norway Basin was followed by a westward shift of axis (Johnson and Heezen, 1967; Talwani and Eldholm, 1977). Spreading about the new Kolbeinsey axis split the Jan Mayen Ridge and its southern extension from the Greenland margin and formed the Iceland Plateau. The aeromagnetic survey of Vogt et al. (1980) revealed that the oldest recognizable anomaly developed on each side of the Kolbeinsey axis is the younger edge of anomaly 7 (Fig. 1).

The following three-stage evolutionary model for the southern Norwegian-Greenland Sea (Nunns, in press) is based on a study of the conspicuous fan-shaped anomaly sequence in the Norway Basin (Talwani and Eldholm, 1977; Nunns and Peacock, in press):

(i) From the time of initial opening until about the end of anomaly 20 time there was regular spreading about the Aegir axis.

(ii) From the end of anomaly 20 time until the end of anomaly 7 time spreading in the Norway Basin took place about a pole situated near the end of the Aegir axis, generating the southward convergent fan-shaped anomaly sequence. Simultaneously, the Jan Mayen block separated from Greenland, with more spreading in the south than in the north. Oceanic crust, generated about the newly formed Kolbeinsey axis, is probably present at the extreme edges of the southern Iceland Plateau: between the Greenland coast and the anomaly 7 lineation shown in Fig. 1, and west of the southern extension of the Jan Mayen Ridge. Anomaly lineations are weakly developed in the former area; Larsen (in prep) has modelled anomalies 7 to 18. A tentative identification of anomaly 7 in the latter area is shown in Fig. 1.

(iii) From anomaly 7 time onwards, no spreading has taken place in the Norway Basin and spreading has been regular about the Kolbeinsey axis.

North of the Jan Mayen Fracture Zones, the evolution of the ocean basins about the Mohns Ridge has been straightforward. The lineations shown in Fig. 1 are based mainly on the detailed U.S. Navy aeromagnetic data presented by Vogt et al. (1978). The evolution of the area north of the Greenland and Senja Fracture Zones has been complex (Vogt et al., 1981 a, b) but is not a main concern here.

## METHOD OF RECONSTRUCTION

In each reconstruction (Figs. 2 to 6) the Eurasian plate is fixed. The Greenland plate and the Jan Mayen "microplate" have been rotated towards Eurasia according to the closure parameters given in the figure captions. The relative positions of the Jan Mayen block and Greenland in the first four reconstructions are determined by the independently derived GR/EU and JM/EU closures. In the anomaly 23 reconstruction the Jan Mayen block has the same position relative to Greenland as it had in the anomaly 20 reconstruction. In each reconstruction, an appropriately partitioned and rotated subset of the outlines and lineations associated with each plate has been plotted. The reconstructed contours have not been adjusted for subsidence and sedimentation and are intended to provide a reference frame rather than an accurate depiction of paleobathymetry. The paleobathymetric reconstructions of Grønlie (1979) and Vogt et al. (1981 a) show that the youngest part of the GSR was probably emergent at each stage of its development.

The GR/EU rotation parameters for anomaly 5 and 13 times are given by Talwani and Eldholm (1977). The same pole latitude (68°) was assumed for closure to the end of anomaly 7 time and the pole longitude was interpolated linearly with respect to time. The amount of rotation was estimated independently. The rotation parameters for GR/EU closure to the end of anomaly 20 time were determined by matching lineations and fracture zones about the Reykjanes and Mohns Ridges. Those for GR/EU closure to anomaly 23 time are averages of the values given by Talwani and Eldholm (1977) and Kristoffersen and Talwani (1977).

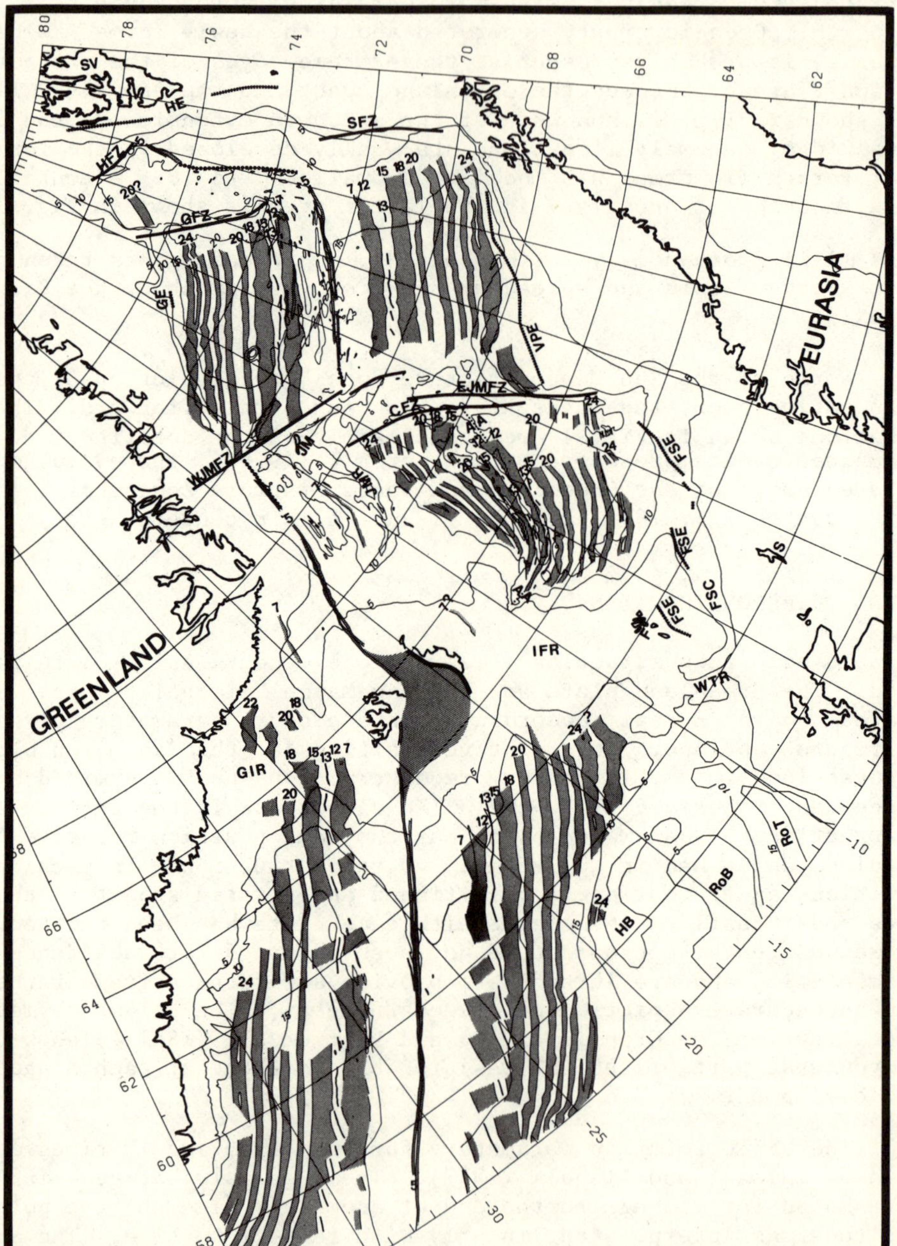


Fig. 2. Anomaly 5 time; -10 Ma.
GR/EU closure pole: lat = 68.0, lon = 137.0, rot = 2.5

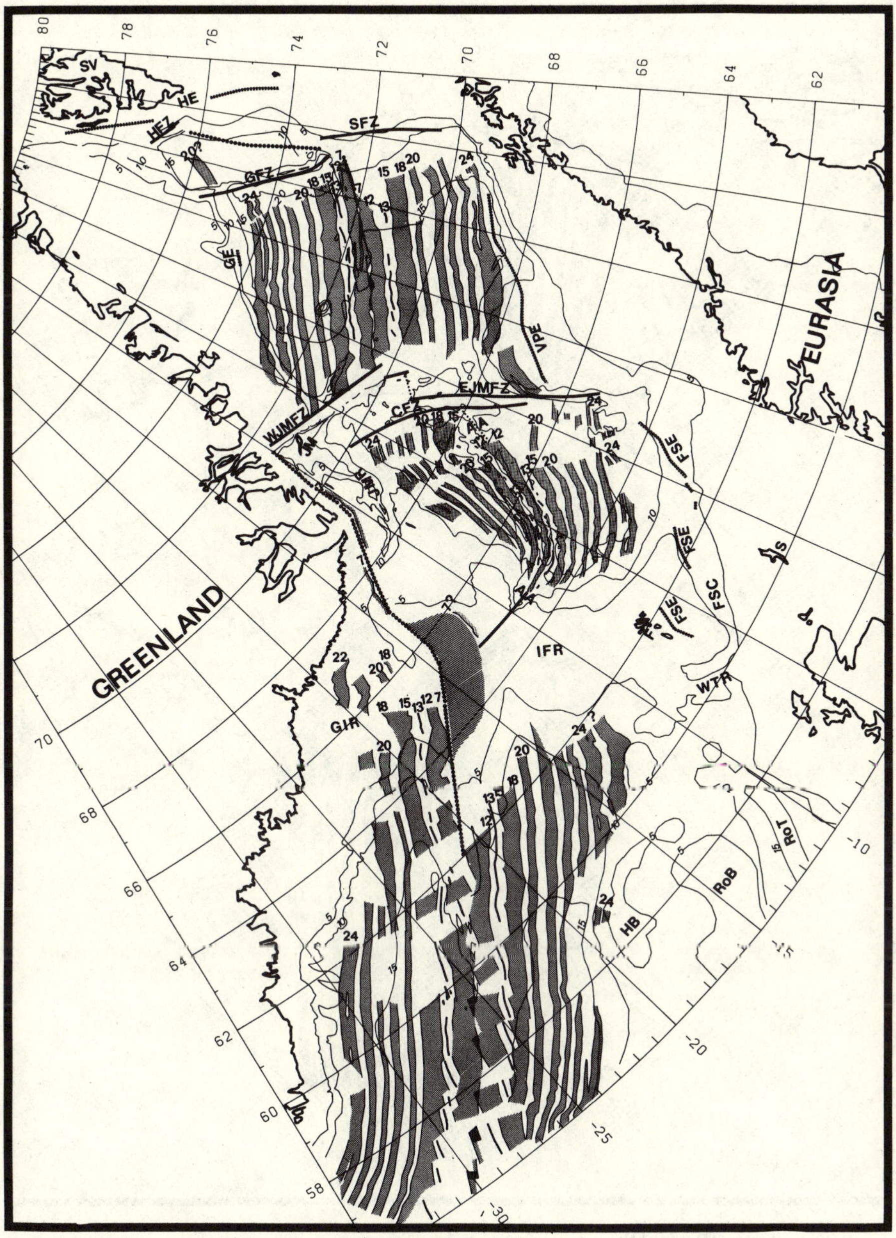


Fig. 3. End of anomaly 7 time; -26 Ma.
GR/EU closure pole: lat = 68.0, lon = 132.2, rot = 5.84

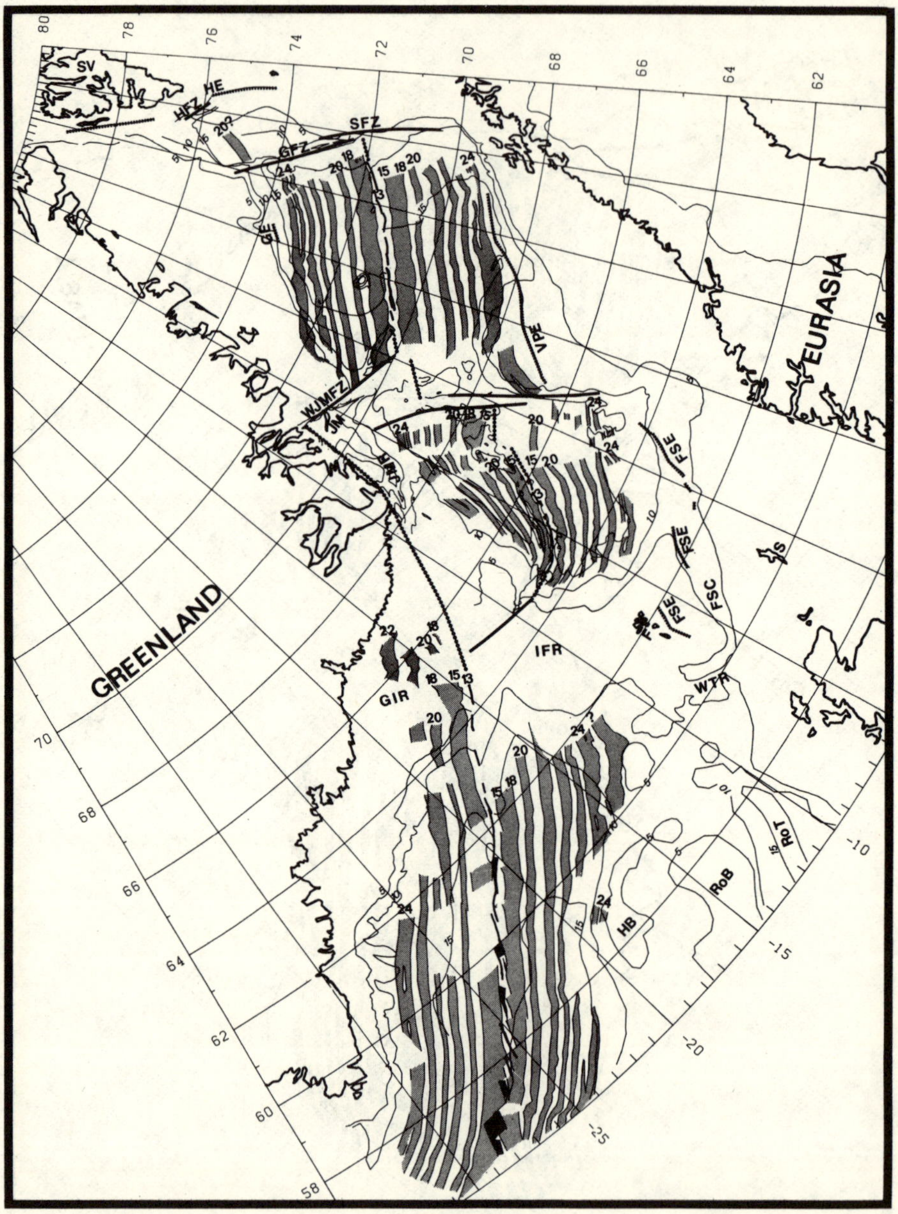


Fig. 4. Anomaly 13 time; -37 Ma.
GR/EU closure pole: lat = 68.0, lon = 129.9, rot = 7.78
JM/EU closure pole: lat = 64.9, lon = -12.3, rot = -8.0

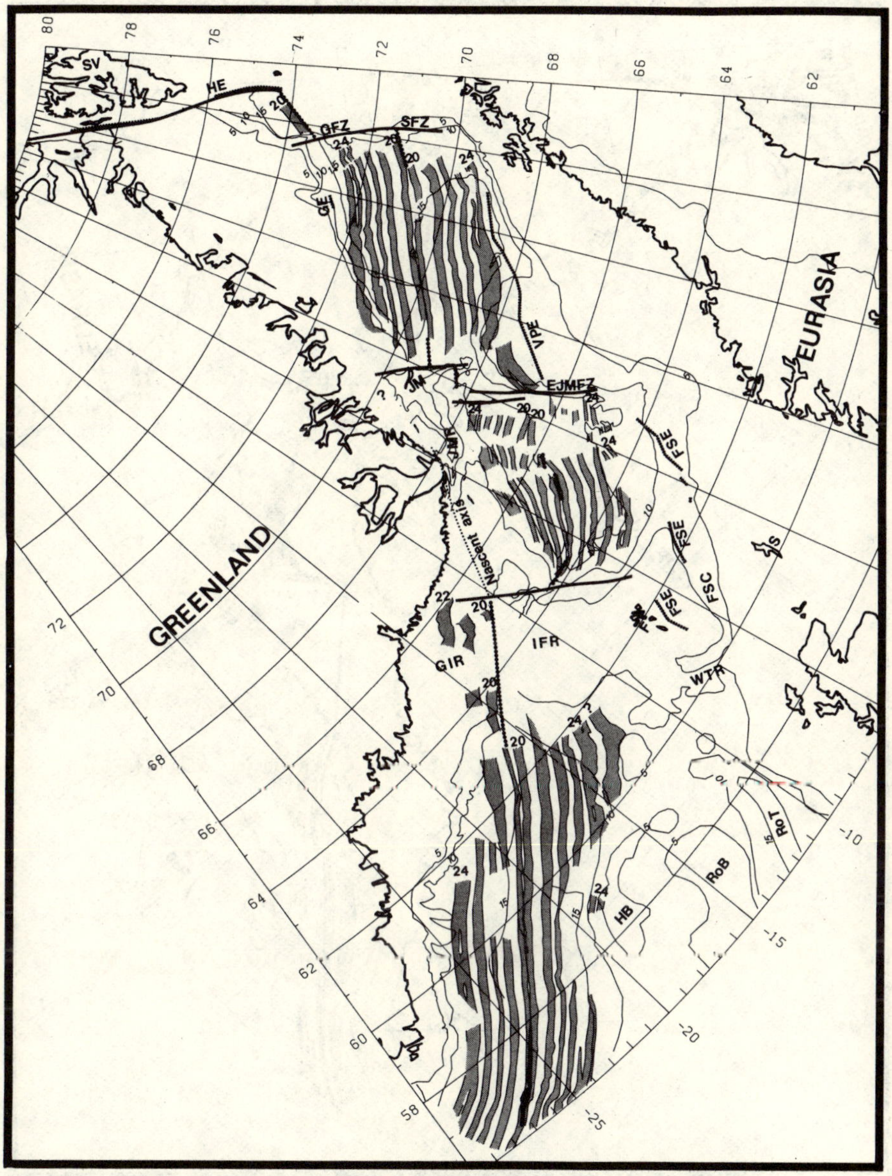


Fig. 5. End of anomaly 20 time; -44 Ma.
GR/EU closure pole: lat = 60.2, lon = 131.6, rot = 8.71
JM/EU closure pole: lat = 64.9, lon = -12.3, rot = -27.4

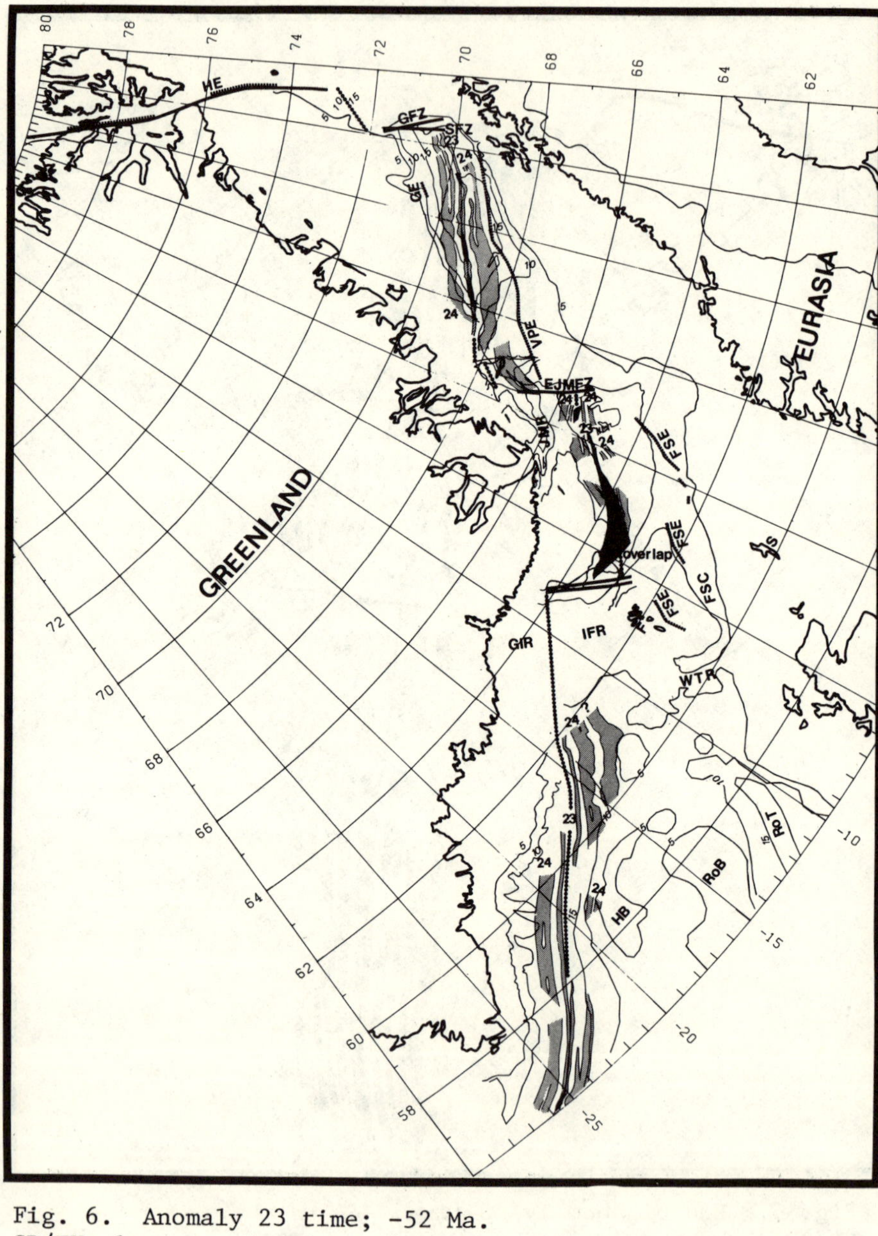


Fig. 6. Anomaly 23 time; -52 Ma.
GR/EU closure pole: lat = 47.3, lon = 128.3, rot = 9.76
JM/EU closure pole: lat = 61.3, lon = -17.2, rot = -28.57

It is assumed that there has been no motion between the Jan Mayen block and Eurasia since anomaly 7 time. The position of the JM/EU closure pole for anomaly 13 time and the end of anomaly 20 time satisfactorily accounts for the configuration of the fan-shaped anomaly sequence and the curvature of the Central Jan Mayen Fracture Zone (Nunns, in press). The western boundary assumed for the Jan Mayen block corresponds to the westerly limit of basement structures which predate the formation of the Norway Basin (Nunns, in press).

## RELATIVE MOTION BETWEEN GREENLAND AND EURASIA

In the anomaly 5 and 7 reconstructions there are good matches between reconstructed lineations about the Reykjanes, Kolbeinsey and Mohns Ridges. The good alignment of the anomaly 7 lineations from the Iceland Plateau supports the contention that there has been no spreading east of the Jan Mayen block since the end of anomaly 7 time. There is a fair correspondence between reconstructed portions of the Western Jan Mayen Fracture Zone in both reconstructions.

In the older reconstructions there are overlaps and gaps between the reconstructed anomalies about the Mohns and Reykjanes Ridges, which cannot be avoided under the assumption of plate rigidity. Parallel to the anomaly trends, control is provided by the Greenland and Senja Fracture Zones (Talwani and Eldholm, 1977) which match well in all the reconstructions.

Talwani and Eldholm (1977) pointed out that the fracture zones in the Norwegian-Greenland Sea exhibit two distinct azimuths: The northwesterly azimuth of the Greenland, Senja and Eastern Jan Mayen Fracture Zones, reflects the early direction of spreading between Greenland and Eurasia; the more westerly azimuth of the Western Jan Mayen and Hovgaard Fracture Zones reflects the new spreading direction after the cessation of spreading in the Labrador Sea. Talwani and Eldholm assumed that the change in spreading direction occurred at about anomaly 13 time. In the present study the predicted direction of opening between anomaly 20 and 13 times is parallel to the motion since anomaly 13 time, rather than to the trend of the old fracture zones. In other words, the main change in plate motion may have taken place earlier than anomaly 13 time. This is consistent with the fact that anomaly 20 is the youngest anomaly developed in the Labrador Sea and with the timing of fracture formation across the Reykjanes Ridge.

The anomaly 23 closure pole of Talwani and Eldholm is such that the opening direction between anomaly 23 and 13 times matches the azimuth of the older fracture zones. The closure pole of Kristoffersen and Talwani (1977) is more consistent with the total openings in the Labrador Sea and the North Atlantic, but does not predict the fracture zone azimuths correctly. An average of the two sets of closure parameters gives a better alignment of the Greenland and Senja Fracture Zones than is achieved with either set. In the present anomaly 23 reconstruction, Greenland is situated about 50 km farther northeast than it is in the Talwani and Eldholm reconstruction. Because the present spreading direction is extrapolated further back in time, Greenland is also situated a similar distance northeast at anomaly 20 time. Thus, the correct early spreading direction is maintained in the present reconstructions. In addition, because of the weight given to the Kristofferson and Talwani parameters, the opening in the Norwegian-Greenland Sea is consistent with the total openings in the Labrador Sea and North Atlantic.

## RELATIVE MOTION OF THE JAN MAYEN BLOCK WITH RESPECT TO EURASIA AND GREENLAND

In Figs. 4 and 5 there is a good match of the reconstructed anomaly 13 and 20 lineations about the Aegir axis. Since anomaly 13 time there has been about 100 km of opening in the northern Norway Basin. With closure from anomaly 7 time to the end of anomaly 20 time, the arcuate Central Jan Mayen Fracture Zone shortens and becomes straight. In Fig. 5 fair parallelism is achieved between the older anomalies about the Aegir axis and also between the Eastern Jan Mayen and Norwegian rises.

In the anomaly 7 reconstruction (Fig. 3) the northern part of the Jan Mayen block is reconstructed close to the Greenland coast. Unless there has been active transverse dislocation of the Jan Mayen block since anomaly 7 time, the Greenland shelf must have prograded dramatically in this area (Vogt et al., 1980). In the anomaly 13 reconstruction, the northern Jan Mayen block is only slightly closer to Greenland, but the southern block is over 100 km closer.

In the anomaly 20 reconstruction the southern Jan Mayen block has moved farther towards Greenland but the northern part has moved outward. This implies consumption of lithosphere between the times of anomalies 20 and 13, which is not an attractive hypothesis.

The second order geometrical problems associated with the relative motion of the Jan Mayen block can be solved in a number of ways if it is assumed that there has been a persistent dislocation between the northern shallow Jan Mayen bank and the Jan Mayen Ridge. The aeromagnetic chart of Navrestad and Jørgensen (1979) shows that there is a clear discontinuity along the projection of the Central Jan Mayen Fracture Zone, separating a disturbed area in the north from a quiet zone in the south. Unfortunately, the areas immediately south of the Jan Mayen Fracture Zones have not been surveyed in detail, so independent reconstructions cannot be made for the Jan Mayen bank. It is clear, however, that the Jan Mayen bank formed opposite the southern part of the Vøring Plateau and it is probably similarly underlain by anomalously elevated oceanic crust.

In the anomaly 23 reconstruction there is a considerable overlap of the lineations in the southern part of the Norway Basin, which perhaps implies that the southern part of the Jan Mayen block began to move away from Greenland prior to the end of anomaly 20 time. It is possible to find an independent JM/EU closure pole based on the best fit of the anomaly 23 lineations in the basin. With this pole, however, an unacceptable GR/EU reconstruction is obtained, with the Jan Mayen block positioned over 100 km farther northeast along the Greenland coast.

## EVOLUTION OF THE GREENLAND-SCOTLAND RIDGE

According to the reconstructions in Figs. 1 to 6, the GSR has evolved in three phases. During the first, which lasted to the end of anomaly 20 time, a major transform fault existed along the northern flank of the GSR, offsetting the mid-oceanic ridge dextrally to the Norway Basin. During the early part of this phase, an area of elevated topography (the southern Jan Mayen Ridge) was situated north of what is now the IFR. Subsequent to anomaly 20 time, a ridge-ridge-transform triple junction formed. Judging from bathymetry, the GSR broadened at this time and the present Icelandic insular platform began to form. During the final stage of spreading since anomaly 7 time, there has been no transform motion between the GSR and the Norway Basin.

In Fig. 2 the reconstructed western anomaly 5 isochron is separated by nearly 200 km from the line of the -10 Ma isochron which has been determined from magnetostratigraphic studies in eastern Iceland (Dagley et al., 1967; McDougall et al., 1976; Ross and Musset, 1976; Watkins and Walker, 1977; Saemundsson, 1979). This separation is too large to be explained by lateral lava flow. In an Icelandic type spreading environment, upper crustal isochrons effectively originate at the edges of a wide accretionary zone (Pálmason, 1973; 1980). Spreading anomalies, if developed, result from polarity contrasts within the lava pile, and are coincident with the upper crustal isochrons (Piper, 1973). Material originating at the center of the accretionary zone is randomly intruded and overlain by later material and follows a steep downward trajectory before passing out of the active zone. This material is not present in the upper crust outside of the active zone, nor is it expected to contribute to the development of recognizable magnetic anomalies. The anomaly 5 reconstruction probably indicates that there was a wide active zone over the GSR at anomaly 5 time, extending eastward of the main line of the mid-oceanic ridge, in a similar fashion to the present active zone in Iceland, which is comprised of a number of individual rift zones (Walker, 1975).

A similarly wide accretionary zone can be inferred for the end of anomaly 7 time, for two reasons: First, the reconstructed anomaly 7 lineation from western Iceland coincides roughly with the -10 Ma isochron in eastern Iceland, so the -26 Ma isochron must lie farther east. Secondly, the asymmetrical configuration of the reconstructed bathymetrical contours suggests that an elevated area lay to the east of the main line of the mid-oceanic ridge. It is difficult to say whether or not a wide accretionary zone existed prior to anomaly 7 time. If the anomaly identifications of Voppel et al. (1979) are correct, the reconstructed anomaly 20 lineations over the GSR match well, implying a narrow accretionary zone.

Two major related problems have not been addressed in the reconstructions presented here because they do not appear to be soluble with the available data. The first of these is the nature of the anomaly 24 lineation immediately south of the GSR. Featherstone et al. (1977) suggested that an early westward shift of axis took place, resulting in a duplication of anomaly 24 northeast of the Hatton Bank and its absence on the Greenland margin. Voppel et al. (1979) considered that a similar shift took place on the GSR; this is consistent with the asymmetrical bathymetry for anomaly 23 time (Fig. 6). To the contrary, Larsen (in prep.) maintains that anomaly 24 is present on the Greenland margin well to the north of its cut-out in Figs. 1 to 6. The second major problem is the location of the ocean-continent boundaries at each end of the GSR, which has been a subject of controversy. It is becoming clear that rapid spreading prior to anomaly 24 time created areas of elevated crust in the Norwegian-Greenland Sea (Mutter, 1981; Mutter et al., 1981). In light of this, detailed seismic data, in addition to magnetic data, will probably be needed to resolve this problem.

## DISCUSSION

The reconstructions which have been presented here differ from those of Talwani and Eldholm (1977) in the treatment of the GSR and the southern Norwegian-Greenland Sea. Vogt et al. (1981a) presented paleobathymetric charts based on the closure parameters of Phillips et al. (in prep.) which are different from those presented here, particularly for the Jan Mayen block. Smoothed paleobathymetric charts based on the present reconstructions would not differ greatly in terms of implications for paleocirculation patterns.

This study was independent of the earlier work of Mirlin (1979) and Mirlin et al. (1980 a, b) who also analyzed the fan-shaped anomaly pattern in the Norway Basin and concluded that the Jan Mayen block separated rotationally from Greenland. According to Mirlin et al. and Larsen (in prep.) the Jan Mayen block began to separate from Greenland at the time of initial rifting. The identifications of anomalies 20 to 24 on the Iceland Plateau (Mirlin et al., 1980b) are not unequivocal, but the possibility of early spreading cannot be discounted. The main phase of separation was probably initiated by the change in spreading direction at the end of anomaly 20 time (Nunns, in press). Before this there was a rectangular configuration of axes and transforms and there was probably not a well defined triple junction over the GSR. However the Reykjanes Ridge may have overshot the GSR (cf. Ramberg and Van Andel, 1977) resulting in a small amount of spreading between Jan Mayen and Greenland.

## ACKNOWLEDGEMENTS

I wish to thank M. Talwani for his advice and for encouraging me to make the computer based reconstructions which have been presented. I gratefully acknowledge the use of bathymetric data digitised by G. Grønlie. I also thank T. Baker and J. L. LaBrecque for providing the coastline data set. Thanks are due to H. C. Larsen, P. R. Vogt and D. Voppel for useful discussions and for providing data and to J. C. Mutter and J. L. LaBrecque for reviewing the manuscript. I thank A. A. Schreider for acquainting me with the Soviet research in the Iceland region. This study is partly based on research carried out at the University of Durham, U. K. and I wish to thank M. H. P. Bott for his continued advice and interest. This study was supported by the NATO Science Advisory Panel for Marine Sciences and Lamont-Doherty Geological Observatory. The manuscript was ably typed (and critically edited) by L. Hayes.

Contribution No. 3237, Lamont-Doherty Geological Observatory.

## REFERENCES

Bott, M. H. P., Browitt, C. W. A., and Stacey, A. P., 1971, The deep structure of the Iceland-Faeroe Ridge, Mar. geophys. Res., 1:328.

Dagley, P., Wilson, R. L., Ade-Hall, J. M., Walker, G. P. L., Haggerty, S. E., Sigurgeirsson, T., Watkins, N. D., Smith, P. J., Edwards, J., and Grasty, R. L., 1967, Geomagnetic polarity zones for Icelandic lavas, Nature, 216:25.

Featherstone, P. S., Bott, M. H. P., and Peacock, J. H., 1977, Structure of the continental margin of south-eastern Greenland, Geophys. J. R. astr. Soc., 48:15.

Fleischer, U., Holzkamm, F., Vollbrecht, K., and Voppel, D., 1974, Die Struktur des Island-Färöer-Rückens aus geophysikalischen Messungen, Dt. hydrogr. Z., 27:97.

Grønlie, G., 1979, Tertiary paleogeography of the Norwegian-Greenland Sea, Norsk Polarinst. Skr., 170:49.

Grønlie, G., and Talwani, M., 1979, Bathymetry of the Norwegian-Greenland Sea, Norsk Polarinst. Skr., 170:3.

Johnson, G. L., and Heezen, B. C., 1967, Morphology and evolution of the Norwegian-Greenland Sea, Deep-Sea Res., 14:755.

Johnson, G. L., Sommerhoff, G., and Egloff, J., 1975, Structure and morphology of the West Reykjanes Basin, Mar. Geol., 18:175.

Kristjansson, L., 1976a, A marine magnetic survey off southern Iceland, Mar. geophys. Res., 2:315.

Kristjansson, L., 1976b, Marine magnetic surveys off the west coast of Iceland, Soc. Sci. Isl. Greinar, 5:23.

Kristofferson, Y., and Talwani, M., 1977, Extinct triple junction south of Greenland and the Tertiary motion of Greenland relative to North America, Geol. Soc. Am. Bull., 88:1037.

Larsen, H. C., and Thorning, L., 1980, Project EASTMAR: acquisition of high sensitivity aeromagnetic data off East Greenland, Rapp. Gronlands geol. Unders, 100:91.

Lowrie, W., and Alvarez, W., 1981, One hundred million years of geomagnetic polarity history, Geology, 9:392.

Luyendyk, B. P., Cann, J. R., et al., 1979, "Initial Reports of the Deep Sea Drilling Project, vol. 49," U.S. Government Printing Office, Washington, D.C.

McDougall, I., Watkins, N. D., Walker, G. P. L., and Kristjansson, L., 1976, Potassium-argon and paleomagnetic analysis of Icelandic lava flows: limits on the age of anomaly 5, J. geophys. Res., 81:1505.

Mirlin, Ye. G., 1979, Movements and deformations of lithosphere plates in the Icelandic region of the North Atlantic and the character of the Faeroes-Greenland threshold, Geotectonics (English Edition), 12:455.

Mirlin, Ye. G., Kostoglodov, V. V., and Suzyumov, A. Ye., 1980a, Cenozoic plate tectonics in the Iceland region of the North Atlantic, Oceanology (English Edition), 20:168.

Mirlin, Ye. G., Popov, K. V., and Finger, D. L., 1980b, Age of the ocean floor in the Icelandic region, Oceanology (English Edition) 19:696

Mutter, J. C., 1981, Layered oceanic basement complex of the Norwegian Margin: a heuristic model, EOS Trans. Am Geophys. Un., 62:407.

Mutter, J. C., Talwani, M., and Stoffa, P. L., 1981, Origin of Subbasement dipping reflection off the Norwegian margin by "subaerial seafloor spreading", submitted to Geology.

Navrestad, T., and Jørgensen, F., 1979, Aeromagnetic investigations on the Jan Mayen Ridge, in: "Norwegian Sea Symposium," Norwegian Petroleum Society, Oslo.

Nunns, A. G., in press, The structure and evolution of the Jan Mayen Ridge and surrounding areas, Am. Ass. Petrol. Geol. Mem.

Nunns, A. G., and Peacock, J. H., in press, Correlation, identification and inversion of magnetic anomalies in the Norway Basin, Earth Evol. Sci.

Pálmason, G., 1973, Kinematics and heat flow in a volcanic rift zone, with application to Iceland, Geophys. J. R. astr. Soc., 33:451.

Pálmason, G., 1980, A continuum model of crustal evolution in Iceland; kinematic aspects, J. Geophys., 47:7.

Piper, J. D. A., 1973, Interpretation of some magnetic anomalies over Iceland, Tectonophysics, 16:163.

Phillips, J. D., Feden., R., Fleming, H. S., and Tapscott, C., in prep., Aeromagnetic studies of the Greenland/Norwegian Sea and Arctic Ocean.

Ramberg, I. B., and Van Andel, T. H., 1977, Morphology and tectonic evolution of the rift valley at lat 36°30', Mid-Atlantic Ridge, Geol. Soc. Am.Bull., 88:577.

Ross, J. G., and Musset, A. E., 1976, $^{40}Ar/^{39}Ar$ dates for spreading rates in eastern Iceland, Nature, 259:36.

Saemundsson, K., 1979, Outline of the geology of Iceland, Jökull, 29:7.

Sundvor, E., and Eldholm, O., 1979, The western and northern margin off Svalbard, Tectonophysics, 59:239.

Talwani, M., and Eldholm, O., 1977, Evolution of the Norwegian-Greenland Sea, Geol. Soc. Am. Bull., 88:969.

Talwani, M., Mutter, J., and Eldholm, O., 1981, The initiation of opening of the Norwegian Sea, Oceanol. Acta, Special Issue (Geology of Continental Margins):23.

Talwani, M., Udinstev, G., et al., 1976, "Initial Reports of the Deep Sea Drilling Project, vol. 38," U.S. Government Printing Office, Washington, D.C.

Talwani, M., Windisch, C. C., and Langseth, M. G., 1971, Reykjanes Ridge crest: a detailed geophysical study, J. geophys. Res., 76:473.

Vogt, P. R., and Avery, O. E., 1974, Detailed magnetic surveys in the north-east Atlantic and Labrador sea, J. geophys. Res., 76:363.

Vogt, P. R., Bernero,C., Kovacs, L., and Taylor, P., 1981a, Structure and plate tectonic evolution of the marine arctic as revealed by aeromagnetics, Oceanol. Acta, Special Issue (Geology of Oceans):25.

Vogt, P. R., Feden, R. H., Eldholm, O., and Sundvor, E., 1978, The ocean crust west and north of the Svalbard Archipelago: synthesis and review of new results, Polarforschung, 48:1.

Vogt, P. R., Johnson, G. L., and Kristjansson, L., 1980, Morphology and magnetic anomalies north of Iceland, J. Geophys., 47:67.

Vogt, P. R., Perry, R. K., Feden, R. H., Fleming, H. S., and Cherkis, N. Z., 1981b, The Greenland-Norwegian Sea and Iceland environment: geology and geophysics, in: "The ocean basins and margins, vol. 5, The Arctic," A. E. M. Nairn, M. Churkin and F. G. Stehli, eds. Plenum Press, N.Y.

Voppel, D., and Rudloff, R., 1980, On the evolution of the Reykjanes Ridge south of 60° N between 40 and 12 million years before present, J. Geophys., 47:61.

Voppel, D., Srivastava, S. P. and Fleischer, U., 1979, Detailed magnetic measurements south of the Iceland-Faeroe Ridge, Dt. hydrogr. Z., 32:154.

Walker, G. P. L., 1975, Excess spreading axes and spreading rate in Iceland, _Nature_, 255:468.
Watkins, N. D., and Walker, G. P. L., 1977, Magnetostratigraphy of eastern Iceland, _Am. J. Sci._, 277:513.

# CRUST AND UPPER MANTLE BENEATH ICELAND

Sveinbjorn Bjornsson

Science Institute
University of Iceland
107 Reykjavik, Iceland

ABSTRACT

Except for greater thickness and more pronounced lateral variations the Icelandic crust is very similar to the crust beneath the Reykjanes Ridge. Layer 3 can be traced without interruption from the 10 m.y. old oceanic crust on the southeastern flank of the ridge through Iceland. It exists also beneath the eastern neovolcanic zone which is supposed to be the present spreading axis. Wide-angle reflection surveys across the active volcanic zones and the adjacent plateau basalts have given a detailed picture of the crust. The cross sections bear a certain resemblance to the crustal structure deduced from geological observations in Eastern Iceland. The dip of refractors and reflectors toward the axial zone and the absence of reflecting horizons below 8 km depth in the central part of the zone, support the predictions of models of crustal accretion mechanism. The episode of crustal rifting in the NE Iceland axial zone, which started in 1975 and still continues, has demonstrated how intimately the rifting is related to episodic magmatic processes at central volcanoes.

The subcrustal structure below Iceland is fundamentally different from the one below the southeastern flank of the Reykjanes Ridge. The Icelandic crust is underlain by low P-wave velocity (7.0 to 7.6 km/s) material down to depths greater than 50 km which is interpreted as mantle material in the state of partial fusion. This anomalous mantle seems to form a diapiric updoming of the astenosphere much broader than is likely beneath mid-ocean ridges proper. Magnetotelluric soundings indicate a low resistivity layer at the base of the crust. The depth to this layer is about 10 km beneath the axial rift zone but increases to about 17 km below the Tertiary basalts east and west. S-waves generally propagate in the crust beneath the axial rift zone to 10 km depth. At greater depth high P- to S-velocity ratio and attenuation of S-waves support the magnetotelluric evidence for partially molten basalt at the base of the crust, not only under the axial rift zone, but under the whole of Iceland.

## INTRODUCTION

In the terms of global tectonics Iceland is a hot spot located near an accreting plate boundary. It is the largest mass of land found on the Mid-Atlantic Ridge and the Iceland transverse ridge is one of the most substantial aseismic ridges of the oceans. The land offers unique conditions to study crustal structure, tectonic features and magmatic processes in far greater detail than ever possible in the submarine parts of the ridge system. On the other hand, Iceland appears in many respects to be anomalous if compared to other parts of the mid-ocean ridge system. It is the summit of a vast regional bulge, has a deep Bouguer gravity low, parallel spreading zones with abundant central volcanoes, different petrochemistry and different crustal and mantle structure. Some of the abnormalities might find their explanation if a mantle plume of low average density and high temperature exists under Iceland.

Considerable research has been conducted on crustal and mantle structure of Iceland and its surroundings in recent years. The status of knowledge in 1974 was presented by Kristjansson (1974), and Palmason and Saemundsson (1974). A review of later progress is given by Saemundsson (1979), Jacoby et al. (1980), and Sanford and Einarsson (1981). This article gives an overview of recent significant data relating to the crust and the upper mantle beneath Iceland. The paper draws heavily on the recent reviews and symposia cited above, to which the reader is referred for further details.

## THE MID-ATLANTIC RIDGE AND THE VOLCANIC ZONES OF ICELAND

### Magnetic Anomalies

Magnetic anomaly lineations indicate that the ocean floor to the north and to the southwest of Iceland has been spreading at a constant overall half-rate of 1 cm/year in both areas since at least 12 m.y. ago, but possibly at a slower rate before this time (Talwani and Eldholm, 1977; Vogt et al., 1980; Voppel and Rudloff, 1980).

The trend of the Charlie-Gibbs fracture zone suggests that the spreading direction in the Iceland region is about N 100°E. South of 56°N the Reykjanes Ridge is orthogonal to the spreading direction but the northern part of the ridge has a trend of N 35°E indicating oblique spreading. Approaching Iceland the crestal region is composed of en échelon crestal segments trending N to NE which are

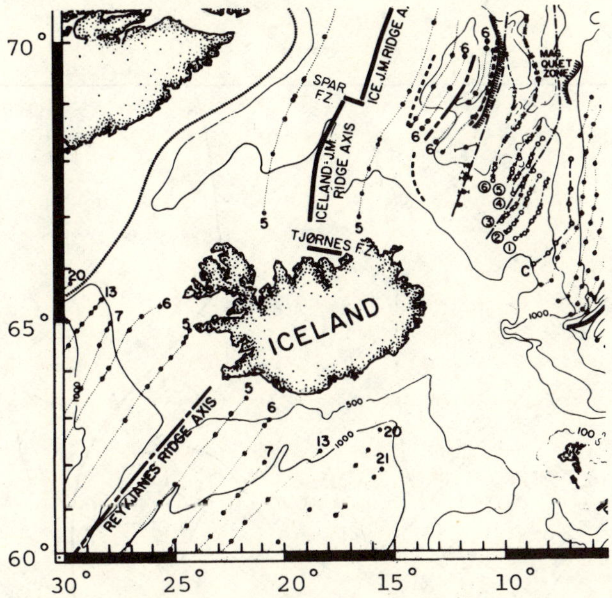


Fig. 1. Some major identified magnetic anomaly lineations in the ocean surrounding Iceland. (After Talwani and Eldholm, 1977)

displaced farther and farther east. (Shih et al., 1978). Where the ridge joins the Reykjanes Peninsula the trend of the crestal zone has become N 80°E (Einarsson, 1979).

Kristjansson (1979) reviewed the available evidence on magnetic anomaly lineations in the shelf area around Iceland. Some major anomaly lineations are easily followed along the Reykjanes Ridge, notably the Anomaly 5 lineations inferred to be of 9-10 m.y. age. West of the ridge axis (Fig.1) this anomaly can be followed to the tip of the Snaefellsnes Peninsula and from there to the NE across the bay north of Snaefellsnes. The eastern Anomaly 5 lineation on the Reykjanes Ridge can be followed as far as 63°N, but only tentatively farther. Continued along its strike it would come ashore between the active volcanic zones in S Iceland where Quaternary rocks are found on the surface. To explain this discrepancy a fracture zone has been postulated in S Iceland. Its existence is supported by seismic evidence and surface fractures indicating right-lateral shear along N-S faults (Einarsson and Bjornsson, 1979). On the Kolbeinsey Ridge north of Iceland magnetic anomaly lineations are very well developed between 67°N and 69°N (Fig. 1). West of the axis Anomaly 5 is known to extend at least south to 66°N and can thus be regarded as essentially continuous through the Iceland area. East of the ridge, on the other hand, the anomaly becomes indistinct south of 67°N. Continuation to the coast meets the active volcanic zone in NE Iceland. The anomaly appears to be offset farther east by the Tjornes Fracture Zone (Saemundsson, 1974; Kristjansson, 1979).

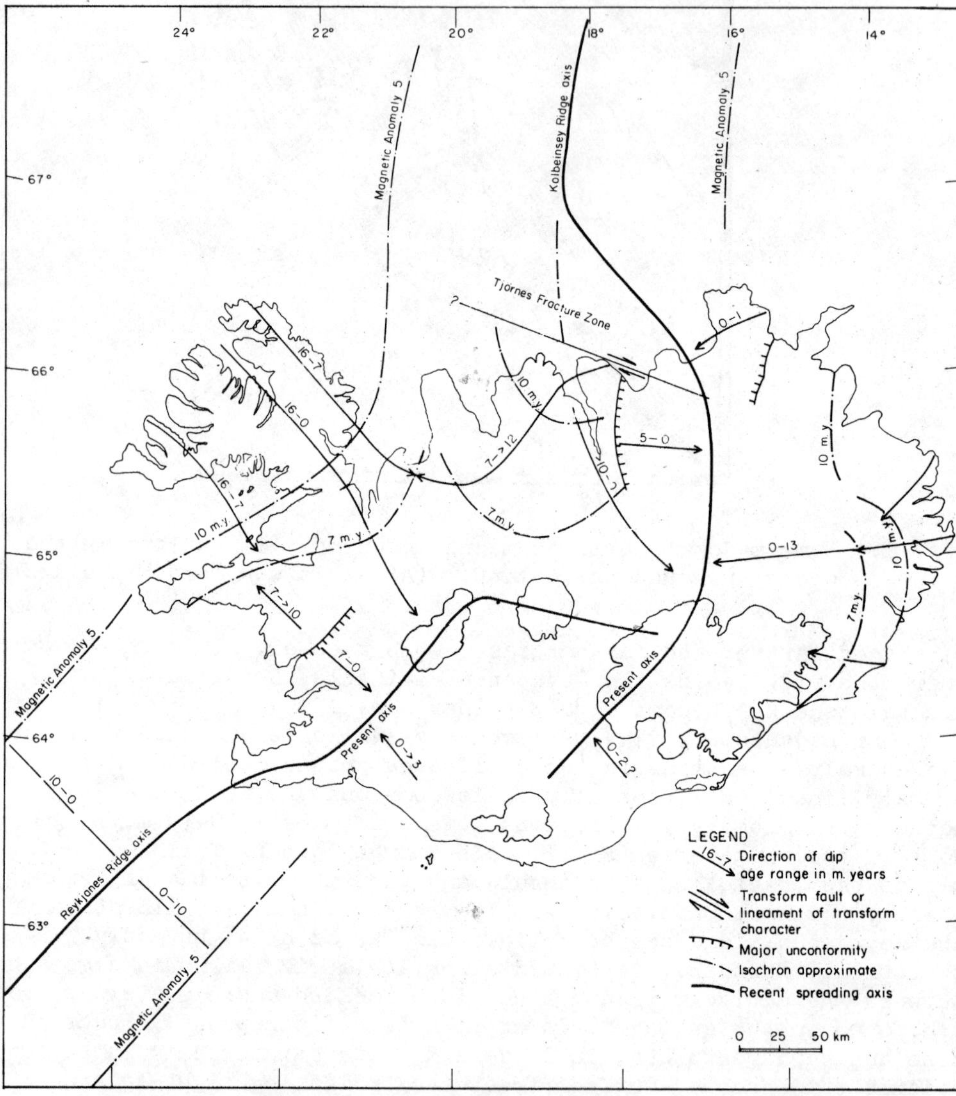


Fig. 2. Development of the present geometry of the tectonically active zones in Iceland. Traces of extinct rift zones are left as structural synclines in the Tertiary areas. Major unconformities have developed between sequences produced within extinct rift zones and those produced from the presently active ones (From Saemundsson, 1979).

## The Lava Pile of Iceland

Magnetic anomaly lineations are not as clear in Iceland as on the ocean bottom. On the other hand, radiometric dating and paleomagnetic reversal patterns observed in the lava sequences of dipping lava piles provide means for stratigraphic mapping of lava groups. Systematic mapping on a regional scale has been initiated in most parts of the Tertiary areas, but it is still far from complete (Einarsson, 1957; McDougall and Wensink, 1966; Einarsson et al., 1967; Dagley et al., 1967; McDougall et al., 1976; McDougall et al., 1977; Watkins and Walker, 1977; Kristjansson et al., 1980; Saemundsson et al., 1980). An outline of the results is given by Saemundsson (1979).

A number of shallow synclines and low anticlines are found in the Tertiary lava pile. The most pronounced syncline is the present axial rift zone formed by crustal extension and downsagging of volcanic products. Other synclines indicate the position of extinct rift zones. The Snaefellsnes rift zone in western Iceland, trending SW-NE in the Snaefellsnes area and N-S in the area west of Skagi (Fig. 2) was active at least from 16 m.y. up to 6.7 m.y. ago. (Johannesson, 1980). The lava pile in western and northwestern Iceland was produced in this zone. A parallel rift zone was probably active during this period in central eastern Iceland where a continuous section spanning the time interval from about 13.4 m.y. up to the Plio-Pleistocene has been found (Fig. 2) (Watkins and Walker, 1977; Musselt et al., 1980).

Studies of dips, magnetic polarity and fragmentary K/Ar ages of lavas indicate that the southern part of the Snaefellsnes rift zone became extinct about 5-6 m.y. ago and a new spreading axis was formed farther east in the present Reykjanes-Langjokull rift zone. The northern part of the Snaefellsnes rift zone became extinct about 3-4 m.y. ago as activity increased in the northern part of the eastern rift zone.

In western Iceland the shift in volcanic activity led to an anticlinal structure of the Tertiary lavas referred to as the Borgarnes anticline (Fig. 2). The ages of its flanks are widely contrasting. The western flank was formed prior to 7 m.y. ago within the now extinct Snaefellsnes rift zone and the eastern flank formed about 7 m.y. ago up to Recent within the Reykjanes-Langjokull rift zone.

In northeastern Iceland the superpositioning of younger lavas has resulted in an age and structural discontinuity within the Tertiary basalts on both sides of the axial rift zone (Fig. 2) (Saemundsson, 1979). Approaching the Vatnajokull region no break is present in the sequence which implies that the axial rift zone in that part of eastern Iceland remained unaffected while activity increased farther north. About 2 m.y. ago rifting and volcanic activity spread southwards southwest of Vatnajokull, leading to the present pattern of axial rift zones (Fig. 3). It is speculated that this southward propagation of the eastern zone may result in

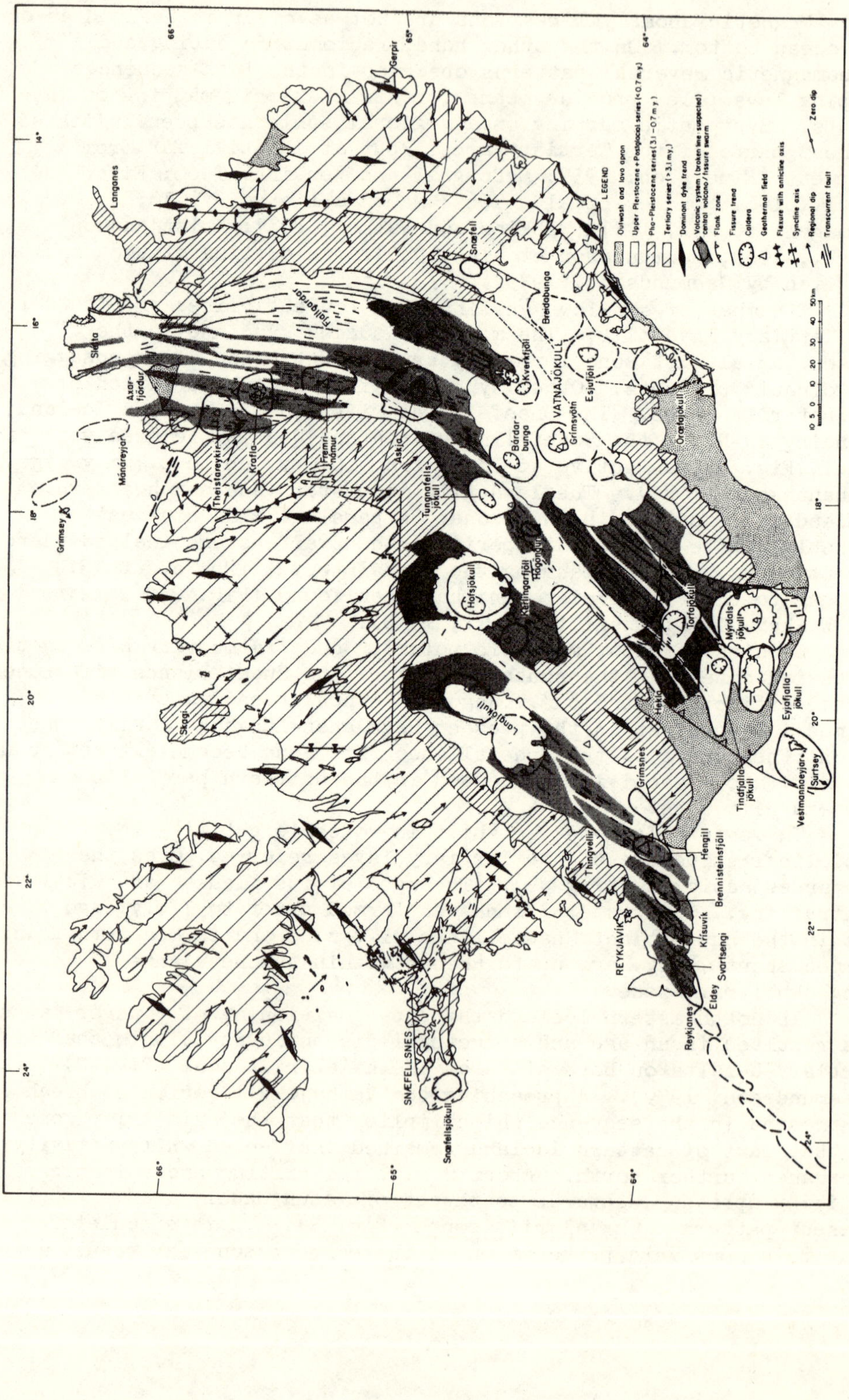


Fig. 3. Tectonic map of Iceland (From Saemundsson, 1979).

extinction of the Reykjanes-Langjokull zone and in a fracture zone joining the Reykjanes ridge and the eastern rift zone. Volcanic activity in the southernmost part of the eastern rift zone and off the southern coast is different from that farther north. The eruptive products are transitional alkali and alkali olivine basalts whereas tholeiitic basalts are erupted in other parts of the eastern zone and in the Reykjanes-Langjokull zone (Jakobsson, 1972). Postglacial lavaflows erupted in the Snaefellsnes Peninsula are also alkalic basalts (Sigurdsson, 1970). Jakobsson (1972) suggested that the tholeiites delineate the crest region of the Mid-Atlantic Ridge as it transects Iceland, and the alkali olivine and the transitional alkali basalts characterize the flank volcanic zones. Saemundsson (1978) proposed that distinction should be made between the axial rift zones that mark the trace of the ridge axis and the lateral rift zones that have developed on the flanks of the ridge axis. This distinction is based on the tectonic structure, the regional stress and the chemistry of volcanic products. The axial rift zones are marked by pronounced linear tensional tectonics. Normal faults, open fissures and eruptive fissures are abundant. They are flanked by volcanic piles which dip and become progressively younger towards them, without major time gaps in the strata. The lateral rift zones are within regions under horizontal shear stress. They have poorly developed features of tensional tectonics and include formations which are distinctly unconformable upon older piles of volcanics.

Superpositioning and the present configuration of the axial rift zones predict that the oldest exposed rocks in Iceland should occur in the farthest northwest, north and east. Radiometric dating of the lowest exposed levels in the east indicate that the oldest rocks in that area are of just over 13 m.y., in the northwest around 16 m.y. and around 12 m.y. in the north (Fig. 2). Southward and eastward displacement of rift zones combined with spreading along two parallel axes is thought to be mainly responsible for the relatively low age of exposed rocks in S Iceland if compared to the age of the sea bottom inferred from magnetic anomaly lineations

## Episodic Rifting of Fissure Swarms

Structural relationships indicate that the lava pile grew as lenticular units along dyke swarms. The dykes run through a central volcanic complex where the volcanic activity is highest. The dyke or fissure swarms are commonly about 10 km wide and their length varies from 30 to over 100 km (Fig. 3).

The fissure swarm and its associated central volcano act as a tectonic unit during magmatic activity. This relationship has recently been clearly demonstrated during a rifting episode in the Krafla fissure swarm in the northern part of the eastern rift zone (Bjornsson et al. 1977, Einarsson 1978, Bjornsson et al. 1979, Brandsdottir and Einarsson 1979, Moller and Ritter 1980, Spickernagel 1980, Torge and Kanngieser 1980, Johnsen et al. 1980,

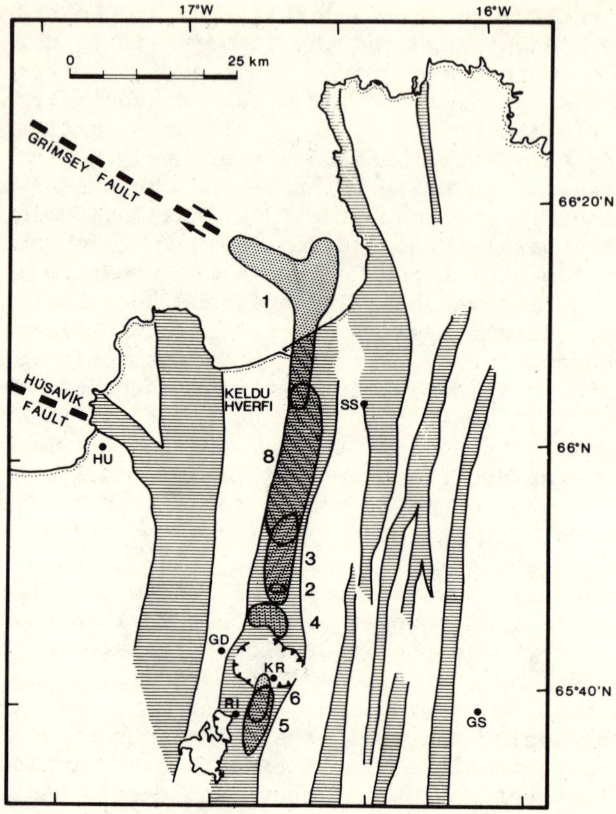


Fig. 4. Fault swarms in the northern volcanic zone, mapped by Kristjan Saemundsson (Bjornsson et al. 1977). The Krafla caldera is located within the Krafla fault swarm. The black dots denote seismic stations. The stippled areas are the zones of maximum earthquake activity during the different deflation events at Krafla. Numbers indicate the sequence of events that occurred in the period Dec. 1975 to Jan. 1978 (From Einarsson and Brandsdottir, 1980).

Tryggvason 1980, Sigurdsson 1980, Einarsson and Brandsdottir 1980).

The center of activity is the Krafla volcanic system. A shallow magma reservoir is receiving constant inflow of about 5 $m^3/s$ of magma. The volcano inflates until the magma pressure in the shallow reservoir reaches the level required to initiate rifting of the associated fissure swarm. As the barrier breaks magma migrates laterally along fissures away from the reservoir and the volcano deflates. Extension of the fault swarm during one deflation event locally exceeds 1 m and the area of maximum rifting generally coincides with the area of maximum earthquake activity. Different segments of the fissure swarm have rifted during individual deflation events (Fig. 4). The extension across the

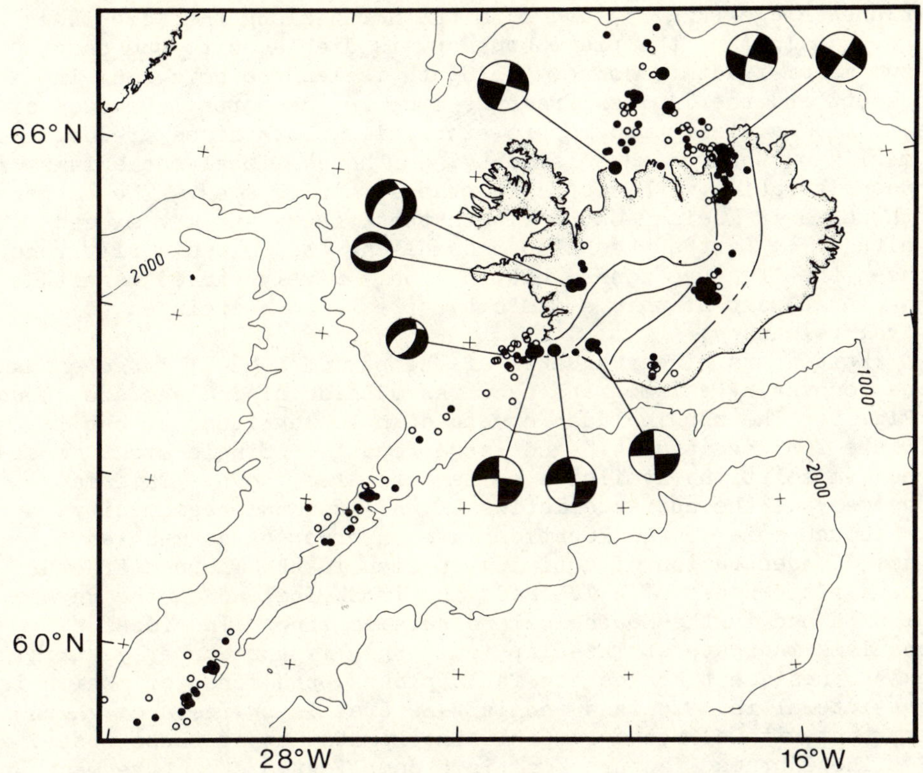


Fig. 5. Epicenters and focal mechanisms of earthquakes in Iceland and the northern part of the Reykjanes Ridge, period 1962-1977. Quadrants containing the least compressive stress axis are shown black (From Einarsson, 1979).

swarm along most of its length amounts already to about 4 m corresponding to about 200 years of tensional stress build up assuming a mean half-rate of spreading of 1 cm/year. Historical accounts indicate that such rifting is episodic, and occurs in northern Iceland every 100-150 years affecting one particular swarm or a part of it at a time.

## Seismicity and Fracture Zones

Einarsson (1979) has described seismicity and earthquake focal mechanisms along the Mid-Atlantic plate boundary between Iceland and the Azores. A more detailed summary of earthquakes in Iceland is given by Einarsson and Bjornsson (1979).

Most of the seismicity of Iceland is related to the Mid-Atlantic plate boundary that crosses the country. The boundary approaches Iceland from the southwest along the crest of the

Reykjanes Ridge (Fig. 5) and from the north along the Kolbeinsey Ridge. In Iceland the plate boundary is displaced to the east by two major fracture zones, the South Iceland seismic zone in the south and the Tjornes Fracture Zone in the north. Because of the lack of a clear topographic expression, both zones are defined primarily by their high seismicity, earthquake focal mechanisms and configuration with respect to the spreading axes. The largest earthquakes in Iceland occur within these zones and may exceed magnitude 7. Earthquakes also occur along the volcanic rift zones between the fracture zones, but they only rarely exceed magnitude 5. A large part of this seismicity appears to be related to central volcanoes.

Epicenters of earthquakes of the period 1962-77 large enough to be located by seismograph networks outside of Iceland are shown in Fig. 5. The map includes events down to magnitude 4, but is complete for magnitude 4.5 and larger events. Single event focal mechanism solutions available so far are also shown. This map shows many of the characteristics of the Icelandic seismicity, even though some of the locations are in error by as much as 40 km. Concentration of activity is seen in the Tjornes Fracture Zone near the coast of N Iceland, and in SW Iceland on the Reykjanes Peninsula and in the South Iceland seismic zone. The focal mechanisms indicate strike-slip faulting. If the easterly striking nodal planes are taken as the fault planes, the sense of motion is right-lateral in N Iceland and left-lateral in SW Iceland, which is consistent with a transform fault interpretation of these zones.

The Tjornes Fracture Zone is a complicated structure made up of both NW-SE trending wrench faults and N-S grabens and volcanic fissure swarms arranged in en échelon pattern (Saemundsson, 1974; Einarsson, 1976; McMaster et al. 1977). Transform motion between the Kolbeinsey Ridge and the northern volcanic zone is taken up by at least two strike-slip fault zones (Fig. 4). Geological evidence points to tens of km right lateral offset on the Husavik faults which are traceable on land for more than 25 km from near the western margin of the axial rift zone to the sea just north of Husavik, juxtaposing rocks of widely contrasting dips and ages. These faults became less active about 1 m.y. ago when the axial rift zone in northern Iceland extended northwards beyond the faults along the Axarfjordur depression (Saemundsson, 1979).

In SW Iceland the Reykjanes Ridge and the eastern volcanic zone are connected by the Reykjanes Peninsula and the South Iceland seismic zone (Fig. 6). On the Reykjanes Peninsula NE-SW trending fissure swarms are arranged along a N 80°E axis so this part of the fracture zone can be viewed as an oblique rift (Klein et al. 1977). A gradual transition from normal faulting in the west to strike slip in the east takes place in the western part of the Peninsula. The connecting segment between the Reykjanes Peninsula and the eastern volcanic zone lies across an eroded pile of Plio-Pleistocene rocks dipping to the northwest. The tectonic pattern indicates sinistral shear which is judged from the great number of N-S and ENE-WSW trending conjugate faults, some of them with lateral offset. Severe earthquakes occur in this zone (Fig. 6) with surface faults arranged en échelon along N-S trending zones indicating right-lateral movement near the surface. The strain in the South Iceland seismic zone is released in sequences of large earthquakes with a recurrence time of 50-100 years. The sequence often starts with a magnitude 7 shock in the eastern part of the zone which is followed by slightly smaller shocks in the western part (Einarsson et al., 1981). Saemundsson (1979) has suggested that this zone might be looked upon as an early stage of opening of an oblique rift axis that eventually will connect the Reykjanes volcanic zone directly to the axial rift zone northeast of the Hekla area.

The stress distribution in the Icelandic crust is still inadequately known. Saemundsson (1978) has summarized the pattern inferred from faults, volcanic axes and focal mechanism solutions (Fig. 7). He suggests that the southern part of the eastern

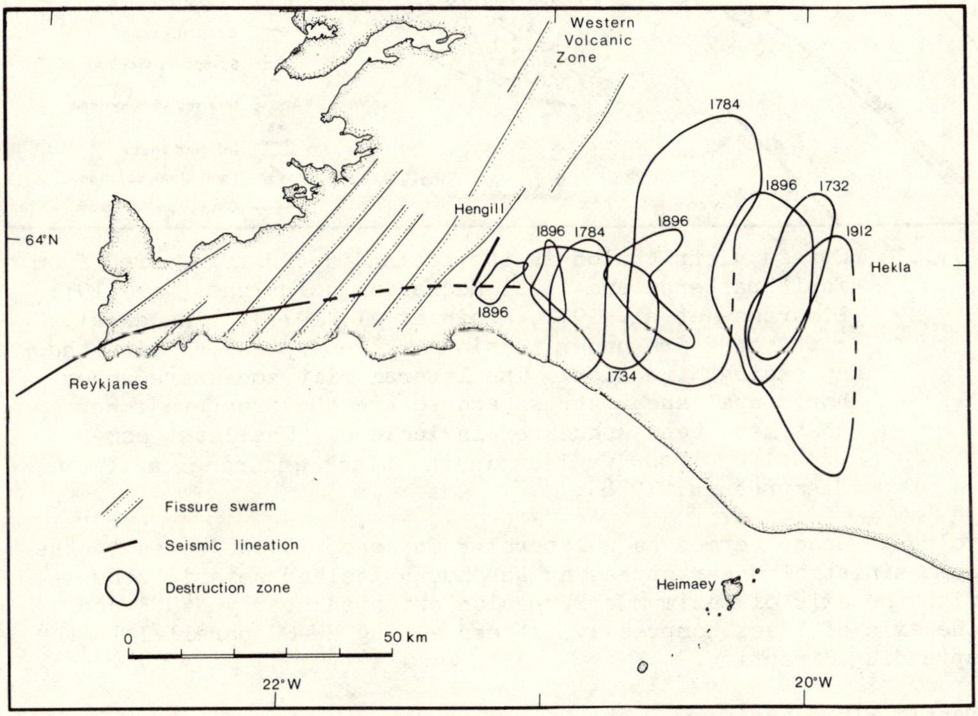


Fig. 6. Map of the seismic zones of SW Iceland, showing some of the tectonic features of the Reykjanes Peninsula and the South Iceland seismic zone. The destruction zones of the historic earthquakes of 1732-34, 1784, 1896 and 1912 are shown. Within these zones more than 50% of houses at each farm were ruined. Corresponding intensity is VIII-IX on the modified Mercalli scale (From Einarsson and Bjornsson, 1979).

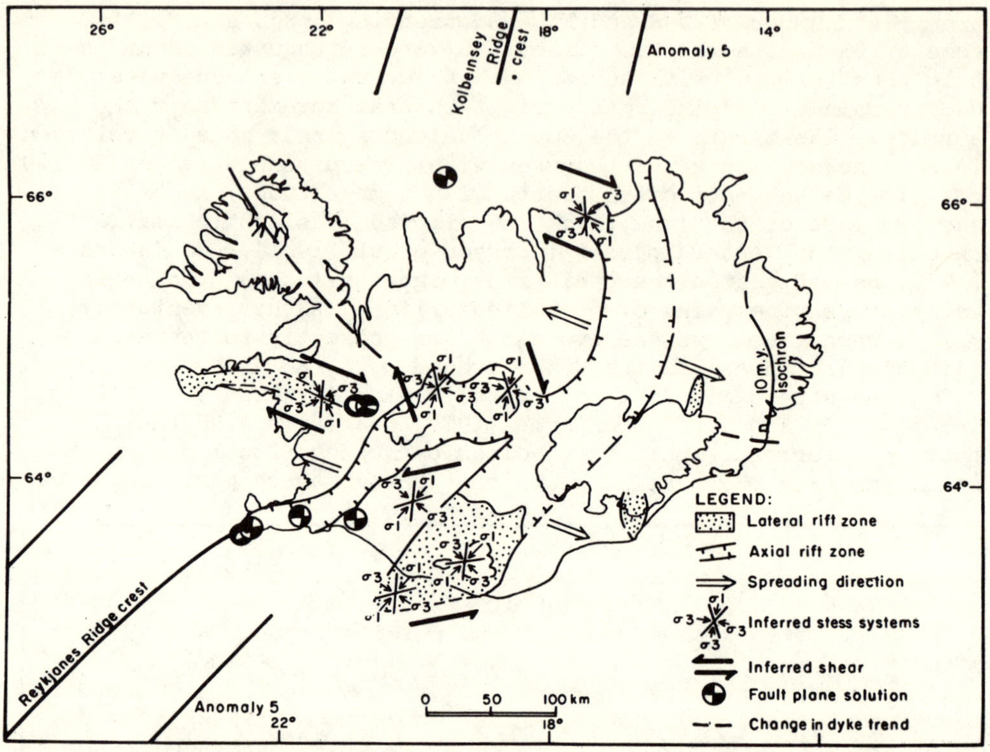


Fig. 7. Stress distribution in the Icelandic crust inferred from fault patterns and focal mechanism solutions (Ward 1971, Einarsson et al. 1977, Klein et al. 1977). The axial rift zones are under tensional stress parallel with the spreading direction. The lateral rift zones are under horizontal shear stress and so are the fracture zones that have been suggested in Iceland. The least compressive stress $\sigma_3$ lies in the black quadrangles (From Saemundsson, 1978).

volcanic zone, termed as a lateral rift zone, is subjected to the same sinistral shear stress as the South Iceland seismic zone with the axis of maximum compression oriented nearly N45°E and the axis of least compressive stress acting NW-SE parallel to the spreading direction.

## Mantle Plume Activity

Several gross features of Icelandic geology seem to find a commom explanation, if one assumes that a mantle plume exists under Iceland (Vogt 1974, Saemundsson 1974, 1978, 1979). These are in particular (1) the elevation of Iceland and the thickness of its oceanic crust, (2) the change in strike of the extensional features across Central Iceland, (3) the decrease in intensity of volcanism along the axial rift zones away from South Central Iceland, (4) region of low seismicity, correlating with the absence of a median rift valley, extending from 57°N through Iceland to the Jan Mayen fracture zone, and (5) the very pronounced offset of the axial rift zones in Iceland relative to the Reykjanes and Kolbeinsey Ridges.

From the trend of linear tectonic features and the eastward offset of the neovolcanic zone relative to the submerged ridges the centre of the plume would be expected to lie under the eastern part of Iceland (Saemundsson, 1974, 1978). The variation of volcanic production and geochemistry of basalts point to a similar location (Jakobsson, 1972; Schilling, 1973; Sigvaldason, 1974).

## KINEMATIC ASPECTS OF CRUSTAL GENERATION IN ICELAND

Palmason (1973, 1980, 1981) has presented a steady-state plate-tectonic kinematic model of crustal accretion in Iceland. The model describes quantitatively the overall time averaged movements of solid crustal elements during the accretion process, and correlates accretion parameters in the axial zone, such as width of lava deposition zone, total lava production rate, width of horizontal strain zone, spreading velocity and normal faulting, with structural properties in the Tertiary lava pile. In the model the discontinuous processes in space and time of lava extrusion, dyke intrusion and tectonic movements within the axial zone are treated as continuous processes representing the average behaviour over a long time. Boundary conditions for the two dimensional flow field of crustal elements are:

(a) At the surface of the crust lava is deposited at a certain average rate which varies with the distance from the axis. For a steady-state process the downward movement of the solid crustal elements must be equal to the deposition rate.

(b) At the axis the horizontal component of the flow must be zero for reasons of symmetry.

(c) The distant lithosphere moves horizontally as a "rigid" body with a constant velocity $V_d$. The movement is assumed to be perpendicular to the axis.

In a simple model which does not account for the effect of normal faulting, the horizontal and vertical velocity components are both assumed independent of depth, but vary with the horizontal coordinate only. The intrusive activity causing horizontal strain across the axial zone of spreading is assumed to have a Gaussian distribution with a standard deviation $\sigma_1$. The deposition rate of extruded lava is also assumed to have a Gaussian distribution with a standard deviation $\sigma_2$. The rate of total lava production q per unit length of the axial zone is integrated across the width of the zone. The four parameters q, $V_d$, $\sigma_1$ and $\sigma_2$ characterize the model. The Tertiary basalt areas on both sides of the axial zone in Iceland provide an opportunity for a comparison with the model. The structural properties which are of particular importance for comparison are (a) the relative dyke volume fraction (b) the regional dip of the lavas and (c) the deposition rate of lavas. Relationships between the structural properties and model parameters were used to place constraints on the parameters on the basis of field observations. Radiometric age data from the Icelandic basalt pile show a remarkably constant rate of deposition of lavas (McDougall et al. 1977, Watkins and Walker 1977). However it was found impossible to infer from measurements of deposition rate and dip about past variations in q. Measurements of dip give primarily information on $\sigma_2$, and measurements of deposition rate information on $\sigma_2/V_d$. Values of 14-25 km for $\sigma_2$ predicted by the model are in rather good agreement with the situation in the present day zone of rifting and volcanism of Iceland, where postglacial lavas are distributed over a zone of some 50-100 km width. These results may indicate a rather uniform intensity of volcanism in Iceland

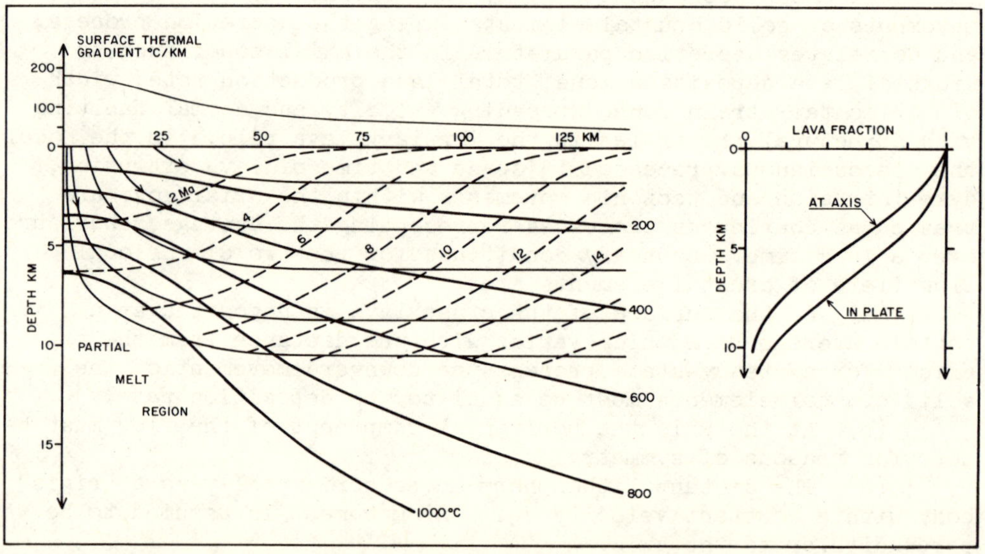


Fig. 8. Section through the lithosphere according to the accretion model of Palmason (1973, 1980). Trajectories (arrows) and isochrons (dashed) of lava elements are shown together with isotherms and surface gradient. The lava fraction at the axis and in the distant lithosphere is shown on the right. The model parameters used are (for definitions see Palmason, 1980): spreading velocity 1 cm/a; lava production rate $4/3 \times 10^{-4} km^2/a$; crustal strain rate, standard deviation 15 km; lava deposition rate, standard deviation 20 km; normal fault parameters, $\varepsilon = 0.75$ and $\eta_o = 1$ (From Palmason, 1981).

for the past 10-15 m.y. The spreading velocity $V_d$ can be estimated with the aid of the model, where measurements of dip and deposition rate have been made at the same level in the Tertiary lava pile. Palmason (1980) obtained a value of 0.84 cm/a for a section in western Iceland and 0.91 cm/a for eastern Iceland.

Dyke observations from eastern Iceland suggest a ratio $\sigma_1/\sigma_2 = 0.69$, if the simple model is applied which assumes that the intrusive activity is independent of depth. Palmason (1980) modified the model to account for the effect of normal faults. The faulting is treated as a continuous process of horizontal strain and an associated subsidence. The horizontal strain rate due to normal faulting is highest at the surface and decreases with increasing depth where dykes and other intrusions take up a larger fraction of the horizontal strain. Both components, faulting and intrusive activity are assumed to be Gaussian distributed across the axial zone. According to this version of the model dyke observations cannot constrain the ratio $\sigma_1/\sigma_2$ more than within the range 0.69-1.0. This ratio and the relative importance of dyke injection and normal faulting in the extensional processes of the axial zone have a pronounced effect on the dyke fraction in the upper part of the crust. The same parameters control the transition from an upper lava-dominated layer to a lower intrusion-dominated layer. It is possible to produce a sharp boundary between lavas and intrusions by a wide extensional zone, where normal faults take up the horizontal strain instead of dykes. In the case of Iceland, however, likely model parameters would give a gradual transition from the lava layer to the intrusion layer.

An example of a section through the lithosphere according to the accretion model is given in Fig. 8. One interesting feature indicated by the trajectories of lava elements is that the eastern Iceland visible lava pile, some 1300 m thick, was formed outside a 50-55 km-wide central part of the Tertiary volcanic zone. Evidence of volcanic and tectonic processes in the central part of the active zone disappears gradually into the deeper part of the distant crust, where it cannot be observed directly. The conspicuous difference in appearance of the two main volcanic regions of Iceland is thus easily understood in terms of the model, and no major change in the behaviour of the volcanic processes in Iceland is required to account for this difference.

The model permits calculation of the thermal state of the axial lithosphere. Isotherms are calculated with the aid of the heat transport equation using a set of parameters deduced from field observations, and using the Parker and Oldenburg (1973) theory as a boundary condition at large distances from the axis. The depth to the 1000°C isotherm at the axis is about 6 km. The model accounts for the cooling effect of extrusive volcanism but it does not incorporate the heat transported to surface by hydrothermal circulation processes in the axial zone. Bodvarsson (1981) discussed terrestrial energy currents through the surface of Iceland and concluded that heat advection by thermal water was equal or even exceeded the time averaged heat advection by surface lava. Deep penetration of water might affect crustal temperatures in rift zones and depress the isotherms to lower levels than indicated by Palmason´s model.

## SEISMIC STRUCTURE OF THE ICELANDIC CRUST

### Upper and Lower Crust

Extensive surveys of crustal structure (Båth, 1960; Tryggvason and Båth, 1961; Palmason 1963; 1971) where interpreted in terms of layers with constant velocity. In this interpretation the Icelandic crust consists of four layers of basaltic material overlying an anomalous mantle with P-velocity 7.2 km/s (Table 1). The crustal thickness is variable between 8 and 14 km. Layer 3 ($V_p$ = 6.5 km/s), often equated to the oceanic layer three, is found at a depth 1-5 km under most of the island, except at the south-eastern coast where the depth increases eastwards to some 10 km. The thickness of the layer is usually 4 to 5 km. The seismic refraction data of Palmason have been reinterpreted by Flovenz (1980) with the aid of synthetic seismograms. The layered model does not satisfy the observed amplitude variation. On the other hand, models assuming continuously increasing velocity with depth fit better with the observed energy distribution in the seismograms. In the model chosen by Flovenz (1980) the P-velocity increases rapidly with depth in the velocity interval 2.0-3.5 km/s followed by an approximately constant gradient of 0.57 $s^{-1}$ down to the 6.5 km/s isovelocity surface, below which the P-velocity becomes nearly constant. In view of this Flovenz (1980) proposed to divide the Icelandic crust into upper crust, where velocities increase continuously with depth, and lower crust, with almost constant velocities ($V_p$ = 6.5 km/s) corresponding to layer 3 of Palmason´s model. The depth to the lower crust is 3-6 km (Fig. 9), except under central volcanoes where the layer may protrude to

Table 1. Layered Seismic Structure of the Iceland Crust after Palmason (1963, 1971) (From Flovenz, 1980).

| Layer No. | P-velocity km/s | S-velocity km/s | Poisson´s ratio | Density g/cm3 |
|---|---|---|---|---|
| 0 | 2.75 | | | 2.1-2.5 |
| 1 | 4.14 | 2.34 | 0.270 | 2.6 |
| 2 | 5.08 | 2.78 | 0.278 | 2.65 |
| 3 | 6.50 | 3.53 | 0.269 | 2.9 |
| 4(Mantle) | 7.20 | | | 3.1 |

levels as shallow as 1 km. The composition of the lower crust is not known. The $V_p/V_s$ ratio is 1.8. Extrapolated temperature gradients (Palmason et al., 1979) suggest that the temperature at the top of the lower crust is 500°C or lower, except in the neovolcanic zones where higher temperatures are considered likely.

Combined reflection-refraction seismic measurements were carried out on profiles crossing the axial rift zones in SW and NE Iceland by Zverev et al. (1980 a,b). In the flood basalts the near surface seismic refractors correspond closely to geological layers and are tilted towards the rift zone (Fig. 10). Within the axial rift zone the refracting boundaries form a slight depression filled with low-velocity formations. Reflecting horizons with an average length of 1-2 km were identified down to depths of more than 15 km. They are generally tilted towards the axial rift zone. The dip is steep, up to 30° at depths of 3 to 7 km. At greater depths the dips decrease to 20° or less. The nature of these deep reflectors is uncertain. One possibility is intrusive sheets as suggested by Walker (1975). In both cross sections volumes of limited lateral extent were identified where no reflectors occurred. These bodies are located under the rift zones. The absence of

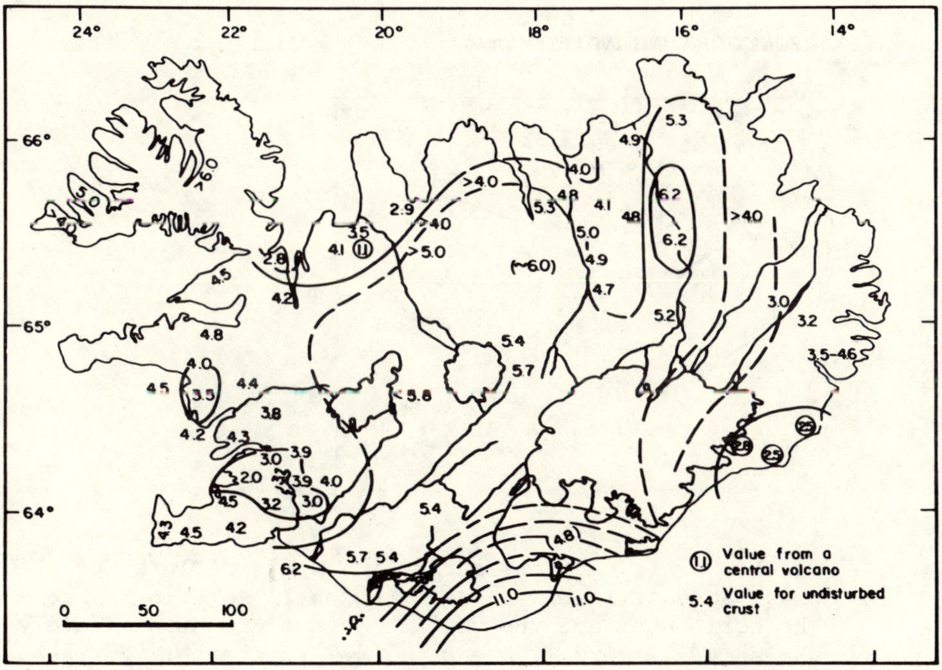


Fig. 9 Depth to the lower crust of Iceland (layer 3, $V_p$ = 6.5 km/s). The velocity in the upper crust is assumed to increase continuously with depth (From Flovenz, 1980).

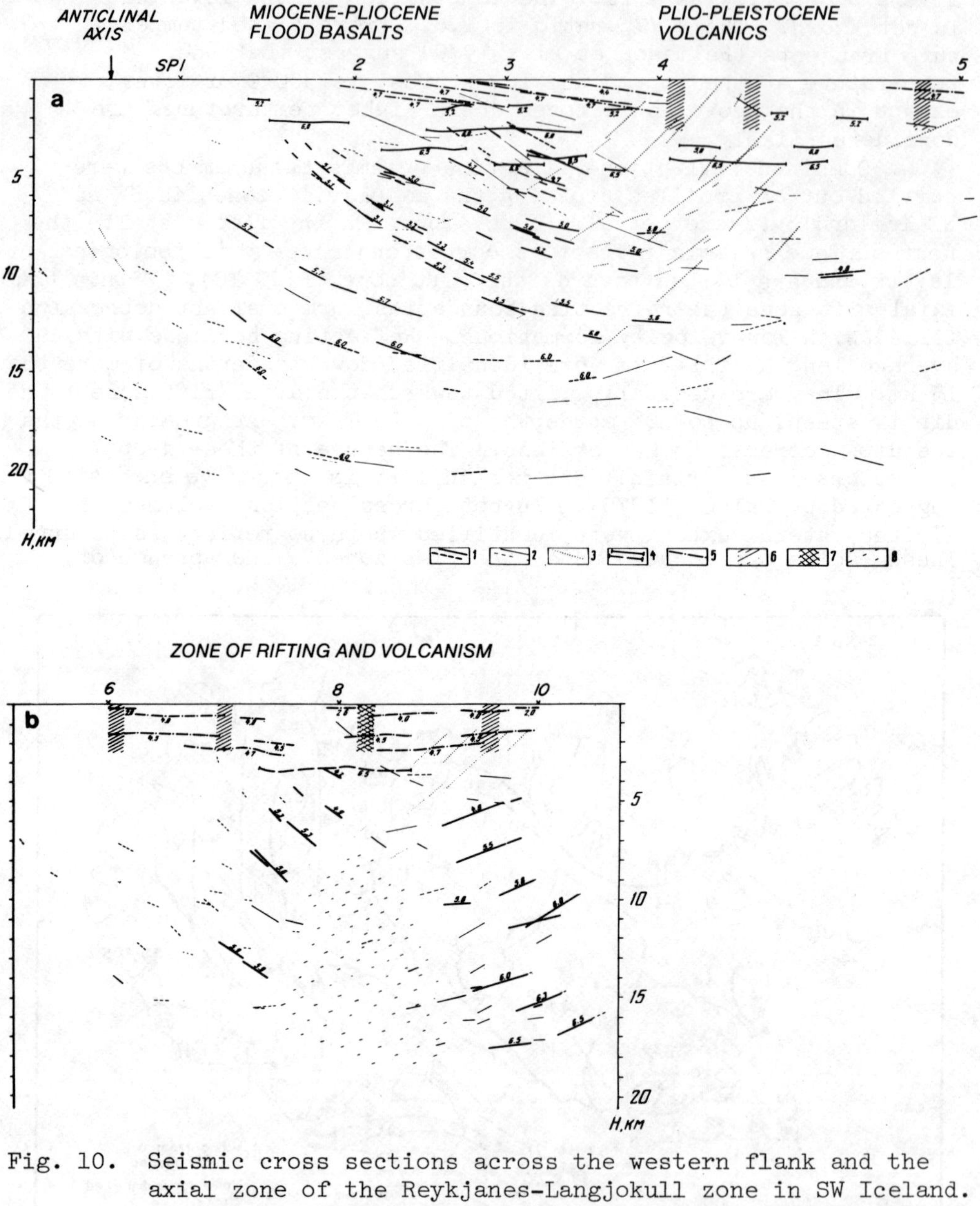


Fig. 10. Seismic cross sections across the western flank and the axial zone of the Reykjanes-Langjokull zone in SW Iceland. The horizontal and the vertical scale are approximately the same. Legend 1, 2 and 3: reflectors, 4: refractors 5: refractors probably connected with gelogical horizons, 6: inactive fracture zones, 7: active fracture zone, 8: body with homogeneous seismic property probably a region of melting temperature (From Zverev et al. 1980a).

reflectors suggests homogeneity in physical properties below 8 km depth. This region may be in the state of partial melting or a magma chamber.

## Crust-Mantle Boundary

The Icelandic crust resembles the oceanic crust in velocity values but the thickness of individual layers is greater. The base of layer 3 has been reached by only few profiles. Palmason (1971) attributed velocities from 7.2 to 7.4 km/s to an anomalous mantle. Combining the seismic data with temperature gradients in boreholes Palmason concluded that the temperature near the crust-mantle boundary is close to the melting range of basalts in SW and W Iceland, where the boundary lies near 8 km depth but somewhat below the melting range in N and SE Iceland, where the crust-mantle boundary is inferred at 14 km depth. Another important conclusion from the seismic data is that the source of the bowl-shaped Bouguer anomaly over Iceland discovered by Tr. Einarsson (1954) is not found in the crust.

Bott and Gunnarsson (1980) have interpreted the results of the North Atlantic Seismic Project from 1972 and earlier short refraction lines and reflection, gravity and magnetic surveys on the Iceland-Faeroe Ridge. The main 5.7 km/s upper crustal layer is locally overlain by lower velocity layers of variable thickness. A 6.7 km/s lower crustal layer underlies the upper crust at a depth of between about 4 and 8 km along the Ridge. This layer is present also beneath the Icelandic shelf but not beneath the Faeroe shelf. A deeper 7.8 km/s refractor interpreted as the Moho occurs at about 30-35 km depth beneath the central and south-eastern part of the Ridge, shallowing by a few kilometres approaching the Icelandic shelf. The Iceland-Faeroe Ridge appears to be underlain by crust of Icelandic type. The Ridge crust is two to three times thicker than the Icelandic crust and about five times thicker than normal oceanic crust.

Combining gravity data from Iceland and the seismic refraction data from NASP 1972, Zverev et al. (1976) concluded that Iceland may be underlain by a very thick crust of continental affinity and normal upper mantle with P-wave velocities of 8 km/s at about 50 km depth. This interpretation is in contrast to seismological observations of teleseismic traveltime residuals (Tryggvason, 1964; Long and Mitchell, 1970) as well as apparent velocities from Mid-Atlantic Ridge earthquakes across Iceland (Francis, 1969). These observations suggested that low velocities might extend down to some 250 km. S-wave travel-time residuals (Francis, 1969, Girardin and Poupinet, 1974) indicate an even more pronounced S- than P-wave velocity anomaly in the upper mantle beneath Iceland. Bott (1965, 1974) suggested that the low velocity mantle and the associated mass deficiency indicated the presence of low density basic magma produced by partial melting within a rising mantle convection current beneath Iceland.

## Results of the Reykjanes Ridge Iceland Seismic Project 1977

A long-range seismic refraction experiment (RRISP) was conducted in 1977 along an 800 km long line across Iceland and along the southeastern flank of the Reykjanes Ridge. The aim of the experiment was to resolve the structure of the crust and upper mantle to greater depth than previously possible and to study the transition from the oceanic to the Icelandic structure (Angenheister et al., 1979; 1980; Gebrande et al., 1980; Goldflam et al., 1980). A generalized crustal and upper mantle cross section obtained on the central part of the main profile is shown in Fig. 11. A more or less normal oceanic crust was found south of Iceland along the 10 m.y. isochron on the eastern flank of the Reykjanes Ridge. In the subcrustal lithosphere P-wave velocities as high as 8.6 km/s were found at depths greater than 20 km. A lithospheric thickness of some 50 km was, somewhat arbitrarily, estimated from the disappearance of the seismic signals at long range. The crustal layers continue into Iceland with increased thickness, whereas the subcrustal oceanic lithosphere seems to be abruptly terminated near the insular shelf margin. The subcrustal velocities under Iceland range between 7.0 and 7.6 km/s down to about 50 km, and no evidence for sharp velocity discontinuities could be found. The transition from the high-velocity oceanic lithosphere to the low-velocity upper mantle beneath Iceland is

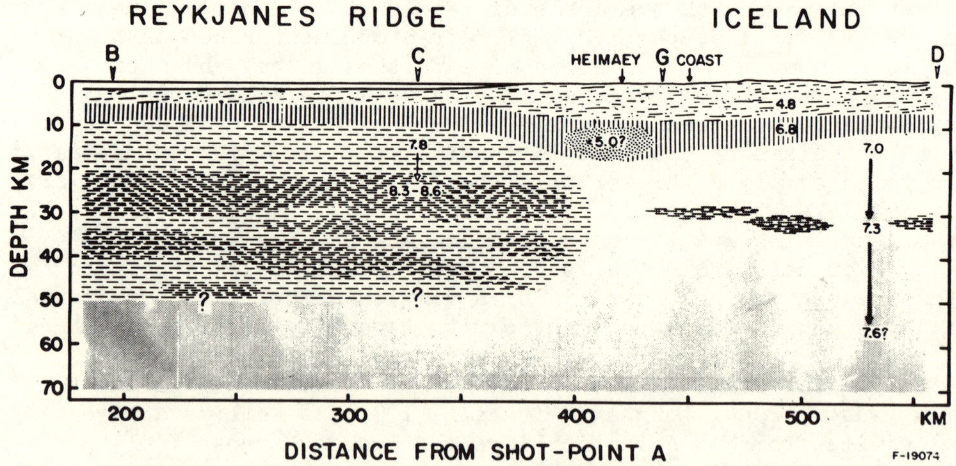


Fig. 11. Generalized crustal and upper-mantle cross section of the central part of RRISP 77 main profile. The profile strikes along 10 m.y. old oceanic crust on the south-eastern flank of the Reykjanes Ridge and along the eastern axial rift zone in Iceland. Letters indicate positions of large shots. Numbers give P-wave velocities in km/s (From Angenheister et al., 1980).

rather abrupt and must occur close to the shelf slope. Other profiles of the experiment across eastern Iceland and a profile of Båth (1960) in the west indicate that the anomalous mantle is not confined to the neovolcanic zone, but extends under all of Iceland. The crustal thickness in Iceland varies between 10 and 15 km and no clear relationship with the axial rift zones was found.

## Propagation of S-waves

Shots at the southern and northeastern coast of Iceland generate considerable S-wave energy. Gebrande et al. (1980) found that S-waves were transmitted through the crust out to a distance of 140 km with a velocity corresponding to a P- to S-velocity ratio of 1.76. Beyond this distance the S-waves became progressively late with respect to the P-waves indicating a velocity ratio approaching a value as high as 2. At the same range the S-waves become attenuated with respect to the P-waves and no S-waves were recorded beyond 250 km. Raytracing calculations show that emergent rays beyond distances of 140 km have penetrated layer 4 with P-velocity 7.0 km/s. Gebrande et al. (1980) therefore concluded that at the base of layer 3 a more fundamental change in physical properties takes place than is indicated by the almost negligible change in P-wave velocity. The disappearance of S-waves was not found to be associated with the axial rift zones in particular, but seemed to be a general feature of the anomalous mantle under Iceland.

Einarsson has analysed high frequency S-waves from earthquakes in Iceland observed at vertical short-period seismograph stations of the Icelandic network (Sanford and Einarsson, 1981). Examples of S-wave rays penetrating under the fissure swarms of the axial rift zone in NE-Iceland are shown in Fig. 12. In most of these cases the S-wave amplitude was larger than that of the P-wave. The numbers shown indicate the depth of penetration of the rays assuming a surface focus of the earthquake. It is therefore a minimum depth. The high frequency S-waves travel through the axial rift zone down to the depth of at least 10 km. If a general continuous zone of magma underlies the axial rift zone, the depth to its top is at least 10 km in NE Iceland. This does not preclude the existence of small magma bodies at higher levels. Zones of high S-wave attenuation have been identified at 3 to 7 km depth under the Krafla volcano (Einarsson, 1978), and gaps in the seismic zone on the Reykjanes Peninsula were interpreted as possible indicators of small magma bodies (Klein et al. 1977). On the Peninsula earthquakes occur from the surface down to 9 km depth, indicating a minimum thickness of the brittle crust near the plate boundary. In the northern volcanic zone earthquakes have also occurred down to similar depth (P. Einarsson, pers. comm.).

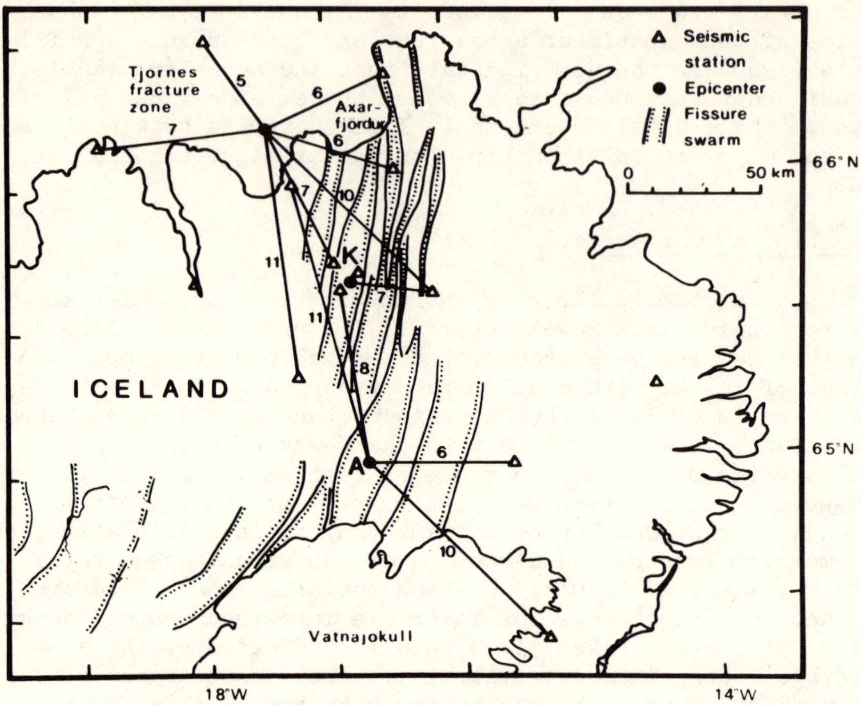


Fig. 12. S-wave propagation across the axial rift zone in NE-Iceland. S-wave paths are shown with lines connecting epicenters and seismograph stations. Numbers indicate the depth of penetration of the respective ray (From Sanford and Einarsson, 1981).

## UPPER MANTLE BENEATH ICELAND

### Electrical Resistivity of the Lower Crust and Mantle

Beblo and Bjornsson (1978, 1980) carried out short period magnetotelluric measurements at 19 sites distributed over the northern volcanic zone and the adjoining older Tertiary flood basalt areas. A two-dimensional model of resistivity distribution beneath an east-west profile across NE-Iceland is shown in Fig. 13. Beneath a 2.5 km surface layer the resistivity is 100 Ω m, except within the active volcanic zone where it is 50 Ω m. Under the whole area a layer of low resistivity is found imbedded between layers of higher resistivity. The resistivity in the layer is about 15 Ω m. The depth to the top of this layer is about 10 km beneath the axial rift zone and the Quaternary areas but increases to about 17 km below the Tertiary basalts east and west. The thickness of the layer is about 5 km on the eastern and western side but reaches about 10 km beneath the Quaternary and neovolcanic rocks. The resistivity below the low-resistivity layers is about 100 Ω m down to at least 100 km. These results are in a general way consistent with earlier magnetotelluric results of Hermance and Grillot (1970, 1974), Hermance (1973) and Hermance et al. (1976) in other parts of Iceland. Thayer et al. (1981) have interpreted magnetotelluric measurements made across the northern volcanic zone in 1973 and confirmed the results obtained by Beblo and Björnsson.

Temperature and Degree of Partial Melting

Comparing their results with temperature-resistivity data of basaltic and ultramafic rocks, Beblo and Bjornsson concluded that the low-resistivity layer most likely consists of partially molten basalt at a temperature of 1000°C-1100°C. This conclusion is supported by observed temperature gradients of 40-60°C/km in the Tertiary rocks east and west and around 100°C/km in the active volcanic zone (Palmason and Saemundsson, 1974; Palmason et al., 1979). The substratum, with a higher resistivity of 70-100 Ω m, was interpreted as the uppermost mantle, consisting of ultramafic rocks. The corresponding temperature range is 950°C-1200°C with

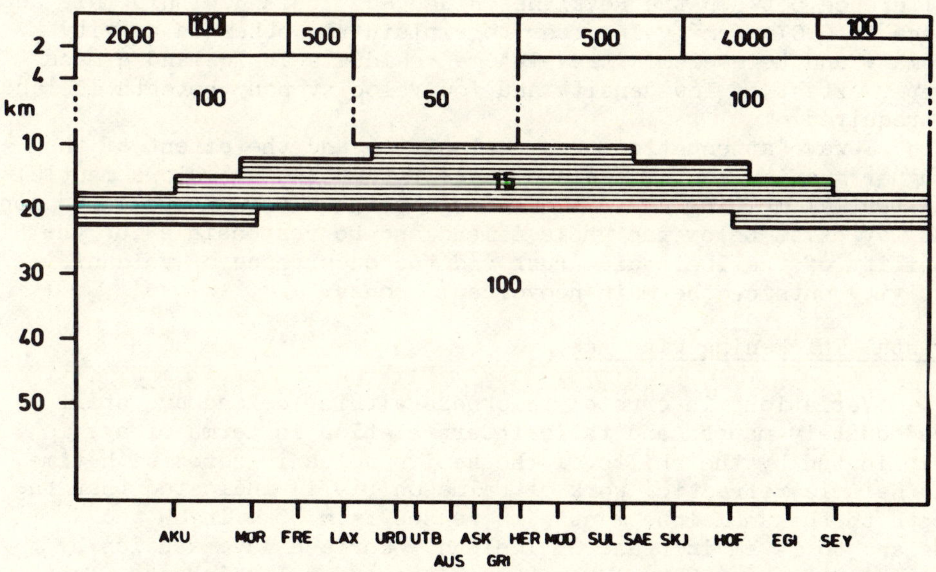


Fig. 13. Two-dimensional model of resistivity distribution beneath an east-west profile across the volcanic zone and adjacent flanks in NE Iceland. The length of the profile is about 200 km, from Tertiary rocks near Akureyri in N Iceland to Tertiary rocks near Seydisfjordur in E Iceland. The numbers indicate assumed resistivities in Ω m. (From Beblo and Bjornsson, 1980).

a most likely value of about 1050°C in agreement with Hermance and Grillot (1974). The nearly constant resistivity down to 100 km depth indicates a very low temperature gradient for the mantle, as already shown by Hermance and Grillot (1974) who inferred a temperature gradient between 0 and 2°C/km in the mantle of SW Iceland.

In this interpretation of the resistivity data the low-resistivity layer of partially molten basalt is supposed to result from upward movement of lighter basaltic melt through the heavier olivine matrix. The ascent causes separation of ultramafic and basaltic material, from which the oceanic crust is generated. This interpretation is in good agreement with theoretical predictions by Bottinga and Allégre (1976).

Gebrande et al. (1980) combined their results on P-wave velocity distribution and P- to S-wave velocity ratio with the temperature estimates of Beblo and Bjornsson (1980) to calculate the amount of partial melt in the anomalous mantle, assuming that it consists of a mixture of the material of the high-velocity layer observed beneath the flank of the Reykjanes Ridge and basaltic melt. The melt content was estimated to be highest (17-23%) at the top of the mantle and to decrease with increasing depth indicating a differentiation process in the upper mantle. Taking a melt content of 13% as the average for the depths from 20 to 50 km the mean density difference between the Reykjanes Ridge and Iceland within this depth range is 0.07 g/cm$^3$. In order to explain the observed gravity anomaly and teleseismic travel-time residuals in Iceland a much deeper extent of low density and low velocity body beneath Iceland is required.

S-wave attenuation, low resistivity and the extent of the Bouguer gravity low of Iceland suggest that the anomalous mantle and partial melting are not confined to the narrow neovolcanic zone but may exist below the whole island and be responsible for the high mobility of the Icelandic crust and the occurrence of volcanic activity outside the main neovolcanic zones.

## Crustal Thickening with Age

Variations in crustal thickness within Iceland are still inadequately mapped and their interpretation in terms of age is complicated by the shifts of the active volcanic zones with time. The seismic refraction work of Palmason (1971) indicated that the depth to the anomalous mantle increases from 8 km in SW and W Iceland to 14 km in N and SE Iceland. Surface wave studies by Evans and Sacks (1979) suggest that the crustal thickness of the Iceland Plateau lying between the Kolbeinsey Ridge and the Jan Mayen Ridge north of Iceland is much greater than that of typical oceanic floors, perhaps exceeding 20 km. Near the axis of the Kolbeinsey Ridge Evans and Sacks (1979) estimated the crustal thickness to be 12-15 km and concluded that the lithosphere structure was similar to that found in studies of structure near the axis of the Reykjanes Ridge (Evans and Sacks, 1980; Keen et al. 1980).

Hermance (1981) has suggested that this evidence indicates crustal thickening with age for Iceland, from a thickness of 8-10 km beneath the modern neovolcanic zone to an average thickness of 20 km for the Icelandic Plateau having an age less than 20 m.y. Hermance (1981) proposed that in addition to creating a 10 km thick crust in the neovolcanic zone through a combination of intrusive and extrusive activity as suggested by the crustal accretion model of Palmason (1973, 1980) there also occurs accretion of new material to the base of the crust beneath regions which at the surface do not exhibit evidence of contemporary volcanism. Lachenbruch and Sass (1978) have proposed an analogous process of crustal underplating associated with extensional processes in the Basin and Range province in the Western United States. The melt accumulation inferred at the base of the crust in Iceland provides support for this model. Jacoby and Girardin (1980) studied Rayleigh wave group velocities measured along paths parallel to the Reykjanes Ridge axis, traversing age slices of lithosphere between 0 and 20 m.y. The data suggested a low velocity zone between 20 and 30 km depth. It was not yet clearly evident at 0-3 m.y. age, distinct at 6-8 m.y. age and appeared to vanish again for older age.

## ACKNOWLEDGEMENTS

The author is obliged to a number of his colleagues for their aid while writing this review, especially Axel Bjornsson, Pall Einarsson, Olafur Flovenz, Leo Kristjansson, Gudmundur Palmason and Kristjan Saemundsson. Elisabet Gudjohnsen and Edda Rikhards-dottir typed the manuscript.

## REFERENCES

Angenheister, G., H. Gebrande, H. Miller, W. Weigel, P. Goldflam, W. Jacoby, G. Palmason, S. Bjornsson, P. Einarsson, S. Zverev, B. Loncarevic, S. Solomon, 1979: First results from the Reykjanes Ridge Iceland Seismic Project 1977. Nature 279, 56-60.

Angenheister G., et al., 1980: Reykjanes Ridge Iceland Seismic Experiment (RRISP 77). J. Geophys. 47, 228-238.

Båth, M., 1960: Crustal structure of Iceland. J. Geophys. Res. 65, 1793-1807.

Beblo, M., and A. Bjornsson, 1978: Magnetotelluric investigation of the lower crust and upper mantle beneath Iceland. J. Geophys. 45, 1-16.

Beblo, M., and A. Bjornsson, 1980: A model of electrical resistivity beneath NE Iceland, correlation with temperature. J. Geophys. 47, 184-190.

Bjornsson, A., K. Saemundsson, P. Einarsson, E. Tryggvason, and K. Gronvold, 1977: Current rifting episode in North Iceland. Nature 266, 318-323.

Bjornsson, A., G. Johnsen, S. Sigurdsson, G. Thorbergsson, and E. Tryggvason, 1979: Rifting of the plate boundary in North Iceland 1975-1978. J. Geophys. Res. 84, 3029-3038.

Bodvarsson, G., 1981. Terrestrial energy currents and transfer in Iceland. Amer. Geophys. Un., Geodynamics Series, in press.

Bott, M.H.P., 1965: The upper mantle beneath Iceland. Geophys. J.R. astron. Soc. 9, 275-277.

Bott, M.H.P., 1974: Deep structure, evolution and origin of the Icelandic transverse ridge. In: Geodynamics of Iceland and the North Atlantic Area, L. Kristjansson, ed.: 33-47, Reidel, Dordrecht.

Bott, M.H.P., and K. Gunnarsson, 1980: Crustal structure of the Iceland-Faeroe Ridge. J. Geophys. 47, 221-227.

Bottinga, Y., and C. Allégre, 1976: Geophysical, petrological and geochemical models of the oceanic lithosphere. Tectonophysics 32, 9-59.

Brandsdottir B., and P. Einarsson, 1979: Seismic activity associated with the September 1977 deflation of the Krafla central volcano in NE Iceland. J. Volcanol. Geothermal Res. 6, 197-212.

Dagley, P., R.L. Wilson, J.M. Ade-Hall, G.P.L. Walker, S.E. Haggerty, T. Sigurgeirsson, N.D. Watkins, P.J. Smith, J. Edwards, and R.L. Grasty, 1967: Geomagnetic polarity zones for Icelandic lavas. Nature 216, 25-29.

Einarsson P., 1976: Relative location of earthquakes within the Tjornes Fracture Zone. Soc. Sci. Islandica Greinar 5, 45-60.

Einarsson P., 1978: S-wave shadows in the Krafla caldera in NE Iceland, evidence for a magma chamber in the crust. Bull. Volcanol. 41, 1-9.

Einarsson, P., 1979: Seismicity and earthquake focal mechanisms along the Mid-Atlantic plate boundary between Iceland and the Azores. Tectonophysics 55, 127-153.

Einarsson, P., F.W. Klein, and S. Bjornsson, 1977: The Borgarfjordur earthquakes in West Iceland 1974. Bull. Seismol. Soc. Am. 67, 187-208.

Einarsson, P., and S. Bjornsson, 1979: Earthquakes in Iceland. Jokull 29, 37-43.

Einarsson, P., and B. Brandsdottir, 1980: Seismological evidence for lateral magma intrusion during the July 1978 deflation of the Krafla volcano in NE Iceland. J. Geophys. 47, 160-165.

Einarsson, P., S. Bjornsson, G. Foulger, R. Stefansson, and Th. Skaftadottir, 1981: Seismicity pattern in the South Iceland Seismic Zone. Amer. Geophys. Un., Maurice Ewing Series, in press.

Einarsson, Th., D.M. Hopkins, and R.R. Doell, 1967: The stratigraphy of Tjornes, northern Iceland and the history of the Bering Land Bridge. In: The Bering Land Bridge, D.M. Hopkins, ed.: 312-325, Stanford Univ. Press.

Einarsson, Tr., 1954: A survey of gravity in Iceland. Soc. Sci. Isl. Rit 30, 22 pp.

Einarsson, Tr., 1957: Magneto-geological mapping in Iceland with the use of a compass. Philos. Mag. (Suppl.) 6, 232-239.

Evans, J.R., and I.S. Sacks, 1979: Deep structure of the Iceland Plateau. J. Geophys. Res. 84, 6859-6866.

Evans, J.R., and I.S. Sacks, 1980: Lithospheric structure in the North Atlantic from observations of Love and Rayleigh waves. J. Geophys. Res. 85, 7175-7182.

Francis, T.J.G., 1969: Upper mantle structure along the axis of the Mid-Atlantic Ridge near Iceland. Geophys. J.R. astron. Soc. 17, 507-520.

Flovenz, O.G., 1980: Seismic structure of the Icelandic crust above layer three and the relation between body wave velocity and the alteration of the basaltic crust. J. Geophys. 47, 211-220.

Gebrande, H., H. Miller, and P. Einarsson, 1980: Seismic structure of Iceland along RRISP-profile I. J. Geophys. 47, 239-249.

Girardin, N., and G. Poupinet, 1974: Teleseismic S travel time delay for Mid-Atlantic Ridge earthquakes. Phys. Earth Planet. Inter. 9, 306-313.

Goldflam, P., W. Weigel, and B.D. Loncarevic, 1980: Seismic structure along RRISP-Profile I on the southeast flank of the Reykjanes Ridge. J. Geophys. 47, 250-260.

Hermance, J.F., 1973: An electrical model for the sub-Icelandic crust. Geophysics 38, 3-13.

Hermance, J.F. 1981: Crustal genesis in Iceland: Geophysical constraints on crustal thickening with age. Geophys. Res. Lett. 8, 203-206.

Hermance, J.F., and L.R. Grillot, 1970: Correlation of magnetotelluric, seismic, and temperature data from Soutwest Iceland. J. Geophys. Res. 75, 6582-6591.

Hermance, J.F., and L.R. Grillot, 1974: Constraints on temperature beneath Iceland from magnetotelluric data. Phys. Earth Planet. Inter. 8, 1-12.

Hermance, J.F., R.E. Thayer, and A. Bjornsson, 1976: The telluric-magnetotelluric method in the regional assessment of geothermal potential. In: Proc. Second U.N. Symposium on the Development and Use of Geothermal Resources, 2, 1037-1048. U.S. Gov. Printing Office, Washington D.C. 20402.

Jakobsson, S.P., 1972: Chemistry and distribution pattern of recent basaltic rocks in Iceland. Lithos 5, 365-386.

Jacoby, W.R., and N. Girardin, 1980: The evolution of the lithosphere at the southeast flank of Reykjanes Ridge from surface wave data. J. Geophys. 47, 271-277.

Jacoby, W.R., A. Bjornsson, and D. Moller, eds, 1980: Iceland: Evolution, Active Tectonics, and Structure. J. Geophys. 47, No. 1-3, 277 pp.

Johannesson, H., 1980: Jardlagaskipun og throun rekbelta a Vesturlandi. (Evolution of rift zones in Western Iceland, in Icelandic with English Summary). Natturufraedingurinn 50, 13-31.

Johnsen, G.V., A. Bjornsson, and S. Sigurdsson, 1980: Gravity and elevation changes caused by magma movement beneath the Krafla caldera, Northeast Iceland, J. Geophys. 47, 132-140.

Keen, C.E., A. Fricker, M.J. Keen, and L. Blinn, 1980: Reykjanes Ridge crest studied by surface waves with an earthquake pair technique. J. Geophys. 47, 265-270.

Klein, F.W., P. Einarsson, and M. Wyss, 1977: The Reykjanes Peninsula, Iceland, earthquake swarm of September 1972 and its tectonic significance. J. Geophys. Res. 82, 865-888.

Kristjansson, L., ed., 1974: Geodynamics of Iceland and the North Atlantic Area, 323 pp., Reidel, Dordrecht.

Kristjansson, L., 1979: The shelf area around Iceland. Jokull 29, 3-6.

Kristjansson, L., I.B. Fridleifsson, and N.D. Watkins, 1980: Stratigraphy and paleomagnetism of the Esja, Eyrarfjall and Akrafjall mountains, SW Iceland. J. Geophys. 47, 31-42.

Lachenbruch, A.H., and J.H. Sass, 1978: Models of an extending lithosphere and heat flow in the Basin and Range Province. Geol. Soc. Am. Memoir 152, eds: R.B. Smith and G.P. Eaton, 209-250.

Long, R.E., and M.G. Mitchell, 1970: Teleseismic P-wave delay time in Iceland. Geophys. J.R. astron. Soc. 20, 41-48.

McDougall, I., H. Wensink, 1966: Paleomagnetism and geochronology of the Plio-Pleistocene lavas in Iceland. Earth Planet. Sci. Lett. 1, 232-236.

McDougall, I., N.D. Watkins and L. Kristjansson, 1976: Geochronology and paleomagnetism of a Miocene-Pliocene lava sequence at Bessastadaá, Eastern Iceland. Am. J. Sci. 276, 1078-1095.

McDougall, I., K. Saemundsson, H. Johannesson, N.D. Watkins, and L. Kristjansson, 1977: Extension of the geomagnetic polarity time scale to 6.5 m.y.: K-Ar dating geological and paleomagnetic study of a 3,500 m lava succession in Western Iceland. Geol. Soc. Amer. Bull. 88, 1-15.

McMaster, R.L., J.-G. Schilling, and P.R. Pinet, 1977: Plate boundary within Tjornes Fracture Zone on Northern Iceland's insular margin. Nature 269, 663-668.

Moller, D., and B. Ritter, 1980: Geodetic measurements and horizontal crustal movements in the rift zone of NE Iceland. J. Geophys. 47, 110-119.

Musselt, A.E., J.G. Ross, and I.L. Gibson, 1980: $^{40}Ar/^{39}Ar$ dates of Eastern Iceland lavas. Geophys. J.R. astr. Soc. 60, 37-52.

Palmason, G., 1963: Seismic refraction investigation of the basalt lavas in Northern and Eastern Iceland. Jokull 13, 40-60.

Palmason, G., 1971: Crustal structure of Iceland from explosion seismology. Soc. Sci. Isl. Rit 40, 187 pp.

Palmason, G., 1973: Kinematics and heat flow in a volcanic rift zone, with application to Iceland. Geophys. J.R. astron.Soc. 33, 451-481.

Palmason, G., 1981: Crustal rifting and related thermo-mechanical processes in the lithosphere beneath Iceland. Geologische Rundschau 70, 244-260.

Palmason, G., and K. Saemundsson, 1974: Iceland in relation to the Mid-Atlantic Ridge. Ann. Review Earth Planet. Sci. 2, 25-50.

Palmason, G., S. Arnorsson, I.B. Fridleifsson, H. Kristmannsdottir, K. Saemundsson, V. Stefansson, B. Steingrimsson, J. Tomasson, and L. Kristjansson, 1979: The Iceland crust: Evidence from drillhole data on structure and processes. In: Deep Drilling Results in the Atlantic Ocean: Ocean Crust, M. Talwani, C.G. Harrison, D.E. Hayes, eds. Amer. Geophys. Un. Maurice Ewing Series 2, 43-65.

Parker, R.L., and D.W. Oldenburg, 1973: Thermal model of ocean ridges. Nature Phys. Sci. 242, 137-139.

Saemundsson, K., 1974: Evolution of the axial rifting zone in Northern Iceland and the Tjornes Fracture Zone. Geol. Soc. Am. Bull. 85, 495-504.

Saemundsson K., 1978: Fissure swarms and central volcanoes of the neovolcanic zones of Iceland. Geol. J. Spec. Iss. 10, 415-432.

Saemundsson, K., 1979: Outline of the geology of Iceland. Jokull 29, 7-28.

Saemundsson, K., L. Kristjansson, I. McDougall, and N.D. Watkins, 1980: K-Ar dating, geological and paleomagnetic study of a 5-km lava succession in Northern Iceland. J. Geophys. Res. 85, 3628-3646.

Sanford, A.R., and P. Einarsson, 1981: Magma chambers in rifts, Amer. Geophys. Un., Geodynamics Series, in press.

Schilling, J.-G., 1973: Iceland mantle plume: Geochemical study of Reykjanes Ridge. Nature 244, 565-571.

Shih, J.S.F., T. Atwater, and M. McNutt, 1978: A near-bottom geophysical traverse of the Reykjanes Ridge. Earth Planet. Sci. Lett. 39, 75-83.

Sigurdsson, H., 1970: Structural origin and plate tectonics of the Snaefellsnes volcanic zone Western Iceland. Earth Planet. Sci. Lett. 10, 129-135.

Sigurdsson, O., 1980: Surface deformation of the Krafla fissure swarm in two rifting events. J. Geophys. 47, 154-159.

Sigvaldason, G.E., 1974. Basalts from the centre of the assumed Icelandic mantle plume. J. Petrol. 15, 497-524.

Spickernagel, H., 1980: Results of height measurements in Northern Iceland 1965/1977. J. Geophys. 47, 120-124.

Talwani, M., and O. Eldholm, 1977: Evolution of the Norwegian-Greenland Sea. Bull. Geol. Soc. Am. 88, 969-999.

Thayer, R.E., A. Bjornsson, L. Alvarez, and J.F. Hermance, 1981: Magma genesis and crustal spreading in the northern neovolcanic zone of Iceland: telluric-magnetotelluric constraints. Geophys. J.R. astr. Soc. 65, 423-442.

Torge, W., and E. Kanngieser, 1980: Gravity and height variations during the present rifting episode in Northern Iceland. J. Geophys. 47, 125-131.

Tryggvason, E., 1964: Arrival times of P-waves and upper mantle structure. Bull. Seismol. Soc. Am. 54, 727-736.

Tryggvason, E., 1980: Subsidence events in the Krafla area. J. Geophys. 47, 141-153.

Tryggvason, E., and M. Båth, 1961: Upper crustal structure of Iceland. J. Geophys. Res. 66, 1913-1925.

Vogt, P.R., 1974: The Iceland phenomenon: Imprints of a hot spot on the ocean crust, and implications for flow below the plates. In: Geodynamics of Iceland and the North Atlantic Area. L. Kristjansson, ed.: 105-126. Reidel, Dordrecht.

Vogt, P.R., G.L. Johnson, and L. Kristjansson, 1980: Morphology and magnetic anomalies north of Iceland. J. Geophys. 47, 67-80.

Voppel, D., and R. Rudloff, 1980: On the evolution of the Reykjanes Ridge south of 60°N between 40 and 12 million years before present. J. Geophys. 47, 61-66.

Walker, G.P.L., 1975: Intrusive sheet swarms and the identity of crustal layer 3 in Iceland. J. Geol. Soc. London 131, 143-161.

Ward, P.L., 1971: New interpretation of the geology of Iceland. Geol. Soc. Amer. Bull. 82, 2991-3012.

Watkins, N.D., and G.P.L. Walker, 1977: Magneto-stratigraphy of Eastern Iceland. Amer. J.Sci. 277, 513-584.

Zverev, S.M., I.P. Kosminskaya, G.A. Krasilschikova, and G.G. Mikhota, 1976: The crustal structure of Iceland and of the Iceland-Faeroes-Scotland region. Soc. Sci. Isl. Greinar 5, 74-96.

Zverev, S.M., I.V. Litvinenko, G. Palmason, G.A. Yaroshevskaya, and N.N. Osokin, 1980a: A seismic crustal study of the axial rift zone in Southwest Iceland. J. Geophys. 47, 202-210.

Zverev, S.M., I.V. Litvinenko, G. Palmason, G.A. Yaroshevskaya, N.N. Osokin and M.A. Akhmetjev, 1980b: A seismic study of the rift zone in Northern Iceland. J. Geophys. 47, 191-201.

# THE CRUST BENEATH THE ICELAND-FAEROE RIDGE

Martin H.P. Bott

Department of Geological Sciences
University of Durham
South Road, Durham DH1 3LE, U.K.

## ABSTRACT

The Iceland-Faeroe Ridge is underlain by exceptionally thick oceanic crust, which otherwise resembles the crust beneath Iceland. The laterally varying upper crust consists of basaltic rocks including lavas, volcanogenic sediments, minor intrusions and the cores of ancient volcanoes. The upper crust is underlain at 3 to 8 km depth by a 6.7 km/s refractor and the Moho occurs at 30-35 km depth at the south-eastern end, shallowing slightly towards Iceland. The Ridge probably formed above sea level between about 55 and 35 My ago as a result of the early activity of the Icelandic hot spot, with the age of the crust younging towards Iceland. The Ridge has subsequently subsided as the underlying lithosphere has cooled. An anomalous passive margin is interpreted as separating the Ridge from the continental Faeroe Block.

## INTRODUCTION

The Iceland-Faeroe Ridge is an elevated bathymetric feature of north-west to south-east trend which joins Iceland to the Faeroe Islands and separates the Norwegian Sea from the Reykjanes Basin (Fig. 1). The ridge is nearly 300 km long and about 200 km wide. It is about 400 m deep along its smooth crest and is separated from the shelves of Iceland and the Faeroes by short, sharp bathymetric scarps which are bordered by sediment-filled troughs at both ends of the Ridge. Sediments are otherwise thin or absent from the crest but they thicken down the flanks into the adjacent deep basins, particularly towards the Norwegian Sea. The Ridge is widely regarded as an anomalous aseismic transverse ridge formed during the earlier stages of the Tertiary opening of the main branch of the north-eastern North Atlantic by sea-floor spreading.

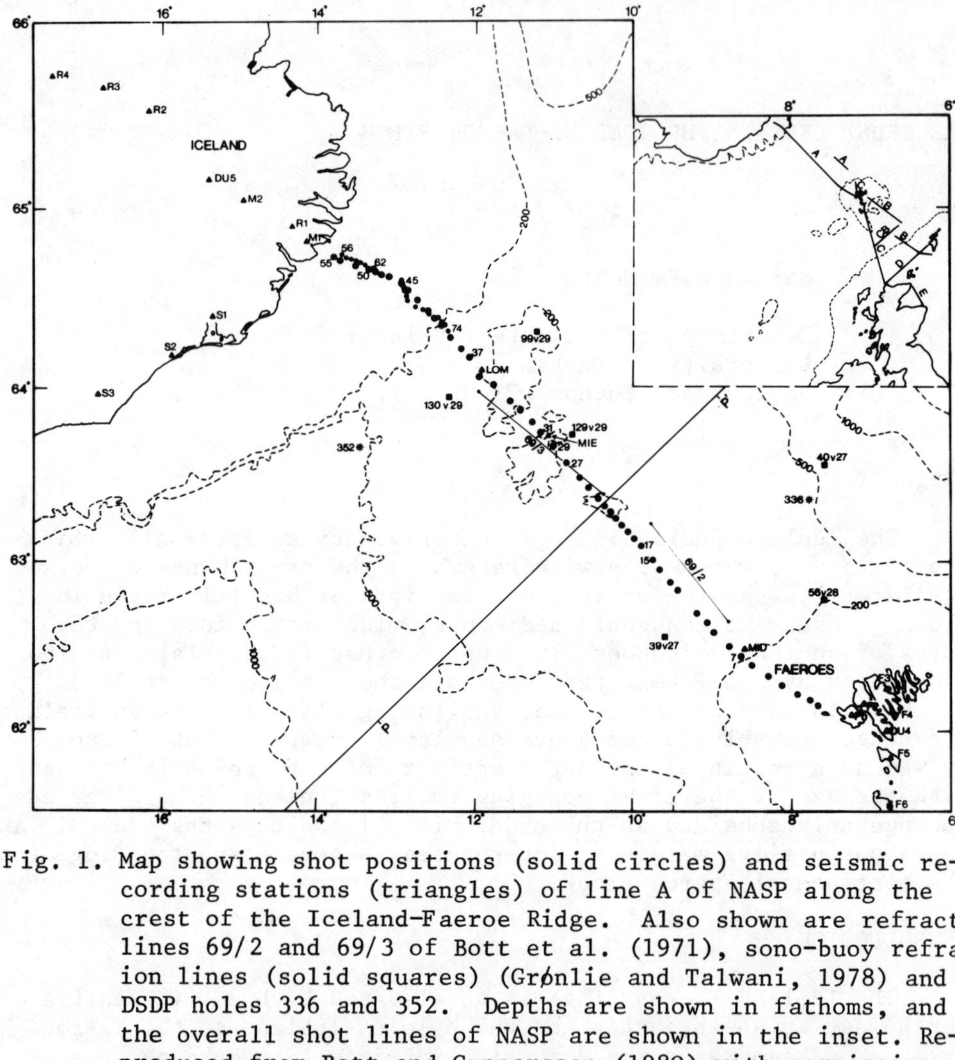

Fig. 1. Map showing shot positions (solid circles) and seismic recording stations (triangles) of line A of NASP along the crest of the Iceland-Faeroe Ridge. Also shown are refraction lines 69/2 and 69/3 of Bott et al. (1971), sono-buoy refraction lines (solid squares) (Grønlie and Talwani, 1978) and DSDP holes 336 and 352. Depths are shown in fathoms, and the overall shot lines of NASP are shown in the inset. Reproduced from Bott and Gunnarsson (1980) with permission.

The earliest crustal seismic investigation consisted of two refraction lines, each about 100 km long, shot along the crest of the ridge in 1969 (Bott et al., 1971). These penetrated to the main 6.7 km/s crustal layer but did not reach the Moho. Knowledge of the deep crustal structure comes from the North Atlantic Seismic Project (NASP) of 1972 (Bott and Gunnarsson, 1980; Zverev et al., 1975), during which the shots along line A were fired between the Faeroes and Iceland along the crest of the ridge, passing near the shorter lines of 1969 (Fig. 1). The line A shots were observed at a series of land recording stations on Iceland and on the Faeroe Islands and by the ships MV Miranda and Mikhail Lomonosov occupying sites along the crest of the ridge. The seismic refraction observations are supplemented by detailed gravity and magnetic surveys (Fleischer, 1971; Fleischer et al., 1974; Bott et al., 1971; Bott and Ingles, 1972), by shallow seismic reflection lines and sonobuoy data (Johnson and Tanner, 1972; Grønlie and Talwani, 1978) and by drilling at DSDP sites 336 and 352 on opposite flanks during Leg 38 (Talwani, Udintsev et al., 1976).

## UPPER CRUSTAL STRUCTURE

Uncompacted sediments are thin or absent on the crest of the ridge, except in local troughs near the ends. This is borne out by the large amplitude and short wavelength magnetic anomalies characteristic of the whole crestal region. These indicate that a strongly magnetic basement probably consisting of basaltic igneous rocks occurs at the seabed or not far below it. A recognisable pattern of dateable magnetic anomaly lineations has not yet been detected but regions of both normal and reverse magnetization appear to be present (Bott and Ingles, 1972). The magnetic anomaly maps (Fleischer et al., 1974; Ingles, 1971) show a series of small, circular-shaped magnetic anomalies which probably reveal the presence of igneous ring intrusions which are the remnants of ancient volcanoes.

DSDP site 336, drilled in 811 m of water on the north-eastern flank of the ridge, penetrated 30 m of oceanic tholeiitic basalt below 515 m of sediment of upper Eocene and later age. This confirms the basaltic nature of the upper crust. These basalts show evidence of sub-aerial erosion indicating that this site must have subsided at least 1350 m since the basalt was erupted. Such subsidence was predicted by Bott et al. (1971) and by Vogt (1972) on the assumption that the ridge is underlain by progressively cooling oceanic lithosphere.

The upper crust is regarded as the region down to a well-defined 6.7 km/s refractor which was originally recognised along line 69/2 (Fig. 1) of Bott et al. (1971) as a 6.8 km/s lower crustal layer, and has been subsequently identified along line A of NASP at between 3 and 8 km depth. Lines 69/2 and 69/3 reveal the presence of one or more uppermost layers with velocities ranging between 3.2 and 4.6 km/s which locally attain a composite thickness of nearly 4 km but elsewhere are thinner or absent. These are underlain by a main upper crustal layer with estimated velocity of between 5.4 and 5.8 km/s. The large lateral variations in uppermost crustal structure produce striking offsets in the travel time curves along lines 69/2 and 69/3. The regions of thick low velocity rock show some correlation with slight bathymetric depressions.

Local variations of free air gravity of up to 40 mgal occur over the ridge crest. An example occurs at the south-eastern end where there are positive free air anomalies of over 70 mgal occupying the north-eastern side of the crest. The steep gravity anomaly gradients indicate that these gravity variations are mostly caused by lateral density variation in the uppermost part of the crust. Along the seismic lines 69/2 and 69/3 (Fig. 1) low gravity values correspond to the regions where thick low velocity rocks occur. Taking into account the shallow magnetic basement, the gravity lows are probably best interpreted in terms of troughs of low density volcanogenic sediments which are probably extensively penetrated by minor intrusions. The gravity highs probably correspond to regions where lavas predominate or where intrusive centres occur.

The uppermost crust of the Iceland-Faeroe Ridge is interpreted in terms of highly magnetic igneous rocks of basaltic composition formed above sea level. Volcanogenic sediments are characteristic of some regions and basaltic lavas of other regions. Several ring complexes representing the cores of ancient volcanoes have been recognised. Apart from the inferred intense pyroclastic activity, conditions of formation were probably similar to those which have more recently occurred in Iceland. After formation, the ridge has undergone sub-aerial erosion and subsidence below sea level as the underlying lithosphere has cooled.

## CRUSTAL STRUCTURE FROM NASP

The 6.7 km/s main crustal refractor was first detected as a horizontal interface at 7.5 km depth along the reversed line 69/3. Further evidence on this refractor and on the depth to the Moho beneath the ridge has come from the North Atlantic Seismic Project data as interpreted by Bott et al. (1976), Zverev et al. (1975) and Bott and Gunnarsson (1980). As the majority of stations observing line A shots were situated on land on different types of crust, the time-term method of analysis has proved useful in interpreting the results. In particular, a variant of this method developed by Mr. K. Gunnarsson which allows for the substantial offset of rays between the refractor and the surface has been useful.

Fig. 2 shows the interpreted depth to the 6.7 km/s refractor as inferred from line A shots along the length of the ridge, assuming a mean velocity of 5.0 km/s for the overlying rocks. It occurs at about 8 km depth at the Faeroe end of the line and shallows northwestwards, reaching a minimum depth of about 3 km just under half way along. It occurs at about 6-7 km depth beneath most of the north-westernmost third of the ridge. The refractor extends beneath the Iceland shelf where it is estimated to be at about 7.5 km depth, but it has not been detected beneath the Faeroe shelf.

The shots along line A gave Moho arrivals at sufficiently distant stations on Iceland and the Faeroe Islands, and most importantly

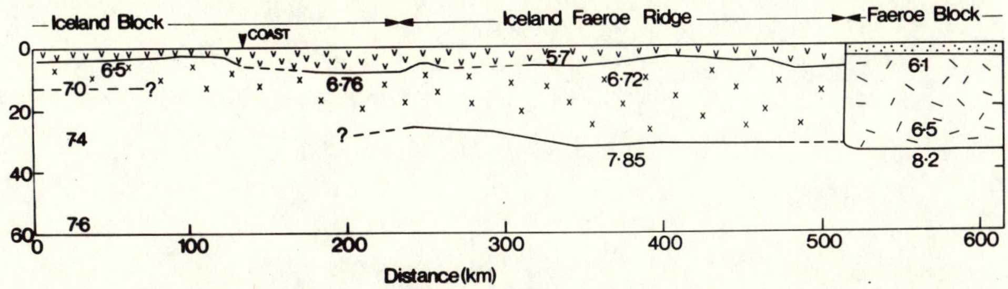


Fig. 2. Interpretation of the crustal structure of the Iceland-Faeroe Ridge and its relationship to the crust beneath the Iceland and Faeroe Blocks. The structure beneath Iceland is after Pálmason (1971) and Angenheister et al. (1979) and that beneath the Faeroe Block interpreted as continental crust is after Bott et al. (1976). Reproduced from Bott and Gunnarsson (1980) with permission.

a few Moho arrivals from shots along the ridge were observed by Mikhail Lomonosov which was situated on the Ridge near its north-western end. A conventional time-term analysis carried out on the data yielded a $P_n$ velocity of 7.78± 0.03 km/s and a mean time-term for the ridge of 2.7 seconds (Bott and Gunnarsson, 1980). Using the modified form of time-term analysis which allows for the ray offset, the velocity was determined to lie between 7.8 and 8.0 km/s. Beneath the south-eastern half of the ridge, the Moho was determined at between 32 and 35 km assuming a 6.73 km/s velocity throughout the lower crust. Less certainly, the Moho appears to shallow northwest-wards to about 25-28 km beneath the junction with the Iceland shelf (Fig. 2).

An interpretation of a free air gravity anomaly profile across the Ridge is shown in Fig. 3. This interpretation is consistent with the seismic refraction line A interpretation at the crossover. It shows that the ridge is isostatically supported, relative to the adjacent Norwegian and Reykjanes Basins, by a root of greatly thickened crust.

## ICELANDIC TYPE CRUST

The crust beneath the Iceland-Faeroe Ridge is similar in thickness to the continental crust of north-west Europe but it is more than four times the thickness of normal oceanic crust. Apart from its thickness, however, this crust does not resemble continental crust, in that the uppermost crust appears to be entirely composed of igneous rocks and a 6.7 km/s layer is systematically present at shallow depths of 3 to 8 km. The closest similarity is with the crust beneath Iceland as described by Pálmason (1971), although

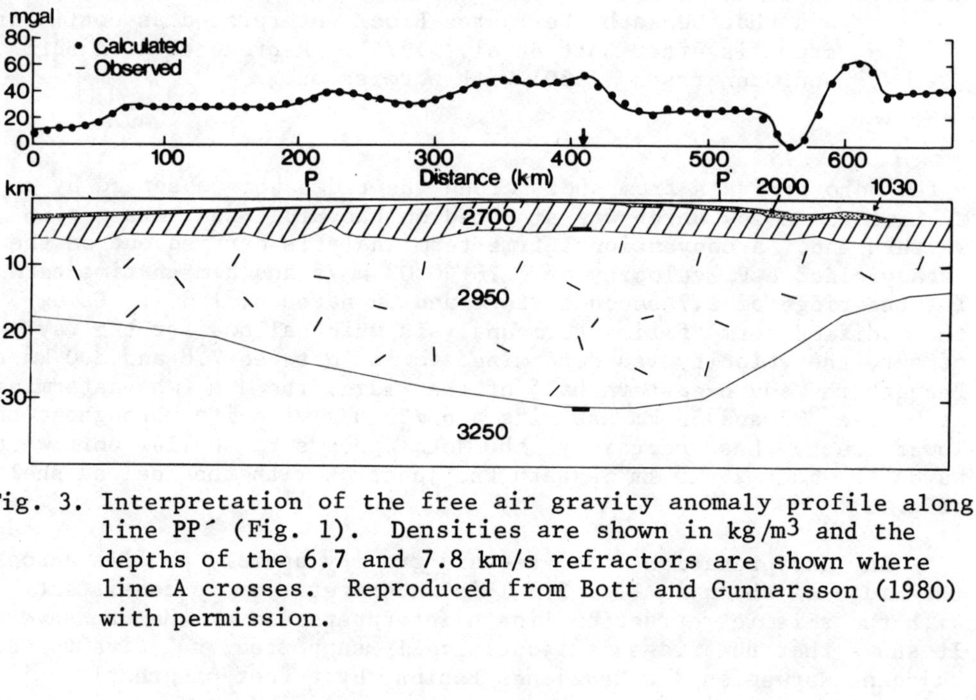


Fig. 3. Interpretation of the free air gravity anomaly profile along line PP´ (Fig. 1). Densities are shown in kg/m$^3$ and the depths of the 6.7 and 7.8 km/s refractors are shown where line A crosses. Reproduced from Bott and Gunnarsson (1980) with permission.

the crust beneath the Ridge is much thicker and the velocities of analagous layers are somewhat higher. The resemblance to Iceland crust is borne out by the continuity of the 6.7 km/s refractor between the Ridge and the Iceland shelf (Fig. 2); it can be equated with the 6.4 km/s lower crustal refractor of Iceland. This type of crust has sometimes been referred to as "Icelandic type" crust (Bott, 1974).

The most satisfactory explanation of Icelandic type crust is that it is anomalously thick oceanic crust formed with its surface above sea level. As indicated by Bott et al. (1971), it is difficult to envisage how the north-eastern North Atlantic could have formed by sea-floor spreading unless the crust of Iceland and the Iceland-Faeroe Ridge was formed by the same basic process. Icelandic type crust could only be regarded as continental if the continents around the North Atlantic Ocean have not significantly drifted apart, as was suggested by Beloussov and Milanovsky (1977).

The region between the 6.7 km/s refractor and the Moho can be regarded as a much thickened version of the normal oceanic layer 3 and the layers above the refractor as an anomalously thick layer 2. The large lateral variations in the structure of the uppermost crust and the apparent absence of clearly dateable magnetic lineations can best be explained by the formation of new crust at a sub-aerial spreading centre of the type now present on Iceland.

Normal oceanic crust forms from the magma fraction produced in the upwelling mantle material below the spreading centre. The explanation of the thick Icelandic type crust is probably that a much larger volume of magma was being produced locally beneath the spreading centre in this region, possibly as a result of upwelling from the deep mantle. The underlying upper mantle would consequently be hotter than normal. Complementary evidence for the occurrence of a hot upper mantle just prior to the 54 My continental split comes from the intense volcanism experienced by the adjacent continental borderlands. The extensive continental volcanism does not support the concept of a narrow plume as envisaged by Morgan (1971) but rather of a more widespread and perhaps irregular region of convective upwelling from the deep mantle over which the spreading axis has continued to be located (Bott, 1973).

## EVOLUTION

The Iceland-Faeroe Ridge has apparently formed by anomalous processes at the spreading centre during the early stages of the evolution of the north-eastern North Atlantic starting about 54 My ago at the time of magnetic anomaly 24. Dateable magnetic lineations cannot be recognised with confidence so that the evolutionary history cannot be inferred by the normal method. This difficulty is accentuated by the contrast in spreading history between the regions to south and to north of the Ridge. To the south, symmetrical spreading about the Reykjanes Ridge has occurred throughout, except perhaps for a westward jump of the spreading axis before the time of anomaly 22. To the north, the history of spreading is much more complicated. Symmetrical spreading in the Norwegian Sea occurred until anomaly 20 time, after which a second contemporaneous spreading axis developed to the west of the Jan Mayen microcontinent. Both axes were active until the time of anomaly 7, after which the Norwegian Sea axis became extinct.

Bott (1974) suggested that the Iceland-Faeroe Ridge formed in conformity with the Norwegian Sea about a now extinct spreading axis at the middle of the ridge. This implies a symmetrical structure, with the oldest crust adjacent to the scarps at both ends of the Ridge and the youngest crust about 43 My old at the middle. It was supposed that the spreading axis subsequently jumped westwards to produce the crust now beneath Denmark Strait. This interpretation was suggested by the apparent lack of a comparable feature to the Ridge between Iceland and Greenland and by the rather slight change of depth along the length of the Ridge.

In reconstructing the evolution of the Norwegian Sea, Nunns (1980) has subsequently shown that the Icelandic transverse ridge formed in conformity with the region to the south without any major westward migration of the spreading axis apart from a possible westward jump before anomaly 22 time. This interpretation implies that the Iceland-Faeroe Ridge becomes progressively younger from the Faeroes margin towards Iceland, forming over about 20 My between the times of magnetic anomalies 24 and 13. This interpretation is supported by the recognition that most of the Greenland shelf west of the Denmark Strait is underlain by oceanic crust formed onwards from anomaly 22(?) rather than by continental crust (Larsen, 1980).

One problem raised by this new interpretation is that the crest of the Iceland-Faeroe Ridge deepens by only about 100 m from north-west to south-east, whereas it should deepen by about 500 m between about 55 and 35 My ages as a result of the progressive cooling of the oceanic lithosphere. This discrepancy can be accounted for, at least qualitatively, by the thinner crust underlying the younger part as indicated by the NASP results.

## TRANSITION BETWEEN THE RIDGE AND ICELAND

Iceland bears the same relationship to the Iceland-Faeroe Ridge as a normal ocean ridge does to the adjacent ocean basin formed earlier at the same spreading centre. Apart from the difference in crustal thickness which is uncharacteristic of oceanic regions, the main difference in the deep structure is that the upper mantle beneath Iceland has anomalously low density and seismic velocity relative to the Ridge.

Bott et al. (1971) showed that the Bouguer anomaly over central Iceland is about 150 mgal lower than over the south-eastern part of the Iceland-Faeroe Ridge. The gradation in the Bouguer anomaly between these regions is gentle. This change is caused by the anomalously low density upper mantle beneath Iceland. If allowance is made for the thinner crust beneath Iceland relative to the Ridge, the negative gravity anomaly caused by the low density mantle beneath Iceland must be at least 250 to 300 mgal. This is a much bigger effect than is normal at ocean ridge crests and it emphasizes that the anomalous upper mantle is much more pronounced beneath Iceland.

The $P_n$ velocity beneath Iceland is about 7.2 km/s in contrast to a value of about 7.8 km/s beneath the Iceland-Faeroe Ridge. This lateral change in the P velocity of the uppermost mantle caused arrivals from Line A NASP shots at stations in Iceland to have progressively larger delay times at increasing distances from the Ridge. The exact nature of the lateral transition of upper mantle velocity is not yet clear, but it probably takes place near the junction between the Ridge and the Iceland Block.

## TRANSITION BETWEEN RIDGE AND FAEROE BLOCK - AN ANOMALOUS PASSIVE MARGIN

The crust beneath the Faeroe Block was determined from NASP data to be about 30 to 35 km thick depending on the crustal velocity distribution which is uncertain at depth (Bott et al., 1974, 1976). The crust beneath the Faeroes does not differ greatly in thickness from that beneath the south-eastern part of the Iceland Faeroe Ridge. However, the NASP data indicate that the crustal layering beneath the Block differs considerably from that beneath the Ridge. The seismic evidence shows that the Faeroe basement underlies between 2.5 and 5 km of lavas with the possibility of some low velocity sediments sandwiched between. The basement velocity lies within the range 5.9 to 6.2 km/s and gives rise to first arrivals up to distances of over 80 km. This arrival is best interpreted as the continental $P_g$ phase; similar velocities are sometimes observed in oceanic crust but they do not propagate as first arrivals over such large distances. There is also no indication of any 6.7 km/s refractor at moderate depth beneath the Faeroe Block although the velocity does appear to increase towards the base of the crust. These results indicate that continental crust probably underlies the Faeroe Block. The alternative view that it is underlain by anomalously thick oceanic crust was proposed by Talwani and Eldholm (1972).

Further indications of a continental origin of the Faeroe Block are as follows: (1) there is lack of evidence of any major subsidence of the Faeroe Islands since the Palaeocene as would be expected if the region were underlain by cooling and subsiding oceanic lithosphere (the Upper Palaeocene intra-basaltic coals now crop out near sea level); (2) a conspicuous negative gravity anomaly of about 40 mgal amplitude occurs on the shelf to the north-west of the Faeroe Islands (Fleischer et al., 1974) and this may possibly represent a continental granite batholith; (3) the continental fit across the North Atlantic, particularly now that it is recognised that the Greenland shelf adjacent to the Denmark Strait is mostly oceanic, is most consistent with a continental Faeroe Block; (4) the new Sr isotope evidence mentioned by Waagstein at the Bressanone meeting indicates an underlying continental crust.

Geophysical observations across the junction between the Iceland Faeroe Ridge and the Faeroe Block also indicate that a fundamental change in the nature of the crust occurs in the vicinity of the bathymetric scarp separating the two regions. Bott et al. (1971) found a steep decrease in the Bouguer anomaly occurs between the Ridge and the Faeroe shelf, the change in level being about 35 to 75 mgal. The gradient is too steep to attribute entirely to change in crustal thickness, and this implies that the mean density of the crust is significantly lower beneath the Faeroe Block than beneath the Ridge. Further evidence comes from the recordings of line A shots on the Ridge at recording stations situated on the Faeroe Islands (Bott et al., 1976). P waves travelling in the 6.7 km/s layer beneath the Ridge were observed to travel through the Faeroe crust with a velocity not significantly greater than $P_g$. This observation indicates that the 6.7 km/s refractor beneath the Ridge terminates at the crustal junction and that the waves travel onwards beyond the junction as $P_g$. A similar phase conversion was recognised with less certainty to affect the wave travelling as $P_n$ beneath the Ridge.

In view of the substantial amount of evidence summarized above, the Faeroe Block is believed to be underlain by continental crust. The junction between the Ridge and the Block approximately marks the location of a passive continental margin of anomalous type with a relatively slight change in crustal thickness across it but a fundamental change in the type and origin of the crust. The crustal transition appears to be much sharper than at a typical passive margin and evidence for a large amount of flexural subsidence of the shelf is missing. The adjoining part of the Ridge has probably undergone thermal subsidence of the order of 2 km since formation about 54 My ago, and this suggests that a major fault (as yet undetected) may occur at the margin.

## CONCLUSIONS

Crustal structure beneath the Iceland-Faeroe Ridge has been investigated down to the Moho by the North Atlantic Seismic Project of 1972 and to shallower depths by other short refraction lines. The main 5.7 km/s upper crustal layer is overlain locally by low velocity layers of variable thickness. Sediments are thin or absent and a strongly magnetic basement interpreted as basalt occurs at shallow depth. The basement shows regions of normal and reverse magnetization but no simple pattern of magnetic anomalies is apparent. The magnetic anomalies reveal the presence of some intrusive ring complexes interpreted as the cores of ancient volcanoes. Gravity anomalies show that large variations of density occur at shallow depth, attributable mainly to thick pockets of pyroclastic deposits and to intrusive centres.

A 6.7 km/s refractor underlies the whole Ridge at 4 to 8 km depth. This extends beneath the adjacent Icelandic shelf but is not present beneath the Faeroe shelf. A 7.8 km/s refractor representing the Moho occurs at 31-35 km depth beneath the south-eastern part of the Ridge and shallows slightly north-westwards towards the Iceland shelf, beneath which it has not been detected. Gravity profiles across the Ridge indicate that it is in approximate isostatic equilibrium as a result of thickened crust in relation to the more normal oceanic regions to the north-east and south-west.

The crust beneath the Iceland-Faeroe Ridge is almost the thickest known example of oceanic crust, being thicker than the continental crust beneath northwestern Scotland. It closely resembles the crust beneath Iceland, except that the layers are thicker and velocities are somewhat higher. The thick crust is attributed to unusually intense differentiation of magma from the underlying mantle during the early stages of opening of the North Atlantic between Greenland and the Faeroe Islands between about 55 and 35 My ago. Complementary crust formed to the west of the spreading centre probably underlies the sediment cover between Iceland and Greenland. The Ridge initially stood well above sea level, accounting for the lateral variation of upper crustal structure, and it has subsequently subsided as the underlying lithosphere has cooled. An anomalous passive margin separates the Ridge from the apparently continental crust beneath the Faeroe Block.

## OUTSTANDING PROBLEMS

Most of the present knowledge of the deep structure and evolution of the Iceland-Faeroe Ridge has been obtained over the last twelve years. Although there has been much progress, there are still many outstanding problems. The following suggested studies are aimed at tackling some of these:

(1) Geochemical and geological sampling along the crest of the Ridge to study the early stages of the evolution of the Icelandic hot spot and to date the formation of the Ridge with better precision;
(2) A series of crustal refraction lines along isochrons to determine the change of crustal thickness and structure with age;
(3) Seismological investigation of the deep crustal and upper mantle transition between the Ridge and Iceland;
(4) Detailed mapping of the lateral variation of upper crustal structure.

An implication of the inferred age structure of the Iceland-Faeroe Ridge, with the crust younging north-westwards towards Iceland, is that a complementary feature formed to the west of the spreading axis must underlie Denmark Strait and the adjacent Greenland shelf, although it is now mainly masked by later sediments. This would be expected to form a mirror image of the Ridge, at least from anomaly 22 time if there was a westward jump before then. The search for a comparable crustal feature west of Iceland is a further item of some priority. There is also need to reconcile the dateable magnetic lineations west of Denmark Strait with the apparently more complex magnetic anomalies of the Iceland-Faeroe Ridge.

## REFERENCES

Angenheister, G., Gebrande, H., Miller, H., Weigel, W., Goldflam, P., Jacoby, W., Palmason, G.G., Björnsson, S., Einarsson, P., Zverev, S., Loncarevic, B., and Solomon, S., 1979, First results from the Reykjanes Ridge Iceland seismic project 1977, Nature, Lond., 279:56.

Beloussov, V.V., and Milanovsky, Y.Y., 1977, On tectonics and tectonic position of Iceland, Tectonophysics, 37:25.

Bott, M.H.P., 1973, The evolution of the Atlantic north of the Faeroe Islands, in: "Implications of continental drift to the Earth sciences," Vol. 1, p.175, D.H. Tarling and S.K. Runcorn, eds, Academic Press, London and New York.

Bott, M.H.P., 1974, Deep structure, evolution and origin of the Icelandic transverse ridge, in: "Geodynamics of Iceland and the North Atlantic area," p. 33, L. Kristjansson, ed., D. Reidel Publishing Company, Dordrecht and Boston.

Bott, M.H.P., Browitt, C.W.A., and Stacey, A.P., 1971, The deep structure of the Iceland-Faeroe Ridge, Marine Geophys. Res., 1:328.

Bott, M.H.P., and Gunnarsson, K., 1980, Crustal structure of the Iceland-Faeroe Ridge, J. Geophys., 47:221.

Bott, M.H.P., and Ingles, A., 1972, Matrix methods for joint interpretation of two-dimensional gravity and magnetic anomalies with application to the Iceland-Faeroe Ridge, Geophys. J. R. astr. Soc., 30:55.

Bott, M.H.P., Nielsen, P.H., and Sunderland, J., 1976, Converted P-waves originating at the continental margin between the Iceland-Faeroe Ridge and the Faeroe Block, Geophys. J. R. astr. Soc., 44:229.

Bott, M.H.P., Sunderland, J., Smith, P.J., Casten, U., and Saxov, S., 1974, Evidence for continental crust beneath the Faeroe Islands, Nature, Lond., 248:202.

Fleischer, U., 1971, Gravity surveys over the Reykjanes Ridge and between Iceland and the Faeroe Islands, Marine Geophys. Res., 1:314.

Fleischer, U., Holzkamm, F., Vollbrecht, K., and Voppel, D., 1974, Die Struktur des Island-Färöer-Rückens aus geophysikalischen Messungen, Dt. hydrogr. Z., 27:97.

Grønlie, G., and Talwani, M., 1978, "Geophysical atlas, Norwegian-Greenland Sea," VEMA Research Series IV, Lamont-Doherty Geological Observatory, Palisades, N.Y.

Ingles, A.D., 1971, The interpretation of magnetic anomalies between Iceland and Scotland, University of Durham Ph.D. thesis.

Johnson, G.L., and Tanner, B., 1972, Geophysical observations on the Iceland-Faeroe Ridge, Jökull, 21:45.

Larsen, H.C., 1980, Geological perspectives of the East Greenland continental margin, Bull. geol. Soc. Denmark, 29:77.

Morgan, W.J., 1971, Convection plumes in the lower mantle, Nature, Lond., 230:42.

Nunns, A.G., 1980, Marine geophysical investigations in the Norwegian-Greenland Sea between the latitudes of 62°N and 74°N, University of Durham Ph.D. thesis.

Pálmason, G., 1971, Crustal structure of Iceland from explosion seismology, Rit. Vísindafj. Isl., 40:1.

Talwani, M., and Eldholm, O, 1972, Continental margin off Norway: a geophysical study, Bull. geol. Soc. Am., 83:3575.

Talwani, M., Udintsev, G., et al., 1976, "Initial Reports of the Deep Sea Drilling Project," Vol. 38, U.S. Government Printing Office, Washington, D.C.

Vogt, P.R., 1972, The Faeroe-Iceland-Greenland aseismic ridge and the western boundary undercurrent, Nature, Lond., 239:79.

Zverev, S.M., Kosminskaya, I.P., Krasilstchikova, G.A., and Mikhota, G.G., 1975, The crustal structure of Iceland and the Iceland-Faeroe-Shetland region, Bull. MOIP Otd. Geol., Vol. L, No. 3:72.

# GEOLOGY AND CRUSTAL STRUCTURE OF THE FAEROE ISLANDS - A REVIEW

P. Hedebol Nielsen

Geological Survey of Denmark, Faeroe Branch
Debesartröd
DK-3800 Thorshavn, Faeroe Islands

## ABSTRACT

The Faeroe Islands consist of plateau basalts extruded during the opening of the Norwegian-Greenland Sea in the early Tertiary. On the shelf outside the 200 m contour line the basalts are overlain by sediments reaching a thickness of 2 km. The shelf is separated from the Iceland-Faeroe Ridge by a scarp with a gravimetric anomaly, and the Faeroese crust has lower velocities (6.2 km/s or less) than the crust of the ridge (6.7 km/s). The Faeroese crust is more than 25 km thick and is probably continental.

## GEOLOGY OF THE ISLANDS

The Faeroe Islands ($62^{\circ}$N $07^{\circ}$W, 1399 $km^2$, coast line 1117 km, population 42 000) consist of an eroded subaerial sequence of plateau basalt lavas ranging from olivine tholeiites to quartz tholeiites. A lava pile of 3 km is exposed on the islands, but no base and no top of the lava pile are found. The lavas were extruded in connection with the opening of the Norwegian-Greenland Sea in the early Tertiary. The topography of the islands is rather rough with heights up to 882 m.

The geology of the Faeroe Islands has been described by Rasmussen and Noe-Nygaard (1969, 1970). The lava pile is divided into three series (Lower, Middle, and Upper, approximately 1 km each). The Middle Series consists of thin pahoehoe flows averaging less than 2 m in thickness and occasionally separated by thin tuffaceous beds. The Upper Series and especially the Lower Series consist of thick flows (Lower Series averaging 20 m, max. 50 m, Upper Series averaging 10 m) separated by tuffaceous beds. In the field the basalts can often be divided into three groups: olivine porphyritic, plagioclase porphyritic, and aphyric, all of which occur in each of the series except that the Lower Series is mainly poor in olivine.

A number of chemical analyses have been published (Noe-Nygaard and Rasmussen, 1968; Rasmussen and Noe-Nygaard, 1969, 1970; Schilling and Noe-Nygaard, 1974; Bollingberg et al., 1975; Waagstein, 1977; Jensen, 1979).

The two horizons separating the three series have been intruded by magma of the same type as the three series and the horizon between Lower Series and Middle Series seems to indicate an interruption of the volcanism giving a variable 10 m clay-coal sequence.

All three series are cut by numerous dikes, some of which can be followed for many kilometres, but the dikes are narrow (<20 m) and their volume is small as compared to the lava flows. The eruption of the Middle Series started explosively, and pyroclastics form a layer of tuff-agglomerate up to 100 m thick.

The three basalt series are tectonically rather undisturbed and with no or only minor unconformities. The lava flows have mostly small dips, often about 2$^\circ$ in an easterly direction (varying from north to south), and although joints and faults are numerous, the motion has been rather small (vertical displacements <10 m are common, but >40 m very rare). The islands seem to be the eastern halves of two large dome structures with centres just west of the islands (Schrøder, 1971; Noe-Nygaard, 1974; Waagstein, 1977) so that the oldest lavas are exposed on Mikines (westernmost island) and on the southwestern part of Suðuroy (southernmost island). The dip of these areas is larger than elsewhere and reaches 16$^\circ$ or more. This does not indicate, however, a generally larger dip of the Lower Series but a spatial variation of the dip. In this connection it can be mentioned that a drill site near Vestmanna (62$^\circ$09'.2N, 07$^\circ$08'.7W, Jarðfrøðisavnið, 1980) gave an almost continuous core down to 657 m (below sea level, starting at 2 m above), and that the horizon between Lower Series and Middle Series was found at 557 m giving a mean easterly dip of this horizon of 2.8°, in agreement with extrapolations from exposures 11-12 km to the west.

Two negative results for the plateau basalts on the islands must be noted: no indication of submarine eruptions and none of central volcanoes have been found.

The Faeroese lavas have been dated by the K/Ar-method to be 50-60 Ma old (Tarling and Gale, 1968), but no correlation between the age of the flows was found, probably because the span of time is rather short and the argon retention varies. Biostratigraphic evidence is consistent with this lower Tertiary age (Rasmussen and Koch, 1963; Laufeld, 1965). Paleomagnetic investigations show that the Upper Series and the Middle Series are both characterized by "reverse" polarity, while the Lower Series shows at least two zones of "normal" polarity separated by a zone of "reverse" polarity (Abrahamsen, 1967; Tarling and Gale, 1968; Løvlie, 1976).

## GEOLOGY OF THE FAEROESE SHELF

A broad shelf surrounds the Faeroe Islands and it was widely glaciated, probably down to the 400 m depth contour in the southeast (Waagstein and Rasmussen, 1975). The present sedimentation is influenced by strong currents: near the islands the tidal current can easily exceed 5 knots (≈9 km/h), and the deep parts of the sea have often strong bottom currents caused by the exchange of water between the Atlantic Ocean and the Norwegian Sea (e.g. Roberts et al., 1979).

The basaltic sea floor around the islands has usually very little or no sedimentary cover, but some depressions contain 50-150 m of sediments. However, outside approximately the 200 m contour line (Figs. 1 and 2) the basalts become overlain by sediments of a presumed Tertiary age (Nielsen et al., 1979). This sedimentary part of the shelf shows the marginal and transverse channels known from many glaciated shelves (Holtedahl, 1970). The depressions on the shelf are often partly filled with transparent sediments of an irregular bedding (Fig. 1). A thin sedimentary cover on Sandoy and Suðuroy Banks and farther east and south is much less transparent and probably includes coarse-grained glacial material. The basalts seem to occur at shallow depth below Nólsoy Bank according to magnetic evidence (Nielsen, 1977), but seismic, gravimetric and magnetic investigations show thick sediments on Sandoy

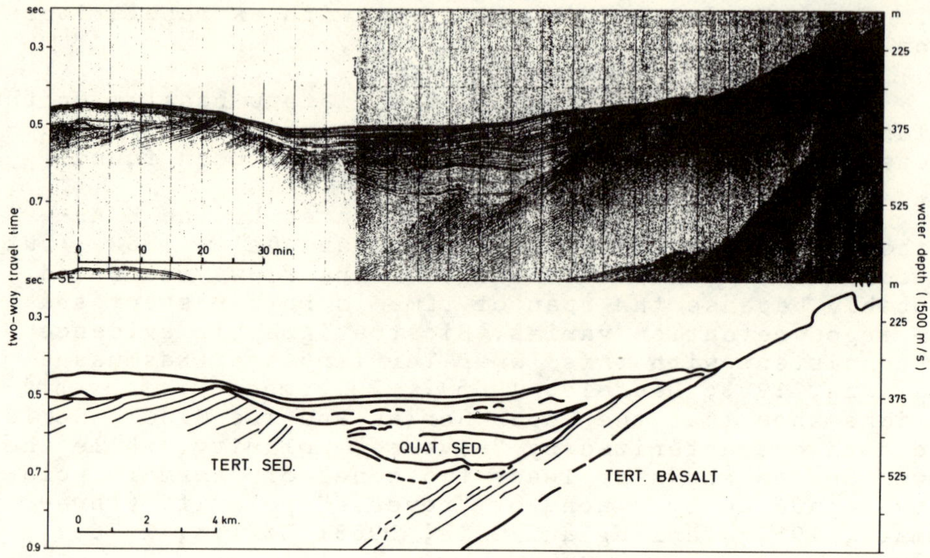


Fig. 1. This seismic record is taken from Nielsen et al. (1979) and is part of a line running in the depression between Sandoy and Suduroy Banks. It shows the transition between the basaltic sea floor of the islands and the sediments of the banks to the east. The record is characteristic in showing that the depressions of the banks are erosional features partly filled with transparent sediments of an irregular bedding. The position of the record is shown in Fig. 2 (line denoted by "Fig").

Suđuroy, and Skeivi Banks (Stride et al., 1967; Bott and Stacey, 1967; Stride et al., 1969; Bott and Watts, 1971; Himsworth, 1973; Korsakov, 1974). These sediments may exceed 2 km in the east, they are well bedded and dip outward at a low angle; the oldest strata overlie the sloping basalt-sediment contact conformably or nearly so with a dip of 2-10$^{\circ}$ (Nielsen et al., 1979) as expected from the dome theory (Schrøder, 1971).

The Iceland-Faeroe Ridge in the northwest consists of subareal basalt lavas with minor sedimentary troughs (Fleischer et al., 1974; Nilsen, 1978; also Talwani et al., 1976), but is separated from the Faeroese shelf by a distinct scarp. The sedimentary boundary in the northwest (Fig. 2) indicates either a trough or the sedimentary cover on the slope down towards the deep Norwegian basin. North and northwest of the islands the basaltic sea floor continues probably to the edge of the shelf, but a conspicuous rough bottom of unknown origin occurs

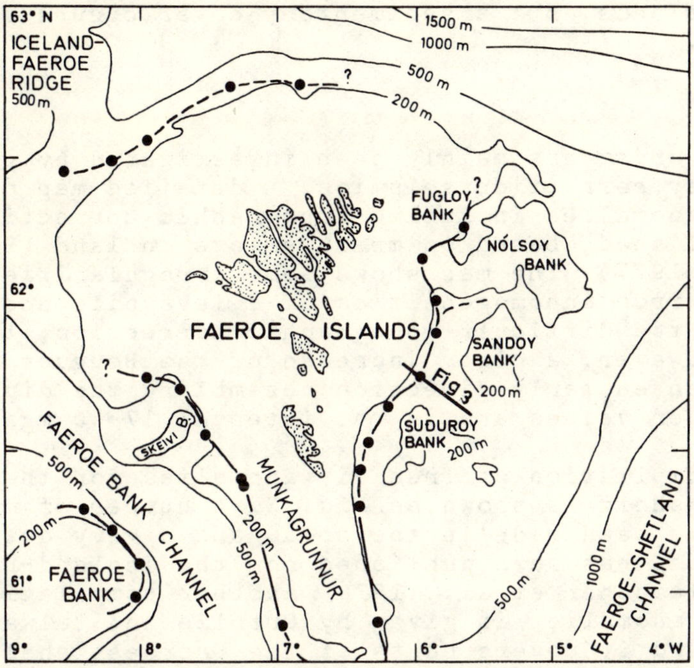


Fig. 2. The map shows the boundary of the Faeroese basalt areas according to the measurements made on board "Dana" in 1979. The heavy dots denote points where seismic profiling and magnetics show that the basalts dip below sediments (from Nielsen et al., 1979). The line denoted by "Fig" refers to Fig. 1 of the present paper.

here on the outer part of the shelf (Fleischer et al., 1974; Nielsen et al., 1979).

The three banks in the southwest (Faeroe, Bill Bailey, and Lousy or outer Bill Bailey Banks) consist probably of plateau basalts much like the Faeroe Islands. The Faeroe Bank has been thoroughly investigated through a magnetic survey (Dobinson, 1970) and through other methods (Stride et al., 1967; Bott and Stacey, 1967; Nielsen et al., 1979). All the banks have been investigated magnetically (Nielsen, 1977). They have also been investigated by dredging followed by chemical and petrographic analyses (Waagstein, 1977). A basaltic sea floor occurs on the banks much like the basaltic shelf of the islands, but a sedimentary cover occurs in the channels.

## THE CRUST

The crust has mainly been investigated by gravimetry and by refraction seismics. A detailed map of the Bouguer anomalies including topographic corrections has been published, based on measurements on land (Saxov, 1969 and 1977). The map shows some irregularities as expected from inhomogeneities of a lava pile and because of the difficult topographic correction; the map shows, however, a clear increase of the Bouguer anomalies in an easterly direction resembling the dip pattern; the Bouguer values are in the interval 17-40 mgal.

A compilation of free air anomalies for the surrounding shelf is shown as Fig. 3. A number of measurements have been made in the area, and a very detailed gravity map has been published for the Iceland-Faeroe Ridge (Fleischer et al., 1974). Another compilation of free air anomalies is given by Grønlie and Talwani (1978), and it covers parts of the Faeroese shelf. Here it can be mentioned that a large number of speculative geophysical profiles have been measured by commercial firms around the Faeroe Islands but these results are unknown to the author.

The free air anomaly map (Fig. 3) shows a pronounced maximum along the scarp between the Faeroe shelf and the Iceland-Faeroe Ridge. The maximum is partly a topographic effect, but it must separate two gravitationally different areas, the ridge with values around 40 mgal and depths around 450 m to be compared with the Faeroese shelf with values around 30 mgal and depths up to 200 m. Although the steep minimum northwest of the islands obscures the situation, a difference in the crust of the two regions is indicated, see, however, Talwani and Eldholm (1977). A number of gravimetric models for the minimum and scarp have been calculated (Bott et al., 1971; Schrøder, 1971; Fleischer, 1971; Casten, 1974; Fleischer et al., 1974), but the origin of the minimum remains speculative.

Another gravitational minimum southeast of the islands on Sandoy and Suduroy Banks is probably connected with nearly 2 km of Tertiary sediments (Bott and Watts, 1971; Korsakov, 1974; Nielsen et al., 1979).

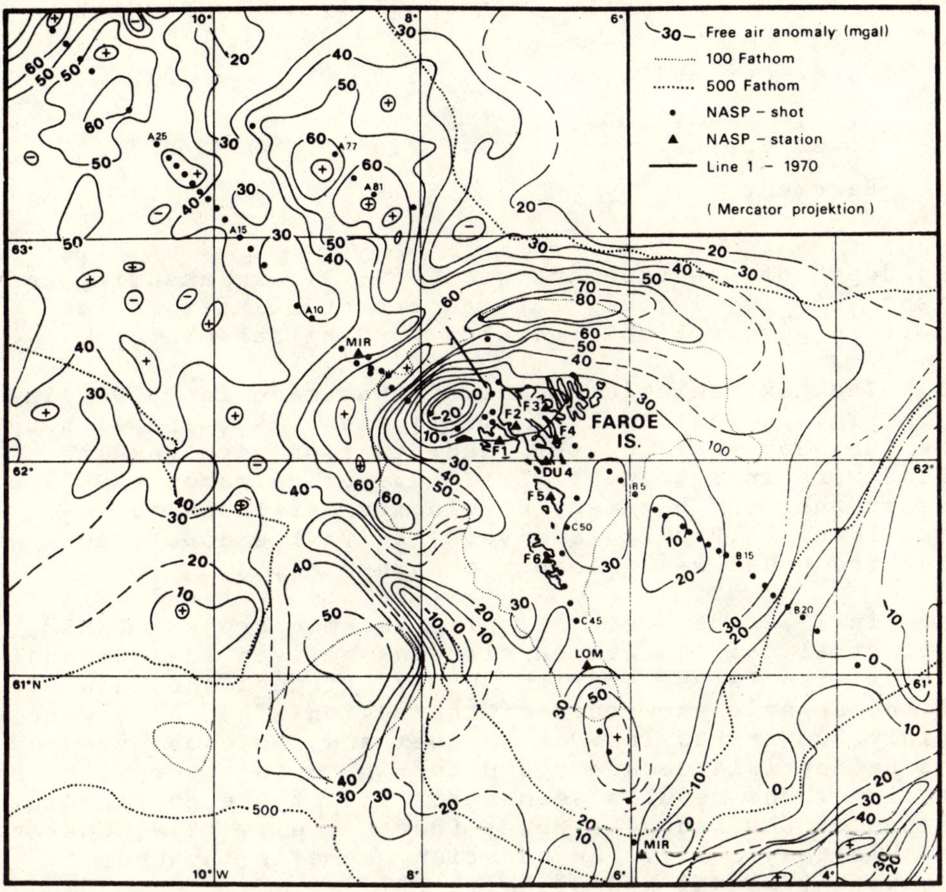


Fig. 3. Free air anomalies around the Faeroe Islands. The map is based mainly on data published by Fleischer et al. (1974) and Bott and Watts (1971). All NASP shots which gave usable first arrivals on the Faeroe Islands are shown. (From Nielsen, 1976).

The first seismic experiment on the Faeroe Islands was made by Pálmason (1965) and gave the following results:

| | P-velocity (km/s) | S-velocity (km/s) |
|---|---|---|
| Upper Series | 3.9 | 2.2 |
| Middle Series and Lower Series | 4.9 | 2.7 |
| Basement | 6.4 | 3.4 |

The depth of the basement was 2.5 - 4.5 km assuming constant velocity layers, but the two profiles were too short to give reliable results for the basement.

The next seismic experiment was made in 1970 (line 1 on Fig. 3, Casten, 1971, 1973, and 1974), and although the sonobuoys failed, the land stations gave results which were in agreement with those of Pálmason (1965) except that the basement P-velocity was found to be much lower (5.9 km/s) and this led to the conclusion that the crust was probably continental.

In 1972 the North Atlantic Seismic Project (NASP, Bott et al., 1974) was carried out between Iceland and the British Isles, and it greatly extended the knowledge of the seismic structure of the region. The shots were mainly 300 or 600 lbs. of Geophex and those observed on the Faeroe Islands are shown on Fig. 3 together with the stations. The results seemed to confirm the continental nature of the Faeroese crust (Bott et al., 1974; Casten and Nielsen, 1975). The question remains, however, disputed (Talwani and Eldholm, 1977).

The NASP data from the shelf of the Faeroe Islands were reported by Nielsen (1976), mainly based on first arrival data and the time term method (Berry and West, 1966). The main results are now described.

The Faeroe Islands and their northwestern shelf have a surface layer with P-velocities close to 5 km/s, but the spacing of shots and of stations was too big to study this basaltic layer in any detail; the results were, however, in agreement with Pálmason's results for the Lower and Middle Series. The only other P-velocity observed before Moho arrivals was close to 6 km/s (5.9-6.2 km/s), and this "basement" gave somewhat scattered time-terms in the interval 0.25-0.54 s corresponding to depths between 2.2 and 4.8 km if a surface layer of constant velocity (4.9 km/s) is assumed. Shot A3 in the middle of the steep gravity low northwest of the islands gave a time-term nearly 0.3 s larger than the neighbouring shots. The 6 km/s basement has probably also been observed 20 km northeast of the islands but with somewhat larger time-terms (see Casten, 1971). However, south and southeast of the islands the 6 km/s basement seems to disappear and give way to lower velocities. Along line B (towards southeast, Fig. 3) an unreversed velocity of approximately 5.2 km/s was found. Along line C (towards south) the land-based stations gave together with the ship station LOM a poorly reversed velocity of 4.9 km/s corresponding to a direct wave in the basalt surface layers south of the Faeroe Islands. The signals received by stations on the Faeroe Islands from shots south of shot C45 were surprisingly weak until more normal signals reappeared at longer distances as Moho arrivals. This points towards a crustal change south of the Faeroe Islands, possibly with basalts underlain by low velocity sediments.

The ship station LOM observed an apparent velocity of 6.6 km/s and time-terms around 0.67 sec for the gravity high south of the Faeroe Islands.

The NASP data gave larger Moho time-terms for the Faeroe Islands than for the Iceland-Faeroe Ridge and the Faeroe-Shetland Channel and indicate a crustal thickness of more than 25 km. However, a time-term analysis (Nielsen, 1976) did not give clear results for Moho arrivals because the time-term method requires a uniform structure within the cone of critical rays or out to a distance which is usually larger than the depth of the refractor. Therefore, an iterative approach should be used for Moho arrivals on the Faeroese shelf, as was done by Bott and Gunnarsson (1980) or Zverev et al. (1975), (both works are based on NASP data).

Bott et al. (1976) found a transition from an anomalous oceanic crust on the Iceland-Faeroe Ridge to a continental crust on the Faeroe shelf. The transition seems to occur a little east of the scarp. This transition seems to be in agreement with the seismic velocities of Zverev et al. (1975) and the structure of the rather inhomogeneous ridge has mainly been confirmed by Bott and Gunnarsson (1980).

Zverev et al. (1975) found decreasing velocities south and southeast of the Faeroe Islands with even smaller velocities for the crust of the Faeroe-Shetland Channel, but the thickness of the crust is much less in the channel than outside (<20 km, but nearly 30 km beneath the Shetland islands and even more beneath the Faeroe shelf). The thin crust of the Faeroe-Shetland Channel corresponds to the rather high Bouguer anomalies here.

## DISCUSSION

It is the author's impression that the shelf of the Faeroe Islands can be considered as a northern continental extension of the Rockall Plateau (Roberts, 1975). This seems consistent with plate tectonic evidence: the oldest magnetic lineation in the Norwegian-Greenland Sea seems to be No. 24, although the structure of the Norway Basin north of the Faeroe Islands is unclear (Talwani and Eldholm, 1977). The same lineation seems to occur just west of the Rockall Plateau (see e.g. Srivastava, 1978). The magnetic survey of Voppel et al. (1979) traces probably lineation 24 just northwest of Lousy Bank (southwest of the Faeroe Islands). This means that the Rockall Trough and its possible northern extension, the Faeroe-Shetland Channel, must be older.

## ACKNOWLEDGEMENT

This review was written mainly during my stay at the Laboratory of Geophysics, Aarhus University. I want to thank the Laboratory and its staff for hospitality and cooperation.

## REFERENCES


Abrahamsen, N., 1967, Some paleomagnetic investigations in the Faeroe Islands, Bull. Geol. Soc. Denmark, 17:371-384.

Berry, M.J., and West, G.F., 1966, An interpretation of the first-arrival data of the Lake Superior experiment by the time-term method, Bull. Seism. Soc. Am., 56:141-171.

Bollingberg, H., Brooks, C.K., and Noe-Nygaard, A., 1975, Trace element variation in the Faeroese basalts and their possible relationships to ocean floor spreading history, Bull. Geol. Soc. Denmark, 24:55-60.

Bott, M.H.P., and Stacey, A.P., 1967, Geophysical evidence on the origin of the Faeroe Bank Channel - II, A gravity and magnetic profile, Deep-Sea Res., 14:7-12.

Bott, M.H.P., and Watts, A.B., 1971, Deep structure of the continental margin adjacent to the British Isles, in: ICSU/SCOR Symposium on East Atlantic Continental Margins 1970, Delany, F.M., ed., Inst. Geol. Sci. Rept., 70/14:89-109.

Bott, M.H.P., Browitt, C.W.A., and Stacey, A.P., 1971, The deep structure of the Iceland-Faeroe Ridge, Marine Geophys. Res., 1:328-351.

Bott, M.H.P., Sunderland, J., Smith, P.J., Casten, U., and Saxov, S., 1974, Evidence for continental crust beneath the Faeroe Islands, Nature, 248:202-204.

Bott, M.H.P., Nielsen, P.H., and Sunderland, J., 1976, Converted P-waves originating at the continental margin between the Iceland-Faeroe Ridge and the Faeroe block, Geophys. J. R. astr. Soc., 44:229-238.

Bott, M.H.P., and Gunnarsson, K., 1980, Crustal structure of the Iceland-Faeroe Ridge, J. Geophys., 47:221-227.

Casten, U., 1971, Seismic measurements on the Faeroe Islands, 1970, Fróðskaparrit, 19:105-116.

Casten, U., 1973, The crust beneath the Faeroe Islands, Nature, 241:83-84.

Casten, U., 1974, Eine Analyse seismischer Registrierungen von den Färöer Inseln, Hamburger Geophys. Einzelschriften, Heft 21.

Casten, U., and Nielsen, P.H., 1975, Faeroe Islands - a microcontinental fragment?, J. Geophys., 41:357-366.

Dobinson, A., 1970, A magnetic survey of the Faeroe Bank, part 2 of Ph.D. thesis, University of Durham.

Fleischer, U., 1971, Gravity surveys over the Reykjanes Ridge and between Iceland and the Faeroe Islands, Marine Geophys. Res., 1:314-327.

Fleischer, U., Holzkamm, F., Vollbrecht, K., and Voppel, D., 1974, Die Struktur des Island-Färöer-Rückens aus geophysikalischen Messungen, Deutsche Hydrograph. Zeitschrift, 27:97-113.

Grønlie, G., and Talwani, M., 1978, Geophysical atlas of the Norwegian-Greenland Sea, VEMA Research Series IV, Lamont-Doherty Geological Observatory, Palisades, N.Y.

Himsworth, E.M., 1973, Marine geophysical studies between North-West Scotland and the Faroe Plateau, Ph.D. thesis, University of Durham.

Holtedahl, O., 1970, On the morphology of the West Greenland shelf with general remarks on the "marginal channel" problem, Marine Geol., 8:155-172.

Jarðfrøðisavnið, 1980, Foreløbig rapport om Vestmannaboringen (preliminary report), addr. Debesartröd, DK-3800 Thorshavn, Faeroe Islands.

Jensen, Aa., 1979, Mineralogical and geochemical variations across three basaltic lava flows from the Faeroe Islands, Bull. Geol. Soc. Denmark, 28:95-114.

Korsakov, O.D., 1974, O tektonike Farero-shetlandskogo zheloba (On the tectonics of the Faeroe-Shetland trough, in Russian), Doklady Akad. Nauk SSSR, 214(3):647-650.

Laufeld, S., 1965, Sporomorphs in Tertiary coal from the Faeroe Islands, Geol. För. Stockholm Förh., 87:231-238.

Løvlie, R., 1976, Post-baking alterations and partial remagnetization in five baked tuff layers from the Faeroe Islands, Geophys. J. R. astr. Soc., 45:219-229.

Nielsen, P.H., 1976, Seismic refraction measurements around the Faeroe Islands, Fróðskaparrit, 24:9-45.

Nielsen, P.H., 1977, A marine geophysical project, Fróðskaparrit, 25:95-100.

Nielsen, P.H., Waagstein, R., Rasmussen, J., and Larsen, B., 1979, Marine seismic investigation of the shelf around the Faeroe Islands, Fróðskaparrit, 27:102-112.

Nilsen, T.H., 1978, Lower Tertiary laterite on the Iceland-Faeroe Ridge and the Thulean land bridge, Nature, 274:786-788.

Noe-Nygaard, A., 1974, Cenozoic to recent volcanism in and around the North Atlantic basin, in: The ocean basins and margins, vol. 2, The North Atlantic, Nairn, A.E.M., and Stehli, F.G., eds., Plenum, NY.

Noe-Nygaard, A., and Rasmussen, J., 1968, Petrology of a 3000 metre sequence of basaltic lavas in the Faeroe Islands, Lithos, 1:286-304.

Pálmason, G., 1965, Seismic refraction measurements of the basalt lavas of the Faeroe Islands, _Tectonophysics_, 2:475-482.

Rasmussen, J., and Koch, B.E., 1963, Fossil Metasequoia from Mikines, Faeroe Islands, _Fróđskaparrit_, 12:83-91 + 2 plates.

Rasmussen, J., and Noe-Nygaard, A., 1969, Beskrivelse til geologisk kort over Færøerne i målestok 1:50 000, _Geological Survey of Denmark, 1st series_, No. 24, p. 370, Copenhagen.

Rasmussen, J., and Noe-Nygaard, A., 1970, Geology of the Faeroe Islands (Pre-Quaternary), _Geological Survey of Denmark, 1st series_, No. 25, p. 142, Copenhagen.

Roberts, D.G., 1975, Marine geology of the Rockall Plateau and Trough, _Phil. Trans. R. Soc. Lond._, ser.A., 278:447-509.

Roberts, D.G., Hunter, P.M., and Laughton, A.S., 1979, Bathymetry of the northeast Atlantic: continental margin around the British Isles, _Deep-Sea Res._, 26A:417-428 + map.

Saxov, S., 1969, Gravimetry in the Faeroe Islands, _Meddelelse No. 43_, Geodetic Institute of Denmark.

Saxov, S., 1977, Complementary gravity observations in the Faeroe Islands, _Fróđskaparrit_, 25:66-83.

Schilling, J.-G., and Noe-Nygaard, A., 1974, Faeroe-Iceland plume: rare earth evidence, _Earth & Planet. Sci. Lett._, 24:1-14.

Schrøder, N.F., 1971, Magnetic anomalies around the Faeroe Islands, _Fróđskaparrit_, 19:20-29.

Srivastava, S.P., 1978, Evolution of the Labrador Sea and its bearing on the early evolution of the North Atlantic, _Geophys. J. R. astr. Soc._, 52:313-357.

Stride, A.H., Belderson, R.H., Curray, J.R., and Moore, D.G., 1967, Geophysical evidence on the origin of the Faeroe Bank Channel - I. Continuous reflection profiles, _Deep-Sea Res._, 14:1-6.

Stride, A.H., Curray, J.R., Moore, D.G., and Belderson, R.H., 1969, Marine geology of the Atlantic continental margin of Europe, _Phil. Trans. R. Soc. Lond._, ser.A, 264:31-75.

Talwani, M., and Eldholm, O., 1977, Evolution of the Norwegian-Greenland Sea, _Bull. Geol. Soc. Am._, 88:969-999.

Talwani, M., Udintsev, G., et al., 1976, Initial reports on the Deep Sea Drilling Project, vol. 38, US Government Printing Office. Also supplement to volumes 38, 39, 40, and 41, 1978.

Tarling, D.H., and Gale, N.H., 1968, Isotopic dating and paleomagnetic polarity in the Faeroe Islands, Nature, 218:1043-1044.

Voppel, D., Srivastava, S.P., and Fleischer, U., 1979, Detailed magnetic measurements south of the Iceland-Faeroe Ridge, Deutsche Hydrograph. Zeitschrift, 32:154-172.

Waagstein, R., 1977, The geology of the Faeroe plateau, unpubl. thesis, Geologisk Centralinstitut, University of Copenhagen.

Waagstein, R., and Rasmussen, J., 1975, Glacial erratics from the sea-floor southeast of the Faeroe Islands and the limit of glaciation, Fróðskaparrit, 23:101-119.

Zverev, S.M., Kosminskaya, I.P., Krasilshchikova, G.A., and Mikhota, G.G., 1975, Glubinnoe stroenie Islandii i Islandsko-Farersko-Shetlandskogo regiona po rezultatam seismicheskikh issledovaniy (NASP-72), Bull. Mosk. o-va ispytateley priroda, T.80, Otd. Geol. T. 50, 99-115.
(Note: the discussion in the present paper is based on a manuscript copy in English).

# ASPECTS OF THE TERTIARY GEOLOGY OF THE FAEROE-SHETLAND CHANNEL

Michael F. Ridd

The British National Oil Corporation
150 St. Vincent Street
Glasgow, G2 5LJ, Scotland

ABSTRACT

Petroleum exploration has moved off the continental shelf and into the deeper waters of the Faeroe-Shetland Channel. This paper describes aspects of this, the southeast flank of the Faeroe Basin, in particular the Tertiary stratigraphy and occurrences of igneous rocks.

A Palaeocene hinge-line can be traced along the basin flank and it is accompanied by listric growth-faults. The Eocene-Oligocene hinge-line is northwest of the Palaeocene one and unaccompanied by growth-faults; it appears to be a prograding regime.

The flank of the Erland Igneous Complex has been drilled (about 100 km NNE of Shetland) and over 400 m of acid and basic volcanic rocks of Campanian to assumed Palaeocene age were penetrated.

A zone of basic igneous intrusions runs at least 250 km from approximately $61^{o}$N Latitude, northeast along the flank of the basin and into Norwegian waters. The intrusions affect sediments of Palaeocene and Cretaceous age and individual igneous bodies range from two metres to over 100 m thick; they appear to be sills rather than dykes. It is likely that their intrusion was closely related to the formation of the basin itself, as was the Axial Opaque Zone - a linear intrusive complex presumed to be related to sea-floor spreading between the Faeroes and Scotland.

## INTRODUCTION

At the time that Cashion (1975) wrote his account of the West Shetland Basin, exploration drilling had not progressed beyond the continental shelf, and the reconnaisance seismic coverage over the deeper-water area of the Faeroe-Shetland Channel was limited and hard to interpret. Ridd's (1981) discussion of the petroleum geology west of the Shetlands considered the Faeroe Basin as a whole; the Devonian to Tertiary stratigraphy of the West Shetland Basin and the Rona Ridge was described and, at late Palaeocene and younger levels, was extrapolated using seismic data far beyond any borehole control.

There have been few new findings of importance in the pre-Tertiary geology since that latter paper and consequently the present account deals largely with the gross features of the Tertiary geology in the Faeroe Basin. In particular it deals with the Tertiary sedimentary and igneous rocks which lie beneath the Upper Palaeocene - Lower Eocene tuffaceous unit* (which is the equivalent of the Rogaland Group of the Northern North Sea; Deegan and Scull, 1977).

Details of the first well west of the Shetlands, Esso's 206/12-1, have now been released by the UK Department of Energy. It was drilled in 1972 on the Rona Ridge (Fig. 1) and encountered Lewisian granite with, incidentally, fractures containing oil, albeit in non-commercial quantities. For the record, the well's stratigraphy is shown on Figure 2. Noteworthy points are the overstepping Campanian - Maastrichtian grey shale unit over 600 m thick, the intra-Palaeocene unconformity which cuts out the Danian, the Palaeocene-Lower Eocene tuffaceous unit with its strong sonic-log response, and the coarsening-upward Palaeogene which comprises the Tertiary to the virtual exclusion of any Neogene.

## SOUTHEAST FLANK OF THE TERTIARY BASIN

Figure 3 shows the Palaeocene at the 206/12-1 borehole (possibly including some Lower Eocene) where it is less than 100 m thick and represented on the seismic section as less than 200 milliseconds two-way-time (TWT). This thickness is typical of the Palaeocene over the Rona Ridge and West Shetland Basin. Northwestwards, out into the Faeroe Basin, it thickens gradually to some 700 milliseconds TWT before thinning over the dome in the northeast part of Quadrangle 205, and thickening again gently toward the axis of the Faeroe Basin.

---

*The tuff has been assigned to the Upper Palaeocene and the Lower Eocene in different wells. By its nature it is unlikely to be diachronous and there do not appear to be two tuff units and the balance of evidence is that there is one, of Lower Eocene age.

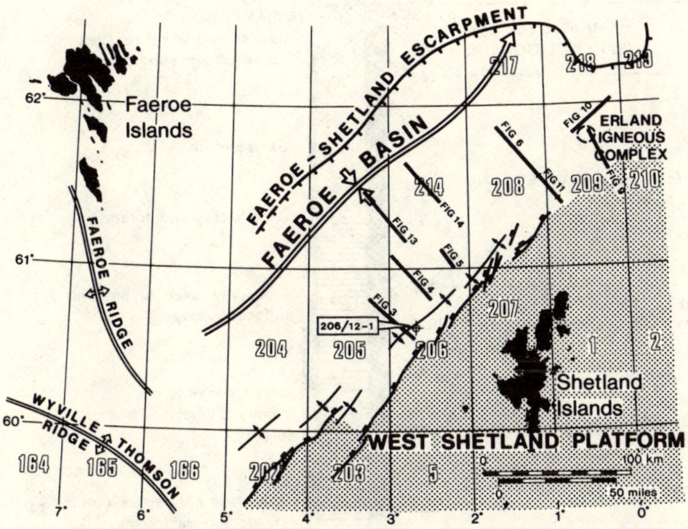


Fig. 1. Tectonic elements of the Faeroe-Shetland Channel, showing well 206/12-1 and the location of cross-sections.

On that traverse of the Faeroe Basin flank the development of the Palaeocene is unremarkable. However, along strike to the northeast the basinward thickening of the Palaeocene is pronounced (Fig. 4). Here, above a basinward thickening wedge of Cretaceous rocks, the Palaeocene maintains a steady thickness of nearly 500 milliseconds as far as a zone of down-to-the-basin faults. These are interpreted as listric or growth faults, and are accompanied by anticlinal roll-over on the downthrown side. The faults are at the hinge-line of the Palaeocene basin and to the northwest the interval rapidly attains a thickness of over 2000+ metres (2.0 seconds) before thinning again toward the Faeroe Basin axis.

Similar listric faulting along the regional strike to the northeast again accompanies a rapid basinward thinning of the Palaeocene (Fig. 5).

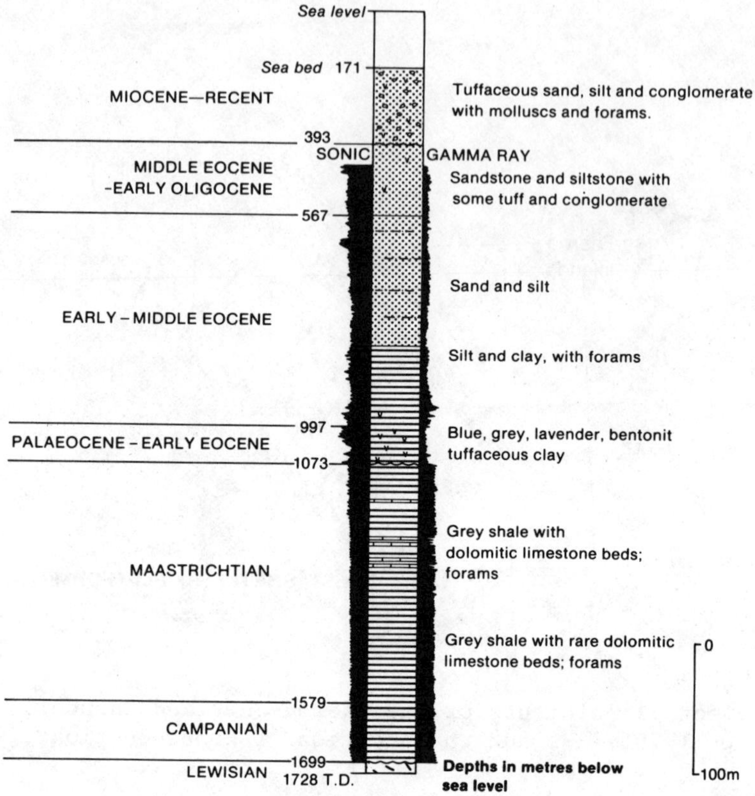


Fig. 2. Well 206/12-1, stratigraphic column. For location see Figure 1.

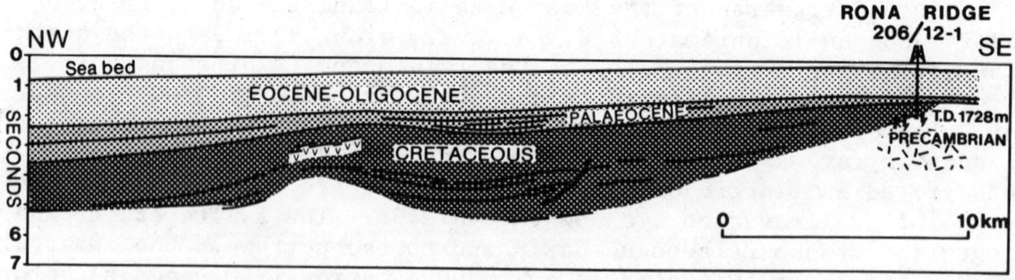


Fig. 3. Geological interpretation of a regional seismic section across the southeast flank of the Faeroe Basin. The v symbol represents possible igneous intrusion. Note listric growth-faulting in the Cretaceous. For location see Figure 1.

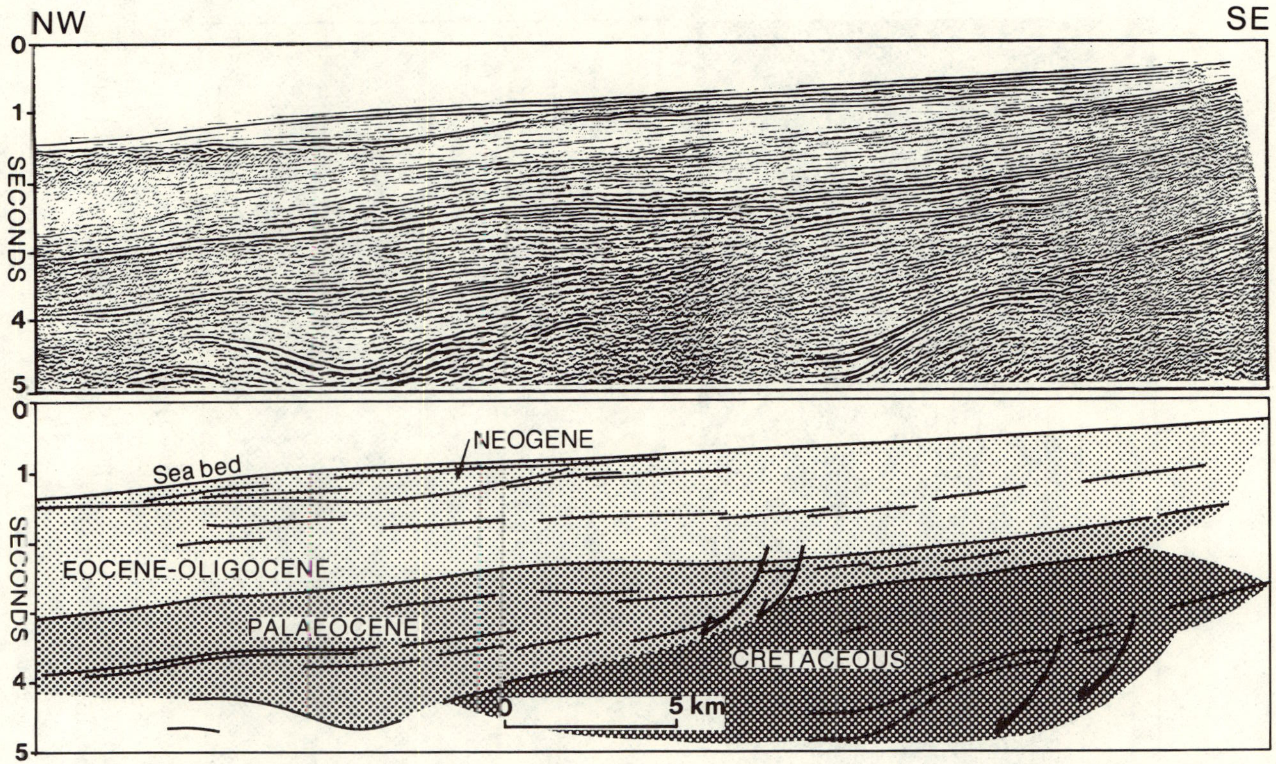


Fig. 4. Seismic section (unmigrated) and geological interpretation across the southeast flank of the Faeroe Basin. Note listric growth-faults in the Cretaceous and Palaeocene, and Neogene (? to Quaternary) fan. For location see Figure 1.

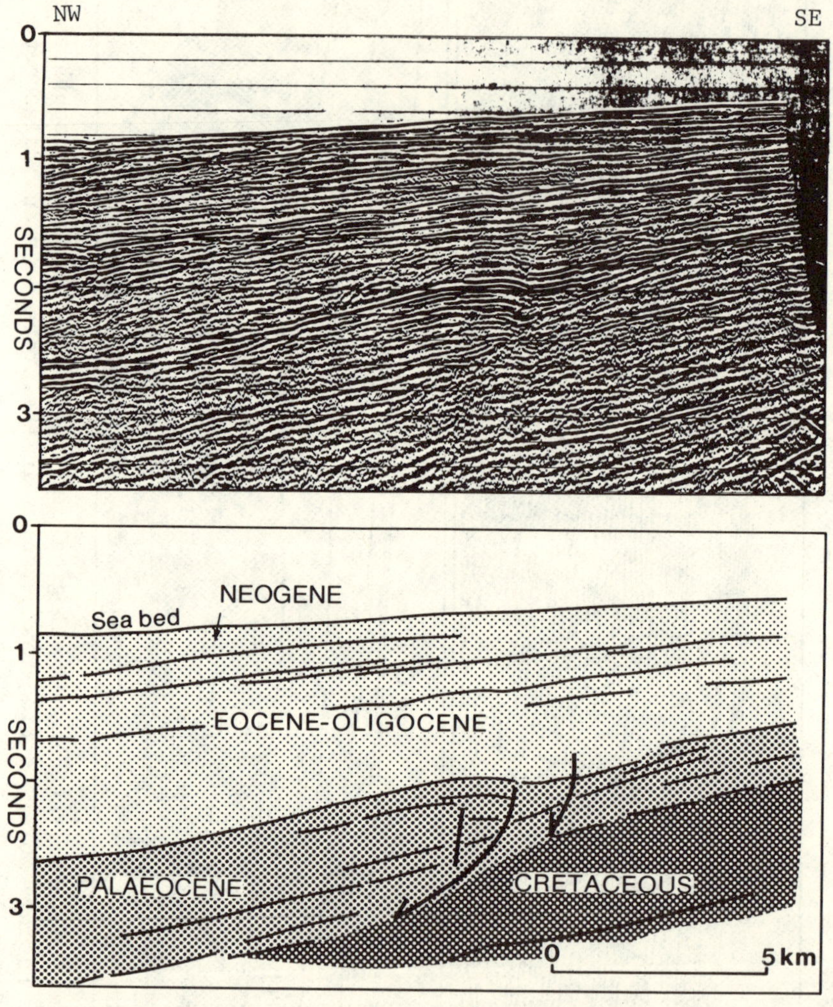


Fig. 5. Seismic section (migrated) and geological interpretation of the Faeroe Basin flank showing listric growth-faulting in the Palaeocene, accompanied by anticlinal roll-over. For location see Figure 1.

Listric faulting and anticlinal roll-over are common phenomena at the tilted margins of rapidly-deposited, thick, marine and deltaic sequences and are a response to gravitational forces acting on the sediment pile. As well as being clearly visible in the Palaeocene section they can also be seen affecting the Cretaceous on Figures 3 and 4, associated with a Cretaceous depocentre which was generally to the southeast of the later, Palaeocene, one.

On the next traverse into the basin (Fig. 6), the character of the basin flank has changed. There is no evidence of listric faulting nor of a Palaeocene hinge-line. However, it is known that the Palaeocene is thick, reaching in excess of 1000 m and including several hundred metres of Danian beds. Basinwards on the same traverse it is the turn of the Eocene-Oligocene succession to thicken rapidly northwestwards. There is no information on its nature in the deeper, thicker, part of the Faeroe Basin although a persistent feature of its seismic character is a jumble of apparently steep dips.

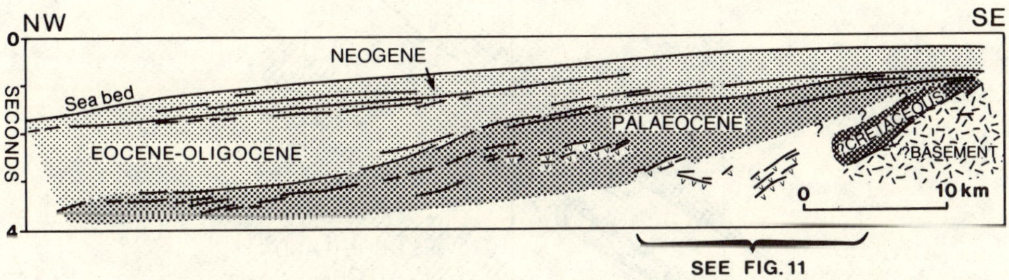


Fig. 6. Geological interpretation of a regional migrated seismic section of the Faeroe Basin flank showing abundant basic igneous intrusions (v symbol) in the Palaeocene and older section. For location see Figure 1.

The Neogene to Quaternary of the Faeroe Basin is thin and, indeed, locally so thin as to be missed in the shallower section of exploration holes before the conductor pipe is set and drill cuttings are obtained. Where it has been identified in wells it is a glauconitic sand, and where its seismic stratigraphy can be seen (e.g. Figs. 4 and 6) it progrades across the basin flank to give a lens, thinning basinwards. In the deeper parts of the Faeroe Basin it occurs as a thin sub-horizontal sequence onlapping against any rises of the mid-Tertiary surface (see Ridd 1981, Fig. 5).

The boundary between the two sedimentary sequences, here designated Eocene-Oligocene and Neogene-Quaternary, is not precisely dated and it could be the equivalent of Roberts' R4 horizon of the Rockall region which he (Roberts 1975) argues instead occurs between the Eocene and Oligocene; that interpretation was favoured by the present writer in his earlier account of the Faeroe Basin (Ridd 1981).

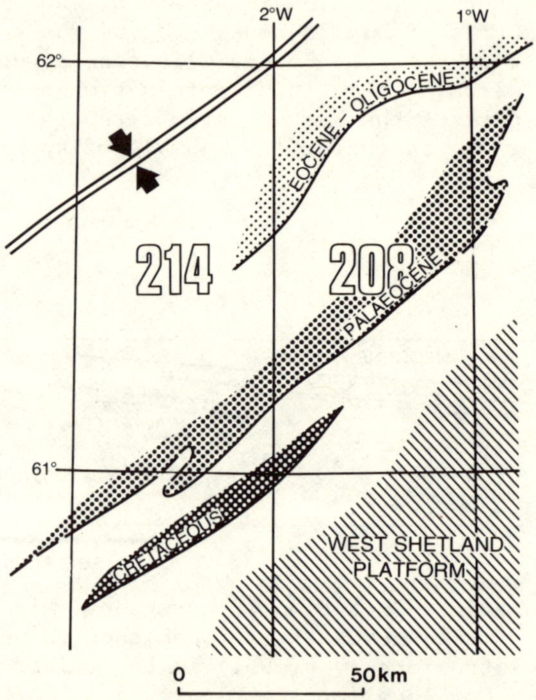


Fig. 7. Map of part of the southeast flank of the Faeroe Basin showing Cretaceous, Palaeocene and Eocene-Oligocene depositional hinge-lines and their migration into the basin.

When the basin-margin hinge-lines (i.e. the lines at which the basinward thickening is most rapid) are plotted (Fig. 7) they can be seen to migrate through the Cretaceous, Palaeocene and Eocene-Oligocene both basinwards and apparently axially along the plunge of the Faeroe Basin.

## IGNEOUS ROCKS

Tuffaceous beds have been encountered in virtually all of the wells drilled in the Faeroe Basin. Close to, or just above, the Palaeocene-Eocene boundary the tuffaceous content is very high and this lithostratigraphic unit, the Rogaland Group of North Sea nomenclature, gives rise to a convenient seismically-mappable horizon over most of the Faeroe Basin. Upward, through the Eocene-Oligocene and even into the Miocene, some wells (for example 206/12-1, Fig. 2) record the continuing presence of tuffaceous material.

Several wells, particularly in the northeast part of the basin have found substantial thicknesses of lava and intrusive rock at various stratigraphic levels beneath the Lower Eocene tuff.

The well 209/9-1, drilled by BNOC 100 km NNE of the Shetlands (Lat. 61°45'44.298"N, Long. 0°20'39.193"W) encountered over 400 m of lavas straddling the Cretaceous-Tertiary boundary (Fig. 8). Chalmers and Western (1979) had earlier predicted the occurrence of a major igneous centre, on gravimetric and magnetic evidence, and the 209/9-1 well and the associated seismic data have gone some way to confirming its existence.

The lower succession of lavas and tuffs is of particular interest, first because microfossils from thin sedimentary interbeds yielded a Cretaceous (Campanian) age and, second, because it is an acid igneous suite of rocks. The following rock-types have been described by J.E. Dixon and B.G.J. Upton (unpublished report) : rhyolite, rhyolitic glass, tholeiitic basalt (possibly a sill), rhyolitic obsidian, pitchstone, pumice, dacite and glassy dacite. A range of complex textures including mixed lava-sediment rocks (peperites) are consistent with explosive interaction of lava with wet, partially consolidated muddy sediment.

The overlying suite of basic extrusive rocks could not be dated palaeontologically. However, they are directly overlain by Lower Eocene claystone and a basal sandy facies containing rhyolite clasts possibly indicates a basal Tertiary unconformity. They are probably, therefore, of Palaeocene to Early Eocene age. They are described by Dixon and Upton as probably quartz tholeiites and, again, texturally they show evidence of having been erupted into, or completely under, water.

On Figure 9 the succession of lavas encountered in well 209/9-1 is plotted onto a NW - SE seismic line which extends over the igneous centre inferred by Chalmers and Western. The acid volcanic sequence cannot be extrapolated far, but the overlying tholeiite sequence can be seen thickening toward the northwest before becoming unmappable in a seismically characterless zone. This latter is interpreted to be the source of the lavas and pyroclastic rocks of this well and, insofar as it does not affect the Eocene and younger succession overlying it, its age is consistent with the Campanian to possibly Early Eocene age ascribed to the volcanic rocks of the well. J. A. Chalmers has recently given the name Erland Igneous Complex to the feature he and Western inferred in 1979 and that name is formally adopted here.

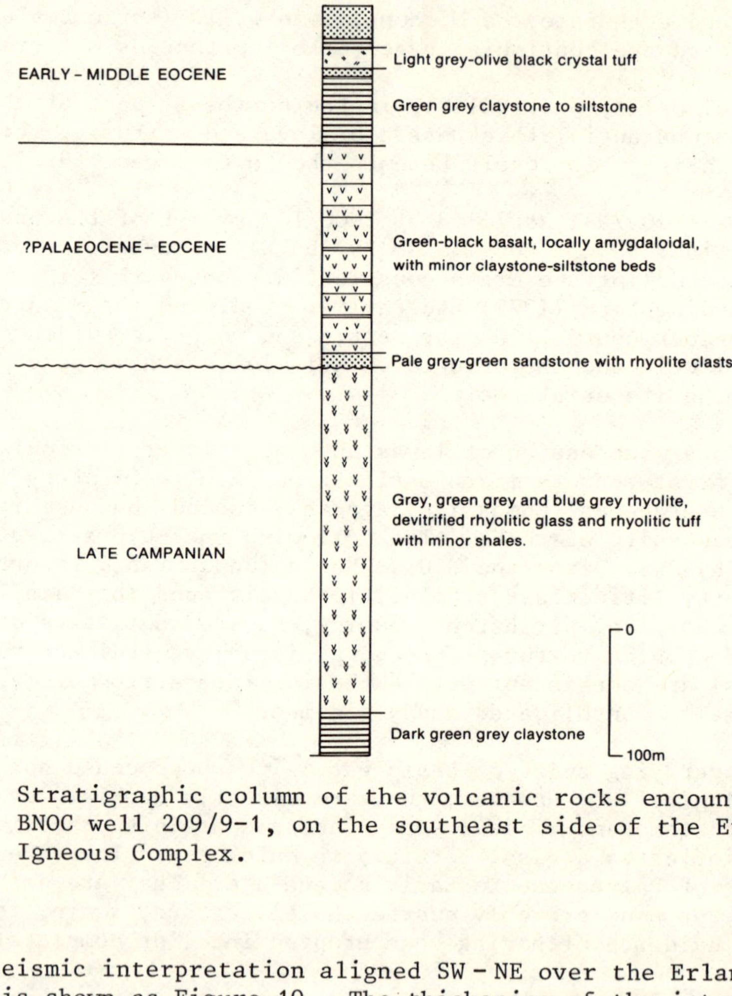


Fig. 8. Stratigraphic column of the volcanic rocks encountered in BNOC well 209/9-1, on the southeast side of the Erland Igneous Complex.

A seismic interpretation aligned SW - NE over the Erland Igneous Complex is shown as Figure 10. The thickening of the interval of lava flows toward the complex can be seen, as can the interfingering of lavas and sediments on the NE side and the progressive advance of sedimentation over the lava pile toward the complex, eventually, in the Early Eocene, overstepping it completely. A gentle drape of the sedimentary cover of the complex can also be seen.

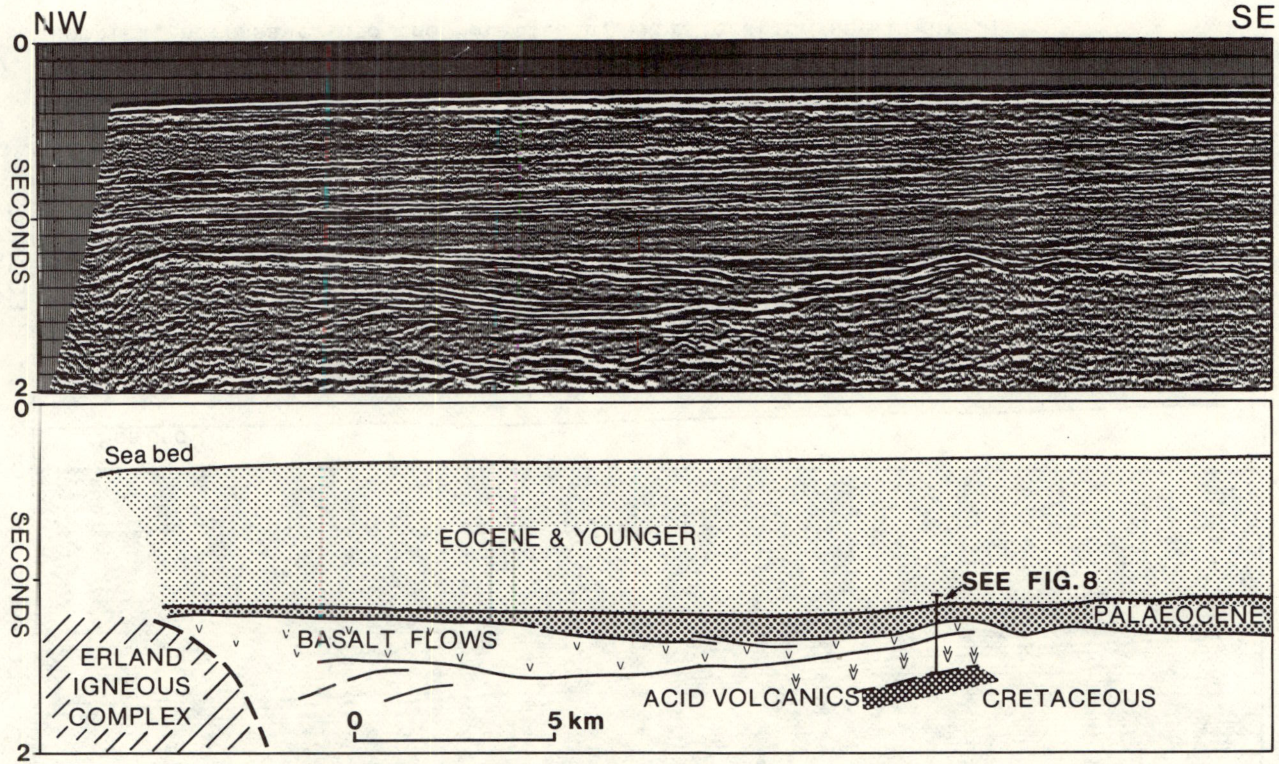


Fig. 9. Seismic section (migrated) and geological interpretation from the Erland Igneous Complex southeastwards through the 209/9-1 well; (the volcanic interval in the well is identified by a vertical bar). For location see Figure 1.

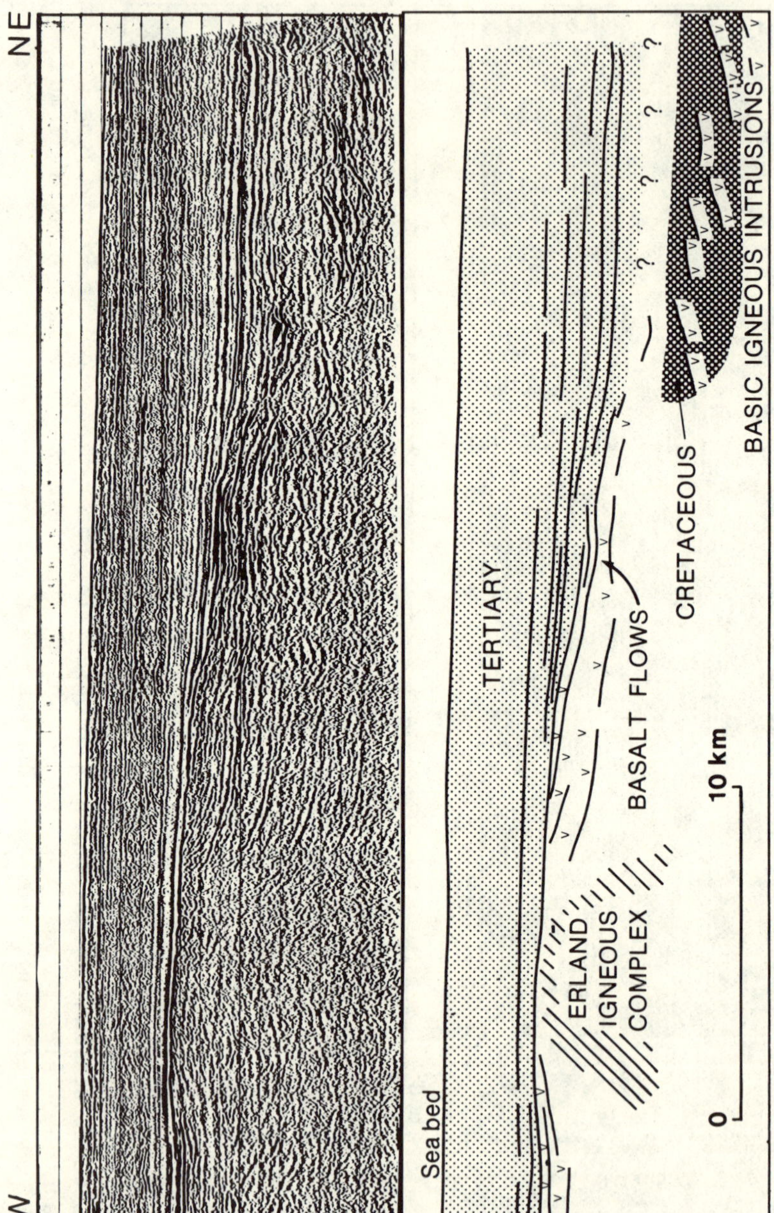


Fig. 10. Geological interpretation of a regional migrated seismic section through the Erland Igneous Complex, northeast into the Møre Basin. For location see Figure 1.

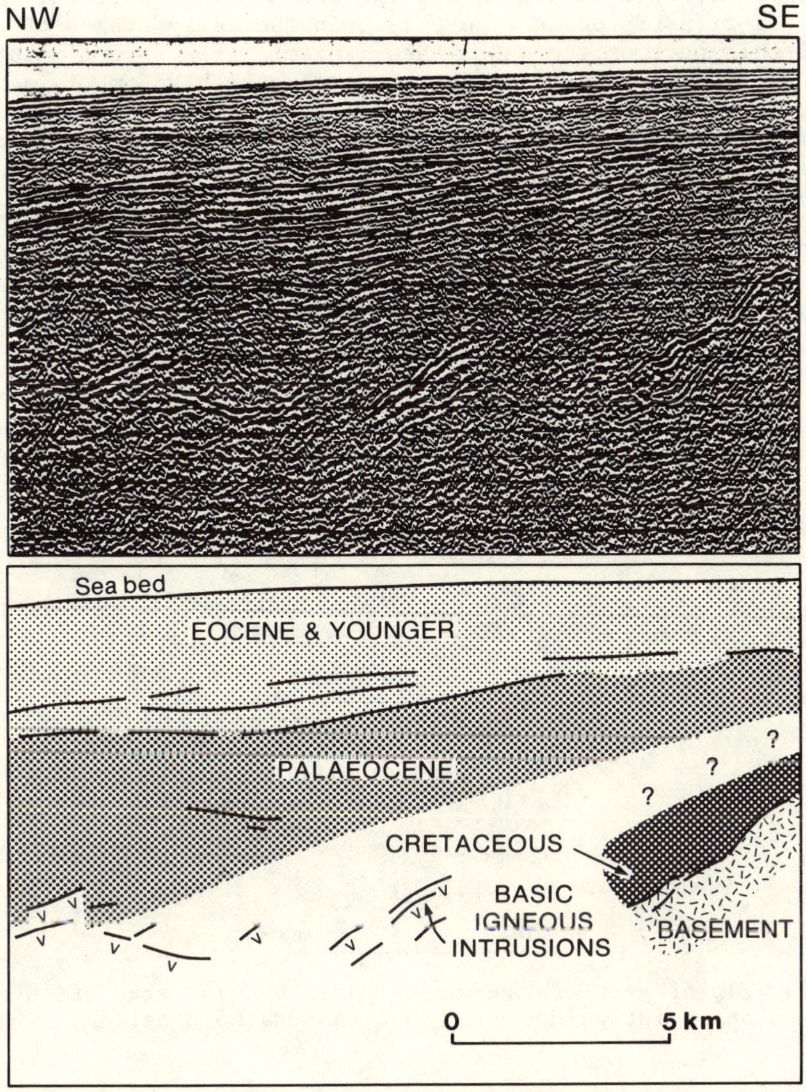


Fig. 11. Seismic section (migrated) and geological interpretation on the flank of the Faeroe Basin showing abundant basic igneous intrusions in the Palaeocene and older succession. For location see Figures 1 and 6.

North of approximately 61° North Latitude, the southeast flank of the Faeroe Basin is characterised by short, generally steeply-dipping, high-amplitude, seismic events which form an indistinct zone on seismic sections in the Palaeocene and older succession. A typical example is illustrated in Figure 11. In the early stages of exploration of this part of the basin these seismic events were interpreted as the tops of deep, tilted fault-blocks, possibly comprising pre-Cretaceous rocks. This geological model was an extrapolation from the North Sea where it has been proved to be commonly applicable. However, drilling has proved that model to be invalid north and northwest of the Shetlands. The short steeply-dipping seismic events are intrusions of basalt, dolerite and gabbro (see, for example, Figs. 6 and 10). In a vertical sequence of sediments say 1000 m thick, six to ten intrusions may be present, ranging in apparent (drilled) thickness from 2 m to more than a hundred metres.

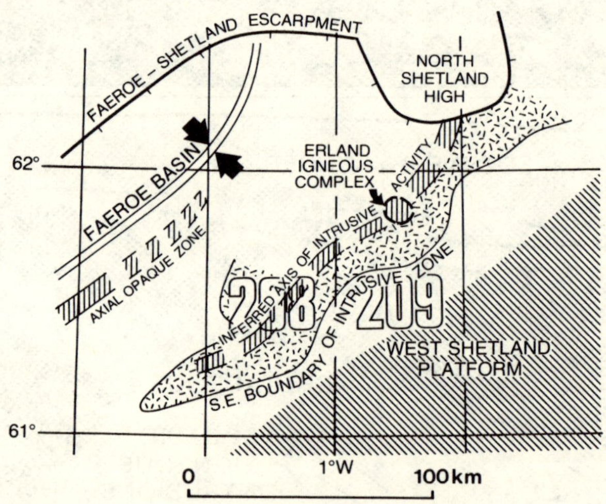


Fig. 12. Map of part of the Faeroe Basin showing area of intense basic intrusions and the Axial Opaque Zone.

The southeast boundary of these intrusions forms a curving line running overall NE to SW, parallel to the basin (Fig. 12), and it appears that their emplacement is genetically related to the formation of the basin itself. It might be speculated that this broad zone of sills (and probably to a lesser extent dykes), only the southeast boundary of which is clear, is related to a NE - SW axis of intrusive activity which runs through the Erland Igneous Complex to the North Shetland High.

En echelon with this inferred axis is a seismically opaque or nearly opaque zone which runs along the deep part of the Faeroe Basin (southeast of the actual axis) and which this writer suggested in his previous paper (Ridd, 1981) is probably a chain of Early Tertiary igneous centres or a linear igneous complex. Interpreted seismic traverses across it are shown as Figures 13 and 14. On both sections there is an anticlinal arching and thinning of the pre-Eocene rocks across the Axial Opaque Zone and commonly this coincides with a gentle anticlinal arching at the level of the base of the Eocene (Fig. 13). As the name implies, the zone is typically seismically opaque in the pre-Eocene succession although on Figure 13 a seismic event can be phantomed through it. On Figure 14 the pre-Eocene seismically opaque zone underlies an Eocene-Oligocene opaque interval, implying some degree of local post-Palaeocene activity along this belt of presumed igneous activity.

No wells have been drilled on or near the Axial Opaque Zone and gravity and magnetic studies are not yet complete. However, this writer's provisional interpretation of the feature is still that it is a linear igneous complex of largely Palaeocene age (but perhaps with some Eocene igneous activity too). In cross-section there are similarities between the Axial Opaque Zone and the Erland Igneous Complex (compare Figs. 13 and 14 with Figs. 9 and 10).

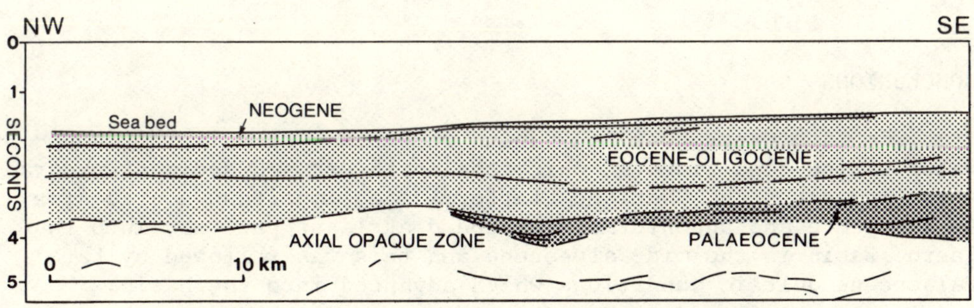


Fig. 13. Geological interpretation of a seismic section across the Axial Opaque Zone, near the middle of the Faeroe Basin. Note the similarity to sections over the Erland Igneous Complex; by analogy, the interval here shown as Palaeocene may include volcanic rocks. Although reflections can be seen at around 5.0 seconds TWT, particularly on the southeast part of the section, their identity is unknown. For location see Figure 1.

One model which will need to be examined against the gravity and magnetic evidence is that the Axial Opaque Zone was an incipient spreading centre in the Palaeocene.

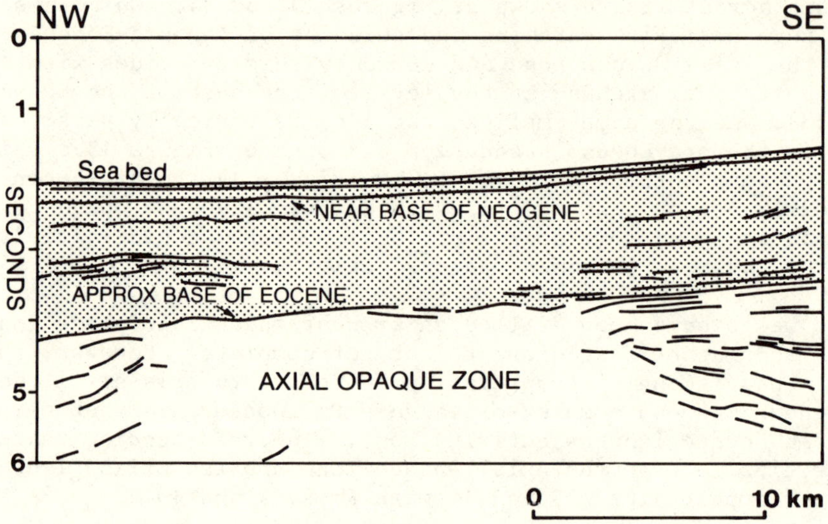


Fig. 14. Geological interpretation of a migrated seismic section over the Axial Opaque Zone.

## CONCLUSIONS

The Late Cretaceous and Danian were times of uniformly deep-water marine conditions in which the northwestward sediment progradation indicates the Shetland Platform as an important source area. An emergence and unconformity extended at least part way into the Faeroe Basin in the mid-Palaeocene and this was followed by Late Palaeocene deltaic conditions which advanced into the basin.

Although some (acid) volcanism occurred as early as the Campanian centred on the Erland Igneous Complex, the acme of volcanism was at the end of the Palaeocene. The Faeroe plateau basalts were erupted and spread over very wide areas of dry land, although locally (as at the Erland Igneous Complex) there is evidence of similar volcanism in an aqueous environment. One can envisage this largely sub-aerial landscape passing laterally into deltaic conditions which might together have provided the land-bridge for the spectacular but brief migration of terrestrial vertebrates between North America and Europe at the end of the Palaeocene (see M. C. McKenna, this volume).

Also at about the end of the Palaeocene there was crustal distension which allowed igneous intrusion along two en echelon NE - SW trending zones, one a dyke and sill swarm running through the Erland Igneous Centre and the other a zone of major intrusions and perhaps extrusions running approximately along the axis of the Faeroe Basin.

This short-lived period of incipient sea-floor spreading (which may in fact have been partially sub-aerial) was no doubt the cause of the relatively thin crust beneath the Faeroe Basin (Bott 1975). As the thermal event waned the basin floor subsided and Eocene marine sediments accumulated, passing laterally into Eocene lavas along the line of the Faeroe-Shetland Escarpment and perhaps locally over the Axial Opaque Zone.

At about the Palaeogene-Neogene boundary a change in sedimentary regime occurred, possibly caused by the subsidence of the Wyville-Thompson Ridge. Such an event would have allowed circulation between the Atlantic and Arctic and the removal by current-action of most of the Neogene sediment which might have entered the basin.

## ACKNOWLEDGEMENTS

This paper is published with the permission of the British National Oil Corporation. Certain seismic sections are reproduced with the kind permission of the Western Geophysical Company. BNOC's Northwest Area exploration group led by Mr. D. D. Taylor were responsible for obtaining much of the data on which this paper is based and the writer is grateful to Mr. A.M.G. Moore, Mr. M.L.B. Miller and Mr. J. A. Chalmers for their assistance. Draughting was done by Mr. Peter Thomson and typing by Ms. G. L. Rusack.

## REFERENCES

Bott, M.H.P., 1975, Structure and evolution of the north Scottish shelf, the Faeroe block and the intervening region, in: "Petroleum and the continental shelf of northwest Europe", Vol 1, A. W. Woodland, ed., Applied Science Publ., England, 105-113.

Cashion, W.W., 1975, The geology of the West Shetland Basin. Proc. Offshore Europe Conf. 1975, Spearhead Publications, 216.1-216.7.

Chalmers, J.A. and Western, P.G., 1979, A Tertiary igneous centre north of the Shetland Islands, Scot. J. Geol., 15, 333-341.

Deegan, C.E. and Scull, B.J., 1977, A standard lithostratigraphic nomenclature for Central and Northern North Sea, Inst. Geol. Sci. Report 77/25.

Ridd, M.F., 1981, Petroleum geology west of the Shetlands, in: "Petroleum Geology of the Continental Shelf of Northwest Europe", L.V. Illing and G. D. Hobson, eds., Heyden, London.

Roberts, D.G., 1975, Marine geology of the Rockall Plateau and Trough, Phil. Trans. Roy. Soc. London, 278A, 447-509.

# FAEROE-SHETLAND ESCARPMENT AND CONTINENTAL MARGIN NORTH OF THE FAEROES

David K. Smythe

Institute of Geological Sciences
19 Grange Terrace, Edinburgh EH9 2LF

## ABSTRACT

The Escarpment is re-interpreted as the buried feather-edge of a thick pile of early Eocene flood basalts, overlying a thinner but more widespread layer of basalts of late Palaeocene age. The Escarpment does not, therefore, define the continent-ocean boundary in the southern Norwegian Sea. In the Faeroe-Shetland Trough and Møre Basin the basalts overlie several kilometres of sediments ranging in age from Palaeocene to at least as old as early Cretaceous, resting in turn on thin crust, with the Moho at around 15 km. The continent-ocean boundary north of the Faeroes underlies a belt of north-dipping 'smooth' intra-basalt reflections seen on multichannel reflection data, which is interpreted as oceanic layer 2, formed during subaerial spreading immediately preceding anomaly 24 time.

## INTRODUCTION

During the last decade the continental shelf and margin north and west of the Shetlands have been commercially surveyed for oil in increasing detail. The resulting data currently include multi-channel seismic reflection profiles, aeromagnetic maps, and more than thirty wells (Figure 1). Regional interpretations, which maintain the confidentiality of sensitive and costly commercial data while providing a stimulus to the exploration industry, are now being published (e.g. Cashion, 1975; Ridd, 1981 and this volume). IGS makes confidential assessments of commercial data for the Department of Energy, and the present regional interpretation is also based on non-confidential IGS and academic information. This account deals briefly with two major geophysical

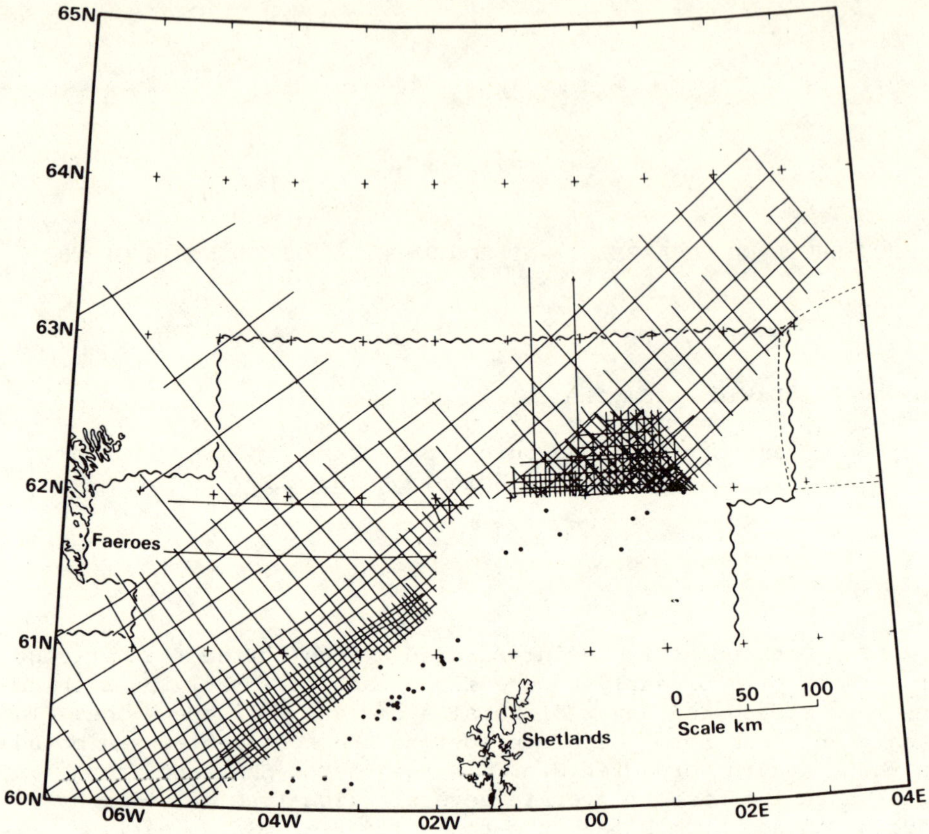


Figure 1. Commercial geophysical data acquired in the Faeroe-Shetland region since c.1972. Solid lines are multichannel reflection profiles (mostly 24-fold); round the Shetlands and south of 62°N into the North Sea the density of lines is too great to be shown. Dots are commercial wells drilled in the West Shetland area and northern Viking Graben. Wavy line denotes the outer limit of commercial aeromagnetic coverage, and dashed line is the SW limit of the published aeromagnetic map of the Norwegian shelf (Åm, 1970).

features of the region which are also relevant to other Atlantic marginal areas and which will be described in more detail elsewhere. These are (1) the nature and origin of the Faeroe-Shetland Escarpment (Smythe *et al.*, in press) and (2) the oceanward-dipping intra-basement reflections observed round the northern and western margins of the Faeroes Plateau (Smythe, in preparation).

FAEROE-SHETLAND ESCARPMENT

The Faeroe-Shetland Escarpment was originally identified by Talwani and Eldholm (1972) as two linear NE-SW trending segments, identified on eight geophysical traverses. The buried SE-facing escarpment feature was interpreted by them as the continent-ocean boundary, at which 'subsided continent', overlain by the sediments of the Faeroe-Shetland Trough and Møre Basin, is downfaulted against elevated early Tertiary oceanic crust on the NW. In 1974 IGS carried out a detailed geophysical survey covering the area between the two segments: 62°-63°N, 1°W-1°E (Figure 2). The results of our re-interpretation of the nature and origin of the escarpment (Smythe *et al.*, in press) are as follows:

(1) The escarpment is a single sinuous feature (Figure 2) caused by the wedging-out of a series of early Tertiary basalts

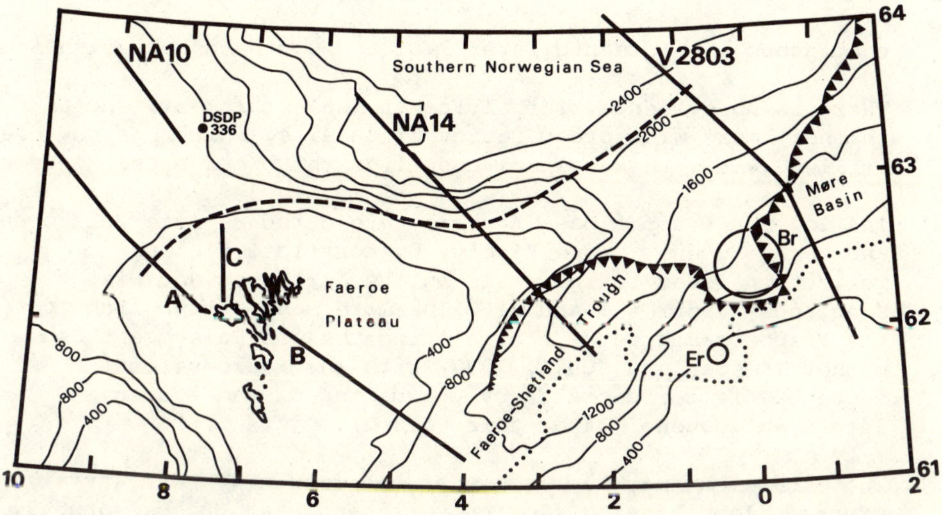


Figure 2. Subcrop of Faeroe-Shetland Escarpment (triangle ornament). Feather-edge of late Palaeocene basalts (see Figure 3) is shown (dotted line). Bathymetry in metres. V2803 - Lamont-Doherty profile (Figure 4). NA14, NA10 - Western Geophysical reflection profiles (Figures 5 and 6 respectively). A,B,C - NASP crustal refraction lines. Dashed line - southern boundary of Tertiary oceanic crust inferred from reflection profiles, gravity and converted P-waves. Round the northern flank of the Faeroes microcontinental block this line thus defines the continent-ocean boundary, but further to the NE it tentatively defines the 'ocean-ocean' contact between crust of different ages (see Figure 4 and text). Br (circle) - location of Brendan pre-Tertiary seamount below basalts. Er (circle) - Erland Tertiary volcanic centre.

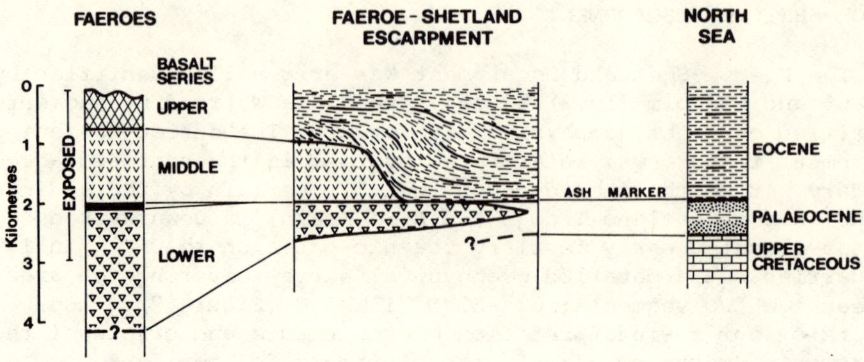


Figure 3. Diagrammatic regional correlation of the Palaeogene, with schematic depth scale. Proportion of succession exposed on the Faeroes is indicated; remainder of Lower Series of basalts is inferred from refraction surveys (Pálmason, 1965). Black band separating Lower and Middle Series represents 10m of coal-bearing sediments, overlain by 100m of tuffs and agglomerate.

contiguous with oceanic layer 2 of the southern Norwegian Sea.

(2) There is no evidence of faulting at the escarpment, and a thinner, more widespread series of basalts, which has now been drilled (Ridd, this volume) underlies the upper basalt series.

(3) The lower and upper basalt series are dated as late Palaeocene and early Eocene, respectively, by correlation along reflection lines to the wells in the North Sea and West Shetlands (Figures 1 and 3). The more speculative link to the Faeroes (Figure 3), correlating the end-Palaeocene ash marker (Deegan and Scull, 1977) with the Faeroese coal sequence, is corroborated by the dating of the latter as latest Palaeocene (Lund, this volume).

(4) The escarpment may have grown as the result of terrestrially-erupted flood basalts flowing south and east, to be abruptly halted at the contemporary shoreline bordering a restricted shallow-water shelf to the SE. The SE-facing protrusion in the subcrop (Figure 2) may thus mark a contemporary peninsula, formed by a slight relative elevation of the crust due to the underlying buried Brendan seamount (Br in Figure 2).

(5) Gravity and magnetic profiles over the escarpment show no consistent change in amplitude, wavelength, regional gradient, etc. ascribable to any major change in crustal structure beneath the escarpment. Figure 4 (bottom) is the interpretation of a 2-D gravity model (Smythe *et al*., in press) along part of Lamont-Doherty profile V2803 (Talwani and Eldholm, 1972; Talwani, 1974), located in Figure 2. Standard densities

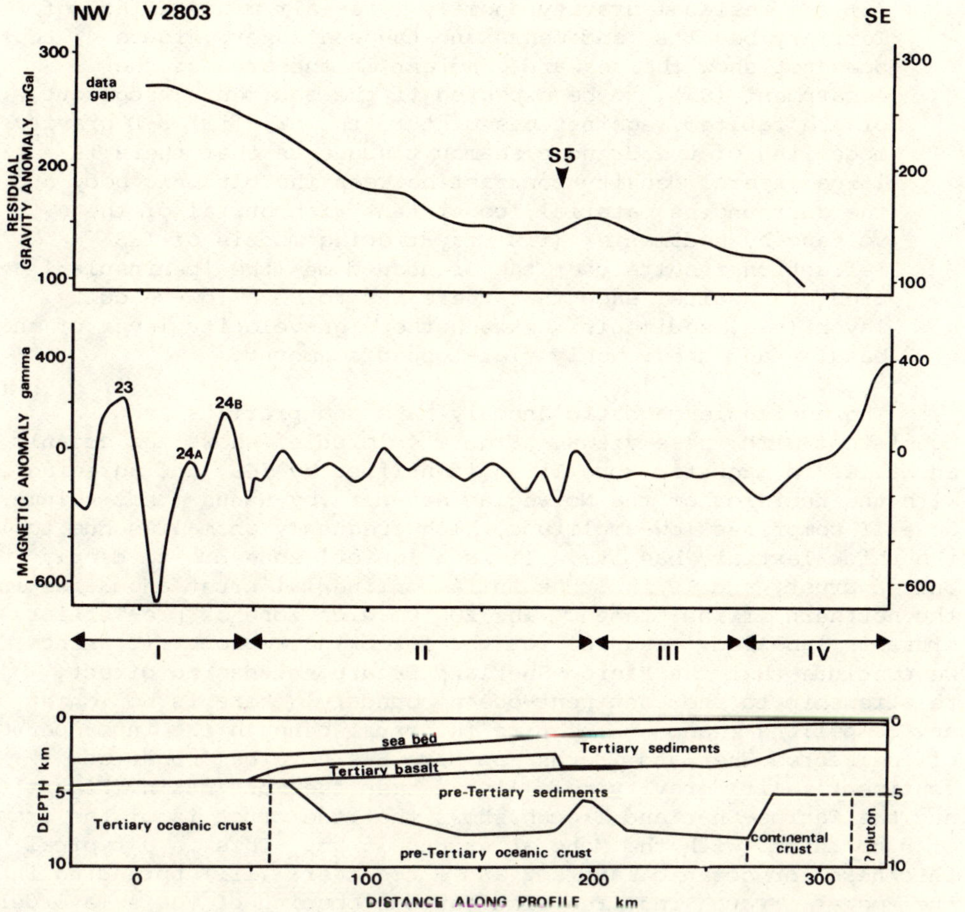


Figure 4. Gravity modelling along LDGO profile V2803 (located in Figure 2). Residual anomaly is free-air after removal of effects of sea layer, Tertiary basalts and Tertiary sediments (bottom cross-section). Note lack of gradient around S5, the location of the Faeroe-Shetland Escarpment. Zones I-IV of magnetic anomaly (middle profile) correlate respectively with Tertiary oceanic crust, Tertiary basalts, deeply buried crust, and continental crust. Airy-type modelling suggests a crustal thickness of 7-8 km for the Tertiary and pre-Tertiary oceanic areas, thickening to 20 km below the Viking Graben at the SE end of the model.

are used (Worzel, 1974). The supra-crustal part of the model is all directly constrained by seismic reflection, except the postulated continuation of the pre-Tertiary sediments beneath the thick basalts. The evidence for their continuation is; (i) the residual gravity anomaly (free-air minus effect of Tertiary basalts, sediments and the sea layer; Figure 4, top) does not show the seaward gradient in the area of the escarpment (S5), to be expected if the sediments wedge out, or are faulted, against basement to the NW; (ii) 3-D gravity modelling of the Brendan seamount suggests that there is a large lateral density contrast between the plutonic body and the surrounding material, consistent with burial of the volcano by sediments; (iii) ray-tracing models of IGS refraction results over the Brendan dome (the 'peninsular' area of basalts) show that there has to be a low-velocity layer (e.g. sediments) between the high-velocity layer of the basalts and the roughly flat-topped seamount.

The available magnetic anomaly maps and profiles are consistent with these views. Figure 4 (middle) shows the oceanic anomalies in magnetic zone I, as identified by IGS, and agreeing with the revision of the Norwegian Sea data by Nunns (this volume). Zone II comprises low-amplitude, high frequency anomalies due to the thick Tertiary basalts, III is a 'quiet' zone due to deeply buried crust, and IV includes intra-continental crust anomalies in the northern Viking Graben. The 200 km wide zone of pre-Tertiary crust is labelled 'oceanic' for the following reasons: (i) since we conclude that the Faeroe-Shetland Escarpment has no direct relationship to the continent-ocean boundary, there is no longer any compelling reason to require the crust beneath the inner parts of the Faeroe-Shetland Trough and Møre Basin to be 'subsided continent'; (ii) gravity modelling across the Møre Basin (Figure 4) and the Faeroe-Shetland Trough shows that the crust is of the order of 8 km thick, with the Moho at around 15 km. This is a typical thickness for oceanic layers 2 and 3 together; (iii) spreading in the Rockall Trough requires simultaneous opening of the same order in the Faeroe-Shetland Trough and eastern Norwegian Sea (Roberts, 1974, p.351; Bott, 1975, p.198). Since individual sections of this ribbon of pre-Tertiary thin crust are unlikely to have evolved by radically different processes from each other, and since the Rockall Trough is generally accepted as oceanic in origin (Scrutton, 1972; Roberts, 1974) it follows that the coeval crust north of the Wyville Thomson Ridge is also oceanic. Available evidence on the crustal structure of these northern regions, in particular, remodelling of Hinz's (1972) crustal refraction line in the eastern Norwegian Sea by (a) simple 'layer-cake' models, or (b) ray-tracing (my unpublished work) is consistent with an orthodox oceanic origin.

The thin crust beneath the Faeroe-Shetland Trough, Møre Basin and Eastern Norwegian Sea is now believed to be overlain by sediments which are at least as old as early Cretaceous in age (e.g. Talwani and Eldholm, 1972; Cashion, 1975; Jørgensen and Navrestad, 1981; Ridd, this volume; Smythe *et al.*, in press). The postulated episode of early Cretaceous sea-floor spreading here (Roberts, 1974) is therefore untenable. The alternative dating of the spreading episode as Stephanian to early Permian (Russell, 1976; Russell and Smythe, 1978) remains valid, as does the conjecture of an intra-continental rifting and subsidence episode during the same period. At present, sea-floor spreading is to be preferred as the mechanism, simply because it leads to clear predictions (as yet unrefuted) about plate geometry, crustal structure, regional magmatism etc., in contrast to the 'subsided continent' hypothesis as applied currently to these areas (Talwani *et al.*, 1981).

## CONTINENTAL MARGIN NORTH OF THE FAEROES

Where is the continent-ocean boundary, if not at the escarpments? It is postulated here that the start of oceanic layer 2 can be defined by the lower end of the "oldest" of the oceanward-dipping series of intra-basement reflectors now commonly observed at passive margins on multichannel reflection lines. The example of the northern Faeroes margin (Figure 5) is particularly good, because it demonstrates the passage northwards from the nearly flat-lying early Eocene terrestrial flood basalts of the Faeroes Plateau into the dipping series, thence into 'normal' oceanic layer 2 which still, however, shows some dipping reflectors. The crucial point is that the dipping series below a smooth top basalt

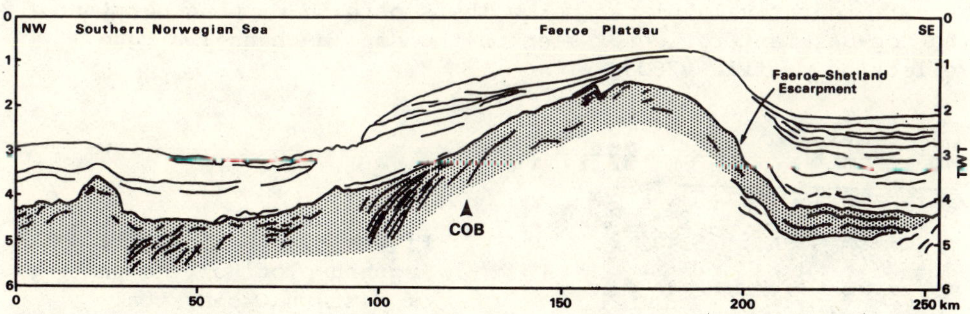


Figure 5. Line-drawing interpretation of Western Geophysical NA14 (located in Figure 2). Strong reflector is top of the basalts (stipple) inferred from reflector amplitude, high interval velocities beneath, magnetic anomalies, and since corroborated by drilling near the Erland pluton (Figure 2 and Ridd, this volume). COB - continent-ocean boundary, defined at onset of dipping reflections (see text). Portion of line 0-175 km has been reprocessed (Smythe, in preparation).

reflector (such as from 90-130 km on Figure 5) does not *overlie* oceanic crust (Talwani *et al.*, 1981), but that it is *part of* the oceanic crust. It may correspond to the mid-ocean ridgeward-dipping flows of the upper part of oceanic layer 2, as postulated by Cann (1974, figure 2). The continuity and length of the reflections, and the smooth upper surface, is considered to be due to subaerial spreading. The abrupt change in character (as at 90 km on Figure 5) from smooth to rough top basement, and from good to poor intra-layer 2 reflections, represents merely the change from subaerial to submarine spreading as the initially elevated mid-ocean ridge subsided below sea-level. Part of line NA14 has been reprocessed, including migration and depth conversion, by Western Geophysical (client IGS), in order to test further the nature of the intra-basalt reflectors. Details of this work will be presented elsewhere (Smythe, in preparation).

The continent-ocean boundary, as inferred from the above criterion and applied to the profiles north of the Faeroes (Figure 1) such as NA14 (Figure 5), is shown in Figure 2. It follows a crescentic free-air gravity high (Casten and Nielsen, 1975), which may be an edge-effect anomaly. The boundary mapped in this way also agrees with an independent determination of the boundary on crustal refraction line NASP A (Figure 2) inferred from converted P-waves (Bott *et al.*, 1976).

In plan, the continental crust beneath the Faeroe Plateau presumably wedges out north-eastwards in the region of 63°N, 2°W (Figure 2), so that pre-Tertiary oceanic crust is in direct contact with Tertiary oceanic crust (Figure 4). There are no multichannel profiles in this region (Figure 1), so the 'ocean-ocean' contact here is tentatively placed below the smooth NW-dipping portion of the top-basement reflector seen on the single channel airgun reflection profile V2803.

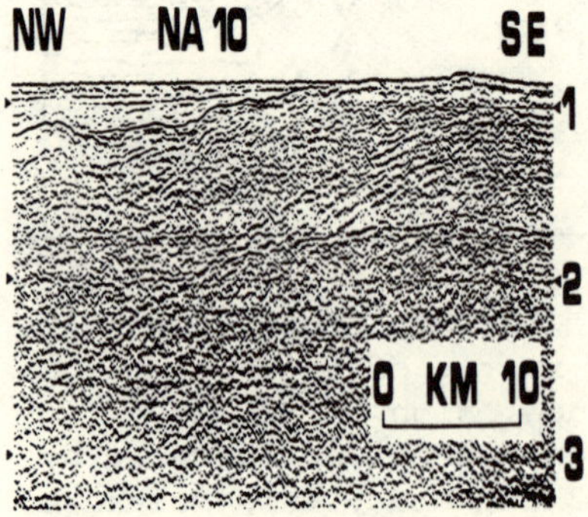


Figure 6. Photographic 'squash-plot' of the part of Western Geophysical reflection line NA10 (located in Figure 2) reprocessed for IGS. Vertical scale is in seconds of two-way time.

Part of reflection profile NA10 (Figure 2) has also been reprocessed by Western Geophysical for IGS. It lies on the NE flank of the Iceland Faeroe Ridge, about 12 km from DSDP 336, which penetrated subaerially weathered Eocene basalt (Talwani *et al.*, 1976). The reprocessed part (Figure 6) shows basalt at the sea bed towards the SE, but buried by up to 300 m of sediment towards the NW. The reprocessing has successfully reduced the amplitude of the strong sea-bottom multiples, and allowed the primary NW-dipping intra-basalt reflectors to stand out. The deepest of these extends to about 1.6 s TWT (∿4 km depth) below the sea bed, in good agreement with the thickness of layer 2 on NASP A (Figure 2) of 5-7 km (Bott and Gunnarsson, 1980). Line NA10 thus provides further corroboration of the subaerial sea-floor spreading origin for dipping reflector sequences.

## IMPLICATIONS FOR THE EARLY OPENING OF THE NORTH ATLANTIC

Preliminary 'pre-drift' reconstructions of the North Atlantic, using the continent-ocean boundary criteria discussed above and applied to various segments of the passive margins, result in a tight reconstruction, not unlike the classic 'Bullard' fit (my unpublished work). The Faeroes block fits tightly against the Blosseville coast of Greenland, interpreted as approximating to the continent-ocean boundary by GGU workers (Thorning, this volume). Hatton Bank appears to have rotated clockwise by several degrees about a pole in the northern Rockall Trough during the early Tertiary, and the nearby Outer Bailey Bank may have moved NW with Greenland (relative to a 'stationary' Europe) during the same interval. Furthermore, Ridd (this volume) suggests that the 'axial opaque zone' of the Faeroe-Shetland Trough may have been an incipient spreading centre in the Palaeocene, implying that the Faeroes block also moved NW relative to Europe by a few kilometres during that time. Irrespective of the earlier development of the Rockall and Faeroe-Shetland Troughs, it is thus apparent that the various microcontinental blocks flanking the Troughs' NW sides moved independently of one another by small, but not insignificant, amounts during the Palaeogene.

## ACKNOWLEDGEMENTS

Western Geophysical kindly gave permission to reproduce the line drawing of their line NA14 in advance of the reprocessing subsequently carried out for IGS. This preliminary account of work involving many of my colleagues, to whom I am indebted, is published with the approval of the Department of Energy and of the Director, Institute of Geological Sciences (NERC).

REFERENCES

Åm, K., 1970, Aeromagnetic investigation on the continental shelf of Norway, Stad-Lofoten (62°-69°N). *Nor.Geol.Unders*. 266: 49-61.

Bott, M.H.P., 1975, Structure and evolution of the Atlantic floor between Northern Scotland and Iceland. *Nor.Geol.Unders*., 316 : 195-199.

Bott, M.H.P., and Gunnarsson, K., 1980, Crustal structure of the Iceland-Faeroe Ridge. *J.Geophys*., 47 : 221-227.

Bott, M.H.P., Nielsen, P.H., and Sunderland, J., 1976, Converted P-waves originating at the continental margin between the Iceland-Faeroe Ridge and the Faeroe Block. *Geophys.J.R. Astron.Soc*., 44 : 229-238.

Cann, J.R., 1974, A model for oceanic crustal structure developed. *Geophys.J.R.Astron.Soc*., 39 : 169-187.

Cashion, W.W., 1975, The geology of the West Shetland Basin. *Proc. Offshore Europe Conf. 1975*, Spearhead, Kingston, 216.1-216.7.

Casten, U. and Nielsen, P.H., 1975, Faeroe Islands - a micro-continental fragment? *J.Geophys*., 41 : 357-366.

Deegan, C.E., and Scull, B.J., 1977, A standard lithostratigraphic nomenclature for the Central and Northern North Sea. *Rep. Inst.Geol.Sci*., No. 77/25.

Hinz, K., 1972, Der Krustenaufbau des Norwegischen Kontinentalrandes (Vøring Plateau) und der Norwegischen Tiefsee zwischen 66° und 68° N nach seismischen Untersuchungen. *Meteor Forsch.-Ergebnisse* C, 10 : 1-16.

Jørgensen, F., and Navrestad, T., 1981, The geology of the Norwegian shelf between 62°N and the Lofoten Islands. *In* Illing, L.V., and Hobson, G.D. (eds) *Petroleum Geology of the Continental Shelf of North-west Europe*. Heyden, London, 407-413.

Pálmason, G., 1965, Seismic refraction measurements of the basalt lavas of the Faeroe Islands. *Tectonophysics*, 2 : 445-482.

Ridd, M.F., 1981, Petroleum geology west of the Shetlands. *In* Illing, L.V., and Hobson, G.D. (eds) *Petroleum Geology of the Continental Shelf of North-west Europe*. Heyden, London, 414-425.

Roberts, D.G., 1974, Structural development of the British Isles, the continental margin, and the Rockall Plateau. *In* Burk, C.A., and Drake, C.L. (eds) *The Geology of Continental Margins*. Springer, Berlin, 343-359.

Russell, M.J., 1976, A possible Lower Permian age for the onset of ocean floor spreading in the northern North Atlantic. *Scott. J.Geol*., 12 : 315-323.

Russell, M.J., and Smythe, D.K., 1978, Evidence for an early Permian oceanic rift in the northern North Atlantic. *In* E.-R. Neumann and I.B. Ramberg (eds) *Petrology and Geochemistry of Continental Rifts*, Reidel, Dordrecht, 173-179.

Scrutton, R.A., 1972, The crustal structure of Rockall Plateau microcontinent. *Geophys.J.R.Astron.Soc.*, 27 : 259-275.

Smythe, D.K., Chalmers, J.A., Skuce, A.G., Dobinson, A., and Mould, A., Early opening history of the North Atlantic: 1. Structure and origin of the Faeroe-Shetland Escarpment. *Geophys.J.R.Astron.Soc.*, (in press).

Talwani, M. (ed), 1974, *Underway Marine Geophysical Data in the North Atlantic, June 1961 - January 1971*. 6 vols + 10 charts. Lamont-Doherty Geological Observatory, New York.

Talwani, M., and Eldholm, O., 1972. Continental margin off Norway: a geophysical study. *Bull.Geol.Soc.Am.*, 83 : 3575-3606.

Talwani, M., Mutter, J., and Eldholm, O., 1981, The initiation of opening of the Norwegian Sea. *Oceanol. Acta* 4 (Supplement: Geology of Continental Margins), 23-30.

Talwani, M., Udintsev, G., *et al.*, 1976, *Initial Reports of the Deep Sea Drilling Project, Volume 38*. U.S. Government Printing Office, Washington D.C., 1256p.

Worzel, J.L., 1974, Standard oceanic and continental structure. *In* Burk, C.A., and Drake, C.L. (eds) *The Geology of Continental Margins*. Springer, Berlin, 59-66.

# OCEAN CONTINENT BOUNDARY UNDER THE NORWEGIAN CONTINENTAL MARGIN

Manik Talwani[1], John Mutter[1] and Karl Hinz[2]

[1]Lamont Doherty Geological Observatory of
Columbia University, Palisades
New York 10964, USA

[2]Bundesanstalt für Geowissenschaften und Rohstoffe
Stillweg 2, Postfach 51 01 53
D-3000 Hannover 51, West-Germany

Some inferences about the early opening history and the location of the ocean-continent boundary for the Iceland-Faeroe section of the Greenland-Scotland Ridge can be drawn from studies of the Norwegian continental margin further north.

A large amount of regional geophysical work including multichannel seismic experiments has been carried out on the Norwegian continental margin. A particular focus of this work has been the Vøring Plateau and Faeroe-Shetland escarpments. Talwani and Eldholm (1972) postulated that these escarpments mark the ocean-continent boundary. While there is no universal agreement about this postulate, the structures on either side of the escarpments are becoming better resolved, and we describe these below. We start by first indicating the location of the escarpments. The Vøring Plateau and Faeroe Shetland Escarpments are shown on the map in Fig. 1. The escarpments are characterized by a sharp increase in the depth to acoustic basement from NW to SE. Basement has not been reached by seismic reflection immediately SE of the escarpments.

The Vøring Plateau Escarpment can be followed from about 66°.5 N to north of 69°N. It is more prominent in the southern part than in the northern part. The Faeroe-Shetland Escarpment is, in general, more poorly defined than the Vøring Plateau Escarpment. Talwani and Eldholm (1972) relying on single channel seismic reflection as well as gravity and magnetic data, concluded that the escarpment consists of two segments. We now believe it may consist of as many as three segments. The eastern segment extends from about 64°N to about 62°.3 N. Smythe (this volume) has pointed out that commercial seismic surveys show a lobate extension of the escarpment to the SE (near 62°.4 N). We have indicated this by a dashed line, noting, however, that the characteristic negative magnetic anomaly associated with the escarpment is not displaced to the SE. Smythe also shows continuity between the two eastern sections of the escarpment. Just east of the Faeroes we have indicated the location of the third segment of the Faeroe-Shetland Escarpment. This location is speculative and is based on the change in character of the magnetic anomalies across this feature (they are much more subdued to the east) and on the existence of a topographic feature here in Nielsen's (this volume) detailed topographic map of the area. The escarpment would coincide with the location of the Faeroe Ridge as mapped by Ridd (this volume) in the southernmost part of the Faeroe's block.

In order to describe the structures associated with the escarpments, it is convenient to divide the area in the vicinity of the escarpments into four zones. The structures in these zones vary, although common features are retained in most of the seismic profiles crossing the escarpments. For illustrative purposes, we have chosen profiles C 165/166 and BFB-1 (Fig. 2); their location is shown in Figs. 1a and 1b.

Landward of the escarpments, lies a deep sedimentary basin in which sediments of Mesozoic and, perhaps, even Paleozoic age might exist. The thickness of sediments varies, but in places it may be even larger than 10 km.

In the vicinity of the escarpment, there exists a distinct seismic horizon lying within the sedimentary column. This seismic horizon which typically extends 30 or 40 km landwards from the escarpment, has a characteristic roughness which manifests itself in a hyperbolated reflection record. We believe that this horizon is composed of basalt flows extruded at a time close to the Paleocene/Eocene boundary. This date is derived from stratigraphic mapping utilizing DSDP hole 338 as well as from the consideration that this layer lies adjacent to and at the same stratigraphic level as a tuff marker horizon, the date for which has been determined in wells in the North Sea. We define Zone I as a zone on the landward side of the escarpment which contains the layer interpreted as basalt flows. Zone I is characterized magnetically as a quiet zone.

Zone II lies adjacent to Zone I immediately seaward of the escarpment. Zone II represents an elevated platform where the layering, when detected, tends to be horizontal or subhorizontal. DSDP holes 338 and 342, which are located in this zone, encountered basalts below the sediments. A very important property of Zone II is the fact that in many instances a horizon, which we have termed K in Fig. 2, appears at a depth generally more than 1 km below the sur-

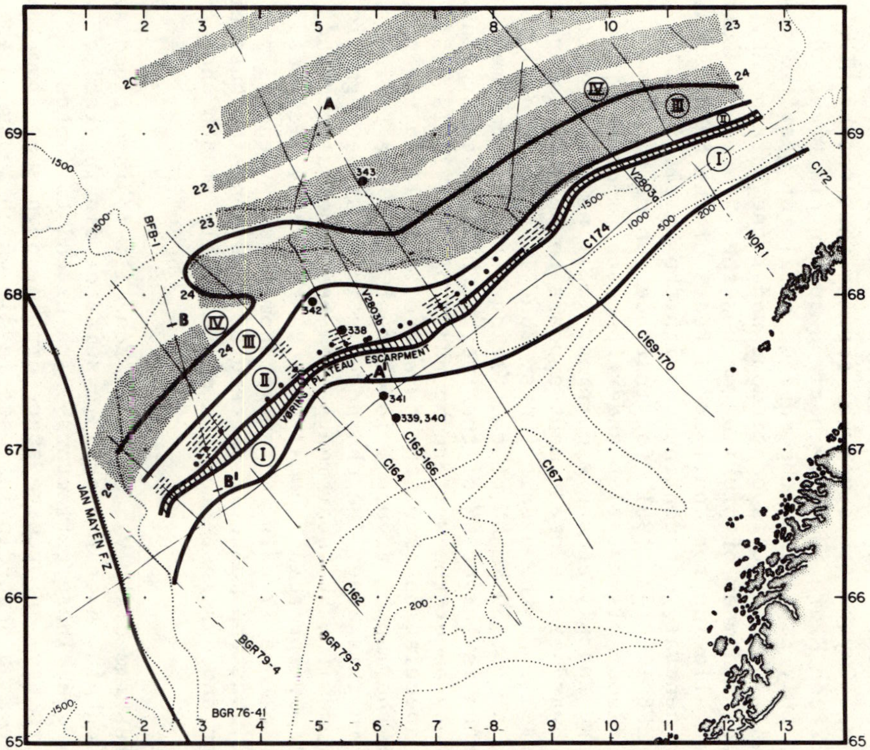


Fig. 1a. Zones I, II, III and IV described in the text are shown in this map of the Norwegian continental Margin. The delineation of these zones is based on an examination of multi-channel seismic records. The preface C refers to lines shot by R/V ROBERT D. CONRAD (Lamont-Doherty); preface BFB or BGR refers to data obtained by Karl Hinz at BGR (Hannover, West Germany) and IGS refers to commercial data reprocessed by Smythe (IGS, Edinburgh). The preface V refers to single-channel Vema lines.

face of the basalt. This horizon, although it has a characteristic appearance, varies considerably from profile to profile. In some profiles, as in C 165/166, it is well developed over a long distance; in other profiles it appears to be very patchy in appearance (see Figs. 1a and 1b), and in still other profiles, it appears to be entirely absent. Its height in the sections is also variable. In some as in C 165/166 and also in C 159 (not shown here) it seems to be nearly continuous with the basalt flows of Zone I near the foot of the escarpment. (As contrasted with the landward basalt flows which are very nearly horizontal, the slope of the K is quite variable). Smythe (this volume) assumes that the basalts of Zone I and horizon K are the same horizon and in his view, then, the escarpment simply represents the edge of a pile of basalt lying on top of a basalt layer that is continuous under the escarpment. On the other hand, a number of profiles across the escarpment demonstrate that, even allowing for "pull-up" effects of the basalt pile, horizon K is, in general, not continuous with the basalt flows of Zone I. Profile BFB 1 (Hinz, 1981) for instance, shows clearly (Fig. 3) that horizon K is not flat under the escarpment but, rather, bends down sharply at the escarpment implying structural control; and landward of the escarpment appears to be at a lower stratigraphic level than the basalt flows. Zone II varies in width from a few km to almost 30 km NW of the Vøring Plateau escarpment. The range of widths is larger NW of the Faeroe-Shetland escarpment, but the uncertainty about the western boundary of Zone II is also greater. If our speculation about the extent of Zone II in the southern part of the map of Figure 1 is correct, then the Faeroe Islands lie within this zone. The characteristic negative magnetic anomaly associated with the escarpments lies near the eastern end of Zone II. Seafloor spreading anomaly No. 24 generally lies west of Zone II. Within Zone II exist magnetic anomalies of generally small amplitudes, but the magnetic pattern of this zone is considerably rougher than and distinct from the magnetically smooth character of Zone I.

Zone III lies immediately seaward of Zone II. It is defined as the area which contains a characteristic suite of dipping reflectors forming a wedge. These reflectors which extend to depths of several kilometers have an arcuate shape dipping down away from the escarpment. There is a down-dip increase in dip which leads to an appearance of downward and seaward divergence of the reflectors. The reflectors individually and the wedge as a whole lack a bottomset aspect. Smythe (this volume) and Mutter et al. (in press) ascribe these reflectors to a seafloor spreading process where the extrusion is subaerial and leads to large flow lengths of the basalt. Talwani et al. (1981) and Mutter et al. (in press) point out that the coincidence of anomaly 24 with the zone containing these reflectors requires that they be created by the process of seafloor spreading. Hinz (1981), who with his co-workers was the first

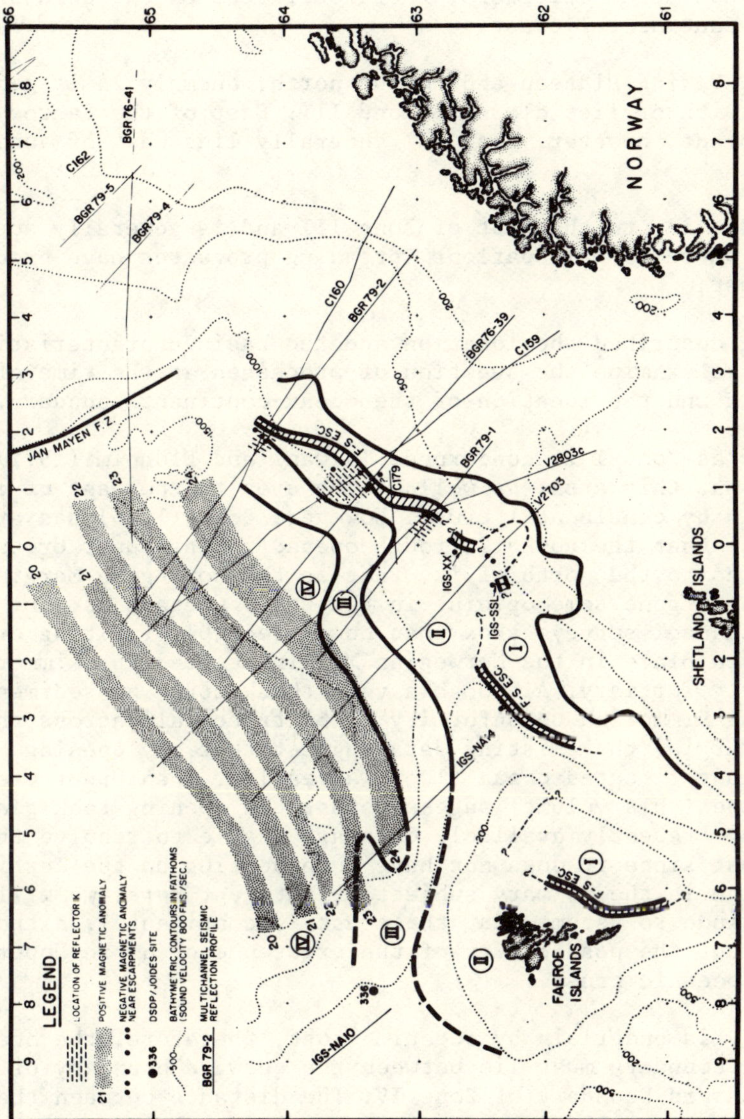


Fig. 1b. See caption for Fig. 1a. The numbering of the magnetic lineations is taken from the work of Alan Nunns.

to observe these reflectors by multi-channel seismic profiling also believes that these reflectors are igneous in nature, but that the volcanism, which produced the lavas and the pyroclastics that make up the layers, took place prior to the initiation of seafloor spreading and these reflectors, therefore, rest on extended and attenuated continental crust.

On the VØring Plateau and to the north, anomaly 24 is generally coincident with or lies close to Zone III. East of the Faeroe-Shetland escarpment, however, Zone III generally lies east of anomaly 24.

Zone IV lies to the west of Zone III and is generally accepted as an area where normal seafloor spreading processes have produced oceanic crust.

Having described the location and the basic characteristics of the zones, we examine the question of processes at the time of initial opening and the location of the ocean-continent boundary.

As far as Zone I is concerned, Talwani and Eldholm (1972) have suggested that this area, as well as the area to the east of it, is underlain by continental crust. However, Bott (1975) has argued on the basis that the Rockall Trough opened in the Upper Cretaceous, that the area to the north of it, that is the Norwegian Margin, may also have undergone some opening in the Upper Cretaceous. The existing multi-channel surveys show that no appreciable faulting or opening has taken place in the Norwegian Margin between the Kimmerian and the Early Tertiary. Although a very thick cover of sediments prevents the Kimmerian unconformity to be traced all across the Norwegian Margin, the existing data suggest that any opening here that would have created ocean floor was earlier than Upper Cretaceous. Smythe (this volume) suggests that the opening took place in the Permian. Presently available data are unable to resolve the question, but since it does not have much bearing on the Tertiary opening which is the primary subject of interest here, we will simply continue to assume that the crust is continental, although not ruling out the possibility of the existence of a pre-Upper Cretaceous oceanic crust.

Zone IV is underlain by oceanic crust, therefore, the ocean-continental boundary must lie between the seaward boundary of Zone I and the landward boundary of Zone IV. The distance between these two boundaries is generally less than 100 km, which then represents the extent of uncertainty in the location of the ocean-continent boundary.

Different workers have differing views on the exact location of the ocean-continent boundary within this area of uncertainty. We briefly describe some of the views that have been expressed and then

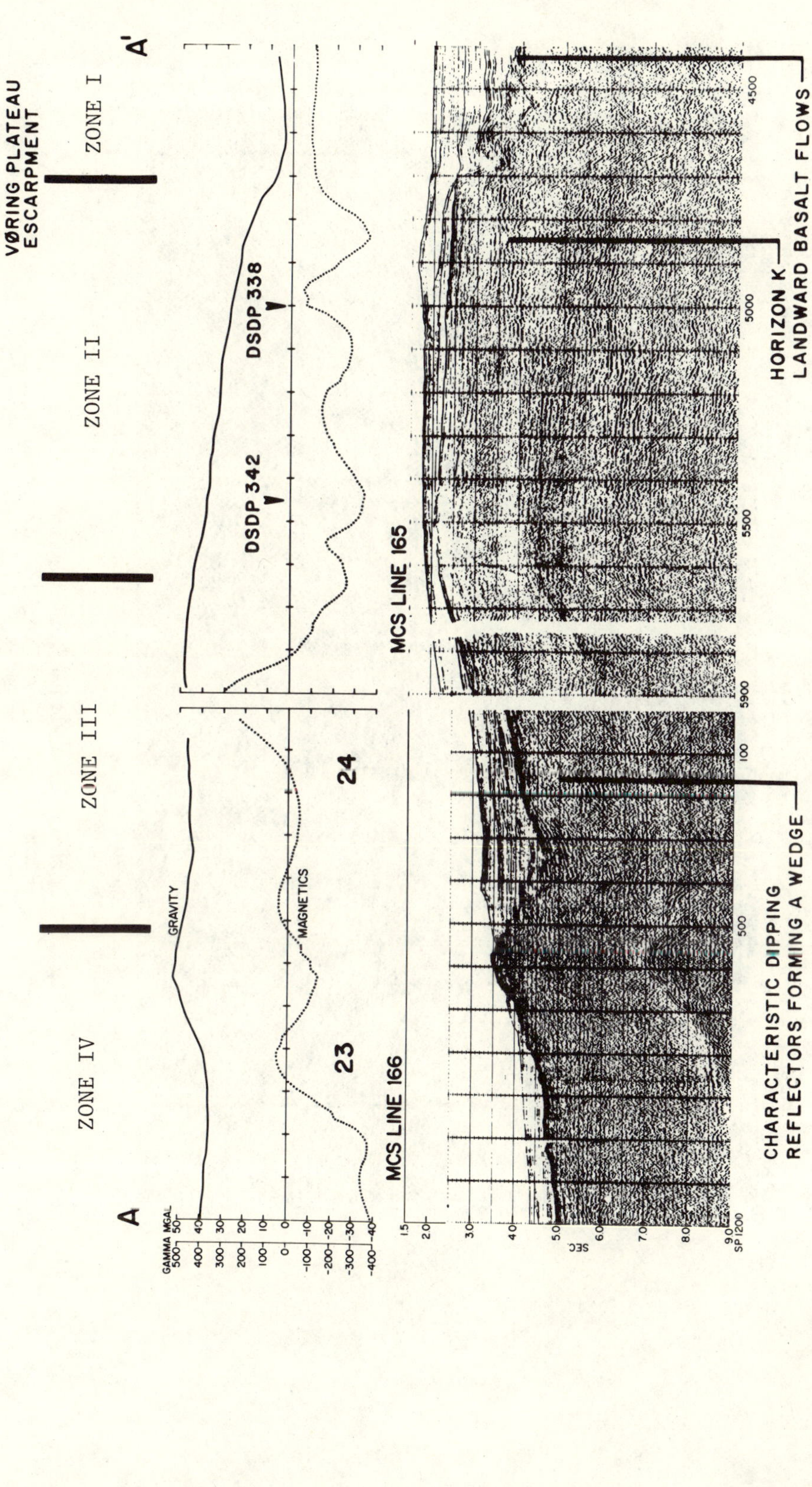


Fig. 2. Portion of multi-channel seismic line C 165/166. See Fig. 1a for location.

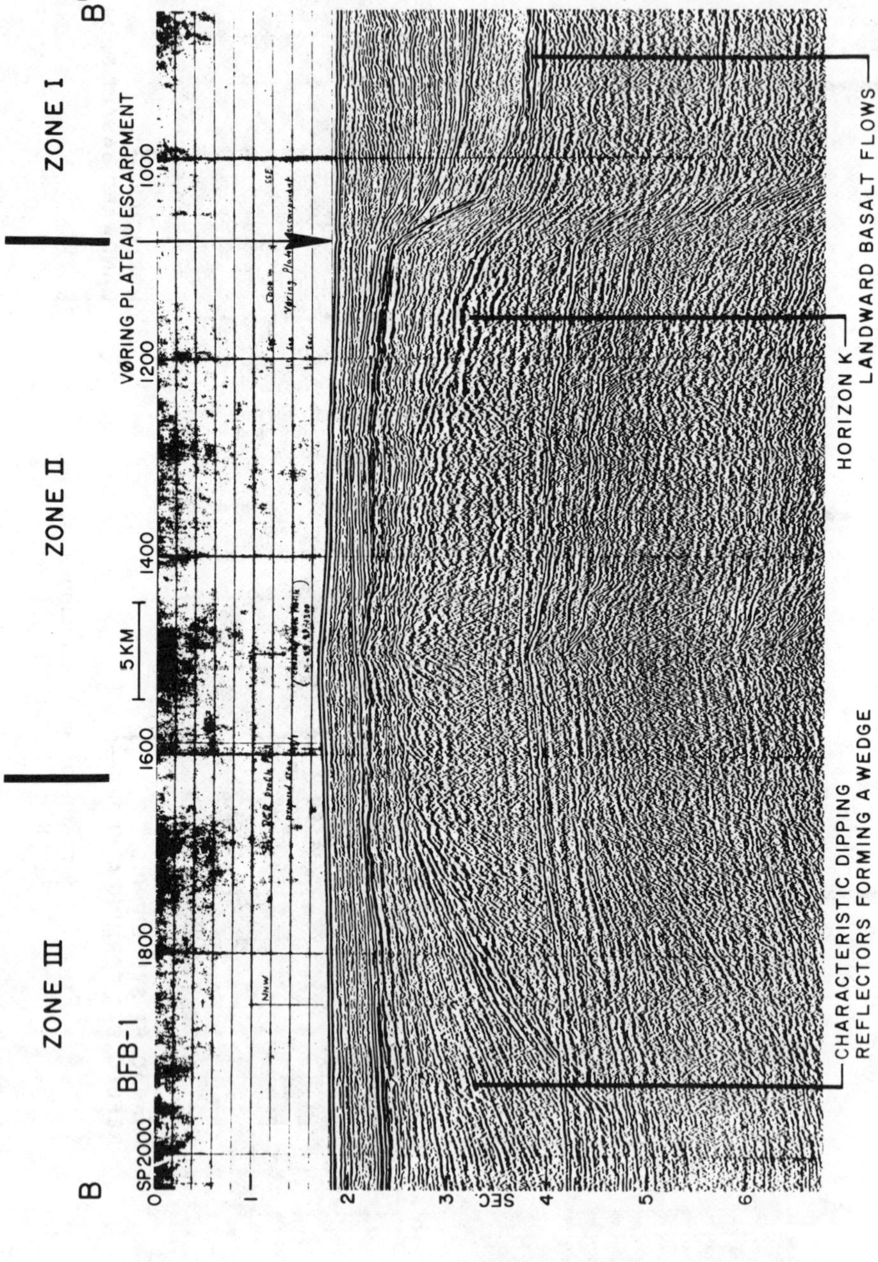


Fig. 3. Portion of multi-channel seismic lines BFB-1 (Hinz, 1981). See Fig. 1b for location.

offer some comments on the problems raised by these views.

Hinz (1981) believes that the layered reflectors in Zone III are composed of lavas and other volcanic products which resulted from an episode of intense volcanism in a sub-aerial and/or shallow water environment, just prior to the initiation of seafloor spreading. This hypothesis supposes Zone III to be in rifted and attenuated continental crust, and the newly created Tertiary seafloor to lie just seaward of Zone III. By implication, the escarpments are associated with the rifting process and do not have a direct bearing on the location of the ocean-continent boundary. Hinz believes that the magnetic lineation identified as the oldest seafloor spreading lineation may usually represent an "edge effect" where the outer edge of the seaward dipping reflectors meets the massive igneous crust of oceanic type. Hinz also believes that since in other areas around the world, the seaward dipping reflectors lie seaward of the oldest seafloor spreading type magnetic lineations, these reflectors may originate in different ways in different areas.

Smythe (this volume) believes (as Mutter and Talwani do) that the characteristic dipping reflectors on Zone III are products of seafloor spreading, but in a subaerial environment. The ocean-continent boundary, according to Smythe, lies at the eastern edge of Zone III; and Zone II consists of continental rocks covered by plateau basalts which were erupted in a period just prior to the initiation of seafloor spreading. A major point that Smythe wants to establish is that the escarpments have no significance with respect to the ocean-continent boundary. Smythe argues that the "landward basalt flows" of Zone I and horizon K of Zone II are in fact the same continuous layer, representing extrusive basalt flows in a continental area. (These flows should, according to Smythe's view, be underlain by Mesozoic sediments in Zone I as well as in Zone II). The escarpment is simply the edge of a pile of basalt emplaced on top of horizon K. Smythe postulates that an ancient shoreline (with deep water to the SE) existed where the escarpments are located now, and the basalt flows originating from the west could not travel very far under subaqueous conditions and piled up to form the escarpments.

In support of his thesis, Smythe cites the lobate outline of the Faeroe-Shetland escarpment near $62^{\circ}.4$ N and the fact that the middle segment of the escarpment appears simply to die out south of $62^{\circ}$N. Smythe states that the above arguments suggest that the escarpments are superficial features and cannot represent deep discontinuities in the crust as the ocean-continent boundary would be expected to entail. There are a number of difficulties which are raised by Smythe's hypothesis. As we have pointed out earlier, horizon K does not in fact continue into the landward basalt flows in all profiles. There is often a difference in level and in some cases K seems to be faulted or folded at the escarpment. There is also a difference in appearance between K and the landward basalt flows; K seems to have a greater amount of low frequency content in the seismic records (although it is not clear how much of this difference can be attributed to filtering by the overlying material). If the boundary between Zone II and Zone III does represent the ocean continent boundary, that is, a boundary with rocks at depth that have greatly different geophysical properties, then the geophysical profiles over the boundary must reflect these differences. However, an examination of a number of gravity profiles reveals no major anomalies at the Zone II - Zone III boundary. In BFB-1 profile, horizon K seems to be continuous seawards with the base of the wedge of dipping reflectors of Zone III, suggesting the lack of a major structural discontinuity between the two zones. Finally, if the escarpments do indeed mark the edge of an ancient shoreline, it is difficult to see why the line of opening should be very nearly parallel to the ancient shoreline and only a short distance away.

Talwani and Mutter believe that the escarpments do locate the position of the ocean-continent boundary. This does not imply that the escarpment itself is a fault which continues deep into the earth. What it does mean is that the escarpment is related to the ocean continent boundary and lies over it. Talwani and Mutter are guided by the consideration that Zone III, which contains the dipping reflectors, was created by seafloor spreading, that there appears to be no major boundary between Zone II and Zone III, but that the boundary between Zone I and Zone II is reflected in geophysical measurements -- there is a characteristic negative magnetic anomaly here and landward of it lies the magnetic quiet zone; a gravity gradient commences here separating an area of low gravity values landward from an area of high gravity values seaward.

A critical question relates to the nature of horizon K and what lies beneath it. Talwani and Mutter have speculated that horizon K might mark the boundary between basalt flows and dikes. Both below and above it would lie igneous material comprising Tertiary oceanic crust. On the other hand if Smythe is correct, Mesozoic sediments underlie horizon K. It seems that it might be necessary to drill to horizon K and penetrate the material below it to resolve the question of the ocean-continent boundary.

Some dates are available for events and rocks associated with the events near the time of initiation of opening. The landward basalt flows are believed (by association with the North Sea tuff marker) to have been emplaced near the boundary between the Paleocene and the Eocene. This is also the date of initiation of seafloor spreading since anomaly 24 is seen in the Norwegian Sea, but anomaly 25 is absent and the Paleocene Eocene boundary lies between these two anomalies. The faunal ages determined for the sediments overlying the basalt at DSDP holes 338 and 342 are also Early Eocene. The uncertainty in all these dates is about a million years or so at the minimum, and the coincidence of dates suggest that all the associated events took place within a short time of each other, but do not allow us to choose between the different views regarding the ocean continent boundary which have been cited above.

With respect to the Iceland-Faeroe Ridge and the initiation of opening there, we believe that the Faeroes appear to be in Zone II (see Fig. 1) and the uncertainties attending the nature of Zone II must apply also to the Faeroes.

## REFERENCES

Bott, M.H.P., 1975, Structure and evolution of the Atlantic floor between northern Scotland and Iceland, Nor. geol. Unders., 316: 195-199.

Hinz, K., 1981, A hypothesis on terrestrial catastrophes. Wedges of very thick oceanward dipping layers beneath passive continental margins. Geologisch. Jahrbuch, Ser.E, 22:3-28.

Mutter, J.C., Talwani, M., and Stoffa, P.L. (in press), Origin of seaward-dipping reflectors in oceanic crust off the Norwegian margin by "sub-aerial seafloor spreading", Geology.

Talwani, M., and Eldholm, O., 1972, Continental margin off Norway: A geophysical study, Geol. Soc. Amer. Bull., 83:3575-3606.

Talwani, M., Mutter, J.C., and Eldholm, O., 1981, The initiation of opening of the Norwegian Sea, Oceanologica Acta, No.Sp.:23-30.

# STRUCTURE AND ORIGIN OF THE WYVILLE-THOMSON RIDGE

D.G. Roberts[1], M.H.P. Bott[2] and C. Uruski[2]

[1]Institute of Oceanographic Sciences
Brook Road, Wormley, Surrey, GU8 5UB, U.K.

[2]Department of Geological Sciences, University of Durham
South Road, Durham, DH1 3LE, U.K.

## INTRODUCTION

The Wyville-Thomson Ridge lies at the northern end of the Rockall Trough and separates the latter from the Faeroe-Shetland Channel. The Ridge forms part of a region of complex topography that includes the Faeroe platform and the prominent seamounts and channels lying to the south and southwest of the platform (Figs 1 and 2).

The region is geologically complex and its history is not yet fully understood. Seismic evidence (Bott et al., 1974, 1976; Nielsen, this volume) suggests that the Palaeocene basalts of the Faeroe Islands are underlain by a microcontinent whose edges lie northwest of the Faeroes and beneath the western part of the Faeroe-Shetland Channel. Reconstruction of the North Atlantic region prior to splitting suggests that this continental crust must extend southwestwards beneath Faeroe, Bailey and Lousy Banks to join up with the Rockall microcontinent, but direct evidence is not yet available. Within the central and southern Rockall Trough, the evidence indicates about a 150 km width of ocean crust of mid to late Cretaceous age. From the evidence of the North Atlantic Seismic Project (NASP) 1972, this crust may trend northwards beneath the Faeroe-Shetland Channel. The great thickness of the Faeroe plateau basalts and the presence of the prominent seamounts, as well as extensive sub-surface lavas, suggests a major and widespread Lower Tertiary volcanic episode in the region which was contemporaneous with that of northwestern Scotland and east Greenland. This igneous activity may have formed the Wyville-Thomson Ridge or

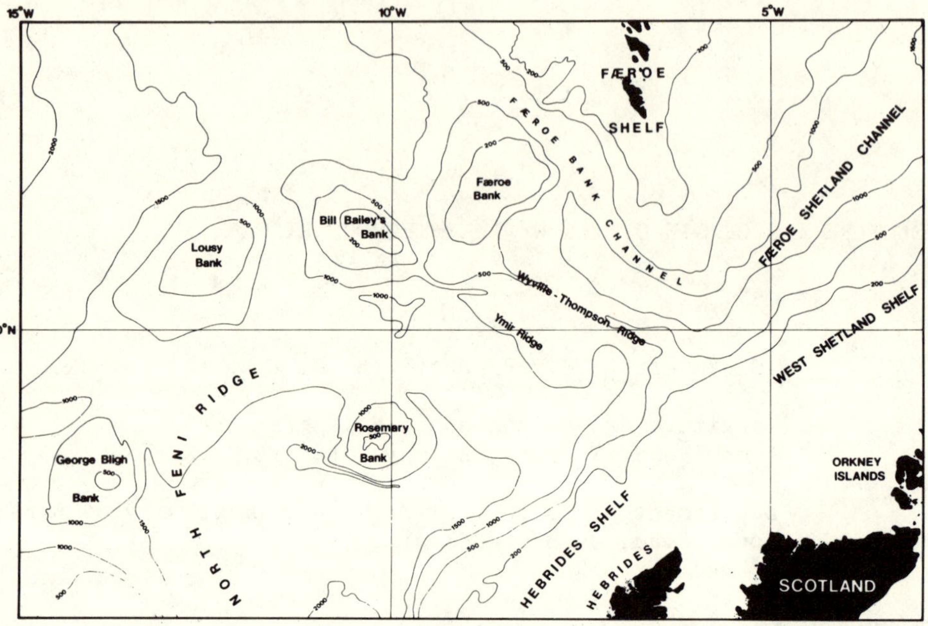


Fig. 1. Location map of the Wyville-Thomson Ridge and neighbouring regions, showing depth contours in metres.

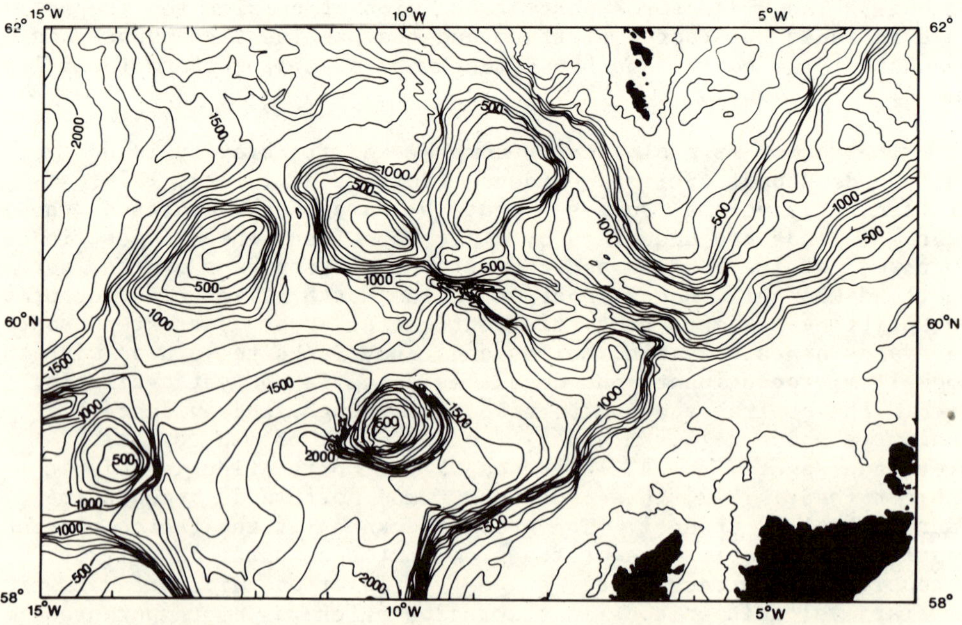


Fig. 2. The bathymetry (in metres) of the Wyville-Thomson Ridge and neighbouring regions. After Roberts et al. (1977).

perhaps modified an older feature associated with the opening of the Rockall Trough. The age and origin of the Wyville-Thomson Ridge is thus relevant to the opening of the Rockall Trough and Faeroe-Shetland Channel, and to the development of Lower Tertiary volcanism in the region.

In a wider context the Ridge and the adjacent seamounts and channels have both palaeontological and oceanographic importance. If the Ridge was sub-aerial during Lower Tertiary time, it may have acted as a land bridge for the migration of land animals from North America to Europe (McKenna, this volume). In an oceanographic context, the Ridge and associated channels have great importance because they act as paths for the passage of overflow water from the Norwegian Sea into the main North Atlantic Ocean. In the area of study, the main path is from the Faeroe-Shetland Channel through the Faeroe Bank Channel (Crease, 1965) but intermittent overflow of water seems to have taken place across the Wyville-Thomson Ridge (Ellett and Roberts, 1973). The date of subsidence of the Ridge and channels to a depth which allowed southward passage to Norwegian Sea water bears directly on the initiation of North Atlantic deep circulation and its profound influence on the dispersal and disposition of sediments in the North Atlantic.

In this paper, we outline the bathymetry of the Wyville-Thomson Ridge and surrounding areas and discuss its origin using seismic reflection data and gravity modelling constrained by seismic refraction and reflection data.

## BATHYMETRY

The bathymetry of the region around the Wyville-Thomson Ridge (Figs 1 and 2) has been previously discussed in detail by Ellett and Roberts (1973), Roberts (1975) and Roberts et al. (1979).

The insular shelf around the Faeroe Islands forms the eastern termination of the Greenland-Iceland-Faeroe Ridge. The Faeroe shelf, which consists of outcropping basalts thinly covered by sediments, is separated from the northwest Scottish shelf by the Faeroe-Shetland Channel and from the Wyville-Thomson Ridge and Faeroe Bank by the Faeroe-Bank Channel. The Faeroe-Shetland and Faeroe Bank Channels comprise a single trough that changes trend from NE-SW to curve northwestwards, debouching from the gap between the Faeroe Bank and Faeroe shelf along the southern flank of the Iceland-Faeroe Ridge. The floors of these channels are generally smooth, although one small knoll composed of Palaeocene lavas occurs near $60^{o}$ 20'N, $7^{o}$ 15'W. These channels are the conduit for Norwegian sea water into the main North Atlantic basin at speeds of about 2 m/s. Long-range sonographs of the floor of the Faeroe Bank Channel show evidence of bedforms below the resolution of the echo-sounder that may have been fashioned by the strong bottom currents (Roberts and

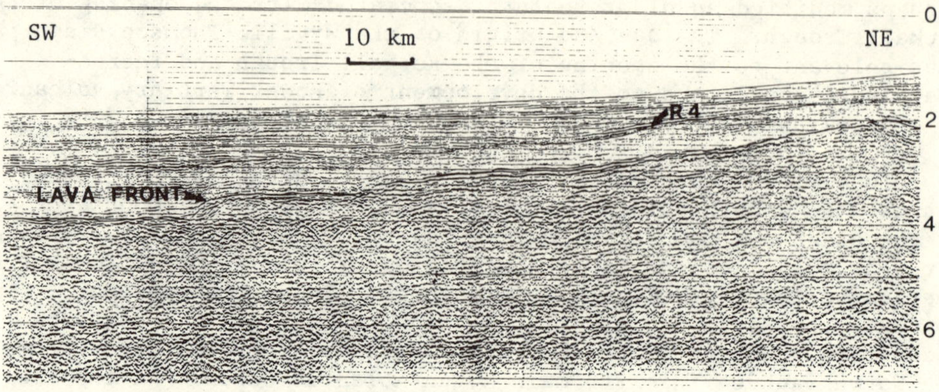


Fig. 3. Seismic section along the sediment drift southeast of Ymir Ridge (part of profile 9, Figs 4 and 5), showing two-way travel time in seconds. Note southward increase in sediment thickness, reflector R4 and contrasting sediments above and below R4, acoustic basement formed by Palaeocene lavas, and step marking the lava front.

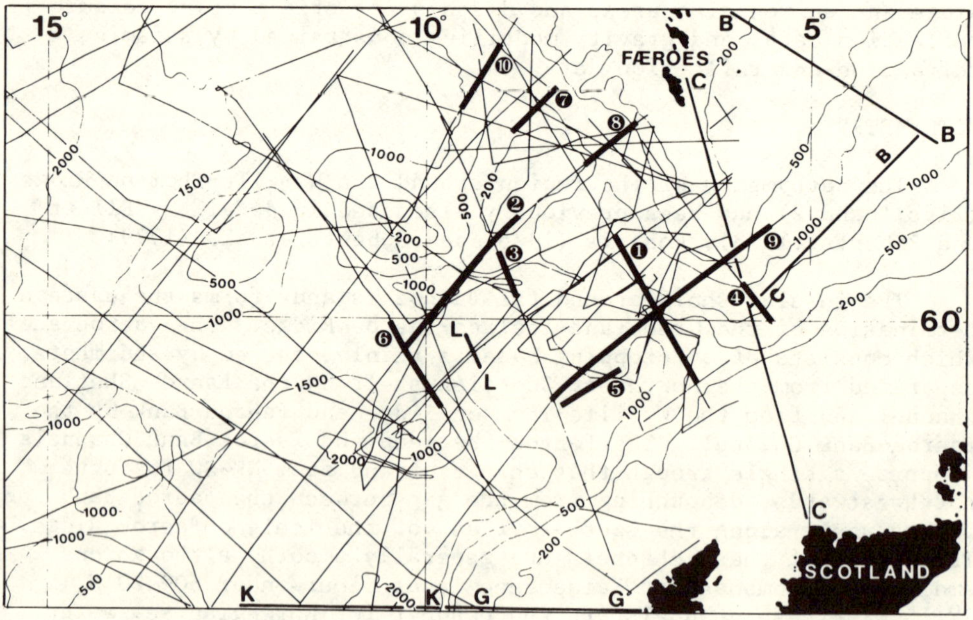


Fig. 4. Location of seismic reflection profiles 1 to 10 shown in Fig. 5, and relevant shot lines of Hebridean Margin Seismic Project (GG, KK, LL) and North Atlantic Seismic Project (BB, CC, CB).

Hunter, in preparation). The complex pattern of channels and ridges observed west of the mouth of the Faeroe Bank Channel (Fig. 2) may also reflect the erosional and depositional effects of the outflow.

The adjoining Wyville-Thomson Ridge system consists of a southern ridge called the Ymir Ridge and a northern ridge called the Wyville-Thomson Ridge (sensu stricto). The northern ridge is elongated in a WNW-ESE direction and consists of a series of knolls of minimum depth 384 m, possibly composed of outcropping basement, and depressions of maximum depth 844 m. The Ymir Ridge parallels the northern ridge in its western part but widens and becomes less steep eastwards. Seismic profiles reported here (Figs 3, 4 and 5) show that this change is related to the outgrowth of a major sediment drift along the southern and eastern part of the Ymir Ridge. The trough between the two Ridges narrows westward and at its most constricted part is deeper than the adjacent floor of the north Rockall Trough (Fig. 2). Ellett and Roberts (1973) suggested that this trough acts as the conduit that channels intermittent overflow across the Wyville-Thomson Ridge into the Rockall Trough and thereby formed the Feni Ridge sediment drift (Jones et al., 1970; Roberts, 1975). Velocities in excess of 0.8 m/s that are competent to erode sediments have been measured in this trough (Gould, personal communication) and the seismic profiles show clear evidence of erosion (Fig. 5, profile 3). The northern flank of the trough trends east-west along the southern margins of Bailey Bank and Faeroe Bank.

Within the northern Rockall Trough the topography is dominated by broad ridges that represent sediment drifts. From west of the channel, one ridge can be followed southward toward Rosemary Bank and is in part a sediment drift that may have accumulated in the lee of a basement high near $60^{\circ}$ 15'N, $9^{\circ}$ 45'W. A prominent sediment drift has also built out southeastwards from the Ymir Ridge (Fig. 3). The GLORIA sonographs show that the ostensibly smooth ground north of Rosemary Bank has also been affected by differential deposition as discrete bedforms are clearly visible on the sonographs (Roberts and Hunter, in preparation). Differential deposition by bottom currents accounts for the development of the moat around Rosemary Bank (Roberts et al., 1974).

## SEISMIC REFLECTION PROFILES

Little has been published hitherto on the regional seismic stratigraphy and structure of the region around the Wyville-Thomson Ridge and southwest of the Faeroe Islands. Stride et al. (1969) reported an early seismic profile across the Faeroe Bank Channel and Roberts (1975) discussed the seismic stratigraphy of the Rockall Trough as far north as Rosemary Bank (Roberts et al., 1974). Seismic profiles obtained by the University of Durham were presented by Himsworth (1973). For this study, unpublished seismic

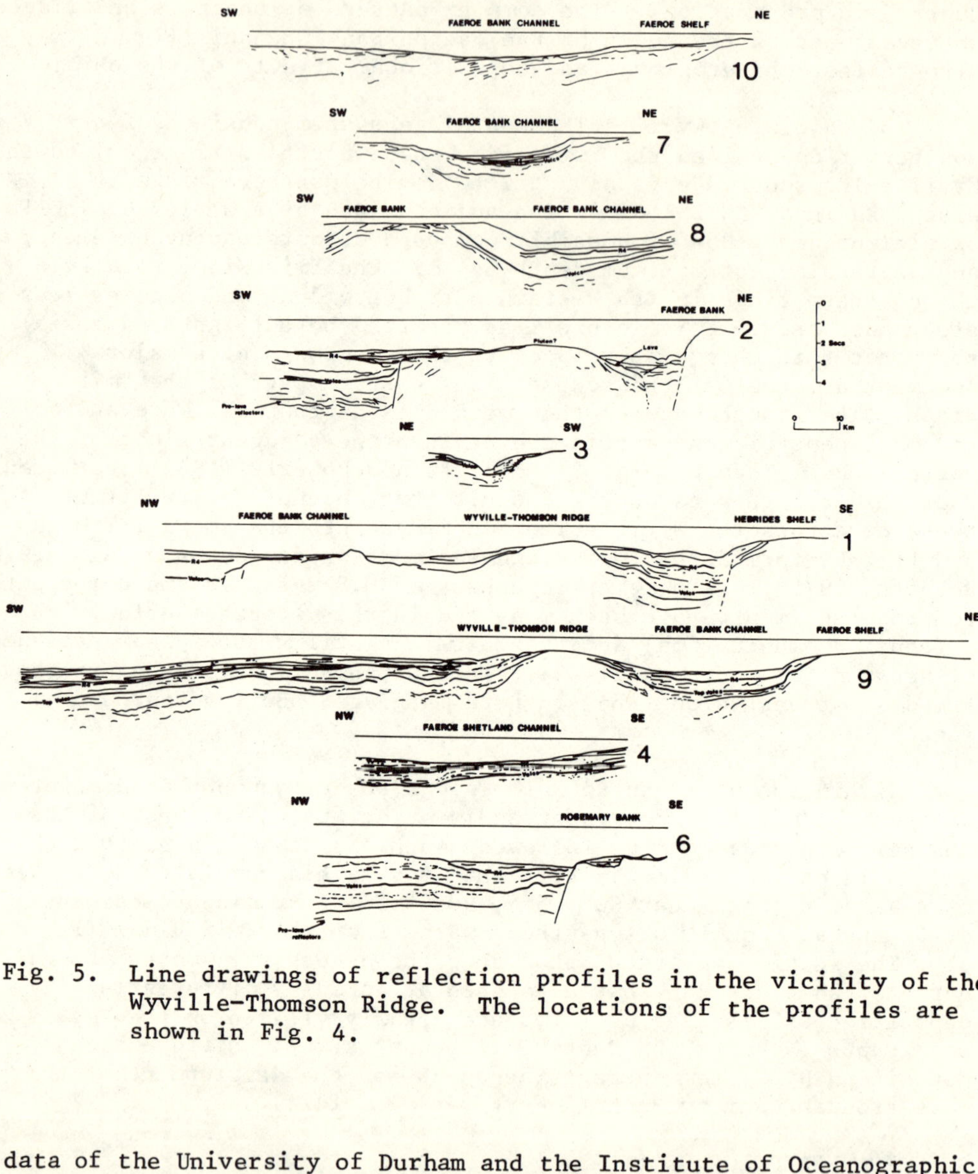


Fig. 5. Line drawings of reflection profiles in the vicinity of the Wyville-Thomson Ridge. The locations of the profiles are shown in Fig. 4.

data of the University of Durham and the Institute of Oceanographic Sciences (IOS) have been integrated with multichannel seismic profiles held by IOS and the single channel seismic profiles obtained by Lamont-Doherty Geological Observatory (Talwani et al., 1974). The adjoining Faeroe-Shetland Channel is not discussed in detail here (see Cashion, 1975; Ridd, 1981 and this volume).

Two prominent seismic reflectors can be identified on almost all profiles in the region (Figs 3, 4 and 5). The uppermost reflector is a regional unconformity that defines the base of a sediment unit characterised by the effects of erosion and differential deposition associated with near-bottom currents (Fig. 3). This reflector is known as R4 and has been assigned an age of Eocene-Oligocene from Deep Sea Drilling results by Roberts (1975). It is likely that a major non-depositional or erosional hiatus may be marked by the reflector. The overlying unit is characterised by numerous unconformities (e.g. Fig. 5, profile 9). In contrast, the underlying unit, above the Palaeocene lavas, is comparatively transparent and does not show comparable evidence of erosion. The strong reflector defining the base of the underlying unit is interpreted as the top of the almost ubiquitous Palaeocene lavas. The Palaeocene lavas, however, probably do not actually represent the true basement in the region of the Faeroe Islands, Faeroe-Shetland Channel and northern Rockall Trough. They may be underlain by either continental or oceanic basement with intervening sediments. There is evidence of a substantial pre-Palaeocene sedimentary section revealed locally in the Rockall Trough as a series of strong laterally impersistent reflectors beneath the lavas. Roberts et al. (1981) considered that these reflectors may represent clastic or volcanoclastic sediments deposited rapidly and in part contemporaneously with the accretion of the oceanic crust (cf Kelts and McKenzie, 1976). The nature and extent of any such sediments beneath the lavas that comprise and flank the Wyville-Thomson Ridge remains unkown. The true underlying basement cannot be mapped with any certainty, and the discussion of the seismic reflection data has thus been confined to the structure and distribution of the Palaeocene lavas and the overlying sediments.

## PALAEOCENE LAVAS

Criteria for the identification of the lavas on the seismic profiles include the distinctive strength and stepped character of the reflections defining the top of the lavas (Figs 3 and 5), high interval velocities and correlation with the outcropping lavas known on the Faeroes. Constraints on their age are offered in the Faeroe-Shetland Channel by their occurrence just below a late Palaeocene-early Eocene tuff (Ridd, 1981), in the Rockall Trough by a longer range correlation with the IPOD wells in Biscay (Roberts et al., 1981), and by the 50 to 60 My dating of the Faeroe Island basalts (Tarling and Gale, 1968). Further information on the subsurface distribution of the lavas and associated plutons has been obtained from marine magnetic maps (Roberts and Jones, 1975) and unpublished aeromagnetic maps.

Isochrons drawn on the top of the lavas (Fig. 6) relative to the sea surface clearly demonstrate that the present-day morphology is closely controlled by the relief of the lava surface. The Palaeocene lavas can be conveniently discussed in terms of (i) the area around the seamounts and the Faeroe shelf and (ii) the Wyville-Thomson Ridge.

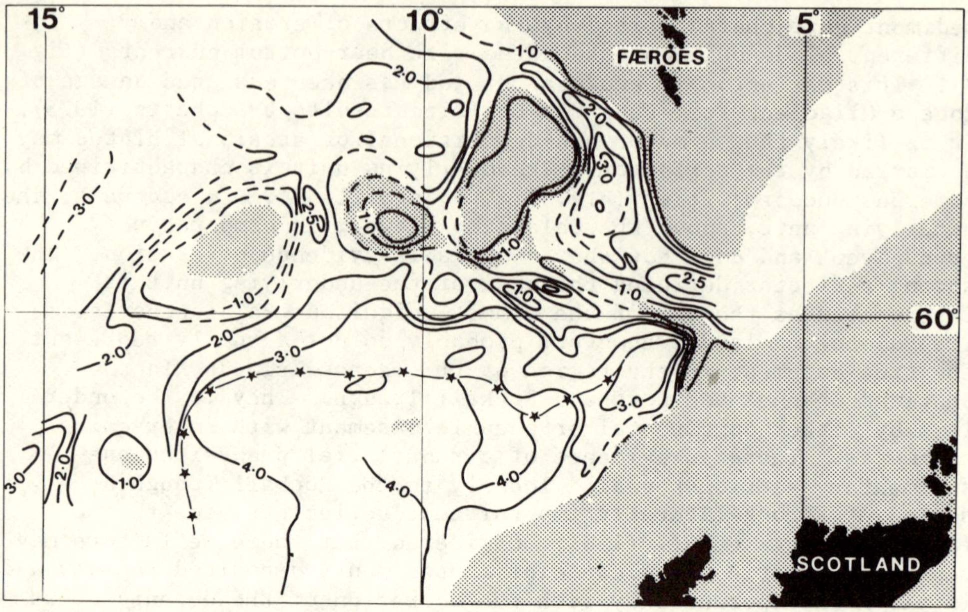


Fig. 6. Isochrons to the surface of the Palaeocene lavas, contoured in seconds of two-way seismic travel time below the sea surface at 0.5 s interval. Shaded areas mark either the Scottish shelf or the outcrop of Palaeocene or earlier rocks at the seabed.

Seismic profiles across the seamounts show evidence of outward dipping reflectors beneath the lava surface (Fig. 5, profile 8). These reflectors are tentatively interpreted as lavas interbedded with pyroclastics. Truncation of these reflectors by the flat top of the seamounts suggests strong subaerial erosion of an original volcanic landform. The present depths of the flat tops of Bailey and Lousy Banks exceeds the post-glacial sea level rise of 200 m suggesting that erosion related to tectonic subsidence may have been the main factor influencing the present topography. Similar reflectors observed beneath the northern flank of the Faeroe Bank Channel may represent lava flows within the flank of the broad anticlinal structure of the Faeroes (Fig. 5, profiles 7 and 10). Evidence for the source of the lavas in these areas is scant although Dobinson (1970) has identified a plutonic centre on Faeroe Bank from magnetic anomalies (Fig. 9). Rasmussen and Noe-Nygaard (1970) considered that the plateau basalts of the Faeroe Islands were erupted from fissures associated with a NW-SE trending system of fractures.

Within the Faeroe Bank Channel, the depth to the top of the

lavas increases progressively southeastwards from 1600 to 3000 milliseconds two-way time which represents about 1.25-2.5 km depth (Fig. 6). The present depth of the lava surface within the Channel is thus between 1 and 2.2 km below that of the adjacent banks. The original relief must have been much greater in view of the evidence of subaerial erosion on the adjacent banks. Towards the Faeroe-Shetland Channel, the lava surface shoals and outcrops on the floor of the channel as a small knoll near 60° 20'N, 7° 15'W (Fig. 5, profile 2). The shoal region is associated with a complex circular pattern of high amplitude magnetic anomalies indicative of an intrusive centre. East of the outcrop, the lava surface deepens towards the Faeroe-Shetland Channel.

The northern Wyville-Thomson Ridge is composed of lava flows that outcrop or are thinly covered by sediments. The flat top of the Ridge suggests sub-aerial erosion. The edge of the northern ridge has a strong WNW-ESE trend but the seismic profiles do not show evidence of post-lava faulting (Fig. 5, profile 9). Within the trough between the Wyville-Thomson and Ymir Ridges, the lavas are synclinal but are distinctively stepped toward the Ymir Ridge. The steps are suggestive of post-lava faulting. The southern flank of the synform is apparently associated with major faulting both in and immediately to the west of the overdeepened channel between Ymir Ridge and Faeroe Bank. The steep slope south of Faeroe Bank is evidently controlled by the fault (Fig. 5, profile 2) and the southern scarp of the Ymir Ridge is associated with a change in lava depth of 1 km in 2 km distance (Fig. 6). Eastwards, however, the Ymir Ridge loses its structural and topographic identity and faulting is not seen in the plunging nose of the Ridge (Fig. 5, profile 9). No major faulting has been observed south of the Ymir Ridge.

South of the Wyville-Thomson and Ymir Ridges, the Palaeocene lava surface progressively increases in depth (Fig. 5, profiles 6 and 9). The southward limit of the thick lavas is shown by a characteristic step on nearly all the seismic profiles associated with a marked change in magnetic gradient (Fig. 5, profile 9). The limit can be followed using the seismic and magnetic data as an irregular boundary from the Wyville-Thomson Ridge as far west as George Bligh Bank where it is clearly shown on the profiles reported by Roberts (1975). The lava limit passes north of Rosemary Bank and bears no obvious relationship to it. A late Cretaceous to early Tertiary age has been inferred for this seamount which may be related either to the Palaeocene igneous episode or more probably to the terminal stages of Cretaceous spreading in the Rockall Trough. North of the limit major changes in the trend and gradient of the isochrons reflects the presence of plutons situated within or below the lava pile (see also gravity map). One such pluton which gives rise to a local gravity anomaly of 60 mgal (Himsworth, 1973) is associated with outcropping basement near 59° 40'N, 8° 40'W (Fig.

5, profile 2). South of the limit of thick lavas, the lavas are apparently thinner and the underlying sediments are more clearly seen (Fig. 5, profile 6).

## SEDIMENTS BETWEEN PALAEOCENE LAVAS AND REFLECTOR R4

An attempt has been made to map the distribution and thickness of the sedimentary succession between the top of the lavas and reflector R4 (Fig. 7). The distribution bears an obvious relationship to sub-surface morphology of the lavas, but the variations in thickness also in part reflect sediment source, supply and dispersal. Post-R4 erosion may also contribute to the thickness variation.

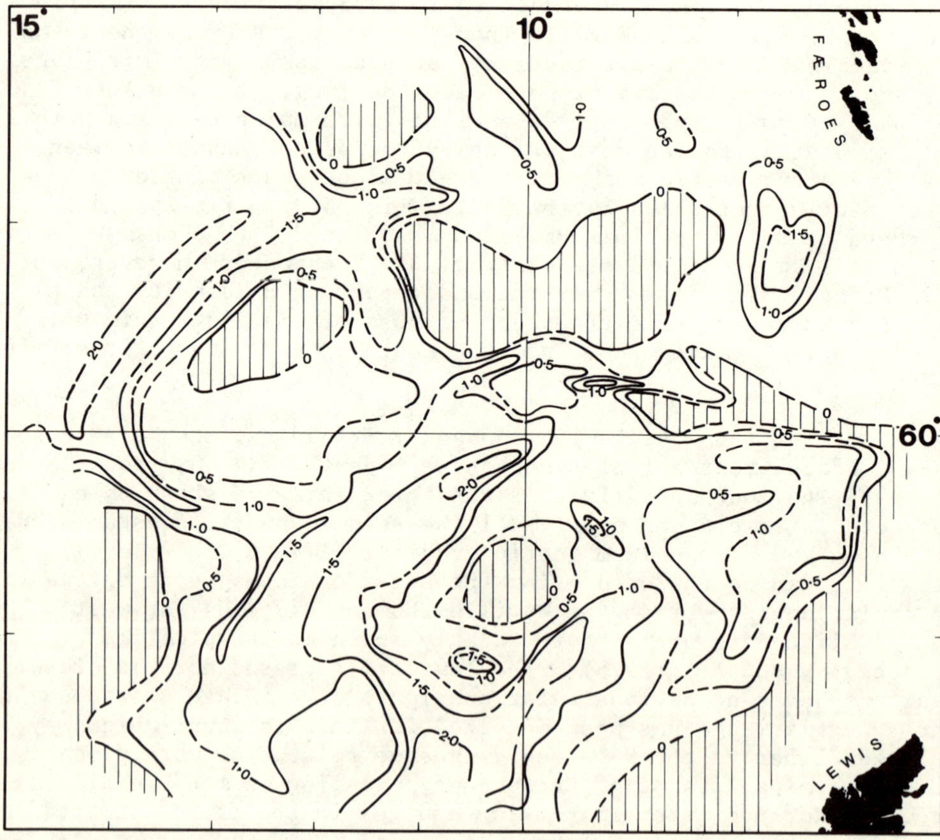


Fig. 7. Isochrons of sediment thickness between the Palaeocene lava surface and the reflector R4, contoured in seconds of two-way travel time at 0.5 s interval.

In the northern Rockall Trough, the succession thins progressively northward to zero as it onlaps the underlying lava surface toward the Wyville-Thomson Ridge. There are two major depocentres in the Trough. One situated south of Bailey Bank comprises a broad ridge of sediments locally exceeding 2000 milliseconds two-way time in thickness that thins westward towards Lousy Bank. A second depocentre is present at the east end of the Wyville-Thomson Ridge and extends southwestwards past Rosemary Bank. These features are tentatively interpreted as deep-sea fans possibly analogous to those of the North Sea (Parker, 1975).

Within the Faeroe Bank Channel, a substantial depocentre occurs southeast of the Faeroe Bank. The sediments below R4 thin drastically northwestward along the channel by more than 1200 milliseconds two-way time. They also thin against the outcropping basement to the southeast of the Channel. Within the Channel, the thickness of the pre-R4 sediments was profoundly influenced by the morphology of the lava surface, but the northwestward thinning may reflect either subsequent erosion by the strong currents or the original pattern of deposition. The location of the depocentre within the channel reflects nearby sources of sediment supply from the adjacent Bank and perhaps from the Faeroe shelf. The same reasoning is equally applicable to the depocentres within the Rockall Trough. Although the adjacent banks are now submerged, the thickness variation suggests that they may have acted as a voluminous source of sediments during the lava-R4 interval when they were probably sub-aerial. Although the thickness variation within the Faeroe Bank Channel may in part be attributed to post R4 erosion, the southeastward thickening suggests the possibility that the highest part of the channel may itself have been sub-aerial so that the depocentre would be partly fed by southeastward longitudinal drainage.

## THE POST-R4 INTERVAL

Seismic profiles across the Faeroe Bank Channel show a westward decrease in thickness of the sediments overlying R4 (Fig. 5, profiles 7, 8 and 10). These are asymmetrically distributed within the Channel both with respect to the present axis and the earlier axis shown by the thickness of the intra-Palaeocene R4 interval (Figs 6, 7 and 8). Within the Channel, major accumulations of post-R4 sediments occur in two areas. Firstly, sediments exceeding 900 milliseconds two-way time in thickness are present in the embayment between the Faeroe Bank and the Wyville-Thomson Ridge, and thin toward the axis of the channel to 200 milliseconds two-way time or less (Fig. 8). In the channel axis erosion revealed by truncation of reflectors at the seabed is evident in the vicinity of the small knoll and elsewhere (Fig. 5, profiles 1, 4 and 7). Secondly, close to the junction of the Faeroe Bank and Faeroe-Shetland Channels, about 500 milliseconds

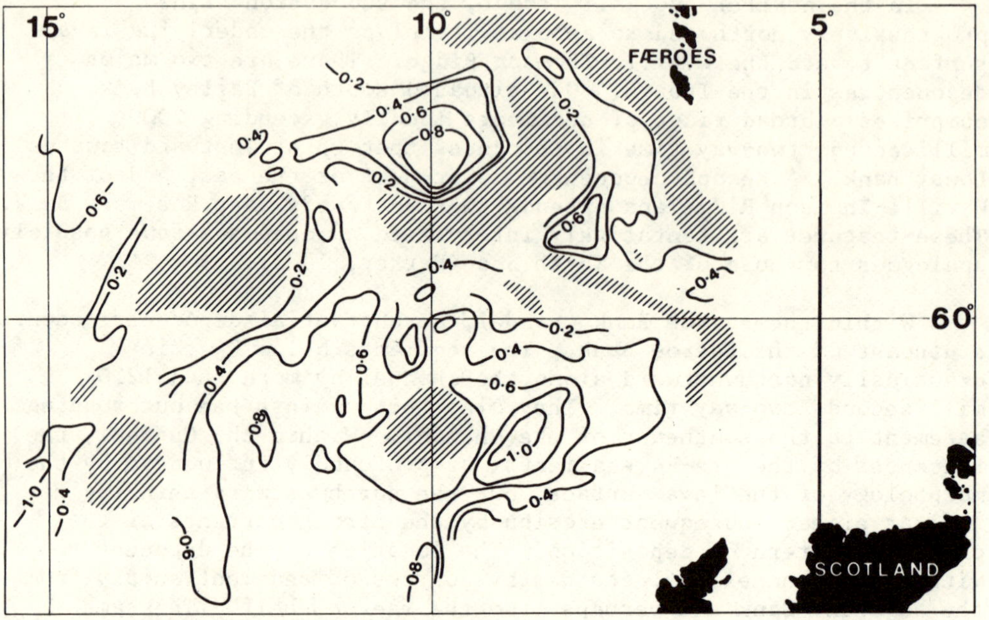


Fig. 8. Isochrons of sediment thickness between reflector R4 and the seabed, contoured in seconds of two-way travel time at 0.2 s interval.

two-way time of post R4 sediments have accummulated on its northern flank. West of the junction, a broad area of exposed basement covering much of the southward slope directly lies beneath the axis of the outflow from the channel indicating the importance of erosion and non-deposition. This region is in contrast bounded to the south by a substantial accumulation of post-R4 sediments whose surface is characterised by small mud waves and channels fashioned by bottom currents. Within the Faeroe-Shetland Channel, erosion down almost to the depth of the R4 reflector is locally evident (Fig. 5, profile 4; Fig. 6 of Ridd, this volume).

South of the Wyville-Thomson Ridge, the post-R4 section is much more uniform in thickness and the seismic profiles show less evidence of erosion although differential deposition has been important. Erosion has, however, been particularly strong in the trough between the Wyville-Thomson Ridge and Ymir Ridge (Fig. 5, profile 3). Within this trough, which acts as the conduit for intermittent overflow across the Wyville-Thomson Ridge, current speeds exceeding 0.8 m/s have been recorded (Gould, personal communication). Such currents may have eroded the 1286 m deep hole at the mouth of the constriction and caused the basement immediately to the southwest to be partly exposed (Fig. 5, profile 2). South of this basement outcrop, the post-R4 sediments thicken rapidly and form a ridge extending south to the edge of the moat surrounding Rosemary Bank. Roberts et al. (1974) ascribed the moat to differential deposition associated with intensified bottom circulation around the seamount. Further south, on the west side of the Rockall Trough, the Feni Ridge sediment drift marks the regional thickening of the post-R4 sediments. The pattern of accumulation of post-R4 sediments cannot be simply ascribed to dispersal and deposition from a single source as a major accumulation of post-R4 sediments is developed above the eastward plunging basalts of the Ymir Ridge away from the trough. Indeed this latter sediment drift, in common with others developed on the southern flank of the Ymir Ridge (Figs 3 and 8), shows an apparent southwestward sense of migration.

The preceding discussion has emphasised the broad aspects of deposition of the post-R4 sediments and may be oversimplified. The presence of several unconformities within the post-R4 sequence demonstrates episodicity in sedimentation. The seismic profiles south of the Wyville-Thomson Ridge show northward onlap of the post-R4 sediments and two unconformities within the post-R4 interval. These unconformities may correspond to those observed elsewhere between the Oligocene and the early Miocene, and between the middle and late Miocene (Montadert, Roberts et al., 1979). In this area, the late Miocene may rest directly on the Eocene. The development of these unconformities undoubtedly reflects variations in the intensity of the near bottom circulation in Neogene time. However, there are as yet insufficient drill and seismic data to document these events in detail.

## DISTRIBUTION OF INTRUSIVE CENTRES

Several prominent intrusive centres occur in the region to the south and southwest of the Faeroe Islands. The distribution of these and other neighbouring centres, together with the limit of the main Palaeocene lava field, are shown in Fig. 9. These have been primarily identified by the large positive gravity anomalies and the associated circular patterns of high amplitude magnetic anomalies. Dobinson (1970) identified the Faeroe Bank centre from the magnetic anomalies and Himsworth (1973) located the centre A in northernmost Rockall Trough from gravity and magnetic anomalies. The Faeroe Channel knoll and centre B have been located during the present work. These centres, apart from that on Faeroe Bank, are associated with shallowing of the Palaeocene 'basement'. They are notably larger in areal extent than the centres of western Scotland and Rockall Bank.

Himsworth (1973) modelled the 60 mgal positive gravity anomaly over centre A in terms of an intrusive body 250 kg/m$^3$ denser than

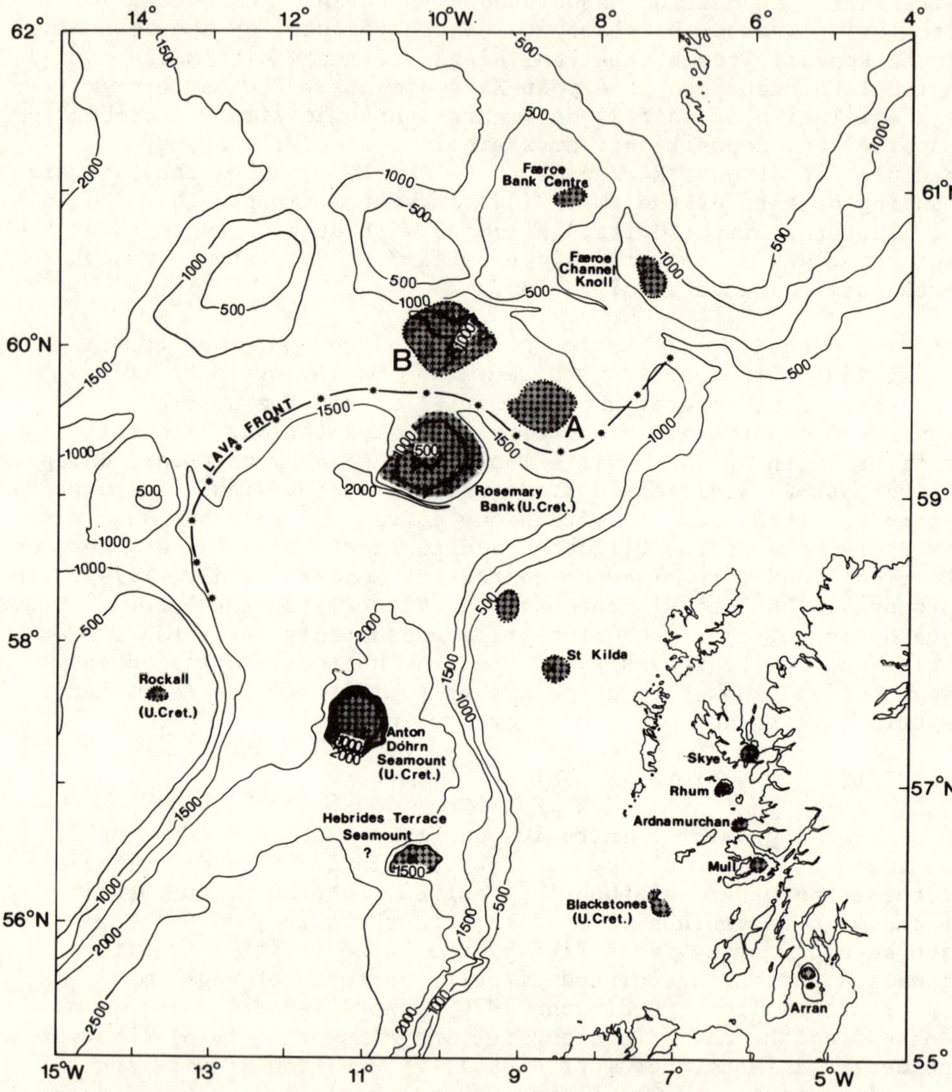


Fig. 9. Distribution of known or inferred Cretaceous to early Tertiary plutonic centres in and around the north Rockall Trough. The lava front south of the Wyville-Thomson Ridge is also shown.

the surrounding rocks and extending down to 13 km depth. The new seismic data show well developed reflectors in the upper part of the Palaeocene 'basement' above centre A, suggesting that the pluton may underlie a large strato volcano. Definite evidence on the age of the centres is lacking.

## GRAVITY ANOMALIES

The free air gravity anomaly map (Fig. 10) shows a linear belt of positive anomalies, peaking at 30 to 60 mgal, overlying the Wyville-Thomson Ridge. A positive anomaly of very large amplitude occurs over Rosemary Bank at the extreme southwestern corner of the map. As previously discussed, other circular, positive anomalies occur at A and B and are interpreted in terms of underlying, dense igneous intrusions which also give rise to complex, high amplitude, circular magnetic anomalies. Further regions of positive anomalies occur over the Faeroe Bank and the Faeroe Ridge. Otherwise the north end of the Rockall Trough and the southwestern end of the Faeroe-Shetland Channel show negative free air anomalies which drop to -40 mgal in the Rockall Trough.

Interpretations of the deep structure beneath the Wyville-Thomson Ridge along line 12/71 (Fig. 10) are shown in Fig. 11. This line crosses the eastern end of the Ridge where the complicating gravity effects of the Faeroe shelf, the igneous intrusions and the Ymir Ridge are avoided. Fig. 11(a) is the original model of Himsworth (1973) which shows that the positive anomaly is smaller than would be expected from the bathymetry and the sediment cover alone. A region of low density rocks must thus occur at depth beneath the Ridge. This model incorporates a rather high density of 2850 kg/m$^3$ for the upper part of the crust and the low density region takes the form of a 10 km root of thickened crust with steep sides.

Although the gravity requires low density rocks below the Ridge, these can partly or wholly occur at shallower depth than the Moho. In view of the seismic evidence for an intervening sedimentary section between the lavas and the underlying oceanic crust further south, this line was re-interpreted to test the hypothesis that the Wyville-Thomson Ridge was formed by the extrusion of a thick lava pile on top of pre-existing oceanic crust. The root would then be caused by downwarping of the oceanic crust in response to the load. In this model the low density region is partly produced by depression of the main crustal layer beneath the lower density lava pile and partly by the corresponding depression of the Moho.

In calculating model (b), model (a) was used as a starting point. The generalised upper crustal density of 2850 kg/m$^3$ was considered too great for a lava pile and so a density of 2700 kg/m$^3$

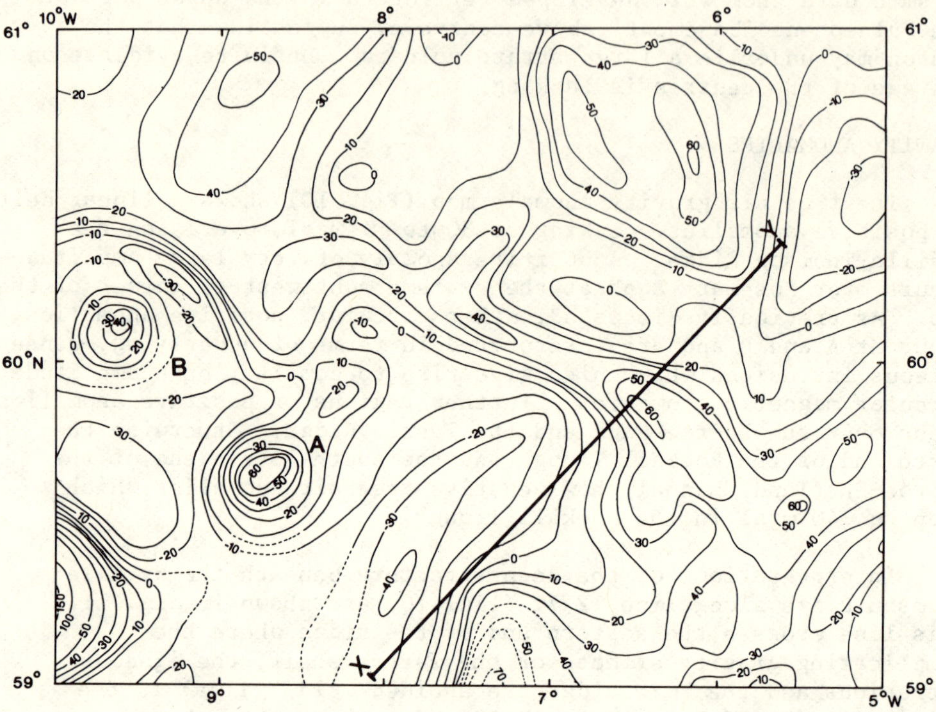


Fig. 10. Free air gravity anomaly map of the Wyville-Thomson Ridge and adjacent regions compiled by Mr. J.E. Lewis at Durham, contoured at 10 mgal interval. A and B mark positive anomalies associated with plutonic centres (Fig. 9), and XY is the line of the gravity profile interpreted in Fig. 11.

was used in the new model (Fig. 11(b)). Oceanic crust, comprising a layer of sediment, a 3 km layer 2 and an 11 km layer 3, was modelled below the basalt. The upper layers, down to the basalt, were well controlled by seismic reflection data. The sub-lava sediment layer was assumed effectively to thin toward the Ridge where it has probably been metamorphosed by the igneous activity and overburden pressure so that it lacks a distinct density contrast. Crustal thickness at the ends of the model was roughly defined by the Hebridean Margin Seismic Project (Bott et al., 1979) for northern Rockall Trough, and the North Atlantic Seismic Project (Smith, 1974) for the southern Faeroe-Shetland Channel. The Moho is gently warped and the oceanic crustal thickness is maintained at a nearly constant value of about 14 km. A fit with errors of less than 5 mgal was obtained. The remaining mis-matches can be explained by variations in thickness of sub-lava sediments or by the presence of lateral density variations within the igneous pile, such as would be produced by intrusions.

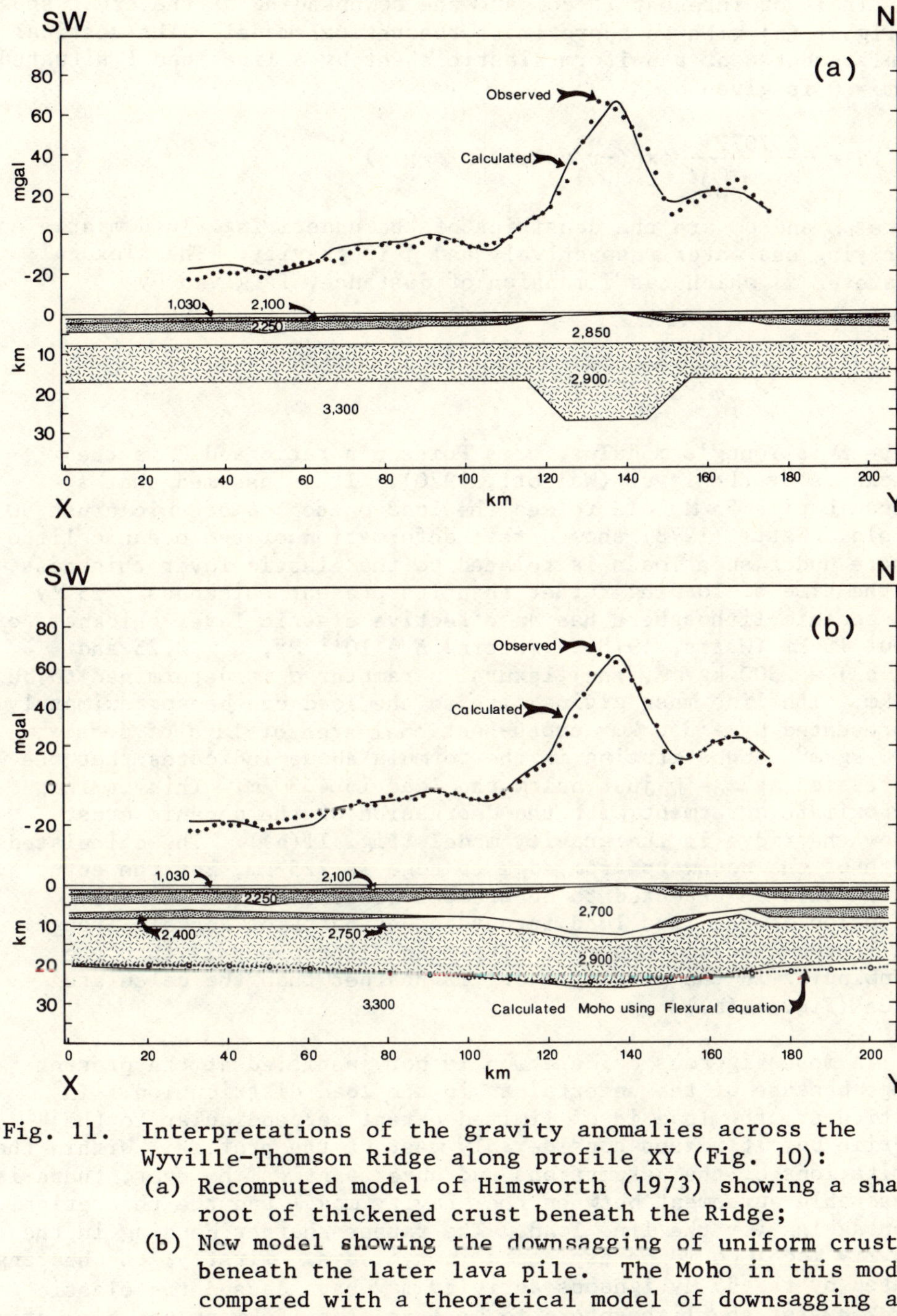


Fig. 11. Interpretations of the gravity anomalies across the Wyville-Thomson Ridge along profile XY (Fig. 10):
(a) Recomputed model of Himsworth (1973) showing a sharp root of thickened crust beneath the Ridge;
(b) New model showing the downsagging of uniform crust beneath the later lava pile. The Moho in this model is compared with a theoretical model of downsagging assuming the load to have the form of a uniform line mass.

It is of interest to compare the downbending of the crust shown in Fig. 11(b) with an approximate theoretical model. The vertical displacement $z$ of a uniform elastic sheet by a line load $P$ situated at $x = 0$ is given by

$$z = \frac{0.707P}{dg(\rho_m - \rho_w)} \exp(-x/d) \cos(x/d - \pi/4)$$

where $\rho_m$ and $\rho_w$ are the densities of the underlying fluid mantle and overlying sea water respectively and $g$ is gravity. The flexural parameter $d$, which has dimension of distance, is given by

$$d^4 = \frac{ET^3}{3g(\rho_m - \rho_w)(1-\sigma^2)}$$

where $E$ is Young's modulus, $\sigma$ is Poisson's ratio and $T$ is the thickness of the layer (Walcott, 1970). It is assumed that a volcanic pile 55 My old formed the load on top of oceanic crust 90 My old. Watts (1978) showed that deformation of the oceanic lithosphere under such loads is related to the elastic layer thickness at the time of loading rather than its present thickness. 35 My old oceanic lithosphere has an effective elastic layer thickness of about 15 km (Watts, 1978). Putting $E = 10^{11}$ Pa, $\sigma = 0.25$ and $(\rho_m - \rho_w) = 2300$ kg/m$^3$, the flexural parameter $d$ is determined to be 48 km. The line mass giving rise to the load can be approximately represented by a 350 km$^2$ cross-sectional area of lava of density 2700 kg/m$^3$. Substituting in the formula above indicates that the depression at $x = 0$ just below the load is 4.3 km. This is in approximate agreement with the depression of the oceanic crust below the ridge in the gravity model (Fig. 11(b)). The calculated width of the downwarp from edge to edge is 225 km, but the actual width would be expected to be greater because of the lateral extension of the lava load into adjacent parts of Rockall Trough and Faeroe-Shetland Channel. The main discrepancy is that the downbending in the gravity model is sharper than the calculated value (Fig. 11(b)).

A more rigorous calculation is not justified at the present stage because of the uncertainty in the load distribution. In particular, the load is of limited extent perpendicular to the profile but it widens considerably west of the profile. Within the limitations of the assumptions and uncertainties, however, there is reasonable agreement between the gravity model and the theoretical downsagging for the line load. The rather sharper bending in the gravity model could be explained in two ways. Firstly, the heating caused by the 55 My igneous activity may have caused the elastic thickness of the lithosphere to be less than 15 km at the time of loading. Secondly, the igneous activity may have caused the old oceanic crust to be thickened by addition of some new igneous material, so that the low density region beneath the Wyville-Thomson Ridge is partly due to loading and partly to crustal thickening. It should be emphasized that only two extreme gravity models are shown in Fig. 11 and that a wide range of intermediate structures could equally well account for the observed gravity profile. The main point is that downsagging under the load appears to be a feasible hypothesis to account for the deep structure, although some degree of crustal thickening may also have occurred by igneous intrusion.

## ORIGIN OF WYVILLE-THOMSON RIDGE

The Wyville-Thomson and Ymir Ridges together form an elevated feature on the sea-floor which, like the nearby seamounts to the west, does not obviously conform to any known type of aseismic ridge. The following four hypotheses suggest possible modes of origin;

(1) If the Rockall Trough and the Faeroe-Shetland Channel have formed by subsidence of continental crust, as supposed by Talwani and Eldholm (1972), the Ridge may represent an upstanding remnant underlain by a root of continental crust and unaffected by subsequent subsidence.

(2) The Ridge may be underlain by anomalously thick oceanic crust formed by hot spot activity at the time of formation of the adjacent Rockall Trough and Faeroe-Shetland Channel by sea-floor spreading. In this case the Ridge could be akin to the later Iceland-Faeroe Ridge.

(3) The Ridge may mark a series of fault blocks along the line of a major transform fault developed either within the older oceanic crust or marking the offset between oceanic crust in the Rockall Trough and continental crust in the Faeroe-Shetland Channel to the north.

(4) The Ridge may have formed as a volcanic pile erupted on top of the earlier sediments and oceanic crust as a result of Palaeocene volcanism. The same volcanic event is assumed to have formed the Faeroe shelf and the nearby Lousy, Bailey and Faeroe seamounts.

Hypothesis (1) implies a continental origin for the adjacent troughs. Hypotheses (2) and (3) imply an oceanic origin, although a variant of hypothesis (3) implies a continental origin for the Faeroe-Shetland Channel. Hypothesis (4) considers the Ridge to be a later feature than the Rockall Trough and Faeroe-Shetland Channel. The evidence discussed below supports this hypothesis, suggesting that the Ridge is the crest of a Palaeocene volcanic pile erupted onto sediments resting on oceanic crust of mid to late Cretaceous age.

Considering hypothesis (1), evidence for oceanic crust in the Rockall Trough is given by: (a) the relatively good fit of the opposite sides using isochrons drawn on basement (Roberts, 1975), (b) the resemblance of the crust in layering and thickness to oceanic rather than continental crust (Bott et al., 1979), and (c) the Trough is much wider and more parallel-sided than many typical continental grabens. Roberts et al. (1981) have recently demonstrated the presence of oceanic magnetic anomalies in the mouth of the Rockall Trough and concluded that about 120 km of oceanic crust was formed in post-Albian, pre-Maestrichtian time. Miles and Roberts (1981) have also suggested that Rosemary Bank seamount is of late Cretaceous to early Tertiary age and was perhaps formed in a terminal volcanic event represented nearby by late Cretaceous centres at Rockall and the Blackstones (Roberts et al., 1974; Durant et al., 1976). This evidence, taken together, is compelling although not quite conclusive.

There is comparatively little direct evidence on the nature of the Faeroe-Shetland Channel. Cashion (1975) and Ridd (1981, this volume) have shown that the eastern margin of the Channel is underlain by faulted blocks formed by mid Cretaceous rifting. Evidence from commercial seismic data (Ridd, 1981) suggests an extinct speading axis may underlie the Channel. Contiguity of the channel with Rockall Trough is supported by the trend of the rift structures and the almost perfect continuity between the east margin of the Rockall Trough and the Faeroe-Shetland Channel. Although the Faeroe-Shetland Channel would appear to be narrower than the Rockall Trough, this can be attributed to the seaward build-up of lavas from the Faeroes during Palaeocene time (Ridd, 1981) over the older sediments and the continental basement high of the original western edge of the Channel (Bott, 1978). This interpretation is supported by results obtained along line B of the North Atlantic Seismic Project which detected an abrupt change in basement seismic velocity from about 6.0 km/s to 5.2 km/s about 50 km southeast of Thorshavn (Bott et al., 1974). The geophysical evidence, although not conclusive, suggests that the Rockall Trough and Faeroe-Shetland Channel form a continuous feature of oceanic origin. The first hypothesis can thus be ruled out as can the variant of hypothesis (3).

The Wyville-Thomson Ridge forms a continuous feature between the Faeroe Bank and the Scottish shelf, but the Ymir Ridge only extends about three-quarters of the way across (Fig. 2). If the ridge system corresponds to thick oceanic crust formed by spreading, it would be expected to be symmetrical. The lack of symmetry is evident. It is also clear from the seismic data that the lavas which comprise the ridge post-date the underlying sediments and thus the true basement below. Although earlier support for hypothesis (2) came from the indication of a sharp root of thickened crust beneath the ridge (Himsworth, 1973), the gravity data can be equally satisfactorily interpreted in terms of downsagging of the crust beneath a later volcanic load.

Considering hypothesis (3), the asymmetry of the ridge system, its lack of continuity and linearity, and absence of the characteristic spreading fabric associated with transform faults (Searle, 1979) argue against a transform origin. Further, the continuity of the eastern margins of the Rockall Trough and Faeroe-Shetland Channel does not support evidence of a margin with transform offset adjacent to the Ridge. In a more regional context, there appears to be insufficient space to accommodate any westward transition of the Rockall Trough by transform faulting. This is because the available space is occupied by the continental basement under the Faeroe Block and no space is available off East Greenland where the Palaeocene continent-ocean transition in the Denmark Strait lies immediately adjacent to the Greenland coast (H.C. Larsen, this N.A.R.I.) The only available space is northward along the Faeroe-Shetland Channel. The Wyville-Thomson Ridge is thus unlikely to have been formed by transform faulting.

The most satisfactory hypothesis, which also explains the adjacent seamounts, is that the Ridge progressively formed as a thick volcanic pile on top of the sediments overlying the pre-existing Cretaceous oceanic crust. This interpretation agrees with the observed volcanic rocks that comprise the Ridge, and occur beneath adjacent regions. It is consistent with our gravity interpretation of the deep structure which involves downsagging of the oceanic lithosphere under the load of the volcanic pile. Our interpretation, which is also consistent with the results of the HMSP (Bott et al., 1979), implies that the pile beneath the Ridge is about 12 km thick. It gradually thins away from the Ridge so that the base rises in both directions along the Trough and Channel. The volcanic activity that produced the Ridge, seamounts and Faeroe basalts was probably contemporaneous and rapid. It was areally widespread and voluminous, the extent of the lavas of both onshore and offshore areas probably being as great as that of the Deccan Traps of India. The volcanism most probably developed in response to hot spot activity that caused rifting and ultimately the spreading apart of Greenland and the Rockall-Faeroe microcontinent. The lavas of the Wyville-Thomson Ridge probably erupted through a series of fissures aligned along the Ridge, possibly marking an underlying line of weakness.

One of the great enigmas of the Wyville-Thomson Ridge has been the gradual decrease in depth of both Rockall Trough and the Faeroe-Shetland Channel toward the Ridge, which represents an abrupt culmination of this otherwise gentle trend. Our interpretation suggests that the trend reflects the thickening and increasing elevation of the lava pile towards the Wyville-Thomson Ridge, rather than a thickening of the underlying oceanic crust.

## A PALAEOCENE LAND BRIDGE?

Prior to the Palaeocene, it has been suggested that the Rockall Trough and Faeroe-Shetland Channel formed a continuous seaway during the late Cretaceous, separating Greenland from Europe. An indication of the depth of the continuous trough at that time can be obtained from the dredging of Maestrichtian foraminifera from the east side of Anton Dohrn seamount between present depths of about 1400 and 900 m (Jones et al., 1974). As the water depth around the seamount is now over 2000 m deep, this suggests that the trough was probably at least 1000 m deep in the late Cretaceous. The bathymetry was then modified radically by the outpouring of Palaeocene lava which formed the Wyville-Thomson Ridge.

The present elevation of the Wyville-Thomson Ridge suggests that it may have provided a Palaeocene land bridge for the migration of land vertebrates. Both the Wyville-Thomson Ridge and the adjacent Faeroe Bank show evidence of sub-aerial erosion, and the thickness variations of the sediments deposited above the lavas prior to reflector R4 also suggest active nearby erosion at that time. Furthermore, if it is assumed that the oceanic crust underlying the Wyville-Thomson Ridge is 90 My old, then thermal subsidence of the cooling oceanic lithosphere will have caused a subsidence of about 1 km during the last 55 My since the lava pile was erupted. If the underlying lithosphere has been significantly re-heated by the Palaeocene igneous activity, then the post-Palaeocene subsidence would be even greater. Both the observations and theory thus suggest that the Wyville-Thomson Ridge stood entirely above sea level after formation in the Palaeocene and has subsequently subsided below sea level as the underlying lithosphere has continued to cool.

As the Faeroe Bank and the Faeroe Bank Channel probably overlie continental crust, such a large amount of post-Palaeocene subsidence may not have applied to them. However, the continental lithosphere beneath them was probably heated to some extent by the Palaeocene igneous activity so that some subsidence would be expected. There is evidence from the erosion that Faeroe Bank stood well above sea level. The situation for Faeroe Bank Channel seems less clear although it may also have been partly emerged. Although these arguments must be treated with caution, they suggest that the Faeroe and Bailey Banks and the Wyville-Thomson Ridge, possibly via the Faeroe Bank Channel, provided a possible route for migration of land vertebrates between Greenland and North Europe during the Palaeocene. The post-Palaeocene subsidence of the Wyville-Thomson Ridge and the Faeroe Bank Channel would have led first to the outflow of water through the Channel and later across the Ridge, although the precise dating of this important event and its relation to onset of overflow across the Iceland-Faeroe Ridge will require further study by drilling.

## CONCLUSIONS

(1) Palaeocene lavas form the acoustic basement of the Wyville-Thomson Ridge, the Faeroe shelf, Faeroe Bank and the adjacent deeper troughs. The lavas occur at the seabed or not far below it on the upstanding regions but are covered by up to 3 km of later sediment in the adjacent deeper regions. Beneath the Ridge, and adjacent parts of Rockall Trough and Faeroe-Shetland Channel, the lavas are probably underlain by mid Cretaceous oceanic crust, with some intervening sediment. Further west, the lavas are underlain by the northern part of the Rockall-Faeroe microcontinent.

(2) The sediments above the Palaeocene lavas are divisible into two units separated by the late Eocene to early Oligocene R4 reflector, which marks a break in deposition with evidence of some erosion. The pre-R4 sediments are thickest in a series of depocentres related to erosion of nearby emergent regions. Post-R4 sediments are conspicuously affected by contour current activity.

(3) Circular gravity and magnetic anomalies reveal the presence of four intrusive centres near the Wyville-Thomson Ridge, two being at the north end of Rockall Trough, one on Faeroe Bank and one in Faeroe Bank Channel.

(4) Gravity anomalies indicate that the Wyville-Thomson Ridge is underlain by a low density root which might represent either crustal thickening or the downsagging of a uniform crust beneath a 12 km thick lava pile.

(5) The available evidence suggests that the Ridge formed by extrusion of a thick Palaeocene lava pile, probably mainly erupted from fissures along the line of the Ridge, on top of mid Cretaceous oceanic crust. The lavas extend outwards along the adjacent Rockall Trough and Faeroe-Shetland Channel, causing the seabed to shallow from both directions towards the Ridge.

(6) The Rockall Trough and Faeroe-Shetland Channel probably formed a continuous seaway at least 1 km deep separating Europe and Greenland from their mid Cretaceous formation until the Palaeocene. The Ridge, formed by Palaeocene volcanic activity, initially acted as a land bridge between the

continental Faeroe Bank and the Scottish shelf (then emergent). Subsequently the Ridge subsided below sea level as the underlying oceanic lithosphere continued to cool, thereby influencing the flow of cold Norwegian Sea water into Rockall Trough.

## ACKNOWLEDGEMENTS

D.G. Roberts acknowledges support from the Department of Energy and M.H.P. Bott acknowledges support from the Natural Environment Research Council (ship time and research grant). Dr. E.M. Himsworth is thanked for use of his earlier data and interpretations and Dr. M.R. Dobson and Dr. R.J. Whittington for allowing use of two Aberystwyth seismic lines.

## REFERENCES

Bott, M.H.P., 1978, The origin and development of the continental margins between the British Isles and southeastern Greenland, in: "Crustal evolution in northwestern Britain and adjacent regions," p. 377, D.R. Bowes and B.E. Leake, eds, Geological Journal Special Issue No. 10, Seel House Press, Liverpool.

Bott, M.H.P., Armour, A.R., Himsworth, E.M., Murphy, T., and Wylie, G., 1979, An explosion seismology investigation of the continental margin west of the Hebrides, Scotland, at 58°N, Tectonophysics, 59:217.

Bott, M.H.P., Nielsen, P.H., and Sunderland, J., 1976, Converted P-waves originating at the continental margin between the Iceland-Faeroe Ridge and the Faeroe Block, Geophys. J. R. astr. Soc., 44:229.

Bott, M.H.P., Sunderland, J., Smith, P.J., Casten, U., and Saxov, S., 1974, Evidence for continental crust beneath the Faeroe Islands, Nature, Lond., 248:202.

Cashion, W.W., 1975, Geology of the West Shetland Basin, Proc. Offshore Europe Conf., Spearhead Publications, Kingston-on-Thames, 216:1.

Crease, J., 1965, The flow of Norwegian Sea water through the Faeroe Bank Channel, Deep-Sea Res., 12:143.

Dobinson, A., 1970, 1. The development of a marine seismic recording system, 2. A magnetic survey of the Faeroe Bank, Ph.D. thesis, University of Durham.

Durant, G.P., Dobson, M.R., Kokelaar, B.P., MacIntyre, R.M., and Rea, W.J., 1976, Preliminary report on the nature and age of the Blackstones Bank igneous centre, western Scotland, Jl. geol. Soc. Lond., 132:319.

Ellett, D.J., and Roberts, D.G., 1973, The overflow of Norwegian Sea deep water across the Wyville-Thomson Ridge, Deep-Sea Res., 20:819.

Himsworth, E.M., 1973, Marine geophysical studies between northwest Scotland and the Faeroe plateau, Ph.D. thesis, University of Durham.

Jones, E.J.W., Ewing, M., Ewing, J.I., and Eittreim, S.L., 1970, Influences of Norwegian Sea overflow water on sedimentation in the northern North Atlantic and Labrador Sea, J. Geophys. Res., 75:1655.

Jones, E.J.W., Ramsay, A.T.S., Preston, N.J., and Smith, A.C.S., 1974, A Cretaceous guyot in the Rockall Trough, Nature, Lond., 251:129.

Kelts, K.R., and McKenzie, J.A., 1976, Cretaceous volcanogenic sediments from the Line Island chain: diagenesis and formation of K-feldspar, DSDP leg 33, hole 315A and site 316, Initial Reports of the Deep Sea Drilling Project, 33:789.

Miles, P.R., and Roberts, D.G., 1981, The magnetisation of Rosemary Bank seamount, Rockall Trough, northeast Atlantic. Earth and planet. Sci. Lett. (Neth.), (in press).

Montadert, L., Roberts, D.G., et al., 1979, "Initial Reports of the Deep Sea Drilling Project," Vol. 48, U.S. Government Printing Office, Washington, D.C.

Parker, J.R., 1975, Lower Tertiary sand development in the central North Sea, in: "Petroleum and the continental shelf of north-west Europe, Vol. 1, Geology," p. 447, A.W. Woodland, ed., Applied Science Publishers, Barking.

Rasmussen, J., and Noe-Nygaard, A., 1970, "Geology of the Faeroe Islands," C.A. Reitzels Forlag, Copenhagen.

Ridd, M.F., 1981, Petroleum geology west of the Shetlands, in: "Petroleum geology of the continental shelf of north-west Europe," L. V. Illing and G.D. Hobson, eds, Heyden, London.

Roberts, D.G., 1975, Marine geology of the Rockall Plateau and Trough, Phil. Trans. R. Soc. Lond., 278A:447.

Roberts, D.G., Flemming, N.C., Harrison, R.K., Binns, P.E., and Snelling, N.J., 1974, Helen's Reef: a microgabbroic intrusion in the Rockall intrusive centre, Marine Geol., 16:M21.

Roberts, D.G., Hogg, N.G., Bishop, D.G., and Flewellen, C.G., 1974, Sediment distribution around moated seamounts in the Rockall Trough, Deep-Sea Res., 21:175.

Roberts, D.G., and Jones, M.T., 1975, Magnetic anomalies in the north east Atlantic, sheets 1 and 2, Institute of Oceanographic Sciences, Wormley.

Roberts, D.G., Hunter, P.M., and Laughton, A.S., 1977, Bathymetry of the northeast Atlantic, sheet 2, Continental margin around the British Isles, Hydrographer of the Navy, Taunton.

Roberts, D.G., Montadert, L., and Searle, R.C., 1979, The western Rockall Plateau: stratigraphy and structural evolution, Initial Reports of the Deep Sea Drilling Project, 48:1061.

Roberts, D.G., Masson, D.G., and Miles, P.R., 1981, Age and structure of the southern Rockall Trough - new evidence, Earth and planet. Sci. Lett. (Neth.), 52:115.

Searle, R.C., 1979, Side-scan sonar studies of North Atlantic fracture zones, Jl. geol. Soc. Lond., 136:283.

Smith, P.J., 1974, A seismic refraction study of crustal structure between the Faeroe Isles and Scotland, Ph.D. thesis, University of Durham.

Stride, A.H., Curray, J.R., Moore, D.G., and Belderson, R.H., 1969, Marine geology of the Atlantic continental margin of Europe, Phil. Trans. R. Soc. Lond., 264A:31.

Talwani, M., (ed.), 1974, Lamont-Doherty survey of the World Ocean: Underway marine geophysical data in the North Atlantic, June 1961-January 1971, part E: Reflection profiles.

Talwani, M., and Eldholm, O., 1972, Continental margin off Norway: a geophysical study, Bull. geol. Soc. Am., 83:3575.

Tarling, D.H., and Gale, N.H., 1968, Isotope dating and palaeomagnetic polarity in the Faeroe Islands, Nature, Lond., 218:1043.

Walcott, R.I., 1970, An isostatic origin for basement uplifts, Can. J. Earth Sci., 7:931.

Watts, A.B., 1978, An analysis of isostasy in the world's oceans 1. Hawaiian-Emperor seamount chain, J. geophys. Res., 83:5989.

# DAVIS STRAIT: STRUCTURES, ORIGIN AND EVOLUTION

S. P. Srivastava

Atlantic Geoscience Centre, Geological Survey of Canada
Bedford Institute of Oceanography
P.O. Box 1000, Dartmouth, Nova Scotia B2Y 4A2

ABSTRACT

A sill with water depths between 400 m and 800 m constitutes the central part of Davis Strait. It divides the deeper waters of Baffin Bay to the north from that of the Labrador Sea to the south. Seismic reflection, gravity and magnetic measurements across the Strait show the presence of a prominent structural high running northeast-southwest. Magnetic anomalies correlatable with seafloor spreading anomalies are not observed over this high. Large quantities of sediments lie to either side.

Seismic refraction measurements over the central part of the sill show a 18 km thick crust with mean velocity of 6.2 km/s overlying mantle of velocity 8.4 km/s. Comparison of crustal velocities under the sill with those observed under other oceanic and continental fragments show similarities to both. Seismic reflection measurements across the sill show presence of strong reflectors parallel to the eastern edge of the high. These reflectors are interpreted as Paleocene or older basaltic flows intermixed with sediments. Similar flows are not observed on the western side of the high.

The offshore extent of the Paleocene basalts which lie onshore in the Cape Dyer-Padloping area of Baffin Island and the Disko-Svartenhuk area of West Greenland has been mapped using geophysical measurements and a limited amount of shallow and deep drilling. The map shows extensive occurence of basalt in the offshore region. Cores of basalt have also been recovered from the structural high in mid-Davis Strait region. Geophysical measurements in the offshore regions and the occurrence of shallow marine deposits on adjacent land at height of 600m or so above sea level indicate that the whole region of Davis Strait was uplifted either due to excessive volcanism or due to the presence of a hot spot sometime in mid Paleocene time.

It is interpreted that the western side of the high probably contains fragments of continental crust which had formed part of south Baffin Island shelf south of Cumberland Sound prior to separation of Greenland from Baffin Island. It's eastern side contains old basaltic flows intermixed with sediments. Sometime in Paleocene or earlier times the portion of the shelf which constitutes the western side of the high got detached from Baffin Island shelf and started to move north, either as part of the Greenland plate or independently, along a leaky transform fault relative to Baffin Island. It stopped moving in early Oligocene time when sea-floor spreading ceased in the Labrador Sea and since has occupied its present position in mid-Davis Strait. Comparison is made between the Iceland-Faeroe-Ridge and Davis Strait Sill in light of the similarities in their crustal structure and geological setting.

## INTRODUCTION

Davis Strait is a constricted passage of water which lies between southern Baffin Island and West Central Greenland (Fig. 1). The sill which forms the main part of the strait separates the deeper water of Baffin Bay to the north from that of the Labrador Sea to the south.

The nature of the crust underlying Davis Strait has been a subject of much controversy. The disagreement centers on whether the region is underlain by oceanic crust formed by sea-floor spreading when Greenland separated from Baffin Island or by continental crust subsequently modified due to the intrusion of basaltic rocks during the Tertiary. Keen and Barrett (1972) on the basis of the velocities observed from seismic refraction measurements favour the former possibility. On the other hand, others (e.g. Kerr, 1967, 1981; Grant, 1975, 1980; and van der Linden, 1975) favour the latter, primarily because the region lacks sea-floor type magnetic anomalies pointing to its true oceanic affinity. Plate tectonic reconstructions (Kristoffersen and Talwani, 1977; Srivastava, 1978) require the generation of some oceanic crust in this region. Furthermore, these reconstructions imply shear motion between Greenland and Baffin Island along a NNE-SSW direction, which may have resulted in much stretching and faulting of the crust. The geological and geophysical measurements made in Davis Strait region are examined here to determine the nature of the crust which lies under this region and to suggest modes of origin which are compatible with the development of the regions north and south of it.

A submarine ridge like the Davis Strait Sill, joins Greenland,

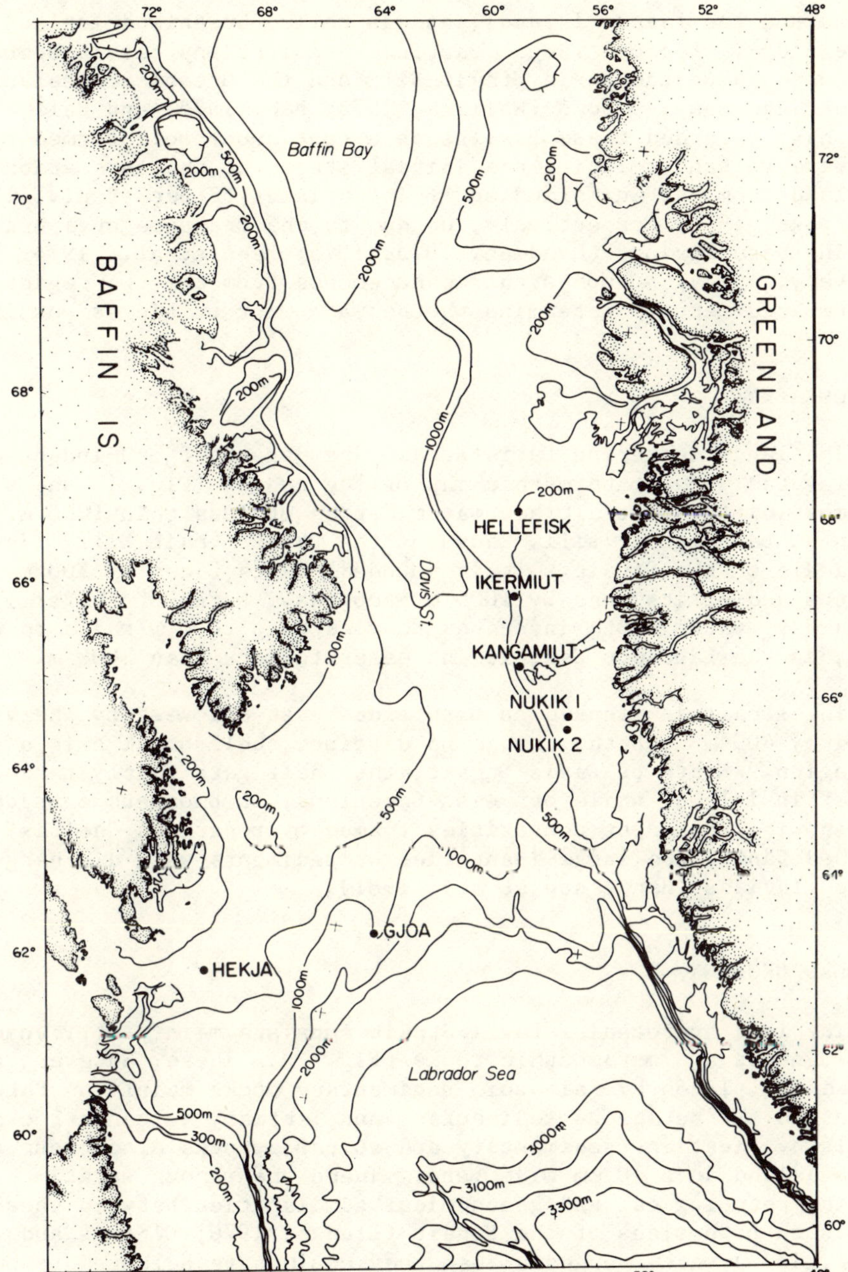


Fig. 1. Generalised bathymetry of Davis Strait region showing prominent physiographic features. Also shown are the locations of deep exploratory wells drilled off Greenland and Baffin Island.

Iceland and the Faeroe Islands, and is called Greenland-Iceland-Faeroe Ridge. Because large quantities of Tertiary basalt occur on land surrounding the Davis Strait Sill and the Greenland-Iceland-Faeroe Ridge many authors (Wallace, 1973; Beh, 1975; and Hall, 1981) have regarded these basalts as a continuous belt formed by extensive volcanism during the initial stages of the separation of Greenland from Baffin Island and British Isles (Soper et al., 1976; and Brooks, 1973) respectively, or due to the presence of a hot spot in these regions (Hyndman, 1973, 1975; Keen et al., 1974; Vogt and Avery, 1974). Geophysical measurements from the two regions are compared here to determine whether a common origin is possible.

## PHYSIOGRAPHY

The lands bordering Davis Strait are jagged, fjord-indented and elevated. The bathymetric map of the region (Fig. 1) shows that most of Davis Strait has water depths of less than 1000 m. A prominent bathymetric sill, known as the Davis Strait Sill, lies in the middle of the Strait. It is bounded on its sides by 500 m isobaths and at its ends by 1000 m isobaths. North of the sill, in Baffin Bay, water depth increases to more than 2000 m while to the south, in the Labrador Sea, it increases to more than 3000 m.

The strait is flanked on both sides east and west by shelves marked by 400 m isobaths though no distinct shelf break exists in the region. North of Davis Strait, the shelf gets fairly narrow off Baffin Island, while off west Greenland, it broadens considerably showing prominent convexities formed by prograded "deltas" and attached fans where large quantities of sediments were discharged by the glaciers (Henderson et al., 1981).

## REGIONAL GEOLOGY

The land surrounding Davis Strait consists mainly of Precambrian igneous and metamorphic rocks (Fig. 2). These, however, are overlain in places by Paleozoic sedimentary rocks mainly on Baffin Island and by Cretaceous sedimentary and Tertiary volcanic rocks on Greenland. Tertiary sedimentary and volcanic rocks also occur on Baffin Island in a 10 km wide zone. There are strong stratigraphic, petrological and geochemical similarities between these basalts on both sides of the Strait (Clarke, 1970; O'Nions and Clarke, 1972) which suggests that they could have belonged to one continuous basalt mass prior to the separation of Greenland from Baffin Island.

A generalised stratigraphic correlation between the volcanic provinces of Baffin Island and West Greenland is shown in Fig. 3, as given by Clarke (1975). The thicknesses of different units

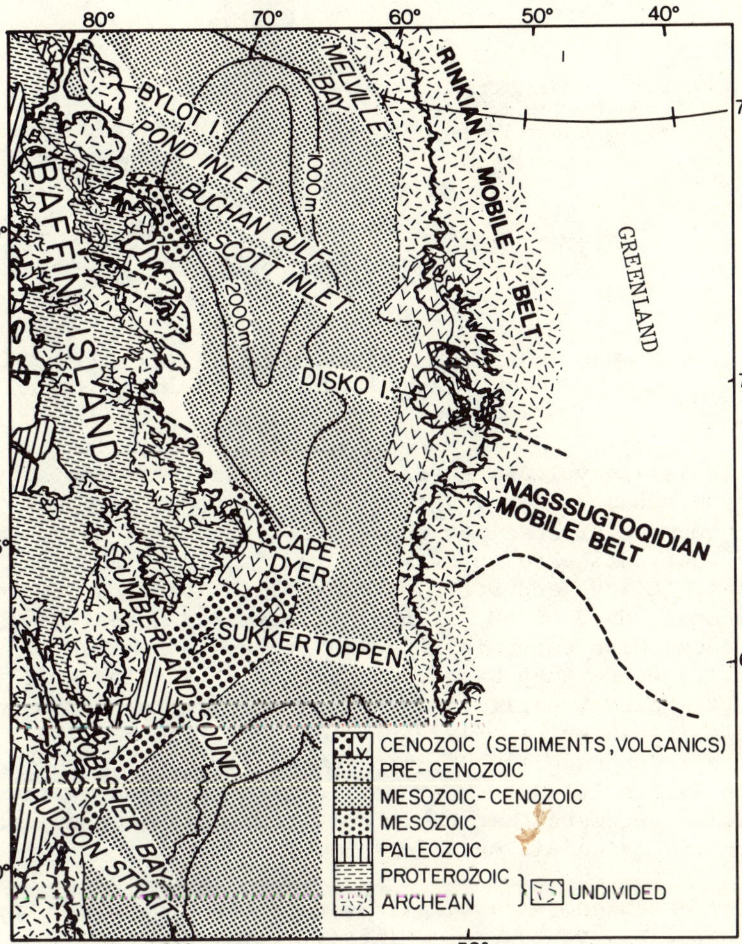


Fig. 2. Generalised bedrock geology of land and shelves surrounding Davis Strait.

shown are only qualitative. The volcanics on Baffin Island are at the most 500 m thick while on West Greenland they may be as thick as 10 km.

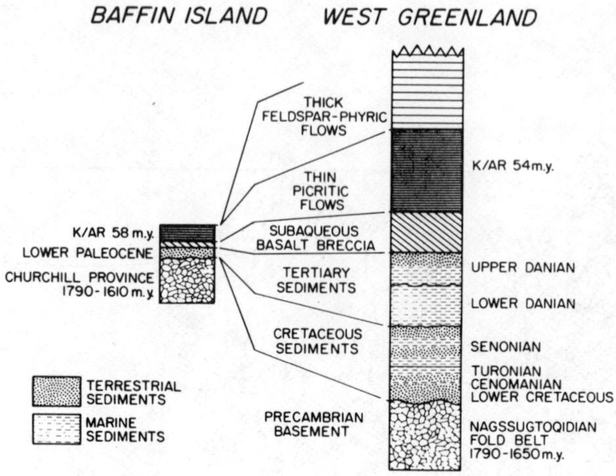


Fig. 3. A comparison of the stratigraphic columns in Baffin Island and West Greenland (After Clarke, 1975).

Baffin Island volcanics (Fig. 3) consist of subaerially erupted flows and subaqueously erupted breccia that in part rest directly on Precambrian crystalline rocks and in part lie on terrestrial Paleocene sediments made up of sandstone, shale, minor coal and conglomerate (Clarke and Upton, 1971). These basalts yielded a potassium-argon date of 58 ± 2 m.y. (Clarke and Upton, 1971). Volcanic rocks also occur in offshore regions southeast of Cape Dyer (Fig. 2) where they have been sampled by shallow drilling and their surface extent outlined by seismic reflection and magnetic profiling (MacLean et al., 1978; MacLean and Falconer, 1979). The basalts sampled offshore are petrogenetically similar to the basalts on Baffin Island. The small amount of glass in the offshore samples suggests that these rocks may have been emplaced under sub-aqueous conditions.

On the Greenland side (Fig. 3), a sequence of Lower Cretaceous (Barremian-Aptian) to Lower Tertiary (Danian) clastic rocks in the Nugssuaq embayment lie on the eroded Precambrian rocks. They contain fluviatile-deltaic sandstone-shale facies that include some coal seams in the south and northward are transitional to prodeltaic marine shales (Henderson et al., 1976; Escher and Watts, 1976). The sediments, though not thick in most of the area, none-theless show that sedimentation continued sporadically for some 60 m.y. before volcanism began. The onset of volcanism is marked by the presence of tuff beds in Danian sediments (Henderson et al., 1976). These were followed by a massive outpouring of subaqueous volcanic breccias which locally attain thicknesses of 600 to 700 m, the source for most of which appears to be located to the west. Emplacement of subaerial, mainly olivine rich, tholeitic basalt followed. These were followed in some places directly and at others after a short hiatus by olivine free, feldspar-phyric basalts (Clarke and Pedersen, 1976). The youngest volcanics in west Greenland are the small, localised occurrence of acidic lavas, pyroclastics, including tuff and ignimbrite and dykes of lamprophyre composition (age 30 to 40 m.y. Parrott and Reynolds, 1975). Dykes are numerous throughout the volcanic pile and have been found to be feeders of the lavas at several localities (Clarke and Pedersen, 1976).

Volcanic rocks also occur in offshore regions off west Greenland and their extent as mapped by Henderson et al. (1981) using aeromagnetic and seismic reflection data is shown in Fig. 2.

Out of the five wells drilled to date on the west Greenland Shelf (Fig. 1), volcanics were encountered at only two (Fig. 4, Greenland Geological Survey 1979). The volcanics from these wells were found to be subaerial in character and yield ages of 93.4 ± 4.7 m.y. in Nukik 2 well (A. J. Ferguson per. comm., 1980) and of 53 to 54 m.y in Hellifisk well (Henderson et al., 1981). Basalts were also encountered in the two deep exploratory wells drilled off Baffin Island (Hekja and Gjoa, Fig. 1). In Hekja the bottom 1000 m of drilled section consists of alkalic volcanics interbedded with chalky clays and ranges in age from Danian to Early Cretaceous (Klose et al. 1981). Gjoa located on the continental slope penetrated 1,300 m of interbedded Paleocene marine shales and alkalic volcanics overlain by 2,700 m of Paleocene to Recent Shales and silts (Klose et al. 1981).

## CRUSTAL STRUCTURE

Extensive geophysical measurements (gravity, magnetic, seismic reflection and refraction) have been made in the Davis Strait region over the past 10 years by government and private agencies. Their findings have recently been summarized by Srivastava et al. (1981) from which the following description of the crustal structure under Davis Strait has largely been extracted.

### Refraction Measurements

Seismic refraction measurements along a limited number of reversed lines and a large number of unreversed lines (expendable sonobuoys) exist for the Davis Strait region (Fig. 5). Figure 6 shows crustal sections from two of the lines from Davis Strait and one from Baffin Bay.

The results show that in the central region of Davis Strait a 2 km thick layer with a velocity of 4.8 km/s underlies a 1.5 km thick sedimentary layer with a velocity of 2.4 km/s. Keen and

Barrett (1972) have associated the layer of velocity between 4.2 and 4.8 km/s with oceanic layer 2 usually observed in the deep oceanic basins (Ludwig et al., 1970) on the basis of other seismic refraction measurements carried out near line D1 (Fig. 5).Grant (1975) on the other hand, has suggested that the above velocity may equally well be ascribed to well-indurated sediments or a wide range of igneous or metamorphic rocks. The main crustal layer under line D1 is about 18 km thick with a velocity of 6.2 km/s and the total depth to 'Moho' is about 22 km (Keen and Barrett 1972).

In contrast to the refraction results obtained over the sill, the results from the reversed line C10 (Keen and Barrett, 1972) which lies in Baffin Bay 300 km north of the Davis Strait sill (Fig. 5) show that the "Moho" lies at a depth of 9 km here compared

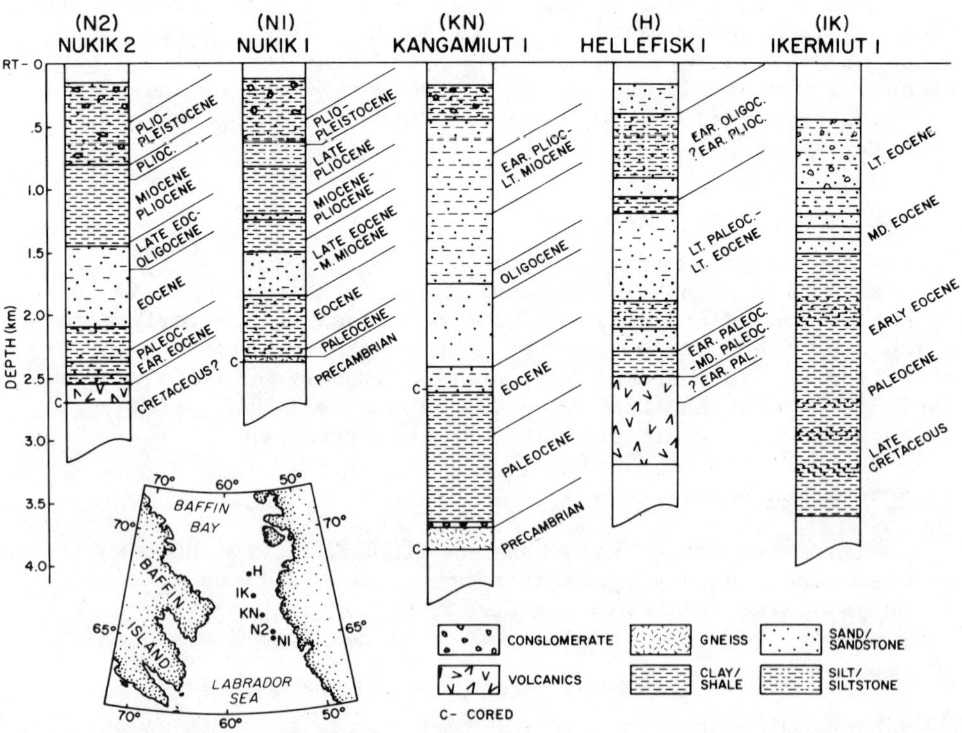


Fig. 4. Lithostratigraphy of West Greenland Shelf wells. (After Greenland Geological Survey, 1979).

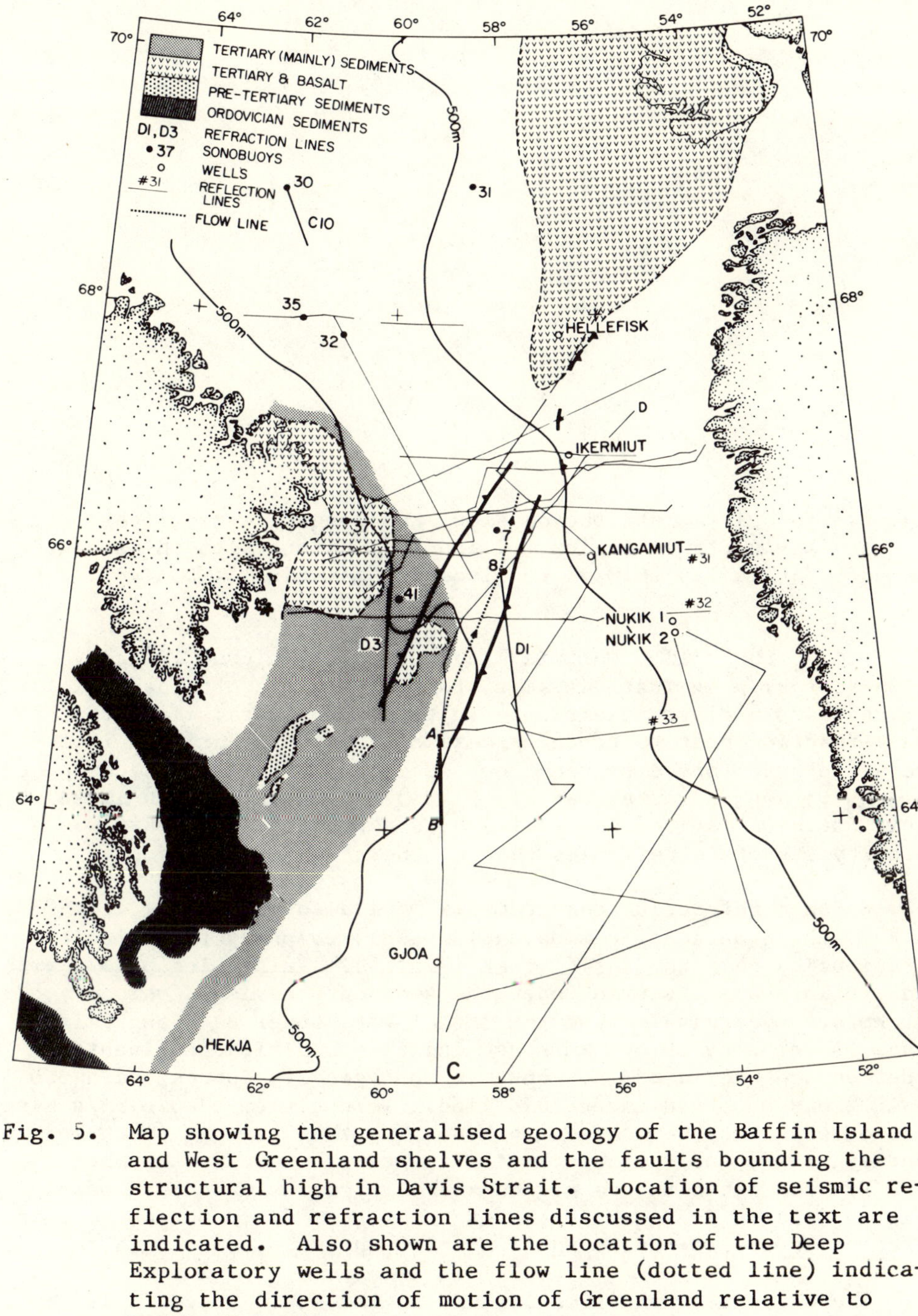


Fig. 5. Map showing the generalised geology of the Baffin Island and West Greenland shelves and the faults bounding the structural high in Davis Strait. Location of seismic reflection and refraction lines discussed in the text are indicated. Also shown are the location of the Deep Exploratory wells and the flow line (dotted line) indicating the direction of motion of Greenland relative to Baffin Island.

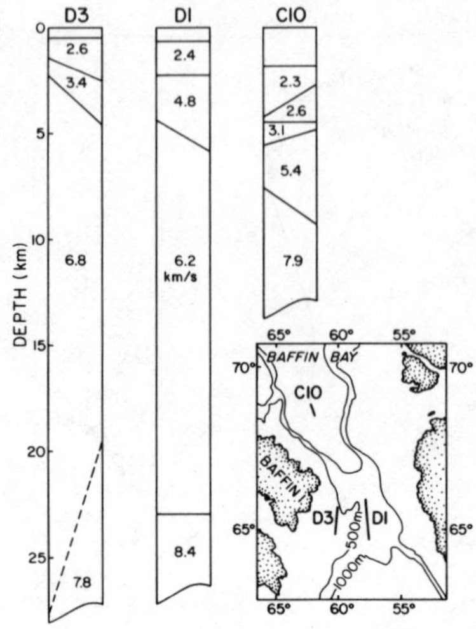


Fig. 6. Velocity distribution along two long reversed lines from Davis Strait region (Dl and D3) and one from the southern Baffin Bay (ClO). For location see Fig. 5.

to 22 km at the sill. On this line a slightly higher velocity value (5.4 km/s against 4.8 km/s) for layer 2 was obtained. This region bears a close resemblance in terms of its seismic reflection and refraction results to deep oceanic basins and has been interpreted to have formed by sea-floor spreading processes (Jackson et al., 1979; and Srivastava et al., 1981). Thus the region immediately north of Davis Strait sill shows a much thinner crust whose velocity structure resembles that of the oceanic basins.

Seismic refraction measurements were also made along line D3 (H.R. Jackson personal communication 1981) to study the velocity distribution near the shelf break (Fig. 5). Velocities of 3.9 km/s and 6.8 km/s are observed under the sediments having a velocity of 2.6 km/s. No arrivals from the 'Moho' were observed along this line. A velocity of 6.8 km/s can indicate modified continental crust or oceanic layer 3 . However, no velocity corresponding to layer 2 was observed under this line. Velcocities of 6.8 km/s have been observed in the transition zones in other regions (Keen and Hyndman, 1979) and the same may be the case here. Furthermore, line D3 lies close to the region where volcanics have been mapped (Fig. 5) and it is possible that the velocity distribution here is affected by the volcanics which overlie the main crustal layers.

There are several other regions in the world which, like the Davis Strait high, exhibit anomalous crustal structures. These structures differ markedly from the relatively simple, three-layer model which applies to most oceanic crust. An analysis of the crustal structures of such plateaus or highs by Carlson et al. (1980) shows that these regions can be divided into two groups: one which are known or thought to be fragments of continental crust (e.g. Rockall Plateau, Lord Howe Rise) and a second which are of oceanic origin (e.g. Shatsky Rise and Broken Ridge). The crustal structure under Davis Strait was then compared with that under other plateaus (Carlson et al., 1980) and under Iceland, the Iceland-Faeroe Ridge (Bott 1974) and Baffin Bay (Fig. 7 and 8) to study the differences between them. Comparison with continental fragments (Fig. 7) shows that (a) the prominent layer with velocity between 6.8 to 7.0 km/s which is present under most continental fragments is not observed under Davis Strait, (b) a much thicker layer of velocity 6.2 km/s is observed under Davis Strait than under other continental fragments (5.9 to 6.4 km/s) and (c) the 'Moho' lies at a shallower depth under Davis Strait than under continental fragments. A similar comparison with oceanic plateaus (Fig. 8) shows that (a) the refractor with a velocity between 7.3 to 7.8 km/s present under most plateaus is not observed under Davis Strait, (b) the main crustal layer under Davis Strait exhibits a velocity of 6.2 km/s which is similar to those observed (5.9 to 6.4 km/s) under some of the plateaus but it is much thicker under Davis Strait and (c) the depth to the 'Moho' under Davis Strait is similar to those of oceanic plateaus.

The above comparison clearly shows significant deviations in the crustal structure observed under Davis Strait from those observed under oceanic and continental plateaus. Strictly speaking the crustal structure neither resembles the structures observed under true oceanic plateaus nor under continental fragments. It is possible that the rocks constituting the Davis Strait sill contain elements of both, a suggestion which was also made by Keen et al (1974).

## Seismic reflection measurements

Examination of a large number of single channel and a limited number of multichannel seismic reflection profiles across Davis Strait shows the presence of a prominent structural high in the middle of the Strait, oriented in a NE-SW direction (Fig. 5). A similar high was also reported by Wallace (1973) and Beh (1975) in this region. Two representative seismic profiles across this high are illustrated in Fig. 9 and 10. Profile 31 (Fig. 9) lies near the northern end of this high where the bedrock is exposed at the sea-floor, while profile 32 lies south of it (Fig. 10). The western end of the high is interpreted to be fault controlled judging by it's steepness. The high shoals to the north. Cores of basalt were obtained from the high, where it outcrops north of profile 31 (MacLean, 1981).

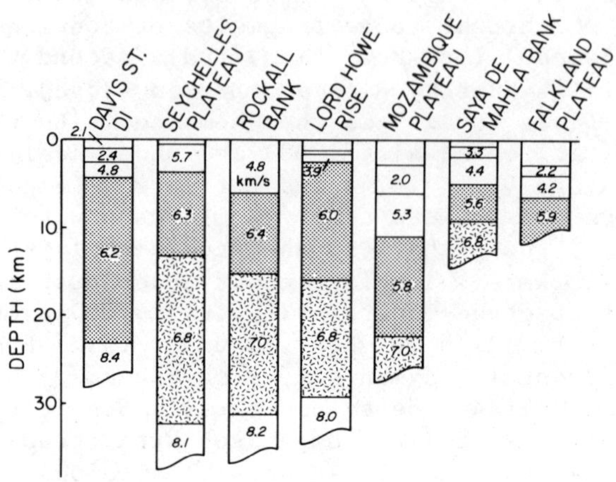


Fig. 7. Crustal structure of continental fragments (after Carlson et al., 1980), and Davis Strait (line D1).

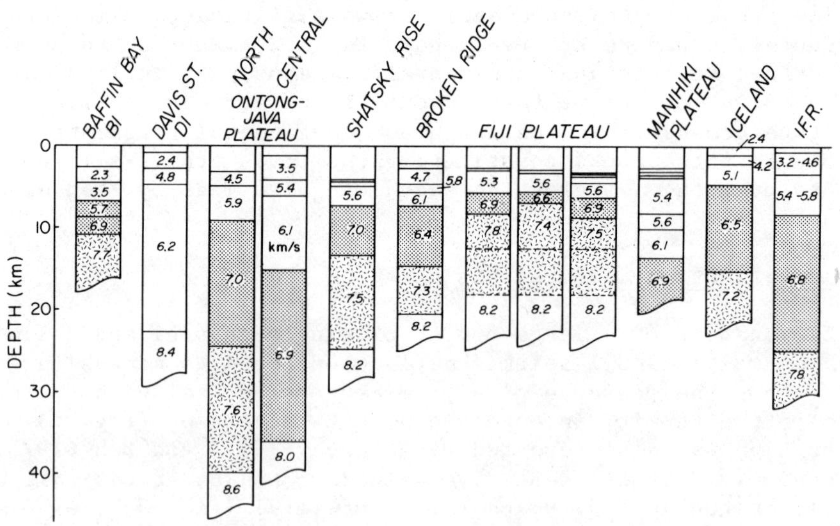


Fig. 8. Crustal structure of oceanic plateaus (after Carlson et al., 1980) Davis Strait, Baffin Bay (line B1), Iceland and Iceland-Faeroe-Ridge (I.F.R.).

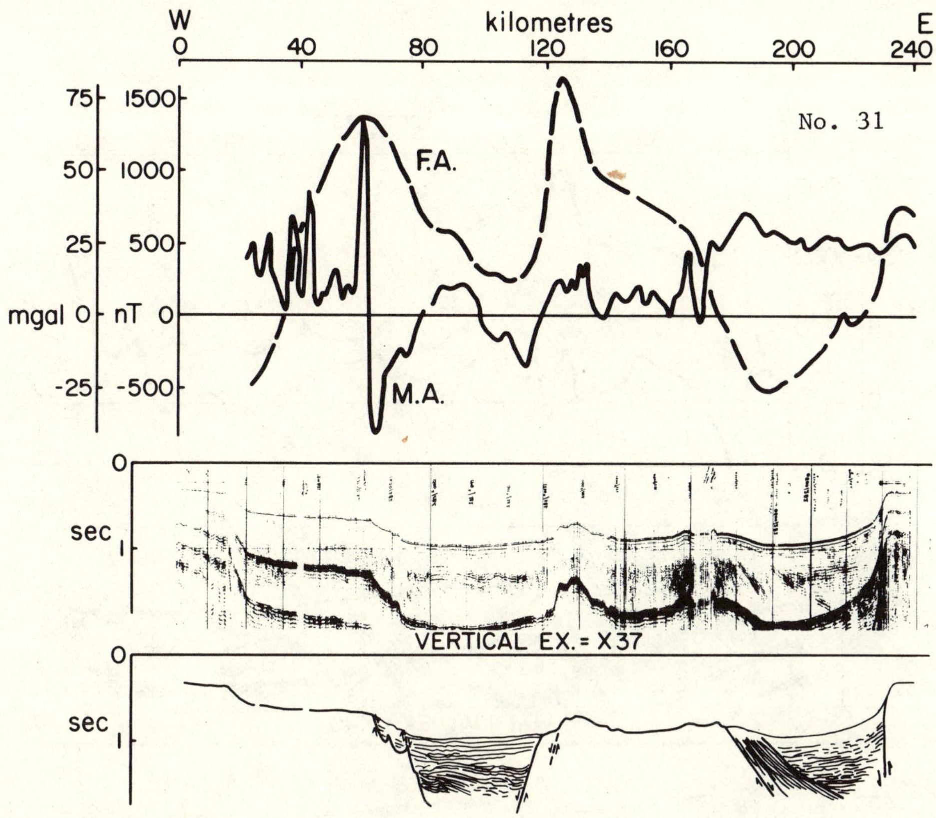


Fig. 9. Photograph and line drawing of seismic reflection profile 31 across Davis Strait (for location see Fig. 5) together with gravity and magnetic profiles.

Variable amounts of sediment overlie this high. A multichannel seismic reflection profile (line CD in Fig. 5) which runs over a considerable portion of this high shows a number of small-fault bounded basins within the high. In places the basement shows the presence of subbasement reflections suggesting presence of basaltic flows intermixed with sediments.

The reflection characteristics over the high obtained from multichannel seismic recordings is different from those observed in the region south of it (Fig. 11). The high shows weak reflections while the region south of it shows presence of strong hyperbolic reflectors resembling oceanic basaltic layers. In some ways the reflection characteristics of the high resemble those observed over basalt outcrops off Baffin Island (at 70 km mark in Fig. 9),indi-

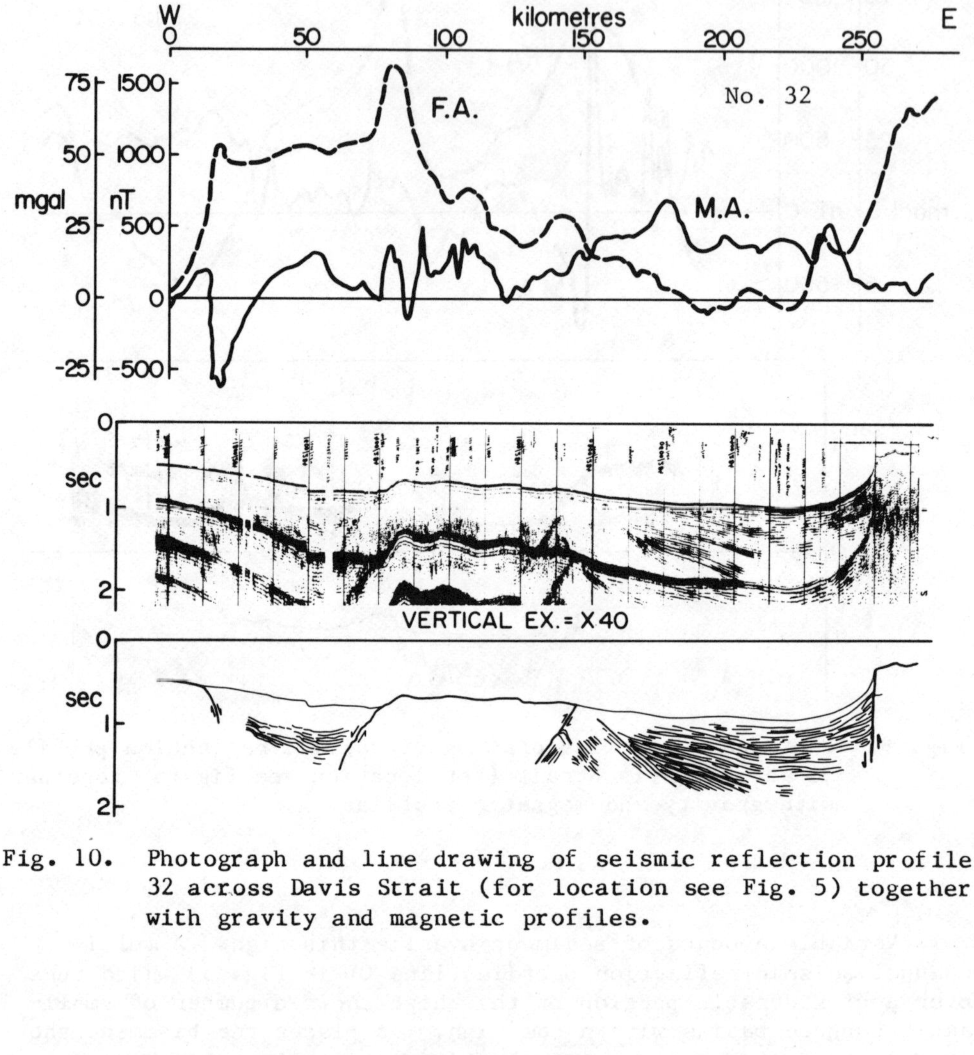


Fig. 10. Photograph and line drawing of seismic reflection profile 32 across Davis Strait (for location see Fig. 5) together with gravity and magnetic profiles.

cating that the top of the high perhaps consists of basalt similar in composition to that lying off Baffin Island.

Several reflectors lie along the eastern side of the high as can be seen at 180 km in Profile 31, Fig. 9. It is difficult to decipher the contact between these reflectors and the bedrock of the high but the absence of any reflectors west of 180 km indicates that it is approximately in the vicinity . The layers within the

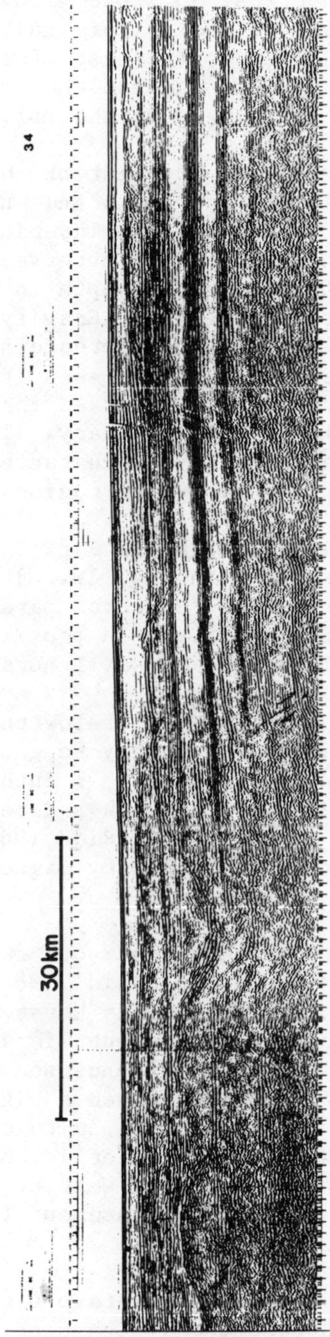


Fig. 11. Photograph of the multichannel seismic reflection profile AB across the Davis Strait high. For location see Fig. 5.

reflectors show a gradual decrease in slope with increase in depth. Furthermore, the eastern end of this unit marks a pronounced unconformity. Sediments to the east of the unit lie horizontally and slightly turn upslope against the unconformity, indicating they were deposited subsequent to the uplift.

On the Baffin Island side, to the west of the high, sediments lie more or less horizontally in the basin and end abruptly against the high. The tilted unit with prominent layering seen at 180 km mark on the east side of the high is not observed on the west side of the high. It may lie at too great a depth to be seen there or perhaps does not exist on the west side. Gravity measurements along profile 31, Fig. 9, show a sharp increase at the west edge of the high. A similar increase can also be seen in Profile 32, Fig. 10, south of line 31. There is also a marked change in the magnetic signature of the anomaly at this boundary. The anomalies east of it are shorter in wavelength and show characteristics similar to those observed over basalt near Cape Dyer region.

The reflector unit seen at 180 km in Fig. 9 seems to continue to the south, e.g. at 150 km mark in Fig. 10. It is not as distinct here as in Fig. 9. A marked reflector parallel to the western edge of the high lies at 150 km in profile 32, Fig. 10. It may be a fault scarp associated with a small horst to the east.

The presence of a set of reflectors below the unconformity only on the east side of the high strongly suggests that they have moved or uplifted together with the high. I interpret these reflectors as basalt flows interlayered with sediments similar to those off Shetland Islands (Smythe 1981, Ridd 1981). This is also supported by the presence of high frequency magnetic anomalies over them.

Compilation of depth to basement data off the west Greenland coast (Henderson et al., 1981; Manderscheid, 1980) shows the presence of several N-S faults giving rise to horst and graben structures (Fig. 12). Large amounts of sediment lie in the grabens and the horsts form ridge like structures (Henderson et al., 1981). One of the wells, Kangamiut, lies over such a high. A major N-S fault runs fairly close to the shelf edge in this region. Farther to the north, large amounts of basalt underlie the shelf where they are in part covered by young sediments. West of Disko Island the basalt overlies a ridge of Precambrian basement (Brett and Zarudzki, 1979; Henderson et al., 1981).

Examination of seismic reflection data off west Greenland shows that the Davis Strait structural high does not continue to the major N-S fault which lies near the shelf break off west Greenland but is separated from it by a small basin.

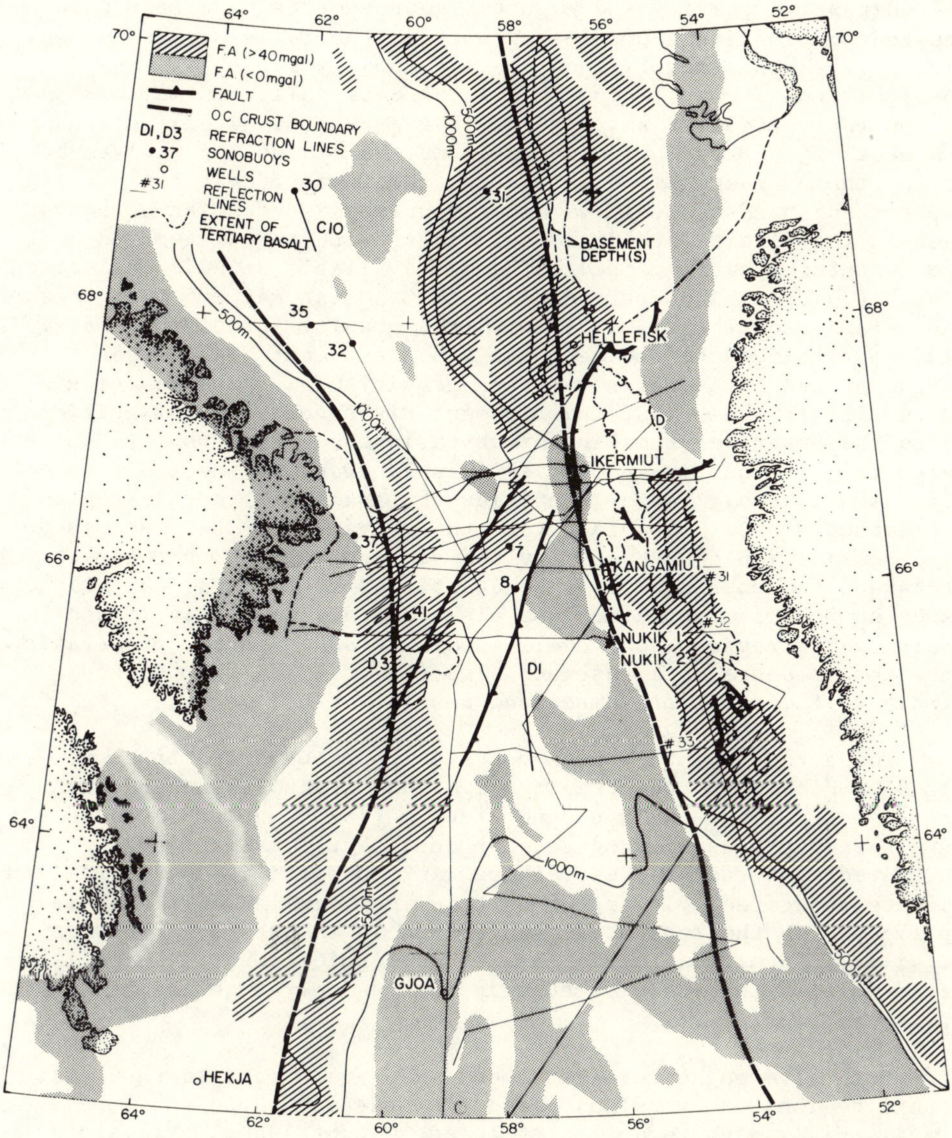


Fig. 12. Map of Davis Strait showing positive and negative gravity anomalies, the structural highs and offshore extent of Tertiary basalts, postulated ocean-continent crust boundary and the location of the deep exploratory wells in the region.

## Gravity and Magnetic measurements

Extensive gravity and magnetic measurements have been made by the Bedford Institute of Oceanography in the Davis Strait region. Figure 12 shows a simplified Free Air anomaly map of this region obtained from a compilation of all the data collected in this region (Srivastava et al., 1981). The map shows several pronounced features: (1) Bands of positive anomalies lie on the shelves bordering the northern Labrador Sea and the Davis Strait region. (2) A prominent band of positive anomaly oriented northeast-southwest lies in the Strait, and it is coincident with the western edge of the structural high, mapped using all available seismic reflection data. (3) Comparison of the depth to basement map off west Greenland (Henderson et al., 1981) with the gravity anomaly map (Fig. 12) shows coincidence between the N-S structural highs (near Kangamiut and Nukik wells) and the gravity high. Such a coincidence clearly shows that the basement highs contribute significantly to the positive anomalies observed in these regions. (4) A very large band of positive anomalies lies landward of the shelf edge off Disko Island. Large quantities of basalt lie in this region (Henderson et al., 1981) and thus the gravity high here arises due to a combination of both namely the structural high of the basement rocks and the presence of large quantities of basalts. (5) Large bands of positive anomalies are also observed on Baffin Island shelf over a region where shallow seismic and magnetic observations show the presence of a basement high (Grant,1975) from Hudson Strait to the mouth of Cumberland Sound.

There thus exists a high correlation between the basement highs or ridges and the gravity highs on Baffin Island and west Greenland shelves. Talwani and Eldholm (1973) showed presence of such basement ridges along several continental margins and suggested that such ridges perhaps mark the position where the continents separated. It is not certain if a similar criterion is applicable in the present instance; but in the absence of other detailed information it can be used as a guide to define the landward limit of the oceanic crust in the region. Such assumed limits are shown in Fig. 12.

Extensive magnetic measurements have been made in the Davis Strait region both from ships and from the air (Hood and Bower, 1975). Fig. 13 shows a plot of selected profiles plotted along tracks. It is quite evident from this plot that except for certain regions on the shelf where large amplitude and high frequency anomalies lie, the anomalies in the rest of Davis Strait region show a highly variable pattern and it is difficult to correlate them from track to track.

On the shelf the high frequency anomalies correlate well with the occurrence of basalt (Fig. 12 and 13) as mapped from seismic reflection profiling and shallow and deep drilling (MacLean et al., 1978). This is clearly visible in Fig. 9 at 60 km near the junction between the basalt off Cape Dyer and the sediments to the east. A similar situation exists on and offshore Disko Island on the west coast of Greenland. Large amplitude and high frequency anomalies are observed in the aeromagnetic map of this region (Greenland Geological Survey, 1978) where basalt is known to occur or has been mapped using other techniques.

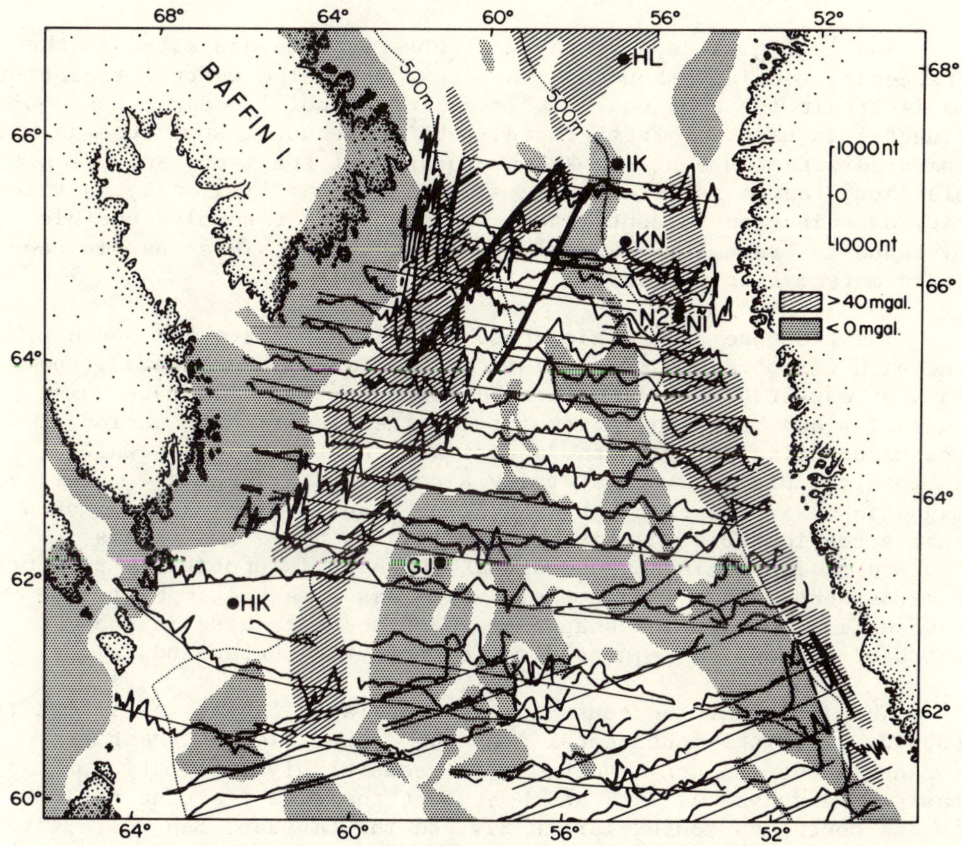


Fig. 13. Map of Davis Strait showing positive and negative gravity anomalies, the structural high, magnetic anomalies plotted along ship's tracks and the location of the deep exploratory wells in the region.

Also shown in Fig. 13 are the positive and negative bands of gravity anomalies. Large amplitude and high frequency magnetic anomalies lie in the region of the gravity high on Baffin Island shelf. A similar situation exists on West Greenland Shelf (between 63°N to 67°N) except that the anomalies here are slightly smaller in amplitude. High frequency and small amplitude anomalies exist only on the western edge of the Davis Strait high except at its northern end where they overlie the entire high (Fig. 9). Similar anomalies are also observed over the outer gravity high off Southern Baffin Island in line with the eastern edge of the Davis Strait high. However, the available seismic reflection data off Southern Baffin Island does not support southward continuation of this high in this region.

## ORIGIN

The geological and geophysical observations described in the preceeding sections show that the crust under the central region of Davis Strait is anomalous. It is neither 'truly' oceanic nor continental in nature. Furthermore, when its crustal structure is compared with those of the known continental fragments and oceanic plateaus, the Davis Strait high shows a greater similarity in velocity distribution to continental fragments but resembles oceanic plateaus in crustal thickness. It is, thus, possible that it contains material from both.

The presence of oceanic crust immediately north and south of the high (as obtained from seismic reflection measurements), the lack of magnetic anomalies correlatable with sea-floor spreading anomalies and the parallelism between the direction of motion of Greenland relative to North America and the fault scraps which enclose the high, favour it being a continental fragment. Gravity, magnetic measurements and the drilling results over the high show that a considerable quantity of basalt lies above and on its eastern side. It is suggested that the high is probably a part of the southern Baffin Island Shelf which was detached from it some time during the initial stages of sea-floor spreading in the Labrador Sea and subsequently moved north with Greenland.

The idea that the central region of Davis Strait was formed by subsidence of the continental crust on either side of the high (Grant, 1980; van der Linden, 1975) seems highly improbable on the basis of geophysical data which support the view that the regions to the north and south, Baffin Bay and the Labrador Sea respectively, were formed by a sea-floor spreading process (Jackson et al., 1979; and Srivastava et al., 1981). If these regions were formed by sea-floor spreading then simple geometrical reconstruction of the plates demands that Davis Strait has to be formed as a consequence of this by a similar mechanism. As mentioned earlier, this does not necessarily imply that the crust under the "high" in the central region of the Strait has to be completely oceanic as it could include a displaced fragment of the continental or older oceanic crust.

On the other hand, if one assumes that Baffin Bay and a large part of the Labrador Sea were formed by crustal attenuation, then the continental crust in Davis Strait has to be attenuated or stretched by an amount compatible with the motion needed between Greenland and Baffin Island. Keen and Pierce (1981) carried out some numerical calculations to examine whether the thin crust observed under Baffin Bay (Keen and Barrett, 1972) could be created by stretching the continental crust. They found no matter which mechanism was chosen to generate the attenuated crust in Baffin Bay, the observed subsidence, the present width and the crustal thickness of the Bay could not be matched satisfactorily. The same situation should apply to the Davis Strait region as it lies south of the Bay. Sea-floor spreading thus seems the only satisfactory mechanism for their formation.

## Davis Strait as an extinct hot spot

The possibility that Davis Strait was the site of a hot spot or mantle plume (Wilson, 1965; Morgan, 1971) has been considered by many workers (Hyndman, 1973, 1975; Keen et al., 1974; Keen and Clarke, 1974). Several independant sources of information have been used to support such an interpretation and we will briefly examine them here. There are several characteristics, common to hot spot regions, which are observed in the Davis Strait region and they are described briefly here.

(1) The hot spot regions have shallow water depths, forming sills or plateau-like features. Davis Strait, as was mentioned earlier, forms a bathymetric sill between the deeper water of Baffin Bay to the north and the Labrador Sea to the south. Such permanence in the elevation of the hot spot regions arises due to the introduction of low density material from the plume into the plate. (2) The regions have extensive outpouring of volcanic lavas at the margins of the plates during the initial openings and at subsequent times when a shift in the spreading takes place. We find the occurrence of Tertiary basalt on either side of Davis Strait, both onshore and offshore. The similarity of basalts in the two regions (Clarke and Upton, 1971) shows they had a common or similar source. This strongly suggests that prior to separation of Greenland from Baffin Island the basalts from these regions may have formed a continuous feature. (3) The basalts from hot spot regions exhibit systematic changes in their chemical composition with distance from the source (e.g. Schilling, 1973). Though there are little data to substantiate this from Davis Strait region, Keen

and Clarke (1974) have shown from their analysis of basalts from Baffin Island, west Greenland and from three dredge hauls located north and south of Disko Island a spatial variation in their chemical constituents from north to south. This, according to them, shows the possibility of an extinct hot spot north of Davis Strait. (4) Because of the excessive elevation under hot spot regions they have anomalously thick crustal structure in comparison with the normal oceanic basins. This we find to be the case for the Davis Strait Region. (5) The depth to basement should increase gradually away from the hot spot along the axis of the ridge (Vogt, 1971; 1976). This seems to be the case on a much larger scale for the Davis Strait region if we include the northern and central part of the Labrador Sea (Hyndman, 1973). On the other hand, the depth to basement increases rapidly towards Baffin Bay, north of the high, as the basement cannot be seen in the seismic reflection data in most of this area (even beyond 3 seconds reflection time).

Presence of sedimentary basins on either side of the Davis Strait high and shallowing of basement from the Labrador Sea towards Davis Strait supports Hyndman's (1975) suggestion that the hot spot must have been active in the Davis Strait region mainly during the second stage of the opening of the Labrador Sea. This is further supported by the ages of the basalt from Baffin Island (58 m.y., Clarke and Upton, 1971), Disko Island (54 m.y.) and from the two deep exploratory wells (Gjoa - 56 m.y., Klose et al., 1981; and Hellefisk - 53 to 54 m.y., Henderson et al., 1981). The youngest volcanics reported so far are the lamprophyre dykes on the west Greenland coast giving an age of 30 to 40 m.y. (Clarke and Pederson, 1976) which roughly corresponds to the time of cessation of sea-floor spreading in the Labrador Sea and perhaps also of the hot spot in Davis Strait.

Volcanic rocks of comparable ages to those which occur on land on either side of Davis Strait occur on either side of Iceland, on the east coast of Greenland (Brooks, 1973) and on the Faeroe Islands as well as in parts of the British Isles. As in the Davis Strait region, the occurrence of these basalts on land bordering the North Atlantic has been related to the onset of sea-floor spreading in the North Atlantic (e.g. Hyndman, 1973; Vogt and Avery, 1974; Soper et al, 1976) as well as to the initiation of a hot spot (Vogt, 1971; Hyndman, 1975) which now lies under Iceland. There are considerable differences in the rock types and intrusive versus extrusive activity between East and West Greenland as well as in their chemical composition (Clarke, 1977) which makes it difficult to relate them to a single source such as a hot spot (Hyndman, 1973). Besides, as pointed out by Clarke (1977), the volcanic rocks in both areas thin towards the inland ice, and perhaps there is no physical connection between them. Clarke (1977) has proposed that the Tertiary volcanics of Davis Strait should be treated as a separate group from those of the Greenland-Iceland-Faeroe group. Hall (1981) on the other hand treats them as a continuous unit.

If Iceland is the present site of the hot spot, then one would assume that during early Tertiary, Davis Strait must have been underlain by low density upper mantle material similar to that which now underlies Iceland. Assuming that isostatic equilibrium prevailed, Davis Strait must have stood about 2 km above its present elevation at that time. Subsequently it subsided as the underlying lithosphere cooled down. Subsidence of the region is supported from the drilling results carried out on Greenland shelf. Shallow marine condition seems to have prevailed on the Greenland shelf during mid Cretaceous time (Gradstein and Srivastava, 1980) but they must have been interrupted for some time during early Tertiary when Davis Strait was uplifted. It is equally possible that a land bridge may have existed for a short period between North America and Europe via Greenland as the uplifted Iceland-Faeroe-Ridge lay between Greenland and Europe at this time. The findings from the five wells drilled on the Greenland Shelf and two off Baffin Island would be most informative to examine if these were the cases.

## EVOLUTION

Even though the geological and geophysical measurements in the Davis Strait region are not conclusive enough to decipher its evolution, nonetheless, sea-floor spreading seems the only plausible mechanism at present which can account for its formation. The two most important source of data on past plate motion are magnetic anomalies and the direction of fracture zones. The former cannot be correlated between tracks with the present density of data and the later cannot be delineated because of the complexity of the structure present in this region. Nonetheless, using the poles of rotation as obtained for the Labrador Sea, a two stage evolutionary pattern is suggested for the region.

Stage 1. During the Late Cretaceous when active sea-floor spreading was taking place in the southern Labrador Sea, Davis Strait was subjected to severe tension. This gave rise to numerous N-S faults along the coast forming graben and half graben structures which were later filled with Tertiary and younger sediments. The evidence for this comes from detailed compilations of basement characteristics using multichannel seismic reflection data on the Greenland shelf (Henderson et al., 1981). By the time active sea-floor spreading had started in the Northern Labrador Sea (Danian) a minor separation had started in the Davis Strait region, separating Greenland from Baffin Island. The onset of this event correlates with the initiation of volcanism in this region as recorded by the presence of tuff layers in Danian sediments on west coast of Greenland.

Stage 2. By early mid Paleocene, a considerable portion of the southern Davis Strait or northern Labrador Sea had evolved and it was at this time that a massive outpouring of basalt on shore and offshore Baffin Island and west Greenland took place. Subsequently, the direction of spreading between Greenland and North America changed drastically because of the commencement of active seafloor spreading between Eurasia and Greenland. Greenland started to move in a NE direction relative to North America, giving rise to a large shear motion between the plates.

The whole region of Davis Strait was uplifted during early-mid Paleocene due to excessive volcanism or to the presence of a "hot spot" east of this region. The direction of movement of Greenland relative to Baffin Island as obtained from the pole of rotation for the Labrador Sea shows that spreading in Davis Strait remained inclined at a shallow angle to the coasts (Fig. 13) thus resulting in large shear motions and only limited spreading.

The history of evolution as described above can account for the formation of Davis Strait but it does not answer the questions; a) Why do we find the NE-SW high in the middle of Davis Strait; and b) Why do we have basaltic layers only parallel to the eastern side of the high? Two possibilities exist; the high may be fragmented continental or old oceanic crust or it may be part of a leaky transform fault. We will consider these two possibilities briefly.

1. Fragmented continental or old oceanic crust

Assuming that the Greenland plate has acted as a rigid plate throughout the development of the Labrador Sea, it is difficult to envisage a solution where the sea-floor spreading takes place on both sides of a continental fragment without displacing it from its original location. Thus, it was assumed displacement of the high occurred. Because no sea-floor spreading type anomalies can be identified in the vicinity of the Davis Strait high its paleogeographic position relative to Baffin Island or Greenland cannot be determined unambiguously.

The presence of reflectors on the eastern side of the high and their absence on the western side as well as the steepness of the western edge of the high indicate that the western edge was probably torn apart from Baffin Island prior to its uplift and movement northward. It was thus decided to treat the high as part of the Greenland plate since anomaly 25 times. The interpretation implies that the reflectors on the eastern edge of the high contain basal-

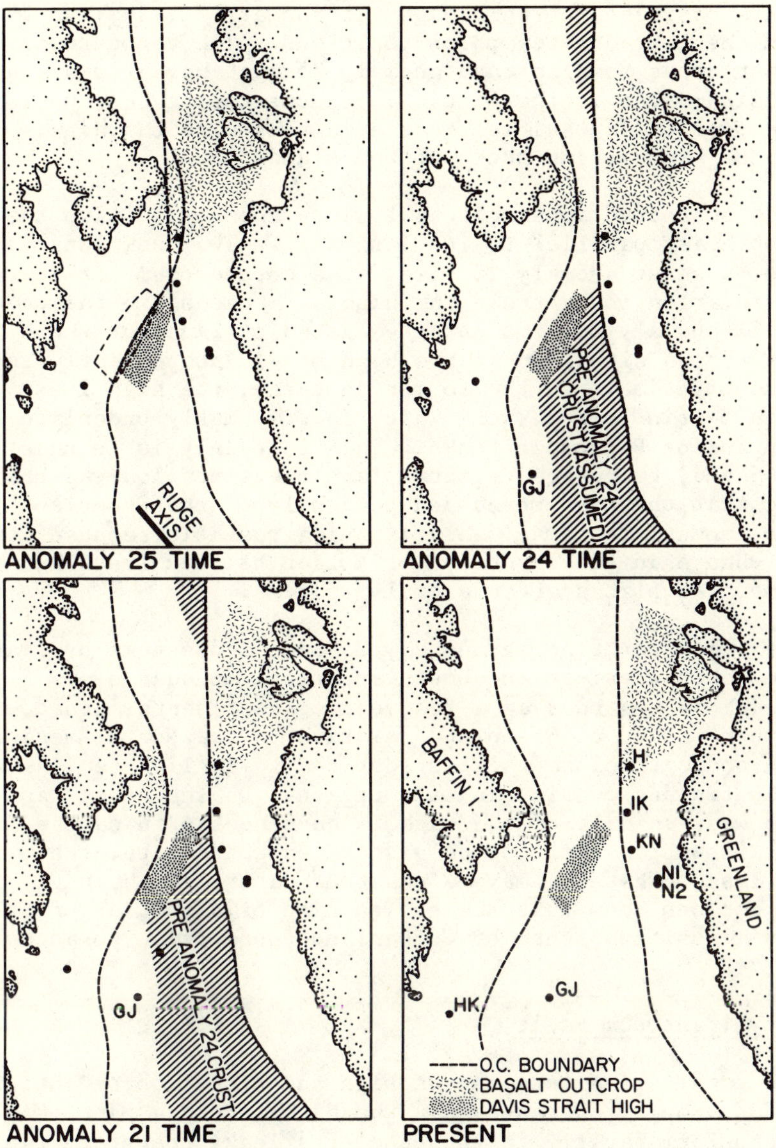


Fig. 14. Paleogeographic positions of Greenland and the Davis Strait high (shown by fine dotted pattern) relative to Baffin Island from Anomaly 25 time to present. The Davis Strait high has been moved as part of Greenland plate except at anomaly 25 time when it is also been treated as an independent plate (shown by dashed lines). Solid lines are the O-C boundaries.

tic flows which are probably older than Paleocene. Figure 14 shows successive positions of this high, at anomaly 21, 24, and 25 times based on the Labrador Sea poles positions. The reconstructions imply that crust younger than anomaly 25 must have formed on the Baffin Island side of the high. To assume the reverse would have made the problem impossible because in that case the high would overlie the Precambrian rocks of Greenland.

The reconstructions show that since anomaly 24 time the high moved northeast parallel to its long axis. Two positions for the high are shown at anomaly 25 time. The dotted position corresponds to the situation if we treat the high as attached to the Greenland plate till anomaly 25 time and the dashed outline for the high corresponds to the situation if the high moved independently from 25 to 24 anomaly time. In the former instance, the high lies in a region on Baffin Island shelf which is presumably underlain by Precambrian and/or Paleozoic rocks. This certainly is an undesirable situation and, therefore, a more plausible situation was derived by assuming that the high moved independently of the Greenland plate prior to anomaly 24 time. This is quite possible because of the drastic change in the direction of motion between Greenland and North American plates after anomaly 25 time.

The reconstruction at anomaly 25 time shows some overlap between the North American and Greenland plate boundaries. If the entire high is regarded as a fragment of continental crust, then any further motion of Greenland relative to the North American plate in order to close the Labrador Sea to earlier periods would result in further overlap in this region. On the other hand, if only the western half of the high is continental in nature, as the basaltic flows parallel to the eastern edge indicate, the overlap will be less. Whatever may be the true nature of the high, the reconstructions show that the entire high or a portion of it had occupied a position south of Cumberland Sound prior to anomaly 25 times.

## 2. Leaky Transform Fault

The parallelism between the high and direction of motion (Fig. 12) strongly suggests that the high formed a part of a leaky transform fault. In the case of a leaky transform fault, some spreading takes place within the fault. Alternatively, the ridge has to jump to different positions. The northern Labrador Sea data shows that the ridge lies more or less in a NW-SE position south of Davis Strait. If prior to anomaly 25 the ridge was oriented in a N-S direction (because of the direction of motion during this time) then a drastic change in direction of motion would tend to reorientate the ridge in a different direction (NW-SE) perhaps resulting in a jump or several jumps to the west. This would imply that as the new crust was formed, it became a part of the Greenland plate and moved with it resulting in a large transform effect represented by the NE-SW high. This solution suffers from the same drawback as the other because it involves too many unnecessary suppositions. In either case the exact pattern of the ridge jumps cannot be ascertained. Further, a portion of this transform fault may contain continental fragments as well giving rise to anomalous crustal structure under Davis Strait.

## ACKNOWLEDGEMENTS

I am grateful to a number of my colleagues from the Bedford Institute of Oceanography for their help in collecting some of the data which has been used in this paper. I am grateful to Compangie Generale de Geophysique for allowing me to reproduce a portion of their multichannel seismic reflection profile in this paper. I thank Drs. R.T. Haworth, J.M. Woodside and Mr. B. MacLean for critically reading the manuscript and for their helpful suggestions. I am also thankful to Dr. J.M. Woodside for carrying out some of the rotations of the plates using his numerical techniques.

## REFERENCES

BEH, R. L., 1975, Evolution and Geology of western Baffin Bay and Davis Strait, Canada. in: "Canada's Continental margins and Offshore petroleum exploraiton", C. J. Yorath, E. R. Parker and D. J. Glass, ed., Canad. Soc. of Petrol. Geol. Memoir, 4: 453-476.

BOTT, M. H. P., 1974, Deep structure, evolution and origin of the Icelandic Transverse Ridge, in: "Geodynamics of Iceland and the North Atlantic", L. Kristjansson, ed., D. Reidel Publ. Co., Boston U.S.A.: 33-47.

BRETT, C.P. and ZARUDZKI, E.F.K., 1979, Project Westmar - a shallow marine geophysical survey, on the west Greenland Continental Shelf in the region 64° to 69° 30' N and a preliminary interpretation of the data, Gronl. Geol. Unders. Rapp., 87: 29.

BROOKS, C. K., 1973, Tertiary of Greenland - a volcanic and plutonic record of continental break-up, in: "Arctic Geology", M. G. Pitcher, ed., Am. Assoc. Petrol. Geol., Mem. 19: 150-160.

CARLSON, R. L., CHRISTENSEN, N.I., and MOORE, R.P., 1980, Anomalous crustal structures in ocean basins: continental fragments and oceanic plateaus, Earth Planet. Sci. Lett., 51:171-180.

CLARKE, D. B., 1970, Tertiary basalts of Baffin Bay: possible primary magma from the mantle, Contr. Mineral. Petrol., 25: 203-224.

CLARKE, D. B., 1975, Tertiary basalts dredged from Baffin Bay, Can. J. Earth Sci., 12: 1396-1405.

CLARKE, D. B., 1977, The Tertiary volcanic province of Baffin Bay, in: "Volcanic Regimes in Canada", W. R. A. Baragar, L. C. Coleman, and J. M. Hall, ed., Geol. Assoc. Can., Special Paper, No. 16.

CLARKE, D. B. and PEDERSEN, A. K., 1976, Tertiary volcanic province of West Greenland, in: "Geology of Greenland", A. Escher and W. S. Watts, ed., Geol. Surv. Greenland, Copenhagen, 364-385.

CLARKE, D. B. and UPTON, B. G. J., 1971, Tertiary basalts of Baffin Island: Field relations and tectonic setting; Can. J. Earth Sci., 8: 248-258.

ESCHER, A. and WATTS, W. S., 1976, Geology of Greenland, Geol. Surv. of Greenland, Copenhagen.

GRADSTEIN, F. M. and SRIVASTAVA, S. P., 1980, Aspects of Cenozoic stratigraphy and paleo-oceanography of the Labrador Sea and Baffin Bay, Paleogeog., Paleoclimat., Paleoecolgy, 30: 261-267.

GRANT, A., 1975, Geophysical results from the continental margin off southern Baffin Island, in: "Canada's continental margins and offshore Petroleum Exploraton", C. J. Yorath, E. R. Parker, and D. J. Glass, ed., Can. Soc. Petrol. Geol., Mem. 4: 411-431.

GRANT, A., 1980, Problems with Plate Tectonics: Labrador Sea, Bull. Can. Petrol. Geol., 28: 252-278.

GREENLAND GEOLOGICAL SURVEY, 1978, Aeromagnetic reconnaissance survey offshore west Greenland for Tenneco Oil and Minerals Ltd., 1970, Contract Geoterrex Ltd., Greenland Geol. Surv. Industry Report 78-1.

GREENLAND GEOLOGICAL SURVEY, 1979. Well data summary sheets, Wells drilled, 1967-1977, Greenland Geol. Surv., Copenhagen, September, 1979, 14 p.

HALL, J. M, 1981, The Thulean volcanic line and its significance in the Tertiary tectonic and sedimentation history of the North Atlantic area, in, "Geology of the Atlantic border lands and basins", J. W. Kerr and A. J. Fergusson, ed., Can. Soc. Petrol. Geol., Mem. 7, (In Press).

HENDERSON, G., ROSENKRANTZ, A., and SCHIENER, E. J., 1976, Cretaceous Tertiary sedimentary rocks of West Greenland, in:"Geology of Greenland", A. Escher and W. S. Watt, ed., Geolog. Surv. of Greenland, Copenhagen,: 340-362.

HENDERSON, G., SCHIENER, E. J., RISSUM, J. B., CROXTON, C. A., 1981, The west Greenland basin, in: "Geology of the Atlantic border lands and basins", J. W. Kerr and A. J. Fergusson, ed., Can. Soc. Petrol. Geol. Mem. 7 (In Press).

HOOD, P., and BOWER, M. E., 1975, Aeromagnetic reconnaissance of Davis Strait and adjacent areas, in: "Canadian continental margins and offshore exploration", C. J. Yorath, E. R. Parker and D. J. Glass, ed., Can. Soc. Petrol. Geol., Mem. 4: 433-451.

HYNDMAN, R. D., 1973, Evolution of the Labrador Sea, Can. J. of Earth Sci., 10: 637-644.

HYNDMAN, R. D., 1975, Marginal basins of the Labrador Sea and the Davis Strait hot spot, Can. J. Earth Sci., 12: 1041-1045.

JACKSON, H. R., KEEN, C. E., FALCONER, R. K. H., and APPLETON, K. P., 1979, New geophysical evidence for sea-floor spreading in central Baffin Bay, Can. J. Earth Sci., 11: 2122-2135.

KEEN, C. E. and BARRETT, D. L., 1972, Seismic refraction studies in Baffin Bay: an example of a developing ocean basin, Geophys. J., Roy. Astron. Soc., 30: 253-271.

KEEN, C. E., and HYNDMAN, R. D., 1979, Geophysical review of the continental margins of eastern and western Canada, Canad. J. Earth Sci., 16: 712-747.

KEEN, C. E., KEEN, M. J., ROSS, D. I. and LACK, M., 1974, Baffin Bay: small ocean basin formed by sea-floor spreading, Bull.Am. Assoc. Petrol. Geol., 58: 1089-1108.

KEEN, M. J. and CLARKE, D. B., 1974, Tertiary basalts of Baffin Bay: Geochemical evidence for a fossil hot spot, in: "Geodynamics of Iceland and the North Atlantic Area", L. Kristjansson, ed., D. Reidel Publ., Co. Boston, U.S.A. 127-137.

KEEN, C. E., and PIERCE, J. W., 1981, The geophysical implications of minimal Tertiary motion along Nares Strait, Greenl. GeoSci. (In Press).

KERR, J. W., 1967, A submerged continental remnant beneath the Labrador Sea, Earth Planet. Sci. Lett., 2: 283-289.

KERR, J. W., 1981, Evolution of the Canadian Arctic Islands--A transition between the Atlantic and Arctic Oceans, in: "The Ocean Basins and Margins, 5, The Arctic Oceans", A. E. M. Nairn, and M. Churkin, Jr, .F. G. Stehli, ed., Plenum Press Pub. Co.

KLOSE, G.W., MALTERRE, E., MCMILLAN, N.J., and ZINKAN, C.G., 1981, Petroleum exploration, offshore northern Labrador Sea, Can. Abstract. Third Internaitonal Sumposium on Arctic Geology, Can. Soc. Petrol. Geolog. Calgary, Alberta. (Abstract).

KRISTOFFERSEN, Y., and TALWANI, M. 1977, Extinct triple junction south of Greenland and the Tertiary motion of Greenland relative to North America, Geol. Soc. Am., Bull., 88: 1037-1049.

LUDWIG, W. J., NAFE, J. E. and DRAKE, C. L., 1970, Seismic refraction, in: "The Sea, vol. IV, pt. one", A. E. Maxwell, ed., Wiley Interscience, N.Y.,: 53-84.

MacLEAN, B., 1981, Cruise report, C.S.S. HUDSON, 80-028, Bedford Institute of Oceanography, Unpublished report.

MacLEAN, B. and FALCONER, R. K. H., 1979, Geological-geophysical studies in Baffin Bay and Scott Inlet-Buchan Gulf and Cape Dyer-Cumberland Sound areas of the Baffin Island Shelf, Current Research, Part B., Geol. Surv. of Can., Paper 79-1B: 231-244.

MacLEAN, B., FALCONER, R. K. H., and CLARKE, D. B., 1978, Tertiary basalts of western Davis Strait; bedrock core samples and geophysical data, Can. J. Earth Sci., 15: 773-780.

MANDERSCHEID, G., 1980. The geology of the offshore sedimentary basin of West Greenland, in: "Facts and principles of world petroleum occurence", A. D. Miall, ed., Can. Soc. of Petrol. Geol., Mem. 6: 951-973.

MORGAN, W. J., 1971, Convection plumes in the lower mantle, Nature 230: 42-43.

O'NIONS, R. K. and CLARKE, D. B., 1972, Comparative trace element geochemistry of Tertiary basalts from Baffin Bay, Earth Planet. Sci. Lett., 10: 235-238.

PARROTT, R. J. E. and REYNOLDS, P. H., 1975, Argon-40/Argon-39 geochronology, age determinations of basalts from the Labrador Sea area, Geol. Soc. Am, Abstract with Prog., 7, no. 6: 835.

RIDD, M. F., 1981, Geology of the Faeroe-Shetland Channel, NATO Advanced Research Institute, Abstract.

SCHILLING, J. G., 1973, Iceland mantle plume: geochemical study of Reykjanes Ridge, Nature 242: 565-571.

SMYTHE, D. K., 1981, Faeroe-Shetland escarpment and continental margin north of the Faeroes, NATO Advanced Research Institute, Abstract.

SOPER, N. J., DOWNIE, C., HIGGINS, A. C. and COSTA, L. I., 1976, Biostratigraphic ages of Tertiary basalts on the east Greenland continental margin and their relationship to plate separation in the Northeast Atlantic, Earth Planet. Sci. Lett., 32: 149-157.

SRIVASTAVA, S. P., 1978, Evolution of the Labrador Sea and its bearing on the early evolution of the North Atlantic, Geophys. J., Roy Astr. Soc., 52: 313-357.

SRIVASTAVA, S. P., FALCONER, R. K. H., MacLEAN, B., 1981, Labrador Sea, Davis Strait, Baffin Bay: Geology and Geophysics--A Review, in: "Geology of the North Atlantic borderlands", J. W. Kerr and A. J. Fergusson, ed., Can. Soc. Petrol. Geol. Mem. 7, (In Press).

TALWANI, M. and ELDHOLM, O, 1973, Boundary between continental and oceanic crust at the margin of rifted continents, Nature, 241: 325-220.

van der LINDEN, W. J. M, 1975, Crustal attenuation and sea-floor spreading in the Labrador Sea, Earth Planet. Sci. Lett. 27: 409-423.

VOGT, P.R., 1971, Asthenosphere motion recorded by the ocean floor south of Iceland, Earth Planet. Sci. Lett., 13: 153.

VOGT, P. R., 1976, Plumes, subaxial pipe flow and topography along the mid-oceanic ridge, Earth Planet. Sci. Lett., 29: 309-324.

VOGT, P. R. and AVERY, O. E., 1974, Detailed magnetic surveys in the Northeast Atlantic and Labrador Sea, J. Geophys. Res., 79: 363-389.

WALLACE, F. K., 1973. Geology of the Davis Strait bathymetric sill and associated sediments, offshore Baffin Island Canada, in: "Canada Arctic Geology", J. D. Aitken and D. J. Glass, ed., Geol. Assoc. Can. and Can. Soc. Petrol. Geol.,: 8198.

WILSON, J. T., 1965, Evidence from oceanic islands suggesting movement in the earth, in: "A Symposium on continental drift", P. M. S. Blackett, E. C. Bullard and S. K. Runcorn, ed., Phil. Trans. Royal Soc. London, Series A. 258: 145-167.

# THE ICELAND MANTLE PLUME: STATUS OF THE HYPOTHESIS AFTER A DECADE OF NEW WORK

Peter R. Vogt

Naval Research Laboratory
Washington, D.C. 20375

## INTRODUCTION

A decade has passed since Morgan (1972) turned Wilson's (1963) hotspots into deep mantle convection plumes, and other workers (e.g., Vogt, 1971, 1974; Schilling, 1973) first applied these ideas to the Iceland phenomenon. Although the basic hypotheses remain unchanged, the data libraries and reprint files have filled out considerably on this subject. This volume on the Greenland-Scotland Ridge (GSR) is an appropriate and timely occasion to review the status of the Iceland mantle plume hypothesis. Not all aspects can be examined in this short review; geochemistry in particular is excluded. We begin by briefly considering whether a deep plume could exist under Iceland; next comes a discussion of hotspot fixity in light of known relative and absolute motions of the Eurasia (EURA) and North America (NOAM) plates, the geometry of the GSR, and the early Tertiary magmatic activity on adjacent continents. Finally we review the various kinds of anomalous "imprints" the Iceland hotspot has left on the ca. $3x10^6 km^2$ oceanic crust affected by this phenomenon. Whatever mechanism is called upon to explain the GSR, which covers only about $4x10^5 km^2$ of ocean floor, must also account for the whole array of anomalous characteristics - topographic, gravity, magnetic, geochemical, seismic, and otherwise. These anomalies extend in some cases 1500km or more from the GSR.

## PHYSICAL ASPECTS: IS THERE A DEEP MANTLE PLUME BELOW ICELAND?

To this day no direct experimental (e.g., seismic) verification of a deep plume under Iceland - or for that matter any other hotspot - has been achieved. Explosion seismology (RRISP, 1980: Bott and Gunnarsson, 1980) shows Iceland and the GSR to be underlain by a thick crust and a low-velocity upper mantle which under much of Iceland must be in a partially molten state. Results from surface wave dispersion, refraction, and a teleseismic delay time of 1.4 sec (Long and Mitchell, 1970) indicate that this low-velocity region must extend to depths of 100-200km or more (RRISP, 1980).

The relatively lower density of the low-velocity, partial melt zone below Iceland helps to provide isostatic compensation for the vast topographic anomaly centered on Iceland (Cochran and Talwani, 1978) but the low viscosity of the compensating mass requires constant inflow of new material, otherwise the topographic dome would decay. The excess mass of this dome cannot be supported by the lithosphere. Thus, the seismic, gravity and topographic data taken together seem to require ascending mantle convection in the Iceland area. Although this convection system must be several thousand

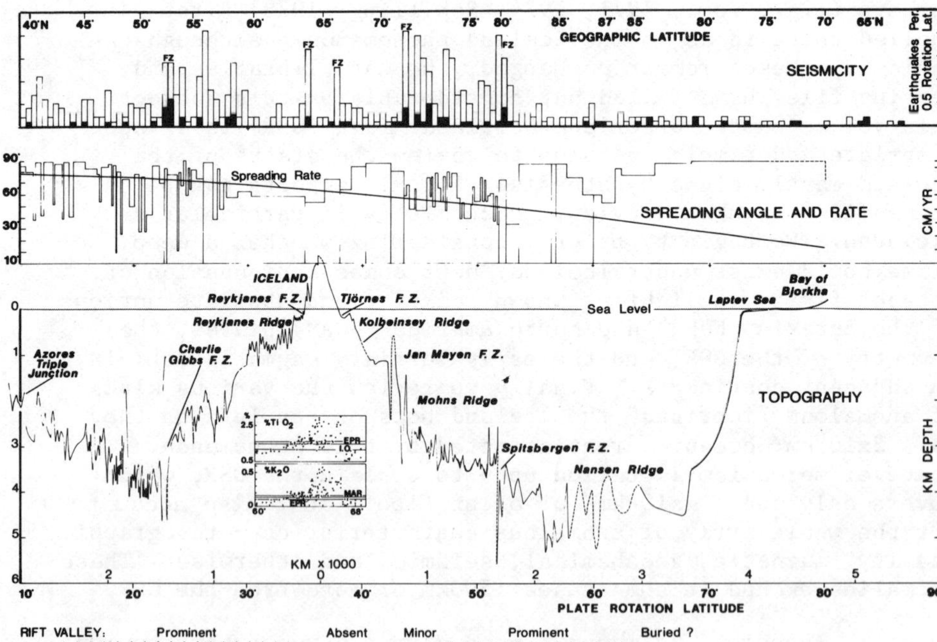


Fig. 1 Longitudinal profiles (along present plate boundary) showing various parameters as a function of distance from Iceland (Vogt et al., 1981a).

kilometers in horizontal extent (Bott, 1973; Vogt, 1974) the various anomalies are sharply peaked in southeast to central Iceland (Fig. 1). The existence of such a narrow (50-100km) center is exactly what would be expected from a narrow plume. The broad extent of the entire anomaly can then be attributed to outward flow in the asthenosphere, particularly under/along the spreading axis (Fig. 2; Vogt, 1974, 1976).

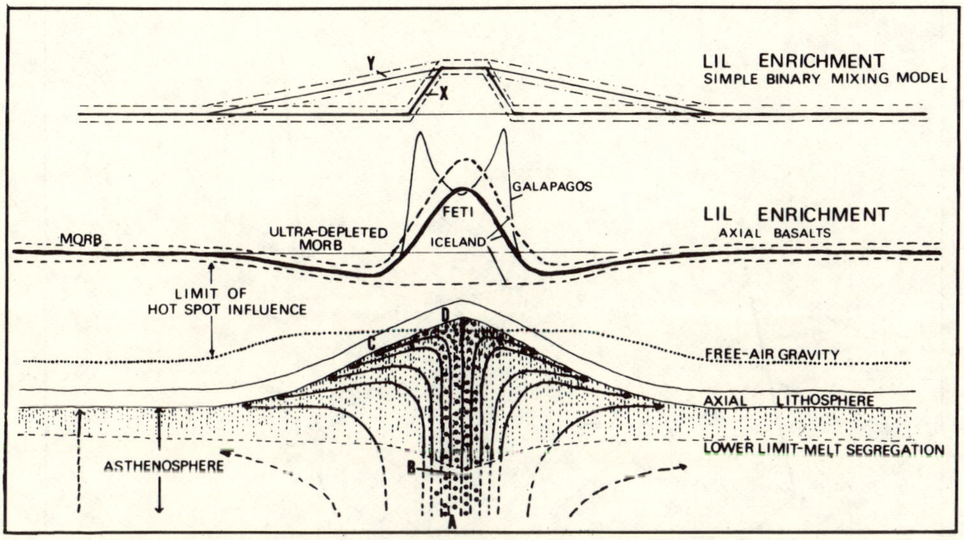


Fig. 2 Schematic longitudinal profile along platè boundary passing over a mantle plume (Compare Fig. 1). Small closed circles depict LIL-enriched pockets rising (A) to begin melting (B) and becoming incorporated in the mantle lithosphere (C-D), while basalt melts derived from these and depleted partial melts rise vertically (fine lines). In some cases (e.g., Galapagos) progressive fractionation produces flanking highs in some LIL-elements (e.g., K, Ti, Fe); the scatter in concentrations (dashed curves) is higher over plume. Postulated "ultra-depleted" basalts on hotspot flanks reflect preferential melting of LIL-enriched pockets near plume (Morgan, 1981) plus further depletion due to LIL-depleted crystal fractionates sinking into the outward flowing asthenosphere. Simple binary mixing (top) would not explain observed data irrespective of whether the plume flow extends only to "X" (e.g., Schilling, 1973) or all the way to "geophysical" limit of plume influence ("Y").

Although the question of existence and origin depth of a deep plume remains experimentally unresolved, several recent studies have strengthened the case for deep mantle plumes: (1) Objections raised against whole-mantle convection in the first

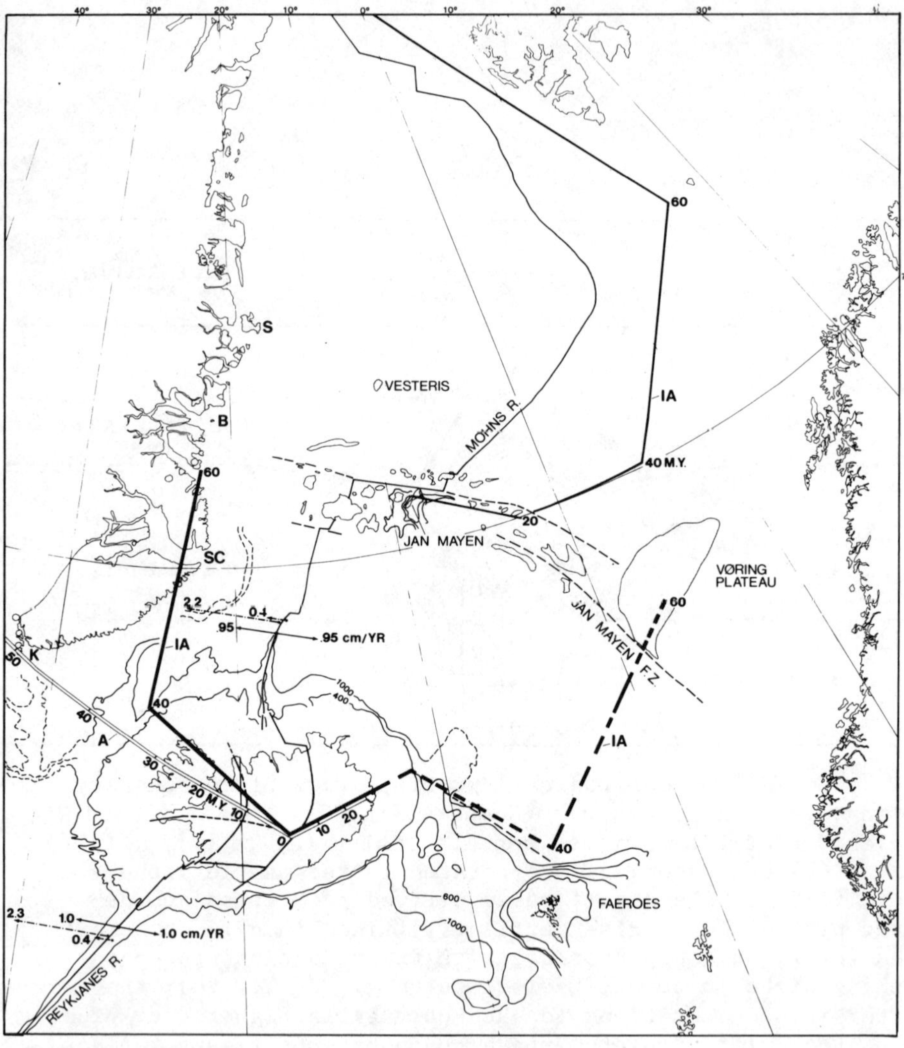


Fig. 3 Elevated ocean floor topography and/or areas of igneous activity; predicted Iceland and Jan Mayen hotspot tracks according to IA and A models (Morgan, 1980, 1981: model A shown for Iceland only); "present" relative and absolute plate motions according to Minster and Jordan (1980). Dashed line approximates trace of EURA points closest to hot spot (under Greenland/NOAM plate) according to IA model. Bathymetry from Perry et al. (1980).

years after Morgan's (1972) publication no longer appear valid (Elsasser et al., 1979; Morgan, 1981). It is likely that the entire mantle participates. (2) Numerical models (Parmentier et al., 1975) suggest that plume-like convection will occur in a base-heated mantle with a strongly temperature-dependent viscosity. (3) Analysis of the low-order harmonics of the geoid and gravity field (such as the broad high over the Iceland-Azores area, and similar features elsewhere) suggest sources in the deep mantle (Chase, 1979) or in some cases even topography on the core-mantle interface (Bowin, 1980). Furthermore, hotspots tend to occur within the long-wavelength geoid highs (Vogt, 1981). (4) Hotspots appear to be fixed relative to one another or move at rates of 3 to 5mm/yr or less (Morgan, 1980, 1981). Such fixity is an important requirement for deep plumes (Morgan, 1972). Unfortunately - in the GSR context - the Iceland hotspot seems less fixed than most.

## KINEMATIC ASPECTS: IS THE GSR THE TRACE OF PLATE MOTION OVER A FIXED HOTSPOT?

In this section we ask whether the GSR could be the trace of NOAM and EURA motion over a fixed Iceland hotspot. When Morgan (1972) first developed the hotspot hypothesis "fixed" meant "a few mm/yr or less", i.e., slow compared to typical rates of plate motion. Subsequently some authors (e.g., Molnar and Francheteau, 1975) concluded that hotspots (including Iceland) wander with respect to each other with speeds of about 1.5cm/yr. Such high speeds are comparable to rates of absolute plate motion in the northeast Atlantic. However, Morgan (1981) verified that relative hotspot motion has indeed averaged no more than 5mm/yr. When only the better known motions of plates surrounding the Atlantic were considered, hotspot motions averaged only about 3mm/yr (Morgan, 1980). These motions appear as random scatter, however, and may be more indicative of errors in the plate reconstructions than actual hotspot motions. Systematic movements of the Iceland hotspot in a Mean Mesospheric Reference Frame (MMRF) cannot yet be proven or disproven. Present data sets lack the resolving power. However rates even as low as 3 to 5mm/yr are not insignificant compared to Cenozoic plate motion rates in the Iceland area (5 to over 20mm/yr relative half-rate, perhaps 0 to 40mm/yr absolute rates). If any hotspot motion was primarily in a NE or SW direction, it would bend the older ends of the GSR SW or NE respectively, with respect to GSR traces predicted by absolute plate motion models. If hotspot motion more nearly paralleled the direction of relative plate motion, the effect would be to change the distance from hotspot to plate boundary as a function of time. This would bear on the time of ridge axis jumps, the distribution of excess crustal thickness between the plates as a function of time, etc. In the following discussion of relative and "absolute" plate motions in the GSR area, the possiblility of hotspot motion should be kept in mind when the geology/geophysics of the GSR is compared with predictions made from absolute motion models: Significant departures could occur even if the Iceland hotspot is as "fixed" as for example the Hawaii hotspot.

## "Present-day" relative and absolute plate motions in the Iceland area

The measured rates of post-anomaly 5 plate separation on the Reykjanes Ridge (2.2cm/yr) and Kolbeinsey Ridge (2.0cm/yr) must bracket the total accretion rate - summed in the direction of EURA-NOAM plate motion - across the entire zone of active tectonism/volcanism on Iceland (Fig. 3). Similarly the direction of EURA-NOAM relative motion at Iceland must be intermediate between the trends of the Charlie-Gibbs transform at 52.5°N(095°T) and the Jan Mayen F.Z. at 71.5°N (113°T). Detailed aeromagnetic data from Kolbeinsey Ridge (Fig. 4; Vogt et al., 1980) suggest these rates have been approximately constant over the last 12 to 14 m.y., and there is no evidence for a change in direction over this time interval.

Several attempts have been made to determine by matrix inversion those present-day plate kinematic parameters which best fit a set of globally distributed plate-boundary measurements such as spreading rates, transform fault azimuths, and earthquake slip vectors (Minster et al., 1974; Minster and Jordan, 1978; Chase, 1978). In model RM-2 of Minster and Jordan (1978, 1980) the mean averaging interval for spreading rates was less than 3 m.y. and for the transform azimuths probably somewhat more, whereas the slip vectors are "instantaneous". RM-2 gives a EURA-NOAM pole at 65.85°N 132.44°E and a rotation rate of 0.231°/m.y.; the 95% confidence ellipse is almost 25° long, however, and its axis follows the extension of Nansen Ridge into Siberia. The pole which best fits the EURA-NOAM boundary data alone is about 5° NNW of the RM-2 pole for these plates. Total opening rates (1.8cm/yr on southern Kolbeinsey Ridge and 2.1 cm/yr on southern Reykjanes Ridge) and azimuths computed from RM-2 are close to observed values. At the presumed hotspot center in southeastern Iceland a EURA-NOAM separation rate of 1.9 cm/yr is predicted.

Global models of "absolute" plate motion have been constructed for different assumptions, e.g., that the Caribbean, African, or Antarctic plates are fixed with respect to the deeper mantle (the MMRF), or that the lithosphere has no net rotation, or that a particular set of hotspots is fixed (the

Wilson-Morgan hypothesis). In most cases the approach is to determine the absolute motion of one plate and then use relative motion models (e.g., RM-2) to add appropriate vectors to find the absolute motions of the other plates. Model AM1-2 (Minster and Jordan, 1978, 1980) was derived from RM-2 by fitting the azimuths of nine hotspots traces and aging rates from five. The hotspot data, pertinent to the last 10 m.y., are dominated by information from the rapidly moving plates in the Pacific. No data from Atlantic hotspots (e.g., Iceland) were used. Model AM1-2 yields absolute motions of 0.38°/m.y. about 0.70°N, 23.19°W for EURA, and 0.247°/m.y. about 58.131°S, 40.67°W for NOAM. In the Iceland area (Fig. 3) the predicted absolute motion of the plates is approximately westward, with a rate of 2.3cm/yr for NOAM and 0.4cm/yr for EURA. (The numerical difference approximately equals the plate separation rate because relative and absolute motions are nearly parallel). The relative errors in the directions of absolute motions for points on slow-moving plates (such as EURA) can exceed 100%, i.e. the directions are not usefully constrained by the data used to derive AM1-2 (Minster and Jordan, 1978, 1980). In fact, the uncertainties are great enough that EURA could be stationary within the hotspot frame. The westward motion of NOAM over a fixed Iceland hotspot (about $2.3 \pm 0.4$cm/yr) is significant, however, and applies at least to the last 10 m.y. The significance of such absolute motions for the generation of the GSR is clear: If the Iceland hotspot has been fixed with respect to the Pacific hotspots for the last 10 m.y., it must have "migrated" eastward 40km with respect to EURA and 230km with respect to NOAM. Another way to view this result is to suppose the hotspot was centered on the accretion axis at 10 m.y.b.p. (just before anomaly 5-time). According to AM1-2, the spot's present position would be 40km east of the EURA 10 m.y. isochron, 230km east of the NOAM 10 m.y. isochron, and provided accretion on Iceland was symmetric - 165km east of the accretion axis.

As one mechanical influence on plate tectonics, hotspots have been suggested to maintain the position of the accretion axis above the hotspot when the axis migrates away from it. This could be accomplished by asymmetrical spreading and/or axis jumps. There is good evidence for two eastward jumps on Iceland (Björnsson, this volume; Saemundsson, 1974), lending support to the inferred absolute motions and to the mechanical effect of the hotspot on the plate boundary. About 6 m.y.b.p. the accretion axis jumped from the Snaefellsnes Peninsula area to the Reykjanes-Langjökull axis, and 4 m.y.b.p. shifted further east into present northeastern Iceland. The eastern axis has expanded southward as far as the Vestmann Islands in the last 2 m.y. This 150-200km total eastward shift of the axis is about the adjustment required for a fixed Iceland hotspot to accommodate 10 m.y. of absolute plate motion. Magnetic anomalies suggest that the young eastward shift of the accretion axis extends offshore north to 67°N. Southwest of Iceland the present Reykjanes Peninsula--Reykjanes Ridge axis south to 63.3°N lies closer to the EURA than to the NOAM anomaly 5. Thus a total of 500 to 600km of the EURA-NOAM boundary has tended to shift eastward in the last few million years.

South of about 63°N the Reykjanes Ridge has been spreading obliquely but symmetrically for at least the last 10 m.y. By contrast the Kolbeinsey axis has developed numerous shortlived fracture zones over the last 10 m.y. (Fig. 4). From 67°N to 68° the present axis lies midway between the 10 m.y. isochrons or slightly to the west. The Spar F.Z. at 68.7°N developed in the last few m.y. largely as a geometrical consequence of an eastward migration of the spreading axis from 68.7°N to 69.5°N. Although these complicated changes in the Kolbeinsey plate boundary probably cannot be attributed to changes in the EURA-NOAM pole, it is not clear whether west-

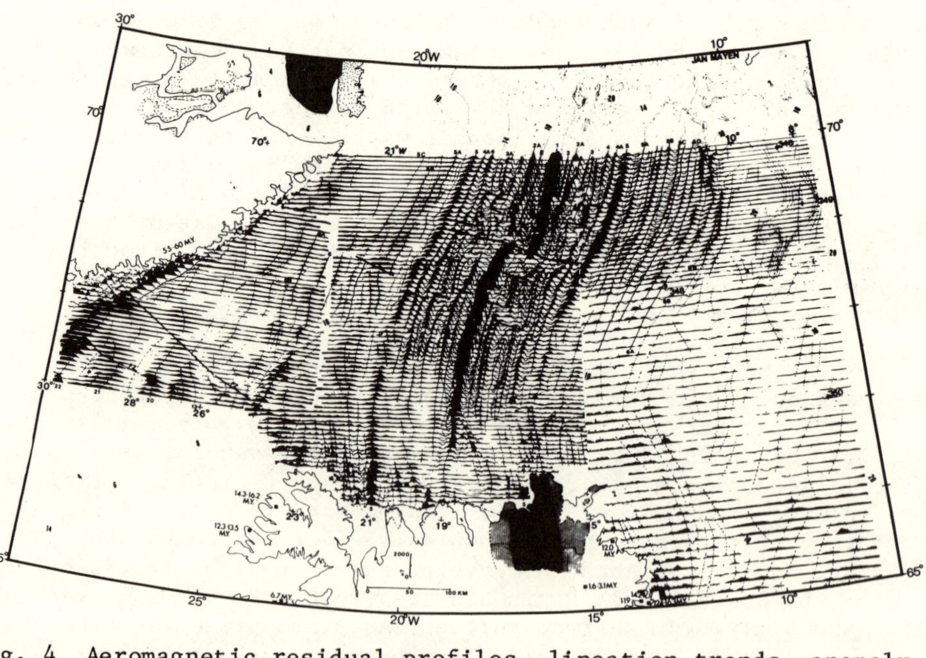


Fig. 4 Aeromagnetic residual profiles, lineation trends, anomaly identifications, bathymetry (thin dotted lines) and selected radiometric ages in the Kolbeinsey Ridge area. From Vogt et al. (1981a).

ward migration of the boundary with respect to the Iceland and Jan Mayen hotspots could be responsible, or, alternatively varying stress regimes related to fluctuating asthenosphere flow from the Iceland plume. Some of the small fractures suggest northward propagating rifts of the type discussed by Hey et al. (1980).

## Absolute and Relative Motions before 10 m.y.b.p.

It is not yet certain that EURA has been relatively stationary compared to NOAM throughout the post-60 m.y.b.p. evolution of the GSR and the associated early Tertiary plutonic/volcanic events on the adjacent continents. However, if this was the case and if the hotspot was located on the plate boundary around anomaly 5 to 6 time, it must have been west of the boundary at earlier times. An initial hotspot location west of the accretion axis may explain the great accumulation of early Tertiary volcanics in east Greenland and its high coastal mountains compared to the EURA plate margins (Figs. 5,6). Perhaps the Iceland platform itself was formed when the westward-migrating accretion axis "caught up" with the Iceland hotspot. (Alternative explanations for the platform include a westward jump of the ridge axis (Talwani and Eldholm, 1977) and an increase in hotspot discharge (Vogt, 1972)). However, the westward jump which separated the Jan Mayen block from Greenland (before 26 m.y.b.p.; Vogt et al., 1980; perhaps as early as 43 m.y., Nunns, 1980) and formed the Kolbeinsey Ridge axis, and the westward jump that occurred before anomaly 20-22 time in the GSR region and south to 61°N (Voppel et al. 1979; Vogt et al., 1980) presumably reflects plume-related events and/or mechanical effects of plate interaction (e.g., Nunns, 1980).

Morgan (1980, 1981) has recently conducted two studies of plate motion over hotspot populations that include Iceland. The technique in both cases involved (1) determining by trial and error a sequence of poles and rotation angles that describe the motion of one plate over its hotspots and reproduce the traces of intra-plate volcanic/epeirogenic lineaments; (2) computing the absolute motion of the other plates by adding appropriate vectors given by plate tectonic data; (3) comparing the predicted hotspot traces on these other plates with available geological data: and (4) returning to step (1), etc., until a satisfactory fit to all the hotspot traces was obtained. In one case (Morgan, 1980) only the plates bordering the Atlantic were considered - since their motions are better known. In the second study (Morgan, 1981) both Atlantic and Indian ocean plates were incorporated into a single model. We

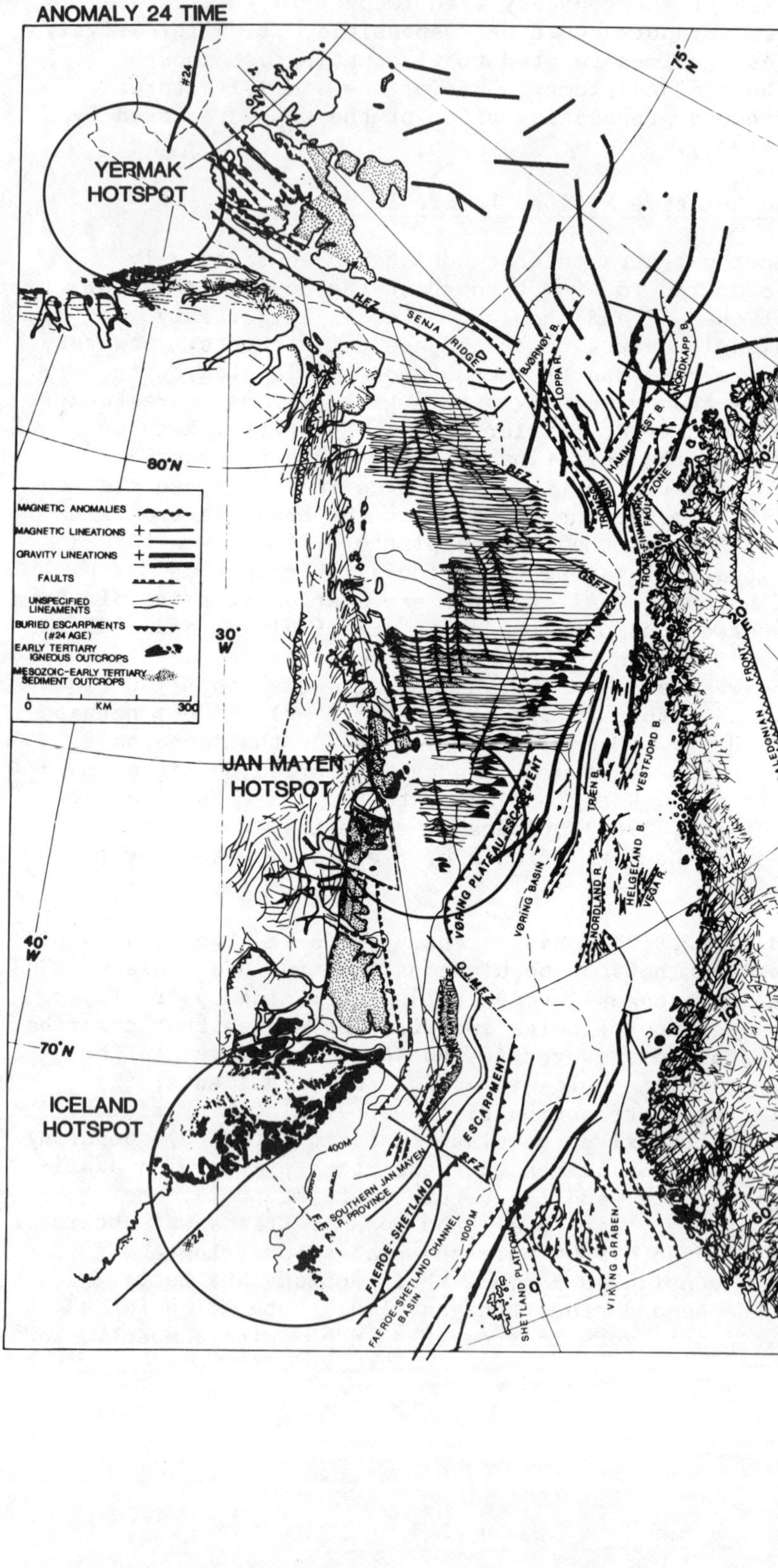


Fig. 5. Reconstruction of Greenland-Norwegian Sea at anomaly 24 time, showing proposed hotspots in relation to early Tertiary igneous provinces and older structures (modified from Vogt et al., 1981b).

shall refer to these models as "A" and "IA" for brevity. The A and IA traces predicted for the Iceland hotspot are shown in Fig. 3. (Note that the traces are not strictly predictions, since data from the Iceland area are embedded in the model). In both models the most rapid motion was in the Paleocene-Eocene, and in both the Iceland hotspot has been under the Greenland (now North America) plate for most of its life, and in terms of absolute motion EURA has been slower than NOAM. The A model predicts a NW trace for the Eocene-Paleocene, versus the more NNW trace predicted by the IA model. Considering the wide, complex, and partly buried or eroded character of the GSR volcanic trace from Iceland to the early Tertiary volcanics in East Greenland, we cannot be certain which predicted line is more nearly correct. Certainly the A trace better fits the overall NW-SE trend of the GSR. The trace intersects Greenland near the Skaergaard intrusions, but this is well south of the center of mass of at least the surviving Tertiary volcanics. The IA trace passes north of the Blosseville coast basalt pile, but if the volcanic and intrusive centers between Bontekoe Is. (B) and Shannon Is. (S) are attributed to the Iceland hotspot, the IA trace terminates in a location centrally located within the entire east Greenland basalt province. Although the IA trace places the hotspot under Greenland in the Paleocene-Eocene, the Vøring Plateau basalt pile was formed not far southeast and is thus explained as a manifestation of the Iceland hotspot. The predicted IA Jan Mayen hotspot trace lies mostly on the Eurasian plate, paralleling the early Mohns Ridge (60-40 m.y.b.p.) and the Spitsbergen-Barents continental margin (60-80 m.y.b.p.). This track might attribute the late Cretaceous-early Tertiary Kap Washington basalts in northeast Greenland to the Jan Mayen hotspot rather than to the separate Yermak hotspot as postulated by Feden et al. (1979). (The late Tertiary manifestations of the Yermak hotspot are however not accounted for by the model). The 60-80 m.y. Jan Mayen hotspot trace (IA) might have pre-weakened the continent, explaining why the intial EURA-Greenland plate boundary veered towards Spitsbergen instead of continuing along the Mohns Ridge-Vøring Escarpment trend. No predicted "A" model trace is shown for the Jan Mayen hotspot. Such a trace would approximately connect the Vøring Plateau with the northern East Greenland volcanics in the reconstruction (Fig. 3).

Whereas the "A" model better fits the azimuths of the Jan Mayen and Iceland hotspot trails, neither model explains why the present hotspot locations are still midway between their early Tertiary manifestations. (The present hotspots should be near the Eurasia continental margin). The observed relationship is only consistent with hotspot fixity provided EURA and NOAM (Greenland) moved over the mantle at similar rates in opposite directions. Yet such motion clearly conflicts with good evidence that other hotspots are fixed, and that NOAM has done most of the "absolute" traveling, not only recently (Minster and Jordan, 1978, 1980) but throughout the Tertiary (Morgan, 1980, 1981). Assuming this problem cannot be solved by acceptable adjustments in plate kinematic parameters, we can either (1) sidestep the problem by increasing the hotspot diameter to say, 1000km (e.g., Bott, 1973; Talwani and Eldholm, 1977; Cochran and Talwani, 1978); (2) give up hotspot fixity by permitting both the Jan Mayen and Iceland hotspots to move northwest at rates of the order 5mm/yr; or (3) allow the GSR to form at a point on the accretion axis that is fed by mantle flow from a fixed hotspot located some distance away from the axis (Morgan, 1978, 1981; Vink and Morgan, 1981).

EMPIRICAL EVIDENCE FOR FLUCTUATING HOTSPOT ACTIVITY (PLUME DISCHARGE?

The Iceland hotspot is associated with anomalies in basement depth, crustal thickness, crustal geochemistry, basement morphology, plate boundary shape, seismicity, upper mantle structure, and gravity and magnetic anomalies (Vogt, 1974). The anomalies extend to different ranges from central Iceland: For example, the depth anomaly extends to the Charlie Gibbs F.Z. 1500km southwest of Iceland, and northeast possibly as much as 2500km to the Nansen Ridge (Figs. 1,6). The lithophile element/isotope anomaly extends only 500 km southwest of Iceland, and terminates off the north coast (Fig. 1; Schilling, 1973). Anomalies in upper mantle velocity structure and crustal thickness may be confined to the GSR/Iceland platform (RRISP, 1980) although surface wave dispersion suggest a thick crust extending north to the Jan Mayen F.Z. (Evans and Sacks, 1979).

To explain the enormous but variable extent of these anomalies by a narrow plume, Vogt (1971, 1974, 1976, 1979) and Schilling (1973) developed the concept of longitudinal flow of plume-generated mantle materials in the pipe-like low-viscosity zone which must exist below accretion axes (Fig. 2). Transform fracture zones would necessarily interrupt this mid-oceanic conduit system to varying degrees and perhaps cause ponding of mantle materials just upstream from the ridge-fracture intersection (Vogt and Johnson, 1975). The concept of a "broad hotspot in the upper mantle" as developed by Talwani and Eldholm (1977) and Cochran and Talwani (1978) is misleading because most anomalies (except gravity) defining the hotspot are sharply peaked and specify a hotspot core area (above the upwelling plume?) no more than 50 or 100km in diameter. In fact, Sigvaldason and Steinthorsson (1974) refer to the Iceland hotspot as the "Kverkfjöll hotspot" because the geochemical anomaly ($K_2O, P_2O_5$, and $TiO_2$) reaches its peak in that area (64.6°N, 16.5°W; Fig. 1). So do the topographic zero-age anomaly (Fig. 1) and the post-glacial lava production rate (Jakobsson, 1972).

Whether or not the Iceland "geosyndrome" is due to a narrow plume, preferential flow under the spreading axis (Fig. 2), and the interactions of this phenomenon with the plate accretion process, the existence of various kinds of imprints on the newly formed oceanic crust makes it possible in principle (Fig. 7) to describe the history of the hotspot's "vigour" by examining the north-south extent and amplitudes of the various anomalies as a function of crustal age. (Another way would be to measure basalt lava (or total crust) production as a function of time. So far there are not enough data to do this except for very young crust (Jakobsson, 1972)). Depth-anomaly is the parameter most easily contoured (Fig. 6) and shows anomalies (1) reasonably constant and large over the last 10-20 m.y., (2) markedly less in amplitude and extent over mid-Tertiary (45-25 m.y. old) crust, and (3) again large on the oldest crust (53-58 m.y.b.p.), although broken into at least three separate maxima in the Vøring Plateau, Faeroes, and Charlie Gibbs F.Z. areas. The late Tertiary depth anomaly maximum corresponds on Iceland itself to the insular platform.

An Iceland mantle plume flowing copiously in the early and late Tertiary - with a slack period between - would "explain" the variation in depth anomaly with crustal age and, if plumes help motivate plates, the roughly parallel fluctuation in spreading rate (Vogt and Avery, 1974). Alternatively, the late Tertiary build-up may simply reflect the overtaking of the hotspot center by the westward-migrating plate boundary (e.g., Saemundsson, 1974, Morgan, 1981) rather than an increase in plume flow. This would not explain the early Tertiary depth anomalies, however. Some may wish to relate the latter to transient "early opening" boundary effects or to the effects of rapid initial spreading. However, if this were so the observed depth anomalies should be universal features of fast spreading and/or first formed oceanic crust adjacent to continental crust. In fact depth anomalies similar to the Iceland-Faeroe Ridge or Vøring Plateau are uncommon features in the first place, and secondly are known to have formed on a pre-existing M.O.R., not only next to a continent (e.g., the J-anomaly Ridge in the western Atlantic and the basement rise south of

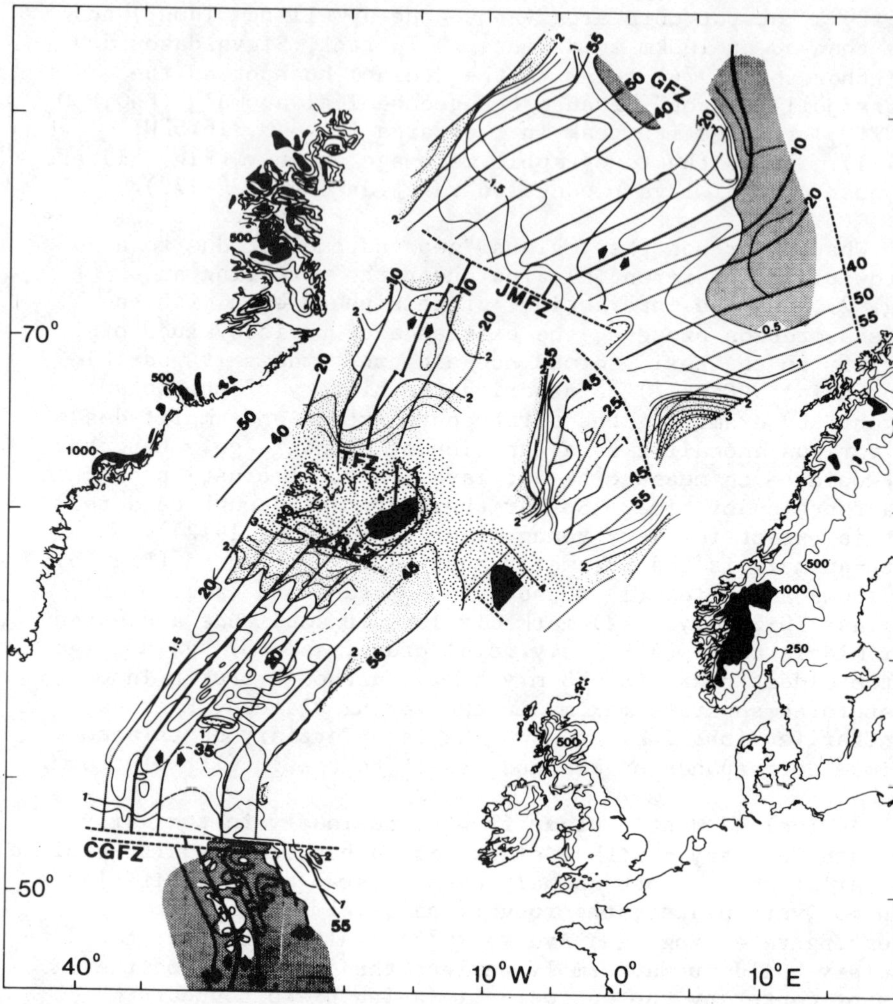


Fig. 6 Depth anomalies (km), isochrons (m.y.b.p.) and coastal elevations (m) in Iceland area. From Vogt et al. (1981a).

the Charlie Gibbs F.Z.; Fig. 6). Fluctuating plume flow activity seems the simplest way to explain the depth anomaly variations. Blocking of this sub-axial flow at transform fracture zones may explain the sharp southern boundary of the Vøring Plateau depth anomaly, the sharp northern boundary of the 52°N depth anomaly, the abrupt termination of the Iceland geochemical anomaly at the Tjörnes F.Z. (Fig. 1); the abrupt southern termination of the late Tertiary depth anomaly at the Charlie Gibbs F.Z. (Fig. 6); the sharp southern termination of the Iceland platform and its thick crust at the "Reykjanes F.Z." (RRISP, 1980); and perhaps the sharp gradient in depth anomaly and crustal thickness across the proposed F.Z. separating the Norway Basin (Aegir Ridge) from the Iceland-Faeroes Ridge with its 30km thick crust (Nunns, 1980; Bott and Gunnarsson, 1980).

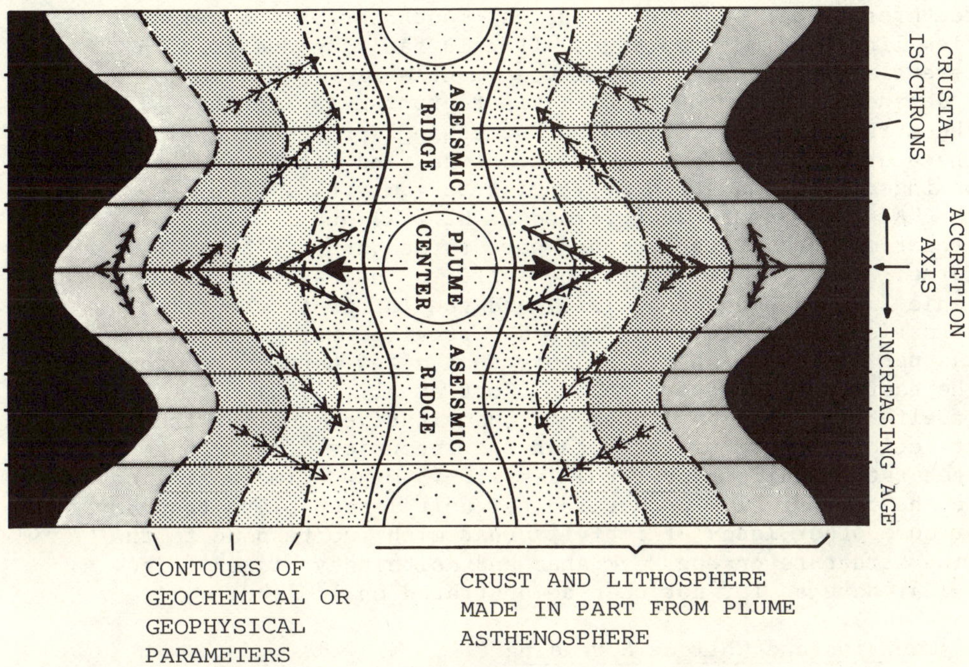


Fig. 7 Generalized anomaly (depth, geochemistry, etc.) contours in relation to crustal isochrons for a hotspot of varying discharge remaining at spreading axis. Herringbone features are time-transgressive ridges (e.g., Fig. 8-10).

Another way in which the history of plume activity might be measured is based on the time-transgressive (V-shaped, diachronous) topographic features found on the Reykjanes Ridge (Figs. 8, 9; Vogt, 1971, 1974) and perhaps also the Kolbeinsey Ridge (Fig. 10). A cross-section of the Reykjanes Ridge along a EURA-NOAM flowline at about 62.7°N ( 200km southwest of the Iceland coast) shows most of the ridges and/or west-dipping escarpments identified by Vogt, plus an older high marked

F and F' (Fig. 8). The most prominent features are the A-A' and E-E' escarpments presumably formed at the spreading axis at 4 and 15 m.y.b.p. at this latitude, earlier closer to Iceland and later to the southwest. Free-air gravity profiles (Fig. 9) show these diachronous structures even more clearly than does the bathymetry. Escarpments corresponding to E-E' were also formed north of Iceland (Vogt et al., 1980). Vogt (1971, 1974) suggested that the escarpments mark the passage of a melt-rich front, representing increased plume activity, traveling southwest from Iceland at rates of the order 10-20cm/yr. Johanson et al. (1981) point out that some of these paired features do not occur on identical isochrons on the same flow line, and may represent off-axis faulting due to differential subsidence, rather than mantle flow. However, independent evidence for off-axis tectonism is lacking. If the structures were formed by "something" propagating along the accretion axis one would expect to find "something" happening on Iceland a few m.y. prior to 4 m.y.b.p. (the time the "A" escarpments were formed at 62.7°N) and 15 m.y.b.p. (the time the "E" escarpments were formed at 62.7°N). The former event may relate to the shift to the Eastern Neovolcanic Zone (Saemundsson, 1974) and the latter to the first formation of an elevated Iceland - provided marine erosion has not moved the shoreline more than a few tens of km from the eastern and western coasts (Vogt et al., 1980). On Iceland itself Pálmason (1980) used his quantitative model of the accretion process to infer that the intensity of volcanism has been rather uniform over the last 10-15 m.y. Time variations in the rate of lava production per unit length plate boundary, or some other index of activity that might correspond to the fine structure on the Reykjanes and Kolbeinsey ridges (Figs. 8-10), have so far not been demonstrated on Iceland.

## Asymmetries and Oblique Spreading

Along the active plate boundary (Fig. 1) the Iceland hotspot exhibits some north-south asymmetry with respect to (1) basalt composition (2) rift valley development, (3) flank morphology, (4) depth, (5) prevalence of fracture zones, and perhaps (6) seismicity. In most cases the anomalous parameter returns to normal more abruptly in a northward direction, perhaps because spreading rates and therefore asthenosphere flow required for mass balance decline in a northward direction. Depth anomalies (Fig. 6) and morphology suggest the north-south asymmetry already existed in mid-Tertiary time. Much of the plate boundary from around Iceland to the Eurasia Basin also exhibits east-west asymmetry, with somewhat higher topography, gravity, and magnetic amplitudes on the NOAM flank (e.g., Fig. 8). This asymmetry, not concentrated in the Iceland area, may relate to the difference in plate/mantle speed rather than to the hotspot's asymmetric location under the ridge crest.

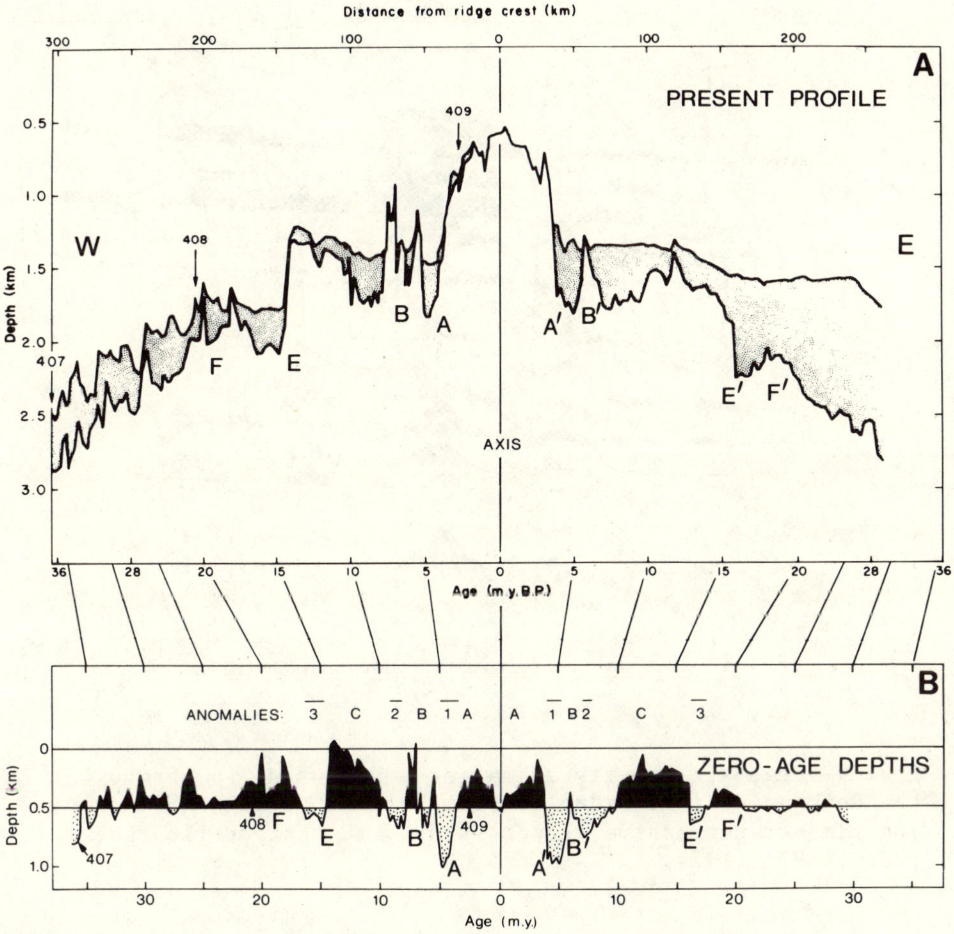


Fig. 8 Seismic reflection (profile A, vs. distance); zero-age depth (B, vs. time) along a flow line crossing Reykjanes Ridge at 63.7°N and connecting DSDP drill sites 407, 408, and 409. Letters A, B, etc. refer to paired bathymetric structures presumed to record fluctuations in Iceland plume flow under Reykjanes Ridge (Vogt, 1974). Adapted from Luyendyk et al. (1979).

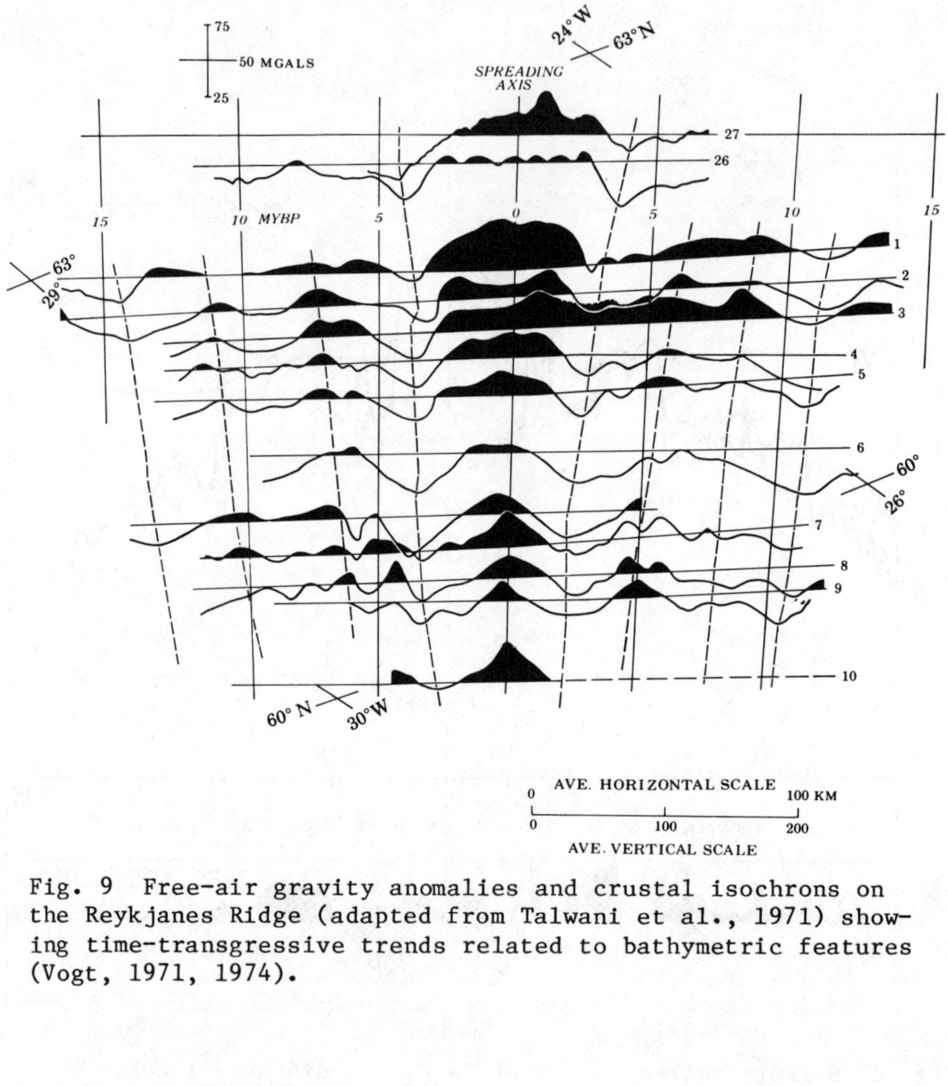


Fig. 9 Free-air gravity anomalies and crustal isochrons on the Reykjanes Ridge (adapted from Talwani et al., 1971) showing time-transgressive trends related to bathymetric features (Vogt, 1971, 1974).

Many segments of the spreading axis are notably oblique (up to 30°-60°) to the transform direction within 1000-2000km of Iceland (Fig. 1). This has been attributed to slow spreading or interplate/intraplate stresses (e.g., Nunns, 1980) but may also reflect stresses due to mantle flow, particularly along the plate boundary (Vogt and Avery, 1974).

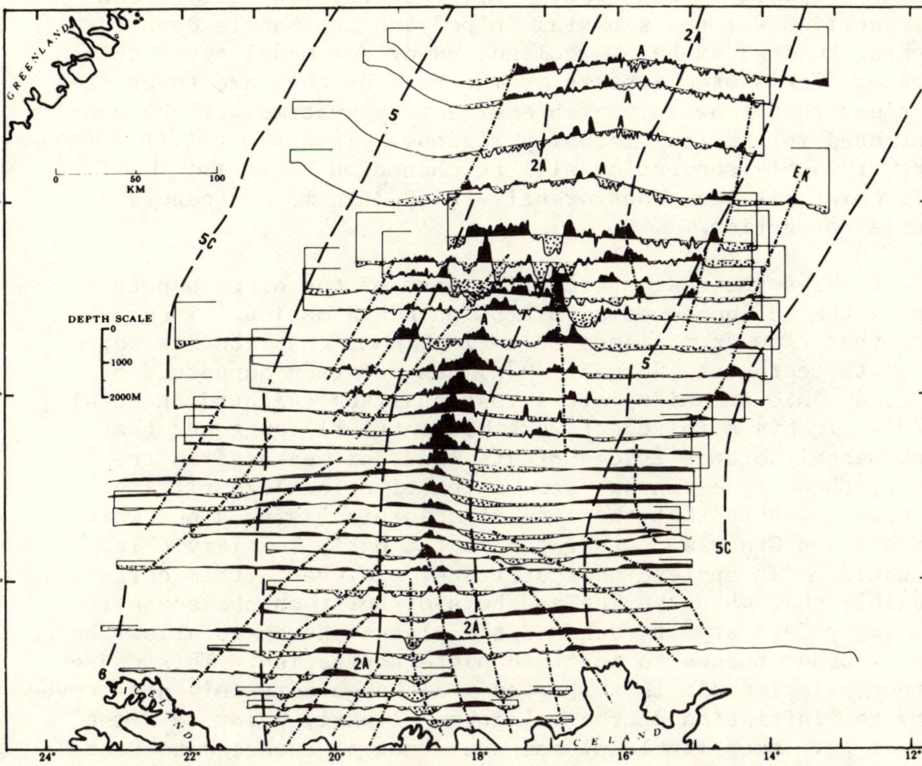


Fig. 10 Bathymetric profiles (solid above smooth curve through each profile), showing trends mainly diachronous with respect to crustal isochrons (dashed, anomalies 2A, 5, and 5C from Fig. 4). From Vogt et al. (1980).

## DISCUSSION

The decade-old Iceland plume hypothesis has not advanced beyond the hypothesis state although some type of upwelling convection is certainly required to perpetuate the anomaly for 55 million years without any sign of systematic decay. The vast topographic/gravity anomaly centered on Iceland represents a mass excess that cannot be supported by the strength of the lithosphere and therefore must be maintained by active mantle flow. And, given both the extra-low viscosity of partial melts below the accretion axis and the longitudinal topographic (hence pressure) gradient (Fig. 1), it is likely that most of the inflow from below is dissipated as subaxial pipe-flow and plate accretion along the Mid-Oceanic Ridge up to 1000-2000km north and south of Iceland (Fig. 2). Until details of the composition and state - particularly viscosity - can be specified for the subaxial "pipe" and the mantle down to several hundred km below Iceland, numerical modeling of this complex flow system appears premature. By the same token it is clear that flow in variable-viscosity systems will be concentrated in the low-viscosity regions. Thus the asthenospheric part of mantle convection will be channeled along the M.O.R. This point has been consistently ignored in modern papers on mantle convection.

Perhaps the most speculative part of the plume hypothesis is the influence of such flow on plate motion. The idea that flow from plumes is an important contributing force to plate tectonics (Morgan, 1972) has not been supported by plate dynamic and mid-plate stress analyses (Richardson et al., 1979). In his more recent work Morgan (1981) suggests that continental hotspot traces create lines of weakness in the crust; these lines may at some later time localize rifting. No one has shown that the line of Mesozoic basins along which Eurasia and Greenland separated in the early Tertiary (Fig. 5) could be an ancient hotspot trace. However, it is certainly possible that when the Iceland hotspot burst on the scene in the early Tertiary it weakened the plates enough to allow the sum of other forces to initiate plate separation. This cause-effect relationship is supported - although certainly not proved - by the initiation of the volcanism slightly prior to first spreading. Deep mantle plumes would not be expected to "care" what kind of lithosphere lies overhead. If this is true, why then did the early Tertiary volcanism develop only within or on the edges of the Mesozoic basins (Fig. 5)? In Morgan's (1981) model of absolute plate motions, the Iceland hotspot was under central Greenland before the early Tertiary. As he has suggested, old cratonic lithosphere may be too thick to be fully penetrated by magma; a slight thermal bulge may be all that marks the cratonic part of a hotspot trace. When the Iceland trace intersected the line of Mesozoic basins, volcanism became possible. This reasoning, however, does not explain the simultaneous development of depth anomalies at pre-existing spreading centers (e.g., at 52°N; Fig. 6), and the near simultaneous eruption of the west Greenland and British Tertiary/ Rockall Plateau volcanics (Bott, 1973) . The synchronism of these distant magmatic events (and the far more distant Deccan Traps; Vogt, 1972) suggests that the initiation of the GSR must be understood in the context of some mantle "event" of at least continental dimension (e.g., a convective overturn as suggested by Bott (1973)). This is true whether or not the Disko, Iceland, Jan Mayen hotspots have their origins deep in the mantle. Most of the intriguing problems concerning the origin of the GSR thus remain unsolved!

ACKNOWLEDGMENTS

I thank Jason Morgan for sharing unpublished manuscripts, Carol Rockelli and Irene Jewett for diligent help, and the Office of Naval Research for financial assistance.

BIBLIOGRAPHY

Bott, M.H.P., 1973, The evolution of the Atlantic north of the Faeroe Islands, in Implications of Continental Drift to the Earth Sciences, Vol., ed. by D.H. Tarling and S.K. Runcorn, Academic Press, London, 175-190.

Bott, M.H.P., and Gunnarsson, K., 1980, Crustal structure of the Iceland-Faeroe Ridge, J. Geophys., 47, 221-227.

Bowin, C., 1980, Why the earth's greatest geoid anomaly is so negative, EOS, Trns. Amer. Geophys. U., 61, 209.

Chase, C.G., 1979, Subduction, the geoid, and lower mantle convection, Nature, 282, 464-468.

Chase, C.G., 1978, Plate kinematics: the Americas, East Africa, and the rest of the world, Earth Planet. Sci. Lett., 37, 355-368.

Cochran, J.R., and Talwani, M., 1978, Gravity anomalies, regional elevations, and the deep structure of the North Atlantic, J. Geophys. Res., 83, 4907-4924.

Elsasser, W.M., Olsen, P. and Marsh, B.D., 1979, The depth of mantle convection, J. Geophys. Res., 84, 147-155.

Evans, J.R., and Sacks, I.S., 1979, Deep structure of the Iceland Plateau, J.Geophy. Res., 84, 6859-6866.

Feden, R.H., Vogt, P.R., and Fleming, H.S., 1979, Magnetic and bathymetric evidence for the "Yermak hotspot" northwest of Svalbard in the Arctic Basin, Earth Planet Sci. Lett., 44, 18-38.

Hey, R., Dunnebier, F.K., and Morgan, W.J., 1980, Propagating rifts on mid-ocean ridges, J. Geophys. Res., 85, 3647-3658.

Jakobsson, S.P., 1972, Chemistry and distribution pattern of recent basaltic rocks in Iceland, Lithos, 5, 365-386.

Johanson, B., Vogt, P., and Eldholm, O. 1981, Further analysis of time-transgressive lineations on the Reykjanes Ridge, Earth Planet. Sci. Lett., in press.

Long, R.E., and Mitchell, M.G., 1970, Teleseismic P-wave delay time in Iceland, Geophys. J.R. Astron. Soc., 20, 41-48.

Luyendyk, B.P., Shor, A., and Cann, J.R., 1979, General implications of the Leg 49 drilling Program for North Atlantic Ocean geology, in Luyendyk, B.P., and Cann, J.R., Initial Reports of the Deep Sea Drilling Project, Vol. 49, U.S. Gov't Printing Office, Washington.

Minster, J.B., and Jordan, T.H., 1978, Present-day plate motions, J. Geophy. Res., 83, 5331-5354.

Minster, J.B., and Jordan, T.H., 1980, Present-day plate motions: Summary, Editons du Centre National de la Recherche Scientifique, Paris, 109-124.

Minster, J.B., Jordan, T.H., Molnar, P., and Haines, E., 1974, Numerical modeling of instantaneous plate tectonics, Geophys. J. Roy. Astron. Soc. 36, 541-576.

Molnar, P., and Francheteau, J., 1975, The relative motion of hotspots in the Atlantic and Indian Oceans during the Cenozoic, Geophys. J. Roy. Ast. Soc., 43, 763-774.

Morgan, W.J., 1981, Hotspot tracks and the opening of the Atlantic and Indian Oceans, Ch. 13, in The Sea, Vol. 7, ed. by C. Emiliani, Wiley and Sons, N.Y., in press.

Morgan, W.J., 1980, Hotspot tracks in the region around the Atlantic, unpublished manuscript.

Morgan, W.J., 1978, Rodriguez, Darwin, Amsterdam - a second type of hotspot island, J. Geophys. Res., 83, 5355-5360.

Morgan, W.J., 1972, Deep mantle convection plumes and plate motions, Am. Assoc. Petrol. Geol. Bull., 56, 203-213.

Nunns, A.G., 1980, Marine geophysical investigations in the Norwegian-Greenland Sea between the latitudes of 62°N and 74°N, PhD Thesis, Durham Univ.

Pálmason, G., 1980, A continuum model of crustal generation in Iceland: kinematic aspects, J. Geophys., 47, 7-18.

Parmentier, E.M., Turcotte, D.L., and Torrance, K.E., 1975, Numerical experiments on the structure of mantle plumes, J. Geophys. Res. , 80, 4417-4424.

Perry, R.K., Fleming, H.S., Cherkis, N.Z., Feden, R.H., and Vogt, P.R., 1980, Bathymetry of the Norwegian-Greenland and western Barents Seas, Chart MC-21, Geol. Soc. Amer.

RRISP Working Group, 1980, Reykjanes Ridge Iceland Seismic Experiment (RRISP 77), J. Geophys., 47, 228-238.

Richardson, R.M., Solomon, S.C., and Sleep, N., 1979, Tectonic stress in plates, Rev. Geophy. Space Phys., 17, 981-1019.

Saemundsson, K., 1974, Evolution of the axial rifting zone in northern Iceland and the Tjornes fracture zone, Bull. Geol. Soc. Am., 85, 495-504.

Schilling, J.G., 1973, Iceland mantle plume, geochemical evidence aljong Reykjanes Ridge, Nature, 242, 565-578.

Sigvaldason, G.E., and Steinthorsson, S., 1974, Chemistry of tholeiitic basalts from Iceland and their relation to the Kverkfjoll hotspot, in Geodynamics of Iceland and the North Atlantic Area, ed. by L. Kristjansson, Dordrecht, Holland, D. Reidel Pub. Co., 155-164.

Talwani, M., and Eldholm, O., 1977, Evolution of the Norwegian-Greenland Sea, Geol. Soc. Am. Bull., 88, 969-999.

Talwani, M., Windisch, C.C. and Langseth, M., 1971, Reykjanes Ridge crest: A detailed geophysical study. J. Geophys. Res., 76, 473-517.

Vink, G.E., and Morgan, W.J., 1981, Iceland-Faeroe Plateau: A hotspot fed ridge, EOS, Trns. AGU, 62, 382.

Vogt, P.R., 1971, Asthenosphere flow recorded by the ocean floor south of Iceland, Earth Planet. Sci. Lett., 13, 153-160.

Vogt, P.R., 1972, Evidence for global synchronism in mantle plume convection, and possible significance for geology, Nature, 240, 338-342.

Vogt, P.R., 1974, The Iceland Phenomenon: Imprints of a hotspot on the ocean crust and implications for flow below the plates, in Geodynamics of Iceland and the North Atlantic Area, ed. by L. Kristjansson, Dordrecht, Holland, D. Reidel Publ. Co., 106-126.

Vogt, P.R., 1976, Plumes, sub-axial pipe flow, and topography along the mid-oceanic ridge, Earth Planet. Sci. Lett., 29, 309-325.

Vogt, P.R., 1979, Global magmatic episodes: New evidence and implications for the steady-state mid-oceanic ridge, Geology, 7, 93-98.

Vogt, P.R., 1981, On the applicability of thermal conduction models to mid-plate volcanism: Comments on a paper by Gass et al., J. Geophys. Res., 86, 950-960.

Vogt, P.R., and Avery, O.E., 1974, Detailed magnetic surveys in the Northeast Atlantic and Labrador Sea, J. Geophys. Res., 79, 363-398.

Vogt, P.R., Perry, R.K., Feden, R.H., Fleming, H.S., and Cherkis, N.Z., 1981a, The Greenland-Norwegian Sea and Iceland environment: Geology and Geophysics, in The Ocean Basins and Margins, v. 5, The Arctic, ed. by A.E.M. Nairn, M. Churkin, and F.G. Stehli, Plenum Press, N.Y., 493-598.

Vogt, P., Bernero, C., Kovacs, L., and Taylor, P., 1981b, Structure and plate tectonic evolution of the marine Arctic as revealed by aeromagnetics, Oceanologica Acta, SP, 25-40.

Vogt, P.R., and Johnson, G.L., 1975, Transform faults and longitudinal flow below the mid-oceanic ridge, J. Geophys., Res., 80, 1399-1428.

Vogt, P.R., Johnson, G.L., and Kristjansson, L., 1980, Morphology and magnetic anomalies north of Iceland, J. Geophys., 47, 67-80.

Voppel, D., Srivastava, S.P., and Fleischer, V., 1979, Detailed magnetic measurements south of the Iceland-Faeroe Ridge, Deut. Hydrogr. Z., 32, 154-172.

Wilson, J.T., 1963, A possible origin of the Hawaiian Islands, Can. J. Phys., 41, 863-870. 24

# CHAPTER 2

# REVIEW OF TECHNIQUES

NEW METHODS IN MARINE GEOLOGY AND GEOPHYSICS THAT HAVE

APPLICATION TO THE STUDY OF GREENLAND-SCOTLAND RIDGE

M. Talwani

Lamont-Doherty Geological Observatory
of Columbia University
Palisades, NY 10964, USA

Present address:

Gulf Research & Development Company
P.O. Drawer 2038
Pittsburgh, PA 15230, USA

It has been difficult to apply the usual methods of deep sea geophysical and geological exploration to the Greenland-Scotland Ridge. Single channel seismic reflection profiling has shown that sediments are either absent or present only in small pockets on the crest of the Iceland-Faeroe Ridge although thicker sediments are present on the flanks. Basalt, which constitutes basement on the ridge, has been mostly impenetrable to single channel reflection profiling and even to conventional multi-channel seismic profiling. The presence of basalt at shallow depths gives rise to high amplitude magnetic anomalies of very short wavelength, anomalies of the kind associated with sea floor spreading are present. Because of shallow water depth and consequent large temporal temperature disturbances, it has been difficult to measure the flow of heat out of the sea bottom on this feature. The Greenland-Scotland Ridge may be particularly suited to the application of electromagnetic methods, but apparently no such experiments have been carried out in this area up to the present time except on Iceland. Fairly detailed gravity surveys exist over the Greenland-Scotland Ridge and have been usefully interpreted in some areas. However, in other areas the absence of complementary information about deep structures makes the unambiguous interpretation of gravity measurements difficult. Important pioneering deep crustal refraction experiments have been carried out on the Iceland-Faeroe Ridge which have yielded interesting results. Costs and problems in logistics make it difficult to carry out more of these experiments or to make substantial improvements. For example, by having very close seismograph spacing much higher resolution could be obtained.

As far as geological sampling is concerned, deep drilling would be very effective. But it has been difficult to carry it out over the crest of the Iceland-Faeroe Ridge for two reasons. The thin sediment cover makes it difficult to spud in and the shallow water depth makes it difficult to position a drill ship such as the GLOMAR CHALLENGER without bending the drill pipe through unacceptably large angles.

We have mentioned the difficulties faced by geological and geophysical methods to point out the new directions that these methods will have to take in the future in order to make further advances. The multi-ship seismic method involving long towed receiving arrays described by Stoffa and the use of ocean bottom seismometers described by Whitmarsh, both in this volume, hold great promise for exploration of the crust of the Iceland-Faeroe Ridge. Better digital filtering techniques will make the existing magnetic data more useful and the thermograd method developed by Langseth and Anderson at Lamont-Doherty for measuring heat flow in shallow water areas can be very profitably applied to studies of the Greenland-Scotland Ridge. Drilling on the crest and flanks of the Greenland-Scotland Ridge can be carried out by the existing GLOMAR CHALLENGER equipment if the drilling sites are located with great care. An intensive drilling program could yield very useful results.

Kullenberg and Zaneveld have stressed the importance of long-term measurements in the bottom boundary layer. The development of such technology is essential in this area where knowledge of the mechanisms of sediment transport are of enormous interest and importance.

# THE SEISMIC REFLECTION/REFRACTION METHOD: WIDE APERTURE DATA OBTAINED IN MULTISHIP EXPERIMENTS

Paul L. Stoffa, Peter Buhl & John B. Diebold

Lamont-Doherty Geological Observatory of Columbia University
Palisades, New York 10964

Multiple ships equipped with multichannel seismic receiving arrays and sound sources can be used in a variety of configurations to acquire wide aperture seismic reflection and refraction data. One experiment, the Expanding Spread Profile, acquires Common Mid-point data as two ships steam on opposite tracks away from the mid-point of the profile. Source-receiver separations of 100 km or greater are achieved with a maximum seismogram separation of 100 m. Source-receiver offsets are accurately measured using Raydist ranging. The density of the recorded data make it possible to transform the data acquired to the domain of intercept time and ray parameter for analysis. This experiment provides detailed velocity-depth information at a fixed geographic location for deep crustal and subcrustal studies. In a second two-ship experiment both ships steam on the same track at a fixed separation selected to track arrivals from target horizons. By alternately firing their sound sources and recording the arrivals from both sources, the two ships can acquire Wide Angle Common Mid-point data at a constant offset. Since the offsets of interest are typically less than 30 km, range can be determined using a Miniranger type ranging unit. These data can also be interpreted in terms of vertical delay time and ray parameter along the track traversed. If the separation between the ships is reduced to the maximum array length employed, no gaps in source-receiver offset will be present. This technique can be extended to include two or more ships. For two ships with 3.6 km arrays the maximum source-receiver offset that can be achieved with no gaps in coverage is 10.8 km, while for three ships offsets of 25.2 km are possible. Wide Aperture CDP profiling of this type will provide detailed crustal velocity-depth information along the track traversed and the data acquired can be processed to provide a normal incidence CDP reflection record section.

## I. INTRODUCTION

The marine seismic reflection/refraction experiment has two fundamental goals: to detect reflected (or refracted) energy from the different subsurface geologic strata encountered along the ship's track; and the more difficult objective of deriving velocity information to determine the thickness of each geologic unit and its depth.

Initially, by rapidly firing air guns and recording the reflected energy with a single detector, vertical (or near vertical) incidence seismic data were displayed as a "cross-sectional" image of the earth's crust. Unfortunately, this cross section is in terms of the time required for the reflected energy to travel from the source to the reflecting horizon and then return to the receiver. In the detection problem it is often adequate to interpret this time section and conversion to depth is not required.

But, even for this limited objective, if the source does not have adequate energy to penetrate the uppermost geologic structure, reflections from the underlying deeper strata will remain undetected. Two solutions to this problem have been proposed. One is to increase the source level. Progress in this area has been rapid and it is quite common to deploy large arrays of air guns whose spectral properties are tuned to detect reflections from various subsurface horizons. Arrays of more than 30 air guns are often deployed and firing pressures of 2000 and more recently 5000 psi are used. There will, however, ultimately be a limit to the amount of energy that can be generated with this method.

A second possibility is to enhance the signal-to-noise ratio by improved reception of the reflected energy. Initially, a single receiving element was used to detect the seismic arrivals. The desire to improve the signal-to-noise ratio and determine velocity-depth information prompted recording the reflected arrivals with an array of detectors. By properly combining the arrivals recorded across a multichannel array,a signal to random noise improvement proportional to the square root of the number of channels is achieved. The success of this approach lies in the ability to predict the X-T trajectories formed by the seismic reflections across the array. For the Common Source geometry (one source and an array of detectors) employed in marine reflection experiments, the effects of dip and topography make it difficult to accurately predict the true X-T trajectories followed by the reflection arrivals, particularly for long array lengths.

An important departure from the Common Source experiment was suggested by Mayne in 1962. In the acquisition mode proposed by Mayne, a Common Source experiment is performed where the locations of the source positions along the track are separated by half the distance between the elements of the multichannel array. During data analysis, traces from different source and receiver positions are collected into an ensemble of seismograms that are spatially focussed, Figure 1. Under ideal acquisition conditions, each seismic event in this Common Depth Point (CDP) gather will have reflected from the identical subsurface location, each with a different angle of incidence. If dip or topography is encountered, the same reflection point will not be common to all the reflections, but the reflected data will have a Common Mid Point (CMP). Even in this case, the spatial focusing inherent in the CMP data acquisition method presents significant advantage over the original Common Source experimental geometry, (Diebold and Stoffa, 1981).

Figure 2 is an example from the Western Pacific of the signal-to-noise improvement possible for 24-fold CDP data obtained with a 2.4 km array. The data at the left are from the near trace of the array and the data at the right were derived by properly combining the 24 traces of the CDP gather. The deep reflections from the M-discontinuity which are quite strong on the 24 fold profile are totally absent on the single channel profile.

In both the Common Source and Common Mid Point experiments, the receivers at different offsets will detect seismic arrivals which have traversed subsurface paths with different angles of incidence at each reflecting horizon. The traveltime of these events as a function of source-receiver offset form the data required for deriving velocity-depth information. As additional data with different angles of incidence are observed, the better the final velocity-depth resolution will be.

It is now common to employ physical multichannel arrays of 3.6 and 4.8 km in length. In shallow water, these arrays will record reflections with many subsurface angles of incidence in the sedimentary horizons. But, for deep crustal studies, derivation of accurate velocity-depth information requires much larger source-receiver offsets to observe seismic reflections having a comparable suite of subsurface angles of incidence. Depending on the water depth and the depth of the horizons of interest, the required source-receiver offsets may range from 5 to 100 km. While it is possible to construct physical arrays of up to 15 km in length, experiments using two or more ships equipped with conventional arrays can be designed to form synthetic apertures of the required length.

## II. WIDE APERTURE SEISMIC DATA ACQUISITION

### 1. Point Velocity-Depth Measurements: Expanding Spread Profiles

To derive velocity-depth information in deep water areas and

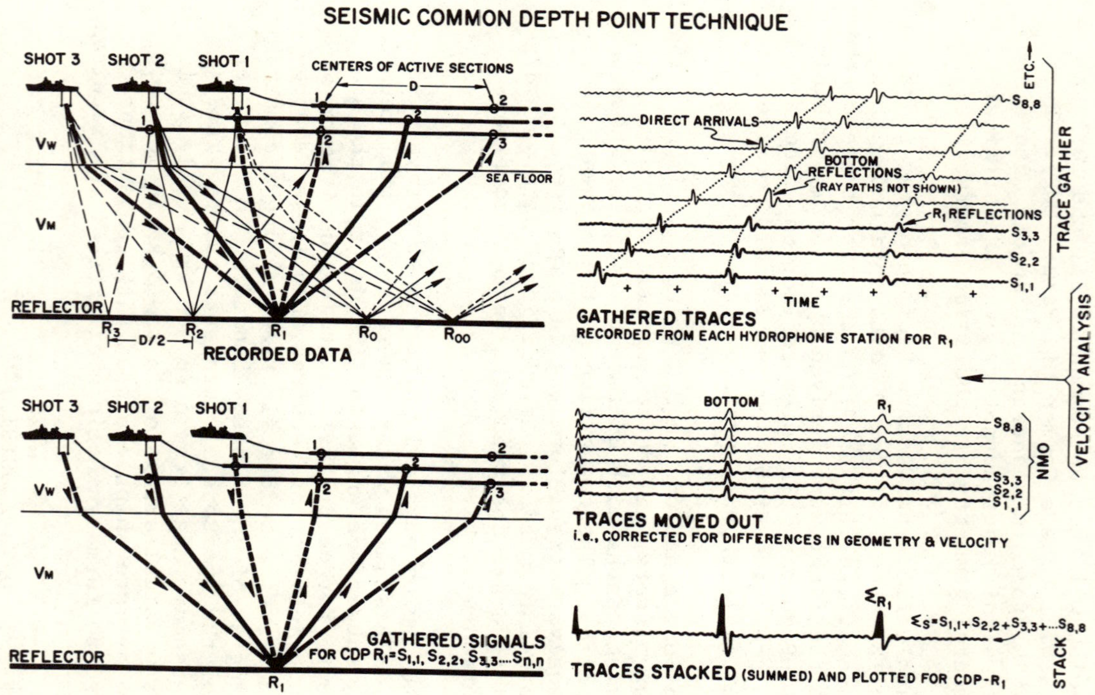


Figure 1. (Left) Common Depth Point data acquisition. Common source experiments (upper left) are performed where each source shot point location advances a distance D/2 from the previous shot point location. The ray paths in the common source geometry (upper left) that contribute to the CDP gather (lower left) are indicated by heavy lines. Under ideal conditions, all the rays in the CDP gather will have reflected from the same ground point. (Right) basic elements of CDP processing. In the upper right figure the reflection arrivals exhibit increasing traveltime with increasing source-receiver offset in the CDP gather. In the center right figure the approximately hyperbolic X-T trajectories of these arrivals are corrected to normal incidence time. Following this "normal moveout correction" these data are summed in the lower right figure to form a stacked trace with improved signal-to-noise ration.

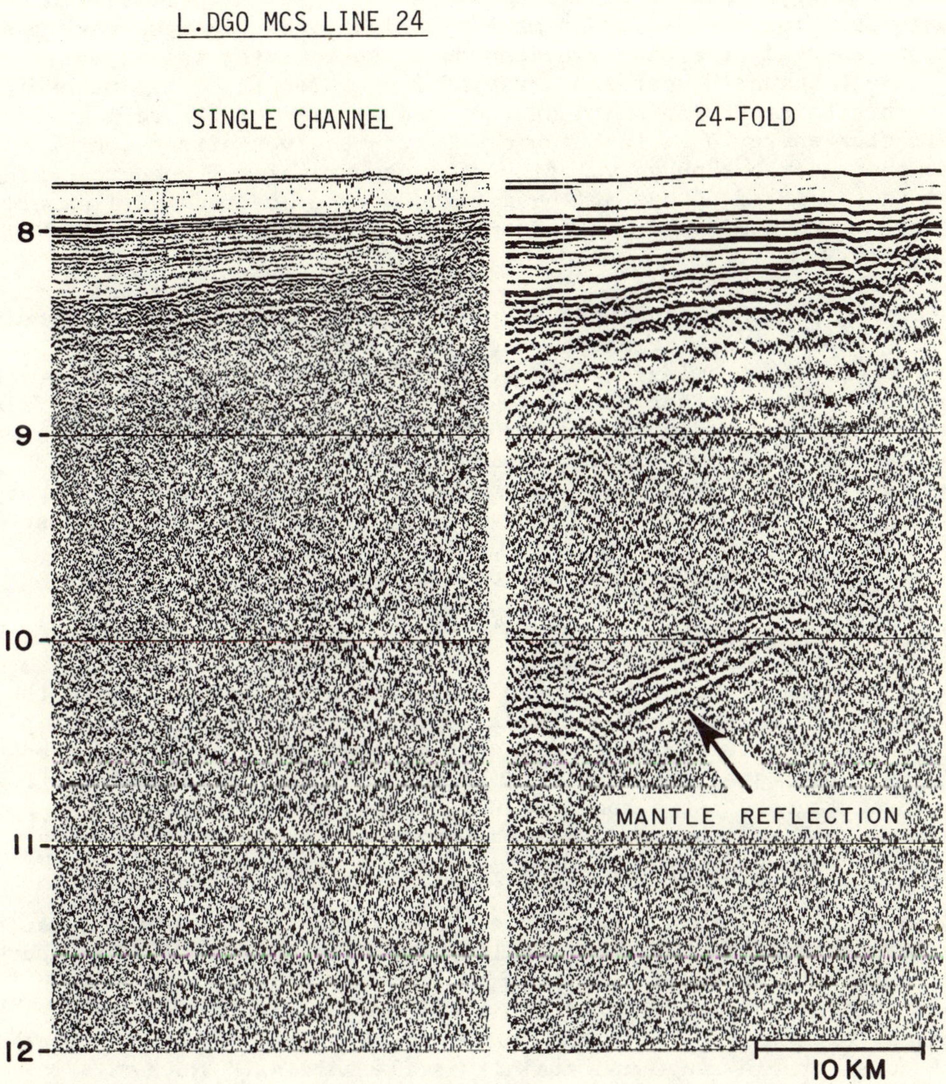


Figure 2. (Left) near trace seismic record section from the Western Pacific in the vicinity of the outer gravity high off Japan. (Right) normal moveout corrected and 24 fold stacked CDP record section for the same data. In this example the improved signal-to-random noise ratio results in the detection of a reflection from the M-discontinuity.

from the deep parts of the continental margins, we initially performed an experiment similar to the two ship refraction type experiment commonly employed in the past, see Ewing et al., 1939. In our implementation of this experiment two significant changes were made. First, instead of a single hydrophone, the receiver was replaced by a multichannel recording array of 2.4 km length. This makes it possible to determine not just the arrival time of the reflected/refracted energy as a function of source-receiver offset, but also the instantaneous phase velocity of each arrival. The second change was that instead of performing a fixed source or (receiver) experiment, a common midpoint experiment was performed, see Stoffa & Buhl, 1979.

During the Expanding Spread Profile, both ships steam on opposite tracks crossing at the midpoint of the profile about one mile abeam. They continue steaming on opposite tracks until the maximum required source-receiver offset is achieved, Figure 3. This type of profile can be extended to 100 km or more if required. The distance between the two ships is monitored by using Raydist ranging and recorded for every shot. Shot instants are detected on the shooting ship by towing a hydrophone and the shot time is transmitted using a touch tone radio link to the recording ship. This signal is then used to start the recording sequence on the receiving ship. Digital recording is used with a variable delay depending on the source-receiver offset and first arrival time.

One source of error is in the shot instant detection due to the uncertainity in estimating the distance between the shot and trip phone on the shooting ship. Since time over the side is recorded and the sinking rate can be estimated, the ship's speed becomes the dominant factor. At a speed of 9 km per hour for small charges (e.g. 1 to 2 kg) the maximum total error is about 5 meters corresponding to a timing error of 3 msec.

In most other wide angle reflection/refraction experiments range is determined from the data itself. For example, in the OBS experiment the direct wave or first and second bottom bounces must be used. This can cause considerable error at ranges beyond 50 km. The use of multiple ocean bottom seismographs reduces this error, but considerable range uncertainity still remains. The use of Raydist ranging provides an independent measure of the distance between the shot and receiving ship. Raydist consists of a master transmitter on the receiving ship and a slave or repeator transmitter on the shooting ship. Raydist is a phase measuring device and an initial calibration is required. Calibration is straightforward and consists of setting the lane count when the ships are at a known separation. The frequency of operation is 3.3 mHz and the discrimination of 1/100th of a cycle yields a relative range accuracy of .45 meters. Our experience with these units in the

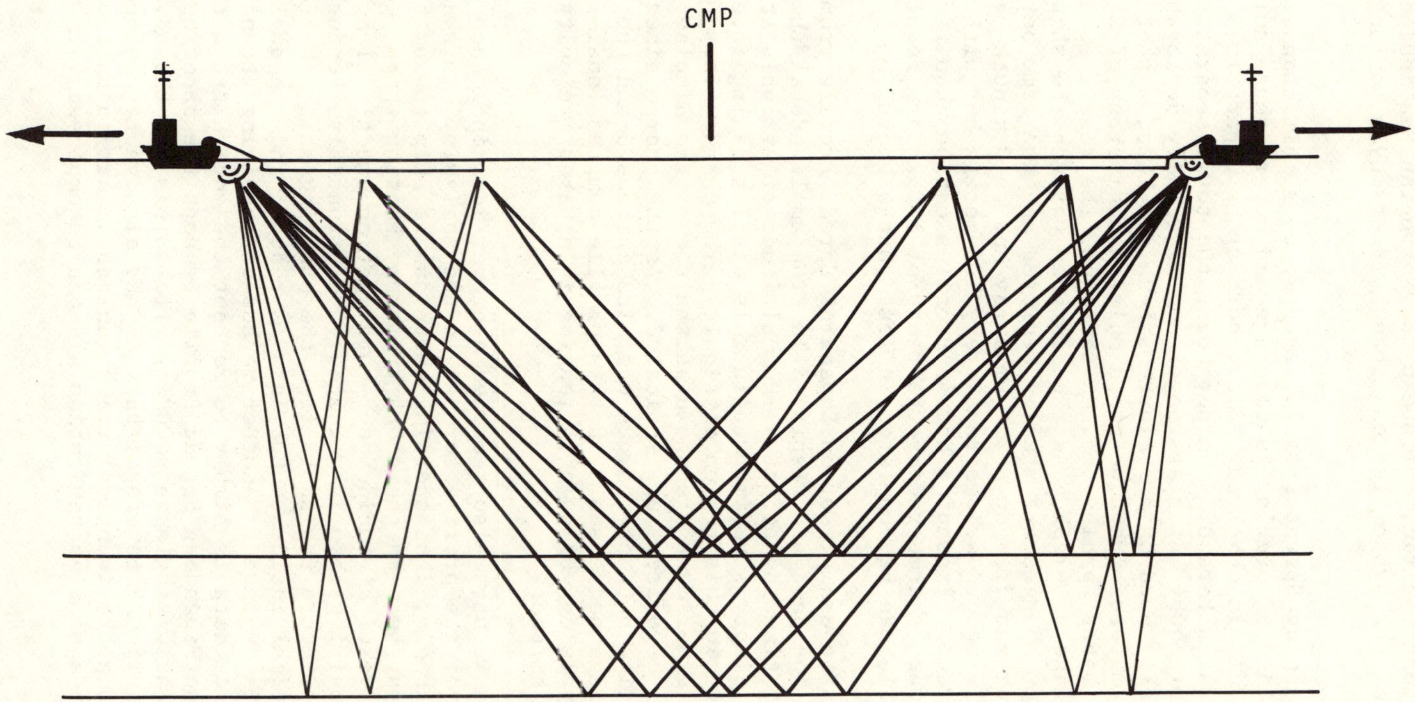


Figure 3. Expanding Spread Profile experimental geometry. In this experiment 1 (or both) ships deploy a multichannel array and depart from a common midpoint to source-receiver offsets of 100 km or greater. As airguns or explosive charges are fired by one ship, the ship towing the multichannel array records the wide angle reflection/refraction arrivals. Range is measured using Raydist ranging. This experiment provides velocity-depth information at the midpoint of the profile.

Western and Eastern Pacific, Caribbean and Norwegian Sea indicate that the error of the system is less than 5 meters even to ranges of 100 km.

In the Expanding Spread Profile, the use of a multichannel array provides a substantial overlap in coverage and a very dense suite of seismic arrivals is observed. Since both ships depart from the common midpoint at a rate of 9 km per hour, the total separation rate is 18 km per hour. At near ranges small, readily deployable charges are usually used on a 1 minute schedule. Thus, a shot is fired every 300 meters. When we use a multichannel array of 2.4 km length with 100 meter spacing between sections, the result is an overlap in coverage of 87.5%. As the distance between the ships increases, larger charges are usually deployed and the time between shots must be increased. For a 2.4 km array and an 8 minute schedule, we can still record seismograms every 100 meters but with no overlap in coverage. Expanding Spread Profiles can be acquired very quickly. Even when source-receiver offsets of 100 km are required, the entire profile can be completed in less than 6 hours.

Figure 4 is an example of an Expanding Spread Profile from the Norwegian Sea. The data are displayed in true amplitude at the recorded arrival time and offset. The multitude of arrivals at the near ranges are simply superimposed in this display making interpretation of data in this region difficult. The charge sizes employed in the experiment are shown at the top of the display and the source-receiver offsets are indicated. As the charge sizes increased, the time between charges was also increased and the overlap in coverage decreases. In this profile compressional wave refraction arrivals can be seen increasing in apparent velocity from 5.8 to 7.1 km/sec.

In the past we have also recorded near vertical incidence data along the track of the profile for use in the analysis. This data is usually acquired by interspersing shots on the ship towing the multichannel array between the shots from the shooting ship. In this way a near vertical incidence record section exactly along the track of the receiving ship is acquired. Unfortunately, because of the necessity to record the shots from the shooting ship, the fold of the near vertical incidence CDP data is reduced and the quality of the profile is degraded. Another method of acquiring the near vertical incidence data is simply to reshoot the entire track traversed during the Expanding Spread Profile by both the shooting and recieving ships as a single ship CDP profile. This has the advantage of maintaining full CDP coverage, but has the disadvantage of possibly not covering the identical track (unless we are in an area where Loran C or a similar navigation aid can be employed).

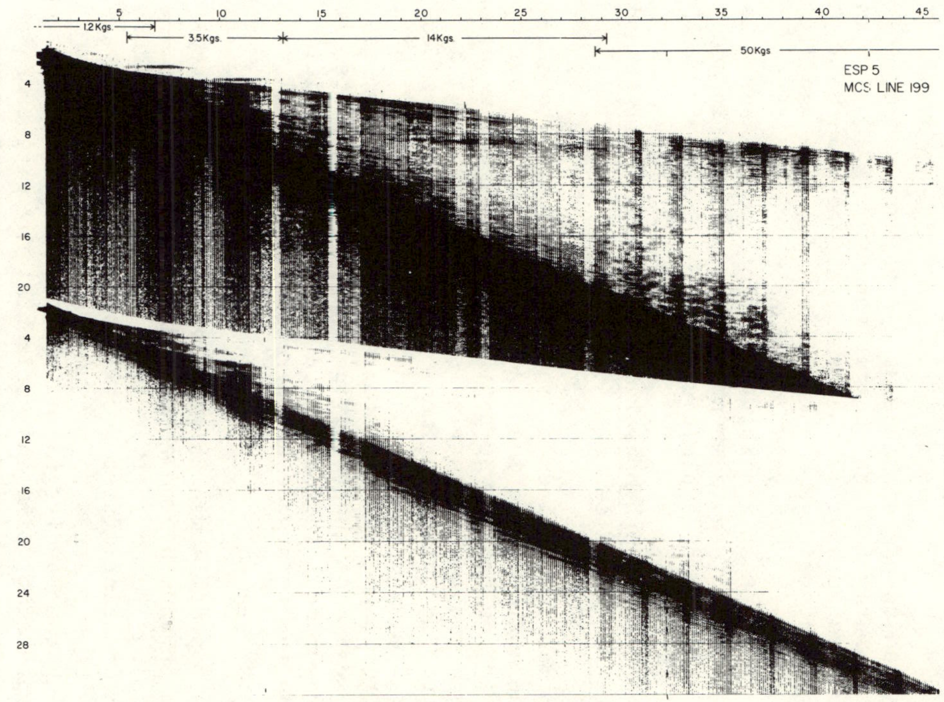


Figure 4. An example of ESP data, Expanding Spread Profile 5, from the Norwegian Sea. The range (in km) and charge sizes used are displayed at the top of the profile. The data are displayed in true amplitude and at their recorded range and arrival time. A high display gain was used in the upper profile and decreased gain in the lower profile. Initially, the overlap in coverage obtained when using a multichannel array results in the recording of many seismic arrivals at nearly the same source-receiver offsets. As the range increases and larger charges are required, the overlap in coverage is less. In the upper section vertical banding occurs when the arrivals from adjacent shots have coincident source-receiver offsets. Unless a shot goes unrecorded the maximum seismogram separation is 100 m.

Since the Expanding Spread Profile is a common midpoint experiment, the effects of dip and topography are minimized (see Diebold and Stoffa, 1981 and below). Our primary objective in this experiment is to determine the velocity-depth structure at the midpoint of the profile. Although the geometry of the experiment minimizes the effect of dip, care must be taken when shooting the Expanding Spread Profile to align the profile in the direction that keeps lateral variability, either in terms of dip, topography, or possible lateral velocity gradients to a minimum.

In the future we plan to use air guns as the primary source of seismic energy (particularly for the shorter ranges, e.g. 0 - 50 km) rather than explosives as we have in the past. Air gun source arrays provide a highly repetitive sound source and can be rapidly fired. Our experience in the analysis of Expanding Spread data (discussed below) suggests that a highly repeatable sound source is needed in the analysis of this data when seismic waveforms in addition to seismic traveltimes are exploited.

2. <u>Continuous Velocity-Depth Measurements, Constant Offset and Wide Aperture CDP Profiles</u>

A second two ship multichannel experiment was performed to continuously record wide angle reflections and refractions. In our initial experiments the objective was to track critical angle reflections from the M-discontinuity and arrivals from oceanic layer 3, a widely found refractor. In this experiment the shooting ship fired explosives on a 1 or 2 minute schedule and the receiving ship, at a distance of from 16 to 25 km, was used to record the arrivals, see Figure 5. In this experiment the receiving ship also fired its air guns to collect near vertical incidence CDP data. The acquisition equipment employed during this mode of operation was identical to that used in the Expanding Spread Profiles.

The constant offset profile was originally designed as a detection experiment, not as a method to observe data suitable for velocity-depth inversion based on traveltime trajectories. Figure 6 is an example of a Constant Offset Profile that recorded arrivals from layer 3. By using the array to measure the phase velocity of the arrivals and using the Raydist range information, both the event vertical delay (or intercept) time and horizontal ray parameter were measured. As a detection experiment this mode of acquisition was highly successful since we were able to determine the time topography.

Recently, Buhl et al. 1981 suggested a modification of the constant offset experiment which can be used when the shooting ship is also equipped with a multichannel receiving array and equivalent sound source. In this acquisition mode, if the arrays are the same

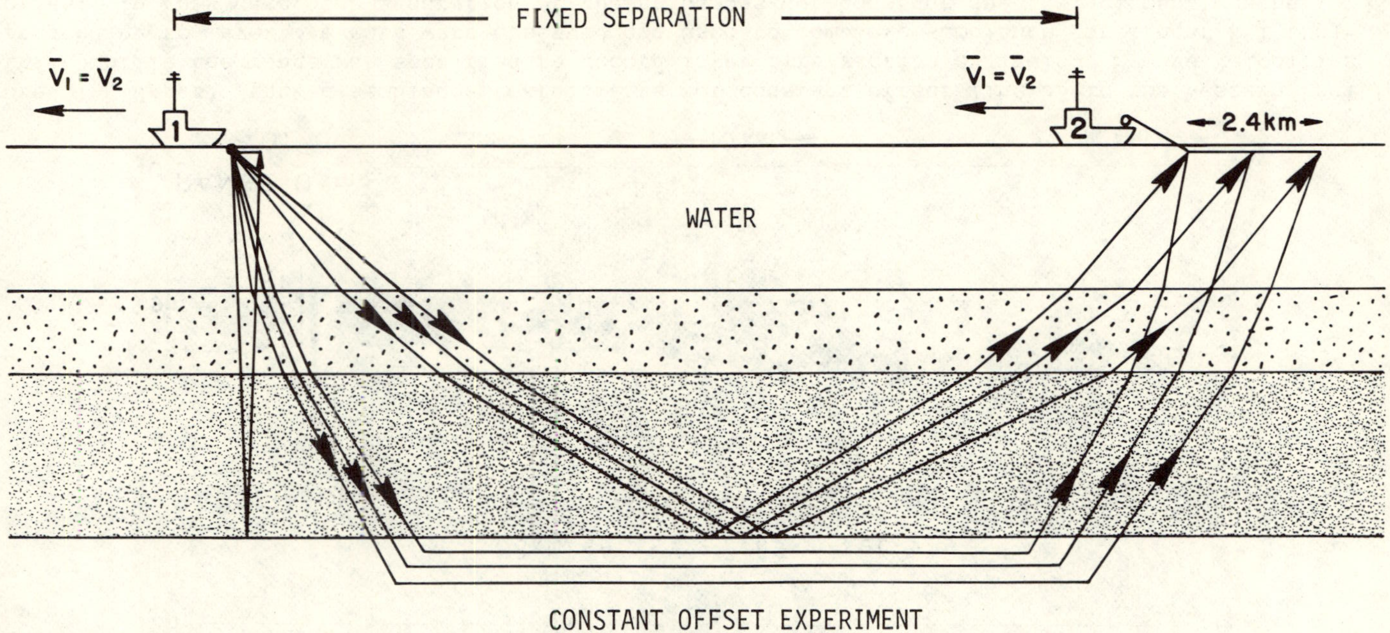


Figure 5. Constant offset profile experimental geometry. Both ships travel at the same speed and in the same direction. The source-receiver separation is continuously monitored by the receiving ship, and speed changes are made when it falls outside the prescribed bounds. Ray paths are shown for reflected and refracted events. The separation between the ships is specified to record either wide angle reflections or refractions from a specified geologic horizon.

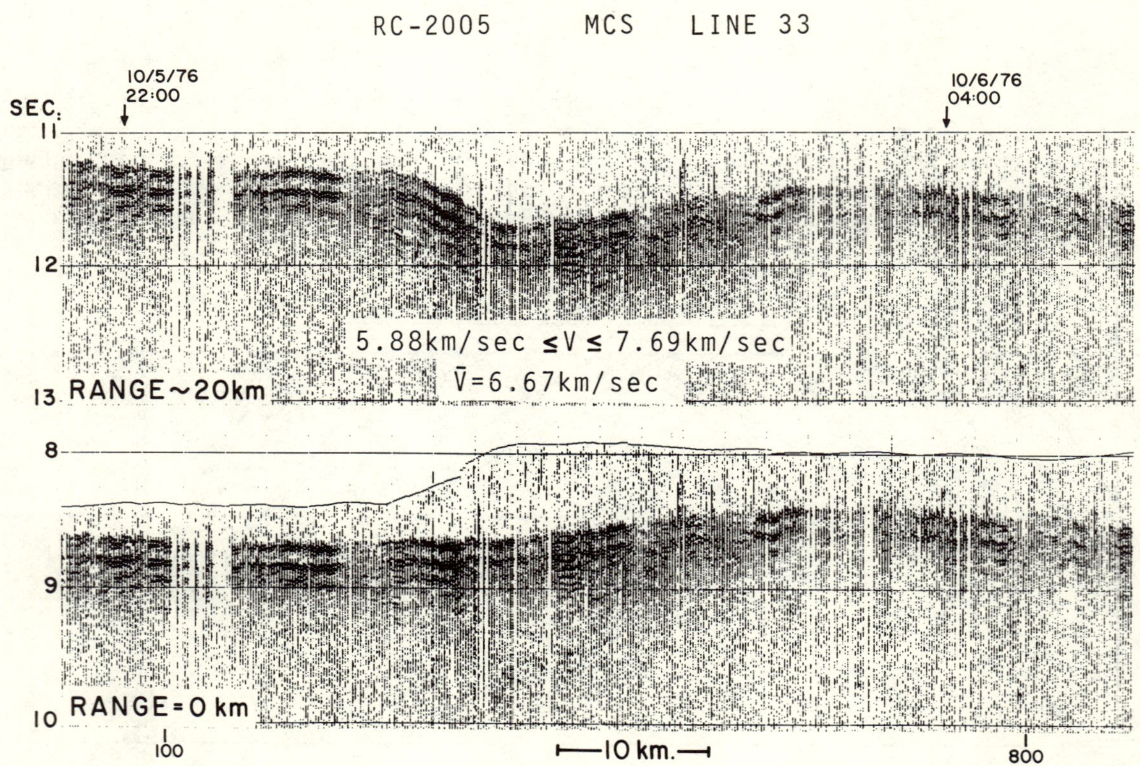


Figure 6. (Upper) Stacked semblance coefficients for constant offset line 33 in the Western Pacific. In this profile the range was specified to record layer 3 refraction arrivals. P wave velocities corresponding to layer 3 events were observed and used to complete semblance for linear X-T trajectories. In this case, the separation of the ships was not constant due to inclement weather. (Lower) Removing the range variation (displaying the same data at intercept time) removes the apparent topographic low in the layer 3 refraction arrival.

length, we separate the second ship a distance equal to an array length from the end of the array of the first ship. By alternately firing a continuous wide angle reflection/refraction experiment with no gaps in source-receiver offsets for up to 3 times the array length will be recorded, see Figure 7. In essence, the physical array lengths are magnified by the separation of the ships. For two ships equipped with 3.6 km arrays, when the lead ship fires it will record arrivals for the source-receiver offsets of 0 - 3.6 km, and the trailing ships will record the offsets of 7.2 - 10.8 km. When the second ship fires, it will again record the source-receiver offsets of 0 - 3.6 km but the lead ship will record the arrivals from 3.6 - 7.2 km, see Figure 8.

By alternately firing and monitoring the distance between the ships accurately, common midpoint data can be constructed just as in the one ship common midpoint experiment. Under ideal circumstances, if each ship fires on a 20 second schedule (40 seconds for a complete cycle) full fold CDP coverage is achieved if the ships are able to maintain a fixed separation and a speed of exactly 5 knots over the ground. Since this is unlikely, the fold of the data along the track will vary. For the case of two ships equipped with 36 channel 3.6 km arrays, 108 fold CDP data can be acquired with source-receiver offsets of 0 to 10.8 km. In this experiment (even under ideal circumstances) the CDP spacing along the ships track will be increased from 50 to 100 (or more) meters. Thus, we have degraded the lateral spatial resolution to acquire wide aperture data with full fold CDP coverage. Alternatively, we could gather the data into 50 m CDP intervals, but gaps in the source-receiver offsets would be present. Wide Aperture CDP Profiling is designed primarily to detect reflections from the oceanic crust in deep water areas and reflections from the deep parts of continental margins. The loss in spatial resolution along the track traversed is a justifiable compromise for the magnified array length possible using this approach.

This technique can be expanded to include more than two ships, Figure 9. In this case, if all array lengths are equal, the third ship should be spaced a distance of three array lengths from the end of the second ships' array. By cyclical firing, and recording the arrivals from each shot, each array is magnified by a factor of 7. That is, for three ships with 3.6 km array lengths, the total spatial aperture becomes 25.2 km. In this mode of acquisition the common depth point spacing along the ships track is (again under ideal circumstances) further increased to 150 meters. Figure 10 indicates how the data from consecutive shots are combined into a Wide Aperture CDP gather for three ships.

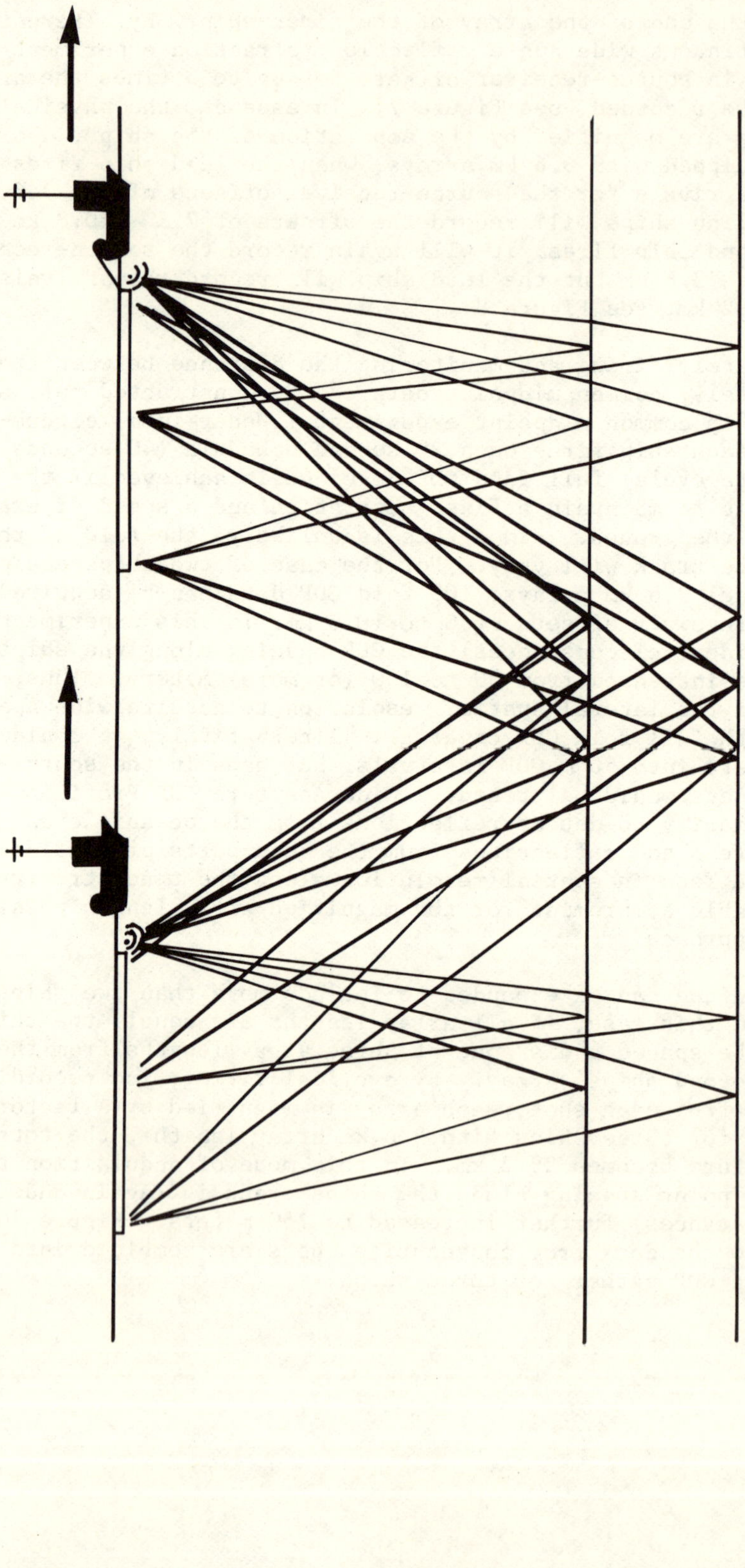

Figure 7. Wide Aperture CDP profiling experimental geometry. In this experiment 2 ships equipped with identical sources and multichannel arrays steam on the same track with a separation equal to one array length. By alternately firing sound sources, source-receiver offsets up to three times the array length will be recorded.

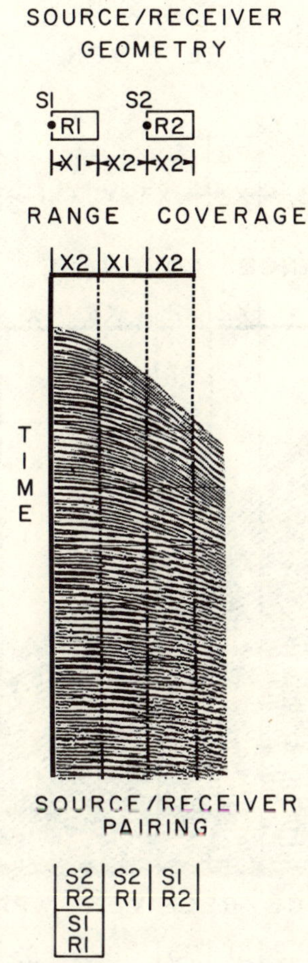


Figure 8. (Top) The spacing of two sources, $S_1$ and $S_2$, and their attached receiving arrays, $R_1$ and $R_2$, of lengths $X_1$ and $X_2$ respectively. The longest array, $R_2$, is shown in the trailing position. To achieve maximum continuous range coverage the distance between the end of the leading receiving array and the trailing sound source is equal to the length of the longest array, $X_2$. (Middle) The range coverage obtained with the spacing shown at top. The coverage is superimposed on a typical time-distance data set. (Bottom) Source-receiver matching which generates true range coverage shown in the middle. Note the duplication at small offsets as each array records the data from its own source ($R_1$ records $S_1$ and $R_2$ records $S_2$). The middle range, $X_2$ to $X_2 + X_1$, is covered by $R_1$ recording $S_2$.

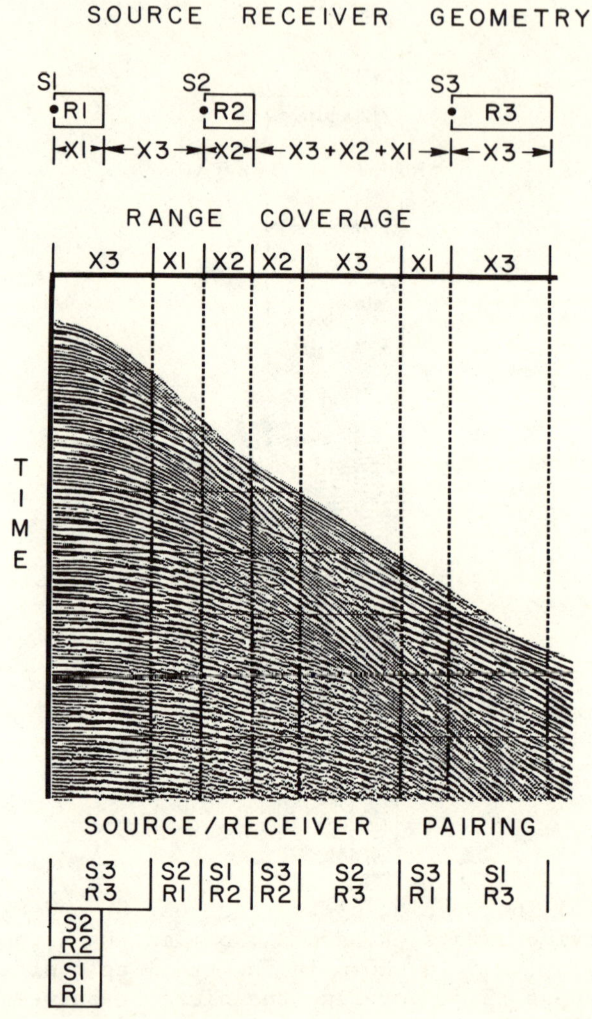


Figure 9. Same as Figure 8, but for three sources and receiving arrays In this illustration the trailing array, $R_3$, is the longest. (Note that $R_3$ can be at either end, but not in the middle.) The other two receiving arrays must be separated by $X_3$. In the source-receiver pairing, $S_i$-$R_j$ is always followed by $S_j$-$R_i$, recovering a continuous range interval $X_i$+$X_j$.

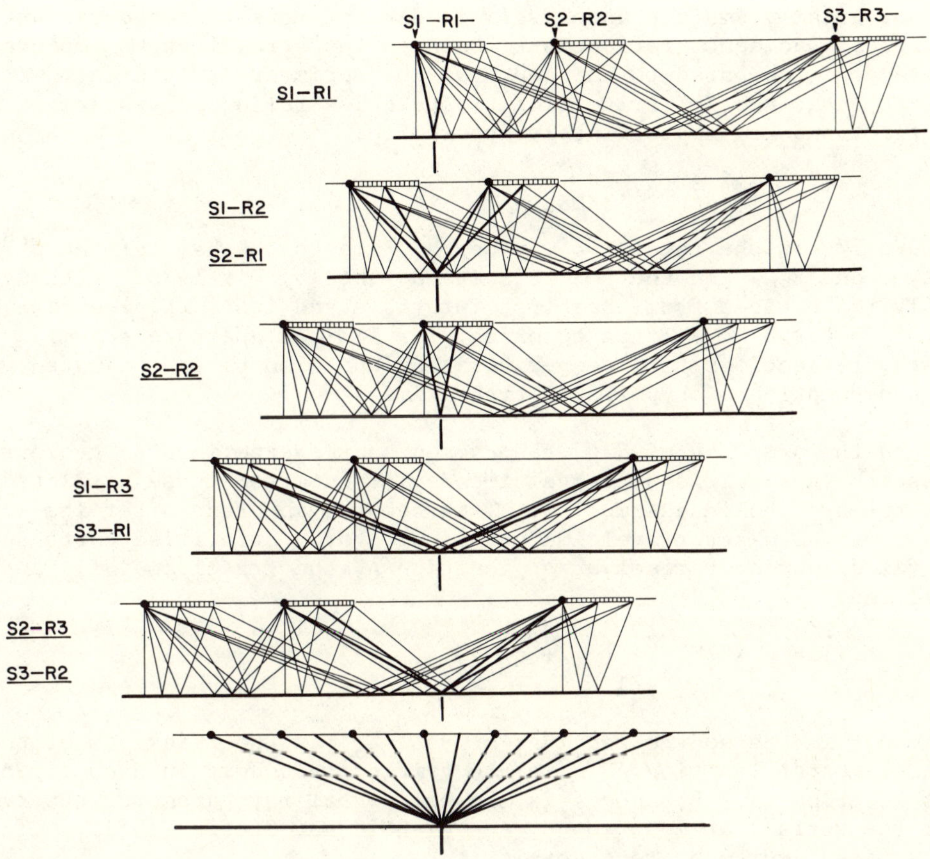


Figure 10. The multiple-source receiving array shot sequence for maximum fold CDP gathering of 3 ship data. ($S_3$-$R_3$ is not shown). The sources and arrays proceed from right to left. The $S_1$-$R_1$ data is obtained first as shown at the top, and is similar to normal CDP gathering. Immediately after the $S_1$-$R_1$ data, both the $S_1$-$R_2$ and $S_2$-$R_1$ data are recorded. Since $S_1$ and $S_2$ are fired at different times, the $S_1$-$R_2$ and $S_2$-$R_1$ bounce points will not in general coincide, smearing the CDP. The heavy ray paths indicate source-receiver pairs which contribute to a single CDP. The light ray paths indicate source-receiver pairs which contribute to other CDPs. Since the complete source-receiver sequence contains three shots, one from each of the ships, the CDP bin size for full fold coverage is three times that of single CDP profiling.

## III. ANALYSIS OF WIDE APERTURE REFLECTION/REFRACTION DATA

Originally, multichannel seismic data primarily recorded near vertical incidence reflections, because the array lengths employed were small compared to the depth of the horizons of interest. In these data, the X-T traveltime trajectories followed by seismic reflections are approximately hyperbolic:

$$T^2 = T\emptyset^2_n + X^2/V^2_{RMS_n} \quad , \qquad (1)$$

where $T\emptyset_n$ is the total two-way normal time to the base of the $n^{th}$ layer and $V_{RMS_n}$ is the RMS velocity defined by Dix 1955. Although only reflections from the first interface and its multiples follow truly hyperbolic X-T trajectories, for limited aperture seismic data, reflections from deeper horizons can also be approximated with the assumption of hyperbolic traveltimes.

In the presence of dip, the common source experimental geometry results in significant traveltime deviations from those predicted by the hyperbolic assumption. This becomes apparent after inspection of the exact traveltime equation. For a stratified earth and a fixed source or receiver at location A, the traveltime, T, can be written:

$$T = p_b X + \Sigma_j Z_{a_j} (q_{a_j} + q_{b_j}) \quad , \qquad (2a)$$

where X is the source-receiver offset, $Z_{a_j}$ is the thickness of the $j^{th}$ layer at location A, q is the vertical slowness in each layer at location A or B, and $p_b$ is the horizontal ray parameter observed at the surface at location B, see Diebold and Stoffa (1981). If the fixed source or receiver is at B,

$$T = p_a X + \Sigma_j Z_{b_j} (q_{a_j} + q_{b_j}) \quad , \qquad (2b)$$

where $Z_{b_j}$ is the thickness of the $j^{th}$ layer at location B, and $p_a$ is the horizontal ray parameter observed at the surface at location A.

With the advent of common depth point data acquisition, the effect of dip on seismic traveltime was observed to be minimal. This can be seen by inspecting the exact form of the CDP (or in the presence of dip common midpoint, CMP) traveltime equation. In this experiment, Diebold and Stoffa (1981), proved that the exact traveltime equation for any source-receiver offset can be found by combining the above fixed source-receiver traveltime equations:

$$T_{CMP} = \frac{(p_a + p_b)}{2} X + \Sigma_j Z_j (q_{a_j} + q_{b_j}) \quad (2c)$$

where Z is the layer thickness at the midpoint of the profile. The minimal effect of dip, and the symmetry of CMP traveltime trajectories is due to the horizontal ray parameter averaging inherent the CMP experiment.

Using the hyperbolic traveltime equation, the original objectives for employing multiple source-receiver offsets can be achieved only for near vertical incidence data. Hyperbolic array velocities and two-way normal times for reflection events can be measured by a two parameter trial and error computer search, see Taner & Koehler 1969. Near vertical incidence, the hyperbolic array velocities found in this search will approach the RMS velocity as defined by Dix (1955):

$$V^2_{RMS_n} = \frac{\sum_{j=1}^{n} v_j^2 t_j}{T\text{Ø}_n} \quad (3)$$

where $v_j$ are the layer velocities and the $t_j$ are their thickness in terms of two-way normal time. Knowledge of the RMS velocities and two-way normal times makes it possible to derive estimates of the interval velocity between reflection events:

$$v_n^2 = \frac{V^2_{RMS_n} T\text{Ø}_n - V^2_{RMS_{n-1}} T\text{Ø}_{n-1}}{T\text{Ø}_n - T\text{Ø}_{n-1}} \quad (4)$$

Thus, using Dix's method, derivation of velocity-depth information, one of the major objectives of recording multiple source-receiver offsets, is achieved.

The second objective, improving the signal to noise ratio, is accomplished by a procedure called "normal moveout correction and stack". In this process, the increase in the arrival time for each reflection with increasing offset is predicted by knowledge of the hyperbolic array velocity. Correcting the arrival times of the events at nonzero source-receiver offsets to two-way normal time is the normal moveout correction, see Figure 1. Once all the reflection events have been corrected for normal moveout, they will lie along horizontal trajectories across a multichannel array or common depth point gather. They can now be combined by simple addition, i.e. "stacked" into a composite trace which will have an improved signal-to-random noise ratio proportional to the square root of the number of channels combined.

Coherent noise is also reduced by the process of normal moveout and stack. For example, multiple arrivals from a shallow interface will usually have a lower hyperbolic array velocity than that of a deep primary reflection, although they may occur at or near the same reflection times. Correcting the data for normal moveout based on the hyperbolic array velocity of the deep primary reflection will not correct the arrival times of the multiples to their two-way normal time. When the data are stacked, the multiples will combine out of phase, while the primary reflection arrivals add constructively. The success of this coherent noise rejection is dependent on the difference in normal moveout between the primary and multiple reflections.

The limitation of analysis methods based on hyperbolic traveltimes immediately became obvious as array lengths increased relative to the depth of the structures in question. If wide angle reflections and refractions are observed the hyperbolic traveltime assumption is no longer appropriate and the goals of increased interval velocity resolution and signal to noise enhancement will not be achieved. In the analysis of the wide aperture data described above, a different analysis method must be found for those arrivals which fall outside the region of applicability for hyperbolic traveltimes.

For wide aperture seismic data we have investigated an analysis method based on the transformation of seismic data from the observational coordinates of source-receiver offset X, and traveltime, T, to the domain of intercept or vertical delay time, $\tau$, and horizontal ray parameter, p. Graphically, this analysis procedure can be described as finding the suite of straight line segments that are present in the observed seismic data rather than finding the best single hyperbolic traveltime trajectory for all the source-receiver offsets recorded. To accomplish this decomposition of the true traveltime trajectory into a series of linear segments, we employ beamforming across a physical array in the Expanding Spread Profile, or across synthetic arrays formed by combining different physical arrays for wide aperture CDP profiling.

The transformation of seismic data into $\tau$-p makes it possible to interpret both pre and postcritical reflection and refraction data simultaneously. Interpretations based on the locus of arrivals in the $\tau$-p plane are facilitated since reflected arrivals terminate at the horizontal ray parameter equal to the slowness just above the reflecting interface and at the intercept time of the headwave refraction along the upper interface (Diebold & Stoffa, 1981).

The parameterization of seismic data by $\tau$ and p appears naturally (see below) and the exact traveltime equations (described above) are all of the form $T = \tau + pX$. For a horizontally stratified medium Schultz, 1976 was able to show that traveltime data parameterized by their traveltime slope, $p = dT/dX$, and intercept time, $\tau = T - pX$, result in $\tau$-p trajectories that are sums of ellipses:

$$\tau_N(p) = \sum_{j=1}^{N} t_j(1 - v_j^2 p^2)^{\frac{1}{2}} = \sum_{j=1}^{N} \Delta\tau_j(p) \qquad (5)$$

In the discussion that follows we show how wide aperture seismic data can be transformed into the domain of intercept time and ray parameter. We then illustrate three methods of deriving velocity-depth information based on equation 5, which predicts the reflection trajectories formed by seismic arrivals in the $\tau$-p plane.

1. Transformation to $\tau$-p

For the one dimensional problem of a stratified medium overlain by a liquid half-space, the reflection response to a $\delta$-function point source excitation at zero depth can be written as:

$$\Phi\,(\omega, X) = \int_0^\infty (-j\omega)\; R\,(\omega, p)\; J_o\,(\omega p X)\; \frac{p}{\eta_1}\, dp,$$

where

$$\eta_1 = \begin{cases} \left(\dfrac{1}{\alpha_1^2} - p^2\right)^{\frac{1}{2}} & p \leq \dfrac{1}{\alpha_1} \\ -j\left(p^2 - \dfrac{1}{\alpha_1^2}\right)^{\frac{1}{2}} & p \geq \dfrac{1}{\alpha_1} \end{cases} \qquad (6)$$

and $R(\omega, p)$ is the plane wave reflection response as a function of horizontal ray parameter, p, and angular frequency $\omega$. $J_0$ is the zero[th] order Bessel function, and $\alpha_1$ is the liquid half-space velocity. $\Phi$ is a potential which can be transformed to pressure by the relation:

$$P\,(\omega, X) = \rho_1 \omega^2 \Phi(\omega, X), \qquad (7)$$

where $\rho_1$ is the density of the half-space. Considering only outward traveling waves and using the standard high frequency approximation for the Bessel function, we can rewrite equation 6 as:

$$\Phi\,(\omega, X) = \frac{1}{\pi\sqrt{2X}} \int_0^\infty \left[\frac{p^{\frac{1}{2}}}{\eta_1}\;(-j\omega)\left(\frac{\pi}{|\omega|}\right)^{\frac{1}{2}} e^{+j\frac{\pi}{4}\frac{\omega}{|\omega|}}\right] \cdot \left[R\,(\omega, p)\; e^{-j\omega p X}\right] dp. \qquad (8)$$

In this expression the reflection response $R(\omega, p)$ includes the time delay required for a plane wave to travel from the first interface to the observation level at depth zero. After an inverse Fourier transform we have:

$$\Phi(T, X) = \frac{1}{\pi\sqrt{2X}} \int_0^\infty \frac{p^{\frac{1}{2}}}{\eta_1} \left[ -\frac{d}{dT} \frac{\theta(-T)}{(-T)^{\frac{1}{2}}} \right] * R(T-pX, p)\,dp, \qquad (9)$$

where $\theta(T)$ is the Heaviside function, and * denotes convolution. The plane wave reflection response can, in theory, be recovered from the seismograms by reversing the above procedure:

$$R(\omega, p) = \frac{\eta_1}{\pi\sqrt{2p}} \int_0^\infty X^{\frac{1}{2}}\, j\omega \left[\frac{\pi}{|\omega|}\right]^{\frac{1}{2}} e^{-j\frac{\pi}{4}\frac{\omega}{|\omega|}}\, \Phi(\omega, X)\, e^{j\omega pX}\, dX \qquad (10)$$

or,

$$R(\tau, p) = \frac{\eta_1}{\pi\sqrt{2p}} \int_0^\infty X^{\frac{1}{2}} \left(\frac{d}{d\tau} \frac{\Phi(\tau)}{\tau^{\frac{1}{2}}}\right) * \Phi(\tau + pX, X)\, dX \qquad (11)$$

These equations show how seismograms are theoretically derived from the plane wave response, and how the plane wave reflection response can be derived from seismograms. (For a detailed theoretical discussion, see Chapman, 1978 and Phinney, et al., 1981).

To generate synthetic seismograms we can simply use a discrete form of equation 9. For a fine ray parameter sampling, the resulting discrete seismograms will closely approximate the analytic response, see Wenzel, et al., 1981. When modeling the seismic response, we are free to choose the ray parameter, spatial and temporal sampling intervals. This freedom makes it possible to trade numerical accuracy against computational effort.

For observational data, we are not usually able to choose the spatial sampling required to implement a discrete form of equation 11 directly. Sampled data are rarely aliased in time, since we can filter the analog data before digitization. But, to avoid aliasing in space requires a spatial sample interval $\Delta X$, less than $V/2f_{max}$, where $f_{max}$ is the highest expected frequency. Thus, the Nyquist ray parameter is equal to $\frac{1}{2}\,\Delta X\, f_{max}$. For marine data with a maximum frequency of 125 Hz, the required spatial sampling is 6 m, which is impossible to employ economically, particularly for long arrays.

Thus, a discrete form of equation 11:

$$R(\tau_i,p_j) = \frac{\eta}{\pi\sqrt{2p_j}} \sum_{k=1}^{N} X_k^{\frac{1}{2}} \Phi(\tau_i + p_j X_k, X_k) , \qquad (12)$$

will be aliased. Another useful function, semblance, has found wide use in detecting coherent arrivals across an array, Taner and Koehler (1969). In our application semblance is defined as:

$$S(\tau_i, p_j) = \frac{\sum_W \left( \sum_{k=1}^{N} \Phi(\tau_i + p_j X_k, X_k) \right)^2}{N \sum_W \sum_{k=1}^{N} \Phi(\tau_i + p_j X_k, X_k)^2} \qquad (13)$$

The time window, W, adds stability to the semblance statistic, but decreases the temporal resolution. Semblance is the ratio of output to input energy along the proposed trajectory, Neidell and Taner (1971), and has the important properties that its value is independent of the amplitude of the arrival to be detected and it is normalized with a maximum value of 1. and a minimum value approaching 1/N.

The use of linear X-T trajectories as defined by equally spaced horizontal ray parameters is a natural selection. For a given bandwidth and a fixed array length the uncertainity in array velocity is characterized by a fixed $\Delta p$, which is inversely proportional to frequency and array length. Figure 11 is an example of stack and semblance for linear X-T trajectories for one shot from ESP 4 in the Caribbean, see Stoffa et al. 1981. The original data are shown on the left followed by semblance and stack, displayed at near trace arrival time for comparison to the original data. The large amplitude arrivals in the wedge shaped region of the stacks are aliases of the sea floor reflector due to the finite spatial sampling and aperture. Note that the corresponding semblance values are a minimum. Although the stack preserves amplitudes and waveforms, it is usually severely aliased and this aliasing often obscures the energy of weak primary arrivals. Semblance yields consistently high amplitudes for arrivals with modest signal-to-noise ratios as long as they are not contaminated by interfering coherent events.

Figure 11 also shows how semblance suppresses aliased energy and thus can be used to derive a window function, W ($\tau$, p) which can be applied to the individual slowness stacks to yield true amplitude, unaliased $\tau$-p stacks for individual subarrays. The

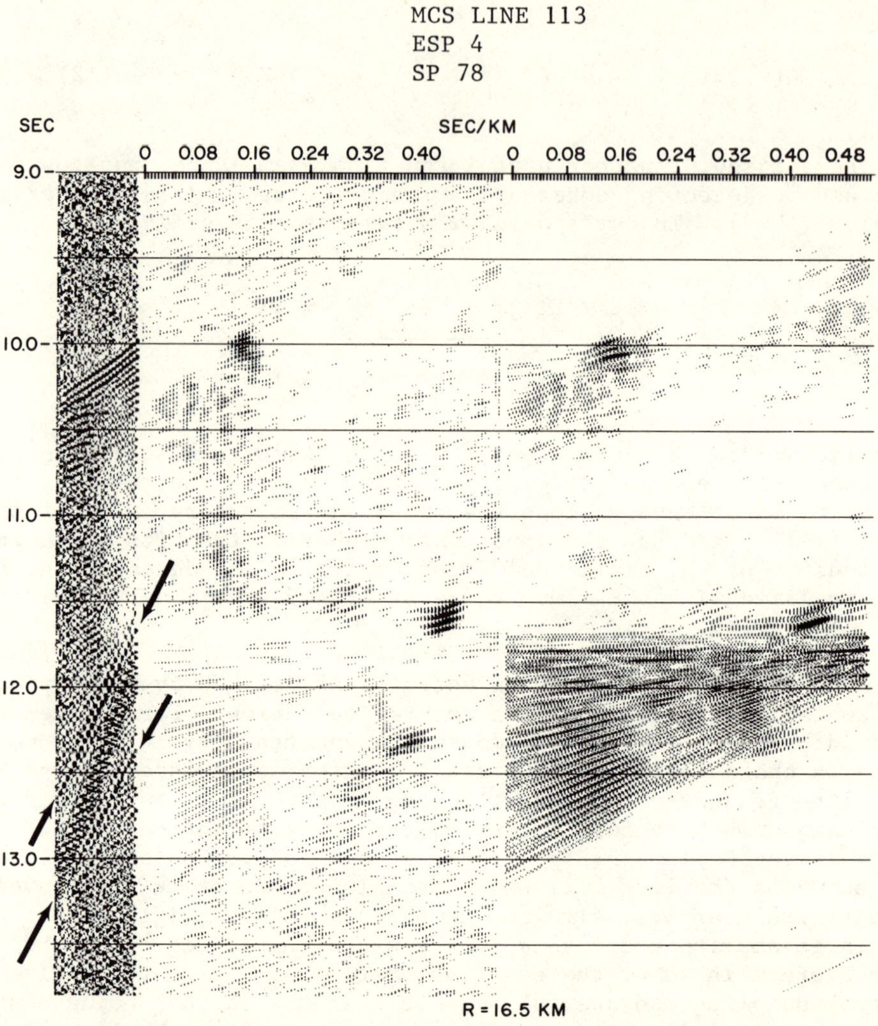


Figure 11. A display of stack and semblance for linear X-T trajectories for one shot of ESP 4 in the Caribbean at a range of 16.5 km. In this case the semblance statistic is far superior to the stack in defining the arrival time and horizontal phase velocity of the second wide-angle reflection event (indicated by the second pair of arrows). At this range, the two wide-angle reflections are 'crossing over' the sea floor reflection. The semblance statistic being insensitive to the relatively high amplitude of the sea floor reflection is unaffected by its presence and readily defines the two reflection events. In contrast, the stack is overwhelmed by the large amplitude of the sea floor reflection event for all velocities scanned, and only the first reflection event is well defined.

$\tau$-p data can then be found by summing the windowed $\tau$-p stacks:

$$R(\tau_i, p_j) \simeq \sum_{\ell=1}^{M} W_\ell(\tau_i, p_j)\, F_\ell(\tau_i, p_j), \qquad (14)$$

where the sum on $\ell$ is over M subarrays. The ideal window has a value of one where there are primary arrivals and zero elsewhere. Stoffa et al. 1981 suggested that the window function be derived by first smoothing the semblance data and then thresholding. For example, after smoothing the semblance data a rectangular window can be defined by defining all semblance values below a specified threshold as zero, and all values above this threshold as unity.

Figure 12 is an example of the $\tau$-p transformation for data from expanding spread profile 4 in the Caribbean using source-receiver offsets up to 22.6 km and using different semblance thresholds. In the mapping shown on the upper left, no window was used in the integration and the $\tau$-p data are severly aliased. In the other examples, the original semblance data was first low-pass filtered from 0-10 Hz and thresholds of .1, .2 and .35 were used to derive a rectangular weighting function. Even the modest semblance threshold of .1 eliminates most of the aliasing. Raising the threshold above .1 eliminates more aliasing but leaves only the principal arrivals that form the monotonically decreasing function of $\tau$ and p.

In this method of transformation the time domain operator, $d/d\tau\,(\tau)/\tau^{\frac{1}{2}}$, is not applied. This operator results from the required integration over all source-receiver offsets for each ray parameter in equation 11. When a weighted integration using overlapping physical or synthetic arrays is used, we automatically isolate the coherent arrivals present at each offset without integrating over all source-receiver offsets. Thus, this time domain operator is not required and the recovered $\tau$-p response will approximate the true plane wave response.

Using the weighted integration method described above, Stoffa et al., 1981, demonstrated that seismic data can be transformed to $\tau$-p without aliasing or the artificial suppression of high frequencies. Figure 13 is an example of this transformation for three wide aperture data sets in varying water depths. Just as the original X-T trajectories of this data can be analyzed to derive velocity-depth functions, we now show how the $\tau$-p trajectories formed by these seismic arrivals can be used. Unlike the hyperbolic traveltime assumption in X-T which fails with increasing offset, velocity-depth information based on the analysis of $\tau$-p trajectories

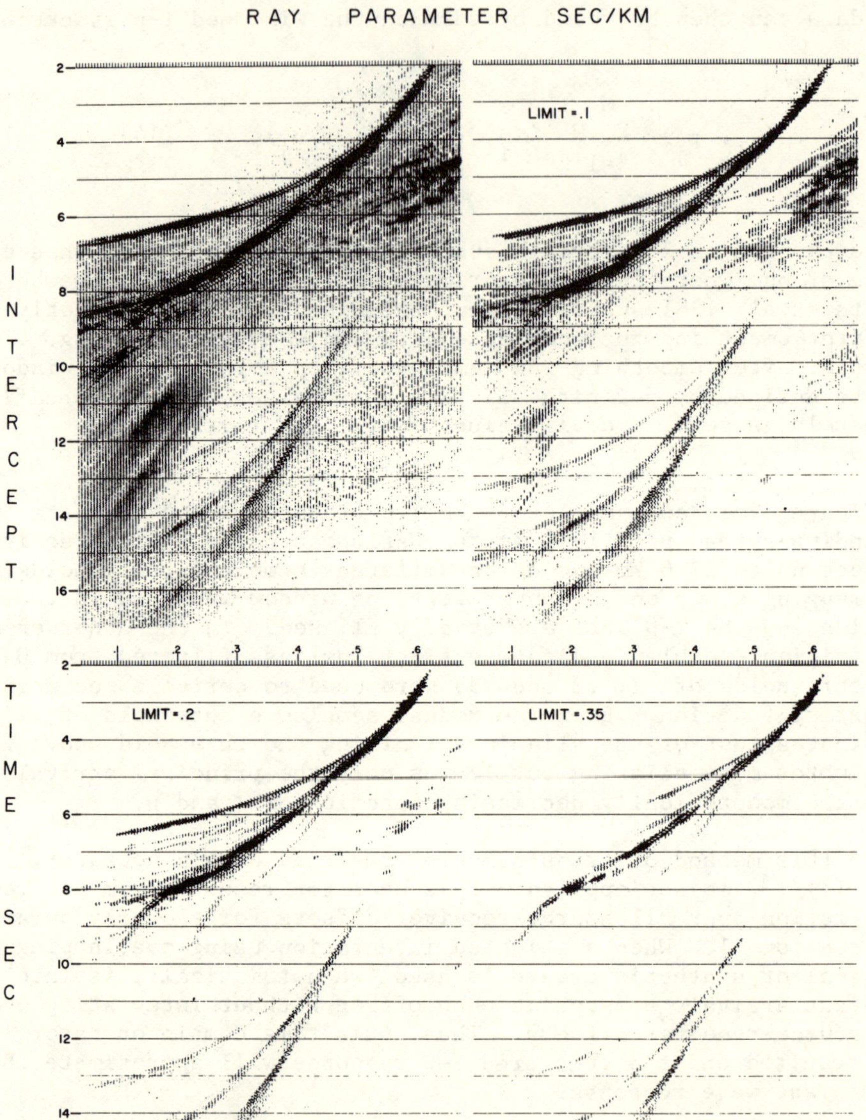


Figure 12. τ-p data for ESP 4 in the Caribbean for the range 0 to 22.61 km; the upper left data were derived without the application of the semblance derived filter and are severely aliased. The other τ-p data were computed using the window function derived from semblance for a pass band of 0 - 10 Hz and semblance limits of .1, .2 and .35, respectively. As the threshold is raised the aliasing decreases but the recovered τ-p response decreases. Only the principal contributions remain at the higher semblance thresholds.

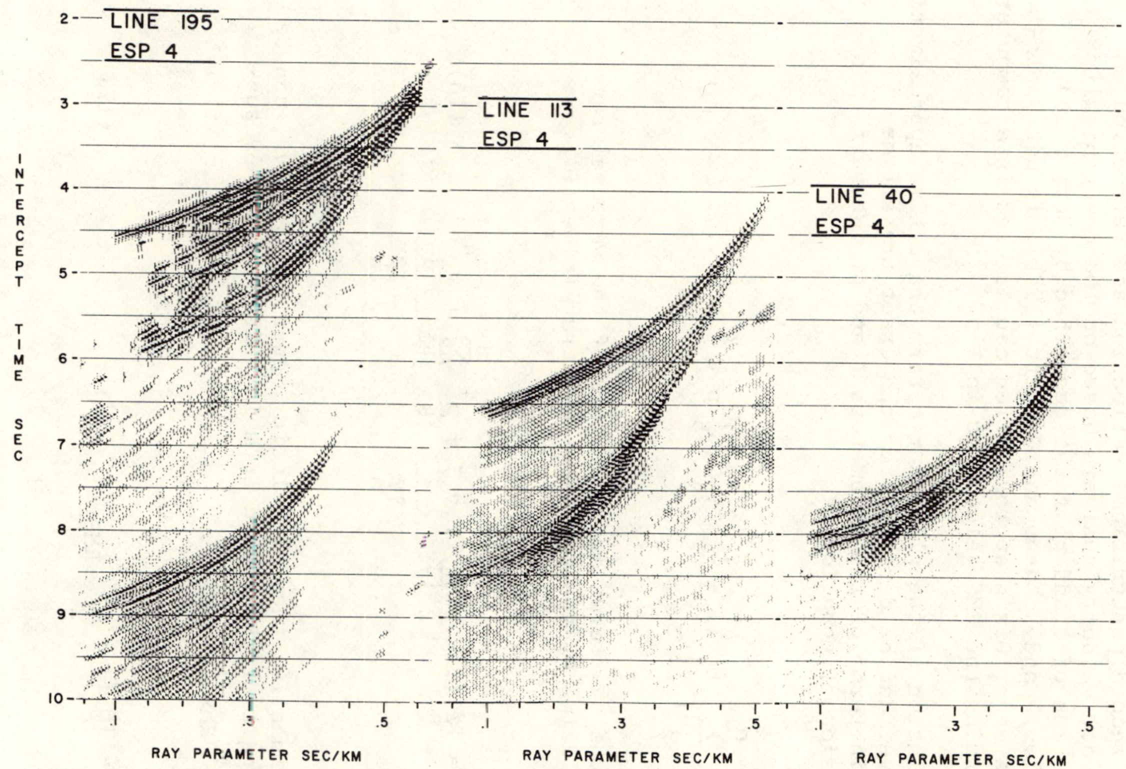


Figure 13. τ-p mapping for 3 Expanding Spread Profiles. In every case, information found at the largest offsets (steepest τ-p trajectories) is present, which could not be properly used with the assumption of hyperbolic X-T trajectories. Strong reflected shear arrivals are seen in the mapping of line 195. Only source-receiver offsets up to 10 km were used in the τ-p mappings. The data on the left are from the Norwegian Sea, in the center from the Caribbean and on the right from the Western Pacific.

will improve as additional ray parameters are observed.

### 2. Best Single τ-p ellipse approximation

Depending on the structure, finite source-receiver offset data may or may not record postcritical reflection and refraction arrivals. If not, velocity analysis based on the monotonically decreasing function of τ and p, first described by Gerver and Markushevich (1967) and Johnson and Gilbert (1972), is not applicable. In this case, Dix's approximation can be used in a computer search to determine the RMS velocities in τ-p.

Dix's formulation of the classic two-term traveltime approximation, equation 1 above, can be derived from the exact traveltime equation cast in terms of the parameters τ and p. We perform a binomial expansion of equation 5:

$$\tau_n(p) \cong \sum_{j=1}^{n} t_j \left(1 - \frac{1}{2} p^2 v_j^{\,2} + \frac{1}{8} p^4 v_j^{\,4} - \ldots\ldots\right), \qquad (15)$$

and note that the $pv_j$ terms are the sines of the vertical ray angles. Normalizing by the total two-way traveltimes $T\emptyset_n$,

$$\tau_n(p) \cong T\emptyset_n \left(1 - \frac{p^2}{2} \sum_{j=1}^{n} \frac{t_j v_j^{\,2}}{T\emptyset_n} + \frac{p^4}{8} \sum_{j=1}^{n} \frac{t_j v_j^{\,4}}{T\emptyset_n} + \ldots\right) \qquad (16)$$

Dix's 1955 formula for the RMS velocity, equation 3 above, can be recognized as part of the second term in this series. By substituting for $V^2_{RMS_n}$:

$$\tau_n(p) \cong T\emptyset_n \left(1 - \frac{p^2}{2} V^2_{RMS_n} + \ldots\right), \qquad (17)$$

we can recognize the resulting two-term series as the binomial expansion of:

$$\tau_n(p) = T\emptyset_n \left(1 - V_{RMS_n}^{\,2} p^2\right)^{\frac{1}{2}} \qquad (18)$$

In this derivation of Dix's approximation, we now see that all terms on the order of $p^4v_j^4$ (sine of the incident angle to the fourth power) or greater have been dropped. Using the definitions of $X=-d\tau/dp$ and $T=\tau+pX$, this single $\tau$-p ellipse which approximates the true sum of individual layer $\tau$-p ellipses can be recognized as the two-term traveltime equation of Dix (1955).

Cutler and Love (1979) first suggested that equation 18 be used in a computer trial and error search to find the total two-way normal time to the base of the $n^{th}$ layer, $T\emptyset_n$, and the RMS velocity, $V_{RMS_n}$. Figure 14 is an example of the best single $\tau$-p ellipse method using semblance as a measure of coherency to detect the RMS velocities and two-way normal times present in the data for trial RMS slowness from .2 to .7 second/km. The points of maximum coherency in the elliptical velocity scan were interpreted in conjunction with the normal incidence data. The two-way normal times and RMS velocities were converted to thickness and interval velocities using Dix's, 1955 formula, see Figure 17.

### 3. $\tau$-sum Inversion

The $\tau$-sum method (Diebold and Stoffa, 1981) for deriving velocity depth data is an inversion procedure that makes use of only the critical and postcritical arrivals. It is equivalent to the standard slope intercept method of Slotnick, 1939, except that both postcritical reflections and refractions are used. The inversion formula is obtained by solving equation 5 for $t_n$:

$$t_n = \frac{\tau_n(p) - \tau_{n-1}(p)}{(1-p^2 v_n^2)^{\frac{1}{2}}} , \qquad (19)$$

the two-way normal time in the $n^{th}$ layer. In this equation, $\tau_n(p)$ is observed, and $\tau_{n-1}(p)$ is known from previous determinations using equation 5.

Diebold and Stoffa (1981) showed that this inversion method is applicable to both reflected and refracted arrivals. It can also be used, if p is densely sampled, when velocity gradients are present. The only requirements are that the experimental aperture be large enough to include the critical reflection/refraction distances and that low velocity zones not be present. In the example of Figure 15 sufficient data for this inversion method are obtained with only 20 km of offset. These data were transformed to $\tau$-p and band-pass filtered (6 - 20 Hz). A Hilbert transform was performed and the resulting analytic trace used to compute the envelope function of the data (Figure 15a).

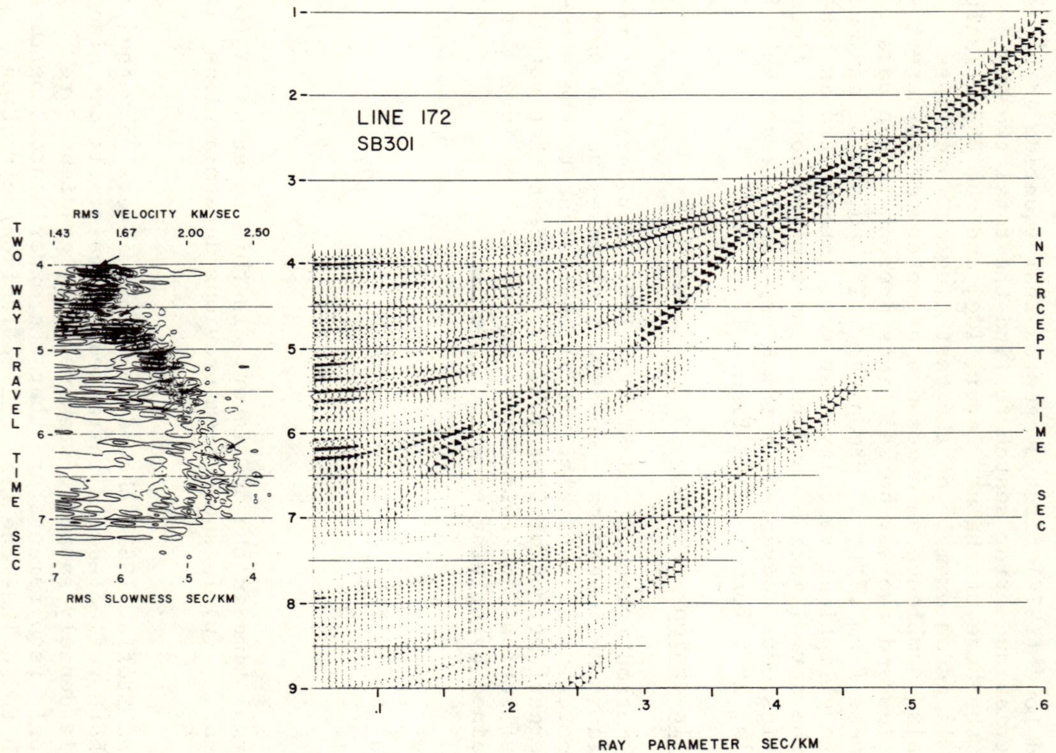


Figure 14. Ten-level contour plot of semblance for elliptical τ-p trajectories (left) for the τ-p data shown on the right. In this example, the τ-p data includes source-receiver offsets for only the first 4.8 km of data. The interpreted two-way normal times and RMS velocities are indicated. The weak events at 4.6, 5.5 and 6.6 seconds of time were identified by correlation with the near normal incidence τ-p data. These data were used to derive an interval velocity depth function using Dix's (1955) formula.

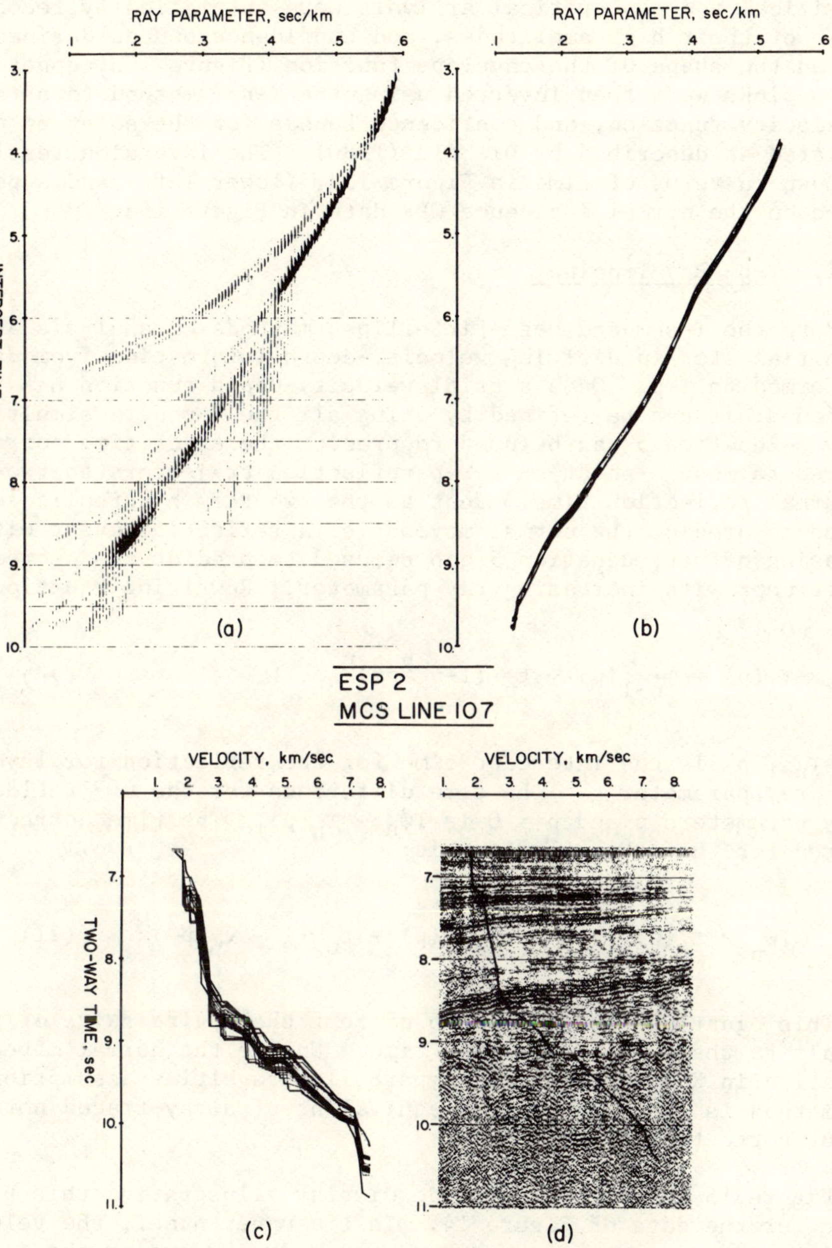


Figure 15. (a) The envelope function of the τ-p data for Caribbean ESP 2. (b) Automatically picked τ-p points used for V(Z) inversion with ± σ confidence bounds. (c) External inversion of the bounds in (b). (d) The most likely V(Z) function plotted on the corresponding normal incidence CDP

The critical and postcritical arrivals were automatically recognized because of their high amplitudes, and confidence bounds defined based on the shape of the envelope function (Figure 15b, upper right). The $\tau$-p picks were then inverted using the $\tau$-sum method to obtain the velocity function, and confidence bounds for the solution were calculated as described by Diebold (1980). The inversion results are shown in terms of time in Figure 15c, (lower left) and superimposed on the normal incidence CDP data in Figure 15d.

4. $\tau$-p "Ray Tracing"

Both the $\tau$-sum and best-fit ellipse methods of analysis are only the initial step in deriving velocity-depth information from data transformed to $\tau$-p. Once a trial velocity-depth function has been found, it can be refined by using all the $\tau$-p data simultaneously. Equation 5 can be used to predict the exact time correction required to move a point on a $\tau$-p reflection trajectory to two-way normal reflection time. Just as the two-term hyperbolic equation is used to predict the normal moveout of a reflection event with increasing offset, equation 5 can be used to predict the decrease in intercept with increasing ray parameter. Rewriting equation 5:

$$\tau_n(p) = \tau_{n-1}(p) + t_n\,(1-v_n^2p^2)^{\frac{1}{2}}\ , \tag{20}$$

where $\tau_{n-1}(p)$ is the intercept time for the reflection for layer n-1 at ray parameter p. The time difference for the $n^{th}$ reflector at ray parameters p and p = 0 is $T\emptyset_n - \tau_n(p)$. The time correction required for the $n^{th}$ reflection is:

$$T\emptyset_n - \tau_n(p) = T\emptyset_n - \tau_{n-1}(p) - t_n\,(1 - v_n^2p^2)^{\frac{1}{2}} \tag{21}$$

This equation can be used to correct the entire suite of $\tau$-p arrivals to their two-way normal time. Unlike the normal moveout correction in X-T based on the hyperbolic traveltime assumption, this method is exact and is the equivalent of a ray-traced normal moveout correction in X-T.

Figure 16 is a true amplitude display illustrating this procedure for the data of Figure 14. In the upper panel, the velocity function used to correct the $\tau$-p data was derived using the $\tau$-sum method of inversion. In the lower panel a refined version of the velocity function obtained from the interpretation of the best single ellipse data was used. The results are quite similar considering that different inversion methods were applied. Semblance

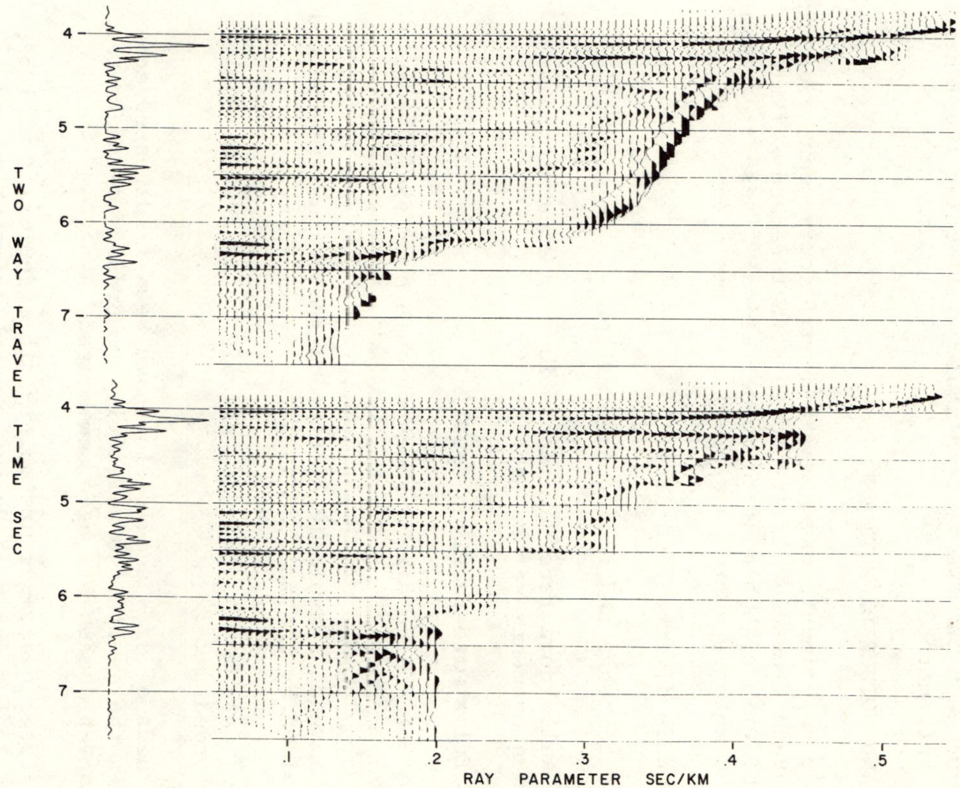


Figure 16. The τ-p data of Figure 14 corrected to two-way normal time by applying a τ-p normal moveout correction based on a velocity function derived from the τ-sum method (upper) and the best τ-p ellipse method (lower). The velocity functions used are displayed in Figure 17. Semblance for the corrected data is shown at the left of each profile, giving a measure of the accuracy of each velocity function.

was also computed for each of the corrected data sets and displayed to the left. In this application semblance is a measure of the success of each velocity model in correcting the τ-p data to normal incidence time. Semblance is different for the two velocity functions, but this is primarily due to the postcritical reflection and refraction arrivals which are included when using the velocity function found by the τ-sum method. Figure 17 is a comparison of the velocity functions used to time correct the τ-p data. The overall agreement between the two is reasonable, reflecting the inherent resolution of both methods.

## IV. SUMMARY AND FUTURE DIRECTIONS

Multi-ship experiments using multichannel arrays can be designed to derive accurate crustal velocity-depth structure and seismic record sections with high signal-to-noise ratios. These experiments have several advantages over those previously employed:

1. True amplitude digital recording.

2. Very accurate range and timing information.

3. Fixed and well-known receiver separation for accurate phase velocity measurements across each array.

4. No difficulties with receiver coupling.

5. Calibration of charge size changes through overlapping receiver positions.

6. A minimum amount of explosives are required because of the redundancy provided by the multichannel receiving array.

7. "Real-time" data monitoring and analysis to insure the quality of the recorded data.

Three methods of data acquisition have been illustrated and our progress in the analysis of digitally recorded Wide Aperture data using the τ-p domain has been summarized.

In the future, Wide Aperture multi-ship experiments will be designed not only for deriving velocity-depth information and a normal incidence time record section, but also to derive the subsurface structure in three dimensions. Analysis methods for the data acquired in these experiments will continue to improve and begin to address problems of lateral velocity variability and structural change. Converted shear arrivals will be routinely used to derive the shear velocity structure; and, in conjunction with compressional wave velocities be used to derive estimates of

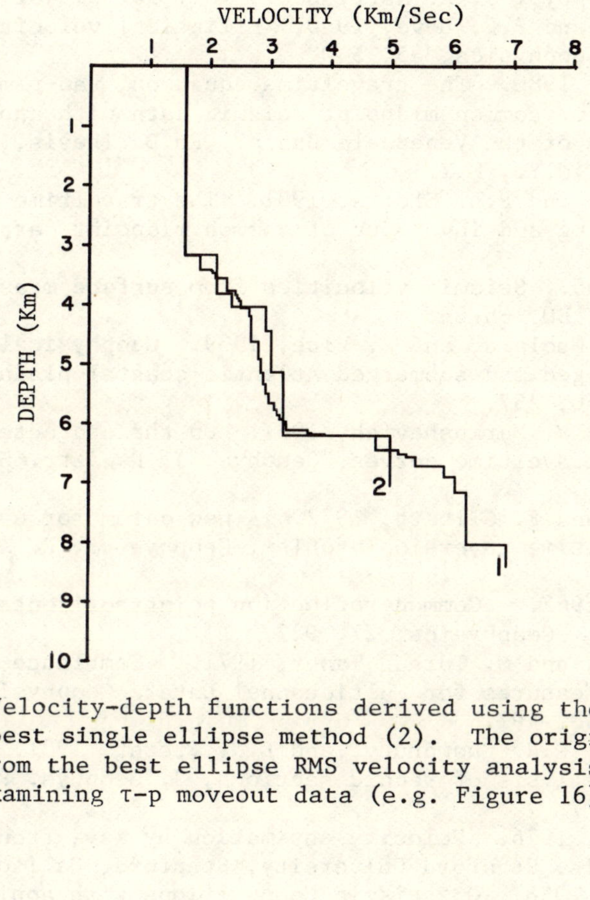


Figure 17. Velocity-depth functions derived using the τ-sum method, (1) and the best single ellipse method (2). The original interval velocities from the best ellipse RMS velocity analysis method were refined by examining τ-p moveout data (e.g. Figure 16).

Poisson's ratio. Inversion methods which exploit the recorded amplitude information will also be developed and used in conjunction with traveltime data to tightly constrain the derived results.

REFERENCES

Buhl, P., J.B. Diebold, and P.L. Stoffa, 1981. Array length magnification through the use of multiple sources and receiving arrays. Submitted to Geophysics.

Chapman, C.H., 1978. A new method for computing synthetic seismograms; Geophys. J. R. Astr. Soc., Vol. 54, p. 481-518.

Cutler, R.T., and P.L. Love, 1980. Elliptical velocity analysis (abstr.), Geophysics, 45, 540.

Diebold, J.B., 1980. The traveltime equation, tau-p mapping and inversion for common midpoint seismic data with applications to the geology of the Venezuela Basin. Ph.D. Thesis, Columbia University, N.Y., N.Y.

Diebold, J.B., and P.L. Stoffa, 1981. The traveltime equation, tau-p mapping and inversion of common midpoint data, Geophysics, 46, 238.

Dix, C.W., 1955. Seismic velocities from surface measurements. Geophysics, 30, 68-86.

Ewing, M., C. Woolard, and A. Vine, 1939. Geophysical investigations in the emerged and submerged Atlantic coastal plain, Bull. Geol. Soc. Am., 50, 257.

Gerver, M. and V. Markushevich, 1967. On the characteristic properties of traveltime curves, Geophys. J. R. Astr. Soc., 13, 241-246.

Johnson, L., and F. Gilbert, 1972. A new datum for use in the body wave traveltime inversion problem, Geophys. J. R. Astr. Soc., 30, 373-380.

Mayne, W.H., 1962. Common reflection point horizontal stacking techniques, Geophysics, 27, 927.

Neidell, N.S., and M. Turhan Taner, 1971. Semblance and Other Coherency Measures for Multichannel Data; Geophysics, Vol. 36, No. 3, p. 482-497.

Phinney, R.A., K.R. Chowdhury, and L.N. Frazer, 1981. Transformation and analysis of record sections, J. Geophys. Res., 86, B1., 359-377.

Schultz, P.S., 1976. Velocity estimation by wave front synthesis, Ph.D. Thesis, Stanford University, Stanford, California.

Slotnick, M., 1936. On seismic Computations with applications, I. Geophysics, 1, 22.

Stoffa, P.L., P. Buhl, J.B. Diebold, and F. Wenzel, 1981. Direct mapping of seismic data to the domain of intercept time and ray parameter: A plane wave decomposition, Geophysics, 46, 255.

Stoffa, P.L., and P. Buhl, 1979. Two-Ship Multichannel Seismic Experiments for Deep Crustal Studies: Expanded Spread and Constant Offset Profiles; Journ. of Geophys. Res., Vol. 84,

p. 7645-7660.
Taner, M.T., and F. Koehler, 1969. Velocity Spectra-Digital Derivation and Applications of Velocity Functions; Geophysics, Vol. 34, No. 6, p. 859-881.
Wenzel, F., P.L. Stoffa, and P. Buhl, 1981. Seismic Modeling in the Domain of Intercept Time and Ray Parameter, Submitted.

# OCEAN-BOTTOM SEISMOGRAPHS

R.B. Whitmarsh, R.C. Lilwall

Institute of Oceanographic Sciences
Wormley, U.K., GU8 5UB

## INTRODUCTION

This paper describes both examples of data obtained with ocean-bottom seismographs and the different sorts of instrument which record such data. The data include recordings of controlled sources (explosions and airguns) and of earthquakes. Instruments which cannot be used in typically oceanic depths are not discussed. Reviews of some early OBS have been written by Ewing and Ewing[1] and by Bradner[2]. In this paper the widely-used acronym OBS stands for all sorts of ocean-bottom seismic sensors whether or not they include seismometers (geophones) or record on the sea-bed. Like many research instruments, OBS are continually being improved and modified and the descriptions provided here, based on the most recent publications, may not correspond exactly to the present configuration of each instrument.

## OBS INSTRUMENTATION

### Historical Development of OBS

The earliest OBS experiments were carried out in 1937 and 1938[3,4]. A set of four bottom charges and four geophones, all connected by electrical cable to a recorder lowered from a ship, were laid out along the seabed and used to generate a refraction profile about 1 km long. Experiments with pop-up recording oscillographs and geophones and free-fall shots with timers followed[4]. These early experiments were then discontinued and it was not until 1951 that OBS work was renewed on a modest scale at the then Lamont Geological Observatory. However, until the late 1950s, little progress was made instrumentally[1]. At that time there was a new requirement for a program to detect and identify clandestine underground and underwater nuclear explosions by seismic techniques which led in 1959 to the setting-up in the U.S.A. of Project VELA-UNIFORM. It was hoped that the ocean floor would provide much quieter sites than on land thereby offering an improved detection threshold if suitable OBS were used.

VELA-UNIFORM provided the impetus for a second generation of OBS incorporating some of the most advanced technology then available, including the rugged Ranger seismometers developed for unmanned lunar landings[5-7,11]. Features of these early instruments included free-fall pop-up spheres, three-component seismometers, triggered digital recording and acoustically commanded ballast release. The principal objective of these OBS was to sample and measure background noise levels on the sea-bed. Most of the noise assessment work was carried out by groups at Lamont Geological Observatory[8], Scripps Institution of Oceanography[2,9-11] and at Texas Instruments Inc.[12,13] who reached the general conclusion that the sea-bed was not quieter than typical sites on land. This negative result halted much of the American effort except at Lamont Geological Observatory[14,15]. Promising results obtained in 1964 from an acoustic-telemetering OBS off Bermuda[16] led to the development of an offshore OBS linked by 100 km of cable to Point Arena near San Francisco. A mass of data was obtained from the second instrument deployed there in 1966 which continued to work for 6.3 years.

Elsewhere in the 1960s limited efforts were also being put into OBS for scientific studies of ocean-bottom noise[17-19], shallow structure of the oceanic crust and sediments[20-22] and natural seismicity[23,24]. Most of these OBS were relatively simple devices lowered on cables from the surface.

A new third generation of mostly free-fall pop-up instruments began to be built in the early 1970s starting in the U.K.[25,26]. Two types of instrument were developed separately for controlled source seismology and for investigating microearthquakes at spreading centres. During the past decade more and more research groups have built OBS to the extent that at least 17 OBS groups have been active within the last three years in the U.K., France, West Germany, Japan, Canada and the U.S.A. A summary of published OBS descriptions is given in Table 1, which also lists papers where further information can be obtained.

## The Pros and Cons of Ocean-Bottom Observations

Usually the only alternative to an OBS today is a sonobuoy. Sonobuoys, unlike OBS are free-floating, usually untethered and their

TABLE 1 - SUMMARY OF OBS SPECIFICATIONS

| FIRST YEAR AT SEA | HOUSING | MAXIMUM DEPTH (m) | WEIGHT AT LAUNCH (Kg) | DEPLOYMENT | SEISMIC SENSORS | METHOD | BANDWIDTH (Hz) | TAPE-SPEED ($cm\ s^{-1}$) | AUTONOMY (hrs) | RECORDING MODE | NOTES | REFERENCES |
|---|---|---|---|---|---|---|---|---|---|---|---|---|
| | | | | | | ← SEA-BED RECORDING SYSTEM | | | | → | | |
| U.S.A. | | | | | | | | | | | | |
| Lamont-Doherty Geological Observatory, New York | | | | | | | | | | | | |
| 1937 | A | | | L | Z | OSC | | | | | | 3 |
| 1939 | A | | | F | Z | OSC | | | | | | 4 |
| 1959 | A | | 300 | L | Z(2Hz) | NONE | | | | | a | 1 |
| 1964 | 2B | | 500 | L | X,Y,Z(15s) | NONE | | | | | a | 14 |
| | | | | | Z(1Hz) | | | | | | | |
| 1965 | 3B | | 750 | L | X,Y,Z(15s) | NONE | | | | | b | 14 |
| | | | | | X,Y,Z(1Hz) | | | | | | | |
| | | | | | SPH,LPH | | | | | | | |
| 1972 | B,L | | 200 | F | X,Z(2Hz) | FM | 2-120 | 2.38 | 12.8 | C | c | 27 |
| | | | | | H | | | | | | | |
| 1974 | B,H | 6000 | 200 | F | X,Z(2Hz),H | DR | 2-80 | 0.064 | 240 | C | c | 28 |
| | | | | | orX,Y,Z(2Hz) | | | | | | | |
| 1976 | A | 5000 | 50 | F | H | DR | 3500 | 2.38 | 3.5 | C,P | c,d | 29 |
| | | | | | | FM | -500 | 2.38 | 3.5 | C,P | | |
| 1977 | A,2H | | | F | X,Z(2Hz),H | DR | 1-30 | 0.0212 | 720 | C | | 30 |
| | | | | | LPZ,LPH | DIG | 0.01-1 | | | | | |
| | | | | | H | DR | 50-150 | | | | | |
| Texas Instruments Inc., Dallas | | | | | | | | | | | | |
| 1961 | K,A | 3000 | | L | X,Y,Z(1Hz) | DIG | 1-10 | | 32.2M | P,E | | 5,7,12,13,31 |
| | | (6100) | | | H | FM | | | 11 | P | | |
| 1964 | C | 7600 | 564 | F | X,Y,Z(1Hz) | DR | 0.5-10 | 0.0191 | 792 | C | | 6,7 |
| Institute of Geophysics and Planetary Physics, La Jolla | | | | | | | | | | | | |
| 1962 | K | 6000 | | | X,Y,Z(3.3-0.5Hz) | FM | 0.1-10 | 4.76 | 1.3 | C,P | | 2,9 |
| Hawaii Institute of Geophysics, Honolulu | | | | | | | | | | | | |
| 1972 | A | | | L | X,Y,Z(1Hz) | NONE | | | | | b | 32,33 |
| | | | | | H | | | | | | | |
| 1973 | A | 10,000 | 320 | L | X,Y,Z(2Hz) | DR | 2-20 | 0.012 | 384 | C | e | 32 |
| | | | | | H | | | | | | | |
| 1975 | A,5E | 10,000 | 150 | F | X,Z(4.5Hz) | DR | 2-20 | 0.012 | 384 | C | | 32 |
| | | | | | H | | | | | | | |
| Scripps Institution of Oceanography, La Jolla | | | | | | | | | | | | |
| 1973 | B | 6000 | 230 | F | Z(1Hz) | DIG | 3-32 | 1370B | 10M | E | | 34 |
| | B | 6000 | 230 | F | Z(1Hz) | DIG | 3-32 | 1370B | 20M | P,E | | 35 |
| 1980 | B | 6000 | | F | X,Y,Z(1Hz),H | DIG | 0.1-30 | 1843B | 159M | P,E | | 36 |
| Oregon State University, Corvallis | | | | | | | | | | | | |
| 1973 | A,L | 6400 | 114 | F | Z(2Hz),H | DR | 2-32 | 0.0254 | 300 | C | | 37,38 |
| | | | | | | | | | (600) | | | |
| University of Washington, Seattle | | | | | | | | | | | | |
| 1973 | I | 5000 | 590 | F) | X,Y,Z(4.5Hz) | DR | 2-100 | 0.1 | 515 | C | c | 39,40 |
| | B | 6000 | 530 | F) | | | | | (300) | | | |
| Woods Hole Oceanographic Institution, Woods Hole | | | | | | | | | | | | |
| 1976 | 2A,4H | 6100 | 325 | F | H | DR | 3-100 | 0.064 | 240 | C | d | 41 |
| Massachusetts Institute of Technology, Cambridge | | | | | | | | | | | | |
| | A,3H | 3400 | 295 | F | X,Y,Z(4.5Hz) | DIG | 2-30 | 1680B | 75.6M | E | | 42,43 |
| Texas University, Galveston | | | | | | | | | | | | |
| 1976 | H | 6400 | 61 | F | Z(4.5Hz) | FM | 4.5-20 | 2.38 | 32 | E | | 44,45 |
| | | | | | Z(8Hz) | FM | 8-40 | | | | | |
| 1978 | H | 6400 | 90 | F | X,Y,Z | DIG | 0.1-10 | | 1.5 | E | f | 46 |
| University of California, Santa Barbara | | | | | | | | | | | | |
| 1978 | A,4F | 6000 | | F | X,Y,Z(4.5Hz) | DIG | 4-32 | 800B | 17.3M | E | | 47 |

TABLE 1 (Continued)

| FIRST YEAR AT SEA | HOUSING | MAXIMUM DEPTH (m) | WEIGHT AT LAUNCH (Kg) | DEPLOYMENT | SEISMIC SENSORS | SEA-BED RECORDING SYSTEM: METHOD | BANDWIDTH (Hz) | TAPE-SPEED (cm $s^{-1}$) | AUTONOMY (hrs) | RECORDING MODE | NOTES | REFERENCES |
|---|---|---|---|---|---|---|---|---|---|---|---|---|
| U.S.S.R. | | | | | | | | | | | | |
| Moscow State University, Moscow | | | | | | | | | | | | |
| 1962 | A | 2200 | 80 | L | Z(3Hz) | DR | 2-15 | 0.12 | 48 | C | | 48 |
| UNITED KINGDOM | | | | | | | | | | | | |
| University of Cambridge, Cambridge | | | | | | | | | | | | |
| 1963 | A | 9100 | | L | Z(4.5Hz),H | OSC | 1-400 | | 0.3 | C | | 20 |
| 1966 | G,E | | 50 | F | Z(4.5Hz),H | DR | 4-500 | 0.635 | 6.6 | C | | 22 |
| Institute of Oceanographic Sciences, Wormley | | | | | | | | | | | | |
| 1968 | D | 5000 | 268 | F | H | FM<br>DR | 2-100<br>60-2000 | 2.38 | 12.8 | C,P | d | 25,49 |
| 1977 | D | 6000 | 284 | F | X,Y,Z(4.5Hz)<br>H | FM/DR | | 1.19<br>0.149 | 25.6<br>200 | C,P<br>C,P | | 50 |
| 1978 | D,L | 6000 | 424 | F | X,Y,Z(2Hz),H | FM | 2-40 | 0.095 | 850 | C | | 51 |
| Institute of Geological Sciences, Blacknest | | | | | | | | | | | | |
| 1970 | 2I | 6000 | 570 | F | X,Y,Z(2Hz)<br>H | FM | 2-40<br>0.2-10 | 0.19 | 213 | C | | 26,52,53 |
| JAPAN | | | | | | | | | | | | |
| Earthquake Research Institute, Tokyo | | | | | | | | | | | | |
| 1963 | A | 8000 | | L | Z(4.5Hz),H | DR<br>FM | 1.5-10<br>2-25 | 0.0145<br>0.145 | 1000<br>100 | C<br>C | | 54-60 |
| | | | | Z | (1Hz)X(4.5Hz) | DR | | 0.0145 | 1000 | C | | |
| Hokkaido University, Sapporo and Tokyo University, Tokyo | | | | | | | | | | | | |
| | A | | | L | Z(2.5Hz) | DR | 0.2-25 | 0.015 | 336 | C | | 61,62 |
| Japanese Meteorological Agency | | | | | | | | | | | | |
| 1978 | A | 4000 | | L | X,Y,Z(3.5Hz) | NONE | | | | | b | 63 |
| FRANCE | | | | | | | | | | | | |
| Centre Oceanologique de Bretagne, Brest | | | | | | | | | | | | |
| 1975 | J | 6000 | 430 | F | X,Y(4.5Hz),<br>Z or H | PWM<br>DIG | 5-25 | 0.595 | 8<br>130 | C,P<br>C | c | 64 |
| Institut de Physique du Globe, Paris | | | | | | | | | | | | |
| 1976 | D | 6000 | | F | Z(2Hz) | FM | | 4.76 | 1.5 | P,E | | 65 |
| CANADA | | | | | | | | | | | | |
| Atlantic Geoscience Centre, Dartmouth | | | | | | | | | | | | |
| 1976 | A,L | 6000 | 175 | F | X,Z(4.5Hz)<br>H | DR | 4-40<br>300-5000 | 0.05-<br>0.02 | 96-<br>240 | C | | 66,67 |
| WEST GERMANY | | | | | | | | | | | | |
| Institut für Geophysik, Hamburg | | | | | | | | | | | | |
| 1978 | A | 4000 | 550 | L | X,Y,Z(1Hz),H | FM | 0.1-1 | | 100 | | e | 68 |

This table is based on published descriptions of OBS. Column (2) A=special tube; B=61 cm O.D. spun aluminium sphere; C=102 cm O.D. spun aluminium sphere; D=71 cm O.D. forged aluminium sphere; glass spheres E=25 cm O.D.; F=33 cm O.D.; G=41 cm O.D.; H=43 cm O.D.; I=71 cm O.D. epoxy resin sphere; J=44 cm O.D steel sphere; K=56 cm O.D. sphere; L=syntactic foam. 3B means 3 spheres used. Column (5), L=lowered to sea-bed; F=free-fall. Column (6) uses X,Y,Z to denote 3 orthogonal geophones and X,Z means 1 horizontal and 1 vertical (natural frequencies in brackets), H denotes hydrophone. SP=short-period, LP=long-period. Column (7) OSC=oscillograph, DR=direct recording, FM=frequency modulated recording; DIG=digital recording, for DIG B=bits/inch, M= megabits. Column (11), C=continuous, P=programmed, E=event recording. Column (12) a=data telemetred acoustically; b=permanent cable to shore; c=uses amplitude compression; d=instrument is tethered above bottom; e=data telemetred by cable to surface; f=strong-motion instrument.

hydrophones are susceptible to noise from waves, near-surface current shear and water flow (tethered sonobuoys can and have been used[70-72] but tend to suffer from even greater flow noise). The changing structure and sea-depth under a drifting sonobuoy often confuse the results of a refraction profile beyond measure and the buoys' drift often precludes profiles over such narrow features as median valleys, fracture zones, linear ridges and seamounts. OBS offer the positive advantages of fixed receivers situated away from the high energy sea-surface environment and, if they include a three-component seismometer, there is the opportunity to positively identify shear-wave arrivals and even to calculate particle motions. An OBS offers more favourable circumstances for observation of upper crustal body waves as first arrivals and allows the determination of sediment elastic properties by using suitable bottom sources. Although free-floating sonobuoys have been used for microseismicity studies[73] they are not practical for prolonged measurements. OBS can observe pure oceanic surface-waves, teleseismic delay times, solid earth tides as well as microseismicity and intraplate events. OBS can be deployed in fixed arrays almost anywhere on the sea-bed. Island stations typically suffer from higher noise-levels (due to surf) and anyway are atypical sites.

The main difficulty in operating OBS is the very hostile environment in which they must work (as low as 2 degrees C and 200 to 600 bars hydrostatic pressure). If seismometers are used problems may arise from the unpredictable and usually unknown coupling to the sea-bed since their emplacement is effectively at random, the instrument being dropped from between 3 and 6 km above the sea-floor. The coupling problem is currently being evaluated by several OBS groups because many OBS seismometers suffer from ringing, after the first arrival, caused by OBS/sea-bed dynamic interactions[14,28,39,44,46,74].

## Aspects of the Design and Operation of OBS

There are only four fundamental classes of OBS:

i) where the recorded data, and usually the sensor package, is brought by line or by its own buoyancy back to the surface,

ii) where the sensor telemeters data acoustically, either continuously or on command, to another device in the sea (buoy, ship, etc.) for onward transmission ashore,

iii) where the sensor telemeters data electrically by cable, otherwise as in ii),

iv) where the sensor is permanently linked by cable to the land.

Most OBS today fall into class i) and will be discussed later. Class ii) OBS were used by Ewing and Ewing[1] and by Sutton et al.[14]. Although many OBS use acoustic telemetry for checking instrument position, and in some cases even its status[14,28,30,35,47], it is not suitable for the high data rates needed for most seismic experiments. Class iii) OBS are expensive and pose problems in deployment and upkeep, but have been used in conjunction with radio telemetry from a surface buoy with back-up internal recording[32]. Class iii) OBS include downhole seismometers as used within the sediments and upper oceanic crust[75,76]. The main advantage of a Class ii) or iii) OBS is the continuous monitoring of data which it can provide during an experiment. A revolutionary Marine Seismic System of this type with a downhole package telemetering via a surface buoy to an orbiting satellite has been proposed and is currently undergoing trials at sea[77]. Class iv) OBS are also expensive but are worthwhile in some circumstances and the cables are perhaps less vulnerable to damage than in Class iii). Offshore OBS have been installed off California[14,15], Hawaii[32,33] and Japan[63]. Bolt[78] has suggested that use could be made of discarded underwater telephone cables for such OBS.

Nearly all OBS in use today are of the Class i) type used in the free-fall/pop-up mode. Some simpler types which are lowered and pulled up on ropes, mostly preceded the present generation of OBS[3,19,20,23,24,60,62,68] or were used only at the prototype stage. A unique use of this type of OBS is the recording of data, from seismometers emplaced down bore-holes, by a recording package which is subsequently retrieved by a pop-up rope[79,80].

The great diversity of OBS designs has resulted from the different skills, ships and experience available to each group as well as from financial considerations and, in some cases, narrowly defined scientific objectives. There have been, however, some common trends and convergence of ideas. Fig. 1 illustrates two representative designs.

Spherical OBS pressure cases are usually of spun or forged aluminium alloy but have also been made from steel[64], glass[44] or epoxy resin[39]. A sphere, while offering the maximum buoyancy/weight ratio is nevertheless a back-breaking entity to handle in a laboratory. Some investigators now prefer cylinders which can be simply detached on deck from the remaining buoyant part of the instrument and brought into the laboratory. Cylinders are also practical with the advent of miniaturised circuity and the use of smaller cassette or cartridge tape-recorders. In the mechanical design careful consideration of the distribution of buoyancy, volume and mass is critical for not only must an OBS present a low-profile to near-bottom currents, liable to create turbulence and flow-noise[39], but also the OBS must be hydrodynamically stable in descent and ascent, be stable at rest on slopes up to a few tens of degrees (and under no circumstances turn over!) and have a reasonable righting moment at the sea-surface.

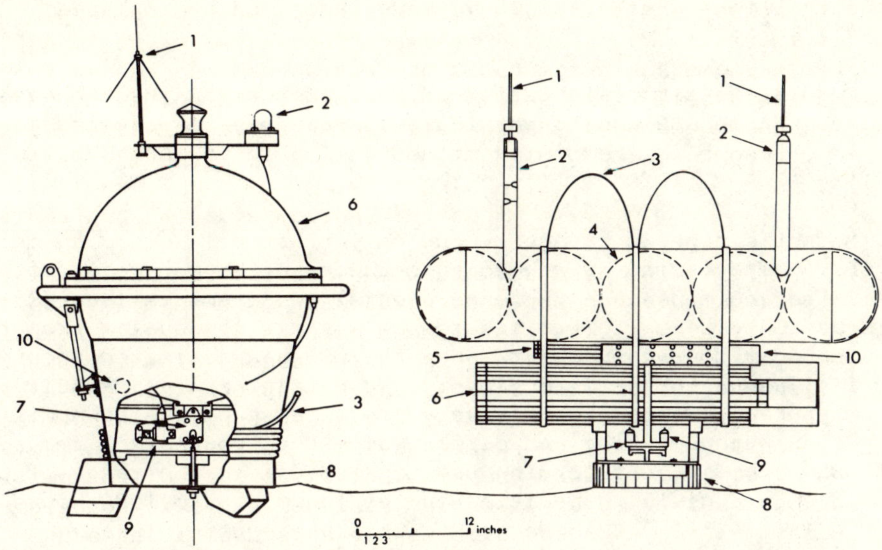


Fig. 1. A comparison of two typical OBS based on a sphere[50] and on a cylindrical pressure housing[32], at the same scale. Note the smoother outline of the spherical OBS but the lower profile of the cylindrical OBS. 1=radio aerial; 2=flashing light; 3=lifting strop; 4=glass spheres; 5=timer; 6=instrument pressure case; 7=release mechanism; 8=ballast weight; 9=release actuator; 10=hydrophone.

The choice of seismic sensors normally lies between short period seismometers (geophones), with natural frequencies in the range 1 to 8 Hz (most often 1, 2 or 4.5 Hz), and hydrophones. Longer-period seismometers have rarely been used in free-fall OBS[2,30] but were incorporated in some more permanent ocean-floor observatories[14]. There are good reasons for designing OBS to be insensitive to signals for several octaves below 2 Hz because of microseismic noise. A variety of levelling arrangements involving free gimbals[19,32,36], "sticky" gimbals[28,38], oil-damped gimbals[39,43,50], self-levelling servoed devices[2,14,30,34], clamped gimbals[52] and no levelling at all[44,64] have been used. The azimuth of horizontal geophones on the sea-bed can be determined roughly by a suitable distribution of shots but some OBS incorporate one of a variety of direction indicators[2,43,46,53].

Hydrophones are invaluable, particularly for controlled source work, because the clear identification of water-waves is usually the key to accurate shot-to-OBS range calculations. Hydrophones have the advantage that their coupling characteristics to the pressure field are site-independent. Some OBS, designed primarily for controlled source seismology, have relied on hydrophones alone[27,41,49].

Regular in-situ calibration of seismometers and hydrophones feature in some OBS and is particularly valuable for amplitude studies or where the recording method is liable to lack amplitude fidelity[2,6,14,28,30,32,35,36,41].

Magnetic tape is presently the only practical recording medium for OBS. The choice of recording method and the mode of recording are linked and depend on the experimental requirements (see (39) for a discussion). For controlled source work it is usual to run the tape-recorder continuously or, when the OBS has a limited recording capacity, at pre-programmed times. To obtain greater capacity there has been a trend towards very low tape-speeds using modified or special purpose decks and direct (amplitude-modulated) recording. Most OBS with direct recording use tape-speeds in the range of 0.015 to 0.15 cm $s^{-1}$ giving capacities in the range 200 to 1000 hours[28,30,32,39,40,41,50,56,66]. However, to obtain adequate signals on replay, tape-speeds of up to several hundred times the record speed must be used. Frequency modulated (FM) recording offers much better low-frequency fidelity but requires higher tape-speeds[27,29,44,49,52,56,65,68]. Attempts to overcome the limited dynamic range of DR and FM recording have used non-linear amplitude compression by applying a square root[39], logarithmic[29] or simple threshold dependent[64] law. Recording on two channels with different gains is also used.

A significant and increasing proportion of OBS now use digital recording techniques, often in association with microprocessor control of the OBS. While offering much greater fidelity and dynamic range than other methods this technique is less efficient in tape usage and therefore is almost always used in a triggered or event-recording mode[30,34,35,36,43,47,64]. Typical sampling rates are 64 or 128 samples/second with 12-bit words. The triggering criterion is simply the ratio of the outputs from a short period and a long period signal integrator. Such schemes involve 12 to 18 second "memories" (shift registers) which capture at least part of the data before the tape-recorder motor switches on. Tape-recorder vibration can be a serious source of noise in some OBS. By their design event-recording OBS produce records of high quality but are susceptible to triggering by non-seismic, probably biological, events and do not reliably trigger on later arrivals.

Whatever the recording scheme used many OBS users now digitise their records for subsequent computer processing and display. This technique is particularly valuable in controlled source seismology for producing record sections.

The means of relocating an OBS on the sea-bed, releasing its ballast and finding it at the surface are normally designed to be effective in all weathers, day or night. Sea-bed relocation, after crudely positioning the ship by dead-reckoning from a fix, is achieved by means of an acoustic pinger or transponder. The ballast can be released by one or more pre-set timers but the most convenient means is by acoustic command. Many release systems have built-in redundancy of power supplies and release actuators. Actuators include explosive bolts[32,34,35,36,47,52,64], small detonators, squibs or pyrotechnic devices[28,50,65], electro-mechanical latches or an electromagnet[39,43,64], parting a wire by electrolysis or heat[29,44,46,66] and a cable cutter[38]. Some systems will also release the ballast on detecting a leak or a substantial drop in supply volts.

Some OBS are tracked acoustically during their ascent. At the surface they are found by radio-direction-finding and by flashing light, frequently using proprietary equipment, or by observation.

## THE USE OF OBS FOR CONTROLLED SOURCE SEISMOLOGY

This section gives examples of how OBS may be exploited in controlled-source experiments.

### Long-Range Refraction Profiles

In order to study the seismic properties of the upper mantle, and particularly the whole oceanic lithosphere, is is necessary to observe either pure oceanic surface-waves or body waves over distances of several hundreds, or even a few thousands of kilometres. Surface wave studies have only about one tenth of the resolution of body wave experiments[81] and in addition it is often difficult to observe surface waves which have travelled over pure oceanic paths. OBS, however, are well suited for long-range seismic refraction experiments in which a relatively small number of large shots are fired over a period of a few days[82-88].

Asada, Shimamura and their colleagues carried out a series of long-range refraction profiles in the northwest Pacific in 1973 and 1974[83]. They recognised the inherent difficulty of using large shots, with bubble-pulse frequencies of 2 to 3 Hz, in that part of the ocean-bottom noise spectrum below 6 Hz in which noise increases quite rapidly as frequency decreases. An initial experiment attempted to record 1.5 ton shots out to 1400 km but useful data beyond 600 km was obtained only subsequently when 5 and 7 ton shots were fired. In spite of these difficulties they were able to construct a record section out to 1250 km (Fig. 2) which demonstrated that a 50 km thick $V_p$=8.2 km $s^{-1}$, $V_s$=4.9 km $s^{-1}$ layer overlay a 20 km thick 8.6 km $s^{-1}$ layer at the base of the lithosphere, which in turn was underlain by a low-velocity layer (8.4 km $s^{-1}$) some 30 km thick[89]. The seismic model is consistent not only with

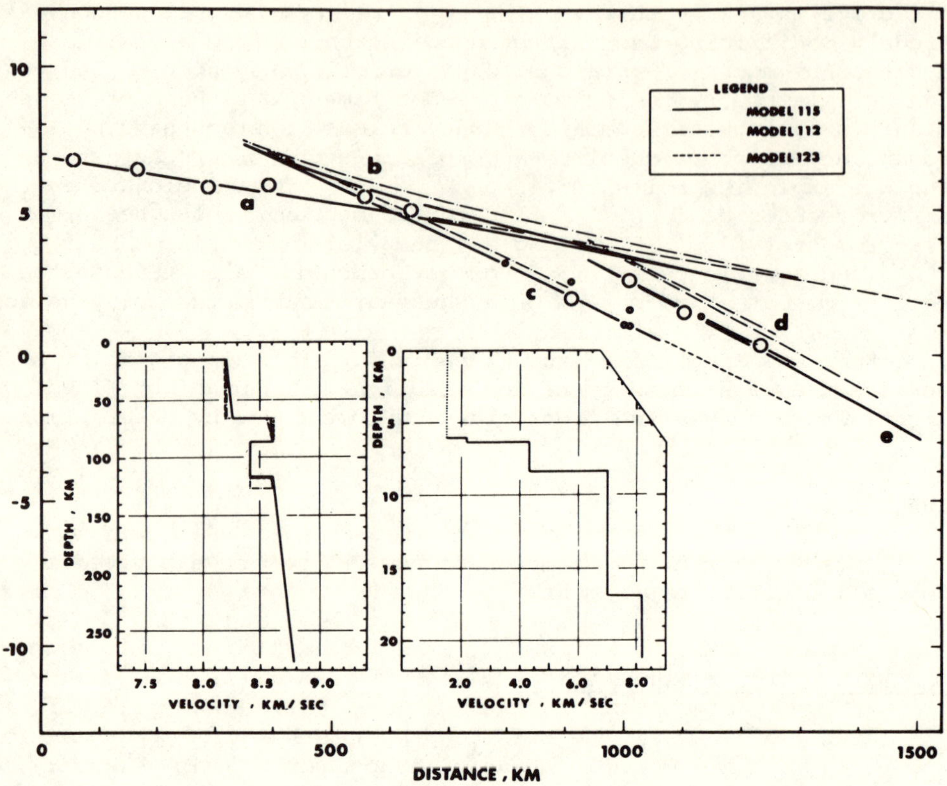


Fig. 2. Hodochrons (lines) fitted to arrival times (circles) observed along a long-range refraction line in the NW Pacific[83] using a range of models shown in the inset.

data from explosions but also with apparent velocities of earthquake arrivals across the OBS array[139]. The combination of data from earthquakes and explosions in long-range profiles can be advantageous, particularly if the earthquakes are well-located, and has been exploited elsewhere[85,86].

More recently long-range profiles have been conducted with greater numbers of OBS and large shots. A further refinement has been the use of dispersed charges[90] whereby the shot is divided into a small number of separate, but electrically interconnected equal packages, detonated together. In this scheme the bubble-pulse frequency is higher, depending only on the size of the unit package, thereby making more efficient use of the explosive. Using this technique in 1976, Orcutt and Dorman[81] shot a 600-km long line with shots of up to 2 tons along the 70 m.y. isochron in the east Pacific.

Using the tau-p inversion technique (which inverts smoothed first arrival travel time/distance data to yield bounds on possible velocity/depth models) they established that beyond 100 km all velocities lay between 8.2 and 8.4 km $s^{-1}$. Considerable velocity "structure" was proposed in the top 60 km of the lithosphere on the basis of the travel-time inversion. Orcutt and Dorman[84] refined the earlier model to demonstrate that the upper lithosphere is characterised by a general tendency for velocity to decrease with depth, except at about 27 and 52 km depth where there are rapid velocity increases perhaps associated with phase or compositional changes.

Long-range refraction profiles are often logistically complex and require the use of OBS which are reliable and have similar or well-known recording characteristics. The scale of operation normally requires inter-laboratory, or even international, collaboration. There is no doubt, however, that such experiments can increase our knowledge of the finer seismic structure of the oceanic lithosphere to an extent not available by any other technique.

## Studies of Spreading-Centres

OBS are also effective for conducting refraction profiles in areas of high bottom relief or where the region of interest is narrow and elongated (such as at spreading centres or fracture zones). OBS have been used in this way to study fracture zones[91-93] and regions close to active spreading-centres[82,88,94-101]).

The Mid-Atlantic Ridge (MAR) has the rough bottom relief characteristic of slow-spreading ridges. Steinmetz et al.[82,95] conducted an experiment close to the MAR axis north of the Azores to study the relatively rapid lateral changes of upper mantle structure which occur in such a region. Although some of the Azores islands were conveniently situated as sites for recording stations the work on the more typical MAR north of the Azores Platform required the use of OBS. Four refraction lines up to 460 km long were shot either parallel or normal to the MAR axis. The quite variable velocity structures, confirmed by synthetic seismogram modelling of the observed record sections, nevertheless agreed in a qualitative manner with velocities deduced from a petrological model of lithospheric development[102]. A profile shot along the 4 m.y. isochron detected a high-velocity layer 3 of 7.6 km $s^{-1}$ overlying an 8.3 km $s^{-1}$ refractor 9 km below the sea-bed. Along the 9 m.y. isochron a velocity of 8.3 km $s^{-1}$ exists only much deeper at about 30 km (Fig. 3). The two 8.3 km $s^{-1}$ refractors represent two different residual peridotite layers, the upper one having undergone hydration (probably including serpentinisation) before it was 9 m.y. old thereby causing the disappearance of the

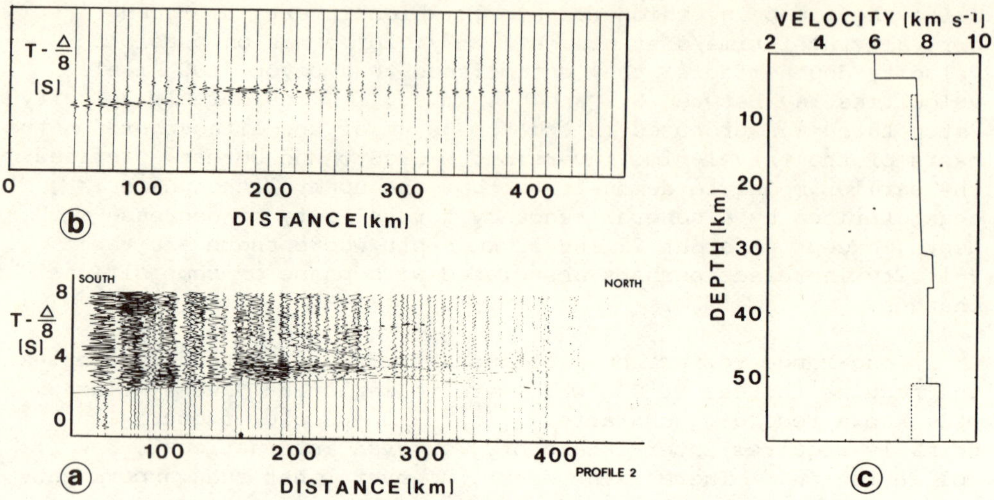


Fig. 3. Observed (a) and synthetic (b) record sections for a refraction line along a 9 m.y. isochron north of the Azores[82]. Velocity model [dotted lines in (c)] was used to calculate the synthetic section.

upper 8.3 km $s^{-1}$ refractor and revealing the deeper 8.3 km $s^{-1}$ refractor to the seismic refraction method. Two of the profiles were shot normal to the ridge axis. These profiles established that although energy was transmitted across the spreading plate-boundary within the crust transmission within the upper mantle was practically undetectable. This was explained by the existence of an extensive negative velocity gradient in the top 20 km of the axial part of the model as predicted by the petrological model.

Different results were obtained by Orcutt et al.[94] from a set of three refraction profiles shot parallel to the faster-spreading East Pacific Rise along its crest and along the 2.9 and 5 m.y. isochrons. The profile along the ridge crest, which consists of a horst only 10 km wide, was interpreted in terms of a crustal 4.9 km $s^{-1}$ low-velocity zone 1.4 km thick, beginning only 2 km below the sea-bed, and underlain by 7.7 km $s^{-1}$ low velocity mantle (Fig. 4). The low-velocity zone was absent from the other two profiles along the 2.9 and 5 m.y. isochrons. Orcutt et al. proposed that the low-velocity zone represented a zone of partial melt or a magma chamber, as required by some petrological models of the formation of oceanic lithosphere. Other systematic changes in seismic structure were also detected by the three profiles, namely the development of a low-velocity upper crust for profiles off the central horst, and the appearance of a "normal" upper mantle velocity (8.1 km $s^{-1}$) only on the oldest 5 m.y. profile.

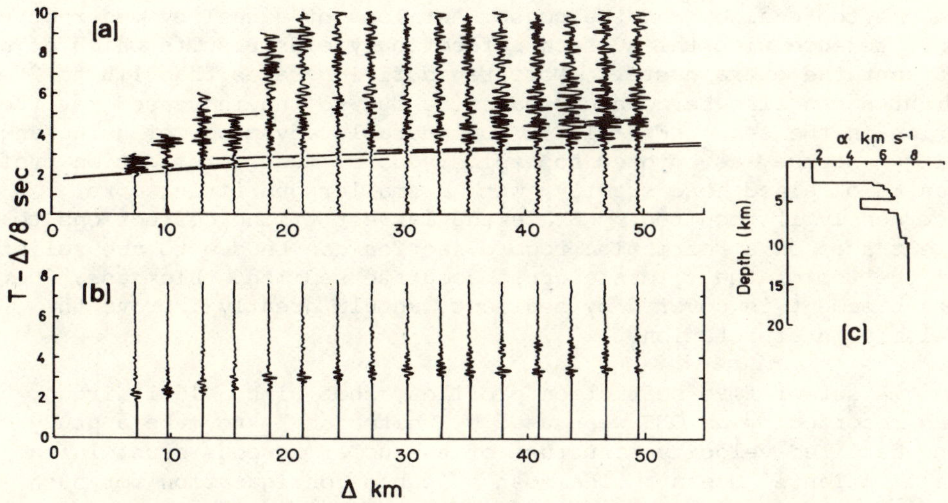


Fig. 4. Observed (a) and synthetic (b) record sections for a refraction line along the E. Pacific Rise crest[94]. The velocity model (c) was used to generate the synthetic section.

The above two examples illustrate how the use of OBS has led to significant discoveries concerning the generation of oceanic lithosphere.

## OBS and Detailed Crustal Studies

One aim of refraction seismologists is to obtain record sections with more traces per km and a better signal to noise ratio so as to improve the resolution of their models either in terms of lateral heterogeneity or, assuming homogeneity, in terms of more exact velocity gradients, layer thicknesses etc. or some combination of the two. This is easily done in two-dimensions (line profiles) and, with more effort, can be achieved in three-dimensions (non-linear arrays), by using OBS and repetitive sound sources. Airguns have been used most commonly with chamber sizes up to 140 litres[64] or with arrays of guns[103] and several examples of airgun record sections obtained with OBS can be found in the literature[44,45,64,66,67,97,98,104,105,106]. Useful signals to a range of about 20 kms are regularly obtained with a 16 l airgun in oceanic depths. Examples of other repetitive sources can be found in Ibrahim et al.[105] (Flexichoc) and Whitmarsh[107] (I.C.I. Aquaflex). An effective arrangement has been to tow one or more 16 l airguns at about 4 knots and to fire them every 1.5 to 2 minutes at 1800 psi. At 4 knots the gun(s) can be maintained at 15m depth thereby generating a 9 Hz signal. Higher speeds have been used[105] but the guns tend to become too shallow and the consequent loss of signal by destructive interference with the surface reflection yields results which do not warrant the extra haste. More rapid firing rates than 1.5 to 2 minutes can also be counterproductive due to the increased reverberation in the sea. Finally, a considerable advantage of using an airgun source, as opposed to explosives, is that a reflection profile can be obtained concurrently, with a smaller gun firing more frequently if required. Since the largest timing corrections to the traces on a refraction record section can be due to the relief of the top of the volcanic crust accurate sediment thickness, where the basement is covered by sediment, should greatly improve the seismic interpretation.

A set of five refraction profiles, shot with a 16 l airgun, and recorded by an OBS were used by Whitmarsh[104] to make a study of the detailed velocity structure of the upper igneous crust in the North Atlantic Ocean. The most suitable configuration for such experiments is to use bottom receivers and bottom shots, nevertheless OBS recordings of near surface shots involves a substantially better geometry than surface recordings of near surface shots in terms of the interference of water-borne energy and the range over which upper crustal body waves can be observed as first arrivals. The five refraction profiles were unreversed and mostly lacked information about sediment thickness variations along them. Travel times were corrected for the varying water-depth only. By modelling the travel times using a) models with only first order and b) models with only second order discontinuities in the velocity/depth profile it was found that a more consistent set of models was obtained from the second set (Fig. 5). This conclusion was supported by using synthetic seismograms as well as being geologically more plausible. Thus the preferred models obtained over young crust (2 and 3.5 m.y.) possess a thin low-velocity upper layer with a strong positive velocity gradient. All models show a gradient of 0.85-1.35 km $s^{-1}$ $km^{-1}$ down to about 2 km below the sea-bed where velocities typical of unfractured basalt (5.75-6.4 km $s^{-1}$) are encountered. The latter gradient has been interpreted in terms of a downward decrease of voids and fractures in the basaltic layer and crude estimates were made from the velocity profiles of the effective porosity at different depths. Such a model for Layer 2 is now widely accepted and remains consistent with other seismic and geological evidence.

There are very few reports in the literature of OBS experiments using bottom sources but such techniques offer the possibility of generating shear waves and even surface-waves and most closely resemble profiles shot on land. The sources are often weak within the bandwidth of an OBS and therefore there may be a navigational problem in siting the sources close enough to the OBS. Such experiments have demonstrated the very low but rapidly increasing

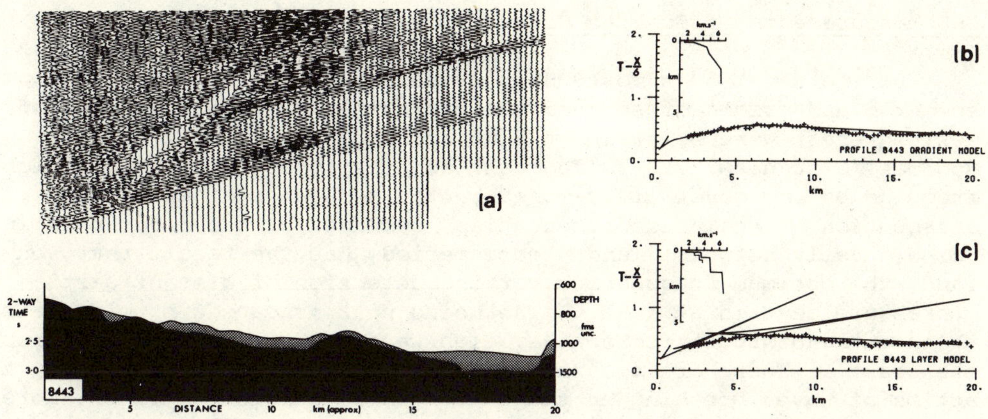


Fig. 5. Record section (a) obtained with a 16 l airgun in the NE Atlantic[104]; (b) and (c) illustrate the different hodochrons obtained with stepped and sloping velocity/depth models.

shear-wave velocity near the seabed in sediments and fresh volcanic rocks[1,21,74,108,109].

## Discussion

OBS are particularly appropriate for carrying out the sorts of studies discussed above and in many cases these experiments would have been impractical with surface receivers. Today refraction lines in other situations are often shot with OBS, not infrequently in conjunction with sonobuoys, simply because they are readily available and convenient to use. Many examples of profiles of this sort exist in the literature but there is not room to quote them here.

## THE USE OF OBS FOR OBSERVING EARTHQUAKES

The detection and location of earthquakes, together with the study of earthquake source processes, is greatly facilitated by observations from points suitably distributed about the hypocentre. Unfortunately most of the earth's seismically active areas lie adjacent to, or are surrounded by, oceans. Without recordings from these oceanic regions the study of many larger earthquakes is far from ideal and the study of smaller events is impossible. However, OBS investigations of ambient noise levels, described below, indicate that the sea bed is a poor environment, compared with land, for monitoring distant earthquakes. Therefore most recent OBS studies have been either close-in microearthquake surveys, where higher predominant frequencies give relatively good signal to noise (S/N) ratios, or in regions where such frequencies propagate to unusually great distances.

## OBS Measurements of Sea-Bed Noise

During the 1960's most OBS work was conducted to assess the spectrum and origin of sea-bed seismic noise. The frequencies of most interest were below about 2 Hz since for relatively distant sources (distance $\Delta$ =20-30 degrees) compressional (P) waves at higher frequencies are generally removed by the mechanism of anelastic attenuation. Anelastic attenuation, combined with a major peak in noise usually between 8 and 4 secs period, has the result that, on land, the optimum frequencies for the detection of distant earthquakes are centred at 1 Hz. The noise peak is caused by microseisms, which on land are retrograde Rayleigh waves propagating inland from the coast. There are two possible modes of generation either by the action of waves breaking on the shore and in shallow water or, more likely, by coupling to the sea-bed of pressure oscillations caused by standing waves in the surface swell further out to sea[111].

Noise levels at OBS in up to 1.2 km of water exceeded those at nearby California land stations[12], especially in deeper water, but compared well with average continental sites. OBS recordings of nearby nuclear tests ($\Delta$=4.5 degrees) also had poorer S/N in the band 1-3 Hz compared with land records. OBS work further out in the Pacific[9,11] established that noise levels in deep water were much higher than on an island site in the band 8 secs to 9 Hz and spectra exhibited peaks corresponding to organ-pipe modes in the water column. The dominant noise frequency was found to be twice that of the local sea swell and, assuming retrograde Rayleigh wave motion, appeared to originate beneath storm centres. Noise measurements in deep water 200 km off the Californian coast[112] showed microseisms moving in from further out to sea and although causally connected with storms they did not necessarily originate at their centres. Latham and Sutton[16] established that for oceanic sediments microseismic motion is prograde and this led to an alternative interpretation[110] which puts the source as standing waves set up by the interaction of opposing swells off nearby islands. Central Pacific noise levels ranged from one micron of ground displacement at 1 Hz to ten microns at the microseismic peak between 0.1 and 0.2 Hz[16].

In the Atlantic, noise measurements on Bermuda and by an OBS in deep water also favoured local interactions between direct and reflected swell as the main source of microseismic noise[16]. Again ground displacement values were always greater at the OBS site.

Theoretical analyses[16] have shown that even if the same total energy flux of microseisms exists on the sea-bed and on land for oceanic structures the flux is concentrated in the sub-oceanic

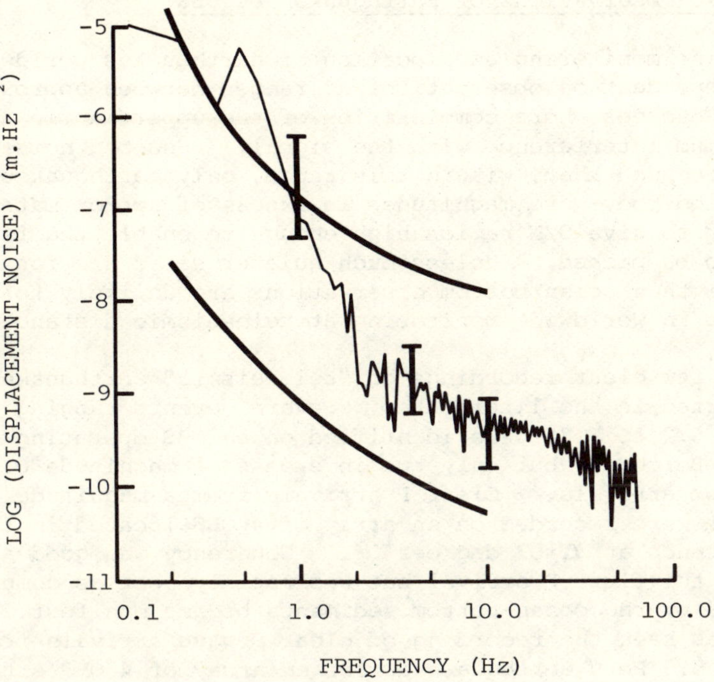


Fig. 6. Ocean-bottom noise spectrum from an OBS in the Middle America Trench after Dorman et al.[114]. Upper and lower curves are for noisy and quiet continental sites respectively. The bars represent the range at other Pacific sites.

sediment layer, whilst for continental structures it is distributed more evenly with depth. In shallow water, or where sediments are thin or absent, microseismic noise levels should tend to be lower and approximate those on coastal regions.

The high noise levels measured in the microseismic band led to a decline in interest in the use of OBS for monitoring distant earthquakes and explosions. Recently, however, deep-sea drilling has led to the possibility of placing instruments in boreholes below the sediments and there has been a reassessment of near sea-bed noise[113,114]. Recent noise-level measurements in the NE Pacific have confirmed earlier results (Fig. 6). Around 1 Hz noise levels fall rapidly with increasing frequency and above about 4 Hz levels become comparable to, or even better than, average land stations (about one millimicron). Similar results above 1 Hz were obtained by Russian researchers in the Black Sea[19].

## Teleseismic Observations of Earthquakes by OBS

Routine monitoring and location of earthquakes worldwide is heavily dependent on observations at ranges between approximately 25 and 90 degrees where complexities of the upper mantle and core give minimum interference with the signal. Sea-bed noise levels at 1 Hz are such that, within this range, only earthquakes with short-period body-wave magnitudes in excess of approximately 6.0 are likely to give S/N ratios high enough to enable the first P arrival to be picked. Unless much quieter sites are found (e.g. boreholes) then ocean bottom observations are unlikely to play a major role in worldwide monitoring at teleseismic distances.

Very few clear recordings of "teleseismic" earthquakes have been reported in the literature. Several events ranging in magnitude from 5.2 to 7.75 were identified on an OBS operating over 8 days near Bermuda[16] but only two in excess of magnitude 6.75 had clear first arrivals. Clear P arrivals from a magnitude 7.2 Tonga-Fiji event were recorded on an array of 4 OBS located in the Middle America trench at $\Delta$=83 degrees[114]. Coherency was good across the array for the first P arrival but was rapidly lost as complexities introduced by the ocean-bottom sediments became manifest. A notable success has been the recording of clear P wave arrivals from a magnitude 5.9 Peru earthquake across an array of 4 OBS situated at the junction of the Vema fracture zone and the Mid-Atlantic Ridge[115]. Arrivals at the junction were delayed by 0.5 secs relative to two OBS about 10 kms away. Such rapid lateral variations cannot be explained solely in terms of variations in crustal thickness and the suggested explanation is in terms of a narrow zone of partial melting at the ridge axis and junction.

Teleseismic events were recorded by ocean-bottom hydrophones (OBH) in the Pacific[116]. Soviet nuclear tests, together with intermediate and deep earthquakes of the same magnitude and distance ($\Delta$=60-90 degrees), can be identified but similar size nuclear tests in Nevada and shallow focus earthquakes are not seen, possibly because Nevada is adjacent to a seismically active low Q region and Q is lower in the upper crust, respectively. Nevertheless, the overall detection threshold is still high at magnitude 6.0.

Phases later than P from teleseismic earthquakes have been seen[16] but not investigated. In the North Atlantic higher-mode Rayleigh waves generated by Mid-Atlantic Ridge earthquakes have been clearly recorded by an OBH northeast of Antigua and on land in Bermuda[117]. As with the fundamental mode microseismic Rayleigh waves, dispersion of the higher mode is very sensitive to the thickness and velocity of ocean-bottom sediments. To obtain agreement with the observations it is necessary to assume that either thick pockets of sediment have a dominant effect or that shear-wave velocities are very low (75m $s^{-1}$) for the top 150 m.

Microearthquake Observations

The installation of seismometers within active seismic zones enables the observation of large numbers of earthquakes. The large proportion of seismic zones under the sea, the relatively large number of events detectable in a short time and the high frequency of the resulting ground motions have made the OBS a powerful tool in earthquake studies. The observations can be conveniently divided into those from accreting plate boundaries (mid-ocean ridges and associated transform faults) and those from subduction zones (principally the trenches of the western Pacific).

Accreting plate-boundaries tend to be remote from large land areas and microearthquake observations are then only feasible using OBS. Data from at least three instruments is required for the unambiguous location of any earthquake, however much has been learnt from deployments of single instruments. Some of the earliest observations were from a land-linked OBS situated off the coast of California south of the Mendocino fracture zone[129,141]. This OBS detected five times as many events from the fracture zone as did an equidistant land station, thus highlighting the disadvantage of observing events across an ocean-continent transition zone. The Medocino fracture zone was confirmed to be a true transform fault, in the sense that microearthquakes are not observed beyond the junctions(s) with spreading axes, in common with other OBS observations at fracture zones. Deployments elsewhere in fracture zones have used OBS with recording periods of up to two weeks. Average earthquake counts have been extremely variable, ranging from 1 to 100 per day.

On the Mid-Atlantic Ridge (MAR) near 45°N single OBS deployments, both on and away from the ridge axis, showed that seismic activity was largely within the median valley[118]. Events also appeared in swarms characterised by repeated shocks of different sizes but with identical seismograms. A similar result was obtained for an array of two OBS on the MAR near 37°N. Here the observed activity was confined to within a few kilometres of the ridge axis or the topographic deep of the fracture zone. High S/P velocity ratios were measured, reflecting the highly fissured nature of the crust[119]. Earthquakes recorded by two OBS deployed on the Reykjanes Ridge[120] may also be confined to a narrow zone less than 5 km across. At the Indian Ocean triple junction, estimated stress drops of microearthquakes recorded by a single OBS are no different from elsewhere on ridge crests[128].

Absolute hypocentre locations of microearthquakes observed within small arrays of three or more OBS are accurate to 1-2 km[142]. For example, microearthquakes located by such an array on the MAR near 45°N show that seismicity was restricted to a 5 km wide zone with most events 8 km below the centre of the median valley[121,122].

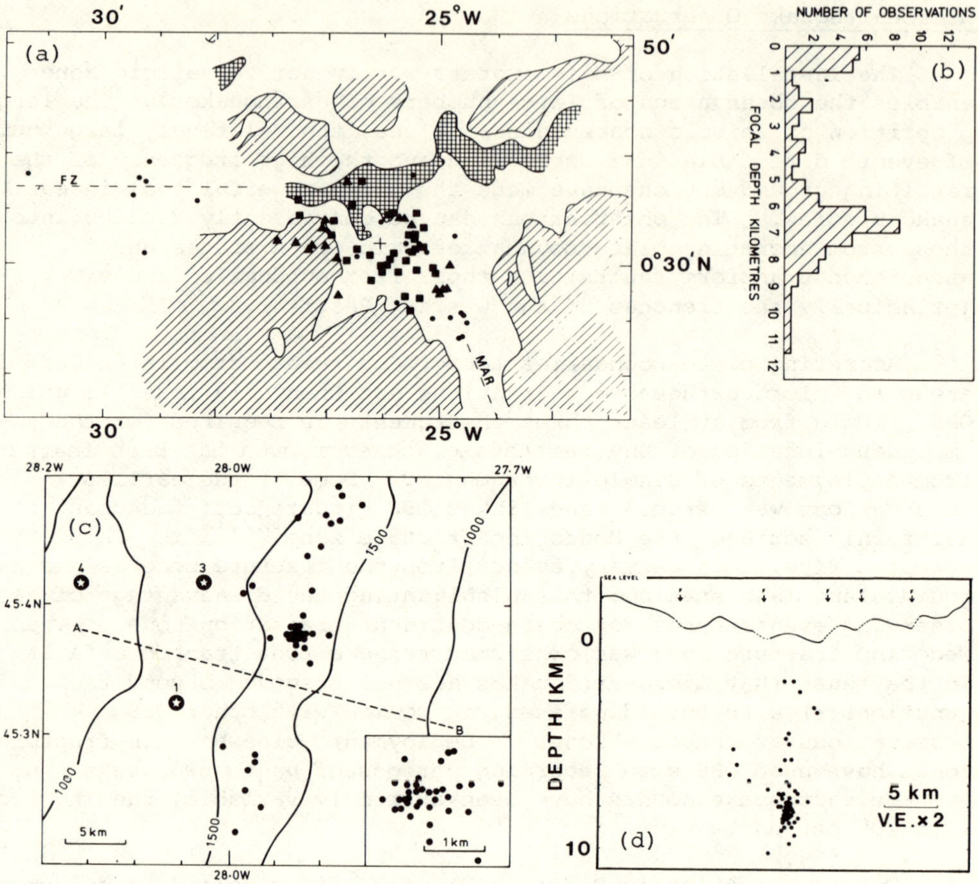


Fig. 7. Spatial and depth distributions of microearthquakes at the MAR/St. Paul's fracture zone intersection (a,b) (ref. 123) and in the median valley of the MAR at 45°N (c,d) (ref. 122).

(Fig. 7). Swarm-like sequences of microearthquakes suggest that these events are linked with magma movement. A four OBS array, deployed at the eastern junction of St. Paul's fracture zone, located many events within a triangular zone filling the corner between the MAR spreading axis and the transform part of the fracture zone[123] (Fig. 7). Focal depths were found to peak at 7 kms and this, combined with the limited distance range over which events are seen by the OBS along the spreading axis, has been interpreted as a measure of the thickness of the lithosphere. This interpretation has been confirmed by waveform modelling of hydrophone records[124].

In the Pacific, tripartite arrays deployed on the East Pacific

Rise (EPR), its junction with the Rivera fracture zone and at the centre of the fracture zone show different microearthquake activities[125]. No events were seen at all on the Rise itself or at the junction, but activity was high on the fracture zone. The fracture zone has a bathymetric trough flanked by ridges and within which the activity was located, but two small offsets in otherwise parallel trends suggest small spreading centres. Several earthquakes, large enough to be recorded by seismic stations elsewhere, were recorded and travel times to these distant stations indicate a slow P wave velocity beneath the fracture zone[126]. Another OBS deployment on the EPR, at its junction with the Wilkes fracture zone, also detected very low activity on the spreading axis and within 10 km of the intersection[127]. It appears that the lithosphere at the rise crest is either too thin, or is not competent, to support sufficient stress to generate even small earthquakes.

Deployments in the trenches of the West Pacific have been made mainly by Japanese workers using tethered OBS. The presence of land nearby enables OBS to be used in conjunction with extra land stations. Observations off Japan and the Kuriles have produced different results concerning the level of microearthquake activity on either side of the trench. OBS deployments on the oceanic side, at the point where the lithosphere flexes downwards, have detected very high activity[57,60]. These results have not been reproduced in other surveys[83,131,132]. It seems likely that such variations in activity result from the relatively short time over which OBS can record at present. Events on the continental side, detected by OBS on the same side, are found beneath the steepest part of the trench wall and appear to coincide with the top boundary of the dipping oceanic slab[130]. The distribution of microearthquakes located by OBS recording for relatively short periods of time delineates the Benioff zone right up to the trench axis[60].

Seismicity surveys have been made elsewhere using single OBS. High rates of activity have been observed in the Marianas trench, especially at the southern termination of the Marianas Ridge, and on the Fiji plateau where swarm-like sequences have been seen[60]. Very high microseismicity has been detected by OBS in the back arc basin of the New Hebrides[105].

Ray paths from deep focus events to OBS situated in a trench often traverse the dipping subducted slab. Travel time measurements along such rays have enabled P-wave velocity estimates of between 8.4-8.5 km $s^{-1}$ for the slab beneath Japan and the Kuriles[57,132]. The attenuation factor Q for this slab has been estimated at over 4000, using OBS recordings of T phases generated within the trenches by deep focus earthquakes[134]. Clear S-P and P-S conversions observed by OBS located on the trench inner wall have been interpreted in terms of very low S wave velocities, suggesting that the sediments there are in a plastic state with low rigidity[133].

Studies of the Oceanic Lithosphere Using Earthquake Sources

OBH earthquake recordings have enabled investigation of the ocean crust and upper mantle in the NW Pacific. High frequency P type arrivals seen on hydrophones at Wake, Midway and Eniwetok have been reported with average velocities ranging from 7.8 to 8.5 km $s^{-1}$ [135]. OBH and OBS recordings have revealed the upper mantle in the NW Pacific to be highly transparent for Pn and Sn paths with effective Q values of 8400 and 19100 respectively[137,138]. Similar high Q values have been found by Japanese researchers[83,134]. Earthquake recordings by OBS on oceanic crust to the east of the Japan trench show simple high frequency waveforms but are surprisingly devoid of low frequencies. The slope of the high frequency part of the spectrum puts Q in excess of 6000. Low frequencies are present on recordings of the same earthquakes made on the continental side of the trench but the lack of high frequencies indicates that Q along these paths is at least ten times less. The absence of low frequencies on the oceanic side is interpreted as evidence for a narrow waveguide in the lithosphere. Apparent velocities of arrivals from these earthquakes, measured using OBS arrays, indicate a velocity of 8.64 km $s^{-1}$ for the lower oceanic lithosphere[139]. Offsets in the travel-time plots suggest an underlying low-velocity layer.

A structure very different from that of the NW Pacific was revealed during an OBS microearthquake survey on the central basin fault of the Phillipine Sea[140]. The fault was seismically quiet but arrivals were seen from many earthquakes on the boundaries of the sea. Travel times and spectra indicate relatively low P wave velocities and relatively low Q values (2000) down to 80 km depth with a rapid rise (to 4000) beneath.

REFERENCES

1. M. Ewing and J. Ewing, A telemetering ocean-bottom seismograph, J. Geophys. Res. 66, 3863 (1961).
2. H. Bradner, Seismic measurements on the ocean bottom, Science, 146, 208 (1964).
3. M. Ewing and A.C. Vine, Deep-sea measurements without wires or cables, Trans. Amer. Geophys. Un., 19th Ann. Mtg. (1938).
4. M. Ewing, G.P. Woollard, A.C. Vine and J.L. Worzel, Recent results in submarine geophysics, Bull. Geol. Soc. Amer., 57, 909 (1946).
5. J.T. Thomson and W. Schneider, An automatic marine seismic monitoring and recording device, Proc. Inst. Radio Eng., 50, 2209 (1962).
6. R.A. Arnett and T.W. Newhouse, Ocean bottom seismograph, Proc. I.E.E.E., 53, 1899 (1965).
7. G.D. Ezell, Historical review of the development of a deep-sea

ocean bottom seismograph system, Mar. Tech. Soc., Wash. D.C., Trans. 3rd Ann. Confer., 147 (1967).

8. D.D. Prentiss and J.I. Ewing, The seismic motion of the deep ocean floor, Bull. Seismol. Soc. Amer., 53, 765 (1963).

9. H. Bradner and J.G. Dodds, Comparative seismic noise on the ocean bottom and on land, J. Geophys. Res., 68, 4339 (1964).

10. H. Bradner, J.G. Dodds and R.E. Foulks, Coherence measurements with time-sampling ocean-bottom seismometers, Proc. I.E.E.E., 53, 1906 (1965).

11. H. Bradner, J.G. Dodds and R.E. Foulks, Investigation of microseism sources with ocean-bottom seismometers, Geophys., 30, 511 (1965).

12. W.A. Schneider and M.M. Backus, Ocean-bottom seismic measurements off the California coast, J. Geophys. Res., 69, 1135 (1964).

13. W.A. Schneider, P.J. Farrell and R.E. Brannian, Collection and analysis of Pacific ocean bottom seismic data, Geophys., 29, 745 (1964).

14. G.H. Sutton, W.G. McDonald, D.D. Prentiss and S.N. Thanos, Ocean-bottom seismic observatories, Proc. I.E.E.E., 53, 1909 (1965).

15. G.H. Sutton, M.E. Odegard, N. Mark and N.J. Letourneau, Research in seismology related to the Columbia ocean-bottom seismograph, Hawaii Inst. Geophys. Rept., 70-12 (1970).

16. G.V. Latham and G.H. Sutton, Seismic measurements on the ocean floor, J. Geophys. Res., 71, 2545 (1966).

17. F.I. Monakhov, Microseisms on the bed of the Black Sea, Izv. Acad. Sci. USSR, Geophys. Ser., 461 (1961).

18. F.I. Monakhov, Microseisms at the bottom of the Baltic Sea and in the northern part of the Atlantic Ocean, Izv. Acad. USSR, Geophys. Ser., 573 (1962).

19. L.N. Rykunov and V.V. Sedov, Seismic noise in the 2-15 cps frequency range on the bottom of the Black Sea, Izv. Phys. of Solid Earth, 7, 443 (1965).

20. J.M. Shorthouse, Seismic measurements on the deep-sea bed, Unpubl. Ph.D. thesis, Univ. of Cambridge (1964).

21. D. Davies, Dispersed Stoneley waves on the ocean bottom, Bull. Seismol. Soc. Amer., 55, 5, 903 (1965).

22. R.B. Whitmarsh, A buoyant seismic recording apparatus for use on the ocean bed, I.E.R.E. Conference Proceedings No. 8, (Electronic Engineering in Oceanography) Suppl. Vol., paper 28, 4 pages (1968).

23. S. Nagumo, H. Kobayashi and S. Koresawa, Construction of ocean bottom seismograph, Bull. Earth. Res. Inst., 43, 671 (1965).

24. S. Nagumo, H. Kobayashi and S. Koresawa, Seabottom seismic observation at Sagami Bay, (i) seismic activity, Bull. Earth. Res. Inst., 46, 877 (1968).

25. R.B. Whitmarsh, An ocean-bottom pop-up seismic recorder, Mar. Geophys. Res., 1, 91 (1970).

26. T.J.G. Francis and I.T. Porter, Microearthquake survey of the Mid-Atlantic Ridge, Nature, 290, 547 (1972).

27. D. Carmichael, G. Carpenter, A. Hubbard, K. McCamy and W. McDonald, A recording ocean bottom seismograph, J. Geophys. Res., 78, 8748 (1973).
28. W.G. McDonald, A.C. Hubbard, R.G. Bookbinder and K. McCamy, Design and shipboard operation of a multipurpose ocean bottom seismograph, Mar. Geophys. Res., 3, 179 (1977).
29. D. Carmichael, W.J. Ludwig, J.I. Ewing, G. Carpenter and A.C. Hubbard, A pop-up ocean bottom hydrophone, Mar. Geophys. Res., 3, 225 (1977).
30. R.G. Bookbinder, A.C. Hubbard, W.G. McDonald and P.W. Pomeroy, Design of an ocean bottom seismometer with response from 25Hz to 100 seconds, Oceans, 178, 510, M.T.S. Wash. D.C. (1978).
31. R.A. Frosch, V.C. Anderson and H. Bradner, Instrumenting the sea-floor, I.E.E.E. Spectrum, 1, 101 (1964).
32. G.H. Sutton, J. Kasahara, W.N. Ichinose and D.A. Byrne, Ocean bottom seismograph development at Hawaii Institute of Geophysics, Mar. Geophys. Res., 3, 153 (1977).
33. J.G. Blackington and M.E. Odegard, An ocean bottom seismograph using digital telemetry and floating-point conversion, I.E.E.E. Trans. on Geoscience Electronics, GE-15, 74 (1977).
34. W.A. Prothero, An ocean-bottom seismometer capsule, Bull. Seismol. Soc. Amer., 64, 1251 (1974).
35. W.A. Prothero, A digital event-recording ocean bottom seismometer capsule, Mar. Geophys. Res., 3, 119 (1977).
36. R.D. Moore, L.M. Dorman, H. Chin-Yen and D.L. Berliner, An ocean bottom, microprocessor based seismometer, in prep.
37. M.D. Cranford, S.H. Johnson, J.E. Bowers, R.A. McAlister and B.T. Brown, A direct-recording ocean-bottom seismograph, Bull. Seismol. Soc. Amer., 66, 607 (1976).
38. S.H. Johnson, M.D. Cranford, B.T. Brown, J.E. Bowers and R.E. McAllister, A free-fall direct-recording ocean bottom seismograph, Mar. Geophys. Res., 3, 103 (1977).
39. C.R.B. Lister and B.T.R. Lewis, An ocean-bottom seismometer suitable for arrays, Deep-Sea Res., 23, 113 (1976).
40. R.V. Johnson, C.R.B. Lister and B.T.R. Lewis, A direct recording ocean bottom seismometer, Mar. Geophys. Res., 3, 65 (1977).
41. D.E. Koelsch and G.M. Purdy, An ocean bottom hydrophone instrument for seismic refraction experiments in the deep ocean, Mar. Geophys. Res., 4, 115 (1979).
42. B.P. Ambuter and S.C. Solomon, An event-recording system for monitoring small earthquakes, Bull. Seismol. Soc. Amer., 64, 1181 (1974).
43. P.J. Mattaboni and S.C. Solomon, MITOBS: a seismometer system for ocean-bottom earthquake studies, Mar. Geophys. Res., 3, 87 (1977).
44. G. Latham, P. Donoho, K. Griffiths, A. Roberts and A.K. Ibrahim, The Texas ocean-bottom seismograph, 1978 Offshore Tech. Conf. 3, 1467 (1978).
45. A.K. Ibrahim and G.V. Latham, A comparison between sonobuoy and ocean bottom seismograph data and crustal structure of the

Texas shelf zone, Geophys., 43, 514 (1978).

46. R.L. Steinmetz, P.L. Donoho, J.D. Murff and G.V. Latham, Soil coupling of a strong motion, ocean bottom seismometer, 1979 Offshore Tech. Conf., 2235 (1979).

47. W.A. Prothero, An operationally optimised ocean-bottom seismometer capsule, Tectonophys., 18, 71 (1979).

48. L.N. Rykunov and V.V. Sedov, An ocean-bottom seismograph, Izv. Phys. of Solid Earth, 8, 537 (1967).

49. J.J. Langford and R.B. Whitmarsh, Pop-up bottom seismic recorder (PUBS) of the Institute of Oceanographic Sciences, U.K., Mar. Geophys. Res., 3, 43 (1977).

50. R.E. Kirk, J.J. Langford and R.B. Whitmarsh, A functional three-component ocean-bottom seismograph for controlled source and earthquake seismology, in prep.

51. T.J.G. Francis, personal communication.

52. T.J.G. Francis, I.T. Porter, R.D. Lane, R.J. Osborne, J.E. Pooley and P.K. Tomkins, Ocean bottom seismograph, Mar. Geophys. Res., 2, 195 (1975).

53. T.J.G. Francis and I.T. Porter, Experience gained with the Blacknest ocean bottom seismograph, Mar. Geophys. Res., 3, 143 (1977).

54. F. Kishinouye, Y. Yamazaki, H. Kobayashi and S. Koresawa, A submarine seismograph; the first paper, Bull. Earth Res. Inst., 41, 819 (1963).

55. S. Nagumo, H. Kobayashi and S. Koresawa, Improvements of ocean-bottom seismograpn-construction of a long-life magnetic tape-recorder, Bull. Earth. Res. Inst., 46, 861 (1968).

56. S. Hasegawa and S. Nagumo, Construction of a long-life magnetic tape recorder and some features of ocean-bottom seismographs, Bull. Earth. Res. Inst., 48, 967 (1970).

57. S. Nagumo, S. Hasegawa, S. Koresawa and H. Kobayashi, Ocean bottom seismographic observation at the off-side of Japan trench near the Erimo seamount - seismic activity of the oceanic lithosphere and velocity structure around the geophysical ocean-continent boundary, Bull. Earth. Res. Inst., 48, 769 (1970).

58. J. Kasahara, S. Nagumo, S. Koresawa and T. Ouchi, Low-power amplifier and tape-recorder for the ocean bottom seismograph, J. Seismol. Soc. Japan, 27, 75 (1974).

59. J. Kasahara, S. Koresawa, T. Ouchi and S. Nagumo, Free-fall pop-up ocean bottom seismograph system, Spec. Bull. Earth. Res. Inst., 13, 1 (1974).

60. S. Nagumo and J. Kasahara, Ocean-bottom seismograph study of the western margin of the Pacific, in Geophysics of the Pacific Ocean basin and its margin, A.G.U. Mono. 19, 155 (1976).

61. H. Shimamura, T. Asada and K. Takano, Measurements of microearthquakes at sea bottom, Bull. Soc. Fr. Jpn. Oceanogr., 8 (1970).

62. H. Shimamura and T. Asada, A cassette recorder for an ocean-bottom seismometer, Geophys. Bull. Hokkaido Univ., 32, 17 (1974).

63. Anonymous, Permanent ocean-bottom seismographic observation system, Tech. Rept. Meteorol. Res. Inst. Japan, No. 4, pp. 233 (1980).

64. F. Avedik, V. Renard, D. Buisine and J-Y. Cornic, Ocean bottom refraction seismograph (OBSR), Mar. Geophys. Res., 3, 357 (1978).

65. L. Steinmetz, personal communication.

66. D.E. Heffler and D.L. Barrett, OBS development at Bedford Institute of Oceanography, Mar. Geophys. Res., 4, 227 (1979).

67. D.L. Barrett, D.E. Heffler and C.E. Keen, Ocean bottom seismometer developments, in Report of Activities, pt. B, Geol. Surv. Canada, paper 77-1B (1977).

68. R. Herber, Ocean-bottom seismograph of the Institut für Geophysik, Hamburg, Mar. Geophys. Res., 4, 247 (1979).

70. C.E. Keen and C. Tramontini, A seismic refraction survey on the Mid-Atlantic Ridge, Geophys. J. R.A.S., 20, 473 (1970).

71. H-W. Kebe, Eine ferngesteuerte Messboje mit Datenspeichrung für refraktionsseismische Untersuchungen auf See, Meteor. Forsch. Ergebn. Reche C, 14-20 (1971).

72. W. Weigel, P. Goldflam and K. Hinz, The crustal structure of the Conception Bank, Mar. Geophys. Res., 3, 381 (1978).

73. I. Reid and K. Macdonald, Microearthquake study of the Mid-Atlantic Ridge near 37°N, using sonobuoys, Nature, 246, 88 (1973).

74. G.H. Sutton, B.T.R. Lewis, J. Ewing, F.K. Duennebier, B. Iwatake and J.D. Tuthill, Lopez Island ocean bottom seismometer intercomparison experiment, Hawaii Inst. Geophys. Rept., 80-4 (1980).

75. R.A. Stephen, The oblique seismic experiment in oceanic crust, equipment and technique, Mar. Geophys. Res., 4, 213 (1979).

76. R.A. Stephen, K.E. Louden and D.H. Matthews, The oblique seismic experiment on DSDP Leg 52, Geophys. J. Roy. astr. Soc., 60, 289 (1980).

77. Anonymous, Marine Seismic System, Gould Inc., Chesapeake Instrument Division, (1980).

78. B.A. Bolt, Ocean bottom seismometry, a new dimension to seismology, Boll. di Geofisica, 19, 107 (1977).

79. F. Duennebier and G. Blackington, A man-made hot spring on the ocean floor, Nature, 284, 338 (1980).

80. R. Von Huene, J. Aubouin et al., Leg 67: The Deep-Sea Drilling Project Mid-America Trench transect off Guatemala, Geol. Soc. Amer. Bull., 91, 421 (1980).

81. J.A. Orcutt and L.M. Dorman, An oceanic long range explosion experiment, J. Geophys., 43, 257 (1977).

82. L. Steinmetz, R.B. Whitmarsh and V. Moreira, Upper mantle structure beneath the Mid-Atlantic Ridge north of the Azores based on observations of compressional waves, Geophys. J. Roy. astr. Soc., 50, 353 (1977).

83. T. Asada and H. Shimamura, Observations of earthquakes and explosions at the bottom of the western Pacific; structure

of oceanic lithosphere revealed by Longshot experiment, in The geophysics of the Pacific Ocean Basin and its margin, A.G.U. Mono. 19, 135 (1976).

84. J.A. Orcutt and L.M. Dorman, Results of Scripps long range profiles in the Pacific, Tectonophysics, 56, 83 (1979).

85. R.B. Whitmarsh, R. Lilwall, M. Donegan, C.E. Keen, B.D. Loncarevic, D. Barrett, L. Steinmetz, J. Tomblin, J. Shepherd and W. Aspinall, Preliminary results from the Lesser Antilles Deep Lithosphere Experiment (LADLE), (Abstract) Geophys. J. Roy. astr. Soc., 61, 209 (1980).

86. R.B. Whitmarsh, R.C. Lilwall, M.J. Donegan, C.E. Keen, B.D. Loncarevic and L. Steinmetz, Progress report on LADLE (Abstract), Geophys. J. Roy. astr. Soc., 65, 263 (1981).

87. S. Nagumo, T. Ouchi, J. Kasahara, S. Koresawa, Y. Tomoda, K. Kobayashi, A.S. Furumoto, M.E. Odegard and G.H. Sutton, Sub-Moho seismic profile in the Mariana Basin - ocean bottom seismograph long range explosion experiment, Earth Planet. Sci. Letts., 53, 93 (1981).

88. RRISP Working Group, Reykjanes Ridge Iceland Seismic Experiment (RRISP 77), J. Geophys., 47, 228 (1980).

89. H. Shimamura, T. Asada and M. Kumazawa, High shear velocity layer in the upper mantle of the Western Pacific, Nature, 269, 680 (1977).

90. A.W.B. Jacob, Dispersed shots at optimum depth - an efficient seismic source for lithospheric studies, J. Geophys., 41, 63 (1975).

91. R.S. Detrick and G.M. Purdy, The crustal structure of the Kane Fracture Zone from seismic refraction studies, J. Geophys. Res., 85, 3759 (1980).

92. P.J. Fox, E. Schreiber, H. Rowlett and K. McCamy, The geology of the Oceanographer Fracture Zone: a model for fracture zones, J. Geophys. Res., 81, 4117 (1976).

93. R.S. White and D.H. Matthews, Variations in oceanic upper crustal structure in a small area of the northeastern Atlantic, Geophys. J. Roy. astr. Soc., 61, 401 (1980).

94. J.A. Orcutt, B.L.N. Kennett and L.M. Dorman, Structure of the East Pacific Rise from an ocean bottom seismometer survey, Geophys. J. Roy. astr. Soc., 45, 305 (1976).

95. L. Steinmetz, A. Hirn, M. Sapin, R.B. Whitmarsh and V. Moreira, Zones of P wave attenuation beneath the crest of Mid-Atlantic Ridge, Bull. Soc. Geol. France, 7$^{e}$ ser., 18, 931 (1976).

96. R.B. Whitmarsh, Median valley refraction line, Mid-Atlantic Ridge at 37$^{o}$N, Nature, 246, 297 (1973).

97. R.B. Whitmarsh, Axial intrusion zone beneath the median valley of the Mid-Atlantic Ridge at 37$^{o}$N detected by explosion seismology, Geophys. J. Roy. astr. Soc., 42, 189 (1975).

98. R.D. Hyndman, G.C. Rogers, M.N. Bone, C.R.B. Lister, U.S. Wade, D.L. Barrett, E.E. Davis, T. Lewis, S. Lynch and D. Seeman, Geophysical measurements in the region of the Explorer ridge off western Canada, Canad. J. Earth Sci., 15, 1508 (1978).

99. B.T.R. Lewis and W.E. Snydsman, Fine structure of the lower oceanic crust on the Cocos plate, Tectonophys., 55, 87 (1979).
100. E.E. Davis, C.R.B. Lister and B.T.R. Lewis, Seismic structure of the Juan de Fuca Ridge: ocean bottom seismometer results from the median valley, J. Geophys. Res., 81, 3541 (1976).
101. C.M.R. Fowler, The Mid-Atlantic Ridge: structure at 45°N, Geophys. J. Roy. astr. Soc., 54, 167 (1978).
102. Y. Bottinga and C. Allegre, Geophysical, Petrological and Geochemical Models of the Oceanic Lithosphere, Tectonophysics, 32, 9 (1976).
103. R.B. Whitmarsh and D.G. Roberts, Seismic studies of the continent-ocean transition in North Biscay, R.R.S. Shackleton cruise 6/79, IOS Cruise Report 80, unpublished manuscript (1979).
104. R.B. Whitmarsh, Seismic refraction studies of the upper igneous crust in the north Atlantic and porosity estimates for Layer 2, Earth Planet. Sci. Letts., 37, 451 (1978).
105. A.K. Ibrahim, B. Pontoise, G. Latham, M. Larue, T. Chen, B. Isacks, J. Recy and R. Louat, Structure of the New Hebrides Arc-Trench system, J. Geophys. Res., 85, 253 (1980).
106. S. Nagumo, T. Ouchi and S. Koresawa, Seismic velocity structure near the extinct spreading centre in the Shikoku Basin, N. Philippine sea, Mar. Geol., 35, 135 (1980).
107. R.B. Whitmarsh, The Owen Basin off the south-east margin of Arabia and the evolution of the Owen Fracture Zone, Geophys. J. Roy. astr. Soc., 58, 441 (1979).
108. H.P. Bucker, J.A. Whitney and D.L. Kerr, Use of Stoneley waves to determine the shear velocity in ocean sediments, J. Acoust. Soc. Amer., 36, 1595 (1979).
109. J.O. Orcutt and K.C. Macdonald, Shallow crustal structure in the Rise hydrothermal field from the Alvin-OBS refraction experiment (abstract), Trans. Amer. Geophys. Un., 60, 863 (1979).
110. H. Bradner and G.V. Latham, Prograde Rayleigh Waves and Microseism Sources, J. Geophys. Res., 77, 6422 (1972).
111. M. Longuet-Higgins, A theory of the origin of microseisms, Phil. Trans. Roy. Soc. London, 243, 1 (1950).
112. G.V. Latham and A.A. Nowroozi, Waves, Weather and Ocean Bottom Microseisms, J. Geophys. Res., 73, 3945 (1968).
113. J.D. Phillips and D.W. McCowan, Ocean Bottom Seismometers for Research: A reassessment Massachusetts Institute of Technology, Lincoln Laboratory Technical Note 1978-1940 (1978).
114. L.M. Dorman, J.A. Orcutt and T.H. Jordan, Seismic Detection and Discrimination using Ocean-Bottom Seismographs, Defence Advanced Research Projects Agency Annaul Technical Report, Contract N F9620-79-00019 (1979).
115. H. Rowlett and D. Forsyth, Teleseismic P-wave Delay times in a Major Oceanic Fracture Zone, Geophys. Res. Letts., 6,

273 (1979).

116. D.A. Walker, Hydrophone Recordings of Underground Nuclear Explosions, Geophys. Res. Lett., 7, 465 (1980).

117. R.S. Anderson and G.V. Latham, Determination of Sediment Properties from First Shear-Mode Rayleigh Waves Recorded on the Ocean Bottom, J. Geophys. Res., 74, 2747 (1969).

118. T.J.G. Francis and I.T. Porter, Median Valley Seismology: The Mid-Atlantic Ridge near 45°N, Geophys. J.Roy. astr. Soc., 34, 279 (1973).

119. T.J.G. Francis, I.T. Porter and J.R. McGrath, Ocean Bottom seismograph observations on the Mid-Atlantic Ridge near Lat. 37°N, Bull. Geol. Soc. Am., 88, 664 (1977).

120. R.C. Lilwall, T.J.G. Francis and I.T. Porter, Some Ocean-Bottom Seismograph Obeservations on the Reykjanes Ridge at 59°N, Geophys. J. Roy. astr. Soc., 62, 321 (1980).

121. R.C. Lilwall, T.J.G. Francis and I.T. Porter, Ocean-Bottom seismograph observations on the Mid-Atlantic Ridge near 45°N, Geophys. J. Roy. astr. Soc., 51, 357 (1977).

122. R.C. Lilwall, T.J.G. Francis and I.T. Porter, Ocean-Bottom seismograph observations on the Mid-Atlantic Ridge near 45°N - Further Results, Geophys. J. Roy. astr. Soc., 55, 255 (1978).

123. T.J.G. Francis, I.T. Porter and R.C. Lilwall, Microearthquakes near the Eastern End of St. Paul's Fracture Zone, Geophys. J. Roy. astr. Soc., 53, 201 (1978).

124. R.C. Lilwall, Fault Mechanisms and Sub-Crustal Seismic Velocities on the Mid-Atlantic Ridge, Geophys. J. Roy. astr. Soc., 60, 245 (1980).

125. W.A. Prothero, I. Reid, M.S. Reichle and J.N. Brune, Ocean Bottom Seismic Measurements on the East Pacific Rise and Rivera Fracture Zone, Nature, 262, 121 (1976).

126. I. Reid and W.A. Prothero, An Earthquake Sequence Studied with Ocean Bottom Seismographs, Geophys. J. Roy. astr. Soc., 64, 381 (1981).

127. R.C. Lilwall, T.J.G. Francis and I.T. Porter, A Microearthquake Survey at the Junction of the East Pacific Rise and the Wilkes (9°S) Fracture Zone, Geophys. J. Roy. astr. Soc., 66 (1981).

128. S.C. Solomon, P.J. Mattaboni and R.I. Hester, Microseismicity Near the Indian Ocean Triple Junction, Geophys. Res. Lett., 4, 597 (1977).

129. B. Auld, G. Latham, A. Nowroozi and L. Seeber, Seismicity off the Coast of Northern California Determined from Ocean Bottom Seismic Measurements, Bull. Seismol. Soc. Amer., 59, 2001 (1969).

130. S. Nagumo, S. Hasegawa, S. Koresawa and H. Kobayashi, Ocean Bottom Seismographic Observation off Sanriku - Aftershock Activity of the 1968 Tokachi-Oki Earthquake and its Relation to the Ocean-Continent Boundary Fault, Bull. Earth. Res. Inst., 48, 793 (1970).

131. C. Frohlich, G.V. Latham, S. Billington, C. Kisslinger, E.R. Engdahl, A. Malahoff and J.G. Caldwell, Ocean Bottom Seismograph Results from the Central Aleutian Subduction Zone, EOS, 60, 878 (1979).

132. J. Kasahara and R.R. Harvey, Seismological Evidence for the High-Velocity Zone in the Kurile Trench Area from Ocean Bottom Seismometer Observations, J. Geophys. Res., 82, 3805 (1977).

133. S. Nagumo, J. Kasahara and S. Koresawa, Large Poissons Ratio and Low S-Wave Velocity Within the Japan Trench Inner Wall Toe, Mar. Geol. 35, 129 (1980).

134. H. Shimamura and T. Asada, T Waves from Deep Earthquakes Generated Exactly at the Bottom of Deep Trenches, Earth. Planet. Sci. Lett., 27, 137 (1975).

135. D.A. Walker and G.H. Sutton, Oceanic Mantle Phases Recorded on Hydrophones in the NW Pacific at distances between 9° and 40°, Bull. Seis. Soc. Amer., 61, 65 (1971).

136. D.A. Walker, High frequency Pn Phases observed in the Pacific at Great Distances, Science, 197, 257 (1977).

137. D.A. Walker, C.S. McCreery, G.H. Sutton and F.K. Duennebier, Spectral Analysis of High-Frequency Pn and Sn Phases Observed at Great Distances in the Western Pacific, Science, 199, 1333 (1978).

138. G.H. Sutton, C.S. McCreery, F.K. Duennebier and D.A. Walker, Spectral Analysis of High-Frequency Pn, Sn phases Recorded on Ocean Bottom Seismographs, Geophys. Res. Lett., 5, 745 (1978).

139. H. Shimamura and T. Asada, Apparent Velocity Measurements on an Oceanic Lithosphere, Phys. Earth. Plan. Int., 13, 15 (1976).

140. H. Shimamura, Y. Tomoda and T. Asada, Seismographic Observation at the bottom of the Central Basin Fault of the Philippine Sea, Nature, 253, 177 (1975).

141. L. Seeber, M. Barazangi and A. Nowroozi, Micro-earthquake Seismicity and Tectonics of Coastal Northern California, Bull. Seism. Soc. Amer., 60, 1669 (1970).

142. R.C. Lilwall and T.J.G. Francis, Hypocentral Resolution of Small Ocean Bottom Seismic Networks, Geophys. J. Roy. astr. Soc., 54, 721 (1978).

# ASPECTS OF BOTTOM BOUNDARY LAYERS IN THE OCEAN

Gunnar Kullenberg
Inst. of Physical Oceanography
University of Copenhagen
Haraldsg.6, 2200 Copenhagen N.

Ronald Zaneveld
School of Oceanography
Oregon State University
Corvallis, Oregon 97331, USA

## 1. BASIC PHYSICAL STRUCTURE AND RELATIONSHIPS

The dynamics of the bottom boundary layer in the ocean is influenced by the rotation of the earth, the mean oceanic circulation being in approximate geostrophic balance. An Ekman boundary layer is developed to match the interior flow, in which the velocity vector changes direction towards the bottom. Seen from above the change is clockwise in the northern hemisphere. Based on a Reynolds number criterion, the oceanic Ekman boundary layer is turbulent, the typical Reynolds number being at least an order of magnitude larger than the critical one (Wimbush and Munk 1971). However, the stability of the density stratification also influences the dynamics of the boundary layer. The static stability may be measured by means of the Brunt-Väisälä frequency

$$N = \left(-\frac{g}{\rho_o}\frac{d\rho}{dz}\right)^{\frac{1}{2}}$$

z being positive upwards and with z = 0 at the boundary. The dynamic stability is measured through the Richardson number Ri,

$$Ri = N^2/\left(\frac{d\bar{U}}{dz}\right)^2$$

having a critical value of 1/4. The flow is unstable for Ri < 1/4.

Ekman boundary layers can exist in the steady state. Further simplifying assumptions are that the flow is statistically homogeneous in the horizontal, that the turbulence is mainly shear-generated and not driven by buoyancy forces, and that the stratification in the lower part of the boundary layer is neutral, i.e. $N = 0$. For the lower part of the boundary layer the following sub-division may then be made

a) Near the boundary the horizontal shear stress $\tau$ changes slowly, giving rise to a constant stress layer in which $\tau$ is almost equal to $\tau(0)$. This defines the friction velocity $u_* = (\tau(0)/\rho)^{\frac{1}{2}}$, which is used as scaling velocity for the whole Ekman layer.

b) Where the boundary is smooth enough a viscous sublayer occurs at the bottom of the constant-stress layer with a thickness of about $12\ \nu/u_*$, where

$$\rho\nu \frac{d\bar{U}}{dz} = \tau(0) \tag{1}$$

The roughness element should have a characteristic scale d less than $3\nu/u_*$.

c) The logarithmic layer takes over above the viscous sublayer. The mean velocity profile is there, using $u = 0$ for $z = z_o$

$$\frac{\bar{U}}{u_*} = \frac{1}{\kappa_o} \ln \frac{z}{z_o} \tag{2}$$

The von Karman constant $\kappa_o = 0.4$, and

$$z_o \approx 0.1\nu/u_* \qquad \text{if} \quad d < 3\ /u_*$$

$$z_o \approx d/30 \qquad \text{if} \quad d > 3\ /u_*$$

for smooth and rough boundaries, respectively (e.g. Wimbush and Munk 1971). The validity of the logarithmic layer beyond the constant stress layer seems to be well established. The thickness of the logarithmic layer has been given as

$$h_{\ln} = \frac{2u_*^2}{fU_g} \qquad \text{(Wimbush and Munk 1971)}$$

or $h_{\ln} = 0.1\ h_E$ (Tennekes 1973).

Defining the constant stress layer as up to the elevation where $\tau \sim 0.8\ \tau(0)$ the thickness is $0.2\ u_*^2/fU_g$ (Monin and Obukhov 1953). The thickness of the Ekman layer, finally, is usually given as $h_E = \kappa_o u_*/f$. The motion in the boundary layer is governed by the external parameters $z_o$, $f$ and $U_g$.

For time dependent and horizontally inhomogeneous conditions these results can be applied provided $\bar{U}/T$ and $U^2/L \ll u_*\bar{U}/h_E$, where

T and L are appropriate scales of time and length for the mean motion $\bar{U}$. We also require that the Ekman layer is developed, i.e. $T > 1/f$ and $L/\bar{U} < 1/f$.

In order to parametrize the effects of turbulence on transfer of momentum and mixing so-called eddy coefficients are often introduced. As yet this is the only practical approach and it is in many cases a reasonably acceptable approach, although unsatisfactory. With an eddy viscosity $K_m$ the stress is

$$\tau = \rho(K_m + \nu)\,\frac{d\bar{U}}{dz} \tag{3}$$

Assuming $K_m \gg \nu$ we may also write

$$K_m = u_*^2\,(\frac{d\bar{U}}{dz}) \tag{4}$$

In the region of overlap of constant stress and logarithmic layers, one finds $K_m = \kappa_o u_* z$. The bottom stress is usually given in the form

$$\tau_b = \rho c_b\ U_g^2 \quad , \quad u_*^2 = c_b\ U_g^2 \tag{5a, b}$$

The friction velocity may also be estimated through the relation

$$u_* = 1/30\ U_g$$

A balance between local dissipation rate and energy production is usually assumed, implying that the net advection of energy is small compared to the dissipation. Then the rate of kinetic energy dissipation per unit mass is

$$\varepsilon = \frac{\tau}{\rho}\,(\frac{d\bar{U}}{dz}) = u_*^2(\frac{d\bar{U}}{dz}) \tag{6}$$

In the constant stress layer we find $\varepsilon = u_*^3/\kappa_o z$. The turbulent energy is mainly dissipated on scales of the order of Kolmogoroff scale $(\nu^3/\varepsilon)^{1/4}$.

A vertical turbulent mixing coefficient $K_z$ may also be introduced as

$$K_z\,\frac{d\bar{\rho}}{dz} = \overline{-w'\rho'} \tag{7}$$

An important parameter is the flux Richardson number Rf giving the ratio between the turbulent energy consumed for vertical mixing against gravity and the total turbulent energy. Thus

$$\mathrm{Rf} = \frac{-K_z \frac{g}{\rho}\frac{d\rho}{dz}}{K_m\left(\frac{d\bar{U}}{dz}\right)^2} = \frac{K_z N^2}{K_m\left(\frac{d\bar{U}}{dz}\right)^2} = \frac{K_z}{K_m}\,\mathrm{Ri} \tag{8}$$

The vertical mixing coefficient can be related to the rate of energy dissipation per unit mass by means of Rf,

$$K_z = \frac{\mathrm{Rf}\cdot\varepsilon}{N^2} \tag{9}$$

Having summarized aspects of the flow conditions let us summarize conditions regarding vertical flux of particles in the boundary layer. First, the assumptions are that the flow is steady and horizontally homogeneous with a concentration of suspended matter which does not influence the density, ρ, of the water. The velocity profile is logarithmic.

The balance equation is

$$F_p = -\,w_p\cdot C - K_p\frac{\partial C}{\partial z} \tag{10}$$

For the overlapping region where we may use $K_m = \kappa_o u_* z$ we find

$$F_p = -\,w_p\,C\;\frac{\kappa_o u_*}{Pr}\cdot z\frac{\partial C}{\partial z} \tag{11}$$

$Pr = Kp/Km$ is the turbulent Prandtl number. Assuming further that the flux $F_p$ is independent of z

$$C = c\cdot z^{-\frac{w_p P_r}{\kappa_o u_*}} - \frac{F_p}{w_p}\quad,\ w_p \neq 0 \tag{12}$$

where c is a constant of integration. For zero settling velocity the profile becomes

$$c - c_1 = \frac{F_p\cdot Pr}{\kappa_o u_*}\,\ell n\,\frac{z}{z_1} \tag{13}$$

where $c(z_1) = c_1$.

Three cases may now be considered.

a) When $\kappa_o u_* \ll w_p$ we have a deposition case and $F_p = -w_p \cdot C$.

b) A balance between gravitational settling and turbulent flux giving $F_p = 0$, and

$$\ell n \frac{C}{C_1} = - \frac{w_p \cdot Pr}{\kappa_o u_*} \ell n \frac{z}{z_1} \tag{14}$$

c) The case of strong erosion giving $w_p \ll \kappa_o u_*$ may be considered as $w_p = 0$.

In the case of fine-grained material a large friction velocity is required for erosion, implying a high turbulence level at the bottom and the possibility of neglecting gravitational settling, i.e. $w_p = 0$ , close to the bottom. The critical friction velocity allowing deposition of fine-grained material is furthermore small, implying a large range of friction velocities over which particles will stay in suspension when once suspended. The balance concentration of particles in suspension will be such that

$$\left| \frac{\kappa_o u_*}{Pr} z \frac{dc}{dz} \right| = \left| w_p c \right| \qquad \text{for } F_p = 0.$$

For friction velocities up to the critical value for erosion the condition of $F_p = 0$ may be expected to hold.

Observations, examples of which are shown in Fig. 1, demonstrate that the bottom boundary layer in the ocean can be thicker than the theoretical expressions yield and that the flow can be very unstable, bursts being important.

## 2. MEASURING TECHNIQUES

Here we will discuss techniques which are used in the deep sea and in the shelf seas to study the properties of the bottom boundary layer. The observations primarily aim at obtaining repeated profiles of velocity, temperature, salinity, suspended matter, certain radioactive isotopes, and fluctuations in space and time of these properties. From the observations characteristics regarding mixing, flux and sedimentation can be deduced. Topographic and sediment characteristics are also needed and are often included in the observations.

Current measurements close to the bottom in the deep ocean are difficult to carry out and a variety of techniques have therefore been attempted. These include the pendulum type instrument measuring the deflection of a plastic ball, the use of dyes, Savonius rotors and propellers. Sternberg (1969) used the technique of sequentially photographing differently colored dye spots ejected at three different levels within 1 m of the ocean floor, mounting

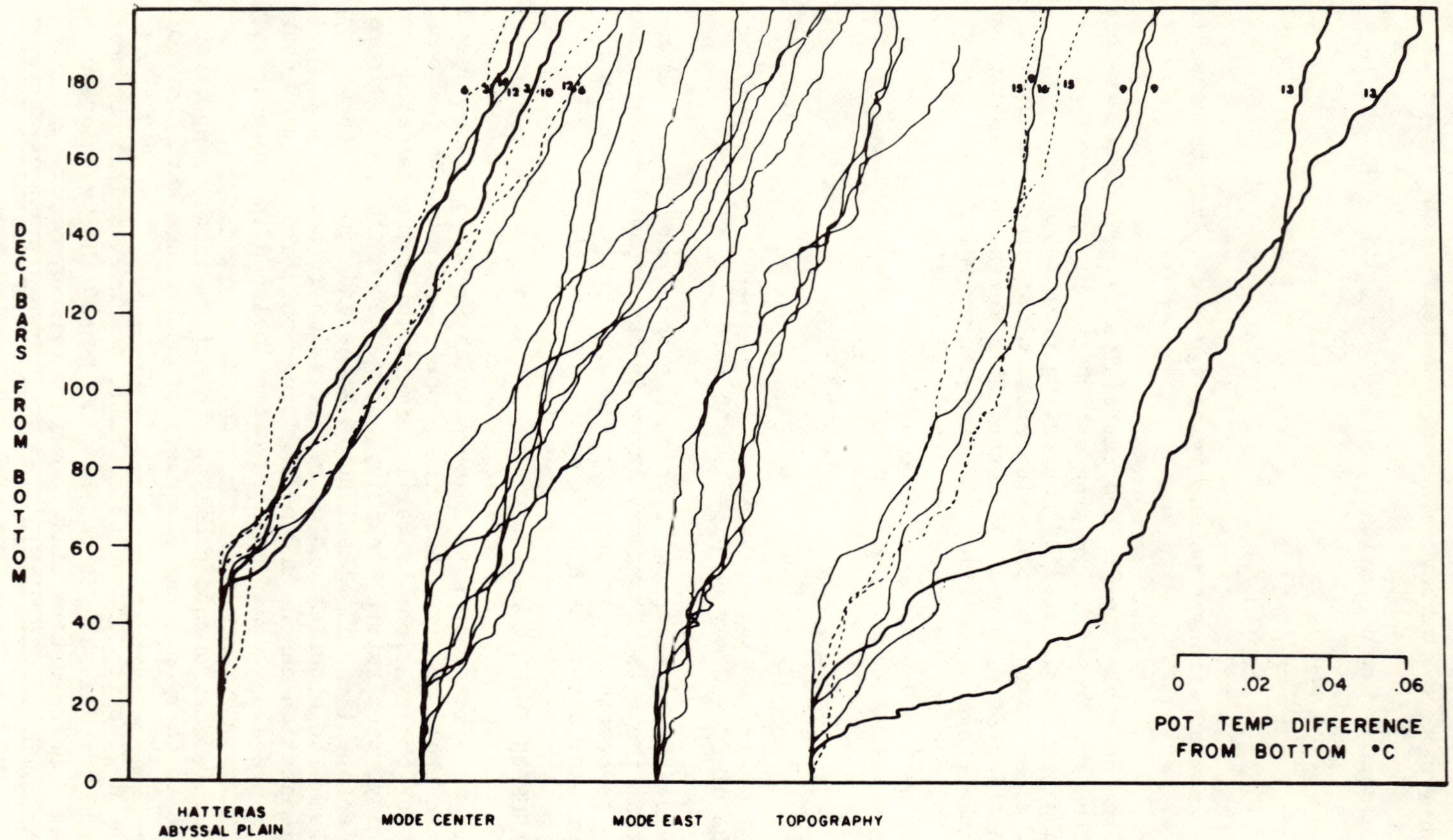


Fig. 1. Observations of bottom boundary layers in the North Atlantic (from Armi and Millard 1976).

the whole package on the bottom. With this device he succeeded in measuring velocities in the cm-range at about 3700 m depth, but only for limited periods (several hours).

Velocity measurements throughout the water column are usually obtained by mounting recording current meters on tight moorings. Problems associated with this technique are the vertical and horizontal oscillations of the mooring. The technique is not well suited to measure the motion in the bottom boundary layer. For such measurements it is much more preferable to mount the sensors on a rig which stands solidly on the bottom. This technique was used by Bowden (1962) in his pioneering studies of the bottom boundary layer in tidal flows. He mounted two electro-magnetic current meters, using different relative arrangements, on a tripod placed on the bottom. However, this was done in shallow water.

One of the most advanced instrument packages so far developed for measurements in the bottom boundary layer of the deep ocean using this approach is the Snodgrass capsule (Snodgrass 1968). The capsule is dropped freely to the ocean floor, records temperature and velocity for periods up to a month, and is recalled by acoustic command from a ship. The sensors are mounted on a frame placed on the bottom with the recording capsule about 15 m above the bottom. Current velocity can be obtained at six levels but both current velocity and direction at two levels only. The temperature can be simultaneously recorded at two levels. The capsule has been used to about 4 km depth (Wimbush and Munk 1971).

A similar approach has been adopted by Williams and Tochko (1977) developing a package to be placed on the bottom equipped with four acoustic velocity sensors for measuring three velocity components each. The sensors are of the Gytre type (Gytre 1975), and they are mounted 50 cm, 120 cm, 250 cm and 600 cm above the bottom.

The approach was also used by Kullenberg (1978) mounting a two-component laser Doppler anemometer (LDA) on a square-formed frame on the bottom for measuring the velocity at one level, 35 cm above the bottom, for water depths to about 100 m (Fig. 2).

Advantages with the LDA are rapid response, no calibration requirements and no sensors introduced in the volume where the flow is measured (Kullenberg and Buckhave 1974; Buckhave 1975).

Rodhe (1973) obtained sections of near-bottom current profiles in the Skagerrak and Kattegat by means of gelatine pendulum current meters attached to a moored string. These meters integrate the current over a 10-20 minute period while the gelatine in a small container coagulates, giving a measure of the velocity by the slope angle of the pendulum imbedded in the gelatine and the direction by

Fig. 2a. Laser Doppler anemometer mounted on frame for placement on the bottom and measurement of the velocity 35 cm above the bottom.

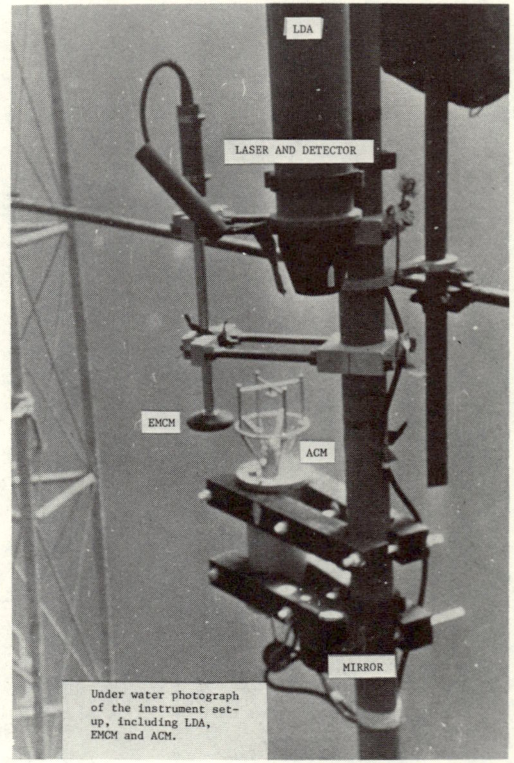


Fig. 2b. Mounting of Laser Doppler anemometer (LDA), electromagnetic current meter (EMCM) and acoustic current meter (ACM) for intercomparison experiment on bottom fixed tower (platform). (From Kullenberg et al., 1975; Jacobsen and Nielsen 1979).

a compass needle attached to the pendulum. The advantage is the quick deployment of the instrument making it possible to obtain several profiles in shelf sea areas across a section over a short period of time.

At greater depths there are considerable difficulties involved in mounting a package on the bottom, and various mooring techniques are often used. In areas where well-defined currents occur the technique has been successfully used (Weatherly 1972, 1977). Weatherly used two moorings, one with nine Savonius rotor current meters from 1 m to 20 m above the bottom and one mooring with 4 film-recording current meters, also using Savonius rotors as sensors, from about 1 m to 33 m above the bottom.

Data from current meter moorings with 4-12 current meters throughout the water column have also been used to determine the properties of the bottom boundary layer in shelf areas, e.g. off north west Africa and Oregon (Kundu 1977).

The acoustic and laser Doppler measuring techniques are considerably more sensitive than the propeller, Savonius rotor or pendulum measuring techniques, being able to detect velocities in the mm $s^{-1}$ range compared to a detection limit of several cm $s^{-1}$ for the others. The electromagnetic sensor falls somewhere in between. An intercomparison experiment between acoustic, LDA and electromagnetic sensors showed these to be relatively well matched (Kullenberg et al 1975; Jacobsen and Nielsen 1980).

Automatically profiling instruments have also been developed whereby a sensor is profiling automatically through the water column at given time intervals, attached to a bottom-mounted rig like the cyclosonde (Van Leer et al 1976). This technique has been used to study the bottom boundary layer on shelfs and continental slopes (Weatherly and Van Leer 1977).

Current measurements in the boundary layer are often combined with temperature and occasionally conductivity recordings. However, continuous profiles of temperature and conductivity can be obtained from ship-board lowered instruments (CTD's). Repeated CTD profiles can be very useful for studying the bottom boundary layer in the deep ocean (Armi and Millard 1976; Armi 1979).

Measurements of the velocity very close to the sea bed have also been carried out (Caldwell and Chriss 1979), from 19 cm above to 2 cm below the bottom, using an automatically traversing heated thermistor, mounted at 199 m water depth. These measurements proved the existence of the viscous sublayer, with properties in good agreement with theory. The surface sediment at the site was indicated to be a silty sand (61% sand, 28% silt, 10% clay) with a mean grain diameter of 0.0042 cm, and no bedforms which could significantly influence the flow were observed at the site.

The existence of near-bottom layers with a relatively high amount of suspended matter, often called nepheloid layers, has been long known (e.g. Jerlov 1953). Suspended matter can be studied by means of optical techniques, measuring total attenuation or scattering properties. Attenuation meters measure the attenuance over pathlengths in the range 25-150 cm, usually in the red part of the spectrum in order to minimize the influence of dissolved substances on the attenuation (e.g. Jerlov 1976). Recently recording attenuation meters have been developed to study the long-term near-bottom fluctuations, by the marine optics groups at OSU (Bartz et al 1978) and at Copenhagen University. The meters are mounted either on moorings or on bottom-attached frames.

Light scattering measurements give a direct measure of the content of suspended matter and can also be used to obtain information of properties of the suspended matter. The scattering can be measured by integrating instruments, giving a measure of the total volume scattering coefficient, or by instruments observing the scattering at a fixed angle (e.g. Jerlov 1976). Instruments measuring the volume scattering function, i.e. the angular distribution of the scattered light, have also been constructed (e.g. Kullenberg 1978a). Usually the scattering in the red part of the spectrum is measured. However, measurements in several wavelengths can yield information about the index of refraction of the scattering particles and thereby about their composition as organic or inorganic material (Zaneveld and Pak 1973).

The naturally occurring isotopes $^{228}$Ra, halflife 5.75 years and $^{222}$Rn, halflife 3.82 days, can be used to study vertical mixing in the deep sea, from the bottom layer, since they both originate in the sediments (Sarmiento et al 1976). There are no known sinks of any significance for either isotope in the deep sea and hence the vertical profiles will be a function of the flux across the sediment-water interface, the radioactive decay and the advective and diffusive transports. This has been used to infer mixing properties of the water column from the bottom several hundred meters upward in the deep ocean (Sarmiento et al 1976).

## 3. SUMMARY OF OBSERVED PROPERTIES OF THE BOTTOM BOUNDARY LAYER

Wimbush and Munk presented results from an analysis of 13 drops of the Snodgrass capsule off Southern California at depths in the range 2-4 km. Typical values of various parameters are given in Table 1, whereby a neutral density stratification was assumed. Current spectra from levels between 20 cm and 640 cm above the bottom showed a rough agreement with the -5/3 slope for wave numbers less than $7 \cdot 10^{-3}$ c.p.cm. However, the slopes of the spectra also indicated the existence of a buoyancy subrange, which would imply a stable stratification. This could however not be substantiated by the direct observations.

Weatherly (1972; 1977) reported results from observations at two sites in the Florida Current at depths of about 350 and 420 m. Characteristic parameters are summarized in Table 1. The values clearly show the expected stronger activity in the Florida Current compared to the deep sea sites of Wimbush and Munk. For comparison, Kundu (1977) determined friction velocities in the range 0.1-1.0 cm $s^{-1}$ from moored current meter observations over the northwest African and Oregon shelfs at depths in the range 40-60 m. Other characteristics determined by Kundu (1977) are given in Table 1.

Caldwell and Chriss (1979) deduced very precise characteristics from their repeated profiles with a vertical resolution of about 1 cm close to and into the bottom, see Table 1. They furthermore determined a shear of 12 $s^{-1}$ giving a stress of 0.18 dyne $cm^{-2}$.

Rodhe (1973) made extensive current measurements, periodically over 3 years, across several sections in the Skagerrak and Kattegat, using pendulum current meters thereby obtaining almost synoptic profiles across the sections. On the basis of these he calculated the flow characteristics using the theory outlined in section 1 above, and some of his results, from one cruise, are summarized in Table 1. The aim of the work was to study the sediment transport in the area towards the Djupa Rännan, an extension of the Norwegian deep trough, near the Swedish west coast. The observations demonstrated the high variability in the flow conditions, with large transporting capacities occurring in the form of episodic events.

Kullenberg (1978) measured the near-bottom horizontal current fluctuations by means of an LDA mounted on a frame on the bottom at depths from 20 to 80 m in the Gulf of Bothnia. The data showed great variability of the flow, with smoothly oscillating flow interrupted by bursts with a duration of about one minute. Some results are summarized in Table 1.

Properties of the bottom boundary layer have also been inferred from observations of the suspended matter distribution, usually in combination with temperature and salinity measurements, in what is often referred to as the nepheloid layer. Such measurements have shown great geographical variations of the properties of the bottom boundary layer, and have also shown large local time variability.

Biscaye and Eittreim (1974) carried out observations including current measurements at 1 m above the bottom, CTD profiles, light scattering profiles, suspended particulate matter and radon concentrations, at sites on the lower slope of the Blake-Bahama Outer Ridge, under the Western Boundary Undercurrent, and on the Hatteras Abyssal Plain. The aim was to compare the properties of the bottom

Table 1. VALUES OF CHARACTERISTIC PARAMETERS DETERMINED BY MEANS OF OBSERVATIONS AT VARIOUS SITES.

| Parameter | Unit | Wimbush and Munk 1971 | Weatherly 1977 | Kundu 1977 Oregon | Kundu 1977 NW Africa | Caldwell and Chriss 1979 | Rodhe 1973 | Kullenberg 1978 |
|---|---|---|---|---|---|---|---|---|
| $u_*$ | cm $s^{-1}$ | 0.1 | 0.4-0.8 | 0.3 | 0.8 | 0.42±0.012 | 0.7-4.0 | 0.1-0.5 |
| Viscous layer | cm | 2 | - | 0.6 | 0.2 | 0.6 | - | - |
| Constant stress layer | cm | 10 | - | - | - | - | - | - |
| Logarithmic layer | cm | 100 | 400 | 200 | 900 | - | - | 30-400 |
| Ekman layer | cm | 500 | 2500 | 1200 | 6000 | - | - | 322-2000 |
| Roughness length | cm | $0.1\nu/u_*$ | 0.03 | - | - | $5.5\cdot10^{-4}$ | 2-0.03 | |
| $\varepsilon$ | $cm^2\ s^{-3}$ | - | - | - | - | 0.013±0.007 | - | $(90-2.5)\cdot10^{-4}$ |
| $c_b$ | - | $2\cdot10^{-3}$ | 0.04 | - | - | $1.1\cdot10^{-3}$ | $(1.2-4.4)10^{-3}$ | |
| Average bottom Current (1-20 m above) | cm $s^{-1}$ | - | 10-40 | 10 | 25 | - | 20-70 | 4-7 |

boundary layer at these different areas and study the variability. They found great differences between the character of the layer at these areas, with a thinner, much less variable layer at the Hatteras Abyssal Plain site than at the undercurrent sites. At the latter area high scattering and particulate concentrations were found, with large variability in the vertical distributions, extending up to about 1500 m above the bottom at a water depth of 3500 m. The bottom (1 m) current velocity was in the range 20-30 cm $s^{-1}$. The radon profiles were not exponential. Vertical diffusion coefficients based on these profiles were of the order of x · 100 $cm^2s^{-1}$. On the contrary at the Abyssal Plain site only a weak nepheloid layer was present, extending at most 900 m above the bottom, at a water depth of 4500 m. The radon profiles were exponential, corresponding to a vertical diffusion coefficient of the order of 10 $cm^2s^{-1}$. Temporal variability was present at both areas although much more pronounced at the undercurrent sites.

Kullenberg (1978a, 1978b) observed light scattering profiles in upwelling areas and in the Southern Ocean (Fig. 3 ). Bottom layers rich in suspended matter were found very markedly separated from the water column above, and with a thickness in the range 1 to 20 m in the upwelling areas and about 50 m in the Southern Ocean. At the slope up towards Cape Horn no increase of suspended matter was observed towards the bottom. This is probably related to the strong currents prevailing in the area. Similar measurements have also been carried out in the Barents Sea area, showing the existence of particle rich layers with different characteristics (Fig. 3 ). Zaneveld and Pak (1981/82) summarized several observations along continental slopes and fronts of the distribution of suspended matter using optical techniques, showing extensions of particle rich layers out from the shelf and slope regions.

A few studies have been mentioned in order to give an idea of the characteristics of the oceanic bottom boundary layer, many more investigations clearly existing. The time- and space-variability is a very essential element. Bursting flows have been found, and periodic deep sea 'storms' may well be responsible for the main part of material motion. In this relation the possible influence of the suspended matter on the stress should be mentioned, studies indicating possible stress reducing effects (e.g. Smith and McLean 1977).

## 4. FURTHER WORK

Detailed, long-term studies of the bottom boundary layer in various regions of the ocean are needed to improve our understanding of the material transport and the processes involved, in particular as regards the variability and the forcing. A good example of an interdisciplinary study of the bottom layer in the deep ocean employing state-of-the-art technology is the High Energy Benthic

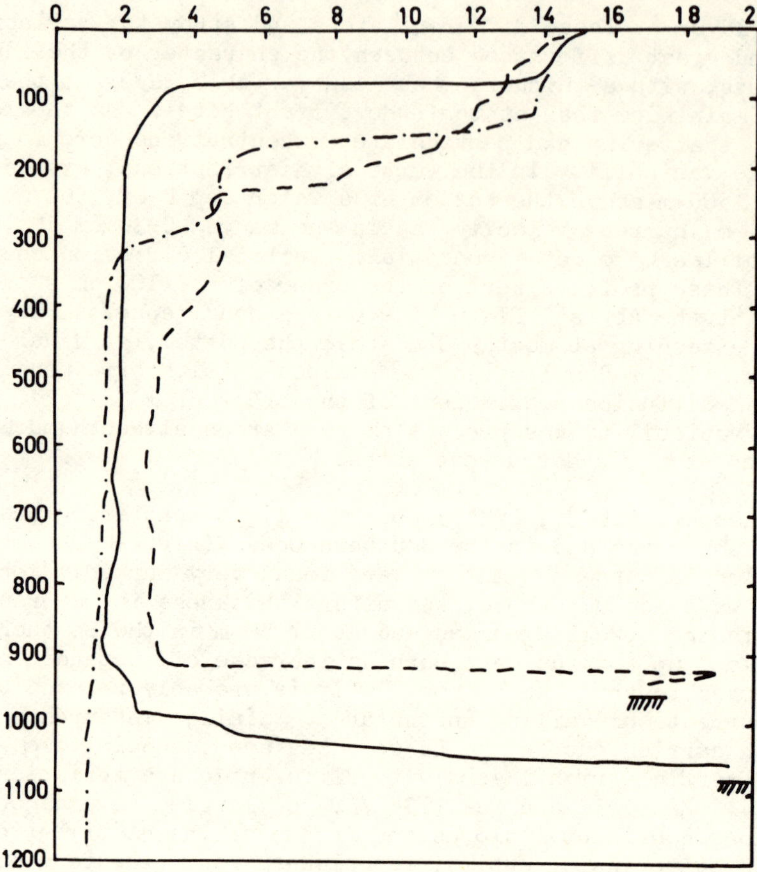

0 2 4 6 8 10 12 14 16 18 20
100
200
300
400
500
600
700
800
900
1000
1100
1200


Fig. 3a

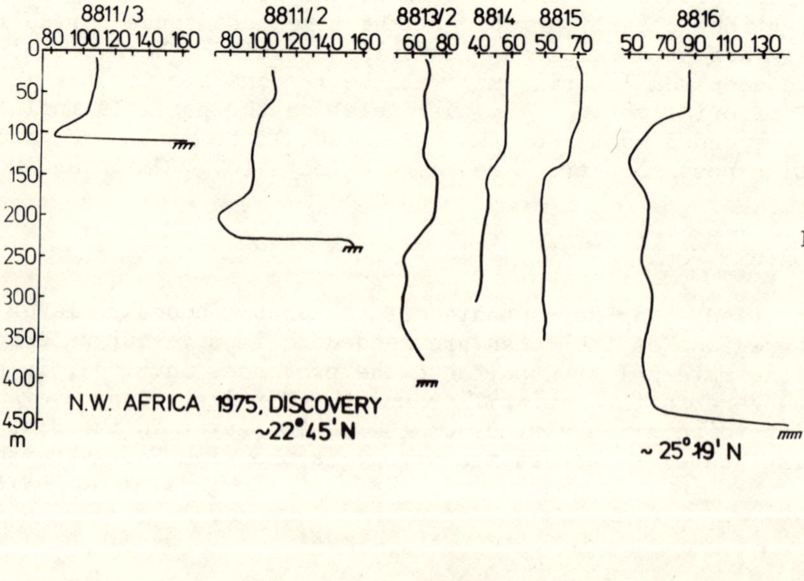

8811/3
80 100 120 140 160
8811/2
80 100 120 140 160
8813/2
60 80
8814
40 60
8815
50 70
8816
50 70 90 110 130
0
50
100
150
200
250
300
350
400
450
m
N.W. AFRICA 1975, DISCOVERY
~22°45'N
~25°19'N


Fig. 3b

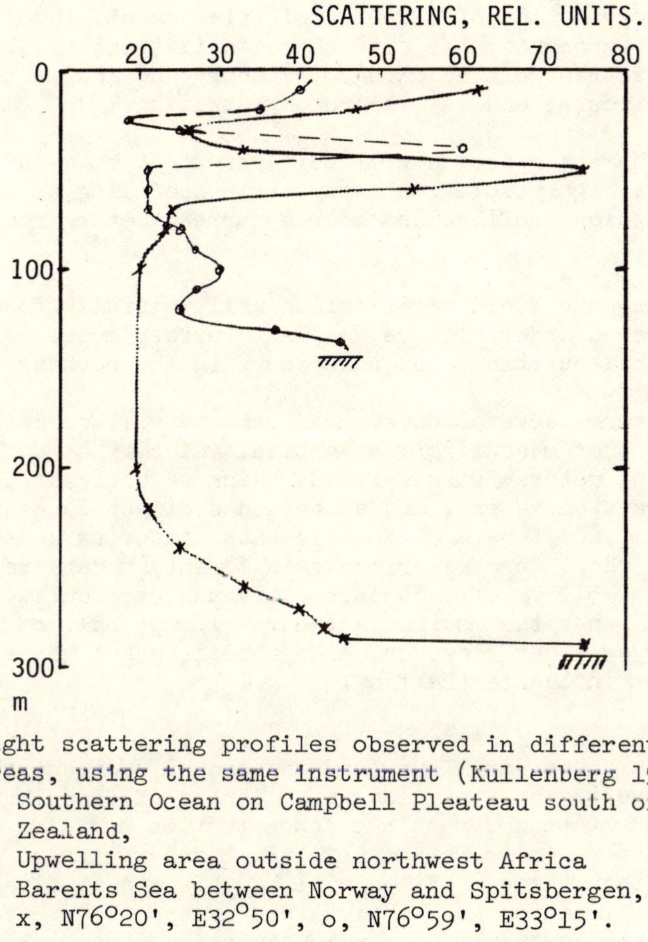


Fig. 3c

Fig. 3. Light scattering profiles observed in different oceanic areas, using the same instrument (Kullenberg 1978a):
a) Southern Ocean on Campbell Pleateau south of New Zealand,
b) Upwelling area outside northwest Africa
c) Barents Sea between Norway and Spitsbergen, x, N76°20', E32°50', o, N76°59', E33°15'.

Boundary Layer Experiment (HEBBLE).

HEBBLE is a comprehensive research program whose aim is the implementation of a boundary layer and sediment transport model and its field verification to predict sediment transport in areas of strong bottom boundary layer flows in the deep ocean. The initial area of experimentation lies at 4-5 km depth on the lower slope region south of Nova Scotia.

The program plan of HEBBLE (Hollister et al, 1980) contains four major components: (1) site characterization; (2) Ekman layer and sediment transport modeling; (3) short and long term boundary layer experiments; and (4) laboratory studies .

Site characterization will be carried out by means of bottom photographs, high-resolution bathymetric profiling using Deep Tow, CTD/transmission profiles and moored current meter/transmissometer arrays.

Modeling and field verification will initially focus on time-averaged second-order closure models. Further modeling will include intermittancy, bursting, and coherency, in the boundary layer.

Short term (several days) measurements will be carried out with the goal of measuring the vertical velocity structure using rotor current meters; stress fields using acoustic stress sensors and hot wire anemometers; and suspended sediment concentrations using transmissometers and acoustic backscattering sensors. The goal of the short term experiments is to obtain data regarding near-bed variability of flow parameters and particulate matter over periods when the motion is the result not only of a simple boundary layer shear flow, but also when internal waves and inertial effects influence the flow.

Long term experiments will study in detail the fluid mechanics of the benthic boundary layer and associated sediment transport. The objective is to obtain long time series of turbulent stress, velocity and suspended particle concentration profiles near the bed, and in the outer region of the boundary layer. It is envisioned that these profiles will be obtained at five locations. At a central location geotechnical and camera observations will also be carried out. The instrumentation is essentially the same as during the short term studies but will also include sediment traps and particle settling velocity meters. As of this writing, measurements have been carried out to obtain shear stress, shear velocity, roughness length and time scales within the benthic boundary layer. Strings of vector-averaging current meters have shown bursts of very strong near-bottom currents (up to 72 cm $s^{-1}$) flowing in a southwesterly direction on the continental rise (Richardson and Wimbush, 1980). The same workers determined that over a long term (i.e. 6 months) the vector averaged flow was west-southwesterly with mean speed of 7-8 cm $s^{-1}$. By using a stack of Savonius rotors (called the chandelier) and an acoustic current array (called BASS: benthis acoustic stress sencor) Williams and Weatherly (1980) determined the Reynolds stress at 1 m above bottom in this area to be 0.28 dynes $cm^{-2}$ ($\pm$25%), and $u_*$ to be 0.4 cm $s^{-1}$. Pak, Zaneveld and Weatherly (1980) obtained profiles of temperature, salinity and light transmission and found that the bottom mixed layer was one of variable thickness, extending anywhere from 15 to 160 meters above the bottom. This information when coupled with the current meter data mentioned above serves to define a filament of colder, fresher, murkier water flowing southwestward at the base of the continental rise (Weatherly, et al. 1980), sometimes at speeds greater than a knot. The consequent "discharge" or volume flow from this filament is estimated at $(2\text{-}6)\times10^6$ $m^3$ $s^{-1}$, a value which agrees well with the volume flow of the Western Boundary Undercurrent. Tidal effects have been shown to be strongly damped near the bottom; this result was obtained by Bird and Weatherly (1980) using a numerical model of the bottom boundary layer.

Changes in particulate concentration and nature as well as bedform morphology have been detected optically, photographically, seismically, and from samples taken with coring devices and sediment traps. Biscaye, et al. (1980) noted the strong variability of particulate concentration in both space and time. "Benthic storms" of extremely turbid water (up to 12000 μg $\ell^{-1}$) with apparent thickness of hundreds of meters and extending hundreds of kilometers horizontally were recorded over several months of near-bottom observation using rephelometers and transmisometers. Tucholke, et al. (1980) used bottom photographs to determine that three sizes of current-produced bedforms were dominant in the hemipelagic muds found in the continental rise off Nova Scotia. The effects of turbidity currents flowing normal to the dip of the rise were observed by Shor and Tucholke (1980). Bottom sediment samples have been studied to learn about biological activity and sedimentary processes (Yingst and Aller, 1980). Sediment trap arrays deployed by Gardner, et al. (1980) indicated that the total particulate flux increased exponentially towards the bottom (from 500 m above bottom, downward) and that the total "apparent" flux (i.e. surface-derived plus resuspended material) increased downslope as the mean current velocity increased. They conclude that in a model characterized by the magnitude of turbulence seen in HEBBLE, resuspension of large particles to elevations of up to 100 meters above bottom must be accounted for. To this end work is being done to determine the hydrodynamic properties of large particles such as fecal pellets (Nowell, et al. 1980). In addition, work is also being done on the development of new in-situ instruments to be used for particle detection in the water column. Of note are the acoustic backscattering system (Orr, et al. 1980), the holographic velocimeter (Carder and Steward, 1980). Eventually an in-situ recirculating seaflume will also be deployed to conduct erosion and deposition experiments. Examples of instrument packages used in HEBBLE are shown in Figures 4 and 5.

The culmination of the HEBBLE project will be a long-term field test of the Ekman layer structure and sediment transport models derived form the work presently being done. From this, it is hoped, a comprehensive statement may be made about the turbulent transfer processes associated with the strong currents in the deep-ocean benthic boundary layer.

## 5. CONCLUSIONS

It can be concluded that the recent technical development has made it possible to study the bottom boundary layer in the deep sea in much more detail than earlier. Studies so far have revealed a variability and complexity not earlier thought of, implying that further research is needed, in different areas. In relation to deep water formation and the outflow from the Norwegian-Greenland seas, studies of the bottom boundary layer across the southern entrances to these areas seem of great interest. Such studies should be planned on the basis of earlier work on the overflow across the ridges (see Meincke, this volume).

## 6. LIST OF NOTATIONS

| | |
|---|---|
| $C(z)$ | particle concentration; $c_b$ friction coefficient |
| $f$ | Coriolis parameter |
| $F_p$ | flux of particles |
| $g$ | acceleration of gravity |
| $h_{\ell n}$, $h_E$ | logarithmic and Ekman boundary layer thickness, respectively |
| $K_m$, $K_z$ | vertical turbulent transfer coefficients for momentum and matter, respectively |
| $Kp$ | vertical turbulent transfer coefficient for particles |
| $N$ | brunt-Väisälä frequency |
| Ri, Rf | Richardson and flux Richardson number, respectively |
| $\bar{U}$ | mean horizontal current velocity |
| $u_*$ | friction velocity |
| $U_g$ | (geostropic) velocity outside boundary layer |
| $w'$ | vertical velocity fluctuations |
| $w_p$ | settling velocity |

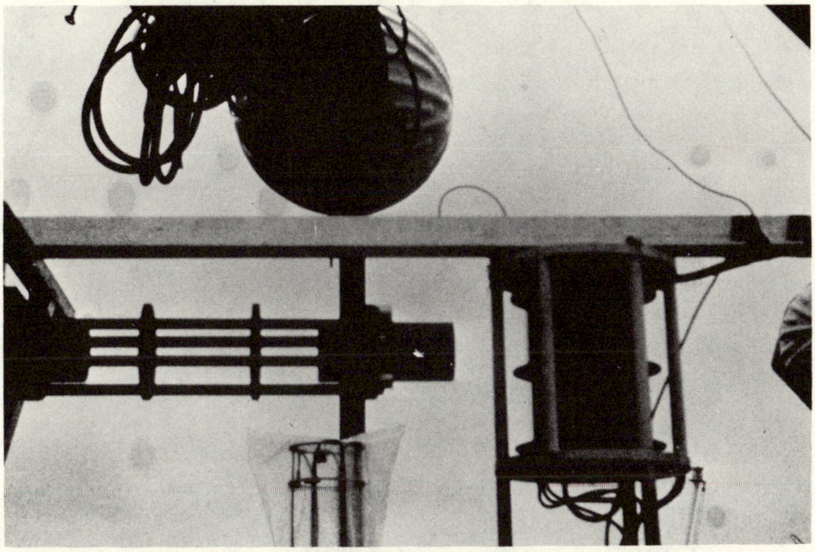

Fig. 4a. Benthic Acoustic Stress Sensor (BASS), showing transmissometer, acoustic current meter and rotor current meter.

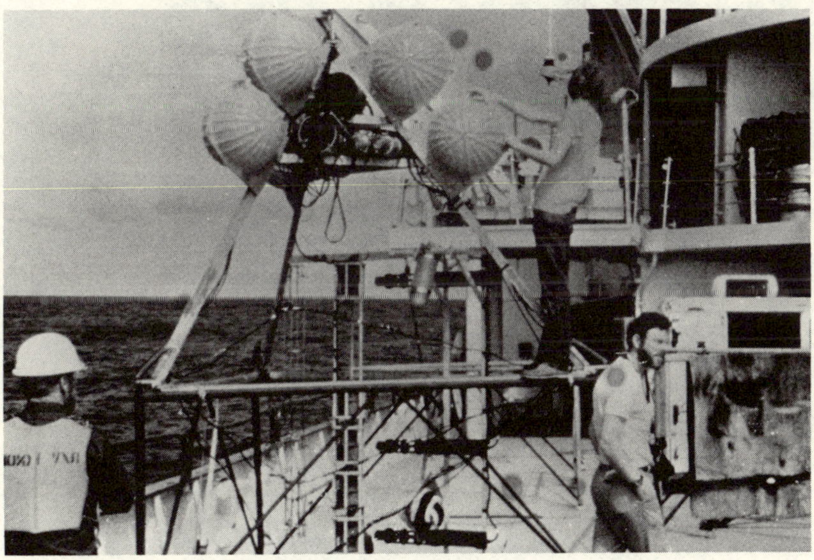

Fig. 4b. Bass mounted for deployment on the bottom showing the three levels of measuring sediment transport. The BASS package was designed by A.J. Williams III, Woods Hole Oceanographic Institution.

Fig. 5a. Profiling instrument package consisting of Lamont nephelometer on top,
Neill Brown CTD with General Oceanics rosette sampler,
Transmissometer, one meter path length, mounted in place of one of the rosette bottles.

| | |
|---|---|
| $z$ | vertical coordinate, positive upwards |
| $\varepsilon$ | rate of kinetic energy dissipation per unit mass |
| $\upsilon$ | viscosity |
| $\kappa_o$ | van Karman constant |
| $\rho$, $\rho_o$ | density and mean density, respectively |
| $\rho'$ | density fluctuations |
| $\tau$, $\tau_b$ | bottom stress |

Fig. 5b. Close up of CTD with rosette sampler and transmissometer

REFERENCES

Armi, L., 1979, Some evidence for boundary mixing in the deep ocean. Journal Geophys. Res., 83(C4):1971.

Armi, L. and Millard Jr., R. C., 1976, The bottom boundary layer of the deep ocean. Journal Geophys. Res. 81:4783-4990.

Bartz, R., Zaneveld, J. R. V. and Pak, H., 1978, A transmissometer for profiling and moored observations in water. Soc. Photo-Optical Instrumentation Engineers, 160, Ocean Optics V:102-108.

Bird, A. A. and Weatherly, G. L., 1980, Numerical simulations of the bottom boundary layer in the HEBBLE area. Abstract, EOS 61(46):1015.

Biscaye, P. E. and Eittreim, S. L., 1974, Variations in benthic boundary layer phenomena: nepheloid layer in the North American Basin, in "Suspended Solids in Water", R. J. Gibbs, ed., Plenum Marine Sc. Ser. 4:227-260, New York (1974).

Bowden, K. F., 1962, Measurements of turbulence near the sea bed in a tidal current. Journal Geophys. Res. 67(8):3181-3186.

Buchhave, P., (editor), 1976, The accuracy of flow measurements by laser-Doppler methods. Proceedings of the LDA Symposium, Copenhagen 1975, Hemisphere Publ. Comp.

Caldwell, D. R. and Chriss, T. M., 1979, The viscous sublayer at the sea floor. Science, 205:1131-1132.

Carder, K. L. and Steward, R. G., 1980, In-situ holographic measurements of the size and settling velocities of ocean particles. Abstract, EOS 61(46):1017.

Gardner, W. D., Hinga, K. R., Richardson, M. J. and Biscaye, P. E., 1980, Large, resuspended particles in sediment traps. Abstract, EOS 61(46):1015.

Gytre, T., 1975, Ultrasonic measurements of ocean currents down to 1 mm $s^{-1}$, Conference Proceeding No. 32 of the IERE Conference on Instrumentation in Oceanography, 23-25 September 1975, University College, N. Wales, Bangor, U.K.,pp69-80.

Hollister, C. D. and McCave, I. N., 1980, HEBBLE, the high energy benthic boundary experiment. Abstract, EOS 61(46):1014.

Jerlov, N. G., 1953, Particle distribution in the ocean. Rep. Swedish Deep- Sea Exped. 3:1-59.

Jerlov, N. G.,1976, Marine Optics. Elsevier Oceanographic Ser. 14, Amsterdam.

Kullenberg, G., 1978, Preliminary results of near bottom current measurements in the Bothnian Sea. Finnish Marine Res. 244: 42-51.

Kullenberg, G., 1978a, Light scattering observations in the northwest African upwelling region. Deep-Sea Res. 25:525-542.

Kullenberg, G., 1978b, Light scattering observations in frontal zones. Journal Geophys. Res. 83(C9):4683-4690.

Kullenberg, G. and Buchhave, P., 1974, An oceanographic in-situ laser Doppler anemometer. ICES C.M. 1974/C:25.

Kullenberg, G., Woods, J. D., Gytre, T., Wiley, R., Buchhave, P., and Crepon, M., 1975, Preliminary report on current meter intercomparisons at STARESO, August 1975, Institute of Physical Oceanography, University of Copenhagen.

Kundu, P. K., 1977, On the importance of friction in two typical continental waters: off Oregon and Spanish Sahara, in: "Botton Turbulence", J. C. J. Nihoul ed., Elsevier Oceanogr. Ser. 19:187-208, Amsterdam, 1977.

Monin, A. S. and Obukhow, A. M., 1953, Dimensionless characteristics of the turbulence in layers of the atmosphere near the earth, Doklady Nat. Acad., USSR 93(12).

Nielsen, P. B. and Jacobsen, T. S., 1980, An intercomparison of acoustic, electromagnetic, and laser Doppler current meters at STARESO 1975. Report No. 41, Institute of Physical Oceanography, University of Copenhagen.

Nowell, A. R. M., Jumars, P. A. and Miller, D. C., 1980, Hydrodynamic characteristics of fecal pellets. Abstract, EOS 61(46):1017.

Orr, M. H., Hess, F. R. and Howard, E. J., 1980, High frequency acoustic detection of suspended particle distributions. Abstract, EOS 61(46):1016.

Pak, H., Zaneveld, J. R. V. and Weatherly, G., 1980, Boundary layer and exchange processes. Abstract EOS 61(46):1015.

Richardson, M. J. and Wimbush, M., 1980, An exceptionally strong near-bottom current on the continental rise of Nova Scotia. Abstract EOS 61(46):1014.

Rodhe, J., 1973, Sediment transport and accumulation at the Skagerrak-Kattegat border. Göteborgs Universitet, Oceanografiska Institutionen, Report No. 8.

Sarmiento, J. L., Feely, H. W., Moore, W. S., Bainbridge, A. E. and Broecker, W. S., 1976, The relationship between vertical eddy diffusion and buoyancy gradient in the deep sea. Earth and Planetary Science Letters 32:357-370.

Shor, A. A. and Tucholke, B. E., 1980, Geologic effects of turbidity currents on the Nova Scotian continental rise. Abstract EOS 61(46):1015.

Snodgrass, F. E., 1968, Deep-sea instrument capsule, Science 162: 78-87.

Smith, J. D. and McLean, S. R., 1977, Boundary layer adjustments to bottom topography and suspended sediment, in "Bottom Turbulence", J. C. J. Nihoul, ed., Elsevier Oceanogr. Ser. 19:123-152, Amsterdam 1977.

Sternberg, R. W., 1969, Camera and dye-pulse system to measure bottom boundary layer flow in the deep sea, Deep-Sea Res. 16:549-554.

Tennekes, H., 1973, The logarithmic wind profile. J. Atm. Sc., 30:234-238.

Tucholke, B. E., Hollister, C. D., Biscaye, P. E. and Gardner, W. D., 1980, Current-produced bedforms in HEBBLE study area, Nova Scotia continental Rise. Abstract EOS 61(46):1014.

Van Leer, J., Düing, W., Erath, R., Kennelly, E. and Speidel, A., 1974, The cyclosonde: an unattended vertical profiler for scalar and vector quantities in the upper ocean, Deep-Sea Res. 21(5)_385-400.

Weatherly, G. L., 1972, A study of the bottom boundary layer of the Florida Current, Journal Phys. Oceanogr., 2:54-72.

Weatherly, G. L., 1977, Bottom boundary layer observations in the Florida Current, in: "Bottom Turbulence", J. C. J. Nihoul, ed., Elsevier Oceanogr. Ser. 19:237-254, Amsterdam, 1977.

Weatherly, G. L. and Van Leer, J. C., 1977, On the importance of stable stratification to the structure of the bottom boundary layer on the Western Florida shelf, in "Bottom Turbulence", J. C. J. Nihoul, ed., Elsevier Oceanogr. Ser. 19:103, Amsterdam , 1977.

Weatherly, G. L., Kelley, E. A., Zaneveld, J. R. V., Pak, H., Richardson, M. J. and Wimbush, M., 1980, A deep, narrow, thin filament of the Western Boundary Undercurrent (WBUC), Abstract EOS 61(46):1016.

Williams 3rd, A. J. and Tochko, J. S., 1977, An acoustic sensor of velocity for benthic boundary layer studies, in: "Bottom Turbulence, J. C. J. Nihoul, ed., Elsevier Oceanogr. Ser. 19:83.

Williams 3rd, A. J. and Weatherley, G., 1980, Reynolds stress and velocity profile at the HEBBLE site 40°N, 63°W. Abstract EOS 61:1015.

Wimbush, M. and Munk, W. H., 1971, The benthic boundary layer, in: "The Sea" 4:731-758, A. Maxwell, J. Wiley and Sons, ed., New York, 1971.

Yingst, J. Y. and Aller, R. C., 1980, Biological activity and associated sedimentary structures in HEBBLE area deposits, Abstract EOS 61(46):1014.

Zaneveld, J. R. V. and Pak, H., 1973, Method for the determination of the index of refraction of particles suspended in the ocean. Jour. Optical Soc. Am. 63:321-324.

Zaneveld, J. R. V., Pak, H., Biscaye, P., Gardner, W. and Richardson, M. J., 1980, The interrelationship of beam transmission, nephels and suspended particulate matter, Abstract EOS 61(46):1017.

Zaneveld, J. R. V. and Pak, H., 1981/82, Processes effecting the distribution of suspended matter in the oceans. Symposium on Transfer Processes, IAPSO, IUGG 17th General Assembly, Canberra 1979, Elsevier Oceanogr. Ser., Amsterdam, in press.

# CHAPTER 3

# REVIEW OF GEOLOGICAL KNOWLEDGE

# OUTSTANDING GEOLOGICAL PROBLEMS OF THE GREENLAND-SCOTLAND RIDGE: AN INTRODUCTION

Jörn Thiede

Department of Geology, University of Oslo
P.O. Box 1047, Blindern
Oslo 3, Norway

The geological contributions of this volume about the Greenland-Scotland Ridge cover a wide range of distinct and controversial subjects most of which have been discussed but not solved during the advanced study institute in Bressanone/Italy during May 1981. The diverse opinions of the individual authors on problems such as for example the nature of the basement under the Faeroe Islands, the origin and type of volcanism building the ridge, the age and nature of the submarine parts of the Greenland-Scotland Ridge, the existence, continuity and break-up of a land bridge along the ridge, its consequences for the exchange of water masses between the Norwegian-Greenland Sea and the main North Atlantic basin etc., vividly illustrate how many unsolved questions about this geological feature exist.

The Greenland-Scotland Ridge (Fig. 1) comprises a major part of the Thulean volcanic province which stretches across the North Atlantic Ocean, which separates the Norwegian-Greenland Sea from the main North Atlantic basins, and which is composed of late Cretaceous-Recent volcanic rocks outcropping on eastern Greenland, Iceland, the Faeroes and the British Islands. However, its regional limits are not well known because the volcanic province seems to continue under the inland ice to western Greenland and across Baffin Bay to the Canadian Arctic islands (Baffin Island). Also the volcanism which is documented north of the ridge along eastern Greenland, which is active on Jan Mayen and which led to the construction of Vøring Plateau off Norway, and its relationship to the thermal events along the Greenland-Scotland Ridge as well as to the episodes of North Atlantic sea floor spreading are poorly understood. The nature of the boundaries of the ridge towards normal oceanic crust is unknown. The large depth anomaly

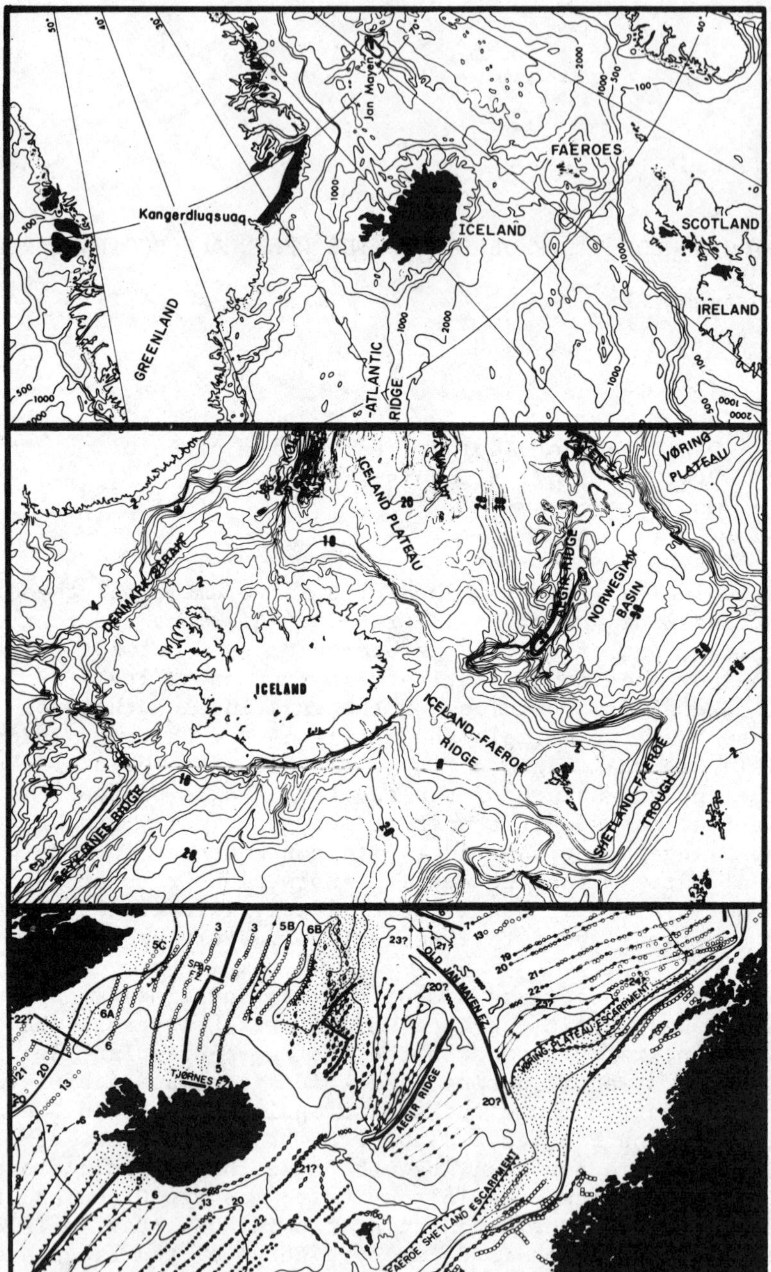


Fig. 1. Morphology of the Greenland-Scotland Ridge, the distribution of its subaerial volcanic terrains, and the magnetic anomalies of the adjacent oceanic crust. Depths in meters. For sources see Thiede and Eldholm (this volume).

generated by the volcanism and by the thermal phenomena along the Greenland-Scotland Ridge since late Cretaceous time can be observed over many hundred kilometers in the adjacent ocean basins south of the ridge, but they seem to be much more pronounced south of the ridge. It is well known that part of the volcanic province which is believed to have been generated by hot-spot volcanism is situated on continental crust, whereas part of it is located on oceanic crust, but the precise position and the nature of the boundaries between the two different types of crust is unknown. For example, based on geophysical evidence some authors believe that the Faeroe Islands are located on oceanic crust, others think on continental crust. New studies of three silicic layers in the volcanic rock pile of the Faeroe Islands suggest that some of the basaltic lavas were contaminated by continental crustal material. In particular, the nature of basement under the Faeroe-Shetland and Faeroe-Bank channels and under several of the smaller banks between Rockall and the Faeroe Islands should be studied in great detail.

The ages of volcanic rocks and intercalated sediments of the subaerial segments of the Greenland-Scotland Ridge which have been analysed by radiometric, magnetic and paleontologic methods, decrease in general towards Iceland, but it is unknown when and where the first volcanism occurred and how old the submarine segments of the ridge are. Therefore, the temporal and spatial sequence of volcanic and thermal events resulting in the construction of the entire volcanic province and the reasons for the episodic production of large volumes of basaltic magma cannot be reconstructed. Lack of these crucial data resulted in great ambiguities how to interprete some anomalies of terrestrial and marine biogeography, of the Cenozoic paleo-oceanography of the North Atlantic Ocean and of the Norwegian-Greenland Sea as well as of the paleoclimate of the northern hemisphere.

Fossils of terrestrial faunas and floras which have been collected on either side of the Greenland-Scotland Ridge or which are preserved in the intrabasaltic sediments of the subaerial segments of the ridge, suggest that a land bridge across the ridge once connected Greenland-North America with northwest Europe. However, it is still unknown when this land bridge first came into existence in late Mesozoic time when probably a shallow seaway connected the Arctic and Atlantic oceans, and how long it functioned as a pathway for terrestrial organisms. Estimates of the timing of the break-up range from Eocene to Miocene, others have suggested that it might have existed until Plio-Pleistocene. Some of the biogeographic changes coincide closely with events of Cenozoic paleoclimate, but with large parts of the ridge unknown it is very difficult to discern between the consequences of climatic and tectono-volcanic processes. Also it is presently impossible to estimate how much of the original volcanic edifice has been removed by erosional processes or altered in its elevation by postvolcanic tectonism. The terrestrial paleoenvironments which developed on the subaerial segments of the Greenland-Scotland Ridge and on the above mentioned land bridge led to the formation of lateritic soils during Paleogene times, to the deposition of limnic deposits in Paleogene and Neogene times, but our knowledge of these paleoenvironments is only very fragmentary. Ice shields covered Iceland, the Faeroe Islands and parts of the British Isles during Glacial periods of the late Cenozoic, but their first appearance and their extent across the present coast lines, possibly across the adjacent shelf regions and the isostatic response of the glaciated areas are largely unknown.

The Greenland-Scotland Ridge is today constraining the water exchange between the main North Atlantic basin and the Norwegian-Greenland Sea. During the existence of an apparently continuous land bridge from northwestern Europe to Greenland this water exchange must have been completely interrupted. However, marine faunal and floral data suggest that connections between both ocean basins existed even in times when terrestrial fossils indicate a complete closure of the sea ways across the ridge. Where these connections have to be sought is open at the present time. Pelagic fossil assemblages indicate that an important water mass exchange between the ocean basins north and south of the ridge began in Eocene/Oligocene times whereas the biogeography of shallow water marine benthic assemblages seems to suggest that the ridge represented an effective barrier between both ocean basins until well into the Neogene. The type of water exchange through the slowly deepening channels across the ridge prior to a climate of the glacial mode on the northern hemisphere, is very difficult to assess and it is unknown when and how these channels have been generated. However, the circulation pattern of shallow and deep oceanic water masses across the ridge after the climatic deterioration during Late Tertiary and Quaternary times is also still an enigma because it is unknown when the overflow of Arctic Deep Water from the Norwegian-Greenland Sea into the North Atlantic began. It is disputed where the limits of a permanent ice cover of the ocean during glacial episodes has to be sought, if the Norwegian-Greenland Sea was completely or partly ice covered during glacials and if large ice shelves developed during glacial stages. Some authors also believe that the Greenland-Scotland Ridge represented a barrier for ice bergs from the north during times of low glacial eustatic sea levels.

The Greenland-Scotland Ridge with its subaerial and submarine segments between northwestern Europe and eastern Greenland provides us with a unique opportunity to study the effect of hotspot volcanism upon continental and oceanic crust. It bears the largest known subaerial segment of an oceanic spreading center whose history can be traced through 70 million years to a time span when the Norwegian-Greenland Sea did not exist. We can see how the structure of continental margins respond to the thermal and volcanic events which probably resulted in the establishment of a land bridge across a wide area which is underlain by oceanic crust. There are few structural highs on our globe which have a more profound influence on the hydrography of the world ocean than the aseismic rises of the Atlantic Ocean of which the Greenland-Scotland Ridge is probably the most important one. The geological investigations proposed by the various authors of this volume will help to solve many of the remaining problems of the nature and origin of the volcanic province which bears the Greenland-Scotland Ridge.

# RARE EARTH GEOCHEMISTRY OF ICELAND BASALTS:

# SPATIAL AND TEMPORAL VARIATIONS

J-G. Schilling, P.S. Meyer and R.H. Kingsley

Graduate School of Oceanography
University of Rhode Island
Kingston, R.I. 02881

## ABSTRACT

There is a progressive secular decline in light-RE enrichment from Tertiary to Present, similar to that previously observed on the Faeroes 50-60mybp. Light-RE depleted oceanic type patterns appear on Iceland only within the last 0.7my. These results suggest pulsating mantle plume activity with time and provide further support to the blob model previously proposed for the origin of the Faeroes-Iceland-Greenland Ridge. Available Sr and Nd isotopic data are consistent with the model.

Regular spatial variations are observed within the SW and SE neovolcanic zones. (La/Sm) decreases and Mg-values increase toward Central Iceland in both zones but at a faster rate in the SE zone. These spatial variations and available isotopic and heat flow data are best explained by a temporal change in melting conditions due to southward propagations of rifts from Central Iceland, starting approximately 6mybp along the SW zone and more recently along the SE zone. The SW zone has now reached thermal maturity, and basalts derived from the upper mantle are composed of a mix of light-RE enriched plume material and depleted asthenospheric material. The SE zone is still in a transient stage of evolution, and lavas from the tip of the propagating rift are derived by melting of the older Late-Miocene lower crust-lithosphere, whereas in the northern part of the zone, lavas are derived from a mantle plume-asthenospheric mix now dominated by the latter type of material. In contrast, there is no geochemical evidence of rift propagation in northern Iceland, as no systematic variations have been observed along either the NE neovolcanic zone or the Skagi paleorift zone.

## INTRODUCTION

There are an increasing number of geophysical, geological and geochemical observations which suggest that Iceland represents an anomalously elevated segment of the Mid-Atlantic Ridge where unusual processes and conditions are superimposed on those associated with simple seafloor spreading.

For example, the Iceland crust is thicker and its structure distinct from that of typical oceanic crust. It is also richer in incompatible trace elements and radiogenic Pb and Sr, reflecting similar anomalies in the underlying mantle, including volatiles which grade outward into the depleted asthenosphere in asymmetrical fashion along the Mid-Atlantic Ridge (1-7). Furthermore, at any fixed Mg-value, basalts of Iceland are Mg, Ca, alkali and notably Ti and Fe-richer and SiO2 poorer than more typical oceanic crust (7). The melting zone beneath Iceland seems to extend to greater depth, volcanism is more intense, and its products greatly more varied than along the submerged part of the Mid-Atlantic Ridge. Iceland's tectonic evolution appears to have been influenced by complex rift jumping and propagation which occurred intermittently throughout its history, sometimes even producing two parallel rifts with activities overlapping in time (8-12), as for instance the SE and SW zones (Fig. 1).

The cause of such unusual geochemical morphologic, tectonic and geophysical anomalies of the Mid-Atlantic Ridge over Iceland remains poorly understood and controversial. The presence of the East Greenland-Iceland-Faroes Ridge, however, suggests that such unusual activity has been long-lived, and apparently variable in intensity.

The purpose of this paper is to present an up-to-date report of rare earth (RE) data we have accumulated on basalts from the Iceland-Faroes-East Greenland region, emphasizing spatial and temporal variations, as well as comparing known Icelandic rift zones at apparently different thermal and tectonic stages of development. For example, the NE and SW zones, which are current landward extensions of the Mid-Atlantic Ridge axis, are considered to be mature rift zones at a quasi-steady state, exhibiting high heat flow and well-developed extensional tectonics (8). On the other hand, the SE zone, a flank zone parallel to the SW rift zone, appears to be a younger rift in the process of propagating southward. This rift zone does not yet exhibit any heat flow anomaly and has only poorly developed extensional features. It appears to be at a transient stage of development. In contrast, the Skagi zone appears to be an ephemeral paleorift which developed about 2.5mybp on the flank of the main rift zone in North Iceland, and subsequently died without ever reaching maturity (14,15). Contrary to the SE zone, there is no petrological and geochemical evidence suggesting that

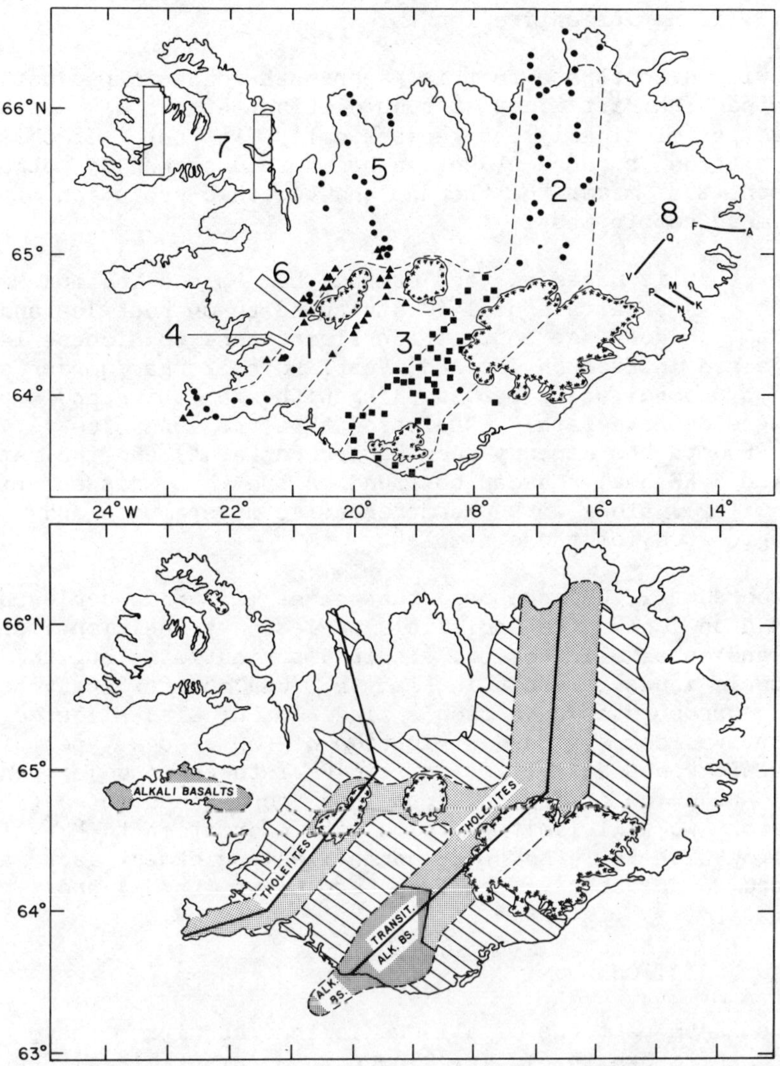


Fig. 1. Iceland maps. Top: location of samples, volcanic zones and regions that have been studied here; 1) SW zone, 2) NE zone, 3) SE zone, 4) West Iceland, 5) Skagi, 6) Borgarfjordur, 7) Vestfirdir, 8) Eastern Iceland. Bottom: Recent volcanic zones (shaded), Plio-Pleistocene formations (oblique lines) and Miocene formations (white) (11). The distribution of petrologic types is after Jakobsson (22,23). The jagged line across the SE zone separates transitional alkali basalts with (La/Sm) > 1.75 from tholeiites with (La/Sm) < 1.75. Central axes through the volcanic zones are based on the orientation of fissure swarms and volcanic systems (23,24).

this ephemeral rift developed by propagation but rather evolved uniformly along its entire length.

We will demonstrate in this paper that volcanic products associated with rift zones at such distinct stages of development are also petrologically and geochemically distinct. If this interpretation is correct, we can begin to decipher by petrological and geochemical means the thermal and tectonic evolution of Iceland in the more remote past.

The study is based on rare earths (La, Sm and Yb) and Mg-values of 334 basalts broadly distributed throughout Iceland (Fig. 1a). A good age control for Pleistocene to Miocene lava piles was obtained in Eastern Iceland, Vestfirdir, Borgarfjordur and West Iceland by coordinating our sampling with age dating and paleomagnetostratigraphic studies of these regions (16-21). Part of the RE data has already been published (4,14) and the remaining unpublished RE analyses can be found in the Ph.D. dissertation of one of us (25) along with a comprehensive mineralogic and petrologic study of these samples.

Jakobsson (22,23) proposed that three main rock series have developed on Iceland; a tholeiitic series, a transitional alkali series and an alkalic series. Their distribution along the neovolcanic zones is shown in Fig. 1b. Our more extensive survey of Iceland corroborates Jakobsson's findings but also allows a more quantitative look at the distribution of these rock types within the neovolcanic zones as well as in Plio-Pleistocene and Tertiary formations, which so far have remained poorly documented (25). In this report we will limit petrological information to Mg-values for comparison with RE. Chondrite normalized enrichment factors for La/Sm and Yb are abbreviated from here on as (La/Sm) and (Yb).

## TEMPORAL VARIATIONS

Rare Earth patterns of all the Iceland basalts we have analyzed are shown in Fig. 2. There is a remarkable difference in the type and range of pattern for Pleistocene-Miocene (13.5my - 0.7my) basalts and basalts from the neovolcanic zones (< 0.7my). Pleistocene-Miocene lavas are consistently enriched in light-RE, whereas Recent lavas exhibit a broad spectrum of RE patterns from light-RE enriched to light-RE depleted patterns more typical of normal Mid-Ocean Ridges.

The difference between Pleistocene-Miocene basalts and Recent basalts is further evident in the (La/Sm) histograms shown in Fig. 3. Pleistocene-Miocene basalts are normally distributed with a (La/Sm) mode around 1.65, whereas Recent basalts are positively skewed with a (La/Sm) mode around 1.35, thus suggesting a broad

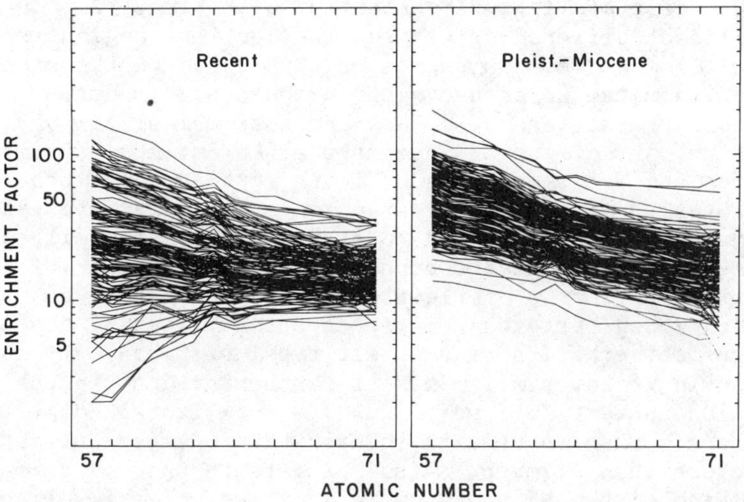


Fig. 2. Rare Earth patterns for Recent basalts and Pleistocene to Miocene basalts. Enrichment factor refers to REE concentration relative to chondrites.

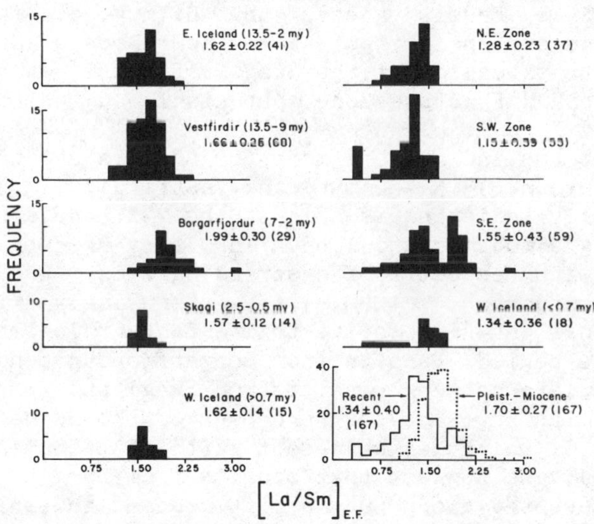


Fig. 3. Histograms illustrating the variation in (La/Sm) between Pleistocene-Miocene basalts (left) and basalts from Recent volcanic zones (right). A comparison between basalt populations older and younger than 0.7mybp is shown in the lower right. Numbers below each heading indicate the mean ± 1 standard deviation and the numbers in parentheses refer to the number of samples.

secular decrease of light-RE enrichment with time. Although there are significant differences between the Tertiary regions studied, their (La/Sm) mode is always greater than 1.5. On the other hand, tholeiites from the three neovolcanic zones have modes systematically less than 1.5. Only transitional alkali basalts from the southern part of the SE zone have a (La/Sm) mode similar to or higher than the Tertiary lavas. Fig. 4 further shows that light-RE depleted basalts in the neovolcanic zones have characteristically high Mg-values, and correspond to either olivine tholeiites or picritic basalts. We have encountered only three picritic basalts in our extensive survey of Pleistocene-Miocene lavas, but these have light-RE enriched patterns. It appears that light-RE depleted basalts reflect eruption of a basalt type which has only developed on Iceland in Recent time. This is further evident in the (La/Sm) distribution in West Iceland (Fig. 3) where light-RE depleted patterns are common in basalts younger than 0.7mybp but absent in basalts older than 0.7mybp. Also, basalts in the now dormant eastern swarm of the NE neovolcanic zone are primarily light-RE enriched, whereas in the currently active western swarm, light-RE depleted patterns are common (see Fig. 6).

The following additional observations can be made on the basis of the histograms in Fig. 3:

1) Rare Earth patterns in the Skagi paleorift zone are remarkably constant (14). Basalts greater than 0.7m.y. old from West Iceland are indistinguishable from those of Skagi and could represent an extension of the Skagi rift zone, south of the Central Iceland Fracture Zone, but displaced westward (Fig. 1b).

2) The RE patterns and (La/Sm) mode for the plateau basalts from Eastern Iceland (13.5 - 2 myr) and Vestfirdir (13.5 - 9myr) are statistically indistinguishable at the 99% confidence level, both having a mode around 1.65. Thus they are apparently symmetric with respect to the spreading axis. On the other hand, Borgarfjordur (7 - 2myr) in western Iceland is distinctly enriched in light-RE relative to its Eastern Iceland counterpart of the same period. Basalts from Borgarfjordur tend, in fact, to be intermediate between basalts from Skagi and transitional alkali basalts from the southern part of the SE neovolcanic zone, both of which may represent rifts at transient states (14). We suggest that basalts from Borgarfjordur may well have erupted along such a rift zone at an early stage of development or along a flank type rift zone.

The secular decrease in light-RE enrichment from Tertiary to present is best illustrated in the Eastern Plateau basalt succession (Fig. 5). A broad regular decrease is readily noted from 13.5my to 2myrs. Note also that extrapolation of this trend below 2mybp to the zero age intercept coincides with the modal value found for the

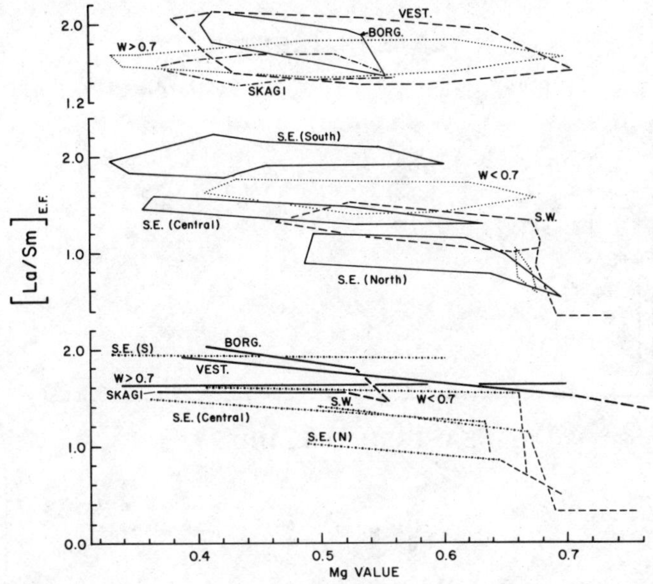


Fig. 4. Covariation between Mg-value and (La/Sm) for individual zones. The upper diagram shows variations in the Pleistocene-Miocene zones; Vestfirdir (VEST.), Borgarfjordur (BORG.), Skagi and West Iceland (W)> 0.7my. The middle diagram shows variations in the Recent zones; Southwest (S.W.), Southeast (S.E.) and West Iceland (W) < 0.7my. The lower diagram represents a composite of the upper two, facilitating the comparison between Recent zones (fine dotted lines) and Pleistocene-Miocene zones (bold solid lines). Note: 1) that older zones exhibit higher (La/Sm) than the younger zones, with the exception of the southern part of the SE Zone where the elevated (La/Sm) is associated with transitional alkali basalts probably derived from remelting of Late-Miocene lower crust and lithosphere (see text); 2) light-RE depleted patterns, (La/Sm) < 1, occur only at high Mg-values in the youngest zones (< 0.7my) and not at all in the Pleistocene-Miocene zones.

NE neovolcanic zone, which is the locus of the present spreading axis. Fig. 5 shows that such a regular secular trend does not appear to be corroborated in western Iceland. This is partly related to our inadequate sampling which does not permit us to take into consideration possible past spatial variations along rift strikes as demonstrated below for the neovolcanic zones, and partly due to complications imposed by rift jumping and ephemeral rifts such as Skagi and Borgarfjordur noted earlier.

Finally, the secular variations shown in Eastern Iceland may not be totally devoid of complexity due to spatial variations. A close look at the broad (La/Sm) secular decreae, may indicate a variation

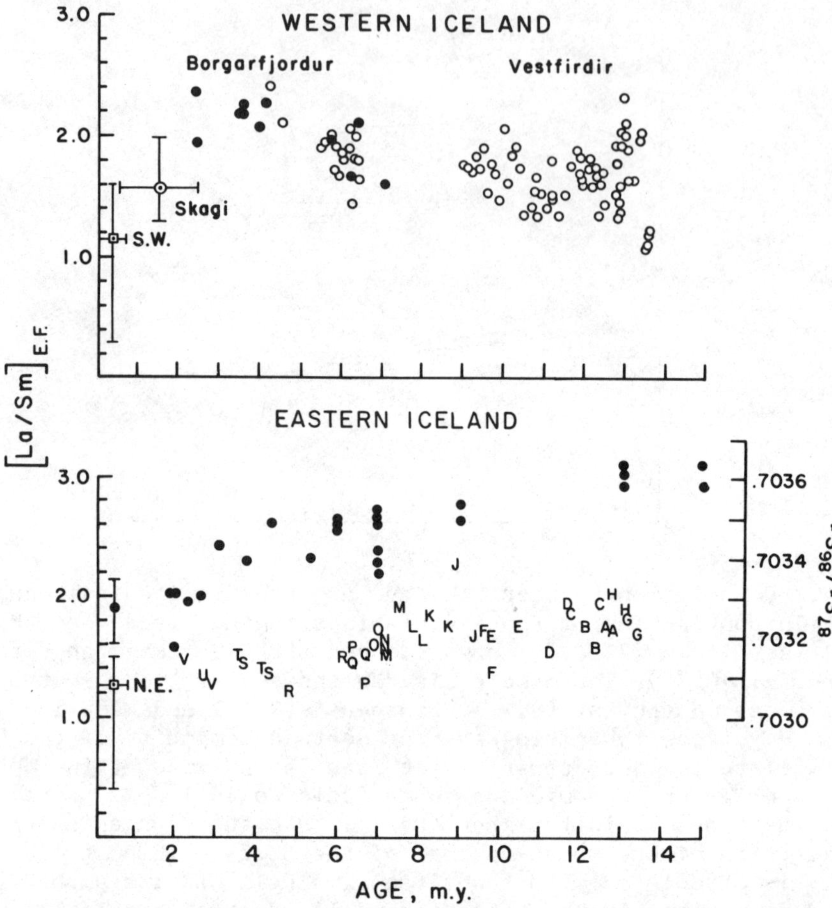


Fig. 5. Secular variation of (La/Sm) through the basalt plateaus in Western and Eastern Iceland. In the Western Iceland sequence, closed circles represent samples dated by the K-Ar method (20) and open circles represent samples whose age has been estimated from their stratigraphic height and the assumed linear relationship between stratigraphic height and age given by McDougall et al. (20) for Borgarfjordur and by McDougall and Kristjansson (in prep.) for Vestfirdir. Ages between 10.5 and 13.5my are based on the average K-Ar ages of several profiles while ages between 10.5 and 9.0my are based entirely on field relations relative to older, dated lavas. In the Eastern Iceland series, letters refer to different profiles (see Fig. 1). The average (La/Sm) for the Southwest (SW), Northeast (NE) and Skagi zones are also shown for comparison, error bars corresponding to the range in age and (La/Sm) observed in each zone. Initial 87Sr/86SR ratios (solid circles) in the Eastern Iceland series are from O'Nions and Pankhurst (31).

of a shorter wavelength, with a maximum in the 7-8my period. The perturbation coincides with a period of high lava extrusion rate (17,18) and the proposed sudden increase in volcanic activity on the Reykjanes Ridge resulting in the initiation of time transgressive V-shaped ridges (26). However, there is also a 60km break in the sample profile at this point, and thus the perturbation could also reflect some small but significant spatial variations existing at the time. Such resolution is not possible at this time.

## SPATIAL VARIATIONS

For practical reasons, our study of spatial variations across Iceland has been limited to known paleo or neovolcanic rift zones. Figure 3 shows that, statistically speaking, tholeiitic basalts from the various Recent volcanic zones (< 0.7my) have (La/Sm) modes (1.2-1.35) which are indistinguishable at the 99% confidence level. However, it is clear that there are some important RE and Mg-value variations between and within the neovolcanic zones which are not revealed in this statistical treatment (difference between two means using the T test). (La/Sm) decreases and Mg-values increase northward along both the SE and the SW zone, but at a faster rate in the SE zone (Fig. 6). On the other hand, 87Sr/86Sr variation is relatively constant along the SW zone, but increases southward along the SE zone (5,28,29). The increase in (La/Sm) along the SE zone is rather stepwise, and the variation essentially reflects the three major petrological zones (22,23) shown in Fig. 1b, whereas this is not the case along the SW zone where basalts are dominantly olivine tholeiites or tholeiites. In contrast, the (La/Sm) and Mg-value variations in the Skagi paleorift zone and the NE neovolcanic zone show no well-developed spatial trends along strike (14,15), though there is a lateral difference in RE patterns between the eastern, dormant and western, active swarms of the NE zone.

Finally, our extensive survey demonstrates that the strongly light-RE depleted picritic basalts previously reported in Central Iceland (14) and the Reykjanes Peninsula (38) are, in fact, confined to these two regions (Fig. 6). The spatial contrast within and between known rift zones of Iceland is further illustrated in the (La/Sm) versus Mg-value diagram (Fig. 4). The (La/Sm) appears insensitive to Mg- value variation in a particular zone, but the (La/Sm) level is distinct from one zone to another.

We conclude that the distinct (La/Sm) levels found between zones reflect similar mantle source distinctions. On the other hand, the different (La/Sm) levels observed within the SE zone, which correspond to three different petrological types, suggest both distinct sources as well as different melting conditions.

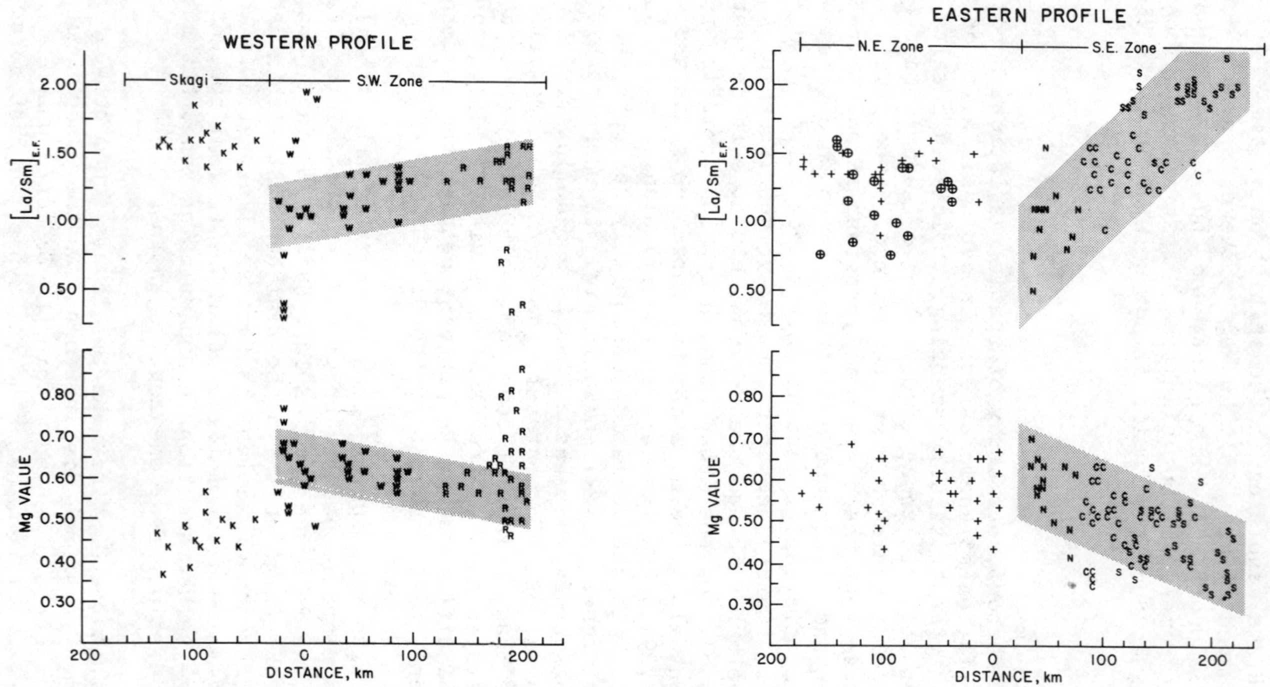


Fig. 6. (La/Sm) and Mg-value variations along the Eastern (SE and NE) and Western (SW and Skagi) volcanic zones. Distance along the X-axis refers to distance away from Central Iceland (64.86°N) along the axes of the volcanic zones (see Fig. 1). Sample locations have been projected normally to zone axes. In the upper diagram, circled crosses and crosses refer, respectively, to samples from the active western and dormant eastern part of the NE zone. Mg-value data for the NE zone (lower diagram) is from Sigvaldason (27).

## MODEL CONSTRAINTS

Interpretation of the above observations in terms of specific operative processes are now briefly considered. In general, processes which have been suggested to affect RE patterns in basalts range from: 1) mantle heterogeneities (of specified or unspecified causes); 2) partial melting conditions in the mantle (eg., different degrees of melting of a periodotite, or eclogite, at the same depth; or same degree of melting of a single peridotite source but at different depths under variable peridotite mineral facies, or any combination of these, etc.); 3) fractional crystallization during magma ascent or during residence in a magma chamber; 4) magma-wall rock interaction during magma ascent; 5) hydrothermal alteration, including zeolitization; 6) mixing of different magmas or reservoirs, etc. This list, of course, is not intended to be complete.

The effect of partial melting and fractional crystallization on RE patterns can be evaluated conveniently on a log-log (La/Sm) versus (Yb) enrichment factor diagram where models and data are readily compared (Fig. 7). High or low pressure fractional crystallization affects the overall enrichment of the rare earths (eg., {Yb}) but has little effect on the (La/Sm) ratio. Most of the fractionation in the Pliocene-Miocene plateau basalt series and localized regions of the neovolcanic zones is along the (Yb) axis. If we were to explain the (Yb) variations of the Pliocene-Miocene group by partial melting of a lherzolite (plagioclase-free or plagioclase-bearing) and chose, for example, a minimum degree of melting of 5%, the range of partial melting would be from 5% to 100%. Clearly such a model would be untenable from a major element point of view. The large (Yb) variation is best explained by low pressure fractional crystallization and crystal accumulation, which is also supported by the abundance of olivine, plagioclase and clinopyroxene phenocrysts, and the large range of Mg-values at relatively constant (La/Sm) shown in Fig. 4.

Variable degrees of melting also has a limited effect on (La/Sm), and thus is also not a viable process for explaining the entire range of (La/Sm) observed on Iceland. For example, variable degrees of partial melting above 10% of a plagioclase lherzolite or a simple lherzolite, which is reasonable from a major element point of view , has essentially no effect on the (La/Sm) ratio which remains essentially identical to that of its source. (La/Sm) is sensitive to the degree of partial melting only over the range of 0-10% in such a model. Although some variation in the degree of partial melting is likely to have taken place in Iceland, distinction of such an effect from that of mantle source heterogeneities cannot be unequivocally established on the basis of our RE data alone.

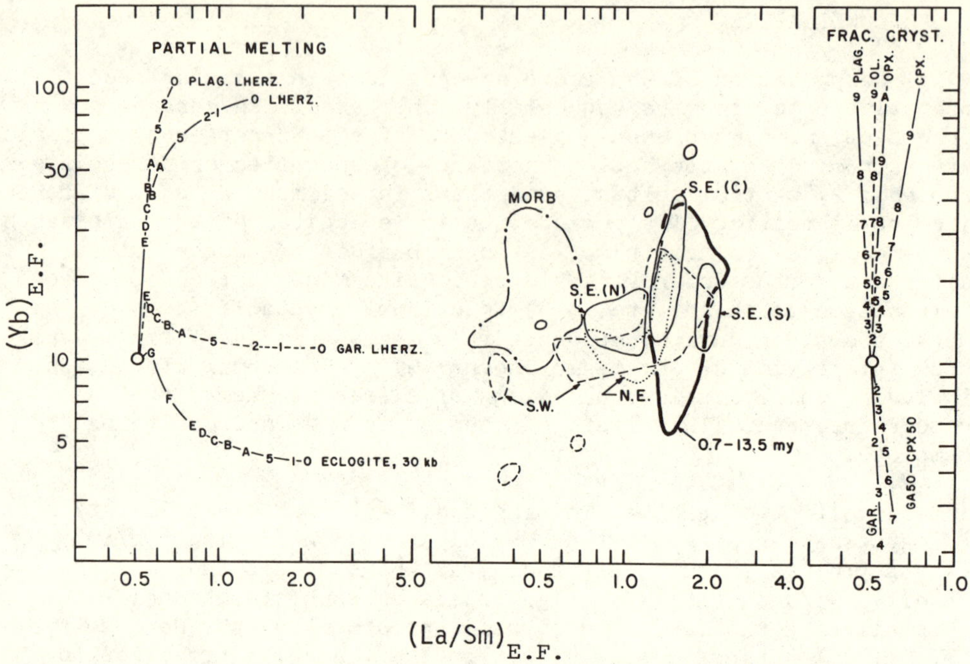


Fig. 7. (Yb) vs (La/Sm) variation for Iceland basalts compared with theoretically derived RE variations generated by different degrees of partial melting (batch melting) (30) of various mantle sources (left) and different degrees of fractional crystallization (right). Partition coefficients used in the models are from Ref. 14. The field for Pleistocene-Miocene basalts (bold line labeled 0.7 - 13.5my) is based on data from the Eastern Iceland Plateau, Vestfirdir, Borgarfjordur and West Iceland basalts > 0.7 my old. SE (N), SE (C) and SE (S) refer to the northern, central and southern part of the Southeast zone, respectively. SW is the Southwest zone and NE is the Northeast zone. MORB refers to normal mid-ocean ridge basalts from the Kolbeinsey and Reykjanes Ridges (7). The starting composition for partial melting and fractional crystallization models was arbitrarily set at (Yb) = 10 and (La/Sm) = 0.5. Curves marked plag. lherz., lherz., gar. lherz., and eclogite are for a plagioclase lherzolite source (20% plagioclase, 48% olivine, 16% orthopyroxene, 16% clinopyroxene), a lherzolite source (60% Ol, 20% Opx, 20% Cpx), a garnet lherzolite source (48% Ol, 16% Opx, 16% Cpx, 20%) and an eclogite source (50% Cpx, 50% Gar). Numbers (0, 2, 5) and letters (A, B, C, D, E, F, G) along the partial melting curves indicate 0, 2, 5, 10, 15, 20, 25, 30, 50 and 77 percent melting. Numbers (2, 3, 4, 5, 6, 7, 8, 9, A) along the fractional crystallization curves indicate 20, 30, 40, 50, 60, 70, 80, 90 and 95 percent crystallization of plagioclase (Plag), olivine (Ol), orthopyroxene (Opx), clinopyroxene (Cpx), garnet (Gar) and an eclogitic mixture (Ga 50-Cpx 50).

There is evidence suggesting that the (La/Sm) in Iceland basalts is likely to be representative of its mantle source, at least to a first order approximation. We have indicated earlier that the distinct (La/Sm) levels independent of Mg-value found in different regions of Iceland are best attributed to mantle sources with distinct (La/Sm) values (Fig. 4). Another indication of mantle heterogeneities is provided by comparing the geographical variation of (La/Sm) and 87Sr/86Sr found along the Mid-Atlantic Ridge and over Iceland (NE and SW neovolcanic zones, Fig. 8). The similarity of the two trends is striking. As it is generally accepted that the 87Sr/86Sr ratio in basalts is unlikely to be affected by partial melting and is, in fact, representative of the isotopic composition of the mantle source, we can safely conclude that, to a first approximation, this is also the case for (La/Sm). The partial melting model shown in Fig. 7 also shows that this is generally the case when the degree of melting is above 15-20% or so, whatever the mantle source mineralogy. Such a range in the degree of melting also makes sense petrologically for the generation of tholeiites or olivine tholeiites, but may be too high for transitional basalts (33-35). Another cause must be sought for the very high (La/Sm) found in transitional basalts on Iceland (eg., the southern part of the SE zone).

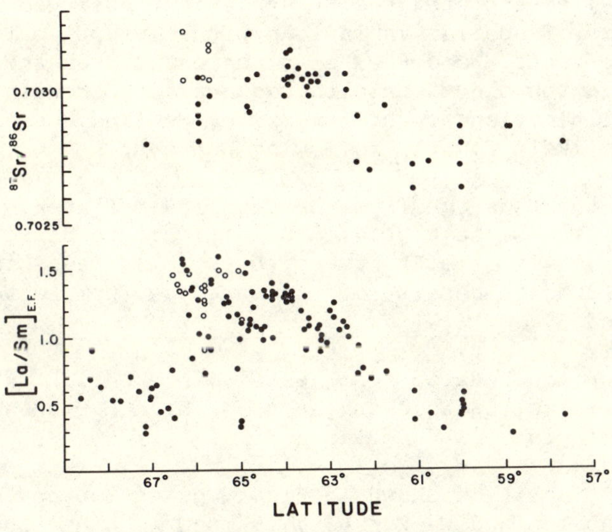


Fig. 8. Variation of 87Sr/86Sr and (La/Sm) along the Kolbeinsey Ridge, the Northeast and Southwest zones and the Reykjanes Ridge. Sr isotope data is from Refs. 5, 31 and 32. O'Nions Sr isotope values have been arbitrarily reduced by 0.00010 to facilitate comparison due to machine bias. Open circles represent basalts from the dormant fissure swarms in the eastern part of the NE zone and filled circles signify basalts from the active fissure swarms in the western part of the NE zone (Dyngjufjoll-Krafla) (24).

## INTEGRATED MODEL(S)

Although numerous models have been proposed to explain trace element and isotopic variations in isolated regions in Iceland (36-38) few have been suggested for the entire Iceland plateau and surrounding ridge segments that account for both spatial and temporal variations in a unified way. These models include the plume and/or blob model (4,39) and Wood's (40) fractional melting model. These models do not address the distinct differences observed between axial rift zones and flank zones. We propose a model for flank volcanism which considers rift propagation onto older crust, resulting in, first, melting of the older large-ion-lithophile (LIL)-enriched lower crust-lithosphere, the enrichment reflecting an earlier stage in plume activity, followed by melting of the upper mantle due to advection (41).

We will now briefly evaluate these and other models on the basis of our new data, discussing first, secular variations; second, spatial variations within mature axial rift zones; and finally, spatial variations along flank zones.

### 1. Secular variation in Eastern Iceland (13.5-2myr).

Wood (42,43) interpreted the RE variations observed in Eastern Iceland in terms of spatial variations within eruptive lenses rather than temporal effects. Wood (42) also suggested that the high (La/Sm) found in the oldest basalts was due to increasing depth of burial and zeolitization of the lava pile. We find Wood's explanations unlikely for the following reasons:

1) Basalts of the same age from the Vestfirdir Plateau (9-13.5myr), which we have studied specifically because they have not been zeolitized, also have high (La/Sm), indistinguishable from that found in basalts of the same age from Eastern Iceland (Figs. 4 and 5).

2) Miocene picritic basalts from Vestfirdir are light-RE enriched and not depleted, as Wood suggests for high MgO basalts from lens margins.

3) The secular variation shown in Fig. 5 using the new dating scale of Watkins and Walker (18) is much smoother than that observed using the scale of Dagley et al. (16). This strengthens our view that the variation is indeed a time-related one, rather than being spatial as suggested by Wood (42,43).

4) Finally, in addition to the secular decrease in (La/Sm), there is a concomitant decrease in 87Sr/86Sr and an increase in 143Nd/144Nd both on Iceland during the last 13my (31,44) and

on the Faeroes 60-50mybp (45). This is too much of a concidence to be attributed to secondary alteration as Wood (42) suggests, since for both Iceland and the Faeroes, the (La/Sm) and isotopic trends were established on totally different sample suites. Furthermore, it is now well established that the radiogenic Nd isotopic ratio is unaffected by weathering and mild metamorphism or metasomatism (46).

We conclude that the secular (La/Sm) variation observed both in Eastern Iceland, and earlier in the Faeroes, reflects a true secular variation in the composition of the mantle source of these basalts. The variation also appears to be episodic.

Wood's (40) model invoking fractional melting is also unlikely to explain the secular variation observed for both the RE and $^{87}Sr/^{86}Sr$ (or $^{143}Nd/^{144}Nd$). It is sufficient to say that the build-up of $^{87}Sr/^{86}Sr$ with time is irreversible. Once a source region high in $^{87}Sr/^{86}Sr$ has been established, successive partial melting of such a source region under equilibrium conditions will not, under any circumstances, decrease this ratio (which has been corrected for any possible isotopic mass fractionation; eg., ref. 5).

Invoking disequilibrium melting of a veined mantle (40) to produce such regular secular trends over a 10my or longer period of seafloor spreading seems most improbable. This would require a mechanical decoupling of the crust generated by spreading and the mantle source from which melts are derived (i.e., decoupling of the crust from the lower part of the lithosphere). This is totally inconsistent with seafloor spreading theory and the distribution of earthquakes. Furthermore, application of the concept of disequilibrium melting to generation of basalts, which seem to have required rather large degrees of melting, does not appear very viable on the basis of Hoffman and Hart's (47) kinetic and scaling analysis.

The binary mixing, asthenosphere-mantle blob (or plume) model (4,39) appears to be a more viable model, as it can explain in a more unified way both the spatial and temporal variation observed during the evolution of the Iceland-Faeroe-Reykjanes Ridge system (Fig. 9). In this model, the RE pattern of a basalt primarily reflects the mixing proportions between a light RE-enriched mantle plume source ($\{La/Sm\} > 1$) and the asthenosphere which is characteristically depleted in light RE ($\{La/Sm\} \leq 0.6$). The degree of mixing depends on the interplay between the rate of mantle plume upwelling and the rate of lithospheric plate divergence above the upwelling region, both of which vary independently through time. Fig. 9 compares the known temporal variation for the Iceland-Faeroe Ridge with the concurrent variation in spreading rate of the Mid-Atlantic Ridge, and a mantle plume flux, modified from Vogt and Avery's (48) curve for rate of basalt discharge.

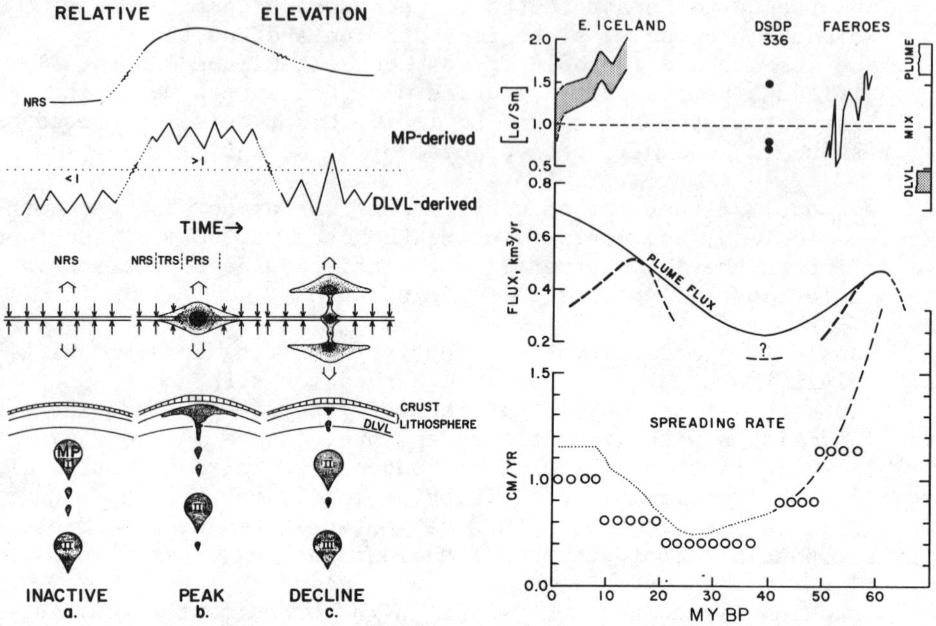


Fig. 9. Left: Model of a mantle plume (MP) ridge system showing variation in plume flux with time and a cycle in plume (blob) activity that is consistent with geochemical variations observed in erupted basalts. Phases: (a) plume is rising beneath an existing spreading ridge (passive spreading); (b) plume flux exceeds the gap created by spreading, resulting in blob flattening and spill along spreading directions (active spreading). Plume-derived material dominates the erupted products which accumulate as plateau basalts; (c) tailing phase of the blob. Plume flux is insufficient to feed the overlying spreading ridge and material from the depleted low velocity layer (DLVL) compensates. Small arrows indicate horizontal component of asthenospheric flow of DLVL material toward the ridge axis. Large arrows indicate spreading directions of the overlying lithospheric plates. NRS, TRS and PRS stand for normal, transitional and plume ridge segments. Right: Reykjanes Ridge spreading rates, Faeroes-Iceland mantle plume flux and (La/Sm) variations with time (modified from Ref. 39). Lower curves are spreading rates from Ref. 48 (dotted-dashed curve) and from Ref. 49 (open circles). The middle curve is the plume flux (solid line) estimated from Vogt and Avery, assuming basalts represent 25% partial melting and, therefore, that basaltic discharge is equal to 1/4 of the total plume flux. The dashed lines indicate our estimate of plume flux for the Iceland-Faeroe plume during the 60-50 (Ref. 39) and 13.5-0my periods, based on (La/Sm) as shown in the upper diagram. The solid circles represent data from DSDP site 336 (Ref. 50).

The (La/Sm) decrease in the 50-60my period cannot be accounted for by a change in spreading rate since spreading would have to increase rather than decrease (39). Therefore, it must reflect a decrease in plume flux. Interpretation of the (La/Sm) secular decline within the last 13mybp in Iceland depends on how reliable the spreading rate estimates are during this period (Fig. 9). Taken at face value, the (La/Sm) decline during the 13-9my period could be accounted for by an increase in spreading rate, but not the decrease in (La/Sm) during the last 9mybp, which again requires a decline in plume flux, particularly severe during the last million years. The blob model further suggests that in Iceland we are currently witnessing the declining phase of mantle blob activity, that is, a rapid change to a more passive mid-ocean ridge type spreading regime. If such a trend is extrapolated into the future, the model predicts either a change from a subaerial to a submarine regime and the possible split of Iceland into two islands located symmetrically about the Mid-Atlantic Ridge, or a new phase of plateau basalt formation, depending on the periodicity and topology of blobs. Such a change of volcano-tectonic regime also occurred in the 50-60mybp period on the Faeroes Plateau, and thus the process appears to be cyclic. The active phase of a blob is on the order of 10-15my and is followed by a rather passive spreading period about twice as long.

It may be suggested that the slightly light-RE enriched basalt, associated with more typical mid-ocean ridge type basalts in DSDP Hole 336 (40-50mybp on the northern flank of the Faeroes-Iceland Ridge), does not support the proposed cycle (Fig. 9). We contend that what happens in the interim between two blobs greatly depends on the topology of the tail end of such blobs, which may be in the form of schlierens. Thus, it is conceivable that the declining phase of blob activity may be characterized by inefficient mixing of this blob material with the light-RE depleted, partially molten asthenosphere. Occasional taping of such schlierens would explain the rapid alternance between light-RE depleted and enriched patterns. Rapid fluctuation between light-RE enriched and depleted patterns is also observed in the upper Faeroe basalt series, and more recently on the Reykjanes Peninsula (38). We suggest that such fluctuation is likely to be characteristic of the tailing phase of blob activity, during which time the mantle may be veined in an irregular fashion (schlierens).

The blob model does not explain how the blobs initially became enriched in incompatible elements and radiogenic isotopes and volatiles. Possible models include a layer instability caused by internal heating from radioactivity (4), oceanic lithosphere recycling (51), or metasomatism (29,38,52,53). Discussion of these possibilities is beyond the scope of this paper.

## 2. Variations along mature axial rift zones.

The first order similarity of the trends exhibited by trends of (La/Sm) and 87Sr/86Sr along the Mid-Atlantic Ridge and its extension across Iceland along the SW and NE neovolcanic zones (Fig. 8) leaves little doubt that it reflects mantle source variations (4-7).

Increasing fractional melting southward, as proposed by Wood (40), to account for the La/Sm gradient along the Reykjanes Ridge was considered by Schilling (4) as a viable model to tailor RE variations, but was found clearly untenable on the basis of the Sr and Pb isotopic evidence (5,6). This is not to say that Wood's fractional melting may not be a viable model in some local cases.

On a finer scale, we have also noted that along the SW zone, (La/Sm) decreases slightly and Mg-value increases slightly towards Central Iceland (Fig. 6b), whereas 87Sr/86Sr stays relatively constant (5). These observations led us to suggest increasing degree of melting toward Central Iceland, or toward the center of the upwelling plume or blob (14). This is further supported by a concomitant increase in volcanic delivery rate, manifested by a widening of the rift zone and a greater abundance of central volcanoes (11,22-24) and a concomitant decrease in 230Th/232Th (54).

Finally, it is important to note that, among Iceland basalts, there is a compositional gap between basalts with (La/Sm) > 0.5 and picritic basalts with (La/Sm) between 0.3 and 0.4 (Fig. 7). Significantly, this gap corresponds to the compositional space occupied by normal ridge segments adjacent to Iceland. These relationships can readily be explained by a combination of the blob model and fractional melting model. The extremely depleted picrites would be derived by unusual fractional melting conditions, while the less depleted basalts from the SE zone would reflect binary mixing between plume-derived and asthenosphere-derived magmas. Clearly, a detailed isotopic study of the kind made by Zindler et al. (38) needs to be undertaken on the light-RE depleted basalts found in Central Iceland.

## 3. Variations along immature flank zones

Interpretation of the large (La/Sm) variation, increase in Mg-value, and change of petrological type observed southward along the SE zone (Fig. 6a) is more problematic. The (Yb) variation shown in Fig. 7 for the northern, central and southern part of the SE zone is clearly related to shallow depth fractional crystallization. However, the (La/Sm) of the least fractionated lavas within the three groups can hardly be accounted for by partial melting of a lherzolite source as the degree of melting would have to vary essentially between 0-100%. Fractional melting rather than batch melting might help, as such melting conditions tend to fractionate the (La/Sm) ratio more (40). This is, however, ruled out as 87Sr/86Sr variation along the zone (although complicated by interlab machine bias) tends to show a southward increase paralleling the (La/Sm) trend (5,28,29,32,36,55).

Oskarsson et al. (56) have proposed a petrological model for rift zones on Iceland which is consistent with the crustal accretion model of Palmason (57). Basalt variations are explained by mixing between mantle-derived olivine-tholeiites and variable amounts of crustal-derived magmas ranging from silicic to nepheline-normative melts. Flank zone volcanism is considered to be dominated by crustal-derived magmas, whereas volcanism within the main axial rift zone is dominated by mantle-derived olivine-tholeiites. A crucial assumption of this model is that the subsiding crust becomes completely hydrated by hydrothermal activity. Studies of DSDP rocks of various ages in the oceans, show that hydration of the oceanic upper crust is very irregular and incomplete, despite the ample abundance of seawater above. It is doubtful that, on Iceland where such an abundant source of fluids is absent, complete hydration would in fact take place. Their model also does not address itself to the temporal variation or spatial variation found along the main axial zones discussed here. However, magmas of crustal derivation in the southern part of the SE zone is an attractive hypothesis but the crust need not be hydrated to the extent indicated.

The difficulty with the SE zone is the fact that the inferred magma source heterogeneity in 87Sr/86Sr and (La/Sm) along the SE zone is geographically of opposite sense to that observed along the SW zone and the Reykjanes Ridge.

A model consistent with the blob model is the southwest propagation of a rift (41). The thermal process is best described in two stages, though in reality it evolves more continuously. We assume first, that crustal tension released at the tip of the rift propagator is sufficient to bring the older and somewhat colder Late-Miocene lower crust (approximately 8my old) into the partial melting stability field, resulting in magmas of crustal origin (and perhaps old lithosphere as well) with Sr and Nd isotopic signatures of Late-Miocene lavas exposed elsewhere on Iceland (28,31,44). RE patterns will also be light-RE enriched, but detailed modeling is required to take into account the possible mineralogical composition of the lower crust (and old lithosphere), degree of melting, and type of lavas generated. As spreading develops and the rift matures, there is a progressive change from primarly crustal-derived products to basalts derived from mantle advection and resulting partial melting. We have seen earlier that the mantle source on Iceland in Recent time, particularly within the last 700,000 years, has become significantly diluted with light-RE depleted asthenospheric material. Thus the basalt erupted in a later stage, far behind the tip of the rift propagator, will also be lower in 87Sr/86Sr, higher in $^{143}Nd/^{144}Nd$ (44), and less light-RE enriched. In other words, the present spatial variation observed along the SE zone is, in fact, the result of a temporal evolution in melting intensity and conditions, due to rift propagation. The tip of the present rift propagator probably lies just south of Surtsey Island which emerged in 1963. Whether such a propagating rift is of the membrane type (58,59) or of Hey's type (60) is unknown.

Finally, as we noted earlier, Skagi represents an ephemeral flank rift zone which developed 2.5-0.7 mybp in northern Iceland. The petrological and geochemical uniformity of basalt erupted along the zone (14,15) precludes the fact that such a zone developed by rift propagation.

## CONCLUSIONS

Our RE survey of basalts from Iceland has allowed us to make the following general observations:

1) There is a secular variation of RE patterns, which is best revealed in Eastern Iceland. The (La/Sm) progressively decreases from about 1.9 at 13.5mybp to 1.1 at 0.7mybp. At this stage light-RE depleted patterns begin to appear in all the neovolcanic zones, though preferentially in the Reykjanes and Tjornes Fracture Zones, and in Central Iceland which may also be a major fracture zone (61). Such a transition period from light-RE enrichment to depletion is not unique to recent times. It also occurred in the late stage of formation of the Faeroes Plateau basalts in the 50-60my period (39) (Fig. 9). In this case the transition was more abrupt and coincided with a major subsidence of the region. Whether this is also the case in Iceland remains to be determined. In both cases, Iceland and the Faeroes, the secular decrease in (La/Sm) is also accompanied by a decrease in $^{87}Sr/^{86}Sr$ and an increase in $^{143}Nd/^{144}Nd$. We conclude that both the 50-60myr and 13.5myr to Present period appear to represent periods of unusually high elevation and constructional volcanism relative to the remaining history of East Greenland-Iceland-Faeroes Ridge. The transition in RE patterns suggest a transition from that of a plume-like regime to that of a more mid-ocean ridge type regime, or from active to passive type spreading (Sacks, 1981 oral presentation). Such a change of regime certainly appears to have been episodic in the lifetime of the Iceland mantle plume. This gives further support to the blob model previously proposed for the evolution of the Faeroes-Iceland-Greenland Ridge (39). If the current geochemical trend continues, the model predicts that Iceland may

split into two islands, or alternatively, a new regime of plateau basalt formation may appear, depending on the periodicity and topology of blobs.

2) Spatial variation of RE patterns and petrological basalt types along the recognized neovolcanic and paleorift zones of Iceland appears to depend on the degree of thermal maturity of the zones and existing mantle source heterogeneities beneath them. The northward progressive decrease in (La/Sm) and the increase in Mg-value observed along both the SW and SE zones, but at a faster rate in the SE zone, are best explained by a temporal change in intensity and conditions of melting resulting from rift propagation. The contrast in 87Sr/86Sr variation along the two zones and heat flow anomalies suggest that the SW zone has approached thermal maturity, whereas the SE flank zone is the result of a rift currently propagating southward. The tip of the propagator would be in the vicinity of Surtsey Island. Melts erupting near the tip of the propagator are primarily derived from the older Late-Miocene lower crust-lithosphere, whereas in the northern part, basalts are derived from a mantle plume-asthenospheric mix, now dominated by the latter type of material. We further suggest that the SW zone may also have developed by rift propagation some 6mybp (62). Recent basalts erupted along the more mature zones are all derived from the upper mantle composed of a blend of LILE-enriched blob and LILE-depleted asthenospheric material, with the degree of melting increasing slightly toward the center of the plume (Central Iceland).

If our assumptions concerning petrologic and geochemical types and degrees of maturity of neovolcanic rift zones are correct and hold true in the more remote past, the following suggestion can be made on the nature of rifting throughout Iceland's history. The Eastern Plateau and NE zone seem to have formed from continuous spreading during the last 13 myrs or so and have been accompanied by flank spreading at least once in the past (Skagi zone 2.5-0.7mybp). So far there is no evidence for rift propagation in northern Iceland. In contrast, the tectonic evolution of southern Iceland has been complicated by recurring southward propagation of rifts of ephemeral duration and simultaneous flank zone volcanism, as evident in the Borgarfjordur region (7-2myrs), the SE zone and the Setberg volcanic region on the Snaefellsnes Peninsula (61). This evolution is generally consistent with the tectonic evolution model proposed for Iceland by Johannesson (62) and Saemundsson (11,12). However, in detail, there are many uncertainties and several discrepancies which cannot be discussed further without more extensive field work in coordination with additional rock dating and geochemical and petrological studies.

ACKNOWLEDGMENTS

This work was partly supported by NSF grants, DES75-03997 and EAR77-23736, to J-G Schilling and H. Sigurdsson. Field work and petrologic studies were carried out in collaboration with H. Sigurdsson. We would like to thank Rannsoknarad Rikisins (The National Research Council of Iceland) for granting us permission to collect samples in Iceland. We would also like to thank Frank DiMeglio, Mike Doyle and staff of the Rhode Island Nuclear Science Center for their assistance in neutron activation analyses, and Gloria Nappi for invaluable assistance.

REFERENCES

1. Moore, J.G. and J-G. Schilling, Contr. Mineral. & Petrol., 41, 105, (1973).
2. Unni, C.K. and J-G. Schilling, Nature, 272, 19, (1978).
3. Rowe, E.C. and J-G. Schilling, Nature, 279, 33, (1979).
4. Schilling, J-G., Nature, 242, 565, (1973).
5. Hart, S.R., J-G. Schilling and J.L. Powell, Nature Phys. Sci., 246, 104, (1973).
6. Sun, S-S., M. Tatsumoto and J-G. Schilling, Science, 190, 143, (1975).
7. Schilling, J-G., M. Zajac, R. Evans, T. Johnston, W. White, J.D. Devine and R. Kingsley, Amer. J. Sci., sub. Feb., (1981).
8. Palmason, G. and K. Saemundsson, Annual Rev. Earth Planet. Sci., 2, 25, (1974).
9. Ward, P.L., Geol. Soc. Amer. Bull., 82, 2991, (1971).
10. Saemundsson, K., Geol. Soc. Amer. Bull., 85, 495, (1974).
11. Saemundsson, K., Geol. J. Spec. Issue 10, 415, (1978).
12. Saemundsson, K. Jokull, 29, 7, (1980).
13. Aronson, J.R. and K. Saemundsson, Earth & Planet. Sci. Lett., 28, 83, (1975).
14. Schilling, J-G., H. Sigurdsson and R.H. Kingsley, J. Geophys. Res., 83, 3983, (1978).
15. Sigurdsson, H., J-G. Schilling and P.S. Meyer, J. Geophys. Res., 83, 3971, (1978).
16. Dagley, P., R.L. Wilson, J.M. Ade-Hall, G.P.L. Walker, S.E. Haggerty, T. Sigurgeirsson, N.D. Watkins, P.J. Smith, J. Edwards and R.L. Grasty, Nature, 216, 25, (1967).
17. McDougall, T., N.D. Watkins and L. Kristjansson, Amer. J. Sci., 276, 1078, (1976).
18. Watkins, N.D. and G.P.L. Walker, Amer. J. Sci., 277, 513, (1977).
19. Ross, T.G. and A.E. Mussett, Nature, 259, 36, (1976).
20. McDougall, I., K. Saemundsson, H. Johannesson, N.D. Watkins and L. Kristjansson, Bull. Soc. Amer., 88, 1, (1977).
21. Wilson, R.L., N.D. Watkins, T. Einarsson, T. Sigureirsson, S.E.Haggerty, P.J. Smith, P. Dagley and A.G. McCormack,

Geophys. J.R. asr. Soc., 29, 459, (1972).
22. Jakobsson, S.P., Lithos, 5, 365, (1972).
23. Jakobsson, S.P., Acta Naturalia Islandica, 26, 1, (1979).
24. Walker, G.P.L., Nature, 255, 468, (1975).
25. Meyer, P.S., Ph.D. Thesis, Univ. of Rhode Island, (1981).
26. Vogt, P.R., Earth & Planet. Sci. Lett., 13, 153, (1971).
27. Sigvaldason, G.E., J. Petrol., 15, 497, (1974).
28. O'Nions, R.K. and K. Gronvold, Earth & Planet. Sci. Lett., 19, 397, (1973).
29. Wood, D.A., J.L. Joran, M. Treuil, N. Norry and J. Tarney, Contr. Mineral. Petrol., 70, 319, (1979).
30. Schilling, J-G., J. Geophys. Res., 80, 1459, (1975).
31. O'Nions, R.K. and R.J. Pankhurst, Earth & Planet. Sci. Lett., 21, 13, (1973).
32. O'Nions, R.K., R.J. Pankhurst and K. Gronvold, J. Petrol., 17, 315, (1976).
33. Reay, A. and P.G. Harris, Bull. Volcanol., 27, 115, (1964).
34. Mysen, B.O. and I. Kushiro, Amn. Mineral., 62, 843, (1977).
35. Scarfe, C.M., B.O. Mysen and C.S. Rai, Carnegie Inst. Yearbook, 78, 498, (1979).
36. Sun, S-S. and B.J. Jahn, Nature, 255, 527, (1975).
37. Bailey, J.C. and A. Noe-Nygaard, Lithos, 9, 185, (1976).
38. Zindler, A., S.R. Hart, S.P. Jakobsson, Earth & Planet. Sci. Lett., 45, 249, (1979).
39. Schilling, J-G. and A. Noe-Nygaard, Earth & Planet. Sci. Lett., 24, 1, (1974).
40. Wood, D.A., Geochim. Cosmochim. Acta, 43, 1031, (1979).
41. Schilling, J-G, P.S. Meyer and R.H. Kingsley, sub. to Nature, (1981).
42. Wood, D.A., J. Geophys. Res., 81, 4353, (1976).
43. Wood, D.A., J. Petrol., 19, 393, (1978).
44. O'Nions, R.K., P.J. Hamilton and N.M. Evensen, Earth & Planet. Sci. Lett., 34, 13, (1977).
45. Gariepy, C., J.N. Ludden, C. Brooks and N. Machado, EOS, Trans. Amer. Geophys. Union, 62, (1981).
46. DePaolo, D.J., EOS, Trans. Amer. Geophys. Union, 72, 137, (1981).
47. Hoffman, A.W. and S.R. Hart, Earth &Planet. Sci. Lett., 8, 44, (1978).
48. Vogt, P.R. and O.E. Avery, J. Geophys. Res., 79, 363, (1974).
49. Herron, E.M. and M. Talwani, Nature, 238, 390, (1972).
50. Schilling, J-G., in: "Initial Reports of the Deep Sea Drilling Project", 38, 741, (1976).
51. Hoffman, A.W. and W.M. White, Annual Report of the director of Terrestrial Magnetism, Carnegie Inst. of Wash. Yr. Book, 79, 477, (1980).
52. Frey, F.A., D.H. Green and S.D. Roy, J. Petrol., 19, 463, (1978).
53. Schilling, J-G., M.B. Bergeron and R. Evans, Phil. Trans. R. Soc. Lond., A297, 147, (1980).

54. Condomines, M., P. Morand, C.J. Allegre and G. Sigvaldason, Earth & Planet. Sci. Lett., in press.
55. O'Nions, R.K., R.J. Pankhurst, I.B. Fridleifsson, and S.P. Jakobsson, Nature, 243, 213, (1973).
56. Oskarsson, N., G. Sigvaldasson and S. Steinthorsson, Nordic Volcanological Inst., 79, 05, (1979).
57. Palmason, G., Geophys. J.R. Astr., 33, 451, (1973).
58. Courtillot, V., P. Tapponnier and J. Varet, Tectonophysics, 24, 317, (1974).
59. McMaster, R.L., J-G. Schilling and P.R. Pinet, Nature, 269, 663, (1977).
60. Hey, R., F.K. Duennebier and W.J. Morgan, J. Geophys. Res., 85, 3647, (1980).
61. Sigurdsson, H., Earth & Planet. Sci. Lett., 10, 129, 1970.
62. Johannesson, H., Natturufraedingurinn, 50, 13, (1980).

# SILICIC BASALTS FROM THE FAEROE ISLANDS:

# EVIDENCE OF CRUSTAL CONTAMINATION

Niels Hald and Regin Waagstein

Geologisk Museum
Øster Voldgade 5-7
DK-1350 København K

Danmarks Geologiske Undersøgelse
Thoravej 31
DK-2400 København NV Denmark

## INTRODUCTION

It is currently discussed whether the Tertiary lava pile of the Faeroe Islands rests on continental or oceanic basement. From geophysical data Bott et al. (1974, 1976) concluded that the Faeroe block is of continental origin and that the boundary towards the Tertiary oceanic crust is located along an escarpment west of the islands. Talwani and Eldholm (1972, 1977), on the other hand, consider the Faeroe block as oceanic with the boundary towards continental crust along its eastern margin and close to the Faeroe-Shetland escarpment.

The Faeroe lavas, with K-Ar ages of ca 55 million years (Tarling and Gale, 1968; Fitch et al., 1978), are tholeiitic basalts (Noe-Nygaard and Rasmussen, 1968; Schilling and Noe-Nygaard, 1974; Bollingberg et al., 1975; Waagstein, 1977). In terms of major and trace elements as well as Sr and Nd isotope ratios the lavas are similar to the tholeiitic basalts of Iceland and East Greenland (Bailey and Noe-Nygaard, 1976; Brooks et al., 1976; Carter et al., 1979). Non-cognate xenoliths, which could provide information on the nature of the substratum, have not been reported from the lavas or from the intercalated pyroclastic deposits.

This short communication calls attention to three silicic basalt lavas found in the volcanic sequence of the Faeroe Islands. We suggest that the unusual composition of these lavas, including anomalously high $^{87}Sr/^{86}Sr$ initial ratios, is a result of contamination of basic magmas by older crustal material.

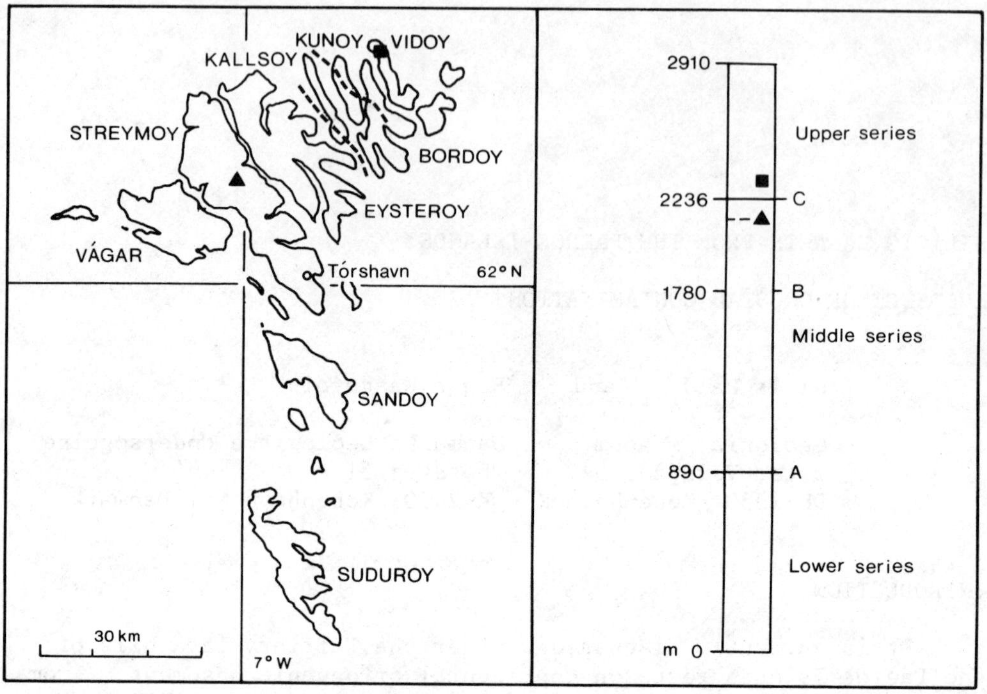


Fig. 1. The Faeroe Islands. Dashed lines: extension of the Klakksvík lava (Hald et al., 1969). Triangle and square: position of the Sneis-14 and Villingadalsfjall-6 lavas in the type profile of Rasmussen and Noe-Nygaard (1969). Right: schematic section through the lava sequence (Rasmussen and Noe-Nygaard, 1969).

## GEOLOGICAL SETTING

The Faeroe Islands are the erosional remnants of a gently warped basalt plateau whose base and top are not seen. The lava sequence is divided into three series and has an exposed thickness of 3 km (Rasmussen and Noe-Nygaard, 1969, 1970) (Fig. 1). The lower series consists of mainly feldspar-phyric and aphyric tholeiites of the aa-type and is capped by a roughly 10 m thick clay-coal sequence. Tuffs and agglomerates with a maximum thickness of 100 m form the lowermost part of the middle series. The rest of the series is dominated by thin pahoehoe flows: olivine-phyric tholeiites overlain by mainly feldspar-phyric tholeiites. The lavas of the upper series appear to have followed without a noticeable break in volcanism. Alternating groups of olivine-phyric and feldspar-phyric tholeiites, mostly of the aa-type, are found in the type profile on Vidoy, while feldspar-phyric tholeiites dominate in the Tőrshavn region and further south.

Table 1. Chemical Analyses of Silicic Basalts, with Comparisons

| | 1 | 2 | 3 | 4 | 5 | 6 | 7 | 8 | 9 | 10 | 11 |
|---|---|---|---|---|---|---|---|---|---|---|---|
| Sample no. | 63135 | 63141 | 69A | X.14 | MM15562 | MM15563 | XI.6 | | | | |
| $SiO_2$ | 53.94 | 53.68 | 53.78 | 54.04 | 53.61 | 53.84 | 52.55 | 46.98 | 48.42 | 47.27 | 53.20 |
| $TiO_2$ | 1.45 | 1.49 | 1.48 | 1.27 | 1.49 | 1.54 | 1.10 | 2.03 | 2.87 | 1.08 | 1.12 |
| $Al_2O_3$ | 12.85 | 12.78 | 12.97 | 10.39 | 13.01 | 13.09 | 15.14 | 12.98 | 14.88 | 14.34 | 14.90 |
| $Fe_2O_3$ | 1.99 | 1.38 | 2.39 | 3.82 | 2.57 | 2.08 | 2.73 | 3.21 | 4.50 | 2.37 | 1.22 |
| FeO | 7.70 | 8.48 | 7.48 | 6.66 | 7.30 | 7.80 | 6.65 | 8.08 | 7.95 | 8.34 | 7.40 |
| MnO | 0.13 | 0.16 | 0.15 | 0.17 | 0.17 | 0.18 | 0.15 | 0.19 | 0.20 | 0.19 | 0.17 |
| MgO | 8.83 | 8.96 | 8.72 | 11.33 | 7.80 | 7.05 | 6.29 | 9.73 | 5.69 | 9.93 | 8.49 |
| CaO | 8.60 | 8.76 | 8.66 | 8.70 | 8.90 | 9.04 | 10.29 | 11.00 | 10.51 | 11.90 | 8.89 |
| $Na_2O$ | 2.35 | 2.43 | 2.52 | 1.95 | 2.38 | 2.55 | 2.51 | 1.95 | 2.52 | 1.87 | 2.16 |
| $K_2O$ | 0.93 | 0.69 | 0.64 | 0.75 | 0.91 | 0.75 | 0.54 | 0.40 | 0.33 | 0.26 | 0.56 |
| $P_2O_5$ | 0.14 | 0.13 | 0.15 | 0.12 | 0.15 | 0.14 | 0.11 | | 0.32 | | 0.15 |
| L.O.I. | 0.90 | 0.78 | 0.99 | 0.92 | 1.32 | 1.08 | 1.09 | | 1.61 | | 1.50 |
| Total | 99.81 | 99.72 | 99.93 | 100.12 | 99.61 | 99.14 | 99.15 | 96.55 | 99.80 | 97.55 | 99.76 |
| FeO*/MgO | 1.07 | 1.09 | 1.10 | 0.89 | 1.23 | 1.37 | 1.45 | 1.13 | 2.11 | 1.05 | 1.00 |
| Rb | 11 | 10 | 14 | 16 | 17 | 20 | 11 | | | | 13 |
| Sr | 227 | 236 | 242 | 193 | 238 | 243 | 177 | 250 | 252 | 118 | 179 |
| Ba | 210 | 230 | 250 | 165 | 270 | 310 | 190 | 68 | 58 | <30 | 255 |
| V | 180 | 220 | 230 | 250 | 250 | 260 | 220 | 235 | 315 | 302 | |
| Cu | 100 | 80 | 103 | 75 | 103 | 115 | 93 | 140 | 192 | 196 | 26 |
| Co | 38 | 47 | 62 | 50 | 53 | 47 | 35 | 65 | 45 | 60 | 37 |
| Ni | 190 | 205 | 210 | 210 | 175 | 150 | 66 | 280 | 98 | 231 | 26 |
| Cr | 415 | 520 | 520 | 780 | 480 | 430 | 195 | 710 | 158 | 426 | 616 |
| Sc | 21 | 26 | 24 | 21 | 38 | 38 | 26 | | | | |
| Zr | 123 | 128 | 140 | 100 | 133 | 170 | 105 | 175 | 202 | 88 | 129 |
| $^{87}Sr/^{86}Sr$ (± 1σ) | 0.7119±2 | 0.7106±2 | | 0.710±1 | 0.7100±2 | | 0.7162±1 | | | | |

1-3: Klakksvík lava (major elements in 1 and 2 from Jensen, 1978). 4-6: Sneis-14. 7: Villingadalsfjall-6. 8: Average of 17 (trace elements 10) olivine-phyric tholeiites from the lower part of the middle series (Waagstein, 1977). 9: Average of 44 (trace elements 9) feldspar-phyric tholeiites from the middle series above the B-horizon (Waagstein, 1977). 10: Average of 22 olivine-phyric tholeiites from the middle and upper series above the B-horizon (Waagstein, 1977). 11: Orthopyroxene-phyric basalt, GGU 113271, from Disko, West Greenland (Clarke & Pedersen, 1976).

Major elements: 1-7 and 9 by XRF, Geol. Surv. Greenland (Sørensen, 1975); 8 and 10 mainly by wet chemistry, Geol. Surv. Greenland; 11 by XRF, Univ. of Leeds. Trace elements: 1-10 (inclusive of Mn in 1 and 2) by optical emission spectrography (Rb and Sr in 1-7 by XRF), Inst. of Petrology, Univ. of Copenhagen; 11 by XRF, Univ. of Leeds. $^{87}Sr/^{86}Sr$: Inst. of Petrology, Univ. of Copenhagen (Eimer and Amend standard: 0.7080)

In general, the basalts are relatively rich in incompatible elements (Table 1, no. 8 and 9). However, the younger olivine-phyric tholeiites, above the B-horizon (Fig. 1) in the middle series, show a marked decrease in the level of these elements (Table 1, no. 10). With La/Sm ratios < 1 they are related to ocean ridge basalts as stated by Schilling and Noe-Nygaard (1974).

## THE SILICIC BASALTS

Two silicic basalts - the Klakksvík lava and the Sneis-14 lava of Rasmussen and Noe-Nygaard (1969) - are located approximately 80 m below the top of the middle series in a sequence of thin-bedded, feldspar-phyric pahoehoe flows. The two lavas were possible erupted during the same volcanic episode, although through different vents.

The Klakksvík lava is easily recognised in the field by its resistance to erosion, light grey to medium grey colour and thick scoriaceous top. It crops out over an area 5-8 km wide and 21 km long from the northernmost end of Kallsoy, over Kunoy to Bordoy where it disappears below sea level (Hald et al., 1969) (Fig. 1). The average thickness is estimated to be 10-12 m, and its volume is at least 1.5 $km^3$. The lava contains phenocrysts of bronzite (Jensen, 1978, 1980) up to 3 mm in size and pseudomorphed olivine up to 1 mm in size set in a fine-grained groundmass with plagioclase, augite, pigeonite, Fe-Ti oxides and small amounts of interstitial sheet silicates. The amount of phenocrysts is up to ca 5%, and dominated by orthopyroxene.

The Sneis-14 lava has only been followed for 0.5 km along the western slopes of the mountain Sneis on northern Streymoy. In the type profile of Rasmussen and Noe-Nygaard (1969) the lava is 20 m thick, and consists of 3 or 4 grey to dark grey flow units separated by scoriaceous horizons. In addition to augite, mostly clustered in glomerophyric aggregates, and bronzite the lava carries a few phenocrysts of plagioclase ($An_{65}$) and pseudomorphed olivine. The amount of phenocrysts varies from 30% (X.14) to 10% (MM 15562 and MM 15563) with a maximum size (orthopyroxene) of 4 mm. The groundmass pyroxenes comprise both augite and pigeonite. MM 15562 has interstitial, partly recrystallised glass.

The third silicic basalt lava, Villingadalsfjall-6 (Rasmussen and Noe-Nygaard, 1969), is located on northern Vidoy 75 m above the base of the upper series. No information on its thickness or areal extent is published. The only available sample is light grey and carries ca. 10% glomerophyric aggregates of plagioclase ($An_{65}$) in crystals up to 1 mm in size, together with a few phenocrysts of augite and pseudomorphed olivine. The groundmass is composed of plagioclase, augite, pigeonite, Fe-Ti oxides and small amounts of interstitial sheet silicates and almost completely recrystallised glass.

## CHEMISTRY

The Klakksvík and Sneis-14 lavas (Table 1, no. 1-6) combine high concentrations of MgO (11.3-7.1%) with a silica content of ca 54%. The Sneis-14 lava, apart from X.14 which is presumably enriched in cumulus phenocrysts, has higher FeO*/MgO ratios and lower Ni than the Klakksvík lava, and thus appears to be slightly more evolved.

Compared with the depleted olivine-phyric tholeiites with similar FeO*/MgO ratios from the middle and upper series (Table 1, no. 10) the Klakksvík and Sneis-14 lavas have higher concentrations of $SiO_2$ and incompatible elements, especially $K_2O$, Sr and Ba. They are depleted in $Al_2O_3$ and CaO, whereas the levels of the refractory elements Ni and Cr are almost identical. Schilling and Noe-Nygaard (1974) quote REE data for Sneis-14 with a normalized La/Sm ratio of 2.4, while the depleted olivine-phyric tholeiites have La/Sm ratios between 0.54 and 0.85.

Of special interest are the $^{87}Sr/^{86}Sr$ initial ratios which vary between 0.7026 and 0.7045 in the basaltic lavas of the Faeroe Islands (Carter et al., 1979; Gariepy et al., 1981). The Klakksvík and Sneis-14 lavas have a much higher proportion of radiogenic Sr with $^{87}Sr/^{86}Sr$ ratios between 0.7100 and 0.7119, comparable to one value given by Gariepy et al. (1981). Considerable variation in $^{87}Sr/^{86}Sr$ is found within the same flow unit (Table 1, no. 1 and 2). There is no evidence that this is due to the incipient alteration.

The feldspar-phyric lava from Villingadalsfjall (Table 1, no. 7) contains 6.3% MgO and has low concentrations of $TiO_2$ (1.10%), $P_2O_5$ (0.11%) and Zr (105 ppm). In terms of the last three elements the lava appears to be related to the depleted olivine-phyric tholeiites and the Klakksvík and Sneis-14 lavas rather than to the surrounding feldspar-phyric tholeiites of the middle and upper series. A relationship between Villingadalsfjall-6 and the Klakksvík and Sneis-14 lavas is also demonstrated by high contents of $SiO_2$, $K_2O$ and Ba and by a $^{87}Sr/^{86}Sr$ initial ratio of 0.7162.

## DISCUSSION AND CONCLUSION

In terms of the high content of $SiO_2$ and MgO and low content of CaO the Klakksvík and Sneis-14 lavas show resemblance to boninites described from the western Pacific subduction zones by Kuroda and Shiraki (1975) and Dietrich et al. (1978). An important difference, however, is the higher concentration of incompatible elements in these Faeroe lavas including a normalised La/Sm ratio of 2.4. For the same reason the two flows are not comparable to the orthopyroxene-phyric lavas of the Troodos complex discussed by Simonian and Gass (1978).

Instead, we call attention to the resemblance between the two MgO-rich silicic basalts and some basic to intermediate lavas and tuffs described by Pedersen (1977, 1978, 1979) from the Tertiary volcanic sequence on Disko and Nûgssuaq, West Greenland (Table 1, no. 11). The composition of these rocks (including high $SiO_2$, incompatible elements and $^{87}Sr/^{86}Sr$ and low CaO) is ascribed to assimilation of shales and sandstones which are present as xenoliths. Feldspar-phyric silica-enriched basalts similar to the lava from Villingadalsfjall are found in the same region.

We conclude that silicic basalts with anomalously high $^{87}Sr/^{86}Sr$ initial ratios were developed on the Faeroe Islands, most likely through wall rock contamination in magma chambers located in sediments or in continental basement rocks (see also Gariepy et al., 1981). Eruption of the contaminated magmas took place at least twice during the formation of the middle and upper series. The magmas parental to the silicic basalts were presumably related to the depleted ocean ridge type tholeiites which are found intercalated with enriched types in the upper part of the lava sequence on the Faeroe Islands.

ACKNOWLEDGEMENTS

We are grateful for chemical analyses from J.C. Bailey, H. Bollingberg, O. Larsen, J. Rønsbo and I. Sørensen. Aa. Jensen kindly provided us with powdered rock samples from the Klakksvík lava. C.K. Brooks, J.G. Larsen, A.K. Pedersen and W.S. Watt are thanked for constructive criticism of the manuscript. XRF (Rb and Sr) and microprobe analyses were carried out on instruments bought with funds from the Danish Natural Science Research Council.

REFERENCES

Bailey, J.C. and Noe-Nygaard, A., 1976, Chemistry of Miocene plume tholeiites from northwest Iceland, Lithos, 9:185-201.

Bollingberg, H., Brooks, C.K. and Noe-Nygaard, A., 1975, Trace element variations in Faeroese basalts and their possible relationships to ocean floor spreading history, Bull. geol. Soc. Denmark, 24:55-60.

Bott, M.H.P., Nielsen, P.H. and Sunderland, J., 1976, Converted P-waves originating at the continental margin between the Iceland-Faeroe Ridge and the Faeroe Block, Geophys. J. R. astr. Soc., 44:229-238

Bott, M.H.P., Sunderland, J., Smith, P.J., Casten, U. and Saxov, S., 1974, Evidence for continental crust beneath the Faeroe Islands, Nature, 248:202-204.

Brooks, C.K., Nielsen, T.F.D. and Petersen, T.S., 1976, The Blosseville Coast basalts of East Greenland: Their occurrence, composition and temporal variations, Contr. Miner. Petrol., 58: 279-292.

Carter, S.R., Evensen, N.M., Hamilton, P.J. and O'Nions, R.K., 1979, Basalt magma sources during the opening of the North Atlantic, Nature, 281:28-30.

Clarke, D.B. and Pedersen, A.K., 1976, Tertiary volcanic province of West Greenland, in: "Geology of Greenland", A. Escher and W.S. Watt, eds., pp. 364-385, The Geological Survey of Greenland, Copenhagen.

Dietrich, V., Emmermann, R., Oberhänsli, R. and Puchelt, H., 1978, Geochemistry of basaltic and gabbroic rocks from the West Mariana Basin and the Mariana Trench, Earth Planet. Sci. Lett., 39:127-144.

Fitch, F.J., Hooker, P.J., Miller, J.A. and Brereton, N.R., 1978, Glauconite dating of Palaeocene-Eocene rocks from East Kent and the time-scale of Palaeogene volcanism in the North Atlantic region, J. geol. Soc. Lond., 135:499-512.

Gariepy, C., Ludden, J.N., Brooks, C. and Machado, N., 1981, Trace element, Sr- and Nd-isotopic constraints on the genesis of the Faeroe lava pile (abstract), Trans. Am. Geophys. Un., 62:424.

Hald, N., Noe-Nygaard, N. and Waagstein, R., 1969, On extrusion forms in plateau basalts. 2. The Klakksvík flow, Faeroe Islands,

Bull. geol. Soc. Denmark, 19:2-7.
Jensen, Aa, 1978, Compositional variations of the pyroxenes from three flows of Faeroe Islands basalts, Bull. geol. Soc. Denmark, 27, Spec. Iss.:63-78.
Jensen, Aa., 1980, Mineralogical and geochemical variations across three basaltic lava flows from the Faeroe Islands, Bull. geol. Soc. Denmark, 28:95-114.
Kuroda, N. and Shiraki, K., 1975, Boninite and related rocks of Chichi-jima, Bonin Islands, Japan, Rep. Fac. Sci. Shizuoka Univ., 10:145-155.
Noe-Nygaard, A. and Rasmussen, J., 1968, Petrology of a 3000 m sequence of basaltic lavas in the Faeroe Islands, Lithos, 1: 286-304.
Pedersen, A.K., 1977, Tertiary volcanic geology of the Mellemfjord area, southwest Disko, Rapp. Grønlands geol. Unders., 81:35-51.
Pedersen, A.K., 1978, Graphitic andesite tuffs resulting from high-Mg tholeiite and sediment interaction; Nûgssuaq, West Greenland, Bull. geol. Soc. Denmark, 27, Spec. Iss.:117-130.
Pedersen, A.K., 1979, Basaltic glass with high-temperature equilibrated immiscible sulphide bodies with native iron from Disko, Central West Greenland, Contr. Miner. Petrol., 69:397-407.
Rasmussen, J. and Noe-Nygaard, A., 1969, Beskrivelse til geologisk kort over Færøerne i målestok 1:50.000, Geol. Surv. Denmark, Ser. 1, 24:370 p.
Rasmussen, J. and Noe-Nygaard, A., 1970, Geology of the Faeroe Islands (Pre-Quaternary), Geol. Surv. Denmark, Ser I, 25:142 p.
Schilling, J.-G. and Noe-Nygaard, A., 1974, Faeroe-Iceland plume: Rare-earth evidence, Earth Planet. Sci. Lett., 24:1-14.
Simonian, K.O. and Gass, I.G., 1978, Arakapas fault belt, Cyprus: A fossil transform fault, Bull. geol. Soc. Am., 89:1220-1230.
Sørensen, I., 1975, X-ray fluorescence spectrometry at GGU, Rapp. Grønlands geol. Unders., 75:16-18.
Talwani, M. and Eldholm, O., 1972, The continental margin off Norway: A geophysical study, Bull. geol. Soc. Am., 83:3575-3606.
Talwani, M. and Eldholm, O., 1977, Evolution of the Norwegian Greenland Sea, Bull. geol. Soc. Am., 88:969-999.
Tarling, D.H. and Gale, N.H., 1968, Isotopic dating and paleomagnetic polarity in the Faeroe Islands, Nature, 218:1043-1044.
Waagstein, R., 1977, The geology of the Faeroe Plateau, Unpubl. thesis, Univ. of Copenhagen.

# CENOZOIC PALEOGEOGRAPHY OF NORTH ATLANTIC LAND BRIDGES

Malcolm C. McKenna

Department of Vertebrate Paleontology
The American Museum of Natural History
New York, N.Y. 10024

Department of Geology
Columbia University
New York, N.Y. 10027

## INTRODUCTION

This paper attempts to synthesize vertebrate paleontological evidence and recently acquired knowledge of the geological history of terrestrial connections across the northern North Atlantic area. In 1975 I published a survey of this subject (McKenna, 1975) but it is appropriate here to bring the synthesis up to date in view of recent geophysical work and in view of further paleontological exploration in the Arctic and elsewhere. As Alfred Wegener (1929) was well aware, the world has had but one history; for that reason geological evidence and biological evidence each provide constraints on any proposed solutions to the ancient geography of an area such as the northern end of the North Atlantic. Although I hope to bring the subject up to date, I intend also to speculate on potential vertebrate paleontological field work that might yield new evidence concerning the details of timing and position of Tertiary land connections between now separated areas.

After a brief section on the biological principles of corridor recognition and an exposition of how Iceland might be coaxed to yield evidence of a biota older than any sediments presently preserved on land there, I will attempt to discuss various geological features around the north end of the Atlantic and in the Arctic relevant to two land bridges that once connected the two sides of the Atlantic. These bridges had somewhat different histories and were affected somewhat differently by whatever climatic gradients existed.

The more northerly of the two bridges (De Geer Route) crossed the high Arctic via the Canadian mainland, Ellesmere Island, Greenland, and Svalbard to Norway. It apparently was continuous as far east as Greenland and as far northwest as Svalbard throughout the Tertiary, but a water gap separated Svalbard from Greenland at least as early as the early Oligocene. An earlier, epicontinental marine connection had existed in the area in the Mesozoic, but presumably the rise of a triple junction at the Yermak hot spot had put an end to that in the Paleocene.

The southern bridge was the well-publicized but poorly known Thulean Bridge. From North America to Greenland there were two western access ramps: one passed northward through Ellesmere Island and then back southward through Greenland coextensive with the west end of the high latitude De Geer Route just mentioned; the other, briefer in duration, connected Labrador and Baffin Island with West Greenland through the Davis Straits area in the late Paleocene and sank down again later in the Tertiary as Baffin Bay reconnected to the Labrador Sea. From Greenland east, proto-Iceland provided elevated terrain from Kangerdlugssuaq Peninsula to the Faeroes. The weakest link in the Thulean Bridge, from the Faeroes to Scotland, broke apart in early Eocene time shortly after forming.

Both bridges were dry land; however, with the advent of a warming climate at the Paleocene-Eocene boundary, the more southerly position of the Thulean Bridge permitted a more extensive mixing of the southwest European and American biotas (and possibly some Asian components as well) than had been the case in the late Paleocene. This mixing was shortlived, however, due to a break in the bridge rather than because of climate. The climate remained warm after the interruption until middle Eocene times before starting a cooling trend that became precipitous at the end of the Eocene as reconnection of the Arctic Ocean with the Greenland-Norwegian Seas occurred. After the earliest Eocene mixing episode the southwestern European and American biotas went their own evolutionary ways. This may not have been so in Fennoscandia, however, because Fennoscandia may well have been an outpost of North America as late as the end of the Eocene (McKenna, 1981). Unfortunately there is no way to test this concept except to prospect the early Cenozoic of Svalbard more intensely than ever. At present, Svalbard's contribution to Cenozoic vertebrate paleontology consists of one specimen of a fish! The right rocks are there, however, and are well known for their flora and invertebrates.

## COMMENTS ON BIOGEOGRAPHICAL PRINCIPLES

### The Corridor-Filter-Sweepstakes Spectrum

From the study of geographical distribution of present-day terrestrial organisms it can be seen that, in general, as distance increases between two samples of the fauna or flora of a large land mass, various indices of faunal or floral similarity decrease. Nevertheless, at about the generic level the resemblance remains high over thousands of km. Simpson (1936, 1940, 1953) denoted large land masses with diverse ecological opportunities by the term "corridor" and noted that dispersal within such areas tends to be geologically instantaneous. In addition, balanced biotas (i.e., with the proper proportions of carnivores to herbivores, large animals to small, etc.) tend to spread as units. Moreover, dispersal in opposite directions within a corridor should occur more or less simultaneously and soon after any initial opportunity to do so arises.

At the other extreme of the spectrum, overwater transport or other unlikely mechanisms become more probable with the passage of time (Simpson, 1952). Such mechanisms may be thought of as a type of sweepstakes in which, once one has become a ticket holder, one may play the game. Thus the presence of the fossil pygmy hippopotamus in Madagascar may be thought of as a result of some coastal African ticket holder somehow having been able to survive an unwelcome 260 km swim across the Pleistocene Mozambique Channel, whereas more arid-country types, such as zebras or giraffes, simply were not ticket holders and thus not able to participate. Presumably, the lucky hippopotamic ticket holder was female, pregnant, and on arrival was successful at making a living after she clambered ashore.

Between these extremes lies a continuous spectrum of filters, allowing only incomplete transfer of organisms. The most obvious of these would be a narrow land bridge on which a narrow range of environments prevail, thus preventing certain organisms from spreading. Variations on the filter theme are endless.

The corridor-filter-sweepstakes spectrum was designed to explain dispersal in a stabilist world, but such dispersal is only one part of the picture as mobilist notions of earth history have made clear. In addition to active dispersal, which is always across pre-existing barriers of some sort, we can also imagine passive dispersal, which is actually not dispersal at all but simply a reordering of the land on which the organisms live.

Plate tectonic theory, with its large-scale Eulerian rotations of rigid plates on the surface of the globe, has provided the theoretical mechanism for sundering of land masses and for the formation of composite terranes through plate collisions. Along with land itself, biotas also must fragment and eventually recombine. I have elsewhere (McKenna, 1973) used the term "Noah's ark" to denote landmasses such as India, Australia, and South America, whose geographic affinities have changed, passively fragmenting a previous biota and then recombining its survivors with the biota of another part of the world with which the landmass makes contact, more or less as a boat would dock. Dispersal after such a docking across the resulting tectonic suture would tend to blur the distinctiveness of a long-docked Noah's ark, but endemic elements should theoretically reveal the composite nature of the biota of a composite landmass because the closest relatives of endemic organisms remaining on the docked ark's side of a tectonic suture would not be among the inhabitants of the terrane across the suture but rather among the inhabitants of the "old country" from whence the ark originally departed. Taking the ark concept a bit further, one can also imagine that exotic terranes now welded to large landmasses might well have carried fossils far from their original main area of distribution, resulting in a situation not unlike a grounded Viking funeral ship (McKenna, 1973). In this case it would be obvious that such fossils would characterize only the exotic terrane component of a composite landmass.

With both Noah's arks and grounded Viking funeral ships it is important to realize that the barriers that separate vicariant biotas arose subsequent to the establishment of the original distribution of the biota on the original land mass. The original biota became passively sundered, with its resulting daughter biotas separated by a widening barrier which for terrestrial organisms is usually water. Provided that the organisms were in existence and living on the parent landmass, active overwater transport mechanisms are not necessary to explain the existence of closely related taxonomic pairs located on opposite sides of a subsequently formed barrier. Such pairs can be explained as the results of evolution in isolation from one another. With the passage of time, however, one may imagine that such primary vicariance would be progressively more completely overprinted by the results of dispersal. Vicariance "signal" would thus eventually be lost in dispersal "noise". At low signal/noise ratios the detection of former vicariant events by means of the study of the sister-group relations of endemic groups would have to involve large numbers of taxon pairs selected without bias. The fragmentation of the Euramerican terrestrial biota in the early Eocene (McKenna, 1975) is such a vicariant event, brought about by the interruption of the Greenland-Scotland Ridge at at least one point, probably between the Faeroe Islands and Scotland.

## Treadmills

Under the mobilist rationale, Iceland can be viewed as a

steady-state phenomenon, spreading from the center, sinking at the coasts, and covering approximately its present land area throughout most of the Cenozoic Era (Vogt, 1972). Early in the Cenozoic, proto-Iceland would have occupied a substantial part of the then much narrower North Atlantic Ocean. At known spreading rates, new crust generated on either side of the accreting plate boundary in central Iceland would have moved laterally toward the northwest and south-east coasts as it does now, sinking as it went and passing beneath the waves in about 20 million years in accordance with logarithmic curves (Sclater, Anderson, and Bell, 1971; Parsons and Sclater, 1977; Cochran and Talwani, 1977; Grønlie, 1979).

The oldest known rocks on Iceland are indeed known from coastal Iceland and are about 16 million years old (Moorbath et al., 1968). To avoid wet feet, the Icelandic biota would hardly have needed to step lively; a pace of about one cm per year toward the center of the island would have sufficed for the inhabitants of either side of the island. The biota would thereby have remained in place atop a moving substrate, more or less as a small child sometimes does on an escalator. From the standpoint of Iceland's biota, the island may be thought of as consisting of two treadmills whose upper ends from which the treads emerge are in contact at the North American-Eurasian plate boundary that crosses the island more or less from southwest to northeast. The lower ends of the two treadmills reach sea level in northwest and southeast Iceland after about 200 km of travel (20 million years at one cm/year) caused by lateral and subsiding motion away from Iceland's central rift.

## THE NORTH ATLANTIC PROBLEM

### Where are the available Paleocene and Eocene Terrestrial Sediments?

Although evidence of Cenozoic sea-floor spreading is everywhere present on the floor of the North Atlantic Ocean, preserved terrestrial sediments capable of producing fossil vertebrates that might tell us about terrestrial faunal exchange are not extensive on land except in Europe. There, as a result of many years of exploration, the ancient biota is now well known over a long time span. In addition, exploration of the known exposures of early Cenozoic continental sediments elsewhere on land around the north end of the Atlantic has yielded Eocene fossil vertebrates on Ellesmere Island. Ellesmere Island is not, strictly speaking, part of the Atlantic coast, of course, but its land continuity with Greenland at the time its known terrestrial Eocene fauna lived is assured and the fauna may be taken as representative. Floral evidence is abundant at most sites throughout the North Atlantic border lands where Paleogene sediments are known.

Great Britain, Ireland, France, Belgium, Netherlands, Switzerland, Germany, Spain, Portugal

Tertiary fossil vertebrates are known from the Thanetian onward in western Europe south of the Baltic Sea area. The literature is enormous and will not be summarized here (see Curry, 1965; Russell, 1968; McKenna, 1975; Tobien, 1976; and Berggren et al., 1978). The earliest known Paleocene terrestrial vertebrates of the area are represented in the rather endemic but faintly "North American" Hainin fauna of Belgium, probably middle Paleocene in age (Godfriaux and Thaler, 1972; Vianey-Liaud, 1979). Later Paleocene terrestrial faunas show moderately increasing similarity to the North American fauna (Russell, 1964; Kurtén, 1966) followed by a sharp increase, which I have elsewhere discussed (McKenna, 1975) as the brief establishment of a terrestrial corridor to North America at the Beginning of Sparnacian time. North of the Artois Uplift (Jacqué and Thouvenin, 1975) along the Franco-Belgian border, most Sparnacian deposits are marine (Argile de Flandres = Ieper Clay, etc.), so that the terrestrial corridor would have extended to the west toward Greenland via the Artois Uplift, Great Britain, the Wyville-Thompson Ridge, Faeroe Bank, and the Faeroes. The northwest British Isles and the Faeroes area were uplifted at the time (Evans et al., 1973; Bott, 1975). After the final phase of graben formation in the central North Sea in the Danian, general subsidence of the North Sea took place (Parker, 1975; Whiteman et al., 1975). Land somewhere to the northwest of the North Sea in the Orkney-Northern Highlands area or beyond supplied much of the coarse clastics locally present in the North Sea Paleocene and Eocene marine deposits, but by the late Eocene this source area was either lower or more distant (Ziegler, 1975, p. 183). Locally, large fans distributed deltaic sands eastward from the Moray Firth area of Scotland in the late Paleocene (Parker, 1975), but during the Eocene these sands were covered by marine shales. Although local tuffaceous deposits are known as early as the Thanetian in the Paleocene (Kent, 1975), the beginning of widespread volcanic ash deposition in the North Sea immediately follows the late Paleocene sands (Jacqué and Thouvenin, 1975; Kent, 1975) during P6 (*Morozovella subbotinae*) time (Berggren, 1972a), which Berggren et al. (1978) interpret to be associated with marine magnetic anomaly 24. Vine (1976) and Berggren et al. (1978) assigned an age of 53 Ma to anomaly 24, which, when corrected to conform with currently accepted IUGS abundance and decay constants for K-Ar dating, would be 54.4 Ma. Ness et al. (1980) assign an age of 55 Ma to anomaly 24. An even earlier date (56 Ma, corrected to 57.5 Ma) has been proposed by Labrecque et al. (1977) but is clearly too old. A tuffaceous sequence about 30 m thick occurs at about the Paleocene/Eocene boundary over a wide area in the North Sea. It probably represents volcanism somewhere to the northwest, although a local source to the east in the Skagerrak may also have contributed.

In the Eocene the Artois Uplift was once more below sea level and was transgressively invaded by marine waters in the Cuisian and early Lutetian. These waters were continuous with those of the Eocene North Sea and they extended as far south as Paris. Southwestward, marine continuity with the Atlantic at this time is also indicated (Pomerol, 1973; Sittler, 1976). Thus, for a time Great Britain and Ireland became isolated from France, more or less as at present, by a narrow seaway that persisted until the Artois Uplift re-exerted itself once more in the middle Lutetian. Therefore at least one interruption existed in the former corridor between France and North America at this time. Land connection between Britain and France then prevailed again from middle Lutetian until Stampian time in the Oligocene.

Other than in the London and Hampshire basins, early Cenozoic terrestrial sediments are rare on land in Great Britain and Ireland. Minor outcrops occur at Petrockstow and Bovey in southwest England. In the northern part of Ireland the lacustrine Lough Neagh Clays are of Paleogene age. These are more than 300 m thick and rest on laterite that followed the uppermost plateau basalts. Eruption of about 2000 $km^3$ of Irish plateau basalts began in the Paleocene but probably continued into the Eocene (Evans et al., 1973). Fossil plant remains are abundant in the Lough Neagh Clays and could be either Paleocene or younger. In southern Ireland presumably early Cenozoic sands, clays, and lignites locally fill topography in Carboniferous limestone near Caher and northwest of Roscrea in County Tipperary (Charlesworth, 1953), but these deposits are not extensive. In Scotland and Ireland additional thin, plant-bearing sediments occur in a few places, interlayered with Brito-Arctic volcanics (Charlesworth, 1953; Deegan, 1976). In northern Ireland extensive laterites were developed between outpourings of plateau lavas that are about 60 Ma in age (Mellor and Mussett, 1975). However, known Paleogene vertebrate fossils of the British Isles are all from the London and Hampshire basins in southern and southeastern England.

## Baltic Epicontinental Marine Connections from the North Sea to Tethys: Poland

Pożaryska and Ciuk (1976) state that where exposures occur in northeastern Poland the Paleogene is primarily marine. Danian deposits thin from north to south and are characterized by small planktonic foraminifers (*Globigerina daubjergensis*). According to these authors the overlying marine Montian of Poland contains the same assemblage of solely benthonic Foraminifera as is found in the Selandian of Denmark and Sweden, and, significantly, also in Lithuania and the Ukraine (see also Anderson, 1976). By the Thanetian, marine Paleocene rocks were no longer being deposited on what is presently land in Poland. Instead, coal-bearing nonmarine sediments were laid down. Nonmarine conditions appear to have continued into the early Eocene except possibly near Szczecin. By Lutetian time marine conditions had returned to northeastern Poland. In contrast, nonmarine conditions, including coal deposition, now prevailed near Szczecin. In the late Eocene a widespread cold-water marine transgression occurred in Poland (Pożaryska, 1978). These waters were connected with those of the Kiev Sea of the Ukraine. In southwestern Poland late Eocene marine waters were also present, but carried warm water Foraminifera including nummulitids derived from Tethys from the south via the Moravian marine portal. Finally, in the Oligocene, marine conditions continued longest in the west of Poland, but gradually changed to a nonmarine environment, starting in the northeastern part of the country. Polish Neogene sediments are all nonmarine except for Paratethyan deposits in the southeast.

From Pożaryska and Ciuk's (1976) report one can conclude that, insofar as the preserved evidence goes, only in the early part of the Paleocene and in the late Eocene was there a marine connection all the way from the North Sea to the Ukraine. From the end of the Montian through the Thanetian, Ypresian, and Lutetian some other marine connection, namely the Moravian portal (Pożaryska, 1978, fig. 2, for location), would have to have operated if terrestrial faunas of Fennoscandia (and also Asia if the Turgai Straits were not a barrier) were not to have contaminated the fauna of terrestrial western Europe. Similarity of the marine teleost fishes of the Paleocene of Turkmenistan with those of the earliest Eocene Mo-clay of Denmark (Bonde, 1974) may be explained by such a connection (Spjeldnaes, 1975).

## Atlantic Coast of Norway

Exposures of Mesozoic and Cenozoic continental sediments in Norway are virtually unknown except in presently subaerial parts of the Svalbard Platform in the far northwest. The Svalbard area is believed to have been continuously connected with "mainland" Norway throughout most of the Cenozoic by a subaerial Barents Shelf (Harland, 1969; Eldholm and Windisch, 1974; Renard and Malod, 1974; Eldholm and Talwani, 1977; Eldholm and Thiede, 1980). South of the Barents Shelf, continental Triassic is known in the northern North Sea, a middle Jurassic flora has been sampled in erratic boulders recovered at about latitude 64° at Beistadfjorden (Os Vigran, 1970; Oftedahl, 1972), and a small patch of Late Jurassic to Early Cretaceous Mesozoic exposures bearing coal and shallow marine fossils is present on Andøy in the Lofoten Islands at about latitude 69° (Ørvig, 1960; Dalland, 1973, 1975; Sturt et al., 1979). Evidently, the sea transgressed northwards during the latest Jurassic in the Andøy area (Ørvig, 1960). Details about the marine Tertiary off Norway still remain little-reported because of the confidentiality of bore-hole data (Ulleberg, 1976), but the geology of the Norwegian continental shelf has been summarized by Ronnevik et al. (1975). Grønlie (1979) has attempted to reconstruct the history of subsidence of the Norwegian-Greenland Sea and believes that the Vøring

Plateau area may have been subaerial until the late Eocene, although not connected with mainland Norway.

Active sea-floor spreading is believed to have begun off the Norwegian coast along the Vøring Escarpment between anomaly 25 and anomaly 24 time (Talwani et al., 1975; Eldholm and Sundvor, 1979; Grønlie, 1979). Those authors, however, accept the Labrecque et al. (1977) time scale, which places anomaly 24 at 56 Ma. It may be that a new and important episode of spreading began at 56 Ma (Grønlie, 1979), but anomaly 24 is younger than 56 Ma (Berggren et al., 1978; Ness et al., 1980). Thus, although graben formation began much earlier, accretion of oceanic crust off the coast of central Norway did not begin until late Paleocene time. Since that time the Norwegian Atlantic Coast has behaved as a trailing-edge coast.

Of the various sites where basaltic basement has been reached in the Norwegian-Greenland Sea area, only DSDP bore-holes 337 and 356 bottomed in <u>submarine</u> lava flows (Kahrin, 1976, 1976).

## Svalbard Archipelago and Barents Shelf

The Tertiary history of Svalbard and the Barents Shelf has recently been summarized by Emelyanov et al. (1970), Lehmann et al. (1978), and by Dypvik and Nagy (1979). According to these authors, more than 2000 m of Tertiary strata are extensively exposed in southern Spitsbergen, where they unconformably overlie Aptian and Albian sediments. The source area for these rocks lay to the north or northeast (Kellogg, 1975). The Tertiary rocks range in age from early Danian to Oligomiocene. Coal-bearing sediments occur in both the Paleocene and Oligomiocene parts of the section. The former are paralic but the latter are limnic. No fossil mammals have yet been reported from Svalbard, but vertebrates (an amioid fish) do occur near Heerodden at a stratigraphically somewhat dubious position in the terrestrial Tertiary of the main island, Spitsbergen (Lehman, 1951). The Basilika Formation, rather uncertainly dated by benthonic Foraminifera and plant remains as Paleocene, contains volcanic ash derived from either northern Svalbard or the Yermak hot spot to the north (Dypvik and Nagy, 1979). The Foraminifera are of interest because mainly textulariines are represented at one locality and only nodosariids are present at another. Dypvik and Nagy (1979) conclude that these foraminifers indicate that the enclosing sediments were deposited in a more or less landlocked shelf sea under unstable or extreme environmental conditions. In the northern part of Svalbard, early Tertiary volcanism including flood basalts has been dated in a range from 60 to 22 Ma (Burov and Zagruzina, 1976; Prestvik, 1977), but whether the tuffs of southern Spitsbergen are related to any of these particular volcanics is not known.

The geology of the continental margin off western Svalbard suggests that some opening of the Greenland Sea had already occurred prior to anomaly 13 time (Schlüter and Hinz, 1978).

The geology of the Barents Shelf has been studied geophysically by Renard and Malod (1974) and by Hinz and Schlüter (1978). Renard and Malod conclude that the Barents Shelf was subaerial during the Tertiary.

Vogt et al. (1978), Sundvor and Eldholm (1979), and various other authors conclude that the spreading phase of the tectonic evolution of both the Greenland-Norwegian Sea and the Eurasia Basin in the Arctic Ocean began during the magnetically reversed interval preceding marine magnetic anomaly 24. Vogt et al. (1978) date the beginning of spreading at 57-58 Ma, evidently on the basis of the Labrecque et al. (1977) time scale. On the basis of the Berggren et al. (1978) and Ness et al. (1980) time scales, the beginning of spreading would have been somewhat later. In any event, active spreading seems to have begun late in the Paleocene, no matter how it is calibrated in years. Vogt and his colleagues calculate, in addition, that Svalbard cleared Greenland at about anomaly 13 time. The manner in which Greenland and Svalbard separated is still unclear (Sundvor and Eldholm, 1979). Talwani's and Eldholm's (1977) model of separation involved primarily shear along the complex Spitsbergen Fracture Zone and Hornsund Fault on the southwest side of Svalbard until anomaly 13 time. That some spreading was also taking place was suggested by Vogt et al. (1978) on the basis of then still unpublished work by J. D. Phillips and his colleagues.

## The Faeroe Islands

The Faeroe Islands are composed of Paleocene and Eocene plateau lavas (Rasmussen and Noe-Nygaard, 1969). They lie on continental crust at the northeastern end of the Rockall-Faeroe microcontinent (Casten, 1973; Bott, 1975; Roberts, 1975). The Faeroe volcanics are separable into three sequences with a total thickness of about 3000 m. Between the lower and middle sequence a thin 15 m lignitic to bituminous coal-bearing unit occurs in the western and southern Faeroes (Rasmussen and Noe-Nygaard, 1970, fig. 2; Laufeld, 1965). Tuffs and clays are also interbedded with the lower basalts, which are about 900 m thick and were produced by fissure eruptions. Above the coal-bearing unit explosive vulcanism began suddenly and gradualy changed to basaltic lava flows. This middle basaltic sequence is about 1350 m thick. After a break, lava production similar to that of the lower volcanic sequence resumed, interspersed with plant-bearing tuffaceous clays. The entire middle and upper sequences are of reversed magnetic polarity, whereas the lower sequence consists of two normal polarity intervals, separated by a reversed interval (Rasmussen and Noe-Nygaard, 1970). Dikes cut all units, including the youngest preserved Faeroe lavas.

Dates from the Faeroe plateau lavas lie between 49 and 62 Ma

(Tarling and Gale, 1968, uncorrected). The aegerine granite that forms Rockall Island is dated at 52±9 Ma (Roberts, 1975). Probably between two and three million $km^2$ of plateau basalt was subaerial in this part of the world before subsiding beneath the Atlantic (Rasmussen and Noe-Nygaard, 1970).

## The Iceland-Faeroe Ridge

The history of submergence of this ridge (Detrick et al., 1977) will be dealt with by virtually every paper at this symposium. Critical to these interpretations is DSDP bore-hole 336, located near the top of the northeast flank of the ridge (Talwani et al., 1975; Nilson and Kerr, 1978). DSDP hole 336 passed through marine Oligocene and late and middle Eocene to bottom in lateritic basaltic rubble derived from subaerial erosion. Indications are that a warm climate prevailed. The microfauna of the Eocene marine sediments suggests "a history of submergence at this site and by extrapolation, for the entire Iceland-Faeroe Ridge" (Talwani et al., 1975). Nevertheless, on the basis of differences in sedimentation and Oligocene faunas from DSDP hole 336 and hole 352 on the other side of the ridge, Talwani and his colleagues concluded that there was no direct connection of the Greenland-Norwegian Sea area with the Atlantic across the Iceland-Faeroe Ridge between holes 336 and 352 until well up in the Tertiary. Vogt (1972) had reached similar conclusions on the basis of topography and sedimentation controlled by the Western Boundary Undercurrent, but calculated that the Faeroe-Shetland Channel would have formed a marine interruption as long ago as 20 Ma, and some evidence of active bottom currents indicated a connection between the Norwegian Sea and the Atlantic as early as the late Eocene. Fossil vertebrates suggest that the marine connection occurred still earlier, in the early Eocene. Such an interruption would have isolated the terrestrial biota of the British Isles (and contiguous land) from that of the western side of the North Atlantic (Grønlie, 1979, fig. 4). That the break had occurred well before 20 Ma is indicated by the vertebrate evidence. Such evidence is not adequate to pinpoint the exact position of the break; rather, it shows that a break existed somewhere. The Faeroe-Shetland area is simply a very weak link in the chain and may have been the site of interruption from early Eocene onward. By late Miocene time (10 Ma) the Iceland-Faeroe Ridge had sunk far enough to have become a northwesterly trending chain of islands. The Faeroe-Shetland Channel also deepened (Grønlie, 1979) at that time.

Strauch (1970; 1972) concluded that the Tertiary Thulean Bridge, of which the Iceland-Faeroe Ridge was a major component, was sufficiently elevated to prevent the marine molluscs of the Greenland-Norwegian Sea (Skandic Sea) and the Atlantic to mix. Strauch believed that the land barrier was actually present as late as the Pleistocene. A subaerial barrier lasting the whole of the Tertiary across the entire Thulean route is out of the question, however, for both geological and terrestrial biogeographic reasons. Evidence from fossil vertebrates indicates that at least one major break in the route must have existed from sometime near the beginning of the Eocene, shortly after the initiation of large-scale plate generation and separation of northern Europe from Greenland (McKenna, 1975).

## Iceland

Latest Miocene terrestrial sediments about 6 Ma in age are interbedded with lavas in eastern Iceland near Bessastadaá (65°03' N, 15°02' W) and Hengifoss (65°05' N, 14°54' W) at an elevation of about 200 m above sea level (McDougall et al., 1976). Plant remains from sediments at Hengifoss and Bessastadaá have been studied by Pflug (1959), who concluded on the basis of spores and pollen from interbedded lignitic layers that a cool temperate late Tertiary environment was represented at Bessastadaá and a cold climate Quaternary environment at Hengifoss. Both sites are latest Miocene in age on the basis of radioisotopic dates and paleomagnetic studies by McDougall et al. (1976). However, Pflug's conclusions about climate are still acceptable.

Vilmundardóttir, 1972 (discussed by McDougall et al., 1976), has reported tillite interbedded in both the Miocene and Pliocene parts of the Bessastadaá section. McDougall et al. are dubious about the identification of these deposits as tillites. The oldest well-authenticated glacial deposit in Iceland, according to McDougall and his colleagues, is about 3.1 Ma in age and occurs at Jökuldalur about 30 km to the north of Bessastadaá (McDougall and Wensink, 1966). Berggren (1972b), on the basis of the earliest ice-rafted detritus in the area collected on DSDP Leg 12, dates the beginning of glaciation near Iceland at about 3 Ma. Thus there is a possibility that the intrabasaltic sediments at Bessastadaá and Hengifoss are preglacial. That assumption accepted, it therefore becomes attractive to investigate these and older intrabasaltic sediments for possible fossil vertebrates and other organisms that had not yet been snuffed out by glaciation in that part of the North Atlantic area.

Although montane glaciation may have begun locally in Iceland at about 3.1 Ma, evidence from northwestern Europe (Zagwijn, 1974) and from DSDP bore-hole data at sites 111, 116, and 410 (Backman, 1980) suggests that the development of large northern hemisphere ice sheets occurred somewhat later, at about 2.5 Ma. Thus any late-surviving Tertiary terrestrial fauna of Iceland might be expected to be recoverable from terrestrial sediments that are older than about 2.5 Ma.

In southeastern Iceland at Hornarfjord (near Hoffells Gletscher) Schwartzbach and Pflug (1957) report possibly Miocene plants from coal-bearing intrabasaltic sediments overlying a "tillite." The pollen from this site indicates a warmer climate than that of

Tjörnes in northern Iceland.

In western Iceland another well-dated and paleomagnetically well-known section about 3500 m thick has been studied by McDougall et al. (1977) in the Borgarfjördur region, north of Reykjavik. Sediments are present in about 15% of this section and locally are even more abundant, especially in the lowest 225 m and again between about 700 m and 1650 m above the base. In terms of time, these intervals are at about 6.5 Ma and about 5.8 to 4.4 Ma, respectively. Glacial sediments are present in the section, notably from 2800 to 3500 m, in rocks that on paleomagnetic grounds belong to the Gauss Epoch (3.32 Ma at its base). Thus another opportunity to sample pre-glacial biotas may exist in the Borgarfjördur section of western Iceland, although McDougall et al. mention no paleontological finds.

Another area where intrabasaltic sediments occur in a probable pre-glacial setting is at Brjanslaekur in northwestern Iceland (Heer, 1868; Friedrich, 1966). Heer believed the Brjanslaekur flora to indicate Miocene age. Several subsequent authors toyed with the idea that this and other Icelandic floras were Eocene in age, but Friedrich (1966) has again opted for the Miocene, and Wolfe (1980) has narrowed the age determination to middle Miocene.

Although no vertebrate remains have been reported, it is important here to mention a thick Plio-Pleistocene sequence of Cenozoic rocks at Tjörnes (lat. 66° N) comprising paralic sandstones and claystones interspersed with volcanics (Strauch, 1963, 1972; Gladenkov, 1981). These rocks crop out on the coast of northern Iceland within the large central graben (Einarsson et al., 1967). The rocks have been divided into three sets of beds separated by lavas: the Tjörnes beds (at the base), the Furuvík beds, and the Breidavík beds (at the top). The lower, Tjörnes beds have in turn been subdivided biostratigraphically (Bardarson, 1925) into the "*Tapes*" zone (at the base), the "*Mactra*" zone, and the "*Cardium*" zone (at the top). The lower two zones of the Tjörnes beds alternate between shallow marine and lacustrine conditions and bear many thin lignitic layers. Pollen from lignite in the "*Mactra*" zone indicates mixed coniferous and hardwood vegetation (Schwarzbach and Pflug, 1957) and suggests a temperate climate. Mollusca from the "*Tapes*" and "*Mactra*" zones are similar to the molluscs from the well-known Coralline Crag beds of England (Durham and MacNeil, 1967) and, like the Coralline Crag molluscs, indicate a somewhat milder climate than that of the present. Already in the "*Tapes*" and "*Mactra*" zones a few molluscs of North Pacific affinities were present (Einarsson et al., 1967), although the North Pacific influence was not so great as in the overlying "*Cardium*" zone (*Serripes* zone of Durham and MacNeil, 1967). Durham and MacNeil believed this minor "Pacific boreal" component of the lower parts of the Tjörnes beds to be inherited from an earlier marine connection with the Pacific across Beringia, but this is no longer considered likely (see below).

The upper part of the Tjörnes beds contrasts with the lower and middle parts in that "Pacific boreal" affinities of the molluscan fauna increase in the "Cardium" zone, especially so between about 150 and 130 m from the top of the Tjörnes sequence where "Pacific boreal" species become very abundant as they are in the Red Crag of England (Hopkins, 1959). Einarsson et al. (1967) date this influx of "Pacific boreal" molluscs in Iceland and England at about 3.0 Ma, slightly before evidence of tillites occurs in the overlying Furuvík and Breidavík beds. On paleomagnetic grounds, Gladenko (1981) dates the "Mactra" zone as lying between 3.0 and 2.1 Ma. Although students of the Icelandic marine Cenozoic at Tjörnes have been interested in the implications of the sudden arrival of "Pacific boreal" molluscs in the Atlantic with regard to the history of inundation of the Bering Land Bridge, their work is also useful in showing that in the late Pliocene a marine biological corridor extended from the Greenland-Norwegian Sea and North Sea area past the Nansen Sill west of Svalbard and across the Arctic Ocean to the Pacific via the Bering Straits area. Clark (1977) and Clark and Kitchell (1979) report that the post-Miocene benthic foraminifers and ostracodes of the Alpha Cordillera of the Amerasia Basin in the Arctic Ocean exhibit a mixture of Atlantic affinities and endemism.

Prior to about 3.5 Ma such North Atlantic-North Pacific marine faunal similarities as existed can also be explained as the result of weak former marine continuity via the Panamic marine portal. Durham and MacNeil's (1967) contention that boreal marine interchange between the North Pacific and the Atlantic began as early as the late Miocene was based on very few species and runs into difficulty when evidence from late Cenozoic terrestrial and marine mammal biogeography is considered (Simpson, 1947; Repenning, 1967; Repenning et al., 1979). Neither terrestrial nor marine fossil vertebrates show any evidence of a late Miocene or early Pliocene marine barrier interrupting the Bering Bridge. Late Miocene molluscs (MacNeil, 1957, 1965; Miller, 1957) from Camden Bay near Point Barrow on the north (Arctic Ocean) side of the Bering Bridge in Alaska show that no marine connection existed at that time to the Pacific across Beringia. By the same token, a marine corridor did exist from the north coast of Beringia through the Arctic Ocean to the Atlantic in the late Miocene as MacNeil (1957) originally showed.

Supposedly Miocene fossil seal and walrus distribution has been used in the past to bolster arguments of late Miocene interruption of the Bering Bridge, with walruses supposedly passing through the Bering Strait area from the Pacific to the Arctic and from there into the Atlantic to reach as far south as the Yorktown Formation of Virginia where *Prorosmarus alleni* occurs (Hopkins, 1967). Similarly, supposedly Miocene phocids were thought to have dispersed from the Atlantic to the North Pacific by the same route. Now, however, with the more recent dating of the Yorktown walrus, found to occur in rocks bearing calcareous nannofossil zones NN13 to NN15

(4.5-3.0 Ma) and the Pacific occurrences of the phocids established as Pliocene and Pleistocene (Repenning et al., 1979), need for an early episode of marine connection across Beringia in the late Miocene has been removed. It is no longer even necessary to invoke overland crossing abilities of pinnipeds (Scheffer, 1967) to achieve the distribution. Either the distribution was achieved late, at about 3.5 Ma when the Bering Strait let the Tjörnes Pacific boreal molluscs escape toward their Atlantic destination (Strauch, 1972), or the Panamic portal was used prior to its Pliocene closure.

Differences in plant species from the eastern to the western sides of Beringia were once used to bolster the idea of a late Miocene marine interruption of terrestrial Beringia (Wolfe and Leopold, 1967), but these might just as well reflect climatic zonation and generally cooler summers rather than a marine filter or barrier. Wolfe (1980, p. 319) has softened his former stand.

Finally, where are trans-Beringian marine Miocene sediments located, if one assumes that such must exist?

Greenland: Wandel Sea area and Peary Land

North of Svalbard and northeast Greenland are two volcanic edifices that were once connected, the Yermak Plateau and the Morris Jessup Rise, respectively. According to Feden et al. (1979), these two elevated areas were produced by a hot spot, the Yermak hot spot, resulting in an Iceland-like lava pile from about 60 Ma or earlier until about 35 Ma (anomaly 12/13). Maximum hot spot activity occurred between 43 and 36 Ma (anomalies 18 and 13).* Presumably the area remained above sea level from some time in the Paleocene until the end of the Eocene. According to Johnson et al. (1978) the Yermak and Morris Jessup lava piles began to separate at 38 Ma. Volcanic ash derived from the area occurs in the early Cenozoic Basilika Formation of Svalbard (Gripp, 1927; Dypvik and Nagy, 1979; Lehmann et al., 1978).

The Kap Washington basalts and basic dikes of northern Peary Land (Dawes, 1976) may be related to the Yermak hot spot, and, if so, they indicate that the hot spot was active in the Paleocene. A Rb-Sr date of 63±2 Ma for the lavas has been obtained (Larsen et al., 1977, 1978) and K-Ar dates are 66.0±6.6 Ma and 72.9±9.0 Ma for dikes (Dawes, 1973, 1976). The Kap Washington basalts had formerly been dated by the K-Ar whole rock method at 32.3±3.2 Ma and 34.9±5.3 Ma (Dawes, 1976). These and several other anomalously young K-Ar

*The Berggren et al. (1978) time scale is used here. On the Heirtzler et al. (1968) scale anomalies 18 and 13 would occur at 46 and 38 Ma. On the Tarling and Mitchell (1976) scale they would be at 39 and 34 Ma.

dates from older rocks in the North Greenland fold belt indicate an early Cenozoic thermal event that reset the K-Ar clock--presumably the Yermak hot spot. It is significant that the Kap Washington basalts are overthrust by folded metasediments. Was the thrusting related to transpression along the Wegener transform fault?

A thermally elevated area centered on a triple junction formed by the Wegener transform fault, Nansen Ridge, and Spitsbergen (-Senja?) Fracture Zone at anomaly 25/24 time would thus have provided a lava bridge between all three present land areas, (1) Ellesmere Island, (2) Greenland, and (3) Svalbard-Barents Shelf-Norway. Such a bridge would have separated the Arctic Ocean from the Greenland-Norwegian Sea and North Sea to the south. By anomaly 13 time the fragments of the bridge sank as the hot spot cooled. Complex transform faulting with many offsets along the Spitsbergen Fracture Zone also pulled apart the area between Greenland and Spitsbergen.

East Greenland

Along the east coast of Greenland from about latitude 66° N (Angmagssalik) to 75° N (Shannon Island) are extensive Paleogene basalts (Noe-Nygaard, 1974, 1976) cut by dikes and other intrusions such as the famous Skaergaard Intrusion (Wager and Deer, 1939). The Skaergaard Intrusion is of reversed polarity (Girdler, 1961) and has been dated by the zircon fission track method as 54.6±1.7 Ma (Brooks and Gleadow, 1977). A paleomagnetic pole position for the time of emplacement of the Skaergard Intrusion has been calculated by Faller and Soper (1979): 180.4° E, 62.9° N. Not all dikes in the region are Tertiary, however. Mitchell (1977), on the basis of K-Ar determinations, has dated some of the dikes near Angmagssalik as Late Cretaceous in age.

Faller and Soper envision a flurry of tholeiitic magmatic activity at about 55.5-52 Ma as Greenland and the Eurasian plate began to separate, just prior to marine magnetic anomaly 24 time. After the main outburst, magmatic activity continued with the injection of a large dike swarm and various intrusions were emplaced under extensional conditions until Oligocene time. At Kap Dalton (70° N, 24° W) on the eastern Blosseville Kust of Kangerdlugssuaq Peninsula about 40 m of Tertiary marine sandstones, shales, and tuffs overlie Tertiary basalt in a down-dropped fault block (Henderson, 1976). The age of these marine beds has been reported variously as Eocene (Ravn, 1933), Eocene to Oligocene (Hassan, 1953), early and middle Eocene (Deer, 1976) and latest Paleocene (Brown and Whitley, 1976). Whatever the age, these sediments place a younger limit on the age of the main tholeiitic outburst. Farther east, at Kap Brewster on the eastern tip of Kangerdlugssuaq Peninsula, about 130 m of sandstones, marl, and conglomerate with a soil zone at the base rest on Tertiary flood basalts. These sediments are referred to the Kap Dalton Formation and are assigned an Eocene to Oligocene age. They are not to be confused with a second set of Tertiary sediments at Kap Brewster comprising 76 m of sandstones and coarse clastics assigned a Miocene age (Hassan, 1953; Birkenmajer, 1972).

The presence of marine rocks at the top of the plateau lava sequence at Kap Dalton indicates that subsidence accompanied lava outpourings and that the area may have remained close to sea level during the whole process (Brown and Whitley, 1976; Deer, 1976; Soper, 1979). These ideas are in strong contrast to earlier ideas of Brooks (1973).

During the early Oligocene magmatic activity continued locally with the emplacement of the Kialineq intrusion (66°45' N, 34° W, at Kialîp Imâ). A Rb-Sr age of 35±2 Ma and a K-Ar age of 35.7±2.6 Ma have been determined by Brown et al. (1977).

Additional evidence of mid-Tertiary magmatic activity in East Greenland has been reported by Rex et al. (1978) for Tertiary intrusions on both sides of Kong Oscars Fjord (lat. 72°-73° N), about 500 km northeast of the Kangerdlugssuaq area. Southwest of Kong Oscars Fjord in the Werner Bjerge is an intrusive complex emplaced in late Paleozoic and Mesozoic sandstones at several different Tertiary episodes. A volcanic extrusive cover is also locally present. The older intrusives have not been dated but the younger ones have been dated by a variety of methods: whole rock Rb-Sr (31±2 Ma), biotite-feldspar Rb-Sr (27.1±2 Ma), and biotite K-Ar (28±1 Ma). Beckinsale et al. (1970) had earlier obtained a biotite K-Ar age of 28.7±2 Ma and Brooks (cited by Rex et al., 1978) obtained a titanite fission track age of 30.3±1.3 Ma. Rex et al. combine these results to indicate a probable age of 30±2 Ma for the intrusives of the Werner Bjerge.

On Traill Island northeast of Kong Oscars Fjord an intrusion on Kap Simpson yields K-Ar dates of 38±2 Ma on both amphibole and biotite. At Kap Parry on the same island two K-Ar runs on an amphibole sample produced dates of 40.0 and 40.6 Ma, averaged to 40.3±1.2 Ma (Rex et al., 1978). The Traill Island dates are not significantly different from each other, but they are significantly older than the 30 Ma date of the Werner Bjerge farther to the southwest.

Still further to the northeast, intrabasaltic sediments near Kap Hartlaub on Lille Pendulum Island (74°40' N, 18°25' W) have produced Paleogene marine invertebrates (Vischer, 1943) and Paleogene plants are found nearby on Sabine Island and the Wollaston Forland.

North of Traill Island, in the Giesecke Bjerge of Gauss Halvø (73°30' N, 22° W), Hald (1977) reports thin coals in a predominantly hyaloclastite section at the base of the plateau lavas. Further northeast, on Hold With Hope and the Wollaston Forland (approximately 74° N, 20° W) Maync (1940) and Vischer (1943) reported early

Tertiary rocks beneath the lavas, but at least some of these (on central Wollaston Forland) have since proven to be Early Cretaceous in age according to F. Surlyk (pers. comm. to Hald, 1977).

Thus, in East Greenland there is abundant igneous evidence of volcanics forming the west end of the Thulean (Greenland-Scotland) bridge in the early Cenozoic. Part of this volcanic pile was subaerially emplaced. Numerous examples of intrabasaltic and pre-basaltic Cenozoic sediments have been reported, and these should be prospected by vertebrate paleontologists in order to study biological aspects of the North Atlantic land bridge problem with high resolution.

## Northeast Greenland

In Kronprins Christian Land 100 m or more of terrestrial early Cenozoic shale deposits with coals occur at the mouth of Danmark Fjord (82° N, 18° W) on Princesse Øer (Schiener, 1976; Dawes, 1976; Harland, 1979) not far from Station Nord.

## Central West Greenland

Between 68° and 73° N (Rosenkrantz and Pulvertaft, 1969; Keen and Clarke, 1974) on the west coast of Greenland early Cenozoic terrestrial sediments are locally preserved beneath and also are interbedded with widespread late Danian and younger tholeiitic basalt lava flows that cover an estimated 55,000 $km^2$ of Baffin Bay and the present Greenland shore (Park et al., 1971; Clarke and Pedersen, 1974, 1976; Zarudzki, 1980). The best exposures of terrestrial sediments are on the southeast coast of Disko Island (Henderson et al., 1976), but extensive prospecting of these by an American Museum of Natural History expedition in 1978 failed to find any fossil vertebrate remains. The lavas on Disko may represent a significant duration of time, but thus far only one radioisotopic date has been obtained, from a flow in southern Disko. This date, obtained from a commercial firm by E. R. Deutsch, has been reported by Athavale and Sharma (1975) to be $70 \pm 4$ Ma (uncorrected for IUGS standards) on the basis of the K-Ar method and is older than would be expected inasmuch as marine upper Danian fossils occur in intrabasaltic sediments on nearby Nûgssuaq Peninsula of mainland Greenland at Marrait Kitdlît (Henderson et al., 1976). East of Marrait Kitdlît lavas rest unconformably upon late Danian marine sediments, then on Cretaceous rocks, and finally on Precambrian basement rocks.

The lavas of northern Disko Island are mainly of magnetically reversed polarity, but magnetically stable normal polarity is observed in 200 m of the middle of the sequence (Athavale and Sharma, 1975) in contrast to the Blosseville volcanics of East Greenland and the other Brito-Arctic volcanics. These latter volcanics are overwhelmingly of reversed polarity. Athavale and Sharma tentatively correlate the 200 m thickness of normally magnetized Disko lavas with marine magnetic anomaly 25.

From the Disko lavas Athavale and Sharma have calculated a mean paleomagnetic pole position at 67.5° N, 165° W and a paleolatitude of 56° N. The declination of the paleomagnetic pole suggests to them that this part of Greenland or all of Greenland has rotated 40° anticlockwise since the lavas were erupted.

## Greenland: Hareøen

Hareøen (70°25' N, 55°00' W) is a small island about 15 km in diameter located at the seaward end of the Vaigat, the strait that separates Disko Island from Nûgssuaq Peninsula in West Greenland. The rocks of Hareøen are mainly flood basalts (Hald, 1971), but intrabasaltic sediments are present as well. Fossil plants (Heer, 1883) have been found in tuffaceous layers on Hareøen correlated with the intrabasaltic Ifsorisok Formation of western Nûgssuaq. The flora of the Ifsorisok (Ivssorigsoq) Formation is reportedly not younger than Eocene (Koch, 1964; Rosenkrantz, 1970; Pedersen, 1976). Wolfe (1980) states that 12 Hareøen species also occur in the late Eocene of Alaska and that the Hareøen occurrence is therefore of that age. No prospecting for fossil vertebrates has been attempted.

Hareøen lies on the seaward and downfaulted side of a fault system that appears to cross the Vaigat, continuing as the northeastward aligned Itivdle Fault of western Nûgssuaq Peninsula of mainland Greenland. As on western Disko, Nûgssuaq, Ubekendt Ejland, and Svartenhuk Peninsula, these westernmost exposed rocks dip generally westward at about 10-30° and are likely to include the youngest lavas and sediments exposed above sea level in western Greenland (Ross and Henderson, 1973).

## Greenland: Nûgssuaq Peninsula

On Nûgssuaq Peninsula (70°15' N, 53° W) both the early Danian Kangilia Formation and the late Danian Agatdal Formation have yielded locally abundant and varied remains of fishes (notably elasmobranchs: Bendix-Almgreen, 1976), coccolithrophores (Perch-Neilsen, 1973), ostracodes (Szczechura, 1971), Foraminifera (Hansen, 1970) and Mollusca and numerous additional kinds of invertebrates (Rosenkrantz, 1951, 1970). A marine late Danian fauna has been found also in intrabasaltic conglomerates at Marrait Kitdlît (70°35' N, 54°15' W). These indicate a temperate marine connection at NP3 time in the Danian with the type Danian of Denmark and also to some extent with the Aquia Formation of Maryland on the American east coast. The Agatdal bivale *Conchocele conradii* also occurs in the Paleocene of Spitsbergen (Schweitzer, 1980). These similarities imply that during at least this part of Danian time waters continuous with those of the Labrador Sea area reached northward past the Davis Straits area as far as Nûgssuaq. Thus, in contrast to the Turonian Late Cretaceous, when marine biological affinities of the Nûgssuaq area were with northern Canada and the North American Mid-Continent Sea (Birkelund, 1965; Rosenkrantz, 1970), Nûgssuaq now was in marine communication with the south. The first southern ammonites occurred in the Campanian, and southern as well as northwestern affinities were evident in the Maastrichtian fauna (Birkelund, 1965). While this marine barrier existed any land connection between North America and Greenland would have had to have lain north of Nûgssuaq.

More importantly, connection of the Dano-Scanian Basin with the Nûgssuaq area via a marine route would have had to have been via the North Sea and then across the present site of the Greenland-Scotland Ridge. An epicontinental marine barrier must have separated Greenland from Scotland throughout most or all of the Danian, prior to the inception of Brito-Arctic rapid sea-floor spreading that began between the times represented by marine magnetic anomalies 25 and 24 at about the end of the Paleocene. There is no evidence that a marine seaway crossed central Greenland from east to west as tentatively proposed by Strauch (1970).

Fossil plant remains are also abundant in the Danian deposits of Nûgssuaq (Koch, 1963; Pedersen, 1976; Wolfe, 1980) but they do not yield paleogeographic information other than that environmental conditions were considerably less extreme than at present. Wolfe (1980) describes the Agatdal Flora as a "broad-leaved evergreen forest that has broad-leaved deciduous and minor coniferous adjuncts." Heer (1883) originally regarded the late Danian Upper Atanikerdluk Flora as Miocene.

Although the period of lava production on Nûgssuaq Peninsula (and Disko Island) is late Danian or younger, nevertheless it is important that volcanic ash was present as early as the two tuff beds deposited during the time of deposition of the "*Thyasira* Member" of the early Danian Kangilia Formation of Nûgssuaq (Rosenkrantz, 1970, fig. 3). Volcanic ash has not been reported from underlying Cretaceous deposits.

## Greenland: Ubekendt Ejland and Svartenhuk Peninsula

Beckinsale et al. (1974) dated plutonic rocks intruding picritic lavas on Ubekendt Ejland, north of Nûgssuaq Peninsula. K-Ar dating yielded 54.0±1.5 and 55.4±1.5 Ma, but Rb-Sr dating gave an older date of 65.2±3.8 Ma. Still younger lamprophyre dikes on Ubekendt Ejland yield dates of 30.6±6.4 and 32.8±6.8 ($2\sigma$) Ma on the basis of the $^{40}A/^{39}A$ dating method (Parrott, 1976).

At Svartenhuk Peninsula $^{40}A/^{39}A$ dating indicates cooling of basic volcanics at 57.9±2.6, 58.1±3.8, and 59.8±3.8 ($2\sigma$) Ma (Parrott, 1976).

Queen Elizabeth Islands, Canada, especially Ellesmere and Axel Heiberg Islands

In the Sverdrup Basin early Cenozoic marine and continental sediments are exposed extensively from Banks Island to Ellesmere Island at the northernmost end of North America (Fortier et al., 1963; Christie, 1964, 1974, 1976; Christie and Rouse, 1976; West et al., 1977; West, 1978; Balkwill, 1978; Miall, 1975, 1979a). These deposits were originally named the Eureka Sound Group (Troelsen, 1950) but, because they have not yet been divided into other than locally mappable rock units, the Eureka Sound is currently regarded as a formation (Tozer, 1963, and later authors). The type section is in central Axel Heiberg Island (Souther, 1963). In the central and western parts of the Sverdrup Basin, deposition of Eureka Sound sediments began as early as Maastrichtian time in the latest Cretaceous (Thorsteinsson and Tozer, 1970; Kerr, 1974; Miall, 1979a) and continued through the Paleocene (Kerr, 1974), but much of the Eureka Sound is Eocene in age (West and Dawson, 1978). Where the unit is well exposed and reaches its greatest thickness (3340 m) on central Ellesmere Island near Strathcona and Bay fiords, fossiliferous paralic marine beds occur in its middle part and fossil plants and terrestrial vertebrates have been obtained from several Eocene levels in the upper part in coal-swamp and fluvial deposits (Dawson et al., 1975, 1976; West, 1977; West et al., 1977; West and Dawson, 1978; McKenna, 1980; this paper, table 1).

In the Lake Hazen area of northeastern Ellesmere Island, feldspathic sediments lacking volcanic cobbles are mapped as the Eureka Sound Formation. These were deposited in a local intermontane basin and are primarily of Eocene age, but they may include rocks as young as Oligocene age (Christie, 1964; Petryk, 1969; Kuc, 1973; Miall, 1979b). Paleocene sediments have not been identified in the Lake Hazen area, a situation that is to be expected in that the Eureka Sound is unconformable on older rocks in that area.

Also in northeastern Ellesmere Island, but lying along Nares Strait near the northeastern end of the Judge Daly Promontory (West and Dawson, 1978, fig. 6), are theoretically very important sediments of a second continental basin mentioned briefly by Miall (1979b, pp. 3, 20). Although these deposits are not as yet reported in detail, nevertheless Miall states that they consist of possibly lacustrine beds and fluvial sediments deposited by northeastward flowing rivers displaying low sinuosity. These Tertiary sediments of the Judge Daly Promontory have remained little deformed since deposition. Significantly, the dominant clasts in these rocks are derived from basic volcanics. In view of the lack of such source rocks in nearby parts of either Greenland or Ellesmere Island in their present positions, where was the source area from which the volcanic clasts came? Has

Table 1. Terrestrial Eocene Vertebrates of Ellesmere Island

| | Horizon[a] | Sister groups in: N.A. | Eu. | As. |
|---|---|---|---|---|
| Fishes | | | | |
| ?*Lepisosteus* sp. (gar) | L | x | x | x |
| *Amia fragosa* (bowfin) | L | x | x | x |
| *Amia* cf. *A. uintensis* (bowfin) | L | x | x | ? |
| *Esox* sp. (pike) | L | x | x | ? |
| Amphibian | | | | |
| *Piceoerpeton* cf. *P. willwoodense* (large salamander) | L | x | | |
| Reptiles | | | | |
| emydine cryptodire (turtle) | L | x | ? | ? |
| emydid cryptodire (turtle) | L | ? | | |
| *Geochelone* sp. (large tortoise) | L & U | x | x | x |
| kinosternine kinosternid (turtle) | L | x | | |
| anosteirine carettochelyid (turtle) | U | | | x |
| *Trionyx* sp. (turtle) | L & U | x | x | x |
| glyptosaurine anguid (lizard) | L | x | x | ? |
| anguid (lizard) | L | ? | | ? |
| varanid (monitor lizard) | U | ? | ? | |
| erycine boid (snake) | L | x | ? | |
| *Allognathosuchus* sp. (alligator) | L & U | x | x | |
| Birds | | | | |
| presbyornithid (?duck-like bird) | L | x | | |
| diatrymatid (flightless bird) | L | x | x | |
| Mammals[b] | | | | |
| *Neoplagiaulax* sp. (multituberculate) | L | x | x | |
| leptictid ("insectivore") | U | x | x | |
| cf. *Pantolestes* or *Chadronia* sp. ("insectivore") | L | x | x | x |
| taeniodont or tillodont | U | x | x | x |
| *Coryphodon* sp. (pantodont) | L | x | x | x |
| hyaenodontid (small carnivore) | L | x | x | |
| *Prolimnocyon* sp. (small carnivore) | ?L | x | x | |
| *Viverravus* sp. (small carnivore) | L | x | x | |
| cf. *Uintacyon* or *Vulpavus* sp. (small carnivore) | L | x | x | |
| ischyromyids (3 species) (rodents) | L & U | x | x | ? |

(Continued)

Table 1. (Continued)

| | Horizon[a] | Sister groups in: N.A. | Eu. | As |
|---|---|---|---|---|
| Mammals[b] (continued) | | | | |
| cf *Ignacius* sp. (2 species) (primates) | L | x | x | |
| plagiomenids (5 species) (colugos) | L | x | ? | x |
| *Hyracotherium* sp. (horse) | L | x | x | |
| *Hyrachyus* sp. (rhinoceros) | L | x | x | x |
| *Lambdotherium* sp. (perissodactyl) | L | x | x | |
| *Manteoceras* sp. (titanothere) | U | x | | x |

[a]L: lower terrestrial vertebrate horizon, approximately Wasatchian (Ypresian) in age. U: upper terrestrial vertebrate horizon, approximately Bridgerian (early Lutetian) in age.

[b]There are some surprising absences in this list (West and Dawson, 1978). Where are the artiodactyls? Why is not *Hyopsodus* present? Where are the omomyids and adapid primates? Marsupials? Phenacodonts? All of these mammals were common to both Europe and North America and both omomyids and *Hyopsodus* are known from Mongolia as well during the early Eocene, implying use of the high latitude Bering Bridge.

it subsided beneath Nares Strait or Kane Basin? Are these sediments younger than the main events of the Eurekan Orogeny? Could palinspastic reconstruction of Greenland have placed any of the volcanics of Peary Land in a suitable position (e.g., Dypvik and Nagy, 1979, fig. 1) to supply the clasts? Did the volcanic clasts come from highlands farther south? Whatever the source, it was evidently not a presently local one. Moreover, subaerial conditions are implied for the Nares Strait area at the time of deposition, presumably in the Eocene or Oligocene. Clearly, the source of these volcanic clasts must be found and the age of both the volcanics and their time of deposition as clasts must be determined.

During deposition of the Cenozoic part of the Eureka Sound Formation in central and northern Ellesmere Island, local renewed uplift and eastward thrusting of the Princess Margaret Arch occurred on neighboring Axel Heiberg Island. Along the Stolz Fault Pennsylvanian strata came to rest on Eureka Sound rocks. Next to the Stolz Fault the Eureka Sound is coarse, even conglomeratic. Following deposition of the Eureka Sound, movements of this orogenic phase, known as the Eurekan Orogeny, compressed, gently folded, and overthrust Eureka Sound sediments. Along about 400 km of the Lake

Hazen Fault Zone in northern Ellesmere Island, early Paleozoic rocks of the Grantland Uplift were thrust southeastward over rocks as young as the Eureka Sound during the late Eocene or Oligocene. Another major fault active during the Eurekan Orogeny was the 140 km Copes Bay Fault, a thrust west of Kane Basin on the east side of Ellesmere Island. Major uplift of the Victoria and Albert Mountains northwest of the Copes Bay Thrust took place at this time. Other major uplifts occurred in the Princess Margaret Range of Axel Heiberg Island and the Sawtooth Range on Fosheim Peninsula of Ellesmere Island. Early Miocene conglomerates were shed eastward over folded Eureka Sound sediments at this time in northeastern Axel Heiberg Island (Balkwill and Bustin, 1980). In the central and western parts of the Sverdrup Basin the effects of the Eurekan Orogeny were less evident, although diapirism played a major role in the deeper parts of the basin (Trettin, 1972). On the basis of Eocene vertebrate fossils in the upper part of the Eureka Sound Formation, much of the folding and thrusting of the Eurekan Orogeny took place after Wasatchian (early Eocene) and Bridgerian (middle Eocene) time (West and Dawson, 1978). The folding was evidently completed before early Miocene time (Balkwill and Bustin, 1980).

Post-Eocene Cenozoic clastics of the Beaufort Formation have been reported from most of the northernmost Queen Elizabeth Islands from Meighen to Banks islands (Kerr, 1980, fig. 10). The Beaufort is famous for its fossil beetles (Matthews, 1979) and virtually unaltered fossil wood. Pollen floras suggest ages ranging from Oligocene to Pleistocene (Hopkins, 1967). Most of these occurrences are along the Arctic coastal plain and dip gently seaward, but in northern Ellesmere Island rocks referred to the Beaufort Formation are overthrust by an upper plate that moved relatively northwest (Wilson, 1976; Kerr, 1980). Kerr believes that dilation and graben formation at this time resulted in the reconnection of the Arctic Ocean with Baffin Bay and the Atlantic to the south as the floor of Parry Channel (including Lancaster Sound) and Nares Strait sank beneath sea level for the first time since the Mesozoic. It is Hopkins' (1967) opinion that the channels separating the Queen Elizabeth Islands from the Canadian mainland developed in the late Pliocene or even after that. Trettin (1972, p. 119) believes that the channels between various of the Queen Elizabeth Islands (except the Eglinton graben between Prince Patrick Island and Melville Island) post-date the deposition of the Beaufort Formation.

At the head of Freemans Cove on southeastern Bathurst Island, limited outcrops of paralic Eureka Sound coal-bearing sediments contain an interbedded andesite flow (Kerr and Temple, 1965; Kerr, 1974) preserved at one locality in a downfaulted block. Isotopic dating of these volcanics would calibrate the lower part of the Eureka Sound Formation. Hills and Wallace (1969) and also Kerr (1974) believed this part of the Eureka Sound to be Maastrichtian in age on palynological and dinoflagellate evidence. Basic dikes cutting Devonian rocks in the southern part of Bathurst Island all give Tertiary K-Ar dates (Kerr, 1974), but these may be unreliable because of alteration. Whether they are the same age as the Eureka Sound flow is not yet known but it is likely.

Pelecypod fragments occur in the Viks Fiord area, Devon Island (Fortier et al., 1963; West and Dawson, 1978; Balkwill and Bustin, 1980, fig. 1). West and Dawson (1978) also report fossil lagomorph remains from mid-Cenozoic lacustrine sediments deposited in Haughton Astrobleme (Frisch and Thorsteinsson, 1978).

On the southwest part of Cornwall Island a small exposure of lower Tertiary sediment occurs (Balkwill and Bustin, 1980, fig. 1).

A limited exposure of lower Tertiary sediment occurs at the northern tip of Somerset Island (Balkwill and Bustin, 1980, fig. 1). A second exposure is preserved in a graben about 150 km to the south (Trettin and Balkwill, 1979, fig. 9).

## Baffin Island and Bylot Island, Baffin Bay, Davis Strait

For about 90 km, flat-lying basalts occur along the east coast of Baffin Island between Cape Searle and Cape Dyer (Wilson and Clarke, 1965; Clarke and Upton, 1971). Grant (1972, 1975) has mapped these basalts offshore to the east where they are downfaulted about one km. Between the lavas and the underlying 1.7 billion year old Precambrian basement of Baffin Island are plant-bearing terrestrial sediments that thicken northward to a total of about 285 m. These sediments consist of unconsolidated quartz sands, impure sandstones, shales, minor coal, and conglomeratic layers typical of cyclical deltaic deposition (Clarke and Upton, 1971). Farrar (1966) reported 10 K-Ar age determinations for the Baffin Island basalts, some of which were obviously in error. Clarke and Upton (1971) give a K-Ar age of 58±2 Ma for the basalt and cite W. A. Bell's estimate that the plant fossils of the underlying sediments are Paleocene in age. Parrott (1976, unpublished) has discussed Farrar's results and on the basis of a weighted mean suggests an age of 59.0±3.7 Ma (1$\sigma$ error) for the lavas (not corrected for new IUGS standards). Parrott's own $^{39}Ar/^{40}Ar$ isochron determinations yield 57.5±16.8 Ma and 50.2±13.4 Ma. There is abundant evidence that some of the Baffin Island lava samples are contaminated with excess argon. Certainly the lavas preserved on land are older than those that would have provided a hot spot track at sea (Burke and Kidd, 1980, fig. 4. For divergent views see Keen and Clarke, 1974; Grant, 1975). Paleomagnetic comparison of the flows on either side of Baffin Bay has been initiated (Athavale and Sharma, 1975), but the number of samples from Baffin Island needs to be increased before comparison of pole positions can be regarded as statistically significant (Kristjansson and Deutsh, 1973).

On the basis of aeromagnetic data, Hood and Bower (1975) suggest that a large fracture zone between two RRF triple points crossed Davis Strait from southwest to northeast, from the entrance of Hudson Strait on the Canadian side to the Itivdle Fault of Nûgssuaq Peninsula just north of Disko Island.

In north-central Baffin Island a very small outcrop of impure freshwater limestone (the "Rimrock Bed") lies on Precambrian granite gneiss at an elevation of 730 m M.S.L. (Andrews et al., 1972). On the basis of palynomorphs ?*Ulmus*, ?*Taxodium*, *Liriodendron*, ?*Carpinus* (or ?*Corylus*), and other plant genera are represented, suggesting a warm-temperature swamp environment and also an early Cenozoic age.

Near the mouth of the Lancaster Sound half-graben or aulacogen (Burke and Dewey, 1973; Burke, 1976) at the northern tip of Baffin Island near Pond Inlet, as well as on the southwest and northern sides of neighboring Bylot Island, a composite thickness of several thousand meters of Mesozoic and Cenozoic rocks has been reported (Wilson and Clarke, 1965; Barrett, 1966; Berkhout, 1973; Beh, 1975). Little is known about these occurrences, but the sequence is mostly continental and begins with sediments as old as Early Cretaceous or Jurassic. Andrews et al. (1972) regarded these beds as wholly Mesozoic in age, but the other authors cited a Cenozoic component. Trettin and Balkwill (1979, fig. 9) show early Cenozoic rocks as well as Cretaceous on southern Bylot Island, but only Cretaceous on northernmost Baffin Island.

Beh (1975) concluded that Baffin Bay opened in the late Eocene, between 46 and 37 Ma. Prior to 46 Ma the Davis Strait area would have been high just prior to rifting (Hyndman, 1973), but whether it was subaerial is not certain. However, Beh's model of separation requires extensive and thus far unconfirmed left-lateral transcurrent faulting across Greenland in order to comply with identification of marine magnetic anomalies. Shallow, nearshore-marine or possible nonmarine conditions prevailed in the Paleocene and Eocene resting on Precambrian at the site of the Kangâmiut bore-hole, 100 km west of the present Greenland Coast and just south of Davis Strait at 66° N (Hinz et al., 1979). At some time between the Paleocene and the late Eocene, however, a block northwest of the Kangâmiut bore-hole was uplifted such that about 4 km of Paleogene sediment was eroded away, presumably subaerially (Risum, 1978). The Oligocene of the Kangâmiut bore-hole is fully marine. About 100 km SE of the Kangâmiut bore-hole, the Nukik 2 bore-hole bottomed in basalt or diabase after passing through a Tertiary section with Paleocene sediments at the base (Hinz et al., 1979).

South of Davis Strait, off Godthaab, Greenland, terrestrially deposited and now subsided early Cenozoic coal has been dredged from a depth of between 404 and 551 m (Johnson et al., 1975). Unlike the truly oceanic Baffin Bay to the north, the Davis Strait Sill has a crustal thickness of more than 20 km, of which about 18 km are basaltic and overlain by 2 km of sediment (Keen and Barrett, 1972). The present depth of the Davis Strait Sill is only about 650 m, whereas Baffin Bay attains about 2400 m.

Grant (1975) concludes that it is possible that Greenland and Baffin Island have moved little relative to each other, thus indirectly supporting arguments by Kerr (1980), Grant (1980), and others that the Nares Strait Lineament (Wegener Transform Fault of mobilists and Robison Megashear of Carey, 1958) has hardly moved at all. On the other hand, Burke (1976), citing evidence of thrusting along the Copes Bay Fault of Ellesmere Island (Trettin, 1972), suggests transpression in the Nares Strait area during the Eurekan Orogeny. Clearly, or perhaps one should say "unclearly", the details of the opening of Baffin Bay and the effects of that activity on the surrounding lands are not well understood.

Although it is still unproven, it is not unreasonable that a terrestrial route for dispersal existed across Davis Strait during some part of the early Tertiary, either because continental crust overlain by a thick and laterally extensive pile of Paleogene volcanics has since foundered there or because the Davis Strait area was kept elevated for a time by hypothetical Davis Strait hot spot. To the south of Davis Strait, much of the subsidence of the Labrador Sea is Miocene or younger (Hinz et al., 1979; Grant, 1980).

## SUMMARY

Events of the following scenario can be visualized with the aid of figures 1-5, modified from Grønlie (1979) and Feden et al. (1979)

### Cretaceous

During Turonian time the Davis Strait area was terrestrial, preventing marine waters at Nûgssuaq on the West Greenland coast from connecting with those of the Labrador Sea and the Atlantic Ocean. Marine biogeographical affinities of Nûgssuaq cephalopods were with northern Canada and the North American Mid-Continent Sea. Thus, part of the Queen Elizabeth Islands area of northeastern Canada was traversed by epicontinental marine waters. By Campanian time Davis Strait was marine, permitting southern ammonites to reach Nûgssuaq. An Arctic-Atlantic barrier separated Greenland from Canada and prevented terrestrial biotic continuity. These conditions continued into the Maastrichtian at the end of the Cretaceous. During the Maastrichtian the climate cooled somewhat.

## Danian

West Greenland's marine fauna at Nûgssuaq shows that marine conditions extended from the Baffin Bay area south across Davis Strait to the Labrador Sea and Atlantic Ocean. Nûgssuaq was apparently isolated from direct access to the Arctic Ocean by uplifted land in the Queen Elizabeth Islands area, but this is not completely certain, although probable. If such a subaerial barrier did exist, Greenland would have been connected with other landmasses at only its northern end, which implies filtering of terrestrial biotas because of climatic considerations. East of Greenland, epicontinental seas extended from the Atlantic across the future site of the Greenland-Scotland Ridge, reaching as far north as Svalbard but evidently not connecting with the Arctic Ocean. Thus a land connection all the way from North America to Fennoscandia via the far northern De Geer Route is likely, but any connection south of there is not. Future collections of fossil vertebrates from the early Cenozoic of Svalbard would test this hypothesis.

Between Fennoscandia and southwestern Europe lay the Danish-Polish trough, a marine barrier for terrestrial animals and plants. The Danish-Polish trough was connected westward with the Atlantic via the North Sea and southeastward with Tethys via eastern Europe. Danian climate was cooler than in most of the Late Cretaceous or in the early and middle Eocene. Brito-Arctic vulcanism began at this time, but rapid plate separation did not start until later.

## Post-Danian Paleocene

Reorganization of major plate movement occurred at this time. Hot spots at Davis Strait, proto-Iceland, and the Yermak triple junction created volcanic edifices sufficiently elevated above sea-level to result in land bridge conditions along both the De Geer and Thulean routes. Terrestrial Paleocene faunas of North America and Europe south of the Danish-Polish trough became generally similar by mixing, but not so similar as they were to become in the early Eocene. Presumably this was because of the filtering effects of a more stringent climatic regime than that of the early and middle Eocene. During the late Paleocene the epicontinental sea between Greenland and Norway was broadened somewhat by rifting. Marine waters between the British Isles, Greenland, and Norway were isolated from both the Atlantic and the Arctic oceans but were connected via the Danish-Polish trough to Tethys. From late Paleocene to some point in the early Eocene the Arctic Ocean would have been completely isolated except at its shallow connection with Tethys via the Turgai Strait. How such isolation of Arctic waters could be compatible with a warm climate is a major unsolved mystery (McKenna, 1980).

Early Eocene

During the early Eocene the previously established Thulean Land Bridge became a corridor as the climate warmed markedly. Eocene climates were to remain warm through the Lutetian before cooling gradually and then precipitously at the end of the Eocene. Broad interchange between the European and American terrestrial biotas occurred during a brief interval in the early part of the early Eocene, but this interchange was suddenly interrupted for geological

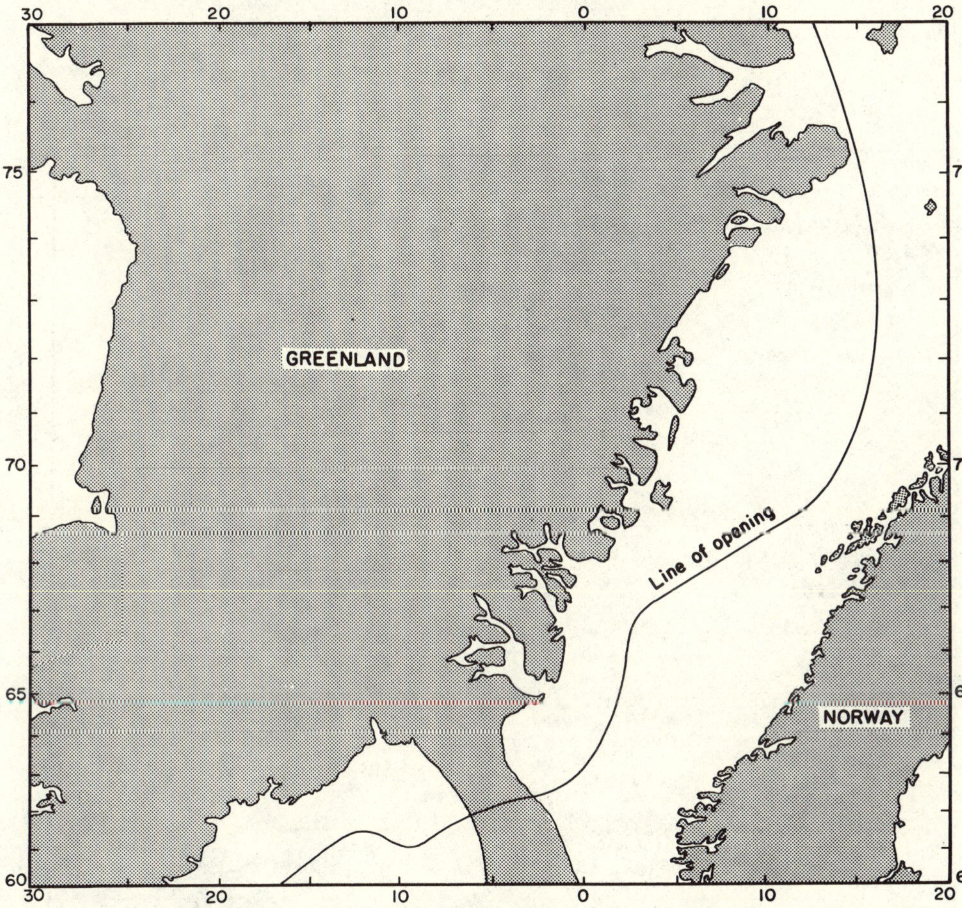


Fig. 1. Paleogeography of the area between Greenland and Norway at the time of initiation of rapid sea floor spreading, between marine magnetic anomalies 25 and 24, approximately 55 million years ago, Paleocene/Eocene boundary. Volcanic edifices connected Greenland with Baffin Island and with Scotland (after Grønlie, 1979, with modifications).

rather than climatic reasons as some part of the Thulean Route became barred suddenly during the early Eocene. This sudden barrier for terrestrial organisms was most likely a marine one and was probably at the site of the Faeroe-Shetland Channel. The Thulean Route never connected to the British Isles again.

At the north end of Greenland (De Geer Route) transpression (a useful term we owe to W. B. Harland) kept Greenland in elevated contact with both Ellesmere Island (North America) and Svalbard (northwestern Eurasia). Fennoscandia therefore probably had a large "North American" terrestrial component of its biota at this time, but no fossils are available except those of Svalbard. Svalbard's fossil plants are similar to those elsewhere in the Arctic early

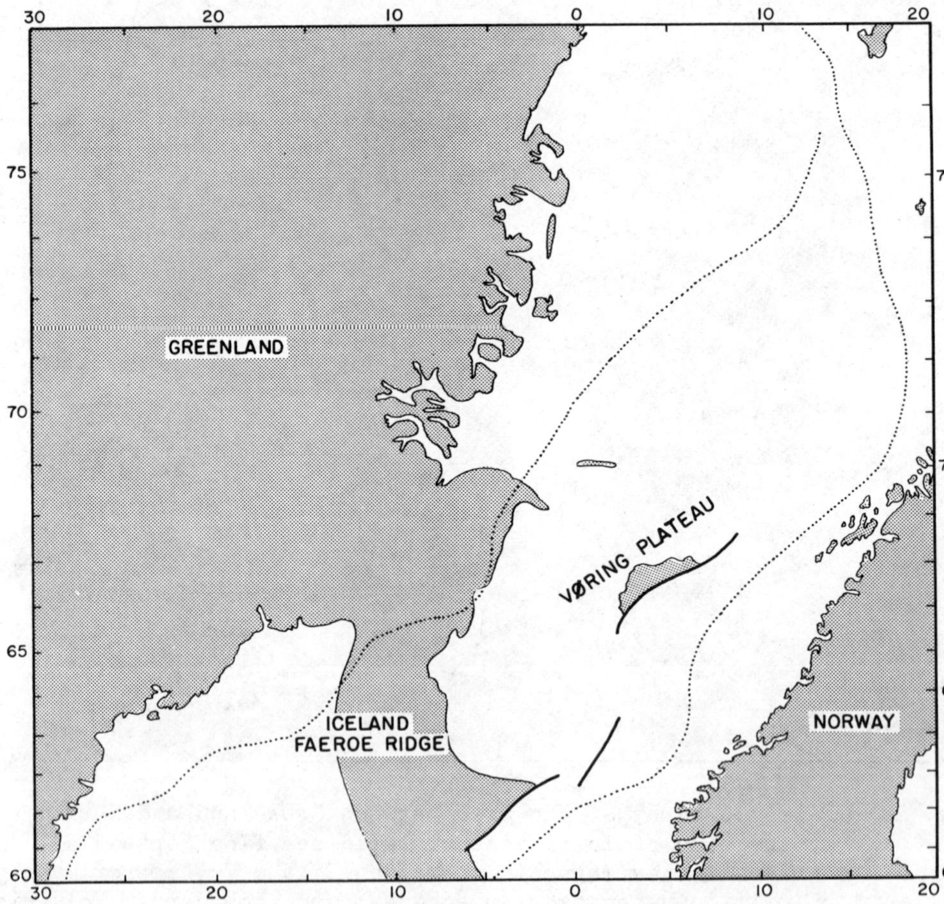


Fig. 2. Paleogeography of the Norwegian-Greenland Sea area at 40 Ma (after Grønlie, 1979, with modifications).

Cenozoic. Fennoscandia was isolated by marine waters from the rest of Europe.

Middle Eocene

Europe south of the Danish-Polish trough continued to be isolated as well as terrestrially fragmented. Terrestrial biotas became markedly endemic. No extensive marine water exchange appears to have occurred across the former east end of the Thulean Bridge, however.

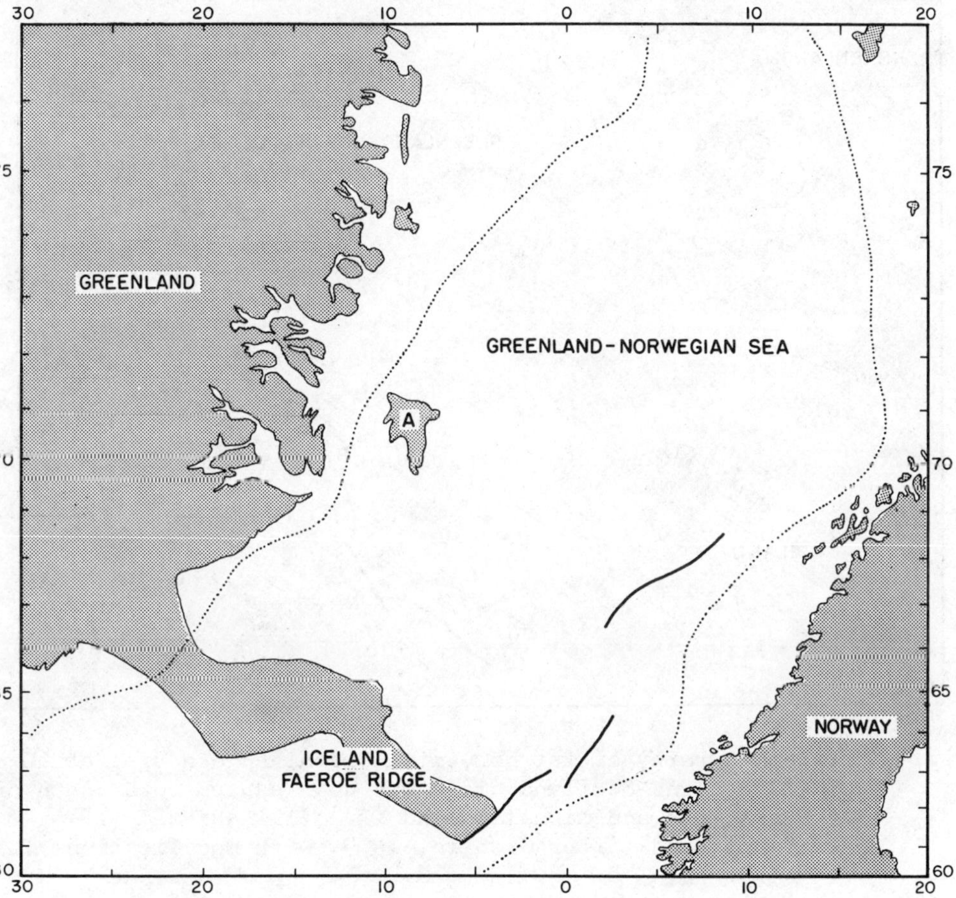


Fig. 3. Paleogeography of the Norwegian-Greenland Sea at 20 Ma. A: former island east of Scoresby Land. Dotted lines: present shelf edges. Solid lines: escarpments at Vøring Plateau and in Faeroe-Shetland Channel area (after Grønlie, 1979, with modifications).

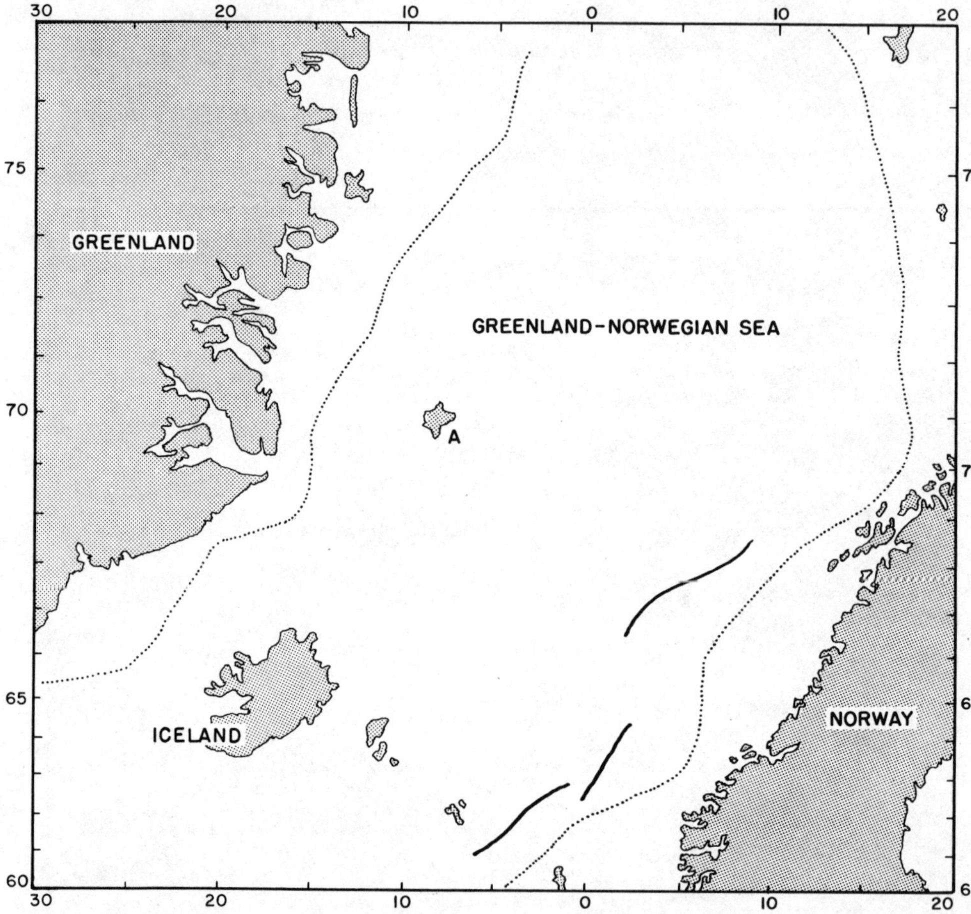


Fig. 4. Paleogeography of the Norwegian-Greenland Sea area at 10 Ma. The Greenland-Scotland Ridge was now reduced to a chain of islands. An additional island (A) still existed off Scoresby Land (after Grønlie, 1979, with modifications).

## Late Eocene

The De Geer Route was presumably still operative but must have been a strong filter because of its northern position and because of a cooling climatic trend. The eastern end of the Thulean Route remained defunct, but marine water exchange remained minimal.

## Early Oligocene

Svalbard cleared Greenland. The last possible terrestrial continuity between Europe and North America via the Atlantic area ceased by this time. Cold, deep water began to flow between the Arctic Ocean and the Greenland-Norwegian Sea in significant quantities. Atlantic temperatures dropped sharply.

## Miocene

The Greenland-Scotland Ridge continued to subside as it lengthened, with low, flat islands remaining locally. A high, volcanic island still formed actively at Iceland. As more and more marine waters traversed the ridge, the Western Boundary Undercurrent became significant in the Atlantic.

## Plio-Pleistocene

Fragmentation and isolation of the Queen Elizabeth Islands of the Canadian high Arctic occurred at this time. Marine flooding of the formerly subaerial Barents Shelf of northwestern Eurasia occurred. Nares Strait was invaded by marine waters, isolating Ellesmere Island from Greenland. The Greenland-Scotland Ridge continued to sink over must of its extent, leaving only the Faeroes, Shetlands, and Iceland as reminders of its former glory.

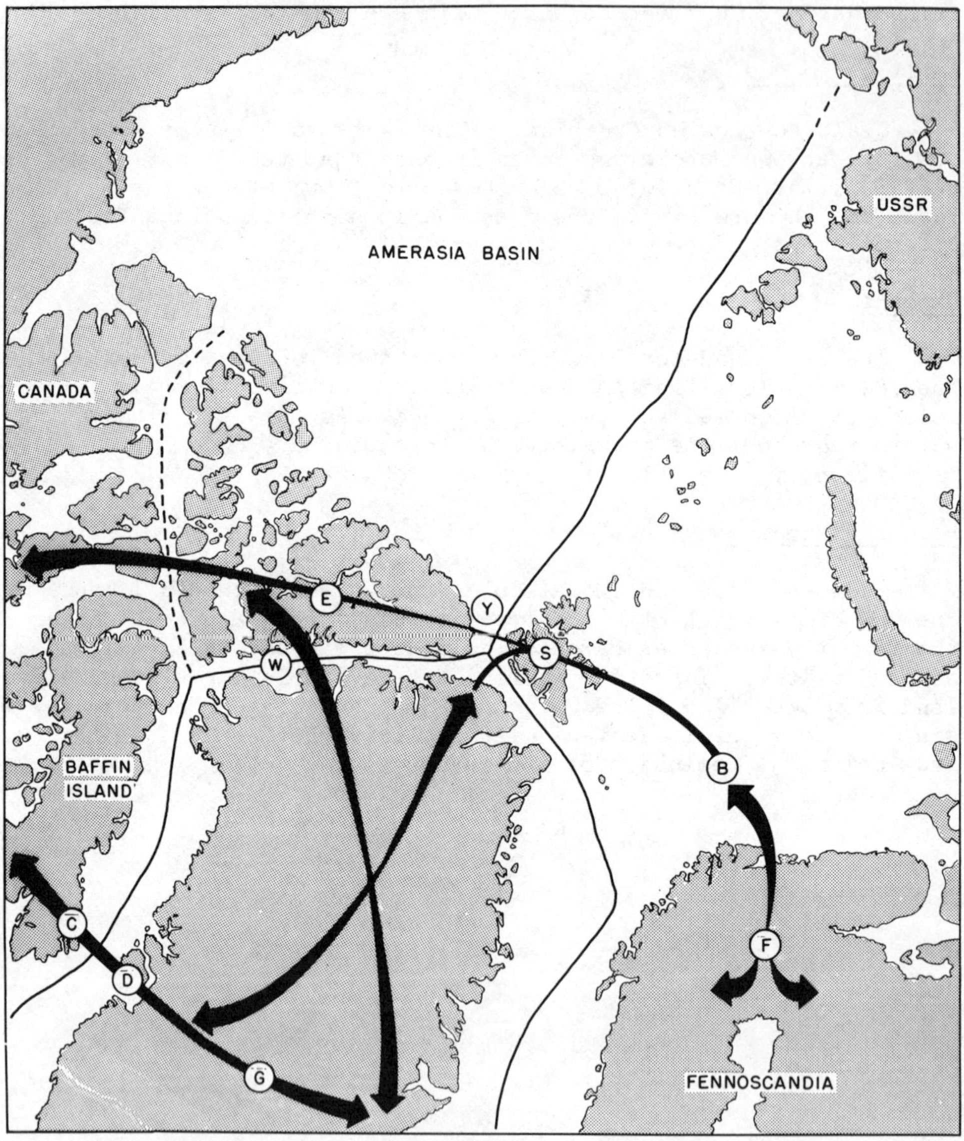
AMERASIA BASIN
USSR
CANADA
BAFFIN ISLAND
FENNOSCANDIA
E
Y
S
W
B
C̄
D̄
F̄
Ḡ

Fig. 5. ⟵ Palinspastic reconstruction of North America, the Queen Elizabeth Islands, Greenland, and northwest Eurasia (after Feden et al., 1979, with various modifications). The western end of the Thulean Route of terrestrial connection with the Greenland-Scotland Ridge passed through Baffin Island, the Disko Island area, and mainland Greenland to Kangerdlugssuaq Peninsula. Off the map, the route continued to Scotland and southwestern Europe. A more northerly placed De Geer Route connected Fennoscandia with North America via the Barents Shelf, Svalbard, the Yermak hot spot triple junction and northern Greenland, Ellesmere Island, Devon Island, Somerset Island, and the Boothia Peninsula of northern Canada. Greenland served as an interchange between the two routes. B: Barents Shelf. C: Cape Dyer, Baffin Island. D: Disko Island, West Greenland. E: Eureka Sound Formation Eocene vertebrate sites on Ellesmere Island in Queen Elizabeth Islands. F: Fennoscandia. G: Greenland. S: Spitsbergen (main island of Svalbard Archipelago). W: Wegener Fault. Y: Yermak hot spot and triple junction.

## REFERENCES

Anderson, H. J., 1976, Stratigraphic and paleogeographic significance of mollusk faunas from the Paleocene of the NW-German Tertiary basin, _in_: "Internatl. Geol. Correlation Programme, Project 124: The Northwest European Tertiary Basin, Rept. No. 1," H. Tobien, ed., pp. 92-95, Dept. Geol., Philipps Univ., Marburg.

Andrews, J. T., Guennel, G. K., Wray, J. L., and Ives, J. D., 1972, An early Tertiary outcrop in north-central Baffin Island, Northwest Territories, Canada: environment and significance, _Can. J. Earth Sci._ 9:233-238.

Athavale, R. N., and Sharma, P. V., 1975, Paleomagnetic results on early Tertiary lava flows from West Greenland and their bearing on the evolution history of the Baffin Bay-Labrador Sea region, Can. J. Earth Sci. 12(1):1-18.

Backman, J., 1980, Pliocene biostratigraphy of DSDP sites 111 and 116 from the North Atlantic Ocean and the age of Northern Hemisphere glaciation, Stockholm Contrib. Geol. 32(3):115-137.

Balkwill, H. R., and Bustin, R. M., 1980, Late Phanerozoic structures, Canadian Arctic Archipelago, Palaeogeogr., Palaeoclimatol., Palaeoecol. 30(3/4):219-227.

Bárdarson, G. G., 1925, A stratigraphical survey of the Pliocene deposits at Tjörnes, in northern Iceland, Kgl. Danske Vidensk. Selskab, Meddel. Biol. 4(5):1-118.

Barrett, D. L., 1966, Lancaster Sound shipborne magnetometer survey, Can. J. Earth Sci. 3:223-235.

Beh, R. L., 1975, Evolution and geology of Western Baffin Bay and Davis Strait, Canada, in: "Canada's Continental Margins," C. J. Yorath, E. R. Parker, and D. G. Glass, eds., Can. Soc. Petrol. Geol. Mem. 4:453-476.

Bendix-Almgreen, S. E., 1976, Palaeovertebrate faunas of Greenland, in: "Geology of Greenland," A. Escher and W. S. Watt, eds., pp. 536-573, Geol. Surv. Greenland, Copenhagen.

Berggren, W. A., 1972a, A Cenozoic time-scale--some implications for regional geology and paleobiogeography, Lethaia 5:195-215.

Berggren, W. A., 1972b, Late Pliocene-Pleistocene glaciation, in: "Initial Reports of the Deep Sea Drilling Project," vol. 12, Univ. California, Scripps Inst. Oceanogr.; A.S. Laughton, W. A. Berggren, R. N. Benson, T. A. Davies, U. Franz, L. F. Musich, K. Perch-Nielsen, A. S. Ruffman, J. E. van Hinte, and R. B. Whitmarsh, pp. 953-963, U.S. Govt. Printing Office, Washington.

Berggren, W. A., McKenna, M. C., Hardenbol, J., and Obradovich, J. D., 1978, Revised Paleogene polarity time scale, J. Geol. 86:67-81.

Berkhout, A. W. J., 1973, Gravity in the Prince of Wales, Somerset, and northern Baffin Islands region, in: "Canadian Arctic Geology," J. D. Aitken and D. J. Glass, eds., pp. 63-79, Geol. Assoc. Can. - Can. Soc. Petrol. Geol., Toronto.

Birkelund, T., 1965, Ammonites from the Upper Cretaceous of West Greenland, Meddel. Grønland 179(7):1-192.

Birkenmajer, K., 1972, Report on investigations of Tertiary sediments at Kap Brewster, Scoresby Sund, East Greenland, Rapp. Grønland geol. Unders. 48:85-91.

Bonde, N., 1974, Palaeoenvironment as indicated by the "Mo-clay Formation" (Lowermost Eocene of Denmark), Tertiary Times 2(1): 29-36.

Bott, M. H. P., 1975, Structure and evolution of the north Scottish shelf, the Faeroe Block and the intervening region, in: "Petroleum and the Continental Shelf of North-West Europe, Vol. 1, Geology," A. W. Woodland, ed., pp. 105-115, John Wiley & Sons, New York.

Brooks, C. K., 1973, Rifting and doming in southern East Greenland,

Nature Phys. Sci. 244:23-25.

Brooks, C. K., and Gleadow, A. J. W., 1977, A fission-track age for the Skaergaard intrusion and the age of the east Greenland basalts, Geology 5:539-540.

Brown, P. E., van Breemen, O., Noble, R. H., and Macintyre, R. M., 1977, Mid-Tertiary igneous activity in East Greenland--The Kialineq Complex, Contrib. Mineral. Petrol. 64(1):109-122.

Brown, P. E., and Whitley, J. E., 1976, Eastern Greenland basalts and their supposed plume origin, Nature 260:232-234.

Burke, K., 1976, Development of graben associated with the initial ruptures of the Atlantic Ocean, Tectonophysics 36:93-112.

Burke, K., and Dewey, J. F., 1973, Plume generated triple junctions: Key indicators in applying plate tectonics to old rocks, J. Geol. 81:406-433.

Burke, K., and Kidd, W. S. F., 1980, Volcanism on Earth through time, in: "The Continental Crust and its Mineral Deposits," D. W. Strangway, ed., Geol. Assoc. Can. Spec. Paper 20:503-522.

Burov, Y. P., and Zagruzina, I. A., 1976, Results of determination of the absolute ages of Cenozoic basic rocks in the northern part of the island of Spitsbergen, in: "Geology of Svalbard" [in Russian], pp. 139, 140, 146, N.I.I.G.A., Leningrad.

Bustin, R. M., 1977 (unpublished), "The Eureka Sound and Beaufort Formations, Axel Heiberg and West-central Ellesmere Islands, District of Franklin," Thesis, Univ. Calgary.

Carey, S. W., 1958, A tectonic approach to continental drift, in: "Continental Drift - A Symposium," S. W. Carey, ed., pp. 177-355, Geol. Dept., Univ. Tasmania, Hobart.

Casten, U., 1973, The crust beneath the Faeroe Islands, Nature Phys. Sci. 241:83-84.

Charlesworth, J. K., 1953, "The Geology of Ireland, An Introduction," Oliver and Boyd, Edinburgh and London.

Christie, R. L., 1964, Geological reconnaissance of northeastern Ellesmere Island, District of Franklin, Geol. Surv. Can. Mem. 331:1-79.

Christie, R. L., 1974, Northeastern Ellesmere Island: Lake Hazen region and Judge Daly Promontory, Geol. Surv. Can. Paper 74-1A:297-299.

Christie, R. L., 1976, Tertiary rocks at Lake Hazen, northern Ellesmere Island, Geol. Surv. Can. Paper 76-1B:259-262.

Christie, R. L., and Rouse, G. E., 1976, Eocene beds at Lake Hazen, northern Ellesmere Island, Geol. Surv. Can. Paper 76-1C:153-156.

Clark, D. L., 1977, Paleontologic response to post-Jurassic crustal plate movements in the Arctic Ocean, in: "Paleontology and Plate Tectonics," R. M. West, ed., Milwaukee Pub. Mus. Spec. Publ. Biol. Geol. 2:55-76.

Clark, D. L., and Kitchell, J. A., 1979, Injection events in ocean history, Nature 278:669.

Clarke, D. B., and Pedersen, A. K., 1976, Tertiary volcanic province of West Greenland, in: "Geology of Greenland," A. Escher and W. S. Watt, eds., pp. 364-385, Geol. Surv. Greenland, Copenhagen.

Clarke, D. B., and Upton, B. G. J., 1971, Tertiary basalts of Baffin Island: field relations and tectonic setting, Can. J. Earth Sci. 8:248-258.

Cochran, J. R., and Talwani, M., 1977, Free-air gravity anomalies in the world's oceans and their relationship to residual elevation, Geophys. J. Roy. Astron. Soc. 50:495-552.

Curry, D., 1965, The Palaeogene beds of South-East England, Proc. Geol. Assoc. 76:151-173.

Daae, H. D., and Rutgers, A. T. C., 1975, Geological History of the Northwest Passage, Bull. Can. Petrol. Geol. 23:84-108.

Dagley, P., Wilson, R. L., Ade-Hall, J. D., Walker, G. P. L., Haggarty, S. E., Sigurgeirsson, T., Watkins, N. D., Smith, P. J., Edwards, J., and Grasty, R. L., 1967, Geomagnetic polarity zones for Icelandic lavas, Nature 216:25-29.

Dalland, A., 1975, The Mesozoic rocks of Andøy, northern Norway, Norges geol. Unders. 316:271-287.

Dawes, P. R., 1973, The North Greenland fold belt: a clue to the history of the Arctic Basin and the Nares Strait lineament, in: "Implications of Continental Drift to the Earth Sciences," D. H. Tarling and S. K. Runcorn, eds., Vol. 2, pp. 925-947, Academic Press, New York,

Dawes, P. R., 1976, Precambrian to Tertiary in northern Greenland, in: "Geology of Greenland," A. Escher and W. S. Watt, eds., pp. 248-303, Geol. Surv. Greenland, Copenhagen.

Dawson, M. R., West, R. M., Langston, W., Jr., and Hutchison, J. H., 1976, Paleogene terrestrial vertebrates: northernmost occurrence, Ellesmere Island, Canada, Science 192:781-782.

Dawson, M. R., West, R. M., Ramaekers, P., and Hutchison, J. H., 1975, New evidence on the palaeobiology of the Eureka Sound Formation, Arctic Canada, Arctic 28(2):110-116.

Deegan, C. E., 1976, The Tertiary of the United Kingdom, in: "Internatl. Geol. Correlation Programme, Project 124: The Northwest European Tertiary Basin, Rept. No. 1," H. Tobien, ed., pp. 26-30, Dept. Geol., Philipps Univ., Marburg.

Deer, W. A., 1976, Tertiary igneous rocks between Scoresby Sund and Kap Gustav Holm, East Greenland, in: "Geology of Greenland," A. Escher and W. S. Watt, eds., pp. 404-429, Geol. Surv. Greenland, Copenhagen.

Detrick, R. S., Schlater, J. G., and Thiede, J., 1977, The subsidence of aseismic ridges, Earth Planet. Sci. Letts. 34(2):185-196.

Durham, J. W., and MacNeil, F. S., 1967, Cenozoic migrations of marine invertebrates through the Bering Strait region, in: "The Bering Land Bridge," D. M. Hopkins, ed., pp. 326-349,, Stanford Univ. Press, Stanford.

Dypvik, H., and Nagy, J., 1979, Early Tertiary bentonites from Svalbard, Geol. Mag. 16(6):457-468.

Einarsson, T., Hopkins, D. M., and Doell, R. R., 1967, The stratigraphy of Tjörnes, northern Iceland, and the history of the Bering Land Bridge, in: "The Bering Land Bridge," D. M. Hopkins, ed., pp. 312-325, Stanford Univ. Press, Stanford.

Eldholm, O., and Sundvor, E., 1979, Geological events during the early formation of a passive margin, Tectonophysics 59:233-237.

Eldholm, O., and Talwani, M., 1977, Sediment distribution and structural framework of the Barents Sea, Geol. Soc. Am. Bull. 88: 1015-1029.

Eldholm, O., and Thiede, J., 1980, Cenozoic continental separation between Europe and Greenland, Palaeogeogr., Palaeoclimatol., Palaeoecol. 30(3/4):243-259.

Eldholm, O., and Windisch, C. C., 1974, Sediment distribution in the Norwegian-Greenland Sea, Geol. Soc. Am. Bull. 85(11):1661-1676.

Emelyanov, E. M., Litvin, V. M., Levchenko, V. A., and Martynova, G. P., 1971, The geology of the Barents Sea, in:"The Geology of the East Atlantic Continental Margin, 2, Europe," F. M. Delany, ed., Great Britain Inst. Geol. Sci., London, Rept. 70/14:1-15.

Estes, R., and Hutchison, J. H., 1980, Eocene lower vertebrates from Ellesmere Island, Canadian Arctic Archipelago, Palaeogeogr., Palaeoclimatol., Palaeoecol. 30(3/4);325-347.

Evans, A. L., Fitch, F. J., and Miller, J. A., 1973, Potassium-argon age determinations on some British Tertiary igneous rocks, J. Geol. Soc. London 129:419-443.

Faller, A. M., and Soper, N. J., 1979, Palaeomagnetic evidence for the origin of the coastal flexure and dyke swarm in central E. Greenland, J. Geol. Soc. London 136(6):737-744.

Farrar, E., 1966 (unpublished), "The extraction and ultra-high vacuum mass spectrometry of argon from rocks," Ph.D. Thesis, Univ. Toronto. [cited by Parrott]

Feden, R. H., Vogt, P. R., and Fleming, H. S., 1979, Magnetic and bathymetric evidence for the "Yermak Hot Spot" northwest of Svalbard in the Arctic Basin, Earth Planet. Sci. Letts. 44:18-38.

Fortier, Y. O., Blackadar, R. G., Glenister, B. F., Greiner, H. R., McLaren, D. J., McMillan, N. J., Norris, A. W., Rootsi, E. F., Souther, J. G., Thorsteinsson, R., and Tozer, E. T., 1963, "Geology of the North-Central Part of the Arctic Archipelago, Northwest Territories (Operation Franklin)," Geol. Surv. Can. Mem. 320:1-671.

Friedrich, W., 1966, Zur Geologie von Brjánslaekur (Nordwest-Island) unter besonderer Berücksichtigung der fossilen Flora, Sonderveroff. Geol. Inst. Univ. Köln 10:1-108.

Frisch, T., and Thorsteinsson, R., 1978, Haughton Astrobleme: a mid-Cenozoic impact crater, Devon Island, Canadian Arctic Archipelago, Arctic 31(2):108-124.

Girdler, R. W., 1961, Some preliminary measurements of anisotropy of magnetic susceptibility of rocks, Geophys. J. Roy. Astron. Soc. 5:197-206.

Gladenkov, Y. B., 1981, Marine Plio-Pleistocene of Iceland and problems of its correlation, Quat. Res. 15:18-23.

Godfriaux, I., and Thaler, L., 1972, Note sur la découverte de dents de mammifères dans le Montien continental du Hainaut (Belgique),

Bull. Acad. Roy. Belgique, classe Sci., 5th Sér. 58:536-541.

Gradstein, F. M., and Srivastava, S. P., 1980, Aspects of Cenozoic stratigraphy and paleoceanography of the Labrador Sea and Baffin Bay, Palaeogeogr., Palaeoclimatol., Palaeoecol. 30(3/4): 261-295.

Grant, A. C., 1972, Geophysical investigations on the continental shelf off southern Baffin Island (Abstr.), Trans. Am. Geophys. Union (EOS) 53:407.

Grant, A. C., 1975, Geophysical results from the continental margin off southern Baffin Island, in: "Canada's Continental Margins," C. J. Yorath, E. R. Parker, and D. J. Glass, eds., Can. Soc. Petrol. Geol. Mem. 4:411-431.

Gripp, K., 1927, Beiträge zur géologie von Spitsbergen, Abh. Naturwis. Ver. Hamburg 21:1-38.

Grønlie, G., 1979, Tertiary paleogeography of the Norwegian-Greenland Sea, Norsk Polarinst. Skrifter 170:49-61.

Hald, N., 1971, An investigation of the igneous rocks of Hareøen and western Nûgssuaq, West Greenland, Rapp. Grønlands geol. Unders. 35:11-13.

Hald, N., 1977, Field investigations of Tertiary basic igneous rocks between 72° and 73°30' N, northern East Greenland, Rapp. Grønlands geol. Unders. 90:109-113.

Hansen, H. J., 1970, Danian Foraminifera from Nûgssuaq, West Greenland, Meddel. Grønland 193(2):1-132.

Harland, W. B., 1969, Contribution of Spitsbergen to understanding of tectonic evolution of North Atlantic region, Mem. Am. Assoc. Petrol. Geol. 12:817-851.

Harland, W. B., 1979, A stratigraphic outline of Greenland, Geol. Mag. 116(2):145-153.

Hassan, M. Y., 1953, Tertiary faunas from Kap Brester, East Greenland, Meddel. Grønland 111(5):1-42.

Heer, O., 1868, "Flora Fossilis Arctica, Vol. 1, Die fossile Flora der Polarlander," J. Wurster & Co., Zurich.

Heer, O., 1883, "Flora Fossilis Arctica, Vol. 7, Flora Fossilis Grönlandica. Die Fossile Flora Grönlands," J. Wurster & Co., Zürich.

Heirtzler, J. R., Dickson, G. O., Herron, E. M., Pitman, W. C., LePichon, X., 1968, Marine magnetic anomalies, geomagnetic field reversals and motions of the ocean floor and continents, J. Geophys. Res. 73:2119-2136.

Henderson, G., 1976, Petroleum geology, in: "Geology of Greenland," A. Escher and W. S. Watt, eds., pp. 488-505, Geol. Surv. Greenland, Copenhagen.

Henderson, G., Rosenkrantz, A., and Schiener, E. J., 1976, Cretaceous-Tertiary sedimentary rocks of West Greenland, in: "Geology of Greenland," A. Escher and W. S. Watt, eds., pp. 340-362, Geol. Surv. Greenland, Copenhagen.

Hills, L. V., and Wallace, S., 1969, Paraalnipollenites, a new form genus from uppermost Cretaceous and Paleocene rocks of Arctic Canada and Russia, Geol. Surv. Can. Bull. 182:139-145.

Hinz, K., and Schlüter, H.-U., 1978, The geological structure of the western Barents Sea, Mar. Geol. 26:199-230.

Hinz, K., Schlüter, H.-U., Grant, A. C., Srivastava, S. P., Umpleby, D., and Woodside, J., 1979, Geophysical transects of the Labrador Sea: Labrador to southwest Greenland, Tectonophysics 59:151-183.

Hood, P., and Bower, M., 1975, Aeromagnetic reconnaissance of Davis Strait and adjacent areas, in: "Canada's Continental Margins," C. J. Yorath, E. R. Parker, and D. J. Glass, eds., Can. Soc. Petrol. Geol. Mem. 4:433-451.

Hopkins, D. M., 1959, Cenozoic history of the Bering Land Bridge, Science 129:1519-1528.

Hopkins, D. M., 1967, The Cenozoic history of Beringia - A Synthesis, in: "The Bering Land Bridge," D. M. Hopkins, ed., pp. 451-484, Stanford Univ. Press, Stanford.

Hyndman, R. D., 1973, Evolution of the Labrador Sea, Can. J. Earth Sci. 10:637-644.

Jacqué, M., and Thouvenin, J., 1975, Lower Tertiary tuffs and volcanic activity in the North Sea, in: "Petroleum and the Continental Shelf of North-West Europe, Vol. 1, Geology," A. W. Woodland, ed., pp. 455-465, John Wiley & Sons, New York.

Johnson, G. L., McMillan, N. J., Rasmussen, M., Campsie, J., and Ditmer, F., 1975, Sedimentary rocks dredged from the southwest Greenland continental margin, in: "Canada's Continental Margins," C. J. Yorath, E. R. Parker, and D. J. Glass, eds., Can. Soc. Petrol. Geol. Mem. 4:391-409.

Johnson, G. L., Taylor, P. T., Vogt, P. R., and Sweeney, J. F., 1978, Arctic Basin morphology, Polarforschung 48(1/2):20-30.

Keen, C. E., and Barrett, D. L., 1972, Seismic refraction studies in Baffin Bay: An example of a developing ocean basin, Geophys. J. Roy. Astron. Soc. 30:253-271.

Keen, M. J., and Clarke, D. B., 1974, Tertiary basalts of Baffin Bay: geochemical evidence for a fossil hot spot, in: "Geodynamics of Iceland and the North Atlantic area, Proc. NATO Advanced Study Inst., Reykjavik, Iceland, 1-7 July, 1974," L. Kristjansson, ed., pp. 127-137, D. Reidel Publ. Co., Dordrecht, Holland.

Kellogg, H. E., 1975, Tertiary stratigraphy and tectonism in Svalbard and continental drift, Bull. Am. Assoc. Petrol. Geol. 59: 465-485.

Kent, P. E., 1975, Review of North Sea basin development, J. Geol. Soc. London 131:435-468.

Kerr, J. W., 1974, Geology of Bathrust Island Group and Byam Martin Island, Arctic Canada, Geol. Surv. Can. Mem. 378:1-152.

Kerr, J. W., 1980, A plate tectonic contest in Arctic Canada, in: "The Continental Crust and its Mineral Deposits," D. W. Strangway, ed., Geol. Assoc. Can. Spec. Paper 20:457-486.

Kerr, J. W., and Temple, P. G., 1965, Stratigraphy and Structure of Bathurst Island, Geol. Surv. Can. Paper 65-1:5-8.

Kharin, G. N., 1976, The petrology of magmatic rocks, DSDP Leg 38, in: "Initial Reports of the Deep Sea Drilling Project," Vol. 38, Univ. California, Scripps Inst. Oceanogr; M. Talwani, G. Udintsev, K. Bjorkland, V. N. D. Caston, R. W. Faas, G. N. Kharin, D. A. Morris, C. Müller, T. H. Nilsen, J. van Hinte, D. A. Warnke, S. M. White, pp. 685-715, U.S. Govt. Printing Office, Washington.

Koch, B. E., 1963, Fossil plants from the lower Paleocene of the Agatdal (Angmârtusut) area, central Nûgssuaq Peninsula, northwest Greenland, Meddel. Grønland 172(5):1-120.

Kristjansson, L. G., and Deutsch, E. R., 1973, Magnetic properties of rock samples from the Baffin Bay coast, in: "Earth Science Symposium on Offshore Eastern Canada," P. J. Hood, ed., Geol. Surv. Can. Paper 71-23:573-598.

Kuc, M., 1973, Plant macrofossils in Tertiary coal and amber from northern Lake Hazen, Ellesmere Island, N. W. T., Geol. Surv. Can. Paper 73-1B:143.

Kurtén, B., 1966, Holarctic land connexions in the early Tertiary, Comm. Biol. 29(5):1-5.

Labrecque, J. L., Kent, D. V., and Cande, S. V., 1977, Revised magnetic polarity time scale for Late Cretaceous and Cenozoic time, Geology 5:330-335.

Larsen, O., Dawes, P. R., and Soper, N. J., 1978, Rb/Sr Age of the Kap Washington Group, Peary Land, North Greenland, and its geotectonic implication, Rapp. Grølands geol. Unders. 90: 115-119.

Laufeld, S., 1965, Sporomorphs in Tertiary coal from the Faeroe Islands, Geol. För. Stockholm Förhandl. 87:231-238.

Lehman, J. P., 1951, Un nouvel amiide de l'Eocene du Spitzberg, Pseudamia heintzi, Tromsø Mus. Arsh. Natur. Avd. 70(3):1-11.

Lehmann, U., Thiedig, F., and Harland, W. B., 1978, Spitzbergen in Tertiär, Polarforschung 48(1/2):120-138.

MacNeil, F. S., 1957, Cenozoic megafossils of northern Alaska, U. S. Geol. Surv. Prof. Paper 294-C:99-126.

MacNeil, F. S., 1965, Evolution and distribution of the genus Mya, and Tertiary migrations of Mollusca, U. S. Geol. Surv. Prof. Paper 483-G:1-51.

Matthews, J. V., Jr., 1979, Fossil beetles and late Cenozoic history of the tundra environment, in: "Historical Biogeography, Plate Tectonics, and the Changing Environment," J. Gray and A. J. Boucot, eds., pp. 371-378, Oregon State Univ. Press, Corvallis.

Maync, W., 1940, Stratigraphie des Küstengebietes von Ostgrønland zwischen 73-75° N Lat. (Östliche Gauss Halvø, Hold-with-Hope, Östliche Clavering Ø, Wollaston Forland, Kuhn Ø, Sabine Ø und Lille Pendulum Ø), Meddel. Grønland 114(5):1-34.

McDougall, I., Saemundsson, K., Johannesson, H., Watkins, N. D., and Kristjansson, L., 1976, Extension of the geomagnetic polarity time scale to 6.5 m.y.: K-Ar dating, geological and paleomagnetic study of a 3,5000 m lava succession in western Iceland, Geol. Soc. Am. Bull. 88:1-15.

McDougall, I., Watkins, N. D., and Kristjansson, L., 1976, Geochronology and paleomagnetism of a Miocene-Pliocene lava sequence at Bessastadaa, eastern Iceland, Am. J. Sci. 276(9):1078-1095.

McDougall, I., and Wensink, H., 1966, Paleomagnetism and geochronology of the Pliocene-Pleistocene lavas in Iceland, Earth Planet. Sci. Letts. 1:232-236.

McKenna, M. C., 1973, Sweepstakes, filters, corridors, Noah's arks, and beached Viking funeral ships in palaeogeography, in: "Implications of Continental Drift to the Earth Sciences," D. H. Tarling and S. K. Runcorn, eds., Vol. 1, pp. 295-308, Academic Press, New York.

McKenna, M. C., 1975, Fossil mammals and early Eocene North Atlantic land continuity, Ann. Missouri Bot. Garden 62(2):335-353.

McKenna, M. C., 1980, Eocene paleolatitude, climate, and mammals of Ellesmere Island, Palaeogeogr., Palaeoclimatol., Palaeoecol. 30(3/4):349-362.

McKenna, M. C., 1981, Discussion [of A. Hallam, Relative importance of plate movements, eustasy, and climate in controlling major biogeographical changes since the early Mesozoic], in: "Vicariance Biogeography, A Critique," G. Nelson and D. E. Rosen, eds., pp. 335-338, Columbia Univ. Press, New York.

Mellor, D. W., and Mussett, A. E., 1975, Evidence for initial $^{36}$Ar in volcanic rocks, and some implications, Earth Planet. Sci. Letts. 26:312-318.

Miall, A. D., 1975, Post-Paleozoic geology of Banks, Prince Patrick, and Eglinton islands, arctic Canada, in: "Canada's Continental Margins," C. J. Yorath, E. R. Parker, and D. J. Glass, eds., Can. Soc. Petrol. Geol. Mem. 4:557-587.

Miall, A. D., 1979a, Mesozoic and Tertiary geology of Banks Island, Arctic Canada: The history of an unstable craton margin, Geol. Surv. Can. Mem. 387:1-235.

Miall, A. D., 1979b, Tertiary fluvial sediments in the Lake Hazen intermontane basin, Ellesmere Island, Arctic Canada, Geol. Surv. Can. Paper 79-9:1-25.

Miller, D. J., 1957, Geology of the southeastern part of the Robinson Mountains, Yakataga District, Alaska, U. S. Geol. Surv. Oil and Gas Inv. Map OM-187.

Mitchell, J. G., 1977, Potassium-argon ages from Phanerozoic basic dykes in South-East Greenland, Rapp. Grønlands geol. Unders. 90:141-146.

Moorbath, S., Sigurdsson, H., and Goodwin, R., 1968, K-Ar ages of the oldest exposed rocks in Iceland, Earth Planet. Sci. Letts. 4:197-205.

Myers, J. S., Bridgwater, D., and Gill, R. C. O., 1980, Mapping of Archaean, Mesozoic, and Tertiary rocks between Watkins Bjerge and Angmagssalik (69°- 65°50' N), Rapp. Grønland geol. Unders. 100:86-90.

Ness, G., Levi, S., and Couch, R., 1980, Marine magnetic anomaly timescales for the Cenozoic and Late Cretaceous: a précis, critique, and synthesis, Rev. Geophys. Space Phys. 18(4): 753-770.

Nilson, T. H., and Kerr, D. R., 1978, Paleoclimatic and paleogeographic implications of a lower Tertiary laterite (latosol) on the Iceland-Faeroe Ridge, North Atlantic region, Geol. Mag. 115(3):153-236.

Noe-Nygaard, A., 1974, Cenozoic to Recent volcanism in and around the North Atlantic basin, in: "The Ocean Basins and Margins, Vol. 2, The North Atlantic," A. E. M. Nairn and F. G. Stehli, eds., pp. 391-443, Plenum Press, New York.

Noe-Nygaard, A., 1976, Tertiary igneous rocks between Shannon and Scoresby Sund, East Greenland, in: "Geology of Greenland," A. Escher and W. S. Watt, eds., pp. 386-402, Geol. Surv. Greenland, Copenhagen.

Oftedahl, C., 1972, A sideritic ironstone of Jurassic age in Beistadfjorden, Trøndelag, Norsk geol. Tidsskr. 52:123-134.

Ørvig, T., 1960, The Jurassic and Cretaceous of Andøya in northern Norway, in: "Geology of Norway," O. Holtedahl, ed., Norges geol. Unders. 208:344-350.

Os Vigran, J., 1970, Fragments of a middle Jurassic flora from northern Trøndelag, Norway, Norsk geol. Tidsskr. 50:193-214.

Park, I., Clarke, D. B., Johnson, J., and Keen, M. J., 1971, Seaward extension of the West Greenland Tertiary volcanic province, Earth Planet. Sci. Letts. 10:235-238.

Parker, J. R., 1975, Lower Tertiary sand development in the central North Sea, in: "Petroleum and the Continental Shelf of North-West Europe, Vol. 1, Geology," A. W. Woodland, ed., pp. 447-453, John Wiley & Sons, New York.

Parrott, R. J. E., 1976 (unpublished), "$^{40}Ar/^{39}Ar$ dating of the Labrador Sea volcanics and their relation to sea-floor spreading," M.S. Thesis, Dalhousie Univ.

Parsons, B., and Sclater, J. G., 1977, An analysis of the variation of ocean floor bathymetry and heat flow with age, J. Geophys. Res. 82(5):803-827.

Pedersen, K. R., 1976, Fossil floras of Greenland, in: "Geology of Greenland," A. Escher and W. S. Watt, eds., pp. 519-535, Geol. Surv. Greenland, Copenhagen.

Perch-Nielsen, K., 1973, Danian and Campanian/Maastrichtian coccoliths from Nûgssuaq, West Greenland, Bull. Geol. Soc. Denmark 22:79-82.

Petryk, A. A., 1969, Mesozoic and Tertiary stratigraphy at Lake Hazen, northern Ellesmere Island, District of Franklin, Geol. Surv. Can. Paper 68-17:1-51.

Pflug, H. D., 1959, Sporenbilder aus Island und ihre stratigraphische Deutung, Neues Jahrb. Geol. Palaont., Abh. 107:141-172.

Pomerol, C., 1973, "Stratigraphie et Paléogéographie. Ere Cenozoïque," Doin, Paris.

Pożaryska, K., 1978, Differences between the late Eocene foraminiferal faunas in western und [sic] eastern Europe, Paläont. Zeitschr. 52(1/2):47-56.

Pożaryska, K., and Ciuk, E., 1976, The Tertiary of Poland (Carpathians excluded), in: "Internatl. Geol. Correlation Programme,

Project 124; The Northwest European Tertiary Basin, Rept. No. 1," H. Tobien, ed., pp. 21-25, Dept. Geol., Philipps Univ., Marburg.

Prestvik, T., 1978, Cenozoic plateau lavas of Spitsbergen: a geochemical study, Norsk Polarinst. Årbok 1977:129-142.

Rasmussen, J., and Noe-Nygaard, A., 1969, Beskrivelse til geologisk kort over Faerøerne i Målestok 1:50.000, Danm. geol. Unders., I Raekke 24:1-370.

Rasmussen, J., and Noe-Nygaard, A., 1970, Geology of the Faeroe Islands (pre-Quaternary), Danm. geol. Unders., I Raekke 25: 1-142.

Ravn, J. P. J., 1933, New investigations of the Tertiary at Cape Dalton, East Greenland, Meddel. Grønland 105(1):1-15.

Renard, V., and Malod, J., 1974, Structure of the Barents Sea from seismic refraction, Earth Planet. Sci. Lett. 24:33-47.

Repenning, C. A., 1967, Palearctic-Nearctic mammalian dispersal in the late Cenozoic, in: "The Bering Land Bridge," D. M. Hopkins, ed., pp. 288-311, Stanford Univ. Press, Stanford.

Repenning, C. A., Ray, C. E., and Grigorescu, D., 1979, Pinniped biogeography, in: "Historical Biogeography, Plate Tectonics, and the Changing Environment," J. Gray and A. J. Boucot, eds., pp. 357-369, Oregon State Univ. Press, Corvallis.

Rex, D. C., Gledhill, A. R., Brooks, C. K., and Steenfelt, A., 1978, Radiometric ages of Tertiary salic intrusions near Kong Oscars Fjord, East Greenland, Rapp. Grønlands geol. Unders. 95:106-109.

Risum, J. B., 1978, Seismic mapping of lower Tertiary sediments offshore West Greenland between 66° and 68° N, Rapp. Grønlands geol. Unders. 95:51-53.

Roberts, D. G., 1975, Tectonic and stratigraphic evolution of the Rockall Plateau and Trough, in: "Petroleum and the Continental Shelf of North-West Europe, Vol. 1, Geology," A. W. Woodland, ed., pp. 77-91, John Wiley & Sons, New York.

Ronnevik, H., Bergsager, E. I., Moe, A., Øvrebø, Ø., Navrestad, T., and Strangenes, J., 1975, The geology of the Norwegian continental shelf, in: "Petroleum and the Continental Shelf of North-West Europe, Vol. 1, Geology," A. W. Woodland, ed., pp. 117-129, John Wiley & Sons, New York.

Rosenkrantz, A., 1951, Oversigt over Kridt og Tertiaerformationens stratigrafiske Forhold i Versgrønland, Meddel. dansk geol. Foren. 12:155-158.

Rosenkrantz, A., 1970, Marine Upper Cretaceous and lowermost Tertiary deposits in West Greenland, Meddel. dansk geol. Foren. 19:406-453.

Rosenkrantz, A., and Pulvertaft, T. C. R., 1969, Cretaceous-Tertiary stratigraphy and tectonics in northern West Greenland, Mem. Am. Assoc. Petrol. Geol. 12:883-898.

Ross, D. I., and Henderson, G., 1973, New geophysical data on the continental shelf of central and northern West Greenland, Can. J. Earth Sci. 10:485-497.

Russell, D. E., 1964, Les mammifères Paléocènes d'Europe, Mém. Mus.

Natl. Hist. Nat., N.S., Sér. C. 13:1-324.
Russell, D. E., 1968, Succession, en Europe, des faunes mammaliennes au debut de Tertiaire, Mém. Bur. Res. Geol. Min. 58:291-296.
Scheffer, V. B., 1967, Marine mammals and the history of Bering Strait, in: "The Bering Land Bridge," D. M. Hopkins, ed., pp. 350-363, Stanford Univ. Press, Stanford.
Schiener, E. J., 1976, Coal geology, in: "Geology of Greenland," A. Escher and W. S. Watt, eds., pp. 506-516, Geol. Surv. Greenland, Copenhagen.
Schluter, H.-U., and Hinz, K., 1978, The continental margin of west Spitsbergen, Polarforschung 48(1/2):151-169.
Schwartzbach, M., and Pflug, H. D., 1957, Beiträge zur Klimageschichte Islands, VI, Das Klima des jüngeren Tertiärs in Island, Neues Jahrb. Geol. Paläont., Abh. 104:279-298.
Schweitzer, H. J., 1980, Environment and climate in the early Tertiary of Spitsbergen, Palaeogeogr., Palaeoclimatol., Palaeoecol. 30(3/4):297-311.
Sclater, J. G., Anderson, R. N., and Bell, M. L., 1971, Elevation of ridges and evolution of the central Eastern Pacific, J. Geophys. Res. 76:7888-7915.
Simpson, G. G., 1936, Data on the relationships of local and continental mammalian faunas, J. Paleont. 10(5):410-414.
Simpson, G. G., 1940, Mammals and land bridges, J. Washington Acad. Sci. 30(4):137-163.
Simpson, G. G., 1947, Holarctic mammalian faunas and continental relationships during the Cenozoic, Geol. Soc. Am. Bull. 58: 613-688.
Simpson, G. G., 1952, Probabilities of dispersal in geologic time, Bull. Am. Mus. Nat. Hist. 99(3):163-176.
Simpson, G. G., 1953, "Evolution and Geography," Condon Lectures, Oregon State System of Higher Educ., Eugene.
Sittler, C., 1976, The Paris Basin in relation to the north-west European Tertiary, in: "Internatl. Geol. Correlation Programme, Project 124: The northwest European Tertiary Basin, Rept. No. 1," H. Tobien, ed., pp. 8-13, Dept. Geol., Philipps-Univ., Marburg.
Soper, N. J., and Costa, L. I., 1976, Palynological evidence for the age of Tertiary basalts and post-basaltic sediments at Kap Dalton, central East Greenland, Rapp. Grønlands geol. Unders. 80:123-127.
Souther, J. G., 1963, Geological traverse across Axel Heiberg Island from Buchanan Lake to Strand Fiord, in: "Geology of the North-central part of the Arctic Archipelago, Northwest Territories (Operation Franklin)," Y. O. Fortier et al., Geol. Surv. Can. Mem. 320:426-448.
Spjeldnaes, N., 1975, Palaeogeography and facies distribution in the Tertiary of Denmark and surrounding areas, in: "Petrol. Geol. & Geology of the North Sea and northeast Atlantic continental margin," A. Whiteman, D. Roberts, and M. A. Sellevoll, eds., Norges geol. Unders. 316:289-311.

Strauch, F., 1963, Zur Geologie von Tjörnes (Nordisland), Sonderveröff. Geol. Inst. Univ. Köln 8:1-129. [English translation available from A. Geol. Inst.]

Strauch, F., 1970, Die Thule-Landbrücke als Wanderweg und Faunenschiede zwischen Atlantik im Tertiär, Geol. Runds. 60(1): 381-417.

Strauch, F., 1972, Phylogenese, Adaptation, und Migration einiger nordischer mariner Molluskengenera (Neptunea, Panomya, Cyrtodaria und Mya), Abh. Senckenberg. Naturf. Ges. 531: 1-211.

Sturt, B. A., Dalland, A., and Mitchell, J. L., 1979, The age of the sub mid-Jurassic tropical weathering profile of Andoya, northern Norway, and the implication for the late Palaeozoic Palaeography [sic] in the North Atlantic Region, Geol. Runds. 68(2):523-542.

Sundvor, E., and Eldholm, O., 1979, The western and northern margin off Svalbard, Tectonophysics 59:239-250.

Szczechura, J., 1971, Paleocene Ostracoda from Nûgssuaq, West Greenland, Meddel. Grønland 193(1):1-42.

Talwani, M., and Eldholm, O., 1977, Evolution of the Norwegian-Greenland Sea, Geol. Soc. Am. Bull. 88:969-999.

Talwani, M., Udintsev, G., Bjørklund, K., Caston, V. N. D., Faas, R. W., Kharin, G. N., Morris, D. A., Mueller, C., Nilson, T. H., van Hinte, J. E., Warnke, D. A., and White, S. M., 1975, Leg 38, Geotimes 20(2):24-26.

Tarling, D. H., and Gale, N. H., 1968, Isotopic dating and paleomagnetic polarity in the Faeroe Islands, Nature 218:1043.

Tarling, D. H., and Mitchell, J. G., 1976, Revised Cenozoic polarity time scale, Geology 4(3):133-136.

Thorsteinsson, R., and Tozer, E. T., 1970, Geology of the Arctic A chipelago, in: "Geology and Economic Minerals of Canada," R. J. W. Douglass, ed., Geol. Surv. Can. Econon. Geol. Rept. 1:547-590.

Tozer, E. T., 1963, Mesozoic and Tertiary Stratigraphy, in: "Geology of the North-Central Part of the Arctic Archipelago, Northwest Territories (Operation Franklin)," Y. O. Fortier et al., Geol. Surv. Can. Mem. 320:74-95.

Trettin, H. P., and Balkwill, H. R., 1979, Contributions to the tectonic history of the Innuitian Province, Arctic Canada, Can. J. Earth Sci. 16:748-769.

Trettin, H. P., Frisch, T. O., Sobczak, L. W., Weber, J. R., Niblett, E. R., Law, L. K., de Laurier, I., and Whitham, K., 1972, The Innuitian province, in: "Variations in Tectonic Styles in Canada," R. A. Price and R. J. W. Douglas, eds., Geol. Assoc. Can. Spec. Paper 11:83-180.

Troelsen, J. C., 1950, Contributions to the geology of northwest Greenland, Ellesmere Island and Axel Heiberg Island, Meddel. Grønland 149(7):1-85.

Ulleberg, K., 1976, The Tertiary of Norway, in: "Internatl. Geol. Correlation Programme, Project 124: The Northwest European Tertiary Basin, Rept. No. 1," H. Tobien, ed., pp. 19-20,

Dept. Geol., Philipps Univ., Marburg.
Vianey-Liaud, M. 1979, Les Mammifères Montiens de Hainin (Paléocène moyen de Belgique), Part I: multitubercules, Palaeovertebrata 9(4):117-131.
Vilmundardóttir, E. G., 1972, "Report on geological investigations at Jokulsa in Fljótsdal Valley, Summer 1970" [in Icelandic], Natl. Energy Authority (Orkustofnun), Reykjavik.
Vine, F. J., 1976, Implications of the proposed geomagnetic reversal time-scales of the early Tertiary and late Cretaceous Period, Trans. Am. Geophys. Union (EOS) 57:674.
Vischer, A., 1943, Die postdevonische Tektonik von Ostgrönland Zwischen 74°und 75°N. Br. Kuhn Ø, Wollaston Forland, Clavering Ø und angrenzende Gebiete, Meddel. Grønland 133(1):1-194.
Vogt, P. R., 1972, The Faeroe-Iceland-Greenland Aseismic Ridge and the Western Boundary Undercurrent, Nature 239(5367):79-81.
Vogt, P. R., Feden, R. H., Eldholm, O., and Sundvor, O., 1978, The ocean crust west and north of the Svalbard Archipelago: Synthesis and review of new results, Polarforschung 48(1/2): 1-19.
Wager, L. R., and Deer, W. A., 1939, Geological investigations in East Greenland, Part III, The petrology of the Skaergaard intrusion, Kangerdlugssuaq, Meddel. Grønland 105:1-352.
Wegener, A., 1929, "Die Entstehung der Kontinente und Ozeane," Friedrich Vieweg und Sohn, Braunschweig, 4th revised edition.
West, R. M., 1977, Ellesmere Island fossils reveal its past, Lore (Milwaukee Pub. Mus.) 27(3):18-26.
West, R. M., 1978, Mammalia, in: "McGraw-Hill Yearbook of Science and Technology," pp. 248-251, McGraw-Hill, New York.
West, R. M., and Dawson, M. R., 1978, Vertebrate Paleontology and the Cenozoic history of the North Atlantic region, Polarforschung 48(1/2):103-119.
West, R. M., Dawson, M. R., and Hutchison, J. H., 1977, Fossils from the Paleogene Eureka Sound Formation, N.W.T., Canada: occurrence, climatic and paleogeographic implications, in: "Paleontology and Plate Tectonics," R. M. West, ed., Milwaukee Pub. Mus. Spec. Publ. Biol. Geol. 2:77-93.
Whiteman, A., Naylor, D., Pegrum, R., and Rees, G., 1975, North Sea troughs and plate tectonics, Tectonophysics 26:39-54.
Wilson, D. G., 1976, Studies of Mesozoic stratigraphy, Tanquary Fiord to Yelverton Pass, northern Ellesmere Island, District of Franklin, Geol. Surv. Can. Paper 76-1A:449-451.
Wilson, J. T., and Clarke, D. B., 1965, Geological Expedition to capes Dyer and Searle, Baffin Island, Canada, Nature 205:349.
Wolfe, J. A., 1980, Tertiary climates and floristic relationships at high latitudes in the Northern Hemisphere, Palaeogeogr., Palaeoclimatol., Palaeoecol. 30(3/4):313-323.
Wolfe, J. A., and Leopold, E. B., 1967, Neogene and early Quaternary vegetation of northwestern North America and northeastern Asia, in: "The Bering Land Bridge," D. M. Hopkins, ed., pp. 193-206, Stanford Univ. Press, Stanford.

Worthington, L. V., 1970, The Norwegian Sea as a mediterranean basin, *Deep-Sea Res.* 17:77-84.

Zagwijn, W. H., 1974, The Pliocene-Pleistocene boundary in western and southern Europe, *Boreas* 3:75-97.

Zarudzki, E. F. K., 1980, Interpretation of shallow seismic profiles over the continental shelf in West Greenland between latitudes 64° and 69°30" N, *Rapp. Grønland geol. Unders.* 100:58-61.

Ziegler, W. H., 1975, Outline of the geological history of the North Sea, *in*: "Petroleum and the Continental Shelf of North-West Europe, Vol. 1, Geology," A. W. Woodland, ed., pp. 165-190, John Wiley & Sons, New York.

# FOSSIL EVIDENCE OF EARLY TERTIARY NORTH ATLANTIC EVENTS VIEWED IN EUROPEAN CONTEXT

Ella Hoch

Geologisk Museum, University of Copenhagen

Ø. Voldgade 5-7, DK-1350 København K., Denmark

## INTRODUCTION

Related forms of organisms in separate geographical areas testify to former interconnection, be it by potential dispersal routes (serving active migration or uniform distribution) or by paths of passive transportation (by water, wind or other organisms, such as man). After fossils have been found and taxonomically determined, a next step towards comprehension of ancient life scenarios is to unravel where the organisms came from and, perhaps, how they got to the site.

On the small scale, dealing with local embedding circumstances, geological conditions around the fossils viewed with common sense and knowledge of living organisms will often permit reliable reconstructions. Fishes in a lake deposit mostly lived in the lake or in waters draining into the lake (there are, however, well-known cases of fishes from the sea periodically fighting their way upstream into fresh waters). Pterosaurs in the same deposit would not have lived submerged in the lake, as is obvious from their shape, but may have hunted prey closely above or even in the water. The stomach content of well preserved fossils can give indications of the preferred living grounds of the animals: marine mollusc shells inside a bird show sea connection, and rounded stones positioned neatly at the skeleton of a dinosaur where its (?)gizzard used to be may permit tracing of the creature's live movements over geologic outcrops.

On the large scale, implying considerations of time and evolution, dispersal, vicariance (see Cracraft, 1975 and Hallam, 1981a) and/or (ante or post mortem) long range transportation, studies of the geographic derivation of fossils meet with difficulties that may be elucidated only by interdisciplinary efforts.

## INTER/INTRA-CONTINENTAL PASSAGEWAYS IN THE NORTH ATLANTIC REGION

Focusing attention on the history of the Cenozoic North Atlantic realm, decades of palaeontologic work have accumulated conspicuous evidence of early Tertiary faunal (e.g. Savage, 1971; McKenna, 1975; Russell, 1975) and floral (e.g. Wolfe, 1975) similarity on the two sides of the present ocean. Land bridges have long been suggested as an explanation, whereas hypotheses of free terrestrial communication via East Europe-Asia find less support in palaeontologic evidence. The epicontinental Turgai Strait, in Late Cretaceous time intersecting Eurasia from the Arctic Ocean to the Tethys (Maack, 1981: Abb. 3), existed also in the Eocene (Heissig, 1979). During the Paleocene it either persisted as a barrier to E-W faunal dispersal or was closed by a southern land corridor (thus forming the Obik Sea) as implied by Heissig (1979) and illustrated by Russell (1975: Fig. 4).

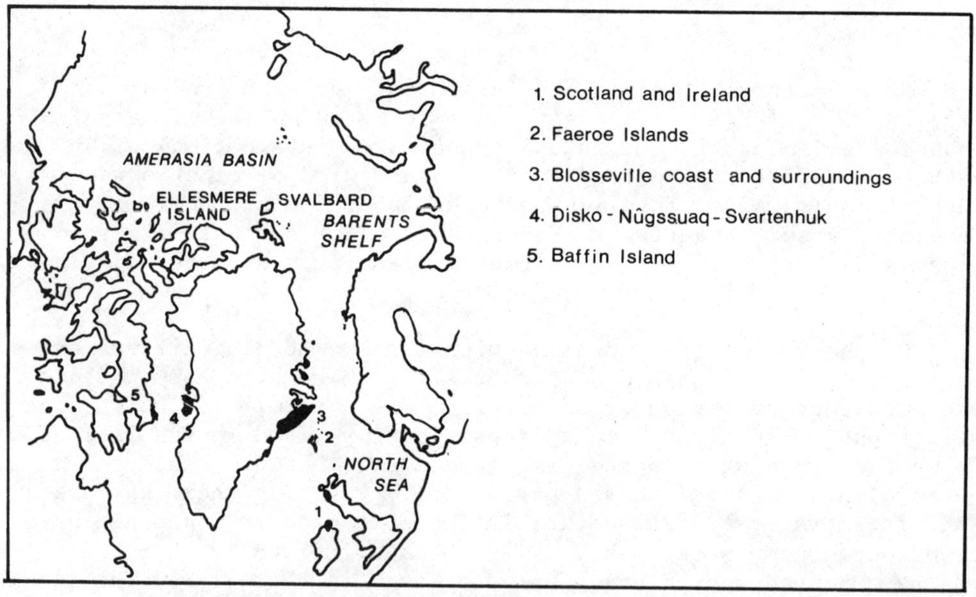


Fig. 1. Reconstruction of the Paleocene North Atlantic region prior to the opening, redrawn from Dypvik & Nagy (1978: Fig. 1). Dark areas show presently subaerial Palaeogene volcanics marking the Thulean route and its probable continuation on the west side of Greenland and Baffin Island. After an extended period of rifting, active sea floor spreading commenced between Greenland and Europe towards anomaly 24 time, i.e. at 55-60 m.y. on the time scale of La Brecque et al. (1977) (Vogt et al., 1981).

Much favoured are the Thulean route over central East Greenland, the Faeroe Islands and Scotland, which was comparatively short in the early Tertiary when rifting commenced, and the De Geer route connecting the Ellesmere Island-North Greenland region with Scandinavia over Svalbard and the Barents Shelf (e.g. McKenna, 1971, this vol.; Strauch, 1970; West & Dawson, 1978).

## EVIDENCE OF TERRESTRIAL LIFE AND MARINE BARRIERS ALONG THE TWO ROUTES

No fossil remains of terrestrial animals have until now been found in the early Tertiary strata which extend over about 10° of latitude from Kap Gustav Holm to Shannon Ø, East Greenland, and which are thought to have been parts of the Thulean route. On the Blosseville Kyst, sediments interbedded with the subaqueous/subaerial Paleocene-Eocene basalt flows are marine and terrestrial (Fawcett et al., 1973; Soper et al., 1976; Nielsen et al., 1981); in the Kangerdlugssuaq Group are molluscs, which have not yet been compared with the Danian molluscs of Nûgssuaq, Denmark and Svalbard (see later). Silicified tree trunks occur in sediments between the early Tertiary lavas (e.g. Orvin, 1931; Fawcett et al., 1966, 1973). A core sample from the NE flank of the Iceland-Faeroe Ridge towards its eastern end contained lateritic palaeosol overlying basalt of mid/late Eocene age and overlain by mid/late Eocene marine strata (Nilsen, 1978; Nilsen & Kerr, 1978); macrofossils have not been found in the approximately 30 m subtropical-tropical terrigenous deposit (T. Nilsen, pers. comm. 1981). No remains of terrestrial faunas are recorded from Palaeogene deposits of the Faeroes or Scotland. Botanical evidence seems presently somewhat better (see also Lund, this vol.). *Aquilapollenites* (angiospermous pollen of unknown botanical affinity) are recorded from Nûgssuaq, central West Greenland, in strata that on the basis of dinoflagellate cysts are referred to Early Paleocene age and a little older (Croxton, 1980). The Nûgssuaq plant assemblages with *Aquilapollenites* are comparable to assemblages known from Alaska and northern Canada. And besides in northern and eastern Asia in the Late Cretaceous-Early Paleocene (Tschudy, 1970; Hao et al., 1979) *Aquilapollenites* occur in Scotland in strata that are most likely between Maastrichtian and Early Eocene in age, and where the European Normapolles flora might otherwise have been expected to occur (Croxton, 1980). Early Tertiary Normapolles also occur in E-North America (Tschudy, 1970).

Birkenmajer's (1972) investigation of the Svalbard Archipelago showed Tertiary deposits only on Spitsbergen and Prins Karls Forland. Four formations of Early Paleocene and two of supposed Late Paleocene or Eocene age were recorded, no post early Palaeogene strata were found (for various views on dating and correlation with East Greenland and Eureka Sound Formation at other arctic sites, see Harland, 1975; Lehmann et al., 1978; Manum & Throndsen, 1978; West & Dawson, 1978; Schweitzer, 1980). Marine invertebrates from the Spitsbergen Tertiary correspond to forms in the Agatdal Formation on Nûgssuaq

(Birkenmajer, 1972; some Spitsbergen species are listed by Lehmann et al., 1978), which fauna shows great similarity on the specific level to the Danish Early Paleocene fauna (Henderson et al., 1976). A seaway connected Nûgssuaq and the Danian Sea in Europe in the early Paleocene, be it south of, across or north of Greenland (Gradstein & Srivastava, 1980: 267 implies a Cretaceous-Cenozoic sea strait linking the Labrador Sea and the Arctic Ocean possibly through the Nares Strait), and south or north of Britain. It seems also to have touched, perhaps crossed Svalbard, at least periodically. The Main Spitsbergen Basin may have been (intermittently) flanked by land to the (south-)west and east: sporadic early Tertiary terrestrial and marine deposits are recorded from North Greenland (e.g. Håkansson, 1979; Croxton et al., 1980), and the major source of pebbles etc. in the basin is thought to be river outlets and coastal beaches of a land which included the northern part of Spitsbergen and the islands to the east of Storfjorden (Birkenmajer, 1972). In the basin, coal seams and fresh water deposits occur from the base upwards interchanging with the marine strata (Ibid.). It shows that periodically the basin floor was subaerially exposed. An (?)Eocene fresh-water fish, *Pseudamia heintzi*, a close relative of the Recent North American bowfins and of the Palaeogene *Amia* in North America (e.g. Estes & Hutchison, 1980) and Europe (e.g. Franzen, 1979), is described from Cap Heer, Spitsbergen (Lehman, 1951).

A boreal nannoplankton species, *Imperiaster obscurus*, is recorded from the Lower Eocene (around zones NP11/12) at the Vøring Plateau, in the Lofoten Basin, "in the central North Sea Basin and to the south in Denmark, Northern Germany, Belgium, and England, to the southeast transported by a cold current via the Ural-Street down to the northern Caucasus" (Schrader et al., 1976, referring Martini). If by the "Ural-Street" is meant a waterway connecting the early Eocene North Sea in a south-easterly direction with the Caucasus, then Scandinavia and Russia were cut off from the rest of Europe at this time, making a De Geer route little useful for North American-(West)European faunal interchange (compare McKenna, this vol.). Bonde (1979) argues for a "fairly direct water connection" to explain similarity in uppermost Paleocene fish faunas in the North Sea Basin and Turkmenistan, but the similarity could be due to cosmopolitan distribution of the oceanic forms in case (N. Bonde, pers. comm., 1981). Odrzywolska-Bieńkowa & Pożaryska (1980) state for the Polish lowlands that "the Upper Paleocene is not developed in marine facies but in the swampy, being dated on the basis of floristic records. The Lower Eocene is also represented by continental deposits". Early Eocene marine connection the North Sea/Lofoten/Caucasus might (alternatively?) be suggested around northern Scandinavia and through the Turgai Strait. The poles were more than 10°C warmer during the Early Eocene than today (Shackleton, 1981; Shackleton & Boersma, 1981), so the early Tertiary Arctic Ocean would hardly have presented a temperature barrier to boreal plankton nor to other North Sea species. The Barents

Shelf has been repeatedly investigated (references are given by Vogt et al., 1981: 529). Renard & Malod (1974), in agreement with others, indicate basinal strata of Mesozoic and Cenozoic age on the south-eastern Barents Shelf, and Tertiary deposits along the western shelf edge and in its marginal depressions. Absence of registered Palaeogene deposits on the Barents Shelf, supporting the idea that the shelf was emergent during the early Tertiary, could perhaps be ascribed to the gross-scale investigation methods and/or to erosion by bottom currents at times of submergence, in the late Cenozoic reinforced by moving ice. Rønnevik & Motland (1979: Fig. 9) indicate Tertiary continental conditions for Iceland and the eastern part of the Barents Shelf. As is obvious regarding Iceland, the palaeogeographic map presents an idealistic picture of conditions through more than 60 m.y. with no indication of transient events. The authors just mentioned refer to Sachs (no reference is given) who suggests an Upper Cretaceous seaway in the Barents Sea towards the Pechora Basin. A similar marine corridor across the Barents Shelf intermittent in the early Tertiary would have periodically invalidated a De Geer route.

*Imperiaster obscurus* could not be found at two sites in the Bay of Biscay (Schrader et al., 1976). Yet there is still no firm evidence that the English Channel was continuously closed during the Early Eocene. King (1981: 113) states that perhaps the Channel was open throughout London Clay times, i.e. upper NP9-mid NP12 zones: latest Paleocene-late Early Eocene. The "Artois axis", often claimed to have connected England and France, was formerly covered by Eocene sediments; until now evidence is found of at least nine transgressive events in the latest Paleocene and Early Eocene on the margin of the North Sea Basin in southern England (King, 1981). But the transgressions supposedly alternated with regressions. The above warns us to keep the mind alert when speculating on ancient corridors.

## PROBABLE TRANS-"NORTH ALTANTIC" TERRESTRIAL CONNECTION

An effective Thulean land bridge between Greenland and the present European continent may have existed, were it only for a relatively very short span of time, in the latest Paleocene-earliest Eocene, corresponding to the peak of registered faunal similarity on the two sides of the North Atlantic Ocean (see West & Dawson, 1978: Fig. 1: Sparnacian is late Paleocene, Cuisian is early Eocene, cf. King, 1981). This would concert with ideas of intermittent regional updoming connected with volcanism (Brooks, 1980) permitting sectional traverse of the "North Atlantic" during intervals of Palaeogene times. On the central East Greenland coast a shallow marine basin occurred through Late Cretaceous-early Paleocene time (Soper et al., 1976), succeeded by late Paleocene, largely terrigenous lavas and sediments (see above). The Greenland-Iceland Ridge was not under water before the Miocene (H.C. Larsen, at the

Bressanone Meeting, 1981), i.e. the early Palaeogene ridge part, the "Proto Iceland", is believed to have been subaerially exposed, as is also argued for the contemporaneous Iceland-Faeroe Ridge (e.g. Schrader et al., 1976; Nilsen & Kerr, 1978). The Faeroe-Shetland Channel may have been dry land for a short time in the Palaeogene judged from the presence of deltaic sediments in the basin (Ridd, at the Bressanone Meeting, 1981). Intermittent Campanian-Paleocene volcanic activity in the area (Ibid.) may have been connected with updoming of "the now-buried Cretaceous and Palaeocene volcanic centre north of Shetland (close to Latitude 62°N)" (Ibid.), such updoming perhaps conditioning terrestrial connection to the British Isles.

## DISPERSAL ACROSS WATER

Terrestrial organisms are prevented from spreading over land areas by environmental conditions to which they are not adapted. Temperature can be as effective a barrier as a mountain range or an ocean. On the other hand, water may be traversed by them. Drifting vegetation islands with trapped animals are well known. And flying land birds blown out of their course during storm may end up on distant atolls (where their descendants in the absence of carnivores may evolve into flightless forms).

Active dispersal across water may occur among land animals that habitually bathe. Swimming elks are sometimes encountered in the northern Baltic Sea (I. Printzlau, pers. comm.). Elephants are reported moving in the sea from one land to another using their trunks as snorkels (a "forced swim" of some 48 km has been observed, shorter distances seem more normal (Johnson, 1980)). Deer, elephant and hippopotamus have populated islands after swimming to them (Sondaar, 1977; Dermitzakis & Sondaar, 1978). Extant sloths may swim in river water to move between hanging tree branches, and the ancestors of the extinct groundsloths on the Greater Antilles probably arrived there by swimming. *Hyracotherium (Eohippus)*, the fox-sized protohorse known from Paleocene-Eocene deposits in Asia (Dashzeveg, 1979), North America (e.g. Morris, 1968; Delson, 1971) and Europe (e.g. Hooker, 1980: the *Hyracotherium* specimens mentioned from below the *Wetzeliella astra* zone are of Late Paleocene age, cf. King, 1981), is often compared adaptationally with tapirs, which occasionally swim, and may have been able to cross narrow marine straits by swimming. Likewise the pantodont *Coryphodon* (for distribution, see later), a possible ecologic analogue of modern *Hippopotamus*, could probably swim across not too wide water gaps.

But taking into consideration the large number of closely related early Tertiary animals on the two sides of the North Atlantic, together with their inferred habitats, terrestrial connection at one time seems obligatory. Marine water gaps, even narrow ones, and especially such with strong currents, do actually constitute rather effective barriers to a large number of land organisms including some birds albeit their flying ability.

EXOTIC ANIMALS IN PALAEOGENE EUROPE

Literature referred above and a review of the Tertiary North Sea Basin fauna (Russell, in prep.) will permit a general idea to be obtained of the traffic of terrestrial animals in the Paleocene-Eocene North Atlantic region. Here shall be mentioned a few fossils and sites relevant to the subject, but not included in the above.

Snakes, crocodiles and mammals of South American ancestry (as far as present evidence goes) occur in the West European Eocene. *Palaeopython*, a boa snake of Boini (South American Boinae) relationship, is recorded from the Middle and Upper Eocene (Rage, 1978). Antunes (1975) described part of the characteristic high snout of a sebecosuchian crocodile, *Iberosuchus macrodon*, from Eocene continental deposits in Portugal; another sebecosuchian, *Bergisuchus dietrichbergi*, is described from the Messel Mid Eocene bituminous shale, West Germany (Berg, 1966); and teeth found at various contemporaneous sites in West Europe may also pertain to the Sebecosuchia. Buffetaut (1980), in a review and interpretation of the Sebecosuchia and their biogeographic history, suggests, in agreement with Antunes (1975), that these presumed principally terrestrial crocodiles spread from South America to Europe via North America and Greenland crossing the North Atlantic area. Abundant remains of ectotherms in the Ellesmere Island Early-Middle Eocene (West & Dawson, 1978; Estes & Hutchison, 1980) and floristic evidence from various high latitude early Palaeogene sites indicate favourable climatic conditions (McKenna, 1980a; Schweitzer, 1980; Wolfe, 1980). Terrestrial passage between the two Americas was possible, at least for a short time, in the Late Cretaceous, whereas there is no evidence of land connection South America-Africa-Europe at relevant times (Rage, 1978; Bonaparte, 1979; Cox, 1980; Hallam, 1981b). Although Notosuchia, ancestral to the Sebecosuchia (Buffetaut, 1980), are recorded from the South American-African Early Cretaceous (Buffetaut & Taquet, 1979; Buffetaut, 1981), no sebecosuchians are known from Africa. Capetta, Jaeger et al. (1978) report remains of various palaeoryctids and some other eutherian mammals from an Early Paleocene deposit in Morocco. Palaeoryctidae are known from the Late Cretaceous and later in East Asia and North America, and from the Late Paleocene in Europe (Romer, 1966). Following Jaeger's (1981) interpretation, the African plate was populated before Early Paleocene by a eutherian stock of Asiatic origin; the Tethys may have been crossed via a "Turkish" plate between the two continents. A similar dispersal route may have been followed by certain plants, as seen from the distribution of e.g. *Aquilapollenites* (Tschudy, 1970; Tschudy & Leopold, 1970). No indications are given to show faunal migration from Africa towards parts of ancient Laurasia. The apparent absence of Sebecosuchia in North America may be ascribed, among other things, to the similarity of sebecosuchian ziphodont teeth with teeth of carnivorous dinosaurs and certain eusuchian crocodiles, the pristichampsines (Buffetaut, 1978). This in combination with general preference of dinosaurs over crocodiles among fossil collectors and investigators may lead to misidentifications (ziphodont teeth, pristichampsines and multitudes of dinosaurs are recorded from North America). Besides, insufficient interest and knowledge can have resulted in secondary burial of undescribed crocodile fossils in museum drawers (see also Buffetaut, 1980).

The only known European xenarthan (edentate), the anteater *Eurotamandua joresi*, also derives from the Middle Eocene lake deposit at Messel (Storch, 1981. For a description of the prolific site, see Hoch, 1980; von Koenigswald, 1980 lists described species and literature). The well preserved skeleton and its state of morphologic specialization leaves no doubt as to its taxonomic relationship. No other Palaeogene or older anteaters are recorded, Miocene and younger ones are known from South America (Paula Couto, 1979). According to Hoffstetter (1981), the Xenarthra constitute among the mammals "el grupo más original de la fauna sudamericana". McKenna (1975) and others agree in the great ancestry of the order. Late Cretaceous mammals in South America were marsupials and placentals (Hoffstetter, 1981); the latter may have comprised early Xenarthra. Hirschfeld (1976) regards it likely that the anteaters formed the earliest radiation of the Xenarthra, thus supporting other workers, such as Hoffstetter (1958), in assigning a high age to the Vermilingua. Fossil non-vermilinguid xenarthrans are known from the Early Paleocene in Argentina (McKenna, 1980b). Edentata reported from the Chinese "Late" Paleocene (Chow & Zheng, 1980) including (or in total?) *Ernanodon antelios* Ding, 1979 may be primitive Xenarthra. And North American Palaeanodonta, although during some time referred to the pangolins (see McKenna, 1980b; Hoffstetter, 1981), may be Palaeogene Xenarthra-relatives. The lack on all continents of clear evidence of pene-contemporaneous close relatives of the anteater that was extracted from the Lutetian Rhine Graben rift zone mud might become a cornerstone in Creationists' argumentation. Others would suggest that anteaters dispersed, perhaps from South America in the late Cretaceous, over North America, and across the North Atlantic area in the early Tertiary. Recent anteaters occur in environments where organisms only exceptionally become fossilized. We should imagine the late Cretaceous-early Tertiary Euramerican warm savannas and humid forests housing populations of anteaters together with multitudes of other creatures (such as sebecosuchian crocodiles) of which we have only extremely sporadic remains compared to the amount of living matter they represent.

Some of the genera in North America and West Europe from the time of little restricted fauna interchange across the "North Atlantic", such as the condylarth *Hyopsodus* (e.g. West, 1979; Hooker,

1979) and the pantodont *Coryphodon* (e.g. West & Dawson, 1978; Hooker, 1980), also occur in the Chinese Early Eocene (e.g. Chow & Zheng, 1980; the coryphodontids seem to be of Asiatic origin (Chow & Wang, 1979)). The filtering effect between East Asia and North America, pronounced in the early Paleocene, diminished towards the Early Eocene (Chow & Zheng, 1979, 1980; see Cox, 1980 for preceding continental connections). In the late Eocene "Upper Level" of China many mammalian forms are comparable to those of the Upper Eocene and Oligocene of Europe (Chow & Zheng, 1980), corresponding to a new pattern of faunal dispersal.

Through the Eocene, West Europe underwent progressive insularization leading to faunal endemism (evolution of e.g. the Palaeotheriidae, perissodactyls well known from the Late Eocene Paris Basin and other places), thus to decreasing similarity with surrounding continents. Land connection to the west seems broken already in the Early Eocene. There is palaeontologic evidence of several fauna provinces in western Europe during the Late Eocene (Rage & Ford, 1980, and works cited therein). And Polish studies indicate a Late Eocene marine corridor between the Baltic and the Tethys region across the Polish lowlands (Odrzywolska-Bieńkowa & Pożaryska, 1980).

Together with many other European animals sebecosuchian crocodiles appear to have become extinct towards the end of the Eocene (Buffetaut, 1980). "La Grande Coupure", a term introduced by Stehlin around 1910 (see Heissig, 1979) and applied in West European literature, designates the results of a sudden invasion in the Early Oligocene from Asia into West Europe, following a Turgai Strait bridging, with Asiatic forms quickly exterminating less effective European ecologic counterparts and occupying their "niches". Clear difference between ante- and post-"Grande Coupure" faunas are seen in e.g. les Phosphorites du Quercy, France (Rage, 1974; de Bonis et al., 1977). Newer studies, however, modify Stehlin's concept. A major climatic deterioration at the end of the Eocene (Wolfe, 1980) may have been fatal to many forms. And, according to Heissig (1979), the "invasion" from Asia was stepwise from Middle Eocene time, partly via a changing pattern of isles and land bridges in the Near Orient and South East Europe, and partly, from the Early Oligocene, across the traditional Turgai land bridge and the Russian platform.

## CONCLUDING REMARKS

Similarity of terrestrial fossils permitted the hypothesis of an early Tertiary "North Atlantic continent", comprising the North American plate including Greenland, the continental Jan Mayen Ridge, the Faeroe-Rockall plate and Europe with shelf areas. Freshwater fishes, such as bowfins and gars *(Lepisosteus (Atractosteus))*, reptiles, such as snakes, crocodiles and alligators, flightless landbirds, as *Diatryma* (Cracraft, 1973; Fischer, 1978; West & Dawson,

1978), and flying ones, as "New World vultures", Cathartidae (see Hoch, 1975 referring Cracraft & Rich, 1972), together with a great variety of mammals are some of the organisms populating the continent (works treating the vertebrates can be found via references given in the present article). Changing epicontinental seas fractured the land, and rifting and drifting finally split the continent apart. Decreasing faunal similarity/diversity ratio on the two sides of the North Atlantic rift zone during the Early Eocene indicates a severing of the Euramerican terrestrial biota at approximately -50 m.y. (West & Dawson, 1978). Times of dispersal possible to land organisms passed into times of vicariance in the North Atlantic region.

The Danish Natural Science Research Council, J.nr. 11-2348, covered the writer's travel expenses Copenhagen-Bressanone.

## REFERENCES

Antunes, M.T., 1975, Iberosuchus, crocodile Sebecosuchien nouveau, l'Eocène ibérique au Nord de la Chaîne Centrale, et l'origine du Canyon de Nazaré, Comunic. Serv. Geol., Port., 59:285-330.

Berg, D.E., 1966, Die Krokodile, insbesondere Asiatosuchus und aff. Sebecus?, aus dem Eozän von Messel bei Darmstadt/Hessen, Abh. hess. L.-Amt Bodenforsch., 52:105 p.

Birkenmajer, K., 1972, Tertiary history of Spitsbergen and continental drift, Acta geol. pol., 22:193-218.

Bonaparte, J.F., 1979, Faunas y paleobiogeografía de los tetrápodos mesozoicos de América del Sur, Ameghiniana, 16:217-238.

Bonde, N., 1979, Palaeoenvironment in the "North Sea" as indicated by the fish bearing mo-clay deposit (Paleocene/Eocene), Denmark, Meded. Werkgr. Tert. Kwart. Geol., 16:3-16.

Bonis, L. de, Crochet, J.-Y., Hartenberger, J.-L., Hoffstetter, R., Rage, J.-C., Sigé, B., Sudre, J. & Vianey-Liaud, M., 1977, Les Phosphorites du Quercy, un laboratoire naturel de l'évolution des vertébrés, Le Courrier du CNRS, 25:6-11.

Brooks, C.K., 1980, Episodic volcanism, epeirogenesis and the formation of the North Atlantic Ocean, Palaeogeogr., Palaeoclimatol., Palaeoecol., 30:229-242.

Buffetaut, E., 1978, Les crocodiles à dents de dinosaure, La Recherche, 89:492-494.

Buffetaut, E., 1980, Histoire biogéographique des Sebecosuchia (Crocodylia, Mesosuchia): un essai d'interprétation, Ann. Paléont., (Vert.), 66:1-18.

Buffetaut, E., 1981, Die biogeographische Geschichte der Krokodilier, mit Beschreibung einer neuen Art, Araripesuchus wegeneri, Geol. Rdsch., Alfred-Wegener-Symposium II, 70:611-624.

Buffetaut, E. & Taquet, P, 1979, An early Cretaceous terrestrial crocodilian and the opening of the South Atlantic, Nature, 280:486-487.

Cappetta, H., Jaeger, J.-J., Sabatier, M., Sigé, B., Sudre, J. & Vianey-Liaud, M., 1978, Découverte dans le Paléocène du Maroc des plus anciens mammifères euthérien d'Afrique, Géobios, 11: 257-263.

Chow, M. & Wang, B., 1979, Relationship between the pantodonts and tillodonts and classification of the order Pantodonta, Vertebr. palasiat., 17:37-48.

Chow, M. & Zheng, J., 1979, The mammal-bearing Palaeocene horizons of China, in: Cretaceous-Tertiary Boundary Events. Symposium, II. Proceedings, addendum: 256-257. W. Kegel Christensen & T. Birkelund, eds. University of Copenhagen.

Chow, M. & Zheng, J., 1980, (with appendix by G.V. Shkurkin), The mammal-bearing early Tertiary horizons of China, PaleoBios, 32:1-7.

Cox, C.B., 1980, An outline of the biogeography of the Mesozoic world, Mém. Soc. géol. Fr., N.S., 139:75-79.

Cracraft, J., 1973, Continental drift, paleoclimatology, and the evolution and biogeography of birds, J. Zool. Lond., 169:455-545.

Cracraft, J., 1975, Historical Biogeography and Earth History: Perspectives for a Future Synthesis, Ann. Missouri Bot. Gard., 62:227-250.

Cracraft, J. & Rich, P.V., 1972, The systematics and evolution of the Cathartidae in the Old World Tertiary, Condor, 74:272-283.

Croxton, C.A., 1980, Aquilapollenites from the Late Cretaceous - Paleocene (?) of central West Greenland, Rapp. Grønlands geol. Unders., 101:5-27.

Croxton, C.A., Dawes, P.R., Soper, N.J. & Thomsen, E., 1980, An occurrence of Tertiary shales from the Harder Fjord Fault, North Greenland fold belt, Peary Land, Rapp. Grønlands geol. Unders., 101:61-64.

Dashzeveg, D., 1979, Hyracotherium from Mongolia, Paleont. J., 3: 364-368.

Delson, E., 1971, Fossil mammals of the early Wasatchian Powder River local fauna, Eocene of northeast Wyoming, Bull. Am. Mus. nat. Hist., New York, 146:305-364.

Dermitzakis, M.D. & Sondaar, P.Y., 1978, The importance of fossil mammals in reconstructing paleogeography with special reference to the Pleistocene Aegean Archipelago, Annls géol. Pays hell., 29:808-840.

Ding, S., 1979, A new edentate from the Paleocene of Guangdong, Vertebr. palasiat., 17:57-64.

Dypvik, H. & Nagy, J., 1978, Early Tertiary Bentonites from Svalbard; a Preliminary Report, Polarforschung, 48:139-150.

Estes, R. & Hutchison, J.H., 1980, Eocene lower vertebrates from Ellesmere Island, Canadian Arctic Archipelago, Palaeogeogr., Palaeoclimatol., Palaeoecol., 30:325-347.

Fawcett, J.J., Brooks, C.K. & Rucklidge, J.C., 1973, Chemical petrology of Tertiary flood basalts from the Scoresby Sund area, Meddr Grønland, 195(6):1-54.

Fawcett, J.J., Rucklidge, J.C. & Brooks, C.K., 1966, Geological

Expedition to the Tertiary Basalt Region of Scoresby Sund, East Greenland, Nature, 212:603-604.

Fischer, K., 1978, Neue Reste des Riesenlaufvogels Diatryma aus dem Eozän des Geiseltales bei Halle (DDR), Mitt. Zool. Mus. Berlin, Suppl. Bd. 54, Ann. Orn., 2: 133-144.

Franzen, J.L., 1979, Senckenberg-Grabungen in der Grube Messel bei Darmstadt, 2. Ergebnisse 1978, Cour. Forsch.-Inst. Senckenberg, 36:1-144.

Gradstein, F.M. & Srivastava, S.P., 1980, Aspects of Cenozoic stratigraphy and paleoceanography of the Labrador Sea and Baffin Bay, Palaeogeogr., Palaeoclimatol., Palaeoecol., 30:261-295.

Håkansson, E., 1979, Carboniferous to Tertiary development of the Wandel Sea Basin, eastern North Greenland, Rapp. Grønlands geol. Unders., 88:73-83.

Hallam, A., 1981a, Methods of palaeobiogeographic analysis, Acta geol. hispan., Concept and method in Paleontology, 16:65-69.

Hallam, A., 1981b, Biogeographic relations between the northern and southern continents during the Mesozoic and Cenozoic, Geol. Rdsch., Alfred-Wegener-Symposium II, 70:583-595.

Hao, Y., Yu, J., Guan, S. & Sun, M., 1979, Some Late Cretaceous and Early Tertiary assemblages of Ostracoda, spores and pollen in China, in: Cretaceous-Tertiary Boundary Events. Symposium, II. Proceedings, addendum: 251-255. W. Kegel Christensen & T. Birkelund, eds. University of Copenhagen.

Harland, W.B., 1975, Palaeogene correlation in and around Svalbard, Geol. Mag., 112:421-429.

Heissig, K., 1979, Die hypothetische Rolle Südosteuropas bei den Säugetierwanderungen im Eozän und Oligozän, N. Jb. Geol. Paläont. Mh., 1979:83-96.

Henderson, G., Rosenkrantz, A. & Schiener, E.J., 1976, Cretaceous-Tertiary sedimentary rocks of West Greenland, in: Geology of Greenland: 340-362, A. Escher & W. Stuart Watt, eds. Geological Survey of Greenland, Copenhagen.

Hirschfeld, S.E., 1976, A new fossil anteater (Edentata, Mammalia) from Columbia, S.A. and evolution of the Vermilingua, J. Paleont., 50:419-432.

Hoch, E., 1975, Amniote remnants from the eastern part of the Lower Eocene North Sea basin, Colloque int. Cent. natn. Rech. scient., 218. Problèmes actuels de paléontologie - Évolution des Vertébrés: 543-562.

Hoch, E., 1980, A new Middle Eocene shorebird (Aves: Charadriiformes, Charadrii) with columboid features, Contr. Sci. Los Angeles, 330:33-49.

Hoffstetter, R., 1958, Xenarthra, in: Traité de Paléontologie, 6,2: 535-636. J. Piveteau, ed.

Hoffstetter, R., 1981, Historia biogeográfica de los mamíferos terrestres sudamericanos: problemas y ensenanzas, Acta geol. hispan., Concept and method in Paleontology, 16:71-88.

Hooker, J.J., 1979, Two new condylarths (Mammalia) from the early Eocene of southern England, Bull. Br. Mus. nat. Hist. (Geol.),

32:43-56.

Hooker, J.J., 1980, The succession of *Hyracotherium* (Perissodactyla, Mammalia) in the English early Eocene, *Bull. Br. Mus. nat. Hist.* (Geol.), 33:101-114.

Jaeger, J.-J., 1981, Les échanges de faunes terrestres entre plaques continentales adjacentes: l'exemple de l'Afrique et de l'Inde au début du Tertiaire, *Terra cognita*, spec. iss., Abstract, P5.

Johnson, D.L., 1980, Problems in the land vertebrate zoogeography of certain islands and the swimming powers of elephants, *J. Biogeogr.*, 7:383-398.

King, C., 1981, The stratigraphy of the London Clay and associated deposits, *Spec. Pap. Tertiary Res.*, 6:158 p.

Koenigswald, W. von, 1980, Fossillagerstätte Messel - Literaturübersicht der Forschungsergebnisse aus den Jahren 1969-1979, *Geol. Jb. Hessen*, 108:23-38.

La Brecque, J.L., Kent, D.V. & Cande, S.C., 1977, Revised magnetic polarity time scale for Late Cretaceous and Cenozoic time, *Geology*, 5:330-335.

Lehman, J.P., 1951, Un nouvel Amiidé de l'Éocène du Spitzberg, *Pseudamia heintzi*, *Tromsø Museums Årshefter, Nat. hist. Avd.*, 39, 70 (1947), 3.

Lehmann, U., Thiedig, F. & Harland, W.B., 1978, Spitzbergen im Tertiär, *Polarforschung*, 48:120-138.

Maack, V., 1981, Zur paläogeographischen Rekonstruktion erdgeschichtlicher Ozeane, *Geol. Rdsch.*, Alfred-Wegener-Symposium I, 70: 78-90.

Manum, S.B. & Throndsen, T., 1978, Rank of coal and dispersed organic matter and its geological bearing in the Spitsbergen Tertiary, *Norsk Polarinstitutt*, Årbok 1977:179-187.

McKenna, M.C., 1971, Fossil mammals and the Eocene demise of the De Geer North Atlantic dispersal route, *Geol. Soc. Am.*, Abstr. with programs, 3:644.

McKenna, M.C., 1975, Fossil mammals and Early Eocene North Atlantic land continuity, *Ann. Missouri Bot. Gard.*, 62:335-353.

McKenna, M.C., 1980a, Eocene paleolatitude, climate, and mammals of Ellesmere Island, *Palaeogeogr., Palaeoclimatol., Palaeoecol.*, 30:349-362.

McKenna, M.C., 1980b, Early History and Biogeography of South America's Extinct Land Mammals, *in*: Evolutionary Biology of the New World Monkeys and Continental Drift: 43-77. R.L. Ciochon & A.B. Chiarelli, eds., Plenum Press, New York and London.

Morris, W.J., 1968, A New Early Tertiary Perissodactyl, *Hyracotherium seekinsi*, from Baja California, *Contr. Sci.* Los Angeles, 151: 1-11.

Nielsen, T.F.D., Soper, N.J., Brooks, C.K., Faller, A.M., Higgins, A.C. & Matthews, D.W., 1981, The pre-basaltic sediments and the lower lavas at Kangerdlugssuaq, East Greenland: their stratigraphy, lithology, palaeomagnetism and petrology, *Meddr Grønland*.

Nilsen, T.H., 1978, Lower Tertiary laterite on the Iceland-Faeroe

Ridge and the Thulean land bridge, Nature, 274:786-788.
Nilsen, T.H. & Kerr, D.R., 1978, Paleoclimatic and paleogeographic implications of a lower Tertiary laterite (latosol) on the Iceland-Faeroe Ridge, North Atlantic region, Geol. Mag., 115: 153-236.
Odrzywolska-Bieńkowa, E. & Pożaryska, K., 1980, Stratigraphy of the Tertiary in the Polish Lowlands, in: International Geological Correlation Programme, Project 124, The Northwest European Tertiary Basin, Report 6:125-131.
Orvin, A.K., 1931, A Fossil River Bed in East Greenland, Norsk geol. Tidsskr., 12:469-474.
Paula Couto, C. de, 1979, Tratado de Paleomastozoologia, Academia Brasileira de Ciências, Rio de Janeiro, 1979:590 p.
Rage, J.-C., 1974, Les serpents des Phosphorites du Quercy, Palæovertebrata, 6:273-303.
Rage, J.-C., 1978, Une connexion continentale entre Amérique du Nord et Amérique du Sud au Crétacé supérieur? l'exemple des Vertébrés continentaux, C.R. somm. Soc. géol. Fr., 6:281-285.
Rage, J.-C. & Ford, R.L.E., 1980, Amphibians and Squamates from the Upper Eocene of the Isle of Wight, Tertiary Res., 3:47-60.
Renard, V. & Malod, J., 1974, Structure of the Barents Sea from seismic refraction, Earth Planet. Sci. Lett., 24:33-47.
Rønnevik, H.C. & Motland, K., 1979, Geology of the Barents Sea, in: Norwegian Sea Symposium, Oslo: Norwegian Petroleum Society, NSS/15:1-34.
Romer, A.S., 1966, Vertebrate Paleontology, 3rd ed. The University of Chicago Press, Chicago and London, 468 p.
Russell, D.E., 1975, Paleoecology of the Paleocene-Eocene Transition in Europe, in: Approaches to Primate Paleobiology. F.S. Szalay, ed., Contr. Primat., 5:28-61.
Russell, D.E. (ed.) (in prep.), Tetrapods of the North western Tertiary basin of Europe (working title, manuscript: 83 p), International Geological Correlation Programme, Project 124.
Savage, D.E., 1971, The Sparnacian-Wasatchian mammalian fauna, Early Eocene, of Europe and North America, Abh. hess. L.-Amt Bodenforsch., 60: Heinz-Tobien-Festschrift: 154-158.
Schrader, H.-J., Bjørklund, K., Manum, S., Martini, E. & Hinte, J. van, 1976, Cenozoic biostratigraphy, physical stratigraphy and paleooceanography in the Norwegian-Greenland Sea, DSDP Leg 38 paleontological synthesis, in: Initial Reports of the Deep Sea Drilling Project, 38:1197-1211, M. Talwani, G. Udintsev et al. eds.
Schweitzer, H.-J., 1980, Environment and climate in the Early Tertiary of Spitsbergen, Palaeogeogr., Palaeoclimatol., Palaeoecol., 30:297-311.
Shackleton, N.J., 1981, Cenozoic climatic variability, Terra cognita, spec. iss., Abstr. N26.
Shackleton, N.J. & Boersma, A., 1981, The climate of the Eocene Ocean, J. geol. Soc. Lond., 138:153-157.
Sondaar, P.Y., 1977, Insularity and its effect on mammal evolution,

in: Major patterns in vertebrate evolution: 671-707, M.K. Hecht, P.C. Goody & B.M. Hecht, eds.

Soper, N.J., Higgins, A.C., Downie, C., Matthews, D.W. & Brown, P.E., 1976, Late Cretaceous-early Tertiary stratigraphy of the Kangerdlugssuaq area, east Greenland, and the age of opening of the north-east Atlantic, J. geol. Soc. Lond., 132:85-104.

Storch, G., 1981, Eurotamandua joresi, ein Myrmecophagide aus dem Eozän der "Grube Messel" bei Darmstadt (Mammalia, Xenarthra), Senckenberg. leth., 61:247-289.

Strauch, F., 1970, Die Thule-Landbrücke als Wanderweg und Faunenscheide zwischen Atlantik und Skandik im Tertiär, Geol. Rdsch., 60:381-417.

Tschudy, R.H., 1970, Palynology of the Cretaceous-Tertiary Boundary in the Northern Rocky Mountains and Mississippi Embayment Regions, Spec. Pap. geol. Soc. Am., 127:65-111.

Tschudy, B.D. & Leopold, E.B., 1970, Aquilapollenites (Rouse) Funkhouser - Selected Rocky Mountain Taxa and Their Stratigraphic Ranges, Spec. Pap. geol. Soc. Am., 127:113-167.

Vogt, P.R., Perry, R.K., Feden, R.H., Fleming, H.S. & Cherkis, N.Z., 1981, The Greenland-Norwegian Sea and Iceland environment: geology and geophysics, in: The Ocean Basins and Margins, 5: 493-598, A.E.M. Nairn, M. Churkin & F.G. Stehli, eds.

West, R.M., 1979, Paleontology and Geology of the Bridger Formation, Southern Green River Basin, Southwestern Wyoming. Part 3. Notes on Hyopsodus, Contr. Biol. Geol., Milwaukee Publ. Mus., 25:1-52.

West, R.M. & Dawson, M.R., 1978, Vertebrate Paleontology and the Cenozoic History of the North Atlantic Region, Polarforschung, 48:103-119.

Wolfe, J.A., 1975, Some aspects of plant geography of the Northern Hemisphere during the late Cretaceous and Tertiary, Ann. Missouri Bot. Gard., 62:264-279.

Wolfe, J.A., 1980, Tertiary climates and floristic relationships at high latitudes in the Northern Hemisphere, Palaeogeogr., Palaeoclimatol., Palaeoecol., 30:313-323.

# BIOSTRATIGRAPHY OF INTERBASALTIC COALS FROM THE FAEROE ISLANDS

Jane Lund

Friedrich-Hoffmann-Strasse 8
D 3101 Wietze
West Germany

## Abstract

Interbasaltic coals from the Faeroe Islands contain a terrestrial microflora of pollen and spores indicating a late Paleocene (Landenian) age. This microflora is younger than the Maastrichtian to Danian microflora known from interbasaltic coals of Mull, Scotland, and is older than the interbasaltic microflora from Iceland.

## Introduction

The Faeroe Islands are located between Scotland and Iceland and form part of the North Atlantic basalt province. For the understanding of the geologic evolution of this province, datings of the individual areas are essential. Biostratigraphic datings are hampered by general absence of fossiliferous marine sediments. However, coals containing terrestrial palynomorphs (pollen and spores) occur interbedded between basalt on Mull (Scotland), Iceland, and the Faeroe Islands.

The main aim of this study is to obtain a palynological dating of the interbasaltic coals on the Faeroe Islands, with the possible secondary aim of establishing a biostratigraphic comparison with Mull, Scotland (and Iceland).

## Previous datings of the Faeroe Islands

No marine fossils are known from the Faeroe Islands, and only one plant macrofossil, *Metasequoia occidentalis* (Newb.) Chaney, has been described, which does not permit any closer dating (Koch in Rasmussen and Koch, 1963). Laufeld (1965) described the microflora from two samples of the coal-bearing sequence and stated that "The percentual composition of the flora combined with the known stratigraphic range of certain types, makes it, however, reasonable to look upon it as Eocene". This dating was in agreement with Rasmussen (1963) who considered the sequence to be of Eocene-Oligocene age based on regional geological considerations. More recent radiometric datings (K/Ar) show an age from 49-62 million years (Tarling and Gale, 1968). This radiometric age implies an early Eocene to early Paleocene age (Curry et al., 1978).

## Geology and location of sample material

The main coal-bearing sequence of the Faeroe Islands occurs between the "lower basalt series" (ca. 900 m thick) and the "middle basalt series" (ca. 1300 m thick) which is overlain by the "upper basalt series" (ca. 800 m thick) (Rasmussen and Noe-Nygaard, 1969). This coal-bearing sequence was sampled at Rókhagi, about 3 km south-east of Hvalba, Suðuroy. Moreover, coal/clay occurring about 20 m below the top of the "lower basalt series" was sampled at Heygsmúli, on the west coast of Suðuroy, about 2 km southwest of Hvalba.

## Preparation and microscopic investigation

The preparation included treatment with cold hydrochloric acid (HCl), and hot hydrofluoric acid (HF), followed by hot HCl. Remaining minerals were removed by gravity separation in zink chloride ($ZnCl_2$) solution with a specific gravity of 1.9. The organic matter was faintly oxidized with "Schulze's solution" for about 1 minute followed by treatment with 5% potassium hydroxide (KOH), and sieving with a 10 µm nylon filter.

The prepared slides were studied under a bright field microscope using x10 eyepieces combined with a x10 objective for searching, and a x100 oil immersion objective for determining the palynomorphs. Six samples from Rókhagi and the single sample from Heygsmúli were selected for quantitative study, with a minimum count of 200 specimens per sample.

## General composition of the microflora

The microflora from Rókhagi comprises thirty-five species and is characterized by *Momipites/Caryapollenites* pollen which in the different samples form 9-29% of the total flora. Disaccate pollen constitute only 2-5%. The frequency of alete pollen is normal for a Tertiary microflora. This group is mainly represented by *Inaperturopollenites hiatus* (Potonié) Thomson and Pflug which comprises 29-48 percent, and *Sequoiapollenites paleocenicus* Stanley comprising 1-12 percent.

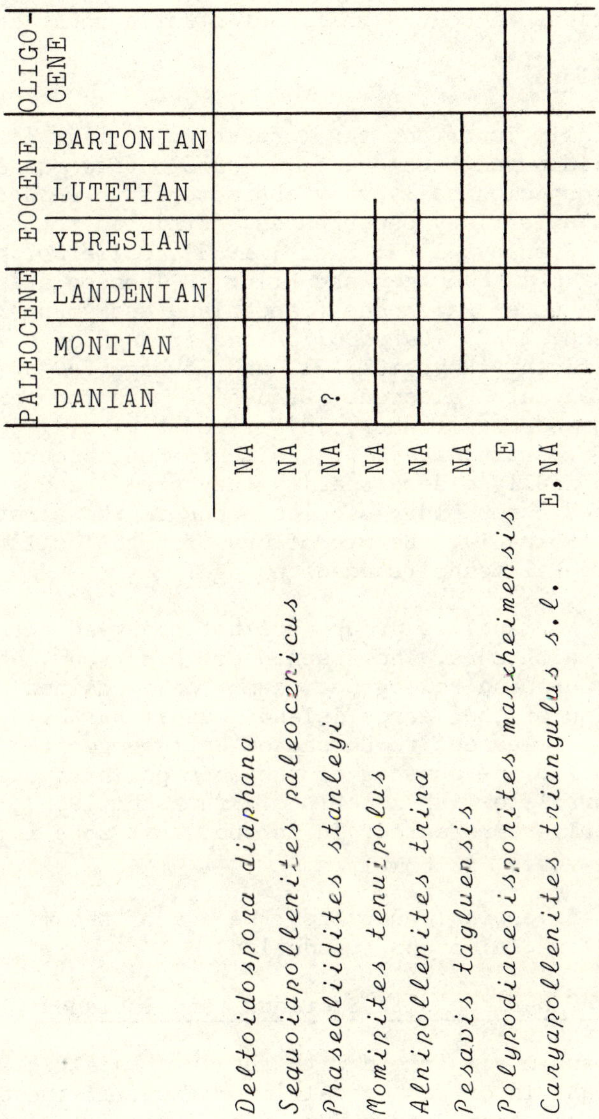


Fig. 1. Selected palynomorphs from the interbasaltic coals on the Faeroe Islands showing previous known range and geographic distribution. NA: North America, E: Europe.

The microflora from Heygsmúli is composed of eighteen species, sixteen of which are common to the Rókhagi microflora. *Momipites/ Caryapollenites* is the major constituent and amounts to 54 percent (23% *Momipites* + 31% *Caryapollenites*), whereas *I. hiatus* only constitutes 26 percent. The differences noted between the palynomorph assemblages from Rókhagi and Heygsmúli probably represent slight ecological variations without significant stratigraphic importance.

## Age of the microflora

Stratigraphically important taxa are shown in fig. 1. *Deltoidospora diaphana* Wilson and Webster, *Sequoiapollenites paleocenicus* Stanley and the morphologically very characteristic *Phaseoliidites stanleyi* Elsik have Tertiary occurrences restricted to the Paleocene of North America (Stanley, 1965; Elsik, 1968), while *Polypodiaceoisporites marxheimensis* (Mürriger and Pflug ex Thomson and Pflug) Krutzsch occurs in upper Paleocene (Landenian) and younger deposits of Europe (Krutzsch, 1967). *Caryapollenites triangulus s.l.* is numerically significant (Heygsmúli 31% and Rókhagi 5-6%) and also indicates an age of late Paleocene (Landenian) and younger (Krutzsch, 1966; Frederiksen and Christopher, 1978; Frederiksen, 1980). For the reasons that *C. triangulus s.l.*, as mentioned, occurs frequently in the stratigraphically older locality (Heygsmúli) and because the first mentioned three older species occur in the stratigraphical younger locality (Rókhagi), the microflora of both localities is indicative of late Paleocene (Landenian) age.

Pollen of the Normapolles group were not observed although they were searched for with care. The absence of this group, which is wellknown in the European Paleogene, cannot be explained by the northernly location of the Faeroe Islands, as it has also been described from the Paleocene to Eocene of Spitsbergen (Manum, 1962). A possible reason for the absence of the Normapolles group is that these sections correlate with zone 10 of Krutzsch (1966), in which this group is poorly represented. In Europe, this zone is restricted to the late Paleocene.

Based on the above considerations, the age of the microflora is very probably late Paleocene (Landenian).

## Biostratigraphic comparison with Scotland (and Iceland)

A biostratigraphic comparison with the interbasaltic coals from Scotland and Iceland is of geological importance. If the ages become progressively younger from Scotland to the Faeroe Islands and to Iceland, then these datings support the sea-floor spreading theory for the origin of the North Atlantic Ocean.

On Iceland, interbasaltic coal-bearing sediments are known from Tröllatunga. The age of this deposit is disputed; Pflug (1956)

<table>
<tr><th colspan="2"></th><th>NW</th><th></th><th>SE</th></tr>
<tr><td colspan="2">PLIOCENE</td><td rowspan="3">ICELAND</td><td rowspan="3"></td><td rowspan="3"></td></tr>
<tr><td colspan="2">MIOCENE</td></tr>
<tr><td colspan="2">OLIGOCENE</td></tr>
<tr><td colspan="2">EOCENE</td><td></td><td></td><td></td></tr>
<tr><td rowspan="3">PALEO-CENE</td><td>LANDENIAN</td><td></td><td>FAEROE ISLANDS</td><td></td></tr>
<tr><td>MONTIAN</td><td></td><td></td><td></td></tr>
<tr><td>DANIAN</td><td rowspan="2"></td><td rowspan="2"></td><td rowspan="2">MULL, SCOTLAND</td></tr>
<tr><td>CRET.</td><td>MAASTR.</td></tr>
</table>

Fig. 2. Palynological correlation of interbasaltic coals from Scotland, Faeroe Islands, and Iceland.

advocated an approximate Eocene age whereas Friedrich and Símonarson (1975) considered interbasaltic plant-bearing localities on Iceland to be of Miocene to Pliocene age. Pflug (l.c.) based his dating mainly on the presence of pollen of the Normapolles group. However, Manum (1962) found by re-investigation of the Tröllatuga material that Normapolles were absent and suggested that Pflug's Normapolles were poorly preserved betuloid pollen. Manum (l.c.) also noted that disaccate pollen occurred frequently making up nearly half of the specimens; this indicates an Oligocene or younger age. *Cicatricosisporites* was noted by Pflug (l.c.) indicating a pre-late Miocene age. An Oligocene to middle Miocene age is thus indicated by the microflora described by Pflug (1956) and Manum (1962). A re-investigation with use of the modern literature would probably yield a more precise age. Whatever the exact age, the Icelandic microflora is younger than that of the Faeroe Islands.

From Scotland, a microflora from interbasaltic coals of the Mull locality was described by Srivastava (1975). It contains several species of *Aquilapollenites*, which was the main criterion for Srivastava to date this microflora as Maastrichtian. It is now known that *Aquilapollenites* also occurs in the Danian (Vakhrameev et al., 1978), which widens the possible age range of the Mull microflora to Maastrichtian to Danian but still definitely older than the interbasaltic coals on the Faeroe Islands.

The biostratigraphic comparisons (fig. 2) thus support the sea-floor spreading theory for the origin of the North Atlantic Ocean.

Acknowledgements

This study was carried out at the Geological Institute, University of Aarhus, Denmark, under supervision of lektor K. Raunsgaaard Pedersen of the mentioned institution, who is thanked for providing the necessary facilities. Dr D. J. Jutson, Deutsche Texaco, Wietze is thanked for linguistic support and Mrs. L. Eklund Jürgensen, Aarhus is thanked for typing the manuscript.

## References

Curry, D., Adams, C. G., Boulter, M. C., Dilley, F. C., Eames, F. E., Funnel, B. M., and Wells, M. K., 1978, A correlation of Tertiary rocks in the British Isles, Geol. Soc. Lond., 12: 1-72.

Elsik, W. C., 1968, Palynology of a Paleocene Rockdale lignite, Milam County, Texas. II. Morphology and taxonomy (end), Pollen et spores, 10(3): 599-664.

Frederiksen, N. O., 1980, Paleogene Sporomorphs from South Carolina and Quantitative Correlations with the Gulf Coast, Palynology, 4: 125-179.

Frederiksen, N. O., and Christopher, R. A., 1978, Taxonomy and Biostratigraphy of Late Cretaceous and Paleogene Triatriate Pollen from South Carolina, Palynology, 2: 113-145.

Friedrich, W. L., and Símonarson, L. A., 1975: Islands Neogene flora, Dansk geol. Foren., Årsskrift for 1975: 41.

Krutzsch, W., 1966, Die sporenstratigraphische Gliederung des älteren Tertiär im nördlichen Mitteleuropa (Paläozän-Mitteloligozän). Methodische Grundlagen und gegenwärtiger Stand der Untersuchungen, Abh. zentr. geol. Inst., 8: 112-149.

Krutzsch, W., 1967, Atlas der mittel- und jungtertiären dispersen Sporen - und Pollen - sowie der Mikroplanktonformen des nördlichen Mitteleuropas. IV-V. Weitere azonotrilete (apiculate, murornate), zonotrilete, monolete und alete Sporenformen, VEB Deutscher Verlag der Wissenschaften, Berlin.

Laufeld, S., 1965, Sporomorphs in Tertiary coal from the Faeroe Islands, Geol. Fören. Stockholm Förhandl., 87(1): 231-238.

Manum, S., 1962, Studies in the Tertiary flora of Spitsbergen, with notes on Tertiary floras of Ellesmere Islands, Greenland, and Iceland, Norsk Polarinst., 125: 1-127.

Pflug, H. D., 1956, Sporen und Pollen von Tröllatunga (Island) und ihre Stellung zu den pollenstratigraphischen Bildern Mitteleuropas, N. Jb. Geol. Paläont., 102(3): 409-430.

Rasmussen, J., 1963, Recent studies in the geology of the Faeroes, in: North Atlantic Biota and their History, A. Löve, and D. Löve, eds., Pergamon Press, Oxford.

Rasmussen, J., and Koch, B. E., 1963, Fossil Metasequoia from Mikines, Faroe Islands, Ann. Soc. Sci. Faeroensis, 12: 83-96.

Rasmussen, J., and Noe-Nygaard, A., 1969, Beskrivelse til Geologisk kort over Færøerne, Dan. geol. Unders., I. Rk. 24: 1-370.

Srivastava, S. K., 1975, Maastrichtian microspore assemblages from the interbasaltic lignites of Mull, Scotland, Palaeontographica Abt. B, 150(5-6): 125-156.

Stanley, E. A., 1965, Upper Cretaceous and Paleocene plant microfossils and Paleocene Dinoflagellates and Hystrichosphaerids from northwestern South Dakota, Bull. Amer. Paleont., 49 (222): 179-384.

Tarling, D. H., and Gale, N. H., 1968, Isotopic dating and palaeomagnetic polarity in the Faroe Islands, Nature 218: 1043-1044.

Vakhrameev, V. A., Dobruskina, I. A., Meyen, S. V., and Zaklinskaja, E. D., 1978, Paläozoische und mesozoische Floren Eurasiens und die Phytogeographie dieser Zeit, VEB Gustav Fischer Verlag, Jena.

# GEOLOGY OF THE GREENLAND-ICELAND RIDGE IN THE DENMARK STRAIT

Birger Larsen

Institute for Applied Geology
Technical University of Denmark
DK 2800 Lyngby, Denmark

## INTRODUCTION

The continental shelf of East Greenland and the insular shelf of Iceland almost merge in the Denmark Strait, forming the Greenland-Iceland Ridge. This is the western part of the Scotland-Iceland-Greenland aseismic ridge, which is an important oceanographic threshold between the Norwegian Sea and the Atlantic Ocean. Roughly half of the inflow of cold arctic water into the Atlantic takes place through the Denmark Strait (Coachman and Aagaard, 1974). This flow is closely related to the warm Norwegian Stream entering the Norwegian Sea south of Iceland. Changes in the cross section in the Denmark Strait may thus imply important changes in oceanography and climate of northern Europe and Greenland (Eldholm and Thiede, 1980). The Scotland-Iceland-Greenland Ridge joins the lower Tertiary volcanic areas of East Greenland, Iceland, the Faeroes and Northwest Scotland (the Brito-Arctic volcanic province). This lead to the suggestion long ago that the ridge was a foundered landbridge of volcanic origin (Holmes, 1918). In the context of seafloor spreading the ridge has been regarded as the trace of the Iceland mantle plume (see Vogt et al., 1981 for references).

In spite of the geological interest of the Denmark Strait area very little data are available. This is chiefly due to the pack ice which covers the East Greenland shelf area for all but a few weeks of the year.

The purpose of this paper is to review the geology of the Denmark Strait area and to report the results of a marine-geological reconnaissance made by the Geological Survey of Greenland in 1974, 1975 and 1979. The magmatic evolution in the East Greenland basalt province and history of seafloor spreading on the East Greenland continental margin are dealt with by H.C.Larsen and Brooks (in preparation).

## BATHYMETRY

The Greenland-Iceland Ridge was discovered by the Danish Ingolf expedition in 1879, and the eastern ice free area was reasonably well sounded before 1900. The general morphology of western Denmark Strait was clarified by Norwegian fishery investigations in 1931-1933 (Iversen, 1936). The Greenland shelf 65°-66° N and the Denmark Strait Channel was throughly sounded by the German R/V "Meteor" in 1956. This provided the base for the German charts no 256 and 257 which are probably the most accurate published charts of the area. On the Greenland shelf the new bathymetric map (Fig. 1) is chiefly based on recent echo-soundings by the University of Durham, the Geological Survey of Greenland and the German Hydrographic Institute along lines positioned by satellite navigation or Loran (Fig. 2). The position of many of the older soundings has proved to be less accurate but these have been included if possible. The contours outside the shelf are based on the international Plotting Sheets nos 2 and 7 from the German Hydrographic Institute.

## MORPHOLOGY

The East Greenland shelf in the Denmark Strait is approximately 100 km wide at 68° N, north of the ridge, and approximately 150 km wide off Angmagssalik, south of the ridge, but broadens to 250 km on the ridge off Kangerdlugssuaq fjord. The shelf is well defined north and south of the ridge, but on the ridge between 66° N and 67° N no major break appears in the bathymetric profile. The water depths on the Greenland shelf are chiefly 200-400 m (Fig. 6, A and B). In contrast to this the depth of the Iceland shelf in the Denmark Strait is only 0-200 m and there is a marked shelf break. The insular shelf of Iceland is thus clearly distinguishable from the Greenland part of the ridge (Kristjansson, 1980). The Greenland-Iceland Ridge is traversed by the 20-30 km wide Denmark Strait Channel.

The morphology of the shelf is in many respects similar to other passive continental margins in previously glaciated areas (Shepard, 1948; Holtedahl and Sellevoll, 1972; Vogt et al., 1981). The most significant features are the very high rocky coasts cut by deep fjords or glacier-filled valleys which continue into deep submarine troughs. The outer shelf is characterized by broad trough like channels - transverse channels - extending almost across the shelf, and by banks. The largest transverse channel is the 350 km long and 50 km wide Kangerdlugssuaq Channel roughly at 31° W. Valleys extending along the coast - marginal channels - are found in the inner part of the shelf, for example northwest of the East Bank (67°30 N 30° W) and the Sermilik Deep south of Angmagssalik.

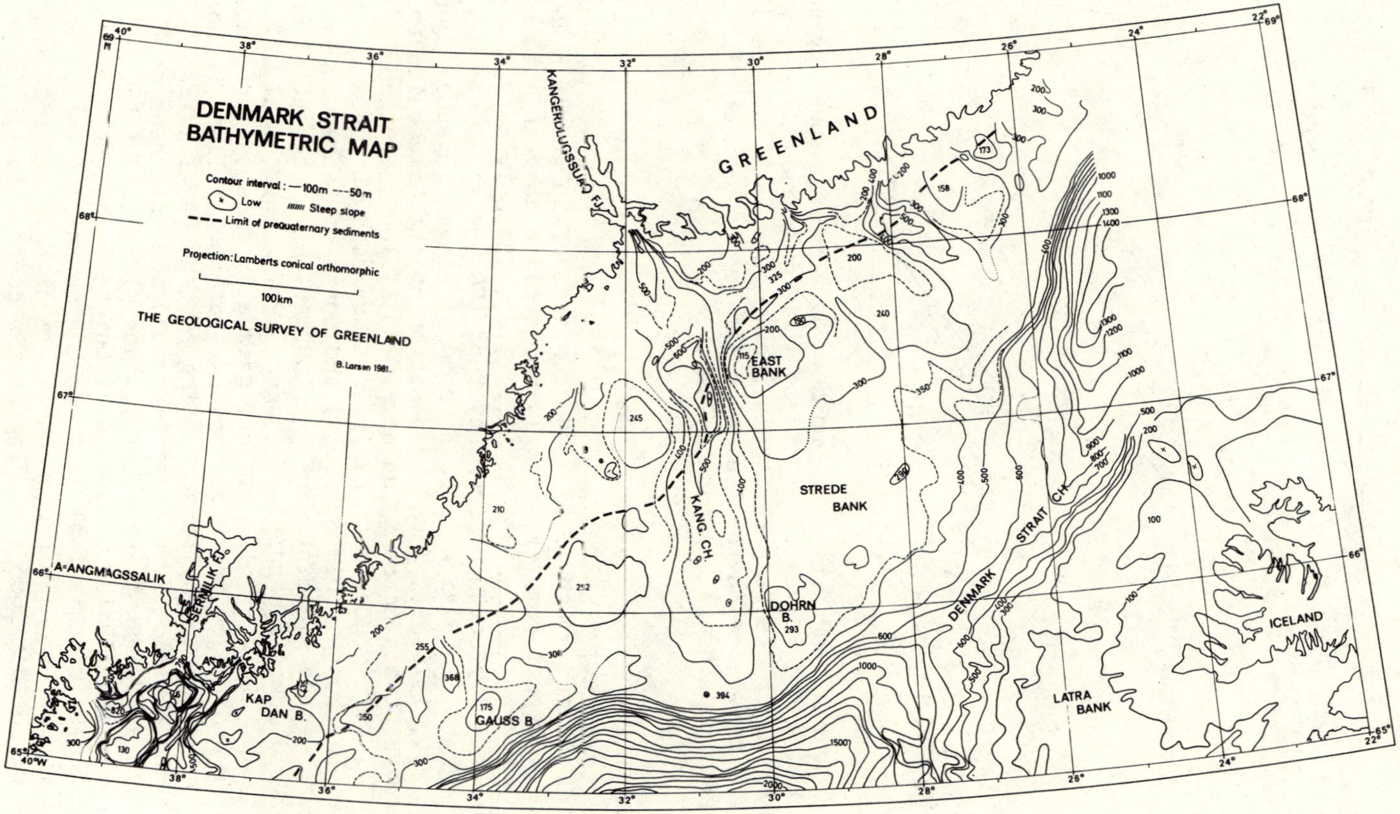


Fig. 1. Bathymetric map of the Denmark Strait area. Sources of data are given in the text.

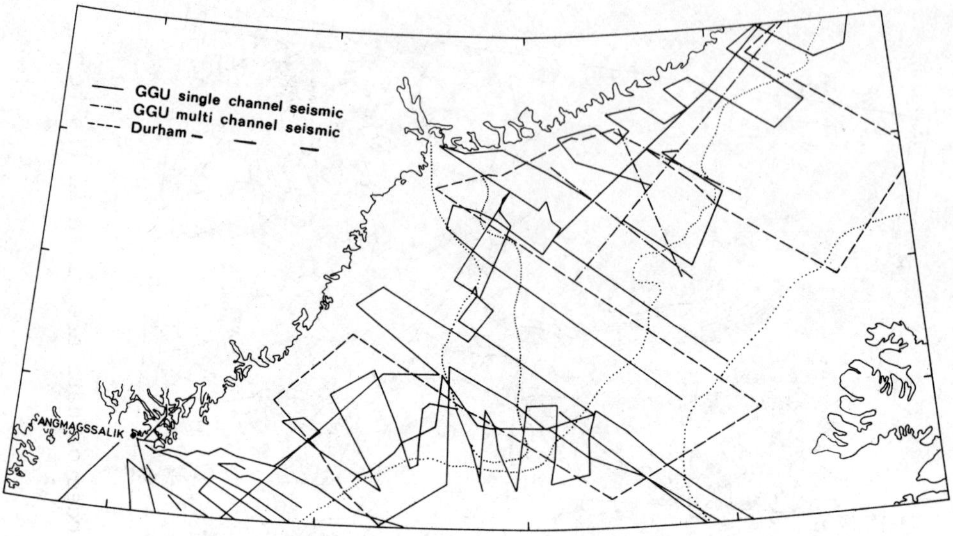


Fig. 2. Tracks of the echograph and seismic lines measured by the Geological Survey of Greenland (GGU) and the University of Durham in the Denmark Strait area. Additional seismic data are available on the Iceland shelf see Johnson et al. (1975) and Johnson and Palmason (1980).

## GEOLOGICAL SETTING

East Greenland south of 68°30 N consists of an Archaean and Proterozoic basement complex, chiefly amphibolite to granulite facies metamorphic rocks ( Bridgwater et al., 1978). The only evidence of Caledonian activity south of Scoresby Sund (70° N) is a deformed alkaline magmatic intrusion near the head of Kangerdlugssuaq fjord (Brooks et al., 1976 a). Thus the position of the Caledonian front at the Denmark Strait coast is unknown. Small local erosional remnants of late Cretaceous to early Tertiary sediments rest on the basement at Kap Gustav Holm (66°30 N) and north of Kangerdlugssuaq fjord. Most of the sediments are shallow marine to terrestial shales, siltstones and arkosic sandstones (Wager, 1947; Soper et al., 1976; Nielsen et al., 1981). The sediments gradually pass up into submarine tuffs and pillow breccias proving that the earliest Paleocene volcanism was submarine. The succeeding plateau basalts are almost exclusively subaerial. Near Kangerdlugssuaq the thickness of the pile has been estimated to be 7 km by Wager (1947) and 5 km by Nielsen and Brooks (1981). The basalts thin towards the north and west (Deer, 1976; Brooks et al., 1976 b). The basalts were erupted in late Paleocene to early Eocene time, immediately prior to the initiation of the seafloor spreading of the north Atlantic (Soper et al., 1976). Subsequently a swarm of coast-parallel dykes arranged

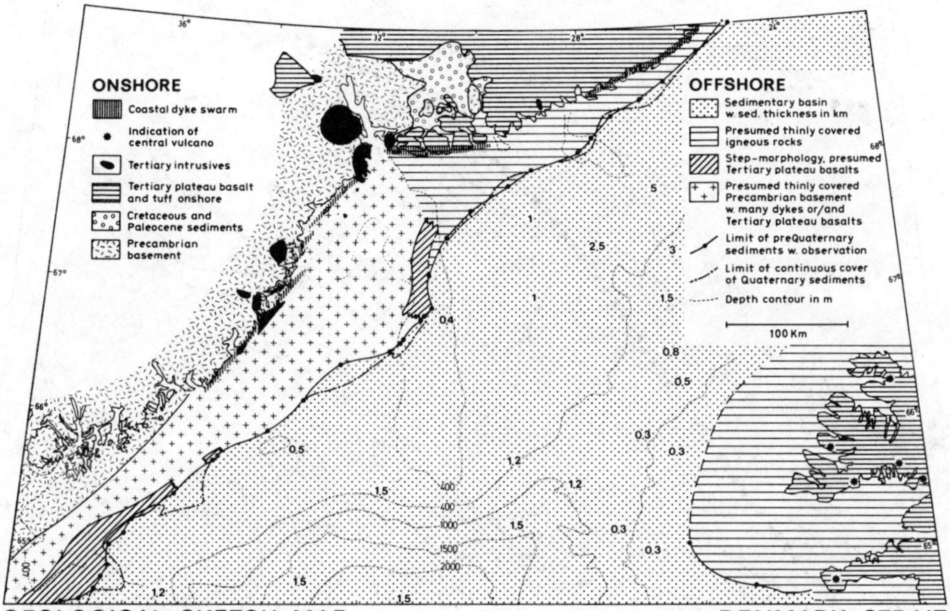


GEOLOGICAL SKETCH MAP DENMARK STRAIT

Fig. 3. Geological sketch map of the Denmark Strait area. Thickness of sediments based on seismic velocity of 2.5 km/s. Sources of information: Onshore Greenland: Bridgewater et al (1978); Iceland; Saemundsson (1980); offshore; B.Larsen (1980) and unpublished GGU data.

en echelon were intruded into the Precambrian basement and the plateau basalts between 66°15 N and 69°45 N (Wager and Deer, 1938; Myers, 1980). The intrusion of chiefly gabbroic and syenitic plutons post dates the dykes. After most of the igneous activity a marked oceanwards tilting of the coastal areas took place producing the so-called coastal flexure. This structure is interpreted as a monoclinal flexure by among others Wager and Deer (1938) and Myers (1980), and as a set of antithetic fault-blocks in a half-graben structure by Nielsen and Brooks (in press) and H.C.Larsen and Jacobsen (in press). The implications of this for the adjacent seafloor are discussed later. A few occurrences of shallow marine sediments of early Eocene and Oligocene age overlying basalts (Soper and Costa, 1976) suggest that the subsidence of the lava pile almost kept pace with the supply of volcanics (Brooks, 1973).

Northwest Iceland adjacent to the Denmark Strait consists of a thick pile of tuffs and plateau basalts extruded subaerially 16-7 m.y. ago (Miocene). The age decreases towards the southeast.

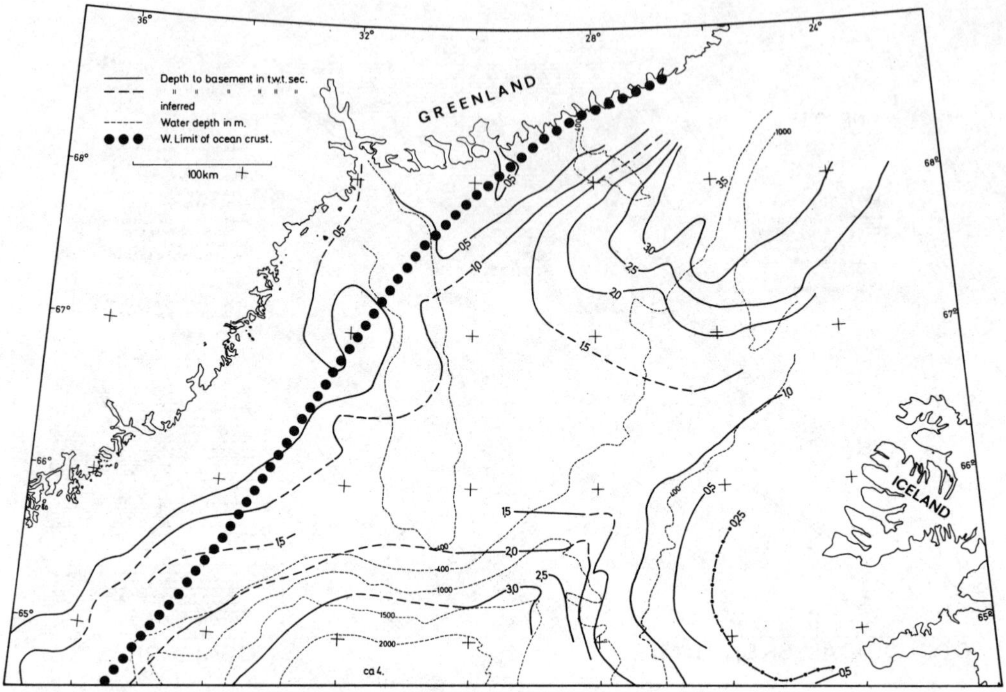


Fig. 4. Map of the depth to acoustic basement in two-way travel time below sealevel based on the GGU lines shown in Fig.2, Eldholm and Windisch (1974), Johnson and Palmason (1980), Hinz and Schlüter (1980) and Featherstone et al. (1977). 1 sec. corresponds roughly to 1 km in depth. The 1 sec contour along the Greenland side is chiefly based on observation at 0.8 sec extrapolated parallel with sediment reflectors in the profiles. In the deepest parts it is not excluded that the surface shown is not the basement and that sediments below this level may be present (Hinz and Schlüter, 1980; Featherstone et al., 1977, and J.Risum, GGU, pers. comm.). The western limit of oceanic crust is based on data in H.C.Larsen and Jacobsen (in press).

Occasionally the sequence includes central volcanoes (Saemundsson, 1980). Sequences of sediments of terrestrial or fresh water facies are interbedded with the basalts. Quaternary volcanic rocks are found at Snaefellsnes; these have not been marked on Fig. 3.

## GENERAL GEOLOGY OF THE RIDGE

The general geology of the Greenland-Iceland Ridge seems to be fairly simple as far as it is revealed by the available seismic sections (Featherstone et al., 1977; Hinz and Schlüter, 1980;

Johnson et al., 1975 a & b; Johnson and Palmason, 1980, B.Larsen, 1975, 1980 and H.C.Larsen, 1978. As far as is known the bedrock of the inner parts of the shelves comprises Precambrian basement or Tertiary igneous rocks (Fig. 3). Due to the uncertain distinction between the two types on the seismic records they will be referred to as "Acoustic basement". Seawards this basement is covered by sediments. Often this sequence is unconformably covered by a sedimentary sequence tentatively designated "Quaternary sediments", but no actual dating is available.

## THE ACOUSTIC BASEMENT

Fig. 4. is a preliminary map of the depth to the acoustic basement in the Denmark Strait region. Seismic data are lacking from the central part of the ridge, but a few depth estimates based on magnetic anomalies indicate 1 - 2 km to basement (H.C.Larsen, 1980, Fig. 1). From the map it is evident that the morphological Greenland-Iceland Ridge has a core in the form of a 1 - 2 km high basement ridge joining the Northwest Iceland with East Greenland between 66° and 68° N, which is south of the East Greenland plateau basalts. Towards north the ridge is limited by a deep sedimentary trough (Hinz and Schlüter, 1980; H.C.Larsen, 1980) while no trough seems to be present immediately to the south (B.Larsen, 1975; Featherstone et al., 1977). The deep sedimentary basins across the ridge, proposed by Henderson (1976) on the basis of very sparse magnetic evidence, have not been confirmed. In general the basement ridge is shaped like a saddle with comparatively flat basement shelves near the coasts, clearly distinguishable from the ridge and the adjoining high coastal areas of Greenland and Iceland. The sparse seismic lines suggest that the southern end of the Denmark Strait Channel is underlain by a graben structure. Possibly it is related to the scarps described by Johnson and Palmason (1980) along the base of the shelf of western Iceland.

It is widely believed that the North Atlantic Ocean was formed over the last 60 m.y. by seafloor spreading. This implies that a great part of the Greenland-Iceland Ridge was also formed by an ocean-floor spreading mechanism (Nunns, this volume) despite its atypical shallow depth (Vogt et al., 1981). Magnetic measurements on the ridge area published by Johnson et al (1975) and H.C.Larsen (1980) reveal a rather complex pattern of large semicircular to oblong anomalies, similar to that of the Faeroe-Iceland Ridge (Voppel et al., 1979). However, according to H.C.Larsen and Jacobsen (in press) it is possible on new high sensitivity data collected by Geological Survey of Greenland to recognize that the large local anomalies are superimposed on a pattern of linear anomalies of ocean floor type. Tentatively anomalies 24 through 20, 13 and 7 have been identified, some of which were also recognized by Voppel et al., (1979) and Vogt et al., (1980). Most of the Greenland-Iceland Ridge was thus formed by seafloor spreading between anomaly 24 and anomaly 6 times (circa 57 m.y. - 16 m.y.). On the Iceland shelf the large local magnetic anomalies are associated with local gravity anomalies. This is also the case on the Iceland-Faeroe Ridge (Kristjansson, 1976). Many major intrusions and volcanoes are known from the landareas adjacent to the Denmark Strait (Fig. 3). The large local magnetic anomalies on the Greenland-Iceland Ridge are therefore probably related to major intrusions and central volcanoes (Johnson et al., 1975 a; Thors and Kristjansson, 1974; Voppel et al., 1979).

Johnson and Palmason (1980) showed that the basement below Latragrund (Fig. 1) has velocities of 3.5 - 4.5 km/s corresponding to "older surface basalts" on Iceland. Thors and Kristjansson (1974) and Kristjansson (1976) suggests that the relative intense central volcano activity in the Snaefellsnes volcanic zone had at some time extended at least 120 km offshore.

It is very probable that the Greenland-Iceland Ridge is characterized by an abnormal thick crust like the better known parts of the Scotland-Iceland-Greenland Ridge (Bott, 1974).

## The inner shelf of East Greenland

The exact location of the transition from oceanic crust to the continental crust of East Greenland is not known. Several proposals have been implied in various models of the seafloor spreading geometry as summarized in H.C.Larsen (1980). The one shown on Fig. 4 is defined as the magnetic low landwards of magnetic anomaly 24 as identified by H.C.Larsen and Jacobsen (in press). This is the nearest to the coast of the proposed positions. According to the models of the coastal flexure between 70° - 67° N of Wager and Deer (1938), Myers (1980), and Nielsen and Brooks (in press) we can expect that the adjacent shelf chiefly consists of plateau basalts. The pattern of magnetic anomalies on the inner shelf at 67° - 68° N is not markedly different from that on the oceanic part of the ridge, which is consistent with a high-magnetic basement. South of 67° N and W of 33° W (see Fig. 3) there is a belt of high-amplitude-short-wavelength magnetic anomalies which Vogt (1970); H.C.Larsen (1978), and Tychsen (1977) interpreted as indications of dyke swarms intruded into a possible metamorphic basement. Parts of the swarm could be considered as a continuation of the coastal dyke swarm (H.C.Larsen, 1978). According to H.C.Larsen the intrusion of the dykes corresponds to a crustal dilatation of the order of 50%.

A characteristic step-like morphology has been observed in two zones just west of the western margin of the sediments on the shelf off Angmargssalik and Kangerdlugssuaq (see map Fig. 3). In these areas the bottom looks like a slightly tilted staircase, with gently seaward sloping plateaus limited by steep scarps towards west (shorewards, Fig. 5). The height of each scarp ranges from 10 to 50 m. No internal reflectors have been observed in this area. Similar steps on the shelf further south have been interpreted as ice- eroded "roches moutonnées" by Sommerhoff (1973), but the ice has to come from the ocean side if this is correct. The steps could possibly reflect a fault pattern activated either in connection

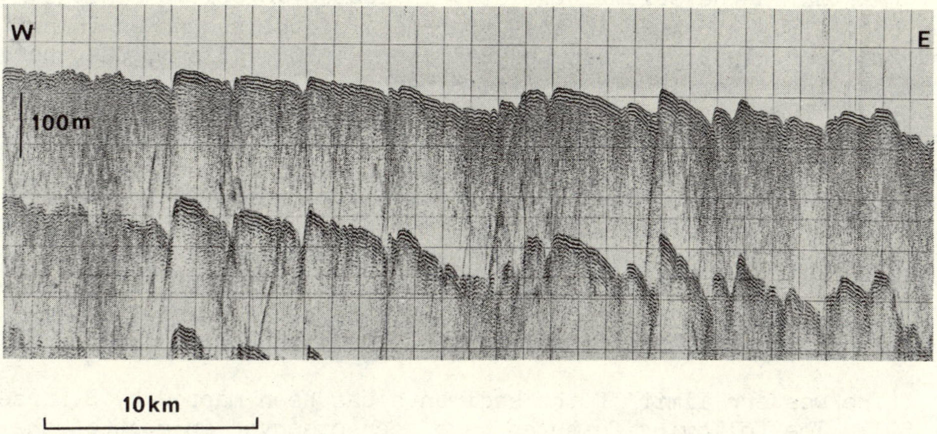


Fig. 5. Seismic profile (10 cu inch Airgun) showing step-morphology at circa 64°30 N, 36°45 N, close to the western limit of sediments. The limit of the step-morphology is seen circa 2 cm from the left of the figure. The repetition of the upper trace is a multiple reflection.

with the inital rifting of the crust or the subsidence of the sedimentary trough. The orientation of the steps is not consistent with normal faults on the western half-graben of a rift. It resembles exactly the surface appearance of the antithetic fault system proposed by Nielsen and Brooks (in press) as the structure of the coastal flexure and the adjoining shelf. However, neither steps nor faulting have been observed beneath or within the adjoining sediments, so faulting is hardly the explanation. Finally the step-morphology could be caused by erosion of a slightly tilted sequence of alternating hard and soft layers. This, combined with the magnetic evidence, suggests that the step-morphology is due to the erosion of a tilted sequence of plateau basalts, which is in agreement with the ideas of H.C.Larsen (1980, Fig. 2). The extension of the possible plateau basalts on the shelf is unknown, but it is smaller than the zone of high-amplitude magnetic anomalies mentioned above. The plateau basalts clearly underlie the sediments. Provided that the position of the transition from oceanic to continental crust on Fig. 4 is correct, the northern occurrence of plateau basalts on the shelf must overlie or be a part of the oceanic crust of Eocene age. This suggests that these basalts are younger than the East Greenland plateau basalts onshore.

## THE SEDIMENTS

The central two-thirds of the ridge is covered by a rather thick sequence of sediments while this cover seems to be thin and discontinuous on the inner parts of the shelves. The sparse information on the seismic velocity indicates velocities in the range of 2.5 - 3.5 km/s in the upper part of the sequences (Hinz and Schlüter, 1979; Johnson and Palmason, 1980; J.Risum, GGU, pers. comm.). Thus the figures in Fig. 3 based on 2.5 km/s are minimum figures. The sequence of sediments on the outer Greenland shelf ($65^{o}30$ N - $67^{o}30$ N) and on the southern flank of the ridge seems to be 1 - 2 km thick. To the north the ridge sediments merge with the up to 8 km thick sequence in the sedimentary trough on the shelf north of $67^{o}30$ N. (Hinz and Schlüter, 1979; unpublished GGU data). The thickness on the insular slope and outer shelf of Northwest Iceland and in the Denmark Strait Channel is of the order of 300 - 600 m, corresponding to a prograding of the shelf edge of 20 km (Palmason, 1974; Johnson and Palmason, 1980; unpublished GGU data).

The western limit of the sediments has been mapped by B.Larsen (1980). The following features have been observed in most of the seismic sections (see Fig. 6). The top of the acoustic basement is gently inclined (1 - $2^{o}$) towards southeast. The basement surface is conformably, or nearly so, overlain by a well bedded sequence of sediments. The western limit is erosional. Faults have not been observed along the margin. No influence of the Greenland-Iceland Ridge on the appearance of the margin has been recognized, except that the base of the sediments is slightly less inclined on the crest of the ridge. It is thus indicated that the margin of the sediments on the ridge has been gently warped downwards at nearly the same rate as the rest of the shelf.

The northern flank of the basement ridge is covered by an on-lapping sequence of sediments, indicating a gradual burial of the flank. (J.Risum and H.C.Larsen, pers.comm.). Not much is known about the deeper structure of the sediments on the ridge. The available data suggest that the sediments were deposited in a succession of lenticular bodies, gradually building the Greenland shelf upwards and outwards (prograding) towards the south and southeast (C.J.Uruski, pers.comm., and Fig. 6 and 7 and unpublished GGU data).

The sedimentation pattern is better known in the southern part of the ridge (S of $66^{o}$ N). The acoustic units a and b rest on a regional unconformity (see Fig. 6 C and D). They can be traced from the shelf into the continental slope. Consequently, the units prove a build-up and a prograding of the shelf. In a zone along the shelf edge b rests on a characteristic belt of strong reflectors in the form of an obtuse angles (Fig. 7). I interpret these as indication of palaeo-shelf breaks. The unit c lacks "topsets". It provides therefore evidence of an episode of erosion or more probably prograding only. Unit c is underlain by unit d, which is of the same type as a and b.

## The age of the sediments

No biostratigraphical dating of the sediments is available. The regional unconformity mentioned above marks an episode of

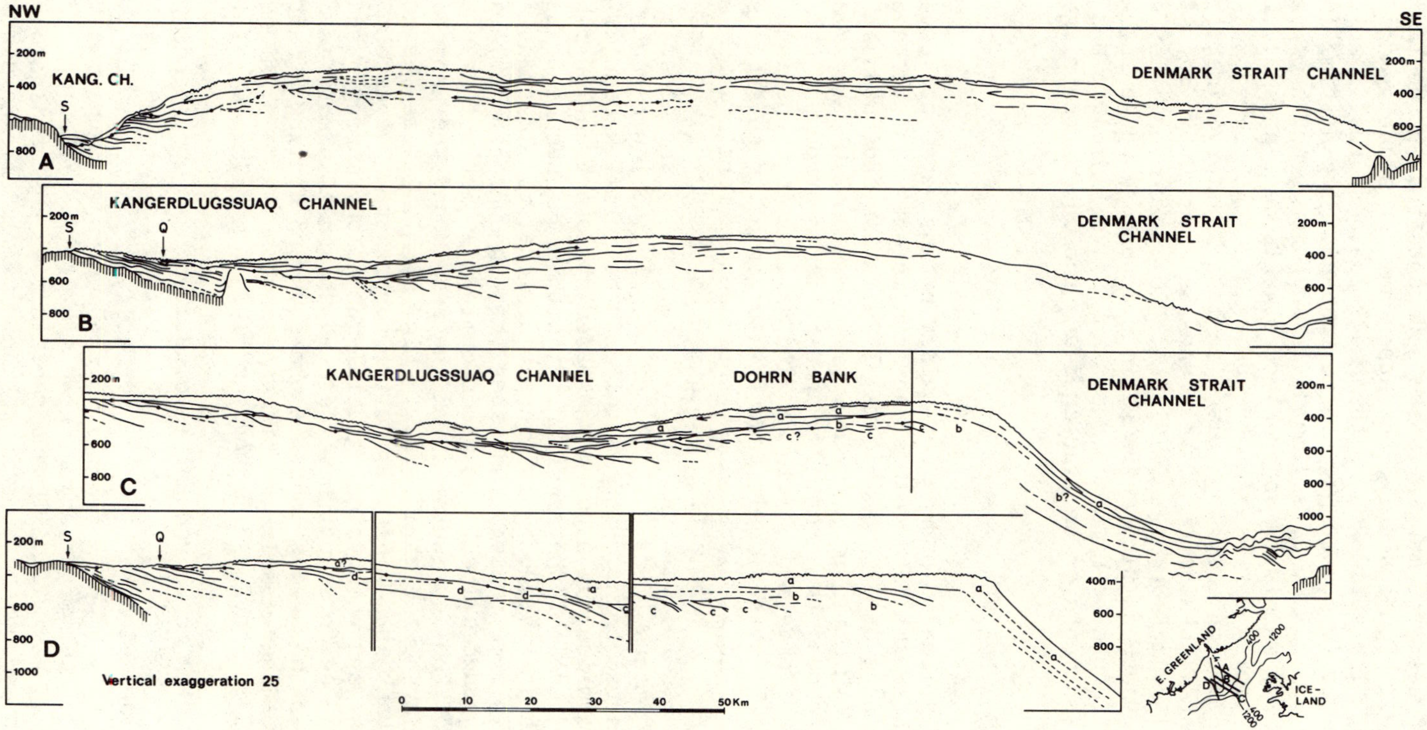


Fig. 6. Single channel seismic section across the East Greenland shelf in the Denmark Strait. The depth axis is based on a sound velocity of 1500 m/s. The acoustic basement is vertically hatched. The representation of the unevenness due to ice-scouring is not exactly to scale. The small letters refer to the seismic unit mentioned in the text.

glacial erosion near the margin of the sediments and in the Kangerdlugssuaq Channel. Consequently, the upper units a and b are of Quaternary age. The sediments on the ridge postdate the oceanic basement which was formed during Eocene to early Miocene time (57 - 16 m.y.). The upper 3 km of sediments in the trough north of the ridge are probably mainly of upper Miocene - Pliocene age (H.C.Larsen, 1980). The onlapping of these on the basement ridge suggests that burial of the ridge chiefly took place in late Tertiary time, following the general subsidence of the area as assumed by Vogt et al. (1980); Bott (1974); Eldholm and Thiede (1980). However, the few occurrences of postbasaltic marine sediments onshore range in age from the early Eocene to Miocene (Soper and Costa, 1976). Thus it is probable that sediments of similar ages are present on the ridge as well.

## GLACIATION OF THE SHELVES

The clearly ice-sculptured coastal mountains were once covered with ice, which in the Kangerdlugssuaq area is estimated to have reached a thickness of 1200 m (Brooks, 1979): this indicates a former extension of the Inland Ice onto the shelf (Weidick, 1976). This ice could have eroded the deep, U-shaped submarine troughs that extend from the fjords and glacier-filled valleys into the inner basement shelf. It is noticable that the larger of these troughs are all situated off the glaciers which, based on the form of the recent ice surface, are the most active outlets from the Greenland ice sheet to the Denmark Strait. The huge transverse channels are clearly eroded down in the sediments (Fig. 6, A, B and C). In the inner parts of the Kangerdlugssuaq transverse channel and the Sermilik Channel ($64^{o}30$ N) the erosion has exposed the basement more than 700 m below sealevel. The level of the unconformity within the sediments has a general seawards rise, but with irregular local culminations and depressions. This morphology indicates that the unconformity is due to erosion by grounded ice. The unconformity is distinct near the western limit of the sediments and in the area round the Kangerdlugssuaq Channel near to the shelf break, (Fig. 6, D). It is less clear in the Strede Bank area, (Fig. 1) to the northeast. The extent of the unconformity and the typical glacial morphology strongly indicate that a grounded ice once must have covered the shelf almost as far as the present shelf break, at least in the southern Denmark Strait and on the shelf off Angmagssalik. This agrees with the conclusions of, among others, Sommerhoff (1973); Brooks (1979), and Vogt et al.(1980).

A 1200 m thick ice cover on the coastal areas corresponds to an extension of the Greenland ice sheet to app. 80 km off the coast, assuming that the height of the ice surface was controlled by steady-state ice flow (cf. Paterson, 1969). This agrees with the maximum extension of the ice margin shown in Weidick (1976). However, the shelf is up to 250 km wide. This discrepancy in the conclusions can be due either to an underestimation of the maximum thickness of ice by Brooks (1979), or to the possibility that the ice sheet on the

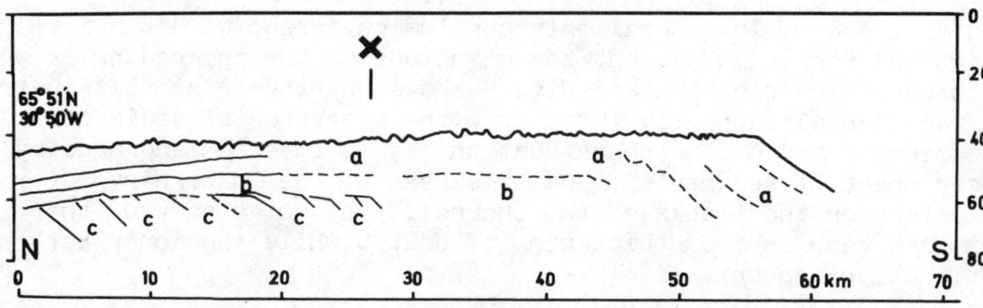


Fig. 7. Seismic section across the mouth of the Kangerdlugssuaq Channel. Note the reflections in form of obtuse angles - unit c - with arms parallel to the shelf surface and the continental slope respectively, interpreted as indications of palaeoshelf-breaks. The shelf has prograded from X to the present shelf-break in Quaternary times.

shelf was to some extent independent of the Greenland Inland Ice. Further investigations are needed on this question.

Recently Vogt et al. (1980) and Vogt et al. (1981) have reviewed the morphology and sedimentation on the Iceland shelf. They suggest that the shelf has been covered with grounded ice out to depths between 200 and 400 m, at least off the deep transverse channels crossing the shelf. Of special importance in this context is the occurrence of a 100 km long, 20 - 30 m high ridge of probably moraine origin on the Iceland shelf on the outer Latra Bank (65 - 66° N) at depths of 200 - 250 m. The ridge is considered to represent the maximum advance of the Weichselian ice sheet in the area (Olafsdottir, 1975). The ridge lies at or very near the shelf break which implies that the underlying sediments predate the glaciation and that very little prograding has taken place since.

## QUATERNARY SEDIMENTS

Often the unconformity on the Greenland shelf is covered by a 100 ± 50 m thick sequence of sediments which must be of Quaternary age. The western limit of the thick and continuous cover is always situated close to the western limit of the underlying sediments (see Fig. 3). This relation has been observed on the East Greenland shelf from Kap Farvel to 72° N (B.Larsen, 1980) and on the Norwegian shelf as well (Holtedahl and Sellevoll, 1972). The significance of this will be discussed elsewhere. The seismic units a and b mentioned above and shown in Fig. 6, C and B belong to the Quaternary cover. Probably the southernmost occurrence of unit c represents the shelf break at the time of the formation of the unconformity (Fig. 7). This palaeo-shelf break has been mapped on the shelf between 33° W and 28° W. From this it is concluded that the shelf has prograded approximately 30 km towards south at the mouth of the Kangerdlugssuaq Channel and approximately 5 km off the Dohrn Bank during the formation of units a and b. In general this conclusion agrees with the ideas of Vogt and Perry (1978), but the magnitude of the prograding is only one third of their estimate. Fig. 6 shows how the erosional Kangerdlugssuaq Channel is partly infilled with a sequence of sediments. In sections B and C it appears that unit a is covered landwards by a younger sheet of sediments. The thickness of the Quaternary cover is greatest on the flanks of the channel. The outer part of the Dohrn Bank results from thickening of unit b while the inner part chiefly comprises unit a.

Unit a and possibly b pass into a sequence of well bedded, acoustically tranparent sediments at the bottom of the Denmark Strait Channel. The surface of the transparent sequence is characterized by large mudwaves. From the subbottom reflections (Fig. 6, C) it appears that the position of the mudwaves is related to the position of the foot of the continental slope. Furthermore, the sedimentation pattern has not changed very much during the sedimentation of the unit a and possibly b. The mudwaves are most probably genetically related to strong bottom currents in the channel (cf. Coachman and Aagaard, 1974; Robert, this volume). Consequently the observations above indicate that strong bottom currents have flowed in Denmark Strait Channel for at least a major part of the Quaternary.

## THE RECENT SHELF

For a summary of the sparse knowledge on the surface sediments in the Denmark Strait area the reader is referred to Eisma and van der Graast (this volume) and the International Quaternary Map of Europa, Sheet 1 (UNESCO and INQUA). Only a few observations are mentioned here. The Iceland Shelf within the 200 m contour is chiefly covered by sand and gravel, while the surface of the Greenland shelf as far as is known mainly comprises glacial-marine tills. "Postglacial" clay deposits, recognized as acoustically transparent accumulations on the glacial surface, are remarkably rare on the Greenland shelf (65-68$^{\circ}$ N). They are found in the fjords and fjord extensions and in a few deep local depressions in the transverse channels. However, much recent sediment may go undetected on the very uneven bottom in most of the area. The surface of the shelf is extensively scoured by icebergs in depths down to 350 m (Fig. 8) and locally down to 650 m depending on the depth conditions in the possible drifting routes of the larger icebergs. Most of the surface of the banks and parts of the channel-floors are completely covered by scour-marks and related features, and the bottom material is more or less disturbed and mixed in the upper 5-10 m of the seafloor. Buried ice-scoured surfaces have not been recognized but they are possibly not seismic reflectors.

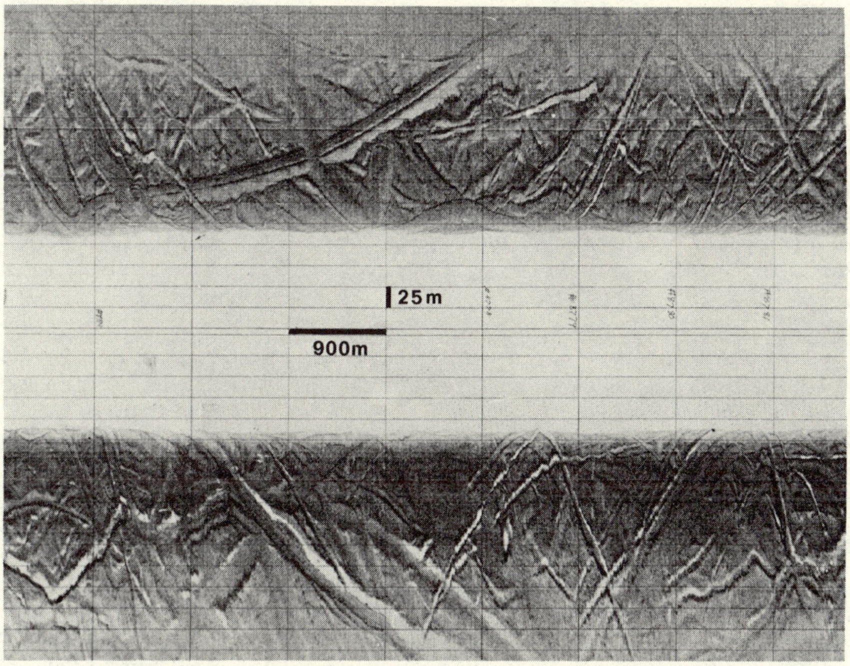


Fig. 8. Side scan record from the Angmagssalik shelf (65° N, 35°30 W) showing extensive ice-scouring. The depth of the bank is approx. 225 m. The side scan fish were towed circa 110 m above the seafloor.

## CONCLUSION

The continental shelf of East Greenland and the insular shelf of Iceland almost merge in the Denmark Strait, forming the Greenland-Iceland Ridge. This anomalously shallow part of the Atlantic Ocean has a core in the form of a 1-2 km high basement ridge, with the crest 1-2 km below sealevel. Some of the magnetic anomalies 24-6 have been traced tentatively across the ridge. Accordingly, the ridge formed by some special type of seafloor accretion in the time interval 57 m.y. - 16 m.y. Local large magnetic anomalies suggest the presence of many intrusive bodies in the ridge. The position of the ocean - continental transition off East Greenland is not known. It may be situated very close or within the coast north of 68° N. It is very likely that igneous rocks in the form of plateau basalts, dykes and plutonic centres are also common on the inner shelf south of the East Greenland basalt province.

In pre-drift reconstructions the southernmost of these volcanics would fit with the Faeroes (cf. Vogt and Avery, 1974, fig. 13).

The central two-thirds of the ridge is covered by 1-2 km of sediments. Most of the sediments are probably of late Miocene to Quaternary age, but older Tertiary sediments may be present. The structure shows that most of these sediments were deposited on the Greenland part of the ridge, while the prograding of the Iceland shelf onto the ridge hardly differs from that in adjacent shelf areas. Vogt and Perry (1978) suggested that the delta-like salient on the East Greenland shelf which forms the ridge is due to a specially high sediment supply to that part of the shelf, especially in Quaternary times. This is to some extent supported by the present data but the effect of spreading the available sediments onto a relatively shallow floor must also be of importance.

On the basis of a regional erosional unconformity and the general morphology it is concluded that a grounded ice sheet covered the shelf almost to the shelf break. The distribution of sediments above the unconformity suggests that an ice cover developed more than once. The icecaps on the shelves and the associated eustatic lowering of the sealevel would cause a very marked decrease of the cross section of the Denmark Strait. These glacial episodes are probably contemporaneous with ice covers to near the shelf break which occurred off Iceland (Vogt et al., 1980), the Faeroes (Waagstein and Rasmussen, 1975) and northern Europe (Andersen, 1979). See Fig. 9. This would seriously impede the oceanographically and climatologically very important exchange of water and heat between

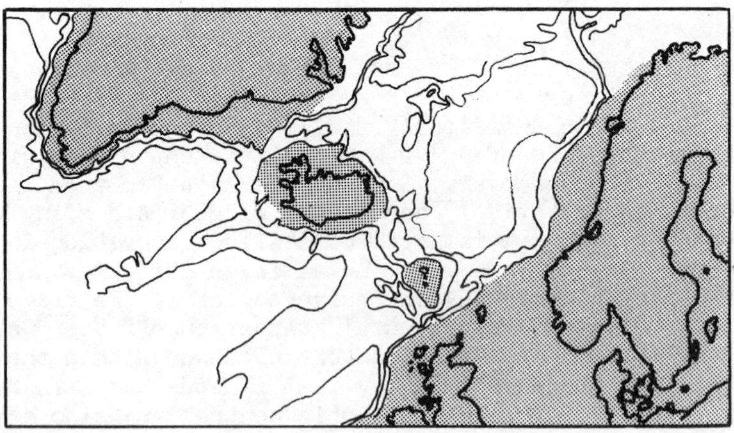

Fig. 9. Sketch of the maximum extension of the ice cover on the Scotland-Greenland Ridge. Depth contours at 500, 1000 and 2000 m. Data from Vogt et al. (1980), Waagstein and Rasmussen (1975) and Andersen (1979).

the Atlantic and the Norwegian Sea. However, some evidence suggests that at least some flow did take place in the deep Denmark Strait Channel contemporaneously with the deposition of most of the Quaternary glacial sediments on the adjacent Greenland shelf. Due to the lack of precise datings the correlations with the palaeo-oceanography of the Norwegian Sea is still an open question.

## ACKNOWLEDGEMENTS

The author wishes to acknowledge for the able assistance of the crews and technical staffs on R/V Tycho Brahe and M/V Dana whose skillful work under rough conditions made this investigations possible. The keen interest and careful assistance of many colleagues at the Geological Survey of Greenland and the Institute for Applied Geology, the Technical University of Denmark, are gratefully acknowledged. Mr. T.C.R.Pulvertaft kindly improved the language. I wish to thank the Danish Natural Science Research Council and the Danish Ministry for Energi for funding the project.

The paper is published with the permission of the Director of the Geological Survey of Greenland.

## REFERENCES

Andersen, B.G., 1979, The deglaciation of Norway 15000-10000 BP, Boreas, 8: 79-87.

Bott, M.H.P., 1974, Deep Structure, evolution and origin of the Icelandic Transverse Ridge, in: "Geodynamics of Iceland and the North Atlantic area", L.Kristjansson, ed., NATO Advanced Study Inst., Ser. C, Math.Phys., Sci. 33-47, D.Reidel Publ. Co., Boston.

Bridgwater, D., Davies, F.B., Gill, R.C.O., Gorman, B.E., Myers, J.S., Pedersen, S., and Taylor, P., 1978, Precambrian and Tertiary geology between Kangerdlugssuaq and Angmagssalik, East Greenland, Rep. Geol. Survey of Greenland, 83: 1-17.

Brooks, C.K., 1973, Rifting and Doming in Southern East Greenland, Nature, 244: 23-25.

Brooks, C.K., 1979: Geomorphical observations at Kangerdlugssuaq, East Greenland, Medd. om Grønland, Geoscience, 1: 1-24.

Brooks, C.K., Fawcett, J.J., and Gittins, J., 1976 a, Caledonian magmatic activity in south-eastern Greenland, Nature, 260: 694-696.

Brooks, C.K., Nielsen, T.F.D., and Petersen, T.S., 1976 b, The Blosseville Coast basalts of East Greenland: Their occurrence, composition and temporal variations, Contr. Miner. and Petrol., 58: 279-292.

Coachman, L.K. and Aagaard, K., 1974, Physical Oceanography of Arctic and Subarctic Seas, in: "Marine Geology and Oceanography of the Arctic Seas", 1-72, Y.Herman, ed., Springer-Verlag, Berlin.

Deer, W.A., 1976, Tertiary igneous rocks between Scoresby Sund and Kap Gustav Holm, East Greenland, in: "Geology of Greenland", A.Escher and W.S.Watt, ed., Geol. Surv. of Greenland, Copenhagen.

Eisma, D. and van der Gaast, S.J., in press, Terrigenous Late Quaternary Sediment components north and south of the Scotland-Greenland Ridge and the Norwegian Sea, this volume.

Eldholm, O. and Thiede, J., 1980, Cenozoic Continental Separation between Europe and Greenland, Palaeogeography, Palaeontology, Palaeoecology, 30: 243-259.

Eldholm, O. and Windisch, C.C., 1974, Sediment Distribution in the Norwegian-Greenland Sea, Geol. Soc. America Bull., 85: 1661-1676.

Featherstone, P.S., Bott, M.H.P., and Peacock, J.H., 1977, Structure of the Continental Margin of South-Eastern Greenland, Geophys.,J.R.Astron. Soc., 48: 15-27.

Henderson, G., 1976: Petroleum Geology, in: "Geology of Greenland", 487-505, A.Escher and W.S.Watt, ed., Geol. Surv. of Greenland, Copenhagen.

Hinz, K.v. and Schlüter, H.-U.,1980, The continental margin of East Greenland, Pro. Tenth. Wld. Petrol. Congr. Bukarest, 2: 405-418.

Holmes, A., 1918, The basaltic rocks of the Arctic region, Mineralog. Mag., 18: 180-223.

Holtedal, H. and Sellevoll, M., 1972, Notes on the influence on glaciation on the Norwegian continental shelf bordering on the Norwegian Sea, Ambio Special Report, 2: 31-38.

Iversen, T., 1936, Sydøstgrønland, Jan Mayen Fiskeriundersøkelser, Fiskeridirektoratets Skrifter Rep. Norwegian Fishery and Marine Investigarions V, no 1.

Johnson, G.L., McMillan, N.J., and Egloff, J., 1975 a, East Greenland Continental Margin, in: "Canada continental margins and offshore petroleum exploration", ed., C.J.Yorath, E.R.Parker and D.J.Glass, Mem.Can. Soc. petrol. Geol., 4: 205-224.

Johnson, G.L., Sommerhoff, G., and Egloff, J., 1975 b, Structure and Morphology of the West Reykjanes Basin and the Southeast Greenland Continental Margin, Marine Geologi, 18: 175-196.

Johnson, G.L. and Palmason, G., 1980, Observations of the Morphology and Structure of the Sea Floor South and West of Iceland, J.Geophys., 47: 23-30.

Kristjansson, L., 1976, Central volcanoes on the Western Icelandic Shelf, Mar. Geophys. Res., 2: 285-289.

Kristjansson, L., 1980, The Shelf area around Iceland, in: "Geology of the european countries", 132-135, Publ. by Comité Nat. Francais de Geologie, 26 Int. Geol. Congr., Graham and Trotman.

Kristjansson, L., Thors, K., and Karlsson, H.R., 1977, Confirmation of central volcanoes off the Icelandic coast, Nature, 268: 325-326.

Larsen, B., 1975, Marine geophysical survey of the East Greenland Shelf south of Angmagssalik, Rep. Geol. Surv. of Greenland, 75: 87-88.

Larsen, B., 1980, A marine geophysical survey of the continental shelf of East Greenland $60^{o}$-$71^{o}$ N - Project DANA 79, Rep. Geol. Surv. of Greenland, 100: 94-98.

Larsen, H.C., 1978, Offshore continuation of East Greenland dyke swarm and North Atlantic Ocean formation, Nature, 274: 220-223.

Larsen, H.C., 1980, Geological perspectives of the East Greenland continental margin, Bull. Geol. Soc. Denmark, 29: 77-101.

Larsen, H.C., and Jacobsen, N., in press, Paleogene Spreading along the Northern Reykjanes Ridge: New evidence from offshore East Greenland, Earth Planet. Sci. Lett.

Myers, J.S., 1980, Structure of the coastal dyke swarm and associated plutonic intrusions of East Greenland, Earth and Planetary Sci. Lett., 46: 407-418.

Nielsen, T.F.D., and Brooks, C.K., 1981, The E.Greenland rifted continental margin: an examination of the coastal flexure, J.Geol. Soc., London, 138, in press.

Nielsen, T.F.D., Soper, N.J., Brooks, C.K., Faller, A.M., Higgins, A.C., and Matthews, D.W., 1981, The pre-baltic sediments and the lower lavas at Kangerdlugssuaq, East Greenland. Their stratigraphy, lithology, paleomagnetism, and petrology, Geol. Surv. of Greenland, in press.

Olafsdottir, P., 1975, Jökulgardur a sjavarbotni ut af Breidafirdi, Natturufrædingurin, 45: 31-36.

Palmason, G., 1974, Insular Margin of Iceland, in: "The Geology of Continental Margins", 375-380, C.A.Burk and C.L.Drake, ed., Springer-Verlag, Berlin.

Paterson, W.S.B., 1969, "The physics of glaciers", London.

Saemundsson, K., 1980, Outline of the Geology of Iceland, in: Geologi of the european countries", publ. by Comité Nat. Francais de Geologie, 26 Int. Geol. Congr., 136-155.

Shepard, F.P., 1948, "Submarine Geology", Harpers, New York.

Sommerhoff, G., 1973, Formenschatz und morphologische Gliederung des südostgrönländischen Schelfgebietes und Kontinental abhanges, Meteor Forsch.-Ergebnisse, C, 15, 1-54

Soper, N.J. and Costa, L.I., 1976, Palynological evidence for the age of Tertiary basalts and post-basaltic sediments at Kap Dalton, central East Greenland, Rep. Geol. Surv. of Greenland, 80: 123-127.

Soper, N.J., Higgins, A.C., Downie, C., Matthews, D.W., and Brown, P.E., 1976, Late cretaceous - early Tertiary Stratigraphy of the Kangerdlugssuaq area, East Greenland and the age of opening of the North Atlantic, J. Geol. Soc. London, 132: 85-104.

Thors, K., and Kristjansson, L., 1974, Westward Extension of the Snaefellsness Volcanic Zone of Iceland, J.Geophys. Res., 79: 413-415.

Tychsen, J., 1977, Interpretation af nogle marin-magnetiske målinger fra Angmagssalik området og Danmarks-strædet. Unpublished Thesis, Geophys. Lab., University of Aarhus, Denmark.

Vogt, P.R., 1970, Magnetized basement outcrops on the south-east Greenland continental shelf, Nature, 226: 743-744.

Vogt, P.R., 1972, Evidence for Global Synchronism in Mantle Plume Convection, and Possible Significance for Geology, Nature, 240: 338-342.

Vogt, P.R., and Avery, O.E., 1974, Tectonic History of the Arctic Bassins: Partial Solutions and Unsolved Mysteries, in: "Marine Geology and Oceanography of the Arctic Seas", 83-117, Y.Herman, ed., Springer-Verlag, Berlin.

Vogt, P.R., and Perry, R., 1978, Post-rifting accretion of continental margin in the Norwegian-Greenland and Labrador Seas: Morphological evidence, EOS, Trans. Am. Geophys. Union, 59: 1204.

Vogt, P.R., Johnson, G.L., and Kristjansson, L., 1980, Morphology and Magnetic Anomalies North of Iceland, Jour. Geophysics, 47: 67-80.

Vogt, P.R., Perry, R.K., Feden, R.H., Fleming, H.S., and Cherkis, N.Z., 1981, The Greenland-Norwegian Sea and Iceland Environment: Geology and Geophysics, in: "The Ocean Basins and Margins", 5, 493-598, Nairn, A.E.M. et al., ed., Plenum Publishing Corporation.

Voppel, D., Srivastava, S.P., and Fleischer, U., 1979, Detailled magnetic measurements south of the Iceland-Faeroe Ridge, Deut. Hydrogr. Z., 32: 154-172.

Wager, L.R., 1947, Geological investigations in East Greenland IV: The Stratigraphy and tectonics of Knud Rasmussen Land and the Kangerdlugssuaq Region, Geol. Surv. of Greenland, 134 (5): 1-63.

Wager, L.R., and Deer, W.A., 1938, A dyke swarm and crustal flexure in East Greenland, Geol. Mag., 75: 49-56.

Waagstein, R., and Rasmussen, J., 1975, Glacial Erratics from the Sea Floor South-East of the Faeroe Islands and the Limit of Glaciation, Frodskaparrit (Annal. Societ. Scient. Færoensis), 23: 101-119.

Weidick, A., 1976, Glaciation and the Quaternary of Greenland, in: "Geology of Greenland", 431-458, A.Escher and W.S.Watt, ed., Geol. Surv. of Greenland, Copenhagen.

# SPECULATIONS ABOUT THE PALEODEPTH OF THE GREENLAND-SCOTLAND RIDGE DURING LATE MESOZOIC AND CENOZOIC TIMES

Jörn Thiede and Olav Eldholm

Department of Geology
University of Oslo
P.O.Box 1047 Blindern
Oslo 3 - NORWAY

Abstract:

The Greenland-Scotland transverse ridge has separated the deep basins of the Norwegian-Greenland Sea from the main North Atlantic Ocean since the early Cenozoic. Originally, it probably formed a large subaerial volcanic ridge which separated the Norwegian-Greenland Sea and the Arctic Ocean almost completely from the North Atlantic. The subsidence of the ridge which can be reconstructed from drilling data and the age-depth relationship of the adjacent oceanic crust suggests that a land bridge existed until Eocene time. However, the main ridge platform did not sink below sea level before the middle Miocene and isolated peaks submerged as late as the Pliocene. Thus, the ridge has allowed exchange of surface water from Eocene-Oligocene times onward. We also suggest that changes in the sea level have to be considered when analysing the ridge as a barrier for surface water circulation.

## INTRODUCTION

The belt of basaltic rock observed on Baffin Island, Greenland, Iceland, the Faeroes and in the northern United Kingdom is commonly described as the North Atlantic or Brito-Arctic volcanic province (Brooks, 1973). It has been generated by volcanism which started in the Late Cretaceous (Campanian) or earlier. Subsequently, the volcanism has resulted in the development of a complex structural high (Fig. 1) presently appearing as the

Greenland-Scotland Ridge (Noe Nygaard, 1974). This elevated region has constrained the mode and quantity of the water exchange between the Norwegian-Greenland Sea and the North Atlantic Ocean since the late Paleocene. Thus, its presence has had important implications for the geologic history of large areas surrounding the ridge, particularly when it existed as a landbridge between Greenland/North America and Europe. The paleobathymetric history of the ridge has been subject to much discussion (Mörner, 1980).

Here, we speculate about the evolution of the ridge morphology, particularly the paleobathymetry of the seaways crossing the ridge. Information of this kind place constraints when assessing how long the Greenland-Scotland Ridge has acted as a land bridge connecting the continents on either side, as well as with respect to the timing of possible water exchanges across ridge channels (Vogt, 1972). A reconstruction of ridge relief in time and space is by no means simple because we do not know when the earliest volcanism in this area occurred and what kind of crust underlies crucial ridge areas such as the Faeroes (Casten, 1973), the Faeroe-Shetland and Faeroe Bank channels and the Denmark Strait (Bott, 1974). However, new geophysical and geological data from the adjacent ocean floor and the Greenland-Scotland Ridge make it possible to constrain the morphologic evolution better than was previously possible.

After a brief description of the morphology of the Greenland-Scotland Ridge (Fig. 1) we review age information obtained from volcanic rocks and their interbedded sediments on the ridge. The rate of subsidence of the ridge due to thermal cooling (Detrick et al., 1977) and the eustatic sea level changes (Vail et al., 1977) are the two most important variables in a reconstruction of the former elevation of the ridge and its transverse channels.

## MORPHOLOGY AND PRESENT DEPTH TO VOLCANIC BASEMENT OF THE GREENLAND-SCOTLAND RIDGE

A simplified morphologic map of the Greenland-Scotland Ridge including the distribution of the main Late Cretaceous to Recent subaerial volcanic terranes is shown in Fig. 1. The sediment cover overlying the volcanic rocks is, except for the Faeroe-Shetland Channel region (Ridd, 1981), very thin or even absent (Miller and Tucholke, this volume). Thus, the present isobaths essentially outline the shape of the volcanic pedestal of the ridge. The areas which are presently above sea level, have been emergent during most of their history as indicated by the occurrence of limnic and subaerially deposited sediments within the volcanic rocks.

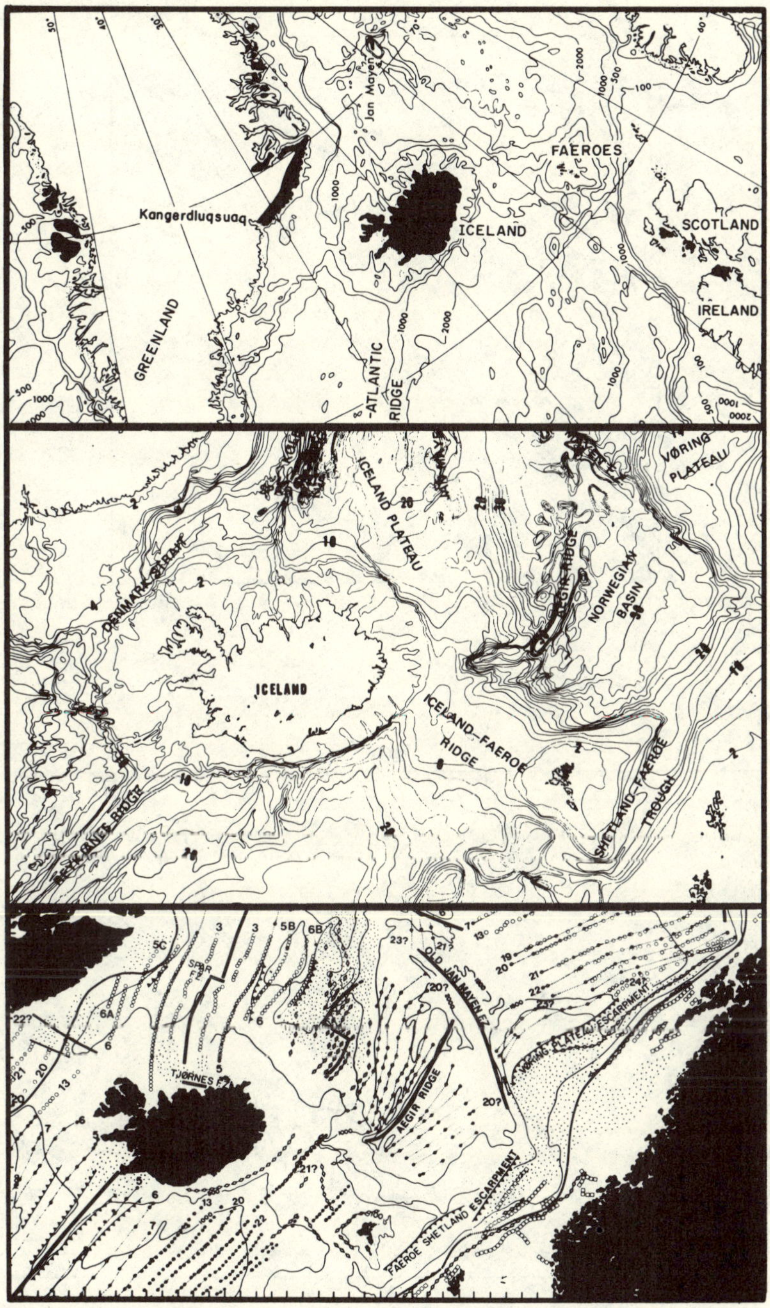


Fig. 1. From top to bottom: Distribution of late Cretaceous to Recent volcanics (Brooks and Jakobson, 1974); topography in meters, magnetic lineations and structural features as compiled by Vogt et al. (1981).

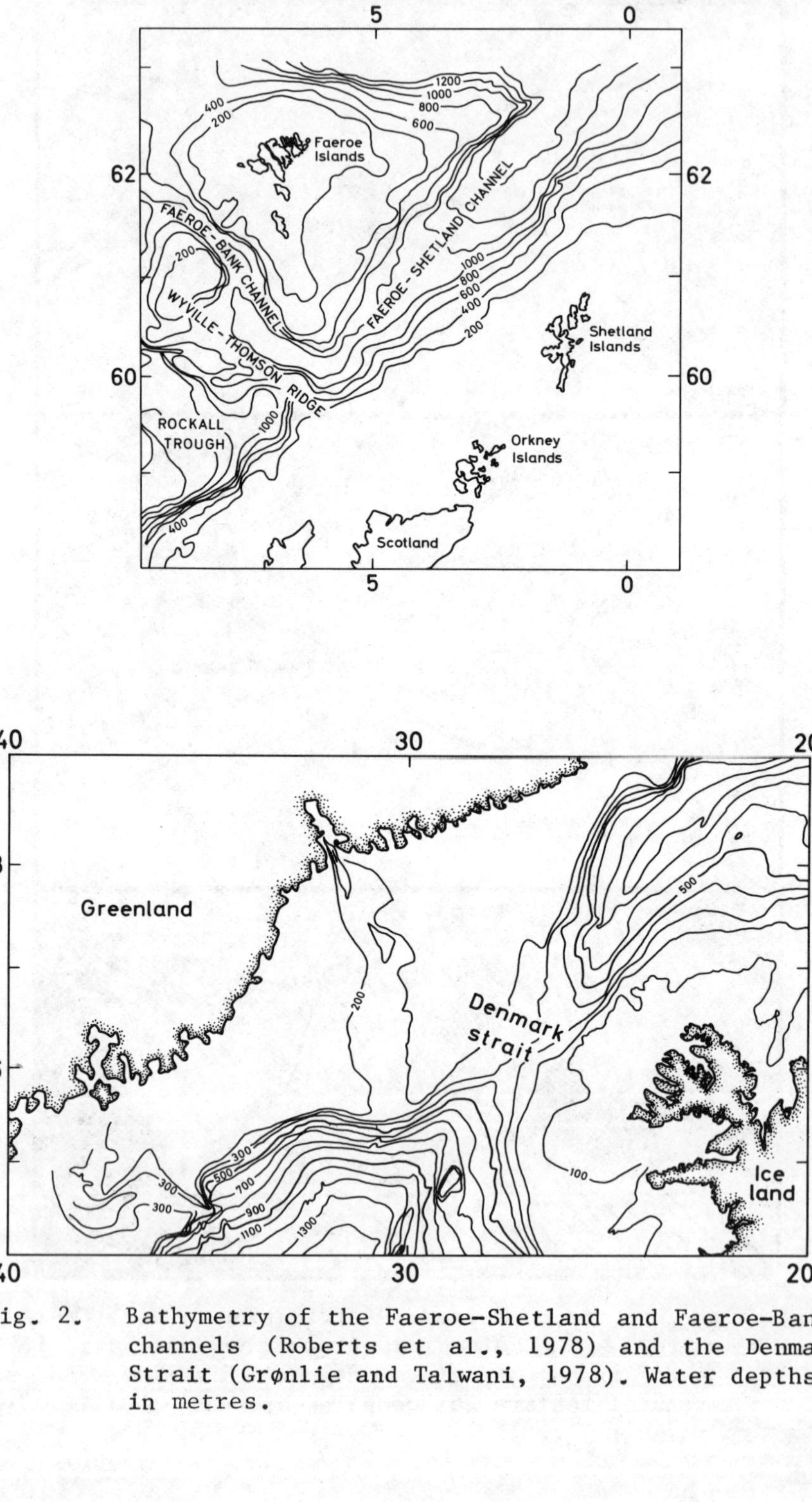


Fig. 2. Bathymetry of the Faeroe-Shetland and Faeroe-Bank channels (Roberts et al., 1978) and the Denmark Strait (Grønlie and Talwani, 1978). Water depths in metres.

It is therefore not necessary to discuss these areas in the context of this study.

The presently submarine parts of the ridge consist of the approximately 900 m deep Faeroe-Shetland Channel (including the Faeroe-Bank Channel), the Iceland-Faeroe Ridge forming the plateau-like 400-600 m deep area between the Faeroes and Iceland, and the almost 600 m deep Denmark Strait between Iceland and the continental margin of eastern Greenland. The two deep channels on either side of the ridge are particularly important with respect to the water exchange between the Norwegian-Greenland Sea and the North Atlantic Ocean (Fig. 2). It is most likely that the first interruption of a land bridge must have occurred in these areas where we expect to find the oldest volcanic rocks (Fig. 1).

## AGE DISTRIBUTION OF VOLCANIC ROCKS AND INTERBEDDED SEDIMENTS

The age distribution of the subaerial parts of the Greenland-Scotland Ridge is reasonably well understood because volcanic rock samples and interbedded fossiliferous sediments can be dated. In general, the region east of the Faeroe-Shetland Channel is of late Cretaceous-Paleocene age, the rocks on the Faeroes of late Paleocene age, the Icelandic rocks yield ages from Miocene to Recent and the basaltic rocks on eastern Greenland are of Paleocene age (Vogt et al., 1981). The age of the submarine parts of the ridge, however, is subject of considerable controversy. The tholeiitic basalts reached at Deep-Sea Drilling Project (DSDP) Site 336 northwest of the Faeroes (Talwani et al., 1976) have provided an age of 43.4 ± 3 my. These rocks are overlain by a late Eocene red clayey soil (Nilsen, 1978).

The other submarine parts of the ridge have not yet been sampled and dated properly and ages can only be estimated by interpolating data from the subaerial parts of the ridge or extending identified sea floor spreading type magnetic anomalies in the adjacent ocean basins onto the ridge (Vogt et al., 1981). The fairly regular increase of ages of the subaerial basaltic rocks away from the present spreading center in Iceland suggests a trend that may be valid also for the submarine segments of the ridge. The deepest part of Denmark Strait is henceforth located over approximately 50 my old oceanic crust (Voppel et al., 1979). The complex pattern of the sea floor spreading type anomalies north of the Iceland-Faeroe Ridge may imply deviations from a continuously increasing age pattern from Iceland to the Faeroes (Talwani and Eldholm, 1977), possibly this part of the ridge may be bounded by fracture zones. In the context of water circulation this may not be of direct importance as this part of the ridge is probably of Miocene or older age representing an emerged region at least until the middle Miocene.

The age and crustal nature of the region between the Faeroe and the Shetland Islands, i.e. the Faeroe-Shetland Channel; the Faeroe-Bank Channel and the Wyville-Thomson Ridge (Fig. 2) are indeed difficult to assess. Paleogeographic reconstructions (Ziegler, 1981) suggest that a trough, which at times was subjected to a marine depositional environment, has existed since late Paleozoic in this area. The rifting which led to the formation of the Rockall Trough occurred during the Albian-Maastrichtian interval (Roberts et al., 1981). Commercial drill sites in the Faeroe Basin west of the Shetland Islands (Ridd, 1981; this volume) encountered Lewisian granites below Campanian-Maastrichtian gray shales deposited in relatively deep waters. Acidic volcanics as old as Campanian have been penetrated at these sites, but the main volcanic event appears to have occurred towards the end of the Paleocene in connection with the eruption of the Faeroe plateau basalts.

However, there are some major unknown features at the easternmost part of the Greenland-Scotland Ridge. Seismic data (Ridd; this volume) indicate the presence of layered volcanic or sedimentary rocks of unknown age under the central part of the Faeroe-Shetland Channel. The age and nature of the basement underlying Wyville-Thomson Ridge, the Faeroe-Bank Channel and the Faeroe Islands (Bott, 1974, Roberts et al., 1981) will have to be determined before the paleobathymetry can be reconstructed with reasonable certainty.

## PALEOBATHYMETRY OF DENMARK STRAIT, THE FAEROE-SHETLAND CHANNEL AND THE ICELAND-FAEROE RIDGE

To construct the paleobathymetry of the three submarine segments of the Greenland-Scotland Ridge it is necessary to take into account the subsidence of the underlying crust, the isostatic response of the crust to the load of sediment and water and the eustatic sea level changes. We shall not deal with the isostatic response of the area caused by the development of the late Weichselian ice shields (Andersen, 1981) on the subaerial parts of the Greenland-Scotland Ridge during the late Tertiary and Quaternary glaciations in the northern hemisphere. We can also neglect corrections for the sediment/water load in the Denmark Strait and the Iceland-Faeroe Ridge because sediments are very thin or absent and water depths have been modest for most of the time. The

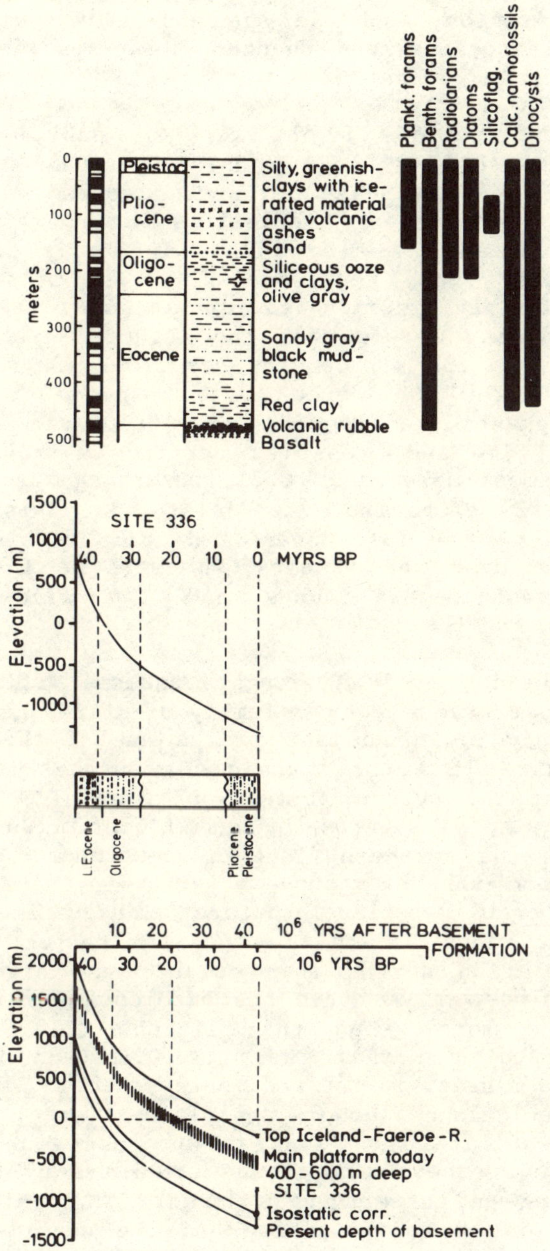


Fig. 3. History of sedimentation and subsidence of the Iceland-Faeroe Ridge during the Cenozoic (Thiede, 1980), based on data from DSDP Site 336 (Talwani et al., 1976).

lacking data about the Paleogene volcanism in the Iceland-Faeroe region, and the age and nature of the rocks under the Faeroe-Shetland Channel, have resulted in great uncertainties in our estimates.

A reconstruction of the rate of subsidence of the Iceland-Faeroe Ridge based on data from DSDP Site 336 is shown in Fig. 3. The basaltic rocks reached at this site resembled mid-ocean ridge tholeiites formed by late Eocene volcanism (Talwani et al., 1976). If it is correct that these volcanic rocks represent the crest of an aseismic ridge originating from volcanism at an accreting plate boundary, we may assume that the structure subsided at a rate similar to the surrounding sea floor (Detrick et al., 1977). Thus, the subsidence curve obtained for site 336 (Fig. 3) suggests that the site sank below sea level during late Eocene time and crossed a neritic environment during Oligocene times. Moreover, it did not reach water depths of around 1000 m before the Plio-Pleistocene. This subsidence is confirmed by the late Eocene red soil covering the volcanic rubble over the basalts, as well as the Oligocene benthic foraminiferal faunas and the Plio-Pleistocene sediments.

The subsidence of an elongate aseismic ridge can only be described completely by a family of different subsidence curves which are dependent upon the age of the underlying oceanic crust. It is not possible to properly assess the ridge age at the present time (Voppel et al., 1979; Vogt, 1981). However, it can be deduced from the subsidence of Site 336 on the northern flank of the ridge that the main ridge platform 600-700 m above the basement of Site 336 did not sink below sea level before middle Miocene times. Although the ridge became "leaky" with respect to surface water circulation during this period; the highest peaks, which lie almost 1 km above the basement at Site 336, did not submerge later than the Pliocene (Fig. 3). It is interesting to note that the late Oligocene to Pliocene deposits are missing in the sedimentary record of both drill sites at the Iceland-Faeroe Ridge (Talwani et al., 1976). It cannot be estimated how long the process generating this hiatus lasted, except that it is bracketed by the ages of the overlying and underlying sediments. The coincidence of the resumption of sedimentation and the subsidence of the highest peaks below sea level suggest that surface water currents were vigorously eroding sediments from the flanks of the ridge during Miocene and early Pliocene times.

The oldest crust under Denmark Strait is most probably close to 50 my old (Fig. 1). A curve describing its paleobathymetric evolution based on the subsidence of the underlying crust (Detrick et al., 1977) and rates of eustatic sea level changes (Vail et al., 1977) is shown in Fig. 4. According to this reconstruction Denmark Strait did not exist as a seaway prior to 15-18 my, but as a result of the combined effects of subsidence and sea level change it had reached about 400 m water depth in the late Miocene and has subsequently deepened slowly.

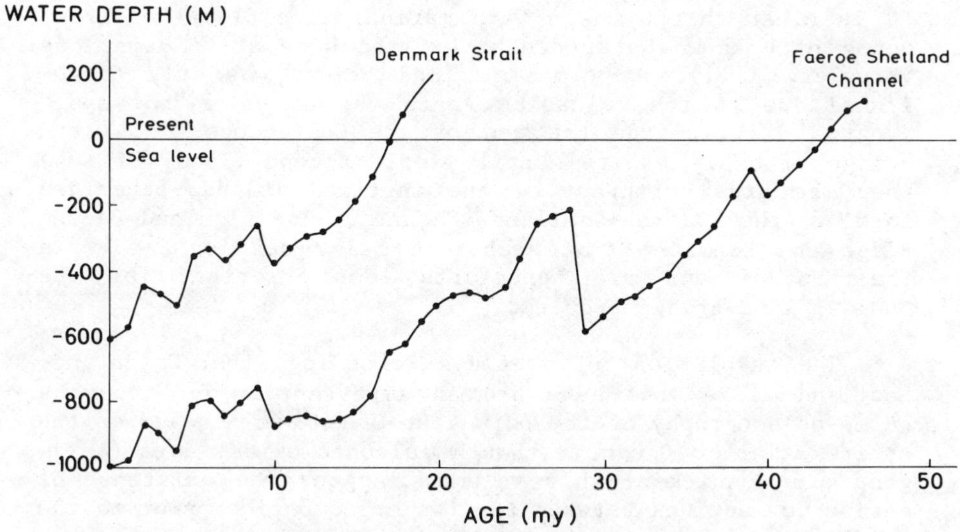


Fig. 4. Evolution of the paleodepth of deposition of the Denmark Strait and the Faeroe-Shetland Channel.

The crust underlying the Faeroe-Shetland Channel probably began its subsidence history approximately 60-70 my ago when the volcanism ceased. We have chosen a maximum age of 70 my to estimate the paleobathymetry. After applying the eustatic sea level changes (Vail et al., 1977) the resulting paleodepths (Fig.4) indicate that the area submerged 40-50 my ago and that it has been submarine since then. However, taking into account the uncertainties and lack of data discussed above it is clear that the resolution in the depth history of the Faeroe-Shetland Channel is poor. In particular, it is possible that the channel fell dry during certain periods in response to sea level fluctuations and the thermal effect of volcanic events in the adjacent regions. On the other hand, it appears that the depth of the Faeroe-Shetland Channel remained modest prior to 25 my and that a substantial deepening beyond 500 m water depth only occurred since the early Miocene.

## CONSEQUENCES FOR THE EXISTENCE OF A LAND BRIDGE AND THE PALEO-OCEANOGRAPHY OF THE ADJACENT OCEAN BASINS

The evolution of the marine seaways crossing the Greenland-Scotland Ridge in the Cenozoic had an important impact on the paleogeography of the entire northern hemisphere, particularly on the paleo-oceanography of the Norwegian-Greenland Sea and the North Atlantic Ocean. Although we do not know much about the pre-Tertiary environment of the shelf between Greenland and the Shetland Islands it is clear that a basin with marine depositional environments of unknown water depths existed here at some periods (Ziegler, 1981). A permanent land bridge was only established due to the volcanic events on top of a "hot spot" (Vogt, 1972) towards the end of the Cretaceous. This land bridge probably existed until middle Eocene times (Fig. 4) when the area northwest of the Shetland Islands submerged forming the Faeroe-Shetland Channel. The Iceland-Faeroe Ridge and Denmark Strait subsided below sea level only in the Miocene and are henceforth less important for the Thulean land bridge problem.

The evolution of seaways crossing the Greenland-Scotland Ridge must have had an important impact upon the paleo-oceanography of the Norwegian-Greenland Sea and of the entire Atlantic Ocean. Although paleontological data (Berggren and Schnitker, this volume) suggest the existence of marine connections between the two ocean basins prior to the middle Eocene (maybe across Europe and through the North Sea), our estimates indicate that a shallow channel did not evolve northwest of Shetland before the middle Eocene (Fig. 4). The channel probably remained shallow until the early Miocene when it started to deepen beyond 500 m water depth. This was approximately the time when seaways also broke across the Iceland-Faeroe Ridge and the Denmark Strait. The estimates we have produced will certainly be refined as new data from key areas such as the Faeroe-Bank Channel and Wyville-Thomson Ridge become available, but we think it is important to have the physical constraints of the seaways in mind if using biogeographic (from terrestrial as well as from marine organisms) or sedimentologic (mainly from oceanic sediment sections) data to infer the existence of a continuous land bridge between Europe and Greenland or seaways crossing the Greenland-Scotland Ridge. Furthermore, the curves in Fig. 4 show quite clearly that the superimposed effects of the sea level curve on the thermal subsidence may play an important role in the detailed understanding of the changes in the water circulation across a trans-oceanic ridge of this kind.

REFERENCES

Andersen, B.G., 1981, Late Weichselian ice sheets in Eurasia and Greenland, p. 3-65, in "The Last Great Ice Sheets", G.H. Denton and T.J. Hughes, eds., Wiley-Interscience, New York.

Berggren, W.A., and Schnitker, D., this volume, Cenozoic marine environments in the North Atlantic and Norwegian-Greenland Sea.

Bott, M.H.P., 1974, Deep Structure, Evolution and Origin of the Iceland Transverse Ridge, in "Geodynamics of Iceland and the North Atlantic Area", L. Kristjansson, ed., D. Reidel Publ. Co., Dordrecht, 33-47.

Brooks, C.K., 1973, Tertiary of Greenland - A volcanic and plutonic record of continental break-up, Am. Assoc. Petrol. Geol. Mem., 19:150-160.

Brooks, C.K., and Jakobson, S.P., 1974, Petrochemistry of the volcanic rocks of the North Atlantic Ridge, p. 139-154, in: "Geodynamics of Iceland and the North Atlantic Area", L. Kristjansson, ed., D. Reidel Publ. Co., Dordrecht.

Casten, U., 1973, The crust beneath the Faeroe Islands, Nature, 241: 83-84. Detrick, R.S., Sclater, J.G. and Thiede, J., 1977, Subsidence of aseismic ridges, Earth Planet. Sci. Lett., 34:185-196.

Eldholm, O., and Thiede, J., 1980, Cenozoic continental separation between Europe and Greenland, Palaeogeogr., Palaeoclimat., Palaeoecol., 30: 243-259.

Grønlie, G., and Talwani, M., 1978, Geophysical atlas Norwegian-Greenland Sea. Lamont-Doherty Geol. Obs., Vema Res. Ser. 4.

Miller, K.G., and Tucholke, B.E., this volume, Development of Cenozoic abyssal circulation south of the Greenland-Scotland Ridge.

Mörner, N.-A., 1980, Earth movements, paleoceanography, paleoclimatology and eustasy: Major Cenozoic events in the North Atlantic, Geol. Fören. Stockholm Förh., 102: 261-268.

Nilsen, T.H., 1978, Lower Tertiary laterite on the Iceland-Faeroe Ridge and the Thulean land bridge, Nature, 274: 786-788.

Noe Nygaard, A., 1974, Cenozoic to Recent volcanism in and around the North Atlantic Basin, p. 391-443, in: "The Ocean Basins and Margins. Vol. 2. The North Atlantic", A.E.M. Nairn and F.G. Stehli, eds., Plenum Press, New York.

Ridd, M.F., 1981, Petroleum geology west of the Shetlands, p. 414-425, in: "Petroleum Geology of the Continental Shelf of North-West Europe", Inst. Petroleum, London.

Ridd, M.F., this volume, Aspects of the Tertiary geology of the Faeroe-Shetland Channel.

Roberts, D.G., Hunter, P.M., and Laughton, A.S., 1978, Bathymetry of the northeast Atlantic. Continental margin around the British Isles, Deep-Sea Res., 26 A:417-428.

Roberts, D.G., Masson, D.G., and Miles, P.R, 1981, Age and structure of the southern Rockall Trough: New evidence, Earth Planet. Sci. Lett., 52:115-128.

Talwani, M., and Eldholm, O., 1977, Evolution of the Norwegian-Greenland Sea, Geol. Soc. Amer. Bull., 88:969-999.

Talwani, M. et al., 1976, Site 336 and 352, Init. Rep. Deep-Sea Drill. Proj., 38:23-116.

Thiede, J., 1980, Palaeo-eceanography, margin stratigraphy and palaeophysiography of the Tertiary North Atlantic and Norwegian-Greenland Seas, Phil. Trans. R. Soc. London, A294:177-185.

Vail, P.R. et al., 1977, Seismic stratigraphy and global changes of sea level, Amer. Assoc. Petrol. Geol. Mem., 26:49-212.

Vogt, P.R., 1972, The Faeroe-Iceland-Greenland aseismic ridge and the western boundary undercurrent, Nature,

Vogt, P.R., Perry, R.K., Feden, R.H., Fleming, H.S., and Cherkis, N.Z., 1981, The Greenland-Norwegian Sea and Iceland environment: Geology and geophysics, p. 493-598, in: "The Ocean Basins and Margins. Vol. 5. The Arctic Ocean", A.E.M. Nairns, M. Churkin and F.G. Stehli, eds., Plenum Press, New York.

Voppel, D., Srivastava, S.P., and Fleischer, U., 1979, Detailed magnetic measurements south of the Iceland-Faeroe Ridge, Dt. Hydrogr. Z., 32:154-172.

Ziegler, P.A., 1981, Evolution of sedimentary basins in North-West Europe, p. 3-39, in: "Petroleum Geology of the Continental Shelf of North-West Europe", Inst. Petrol., London.

# INFLUENCE OF THE GREENLAND-SCOTLAND RIDGE ON THE GEOLOGICAL HISTORY OF THE NORTH ATLANTIC AND NORWEGIAN-GREENLAND SEA AREAS

Tor H. Nilsen

U.S. Geological Survey
345 Middlefield Road
Menlo Park, California, USA 94025

ABSTRACT

The Greenland-Scotland Ridge was a controlling factor in the development of many aspects of the geologic history of the North Atlantic and Norwegian-Greenland Sea areas. It grew and evolved as a transverse ridge during the separation of the North-American-Greenland and North European plates by seafloor spreading processes. The ridge consists from northwest to southeast of the Denmark Strait, the volcanically active Icelandic Plateau, the volcanically inactive and aseismic Iceland-Faeroe Ridge, the Faeroe Islands block, a series of smaller banks and ridges that include Faeroe Bank, Bill Bailey Bank, Lousy Bank, and the Wyville-Thomson and Ymir Ridges, and the Faeroe-Scotland and Faeroe Bank Channels. The Greenland-Scotland Ridge is thus formed of an interconnected assemblage of subaerial and submarine volcanic plateaus, uplifted fragments of oceanic crust, rifted microcontinental blocks, narrow subsided channels, and sediment drifts.

The Greenland-Scotland Ridge devoloped in response to mantle-plume-generated hot-spot volcanism. Uplift and doming of the developing rift in Paleocene time resulted in widespread extrusion of basalt on both the European and East Greenland continental margins. Continued rifting and hot-spot activity yielded development of the Iceland-Faeroe Ridge between 55 and 30 m.y. ago. This ridge, originally a subaerial volcanic plateau constructed on oceanic crust, later subsided below sea level. Subsequent migration of the locus of hot-spot volcanism resulted in the development of the younger and still active Icelandic Plateau to the west.

The Greenland-Scotland Ridge has strongly influenced the

migration of marine microfauna and microflora as well as terrestrial fauna and flora, the development of worldwide patterns of oceanic circulation, and the evolution of Tertiary and Quaternary climates. Apparent continuity of the ridge across the developing ocean in early Eocene time permitted the exchange of land vertebrate faunas and establishment of nearly identical assemblages in North America and Europe. Subsequent subsidence of at least part of the ridge below sea level later in the early Eocene permitted northward migration of marine microfauna and microflora from the North Atlantic to the widening Norwegian-Greenland Sea. This movement of water masses may have helped maintain tropical climates during most of Eocene time in the northern latitudes.

The establishment of marine connections between the Arctic Ocean and Norwegian-Greenland Sea during the interval of time near the Eocene/Oligocene boundary may have resulted in the development of cold bottom water in the Norwegian-Greenland Sea. Southward movement of fast-moving cold bottom water into the North Atlantic initially formed widespread submarine unconformities. Subsidence of the main part of the Iceland-Faeroe Ridge below sea level later in Oligocene time permitted larger quantities of sediment-laden Norwegian Sea bottom water to flow southward and deposit large sediment drifts in the North Atlantic during Neogene time.

## INTRODUCTION

The Greenland-Scotland Ridge, also known as the Icelandic transverse ridge and Thulean land bridge, has been the subject of many and varied studies during the past 200 years. The ridge extends from the southeast coast of Greenland to Scotland, trending almost perpendicular to the presently active midocean spreading ridges (fig.1). The ridge is complex, consisting of elevated ocean crust, subaerial and submarine volcanic plateaus, continental microfragments, sediment drift ridges, and channels with various orientations that were formed by episodes of rifting, subaerial erosion, and submarine erosion. The complex nature of the ridge results from a variety of geologic processes that have constructed, shaped, eroded, deposited and deformed it during Cenozoic time.

The Greenland-Scotland Ridge is critically important to understanding a number of significant aspects of the geologic history of the North Atlantic region. Among the most important aspects and features that require study are:

1. The structure, geochemistry, and evolution of oceanic plateaus that develop over mantle-plume-generated hot spots.

2. The structure and evolution of rifted continental fragments.

Fig. 1. Index map of the Norwegian-Greenland Sea, Artic Ocean and North Atlantic Ocean showing location of Greenland-Scotland Ridge (modified from Vogt and Avery, 1974). Light line represent 1,000 m depth contour; heavy lines, spreading ridges, dashed where extinct; intermediate lines, transform faults; darkened areas, outcrpos of basalt of the early Tertiary North Atlantic igneous province; numbered dots, selected DSDP sites; arrows, bottom flow of Norwegian-Greenland Sea overflow water, modified from Jones et al. (1970).

3. The subsidence history and erosional regime of aseismic ridges.

4. The influence of transverse ridges on oceanic circulation patterns.

5. The influence of transverse ridges on migration patterns and evolution of marine microfauna and microflora.

6. The evolution of transverse ridges as land bridges and their influence on migration patterns and evolution of terrestrial fauna and flora.

7. The influence of transverse ridges on climate.

8. The influence of transverse ridges on oceanic and continental margin sedimentation and erosion, particularly the development of anoxic conditions suitable for generation of hydrocarbons.

The Greenland-Scotland Ridge is a suitable field laboratory close to the research centers of western Europe and eastern North America for the study of many aspects and features related to the processes of continental rifting and separation.

As early as the 1800's, the Greenland-Scotland Ridge was thought to have been a possible land bridge, the Thulcan land bridge, that connected North America and western Europe in early Tertiary time. A second land bridge, the De Geer route, was later proposed across Spitsbergen and northeastern Greenland. The remarkable similarities between the land vertebrates of North America and western Europe during the early Tertiary was the chief reason for speculations on the land bridges (see review in Lindroth, 1957; McKenna, 1972 and this volume). Recognition of early Tertiary paleosols at the bottom of DSDP site 336, at present at a depth of 1300 m below sea level, provided additional new support for this concept (fig. 1).

Flow of marine waters across the Greenland-Scotland Ridge in the early Tertiary also appears to be necessary, because of similarities between marine microfauna north and south of the ridge (Berggren and Schnitker, this volume). The northward migration of warm waters into the newly opening Norwegian-Greenland Sea may have provided a climatic warming for the entire polar region during the early Tertiary.

Southward flow of colder water over the Greenland-Scotland Ridge and into the North Atlantic apparently began no later than Oligocene time (Miller and Tucholke, this volume). This southward flow apparently initiated widespread submarine erosion of the North

Atlantic at the beginning of Oligocene time. Subsequently, the southward-flowing, sediment-laden, colder waters deposited major sediment drifts or contourite ridges as far south as the Blake-Bahamas plateau region. As a result of this new movement of water masses over the Greenland-Scotland Ridge, bottom water circulation patterns for the entire world ocean apparently changed drastically and worldwide climatic patterns may have also altered significantly (Schnitker, 1980).

A great deal of new information about the Greenland-Scotland Ridge and its history was presented at the NATO meeting in Bressanone, Italy, in May 1981. Much of the new data, in widely divergent fields of interest, reflected attempts by various specialists to understand the history and structure of the ridge in relation to important problems in their special fields. Discussions of geophysical, geochemical, geological, paleontological, petrological, paleogeographical, paleo-oceanographical, glaciological and sedimentological studies gradually made the interconnected nature of all the studies clearer. The purpose of this paper is to summarize various critical aspects of the more specialized studies within the region that relate the development and subsidence of the Greenland-Scotland Ridge to processes of plate tectonics. The paper builds on previous attempts to reconstruct the history of one part of the ridge, the Iceland-Faeroe Ridge, and the effects of this segment on sedimentation to the north and south (Nilsen, 1976, 1978a,b; Nilsen and Kerr, 1978).

## GEOGRAPHIC FRAMEWORK

The Greenland-Scotland Ridge extends from the southeastern continental margin of Greenland to the northwestern continental margin of Scotland (fig. 1). It forms a major transverse ridge that separates the North Atlantic Ocean from the Norwegian-Greenland Sea. The ridge is almost 4,000 km long and 1500 km wide at its widest point on the Icelandic Plateau (figs. 2 and 3). The Icelandic Plateau and the Faeroe Islands are the only parts of the ridge that are presently above sea level.

The Greenland-Scotland Ridge can be divided into six parts that are both geographically and geologically distinct. The Denmark Strait forms the westernmost part of the ridge. The strait is 1100 km wide and as deep as 620 m (figs. 1 and 2). It forms a marine passageway between East Greenland and Iceland, permitting southward flow of cold Norwegian-Greenland Sea bottom water into the North Atlantic. This bottom water joins that which flows over the eastern end of the Greenland-Scotland Ridge along the southeastern margin of Greenland, after the latter has passed through the Charlie Gibbs Fracture zone. The Denmark Strait, poorly understood until recently, is now known to be underlain by a thick

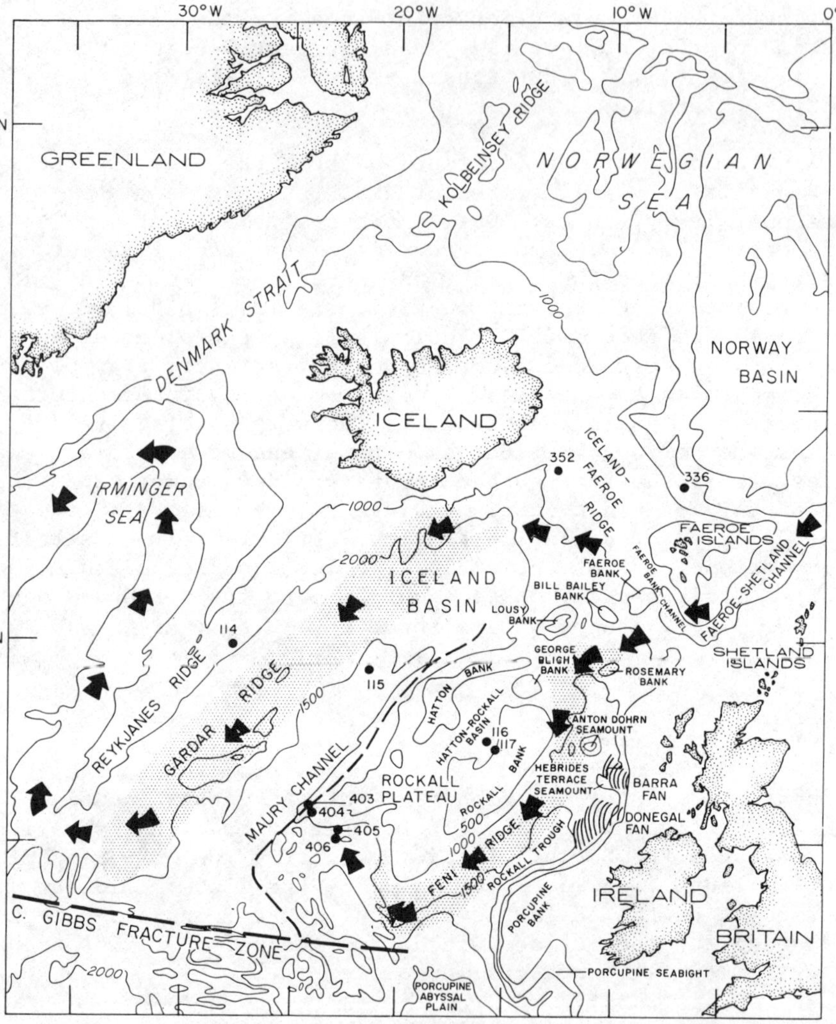


Fig. 2. Index map of the Northeast Atlantic Ocean and adjacent areas showing location of Greenland-Scotland Ridge and selected DSDP sites (modified from Roberts, 1975a and Davies and Laughton, 1972). Small arrows indicate flow directions of Norwegian Sea bottom water. Bottom contours in meters.

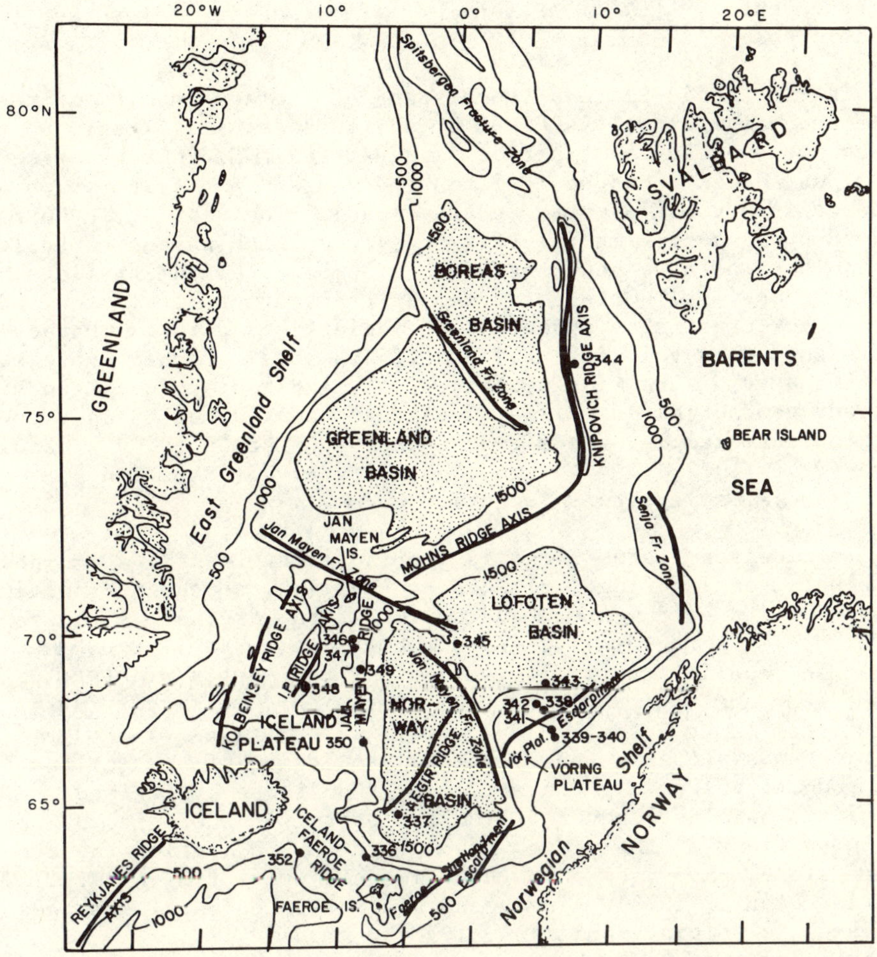


Fig. 3. Index map of the Norwegian-Greenland Sea and adjacent areas showing location of Greenland-Scotland Ridge and selected DSDP sites (modified from Talwani et al., 1976). Bottom contours in nominal fathoms (sound velocity 800 fms/sec).

prograding wedge of sediment. This sediment wedge, derived from erosion of source areas in Greenland, combined with sediment drifts on the east flank of the strait derived from the south-flowing Norwegian-Greenland Sea bottom water, have almost closed off part of the strait (H.C. Larsen, this NARI; B. Larsen, this volume). Scour to depths of 300 m or more by large icebergs helps keep the strait open.

The Icelandic Plateau, consisting of Iceland and its contiguous shelf and marginal slopes, is a broad intraoceanic volcanic plateau. It has been volcanically active since at least Miocene time, because the oldest dated rocks on Iceland have yielded potassium-argon ages of about 16 m.y. (Moorbath, et al., 1968). The plateau is characterized by an active medial rift zone from which extensive tholeiitic plateau basalts have been extruded, a number of large rhyolite domes, and several northwest-trending transform faults. Iceland has a thick crust that appears to be chiefly oceanic in character (Björnsson, this volume). Because of its physiographic expression, location astride the Mid-Atlantic Ridge, and geophysical, geochemical, and geothermal anomalies, Iceland is considered by many to be an excellent example of a mid-ocean hot spot produced by a mantle plume (Vogt, 1972, this volume; Schilling, 1973; Schilling, Meyer, and Kingsley, this volume).

The Iceland-Faeroe Ridge is a broad, flat-topped, 400-m-deep, northwest-trending ridge. It has two shallow, southwest-trending troughs on its northwestern and southeastern margins. Although it is complex structurally, it is thought to be underlain by thickened oceanic crust. It resembles the Icelandic Plateau except that the crustal layers are thicker and seismic velocities higher (Bott, et al., 1971; Bott, 1974; Bott, this volume). The ridge was probably the Paleogene expression of the Iceland hot spot and its volcanic rocks are older than those of Iceland. Tholeiitic basalts recovered from the bottom of DSDP site 336 on the north flank of the ridge, at a depth of 900 m below the crest of the ridge, have yielded potassium-argon ages of 37-47 m.y. (Kharin, et al., 1976). The geochemistry of the basalts suggest a mantle-plume origin (Schilling, 1976).

The Faeroe Islands block is a triangular-shaped platform capped by 3,000 m of Paleocene to Eocene tholeiitic basalts that crop out in the Faeroe Islands (Rasmussen and Noe-Nygaard, 1970). The exposed basalts yield potassium-argon ages of 49-60 m.y. (Tarling and Gale, 1968). The crust beneath the Faeroe Islands block, as indicated by various types of geophysical data, appears to be continental (Palmason, 1965; Bott and Watts, 1971; Casten, 1973; Bott, et al., 1974; P.H. Nielsen, this volume). The geochemistry of the basalts suggests a mantle-plume origin for the lower basalts and a mid-ocean ridge origin for the upper basalts (Schilling and Noe-Nygaard, 1974). The Faeroe Islands block is flanked on all sides by escarpments that slope away from it.

The relatively narrow marine channel that flanks the Faeroe Islands block to the southeast and southwest forms the fifth part of the Greenland-Scotland Ridge. To the southeast it is called the Faeroe-Shetland Channel and to the southwest the Faeroe Bank Channel. The Faeroe-Shetland Channel, being intensively drilled at present for hydrocarbons, is filled at least in part by a thick sequence of Cretaceous and Paleogene sedimentary strata (Roberts, 1975; Ridd, this volume). The channel may also contain some volcanic and plutonic rocks of this age. The Faeroe-Shetland and Faeroe Bank Channel system is locally as deep as 1200 m and forms a passageway for the transport of cold Norwegian Sea bottom water southward into Rockall Trough and the North Atlantic (Crease, 1965; Jones, et al., 1970; Ellett and Roberts, 1973; Roberts, 1975a).

The most southeasterly part of the Greenland-Scotland Ridge consists of a series of submarine banks and ridges of variable size and shape located southwest of the Faeroe Bank Channel. Faeroe Bank, Bill Bailey Bank, and Lousy Bank are flat-topped, roughly equidimensional platforms thought to be continental fragments capped by basalts, similar to the Faeroe Islands block. The west-trending Wyville-Thomson and Ymir Ridges, located southeast of Faeroe Bank and Bill Bailey Bank, respectively, appear to consist of elongate sediment drifts deposited on basalt. The sediment drifts are thought to have been deposited by southwest-flowing Norwegian Sea bottom water as it passes through the Faeroe-Shetland Channel and travels southward into Rockall Trough and southwestward into the North Atlantic (Roberts, 1975a, b; Roberts and others, 1979). The underlying basalts can be traced continuously in seismic reflection profiles under the Faeroe-Shetland Channel to the Faeroe Islands block (D.G. Roberts, this NARI).

The Greenland-Scotland Ridge thus consists of a number of diverse elements. Two major volcanic plateaus, thought to have been constructed on oceanic crust by hot-spot magmatism along an active spreading ridge, form the central part of the ridge. The southeastern part of the ridge consists of isolated continental fragments, the Faeroe Islands block and adjacent smaller blocks to the south, separated from the European continent by a major curvilinear deep basin beneath the Faeroe-Shetland and Faeroe Bank Channels. The northwestern part of the ridge, the Denmark Strait, consists of ocean floor that is probably uplifted and covered by a thick sequence of marine sediments derived from Greenland. Present pathways for southward flow of Norwegian-Greenland Sea bottom water include the Denmark Strait, Iceland-Faeroe Ridge and its marginal troughs, and the Faeroe-Shetland and Faeroe Bank Channels.

## HOT-SPOT TECTONICS AND SUBSIDENCE

On the basis of a variety of geophysical, geochemical, and geological criteria, Iceland is considered to be a prime example of hot-spot activity (see review by Vogt, this volume). Even though its oldest exposed rocks are 16 m.y. old, Iceland is thought to have been an active site of volcanism for the past 30 m.y., concomitant with seafloor spreading along Reykjanes and Kolbeinsey Ridges (figs. 1-3). Other parts of the North Atlantic and Arctic regions that may record hot-spot volcanism include the early Tertiary volcanic province of northern Britain and Ireland, the Faeroe Islands block, the Iceland-Faeroe Ridge, East Greenland, the Vöring Plateau, Jan Mayen Island, the Davis Strait area (Srivastava, this volume), and the Yermak Plateau, located northwest of Spitsbergen in the southeastern corner of the Arctic Ocean. The temporal and spatial distribution of hot-spot volcanism on the Greenland-Scotland Ridge provides the keys to understanding the tectonic development of the entire region and the subsidence history of the ridge.

The pattern and history of seafloor spreading in the region of the Greenland-Scotland Ridge have been complex. Areas to the north and south of the ridge have undergone different histories (see reviews by Nunns, this volume, and Eldholm and Thiede, this volume). Early pre-Tertiary rifting took place in Rockall Trough and the Faeroe-Shetland Channel, possibly extending as far north as the Inner Vöring Basin (see reconstruction, fig. 4). Later early Tertiary seafloor spreading shifted to the Reykjanes Ridge and Aegir Ridge, an extinct spreading ridge in the Norwegian Sea. The early Tertiary rifting was associated with voluminous outpourings of subaerial basalts, with locally associated intrusives, in northern Britain, Northern Ireland, Rockall Plateau, Faeroe Islands, and East Greenland (Roberts, 1969; Rasmussen and Noe-Nygaard, 1970; Purdy, et al., 1972; Evans, et al., 1973; Brooks, 1973a,b; Schilling and Noe-Nygaard, 1974; Brown and Mussett, 1976; and Brown and Whitley, 1976).

This early Tertiary volcanism, centered roughly between the then-contiguous borders of East Greenland and the Faeroe Islands, produced broad doming of the region, basaltic plateaus, eventual rifting, and finally true seafloor spreading along Reykjanes and Aegir Ridges (fig. 4). The hot-spot had a southward-extending zone of rifting and active volcanism into northern Britain and Northern Ireland; this rift may represent the failed arm, or aulocogen, for the opening of the Norwegian-Greenland Sea. Earlier, late Paleozoic and Mesozoic rift systems in northern Europe, which include the Oslo and Rhine grabens, developed during unsuccessful phases of doming, rifting, and volcanism that did not culminate with seafloor spreading and continental separation (Pegrum, et al., 1975; Whiteman, 1975). It is possible that only when the hot-spot activity encountered the ancient Paleozoic Caledonian suture

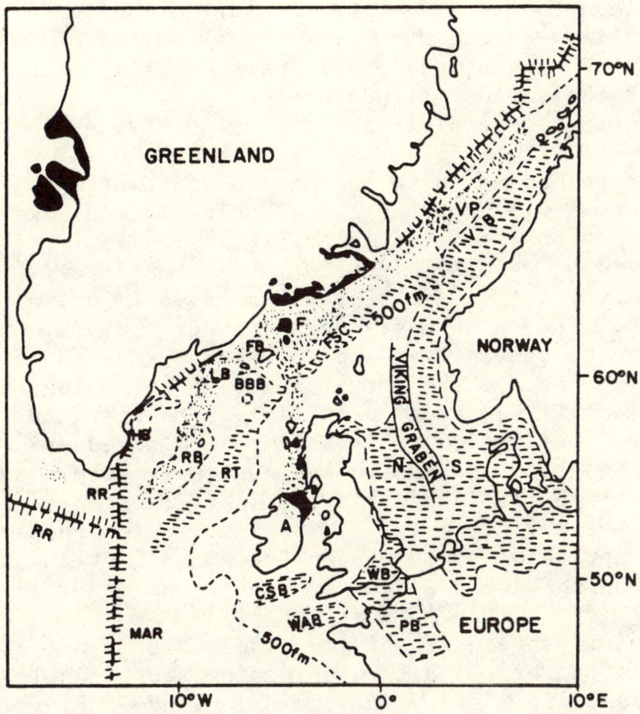


Fig. 4. Plate reconstruction for 60 m.y. ago of North Atlantic and Norwegian-Greenland Sea areas, showing locations of lower Tertiary plateau basalts (modified from Schilling and Noe-Nygaard, 1974, and Vogt and Avery, 1974). Blackened areas indicate outcropping basalts; dotted areas indicate known and inferred submarine lower Tertiary plateau basalts; horizontally dashed areas indicate Early Cretaceous basins. Dot-dash line marks inferred position of rift 60 m.y. ago, with adjacent spreading-ridge basalts that are approximately 60 m.y. old shown by lined pattern. Abbreviations, from north to south: VP, Vöring Plateau; IVB, Inner Vöring Basin; F, Faeroe Islands; FSC, Faeroe-Shetland Channel; FB, Faeroe Bank Channel; LB, Lousy Bank; BBB, Bill Bailey Bank; HB, Hatton Bank; RB, Rockall Bank; RT, Rockall Trough; NS, North Sea Basin; RR', Reykjanes Ridge; RR, Ran Ridge; A, Antrim Basalts; WB, Weald Basin; CSB, Celtic Sea Basin; WAB, Western Approaches Basin; PB, Paris Basin; MAR, Mid-Atlantic Ridge.

between East Greenland and northwestern Europe was true continental separation effected.

The Iceland-Faeroe Ridge grew rapidly as an oceanic plateau in the early Tertiary as the Norwegian-Greenland Sea opened. Volcanism on the ridge took place possibly in part coevally with that in East Greenland and the Faeroe Islands block, but also continued into the middle Tertiary, possibly as late as 30 m.y. ago.

Seafloor spreading north of the Iceland-Faeroe Ridge shifted westward about 30 m.y ago to the Icelandic Plateau Ridge Axis, an extinct ridge northeast of Iceland (fig. 3). This shift in the locus of spreading resulted in the rifting away from the East Greenland continental margin of the Jan Mayen Ridge, a continental microfragment. The spreading ridge axis north of the Greenland-Scotland Ridge subsequently jumped westward a second time, as Kolbeinsey Ridge become the locus of active spreading and has continued to be active to the present time. The westward shifts of the spreading ridge axis were probably accompanied by westward movement of the locus of hot-spot volcanism from the Iceland-Faeroe Ridge to the Icelandic Plateau. Hot-spot volcanism may have also shifted westward from the Vöring Plateau to Jan Mayen Island. The locus of spreading south of the Greenland-Scotland Ridge remained over the Reykjanes Ridge throughout the Cenozoic.

The complex pattern of shifting loci of hot-spot volcanism and shifting positions of the active spreading ridge resulted in an irregular subsidence history for the Greenland-Scotland Ridge. The Faeroe-Shetland and Faeroe Bank Channels apparently remained open through the remainder of Cenozoic time, probably in part because of the erosive strength of the southward-flowing Norwegian Sea bottom water. The Faeroe Islands block, as well as Faeroe Bank, Bill Bailey Bank and Lousy Bank, apparently remained high-standing platforms at or above sea level.

The Iceland-Faeroe Ridge subsided below sea level, in part as a result of the shift westward of hot-spot volcanism to the Icelandic Plateau. The subsidence history of the ridge is constrained by paleobathymetric data from DSDP sites 336 and 352 on the flanks of the ridge (Nilsen, 1978a). The ridge subsided at an average rate of 30 m/m.y. (Fig. 5) and its crest probably subsided below sea level in Oligocene time. Its rate of subsidence is similar to that of other aseismic ridges and normal ocean crust in the North Atlantic region, although its amount of subsidence is less because of its original position high above sea level (fig. 6).

Iceland has been a subaerial volcanic plateau for at least the last 15 m.y. and may have developed as a subaerial feature prior to that. The Denmark Strait has remained a narrow seaway underlain by seafloor 55 m.y. old or younger throughout most of the

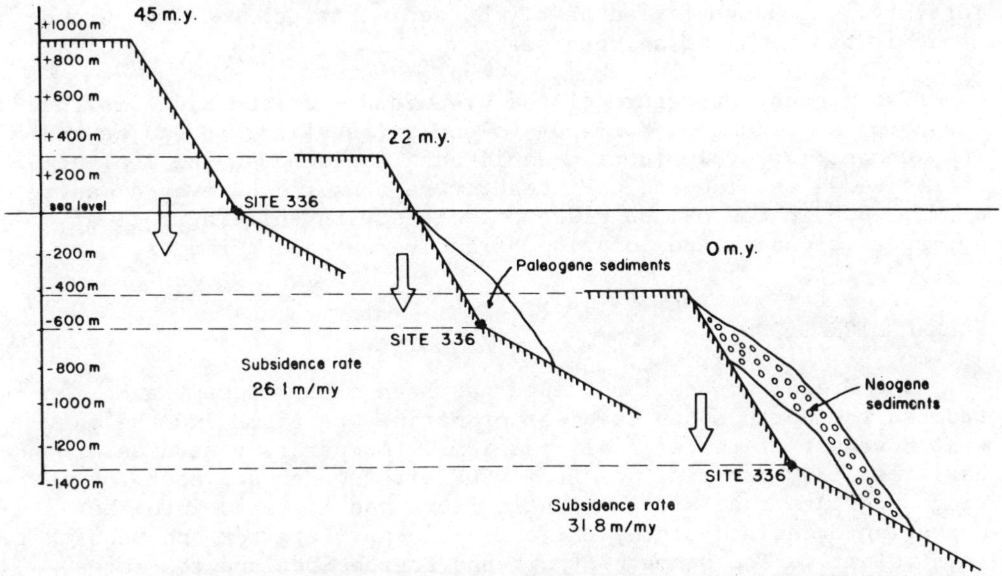


Fig. 5. Diagram showing possible subsidence history of DSDP site 336 and top of Iceland-Faeroe Ridge from 45 m.y. ago to present (from Nilsen and Kerr, 1978).

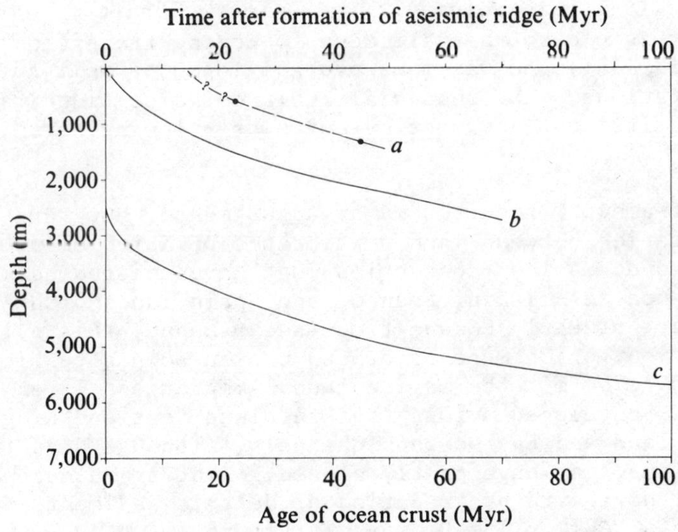


Fig. 6. Subsidence curves for a, the Iceland-Faeroe Ridge, DSDP site 336; b, normal aseismic ridges (from Detrick et al., 1977); and c, North Atlantic normal ocean crust (from Parson and Sclater, 1977).

Tertiary. Although filled partly by sediments derived from Greenland, it has remained an open seaway.

The present structure of the Greenland-Scotland Ridge is thus the complex product of hot-spot volcanism, seafloor spreading and subsidence. Two voluminous oceanic basalt plateaus, the Iceland-Faeroe Ridge and Icelandic Plateau, were constructed over a mantle plume. Subsidence of the ridge coincided with shifting loci of hot-spot volcanism and location of active spreading ridges.

## FAUNAL MIGRATIONS

The Greenland-Scotland Ridge has been a key element that controlled the north-south movement of marine organisms and the east-west movement of terrestrial organisms. Comparative studies of early Tertiary Foraminifera from DSDP sites north and south of the ridge indicate that North Atlantic fauna had migrated into the Norwegian-Greenland Sea by the early Eocene (Berggren and Schnitker, this volume). The Denmark Strait and Faeroe-Shetland to Faeroe Bank Channels were the only possible routes for this migration because the Icelandic Plateau had not been constructed yet and the Iceland-Faeroe Ridge did not subside below sea level until Oligocene time (Nilsen and Kerr, 1978). Pathways through the North Sea were blocked by uplifted land masses. Interpretations of seismic profiles suggest that the Faeroe-Shetland and Faeroe Bank Channels were deep troughs as early as Paleocene time (D.G. Roberts, this NARI; Ridd, this volume). The Denmark Strait is underlain by seafloor as old as the Paleocene to Eocene transition (H.C. Larsen, this NARI) and was thus probably open as well to marine faunal migrations. The subaerial Iceland-Faeroe Ridge was thus bypassed to its east and west by northward-flowing North Atlantic water.

This framework requires major adjustments, however, to explain the similarities between land vertebrates of North America and western Europe. A land connection must have existed between parts of Europe such as Britain, France, and Spain, and North America in at least part of Early Eocene time (see McKenna, this volume). The connection, which must have been broken soon after it was established to permit the marine faunal migrations later in early Eocene time, developed across the Greenland-Scotland Ridge. The Faeroe-Shetland and Faeroe Bank Channels, although deep troughs in Paleocene time, may have partly closed in latest Paleocene time as a result of northward progradation of deltaic sediments (D.G. Roberts, this NARI). The Denmark Strait, because it opened during the Paleocene to Eocene transition, permits the possibility of land connections between the Faeroe Islands block and East Greenland in latest Paleocene time.

The marine and terrestrial faunal migration patterns thus require a land bridge between western Europe and North America-Greenland near the Paleocene-Eocene boundary and break-up of the land bridge in early Eocene time. This framework seems possible within our present understanding of the subsidence history of the ridge, but demands a number of precise and well-timed events.

## OCEANIC PALEOCIRCULATION PATTERNS AND PALEOCLIMATES

The northward and southward flow of marine water over the Greenland-Scotland Ridge is an important factor in understanding Cenozoic ocean circulation patterns, sedimentation, erosion, and climates. Cold Norwegian Sea bottom water presently overflows through the Faeroe-Shetland and Faeroe Bank Channels into Rockall Trough and the northeastern Atlantic Ocean (figs. 1 and 2). Additional Norwegian Sea water overflows across the Iceland-Faeroe Ridge into the Atlantic. Cold Greenland Sea bottom water overflows through the Denmark Strait into the North Atlantic.

The cold, sediment-laden Norwegian and Greenland Sea bottom waters have and continue to deposit sediment drifts or contourite ridges that are locally more than 1,000 m thick in many parts of the Atlantic. Gardar Ridge on the east flank of Reykjanes Ridge and Feni Ridge in western Rockall Trough are examples of sediment drifts (fig. 2). Continued southward flow of the Norwegian-Greenland Sea bottom water, joined with bottom water from the Davis Strait region, results in deposition of sediment drifts along the western margin of the North Atlantic as far south as the Blake-Bahama outer ridge. The combined influence of the southward flow of cold bottom water and northward flow of warm surface water of the Gulf Stream over the Greenland-Scotland Ridge causes major climatic effects. The subsidence history of the ridge is the critical and controlling factor in these circulation changes.

Miller and Tucholke (this volume) carefully document the Cenozoic development of North Atlantic bottom circulation, concluding that the Faeroe-Shetland and Faeroe Bank Channels provided the major pathway for southward flow of cold water. Additional input from the Denmark Strait and Davis Strait regions augmented the southward flow (fig.1). They also demonstrate that the initiation of southward flow commenced near the Eocene-Oligocene boundary, coincident with the final separation of Greenland and Spitsbergen (Talwani and Eldholm, 1977) and probable initial spilling of colder water of the Arctic Ocean into the Norwegian-Greenland Sea. The southward movement of the cold water over the Greenland-Scotland Ridge into the Atlantic is thought to have resulted initially in widespread submarine unconformities in early Oligocene time. Later initiation of major contourite sedimentation in the North Atlantic may coincide with final subsidence of the crest of the

Iceland-Faeroe Ridge below sea level, massive southward overflow, and transport of water containing larger amounts of suspended sediment.

The record of water overflow and sedimentation corresponds well with the subsidence history of the Greenland-Scotland Ridge. The northward movement of warmer North Atlantic water into the Norwegian-Greenland Sea at present warms the northern climate and keeps much of the sea area ice-free. In the early Tertiary, climatic amelioration in the northern latitudes is indicated by the presence of tropical vertebrates such as alligatorids and sea turtles in Ellesmere Island (Dawson, et al., 1976), tropical floras in southwestern England (Davis and Elliott, 1957), warm to temperate floras in West Greenland (Koch,1963), Baffin Island (Andrews, et al., 1972), and Ellesmere Island (Christie, 1964), and warm-water marine phytoplankton in the Arctic Ocean (Clark, 1974). This early Tertiary warming of the Norwegian-Greenland Sea and Arctic areas may have been produced, at least in part, by the northward movement of warmer Atlantic water over the Greenland-Scotland Ridge, as suggested by Berggren and Schnitker (this volume).

Reddish, hematitic weathered zones of early Tertiary age that are at least partly lateritic have been described from Northern Ireland, where bauxite is also present (Wilson, 1972), Scotland (Richey, et al., 1961, p. 44-47), Faeroe Islands (Lövlie,1975), the Iceland-Faeroe Ridge (Nilsen and Kerr, 1978), and West Greenland (Deutsch and Kristjansson, 1974). These paleosols also suggest widespread warming of land climates in northern latitudes, probably related to oceanic circulation patterns. Other paleoclimatic data suggest progressive cooling of the North Atlantic and Norwegian-Greenland Sea regions in middle and late Cenozoic time (Schwarzbach, 1961; Berggren and Hollister, 1974).

## SUMMARY AND RECOMMENDATIONS FOR FUTURE RESEARCH

The Greenland-Scotland Ridge has evolved in a complex rifted ocean basin setting and has played a critical role in the tectonic and paleogeographic development of the North Atlantic and Norwegian-Greenland Sea area. By controlling the northward flow of warm water and southward flow of cold water, it strongly influenced migration patterns of marine microfauna and microflora and affected climates over a very large region. The ridge formed the long-sought-for Thulean land bridge in the early Tertiary between North America-Greenland and western Europe.

Because the ridge developed over a hot-spot whose locus of activity shifted through Cenozoic time, its principal components are an inactive, aseismic, subsiding, submarine oceanic basalt plateau, the Iceland-Faeroe Ridge, and an active, seismic, uplifting, subaerial oceanic basalt plateau, Iceland. Shifting seafloor spreading ridges north and south of the ridge yielded completely different seafloor magnetic anomaly records. Paleogene doming around the hot-spot resulted in separation of Greenland and northern Europe along the earlier Paleozic suture that marked the Caledonian orogeny. A failed arm or aulocogen developed in northern Britain and Ireland during the early Tertiary.

Although much is known about the structure of the Greenland-Scotland Ridge and its influence on the geologic history of adjacent regions, much further work is needed. The following is a list of items which need the attention of various earth scientists working in the region:

1. The age, character, and thickness of sediments deposited in the Denmark Strait, Faeroe-Shetland Channel and Faeroe Bank Channel, on the top and flanks of the Iceland-Faeroe Ridge, and on the Wyville-Thomson and Ymir Ridges need to be determined from seismic profiling and drilling.

2. The biostratigraphy, faunal provinces, paleobathymetry, and geohistory of these channels, plateaus and ridges need to be better understood.

3. The age and character of the ocean floor around the margins of the Greenland-Scotland Ridge need to be more precisely determined.

4. The structure of the Greenland-Scotland Ridge, and particularly the Iceland-Faeroe Ridge, needs to be better defined, by deeper penetration of the basaltic basement by drilling if possible.

5. The subsidence history of the Greenland-Scotland Ridge must be better determined, preferably by drilling in the Denmark Strait, on top of the Iceland-Faeroe Ridge, and in the Faeroe-Shetland and Faeroe Bank Channels.

6. Models for regional paleocirculation and paleogeography need to be carefully examined and improved upon with new studies.

7. The history of and reasons for westward migration of hot-spot activity and shifting of seafloor spreading ridges need to be better defined and understood. The westward jump in spreading from Aegir Ridge to the Icelandic Plateau Ridge Axis requires more thorough documentation, as does the rifting of Jan Mayen Ridge from Greenland.

8. The relation of the proposed Vöring Plateau-Jan Mayen Island hot spot to the Icelandic and proposed Yermak Plateau hot spots needs to be clarified.

9. The age of the oldest volcanic rocks on the Icelandic Plateau need to be determined, probably by drilling, to better understand the history of the plateau and its relation to seafloor spreading.

10. The age of the youngest volcanic rocks on the Iceland-Faeroe Ridge needs to be determined, as well as ages of volcanic rocks older than those recovered from DSDP site 336.

11. An integrated geochemical and petrologic analysis of early Tertiary basalts from the Davis Strait, East Greenland, North Atlantic, Norwegian Sea, Iceland-Faeroe Ridge, Faeroe Islands, Rockall Plateau, northern Britain, Northern Ireland, and Vöring Plateau areas should be made to determine the nature of volcanic and magmatic processes and regional variations in the processes.

12. The failed-arm or aulocogen hypothesis for the northern British Isles volcanic province should be rigorously tested.

13. Regional and worldwide climatic changes based on movement of water masses over the Greenland-Scotland Ridge should be carefully modeled, and the proposed physical processes and changes quantitatively analysed.

14. The models proposed for the origin of sediment drifts and Oligocene unconformities in this volume need to be examined in a more quantitative manner and rigorously tested.

15. Early Tertiary laterites in the North Atlantic-Arctic region need to be studied in an integrated manner to provide more control on their possible climatic significance.

16. More fossil localities, particularly of early Tertiary terrestrial vertebrates, need to be located in the North Atlantic-Arctic region to test the analyses of migration patterns presented in this volume.

ACKNOWLEDGMENTS

I am indebted to the organizers of the NATO Advanced Research Institute on "The structure and development of the Greenland-Scotland Ridge -- New methods and concepts" in Bressanone, Italy, May 11-15, 1981, for travel support and the opportunity to present parts of this paper. Many participants at the meeting provided new ideas and valuable concepts regarding the history of the region, and some of these are incorporated into this paper. I thank my shipboard colleagues from DSDP Leg 38 for many stimulating discussions about the geological history of the Greenland-Scotland Ridge. Bill Patton and Betsy Moll of the U.S. Geological Survey provided helpful reviews of the paper.

## REFERENCES

Andrews, J.T., Guennel. G.K., Wray, J.L., and Ives, J.D., 1972, An early Tertiary outcrop in north-central Baffin Island, Northwest Territories, Canada -- environment and significance, Can. J. Earth Sci., 9:233-238.

Berggren, W.A., and Hollister, C.D., 1974, Paleogeography, paleobiogeography and the history of circulation in the Atlantic Ocean, in "Studies in paleo-oceanography," Hay, W.W., ed., Soc. Econ. Paleont. Miner. Spec. Publ., 20:126-186.

Bott, M.H.P., 1974, Deep structure, evolution and origin of the Icelandic transverse ridge, in "Geodynamics of Iceland and the North Atlantic Area", L. Kristjansson, ed., D. Reidel, Dordrecht, 33-48.

Bott, M.H.P., Browitt, C.W.A., and Stacey, A.P., 1971, The Deep structure of the Iceland-Faeroe Ridge, Marine Geophys. Res., 1:328-351.

Bott, M.H.P., Sunderland, J., Smith, P.J., Casten, U., and Saxov, S., 1974, Evidence for continental crust beneath the Faeroe Islands, Nature, 248:202-204.

Bott, M.H.P., and Watts, A.B., 1971, Deep structure of the continental margin adjacent to the British Isles, Inst. Geol. Sci. Rept. 70/14:92-109.

Brooks, C.K., 1973a, Tertiary of Greenland -- a volcanic and plutonic record of continental breakup, Am. Assoc. Petrol. Geol. Mem., 19:150-160.

Brooks, C.K., 1973b, Rifting and doming in southern East Greenland, Nature, 244:23-25.

Brown, C.W., and Mussett, A.E., 1976, Evidence for two discrete centers in Skye, Nature, 261:218-220.

Brown, P.W., and Whitley, J.E., 1976, Eastern Greenland basalts and their supposed plume origin, Nature, 260:232-234.

Casten, U., 1973, The crust beneath the Faeroe Islands, Nature Phys. Sci., 241(108):83-84.

Christie, R.L., 1964, Geological reconnaissance of northeastern Ellesmere Island, District of Franklin, Geol. Surv. Canada Mem. 331: 79 p.

Clark, D.L., 1974, Late Mesozoic and Cenozoic sediment cores from the Arctic Ocean, Geology, 2:41-44.

Crease, J., 1965, The flow of Norwegian Sea water through the Faeroe Bank channel, Deep-Sea Res., 12:143-150.

Davis, A.G., and Elliot, A.F., 1957, The paleogeography of the London Sea Clay, Geol. Assoc. Proc., 68:255-277.

Dawson, M.R., West, R.M., Langston, W., Jr., and Hutchinson, J.H., 1976, Paleogene terrestrial vertebrates -- northernmost occurrence, Ellesmere Island, Canad. Sci., 192:781-782.

Detrick, R.S., Sclater, J.G., and Thiede, J., 1977, The subsidence of aseismic ridges, Earth Planet. Sci. Lett., 34:185-196.

Deutsch, E.R., and Kristjansson, L.G., 1974, Palaeomagnetism of Late Cretaceous-Tertiary volcanics from Disko Island, West Greenland, Geophys. J. Roy. Astron. Soc., 39:343-360.

Ellett, D.J., and Roberts, D.G., 1973, The overflow of Norwegian Sea deep water across the Wyville Thomson Ridge, Deep-Sea Res., 20:819-835.

Jones, E.J.W., Ewing, J.I., and Eittrem, S.L., 1970, Influences of Norwegian Sea overflow water on sedimentation in the northern Atlantic and Labrador Sea, J. Geophys. Res., 75:1655-1680.

Kharin, G.N., Udintsev, G.B., Bogatikov, J.I., Dmitriev, M., Raschka, H., Kreuzer, M., Mohr, M., Moore, W., and Eckhardt, F.J., 1976, K-Ar ages of the basalts of the Norwegian-Greenland Sea, DSDP Leg 38, in "Initial Reports of the Deep Sea Drilling Project, Volume 38", Talwani, M., Udintsev, G., et al., U.S. Govt. Printing Office, Washington, 755-759.

Koch, B.E., 1963, Fossil plants from lower Paleocene of the Agatdalen (Angmartussut) area, central Nugssuag Peninsula, northwest Greenland, Medd. Groenl., 172.

Lindroth, C.H., 1957, "The faunal connections between Europe and North America", John Wiley, New York.

Lövlie, R., 1975, The oxidation state of some Tertiary rocks from the Faeroe Islands and its implications for paleomagnetism, Geophys., J. Roy. Astron. Soc., 40:55-65.

McKenna, M.C., 1972, Eocene final separation of the Eurasian and Greenland North American landmasses, Proc. 24th Int. Geol. Cong., 7:275-281.

Moorbath, S., Sigurdsson, J., and Goodwin, R., 1968, K-Ar ages of the oldest exposed rocks in Iceland, Earth Planet. Sci. Lett., 4:197-205.

Nilsen, T.H., 1976, Turbidites, redbeds, sedimentary structures, and trace fossils observed in DSDP Leg 38 cores and the sedimentary history of the Norwegian-Greenland Sea, in "Initial Reports of the Deep Sea Drilling Project, Supplement to Volume 38", Talwani, M., Udintsev, G., et al., U.S. Govt. Printing Office, Washington:259-288.

Nilsen, T.H., 1976a, Lower Tertiary laterite on the Iceland-Faeroe Ridge and the Thulean landbridge, Nature, 274:786-788.

Nilsen, T.H., 1976b, Sedimentation in the Northeast Atlantic Ocean and Norwegian Sea, in: "Crustal evolution in northwestern Britain and adjacent regions", Bowes, D.R., and Leake, B.E., eds., Seel House Press, Liverpool:433-454.

Nilsen, T.H., and Kerr, D.R., 1978, Paleoclimatic and paleogeographic implications of a lower Tertiary laterite (latosol) on the Iceland-Faeroe Ridge, North Atlantic region, Geol. Mag., 115:153-182.

Palmason, G., 1965, Seismic refraction measurements of the basalt lavas of the Faeroe Islands, Tectonophysics, 2:475-482.

Parson, B., and Sclater, J.G., 1977, An analysis of the variation of oceanfloor bathymetry and heat flow with age, J. Geophys. Res., 82:803-827.

Pegrum, R.M., Rees, G., and Naylor, D., 1975, "Geology of the North-West European Continental Shelf, Vol. 2, The North Sea", Graham Trotman Dudley, London: 225 p.

Purdy, J.W., Mussett, A.E., Charlton, S.R., Sackford, M.J., and English, H.N. 1972, The British Tertiary igneous province -- potassium-argon ages of the Antrim basalts, Geophys. J. Roy. Astron. Soc., 27:327-335.

Rasmussen, J., and Noe-Nygaard, A., 1970, Geology of the Faeroe Islands (pre-Quaternary), Danmarks Geol. Unders. 25:1-142.

Richey, J.E., MacGregor, A.E., and Anderson, F.W., 1961, "British Regional Geology -- Scotland, the Tertiary Volcanic Districts", Institute of Geological Sciences, H.M.S.O., Edinburgh.

Roberts, D.G., 1969, New Tertiary Volcanic centre on the Rockall Bank, eastern North Atlantic Ocean, Nature, 223:819-820.

Roberts, D.G., 1975a, Marine geology of the Rockall Plateau and Trough, Phil. Trans. R. Soc. London, Ser. A, 278:447-509.

Roberts, D.G., 1975b, Tectonic and stratigraphic evolution of the Rockall Plateau and Trough, in "Petroleum and the continental shelf of North-West Europe, 1, Geology", Woodland, A.W., ed., Applied Science Publishers, Barking:77-91.

Roberts, D.G., Montadert, L., and Searle, R.C., 1979, The western Rockall Plateau -- stratigraphy and crustal evolution, in "Initial Reports of the Deep Sea Drilling Project, Volume 48", Montadert, L., Roberts, D.G., et al., U.S. Govt. Printing Office, Washington:1061-1088.

Schilling, J.-G., 1973, Iceland mantle plume, Nature, 246:141-143.

Schilling, J.-G., 1976, Rare-earth, Sc, Cr, Fe, Co, and Na abundances in DSDP Leg 38 basement basalts -- some additional evidence on the evolution of the Thulean Volcanic Province, in "Initial Reports of the Deep Sea Drilling Project, Volume 38", Talwani, M., Udintsev, G., et al., U.S. Govt. Printing Office, Washington:741-751.

Schilling, J.-G., and Noe-Nygaard, A., 1974, Faeroe-Iceland plume -- rare-earth evidence, Earth Planet. Sci. Lett. 24:1-14.

Schnitker, D., 1980, North Atlantic oceanography as a possible cause of Antarctic glaciation and eutrophication, Nature, 284: 615-616.

Schwarzbach, M., 1961, The climatic history of Europe and North America, in "Descriptive Paleoclimatology", Nairn, A.E.M., ed., Interscience, London:255-291.

Talwani, M., and Eldholm, O., 1977, Evolution of the Norwegian-Greenland Sea, Geol. Soc. Am. Bull., 88:969-999.

Talwani, M., Udintsev, G., et al., 1976, "Initial Reports of the Deep Sea Drilling Project, Volume 38", U.S. Govt. Printing Office, Washington: 1312 p.

Tarling, D.H., and Gale, N.H., 1968, Isotopic dating and palaeomagnetic polarity in the Faeroe Islands, Nature, 218:1043-1044.

Vogt, P.R., 1972, Evidence for global synchronism in mantle plume convection, and possible significance for geology, Nature, 240:338-342.

Vogt, P.R., and Avery, D.E., 1974, Tectonic history of the Arctic basins -- partial solutions and unsolved mysteries, in "Marine geology and Oceanography of the Arctic seas", Herman, Y., ed., Springer-Verlag, New York:83-117.

Whiteman, A.J., Rees, G., Naylor, D., and Pegrum, R.M., 1975, North Sea troughs and plate tectonics, in "Petroleum geology and geology of the North Sea and northeast Atlantic continental margin", Whiteman, A.J., Roberts, D., and Sellevoll, M.A., eds. Norges Geol. Unders. 316:137-161.

Wilson, H.E., 1972, "Regional geology of Northern Ireland", Geological Survey of Northern Ireland, H.M.S.O., Belfast: 103 p.

# SEDIMENT TRANSPORT MECHANISMS AT NE ATLANTIC MARGINS: EVIDENCE FROM THE MAGNETIC ANISOTROPY OF IPOD CORES

Ernest A. Hailwood and William O. Sayre*

Oceanography Department, Southampton University
SO9 5NH, U.K.

## ABSTRACT

Results of magnetic fabric measurements on sediments drilled during Legs 47 and 48 of the International Phase of Ocean Drilling (IPOD) in the NE Atlantic are believed to define primary grain alignments, produced by the action of sediment transport processes at the time of deposition. The use of palaeomagnetic data from the same samples allows these directions of alignment to be specified with respect to magnetic north, so that the inferred sediment transport axes can be related to the palaeogeography and bathymetry at that time. At Site 403, located on the rifted westerly margin of the Rockall plateau, the transport axis deduced for Palaeogene tuffs and mudstones is aligned NW-SE, ie perpendicular to the incipient continental margin. This is consistent with deposition in a deltaic system which drained the adjacent continental areas during the early stages of rifting of Greenland from Rockall. At Site 405, located close to an E-W trending transform-fault which forms part of the southern margin of the Rockall plateau, the transport axis for Palaeogene sediments is aligned perpendicular to the fault scarp. This supports the view that these sediments were deposited in an extensive submarine fan, which prograded southwards from the transform fault scarp. Conversely at Site 398 on the western Portuguese margin the transport axis in Palaeogene sediments is closely parallel with the average local direction of the margin, and this may indicate deposition from contour-following currents. Finally, at Site 400A at the foot of the NE Biscay margin, a sequence of Aptian-Albian black shales

*Now at Department of Earth Sciences, Iowa State University, Ames, Iowa 50011, USA.

are characterised by a weak magnetic lineation, indicative of a transport direction which is perpendicular to the local margin. This supports the proposal that the high organic carbon content of these black shales is due to rapid input of terrigenous organic material, and active downslope transport, rather than deposition in a stagnant oxygen-depleted basin.

## INTRODUCTION

The alignment of individual particles within marine sediments is normally governed by the balance between gravitational and hydraulic forces at the time of deposition (eg Potter and Pettijohn, 1977). In general the gravitational couple will tend to cause the particles to lie with their long-axes within or close to the horizontal plane, whilst the hydraulic couple, applied during deposition from a bottom current, will normally cause the long axes to be aligned parallel to the local direction of flow at the instant of deposition. In the case of particles of ferromagnetic material such as magnetite there will be a third couple due to the geomagnetic field, but in general this will be weak compared with the gravitational and hydraulic couples, except for very small magnetic grains having diameters less than about 20 microns (Rees and Woodall, 1975).

Studies of preferred orientation of mineral grains in deep sea sediments can provide important information on the strength and direction of sediment transport agents, which in turn can help in the development of palaeogeographic and palaeoenvironmental models. However work in this field has been severely inhibited by the tedious and time consuming nature of conventional techniques for grain orientation measurement. In recent years it has been demonstrated that the net alignment of grains within a sedimentary sample can be determined with rapidity and accuracy from measurements of the magnetic susceptibility anisotropy (eg Hamilton and Rees, 1970). The present paper describes results obtained from the application of this magnetic fabric technique to sediment samples from the NE Atlantic margins cored during Legs 47 and 48 of the International Phase of Ocean Drilling (IPOD).

## PRODUCTION AND SPECIFICATION OF MAGNETIC ANISOTROPY IN SEDIMENTS

The magnetic susceptibility, K, represents the relationship between the induced magnetisation, J, and the external magnetic field, H, applied to a material.

$$\text{ie} \quad K = J/H \qquad \text{------------------------------------} \quad (1)$$

In the case of an ellipsoidal grain of magnetic material the magnetic susceptibility will vary with direction, and this phenomenon is known as <u>magnetic anisotropy</u>. For such a grain the maximum susceptibility direction ($K_{max}$) will correspond with the grain long axis and the minimum susceptibility direction ($K_{min}$) with the short axis.

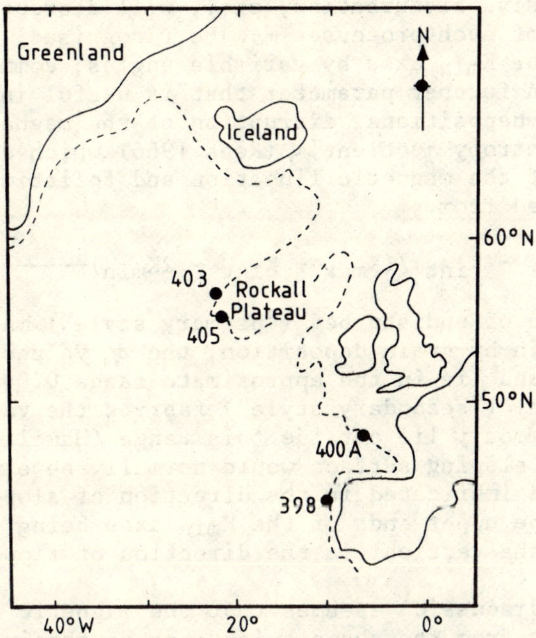


Figure 1. Location of IPOD drill sites for which magnetic anisotropy data are presented in this paper.

A third, intermediate, susceptibility axis ($K_{int}$) will complete the cartesian set. Similarly, for any irregular-shaped grain with a small degree of elongation the variation of susceptibility with direction may be approximated by a triaxial ellipsoid, defined by the three orthogonal principal axes, of length $K_{max}$, $K_{int}$ and $K_{min}$. In a sediment sample comprising an assemblage of such grains the $K_{max}$ axis will correspond with the direction of preferred orientation of the individual grain long dimensions, and the $K_{min}$ axis with that of the short dimensions. Thus for a sediment deposited in the presence of a bottom current the $K_{max}$ axes for different samples would be expected to group close to the azimuth of flow, whilst the $K_{min}$ axes will group close to the vertical. Any subsequent disturbance of the sediment, due to slumping, extensive bioturbation, etc., will disrupt this alignment, and the effect of such processes may be recognised, for example, by departures of the $K_{min}$ axes by variable angles, commonly $\geqslant 20^{\circ}$ from the vertical. A further parameter that is useful in identifying the effects of post-depositional disruption of the magnetic fabric is the "Azimuthal anisotropy quotient" $q$ (Rees 1966) which defines the relative strength of the magnetic lineation and foliation. This parameter may be calculated from

$$q = 2(K_{max} - K_{int})/(K_{max} + K_{int} - 2K_{min}) \qquad (2)$$

In the case of undisturbed ("primary style") magnetic fabrics, produced by grain-by-grain deposition, the $q$ values are normally rather uniform and lie in the approximate range 0.06 to 0.67, whereas for deformational ("secondary style") fabrics the values become erratic, and commonly lie outside this range (Hamilton & Rees, 1970). Deposition on a sloping surface would normally be expected to cause the grains to be imbricated in the direction of slope. This may be recognised by the upper ends of the $K_{min}$ axes being systematically displaced from the vertical in the direction of slope.

The above discussion assumes that the magnetic fabric of the sediment results from the shape anisotropy of the individual grains. Although other types of magnetic anisotropy are recognised (notably magnetocrystalline anisotropy) these are unlikely to be important in the case of sediment in which the dominant magnetic material is detrital magnetite. Furthermore, it is assumed that the orientation of the magnetic particles is essentially the same as that of the non-magnetic quartz, felspar, clay-minerals, etc. Comparative studies (eg Taira and Lienert, 1979) suggest that this will normally be the case. In the present study all magnetic anisotropy measurements were performed by means of a low-field torque magnetometer (King & Rees, 1962), using an applied field of approximately 100 Oe RMS.

## CORE ORIENTATION

One of the difficulties with grain orientation studies of IPOD sediment samples is that the cores are not orientated azimuthally

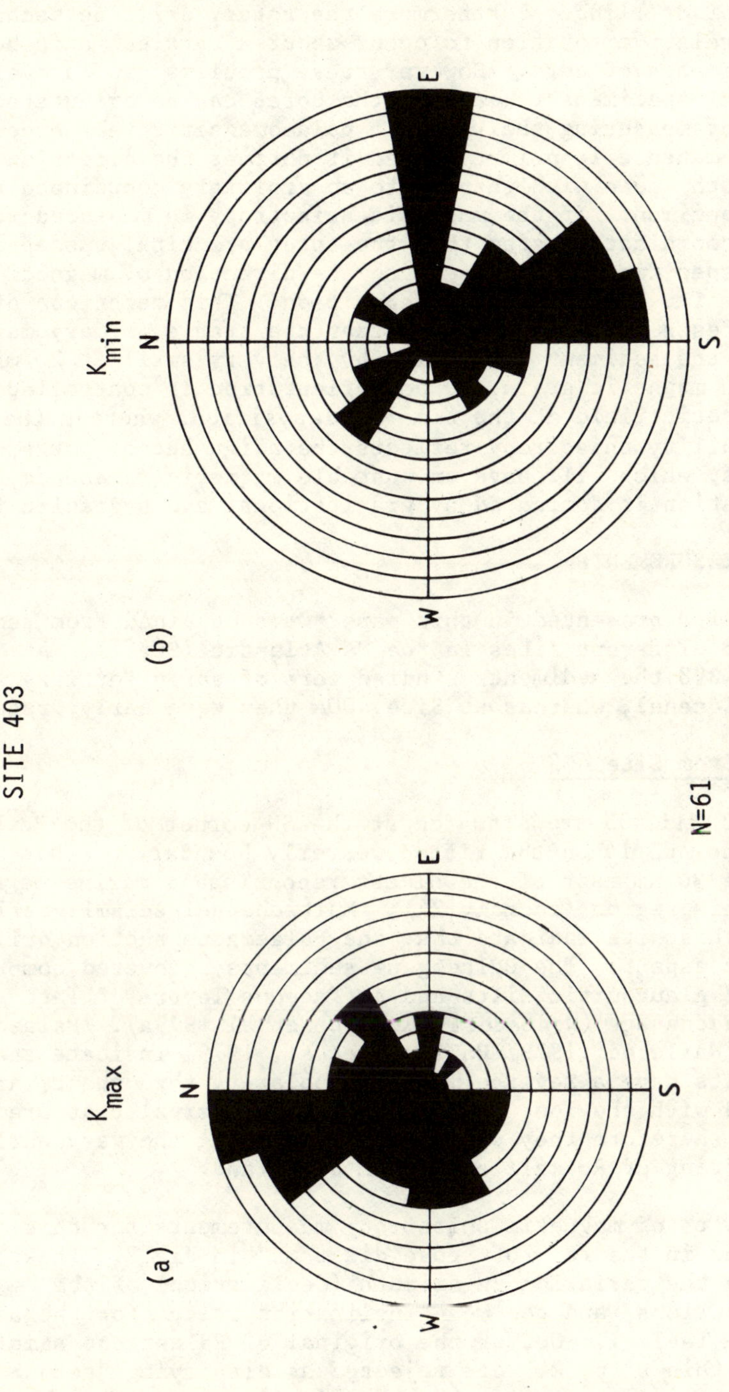


Figure 2. Site 403 Palaeogene sediments. Azimuthal distribution of upper ends of $K_{max}$ and $K_{min}$ axes for samples exhibiting "primary style" magnetic fabrics.

at the time of drilling. Furthermore the rotary drilling technique used causes relative rotation to occur about a vertical axis between different segments of core. However these problems can be overcome and individual specimens taken from the cores can be orientated azimuthally by measuring their stable palaeomagnetic remanence. Provided this remanence is primary, then it defines the direction of the North (or South) pole with respect to an arbitrary coordinate system within the specimen. If the magnetic anisotropy is measured relative to the same coordinate system then the three principal susceptibility axes can be specified with respect to the direction of magnetic north at the drill site at the time of deposition. This technique of core orientation rests on the assumption that the stable primary magnetic remanence of the sediment is carried by the very small (< 20 microns) single-domain magnetic grains, whose orientation is controlled by the local geomagnetic field at the time of deposition, whereas the magnetic susceptibility anisotropy reflects the alignment of larger multidomain grains, which will have an unstable magnetic remanence, but whose orientation is dominated by gravitational and hydraulic forces.

## RESULTS OF MEASUREMENTS

The results presented in this paper were obtained from sediments cored at four different sites in the NE Atlantic (Fig.1). At Sites 403, 405 and 398 the sediments studied were of early Tertiary age (Palaeocene-Eocene), whereas at Site 400A they were early Cretaceous.

### (i) Results from Site 403

Sites 403 and 405 are situated at the SW corner of the Rockall plateau. Site 403 is on the rifted westerly boundary of this plateau, some 30 Km east of the oldest recognisable marine magnetic anomaly in this region (anomaly 24). Multichannel seismic reflection profiles at this site indicate that the Palaeogene section drilled is of deltaic aspect. The Palaeogene sediments recovered comprise 270 metres of glauconitic silts and tuffaceous layers of late Palaeocene-early Eocene age (Montadert, Roberts et al 1979a). Palaeomagnetic studies (Hailwood, 1979; Hailwood et al, 1979) indicate that these sediments have a reverse magnetic polarity throughout, and may be correlated with the long reverse polarity interval that preceded anomaly 24. Therefore they were deposited during the very early stages of rifting of Rockall away from Greenland.

The results of magnetic anisotropy measurements for this site are summarised in the form of "rose diagrams" in Fig.2. These diagrams display the variation in azimuth (declination) of the $K_{max}$ and $K_{min}$ directions, and the mean inclination values for the section are listed in Table 1. Out of the original 82 Palaeogene samples measured for this site, 21 were rejected as displaying "secondary style" characteristics, using the criteria discussed in an earlier section. Although a small degree of bioturbation is present in these sediments the principal cause of the deformed fabric in the rejected samples is believed to be drilling disturbance. The $K_{min}$ axes for the remaining 61 samples are grouped closely around the vertical, with a mean inclination of $-87^{o} \pm 4^{o}$ for their upward ends. Similarly the $K_{max}$ axes are grouped close to the horizontal (Table 1). However there appears to be a clear concentration of the upward ends of the $K_{max}$ axes in the NW quadrant*(Fig.2a), indicating a preferred orientation of grain long axes along a NNW-SSE trend. If this alignment were due to bottom currents then the direction of flow would have been towards the NNW, since the grains are imbricated upwards in this sense.

Similarly, the orthogonal $K_{min}$ axes are systematically displaced from the vertical by a small angle into the SE quadrant. Consequently these results are consistent with deposition on a near flat bed from bottom currents having an average flow direction towards the NNW. Alternatively they may be accounted for by down slope transport on a margin having a persistent dip towards the SSE, but in view of the low dips measured in the Site 403 section this is regarded as less likely. When plotted on a palaeomagnetically determined reconstruction of the NE Atlantic region for early Palaeogene times this NNW-SSE axis is seen to be perpendicular to the incipient continental margin (Fig.6a). Therefore the magnetic fabric results support the view that the early Palaeogene tuffs and mudstones cored at Site 403 were deposited in a shallow water delta system which drained the adjacent continental areas. The relatively wide dispersion of the axes apparent in Fig.2 may be partly due to the complex flow conditions to be anticipated in such a deltaic environment. However some of the variability may be attributed to inaccuracies in the palaeomagnetic orientation method, since the relatively high sedimentation rates at this site (~ 80m/m.y.) result in secular variation of the geomagnetic field being only partially averaged out in a single 2.5 cm cylindrical sample. Scatter due to this source should be random, so that the overall mean $K_{max}$ direction will still correspond with the mean direction of flow.

(ii) Results from Site 405

Site 405 is located on the transform fault-controlled southerly margin of the Rockall plateau (Fig.1). In the lower part of the drill hole a thick sequence of Eocene calcareous and siliceous mudstones were encountered. Multichannel seismic reflection profiles suggest that these sediments form part of an extensive submarine fan system which appears to have been built outwards by southward progradation from the transform fault scarp (Montadert, Roberts et al., 1979b).

*This direction is in the opposite sense to that cited in the preliminary report by Hailwood & Sayre (1979) due to a palaeomagnetic orientation error.

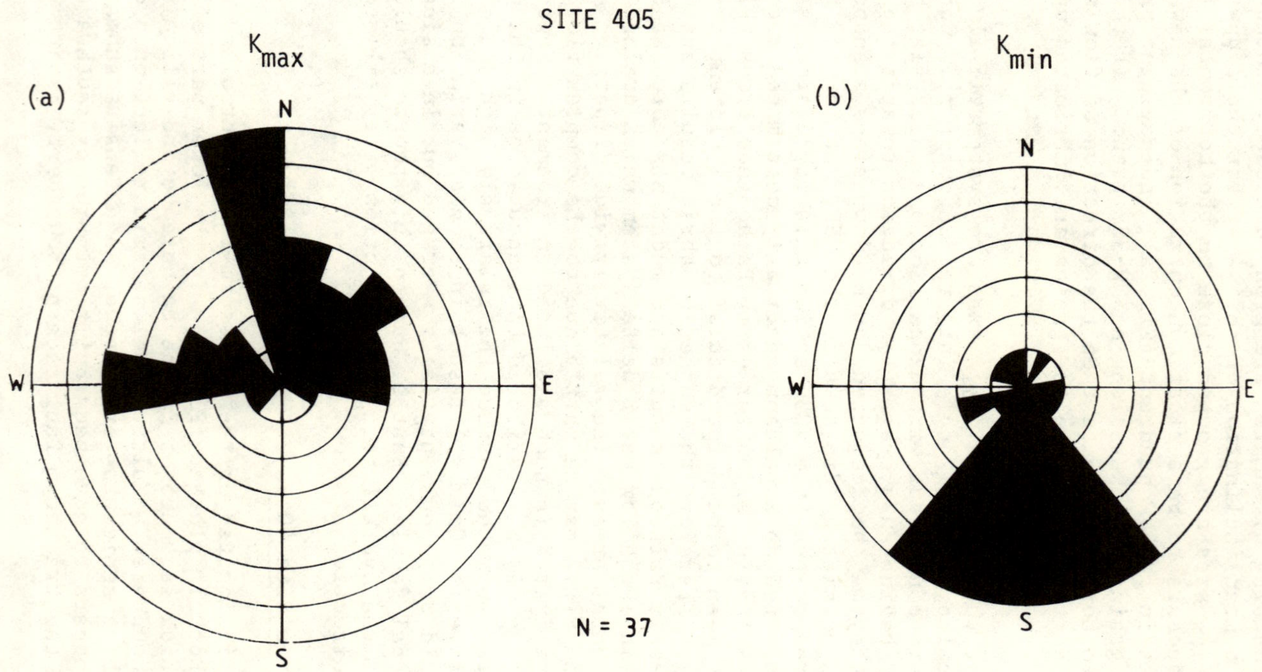


Figure 3. Site 405 Palaeogene sediments. Azimuthal distribution of upper ends of $K_{max}$ and $K_{min}$ axes for samples exhibiting "primary style" magnetic fabrics.

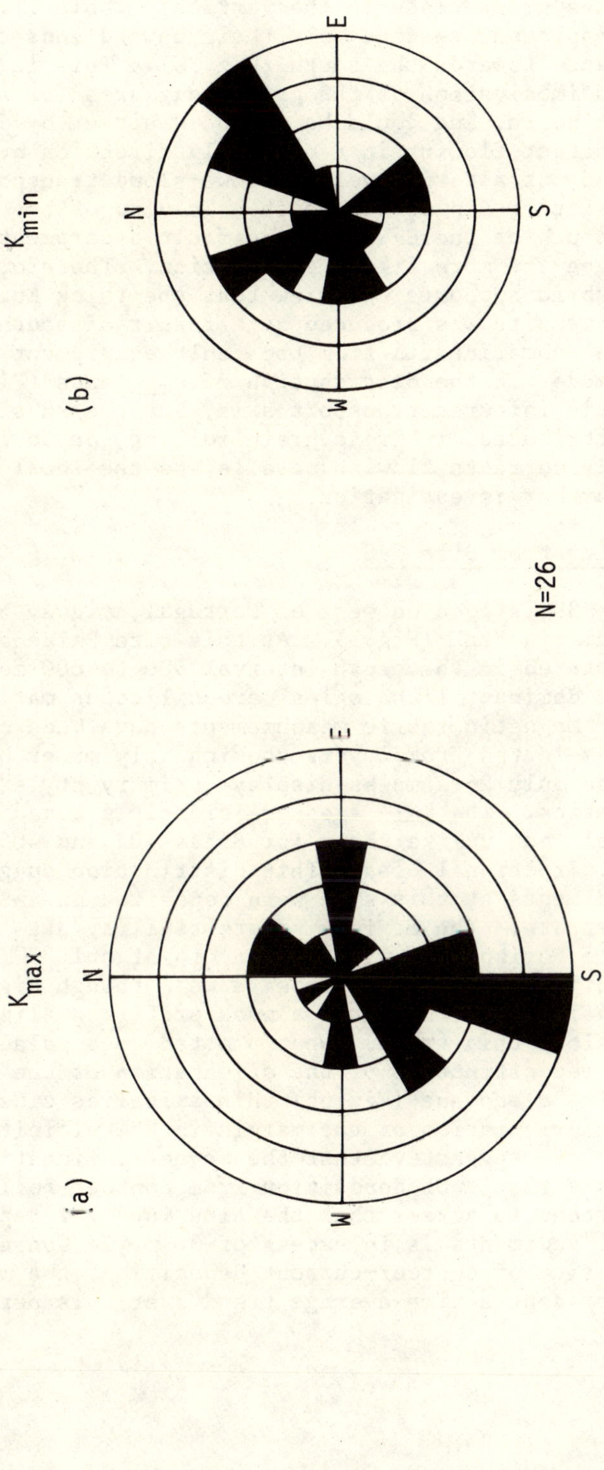


Figure 4. Site 398 Palaeogene sediments. Azimuthal distribution of upper ends of $K_{max}$ and $K_{min}$ axes for samples exhibiting "primary style" magnetic fabrics.

The principal magnetic fabric results for this site are summarised in Fig.3. Approximately 85% of the samples measured display "primary style" fabric characteristics. Although the $K_{min}$ axes for these samples group close to the vertical (Table 1), there is a clearly conspicuous tendency for their upward ends to be systematically displaced towards the south (Fig.3b). This indicates an overall upwards imbrication of the grain long axes (or flat surfaces) towards the north, and could be produced either by deposition from a bottom current flowing in a northerly direction over an essentially flat sea bed, or alternatively by down-slope transport along a bed with a persistent dip to the south. In view of the location of this site just south of the E-W trending fault escarpment the latter appears to be the more likely explanation. Therefore the observed magnetic fabric supports the view that the thick Eocene sediment wedge at this site was produced as a result of southward progradation of a submarine fan from the fault escarpment (Fig 6a). The northerly mode in the distribution of $K_{max}$ axes (Fig.3a) is parallel with this inferred transport axis, but the subsidiary E-W mode could be attributed either to grain rolling, or to intermittent transport by currents flowing parallel to the local slope. This requires further investigation.

(iii) Results from Site 398

Site 398 is located west of Portugal, midway between the mainland and Galicia Bank (Fig.1). At this site Palaeogene sediments were encountered in the depth interval 500 to 800 metres below sea floor. The dominant lithologies were siliceous marly chalks and mudstones. Magnetic fabric measurements have been carried out on 46 samples selected from layers showing only minor bioturbation, but of these only 26 samples display "primary style" magnetic fabric characteristics. The $K_{min}$ axes are closely grouped around the vertical (Fig.4b), but unlike those for Sites 403 and 405, they show little systematic directional bias. This distribution suggests that the Palaeogene sediments at this site were deposited on an essentially horizontal substrate. The maximum susceptibility axes themselves lie close to the horizontal, with a mean dip of only $3^{o}$ (Table 1), but their azimuthal distribution shows a weak though significant S to SSW mode (Fig.4a). This indicates a weak preferred alignment of grain long axes along this trend. When plotted on a palaeomagnetically determined reconstruction of the orientation of the Portuguese margin in the early Palaeogene (Fig.6b) this axis lies closely parallel to the average orientation of the margin in the vicinity of Site 398. The results are suggestive that the magnetic lineation may have been produced as a result of deposition from contour-following currents. It is important to stress that the time interval represented by this sequence of sediments is in excess of 30 m.y. Consequently, if this has been a site of contour-current deposition the magnetic fabric results represent a time-average trend over this period.

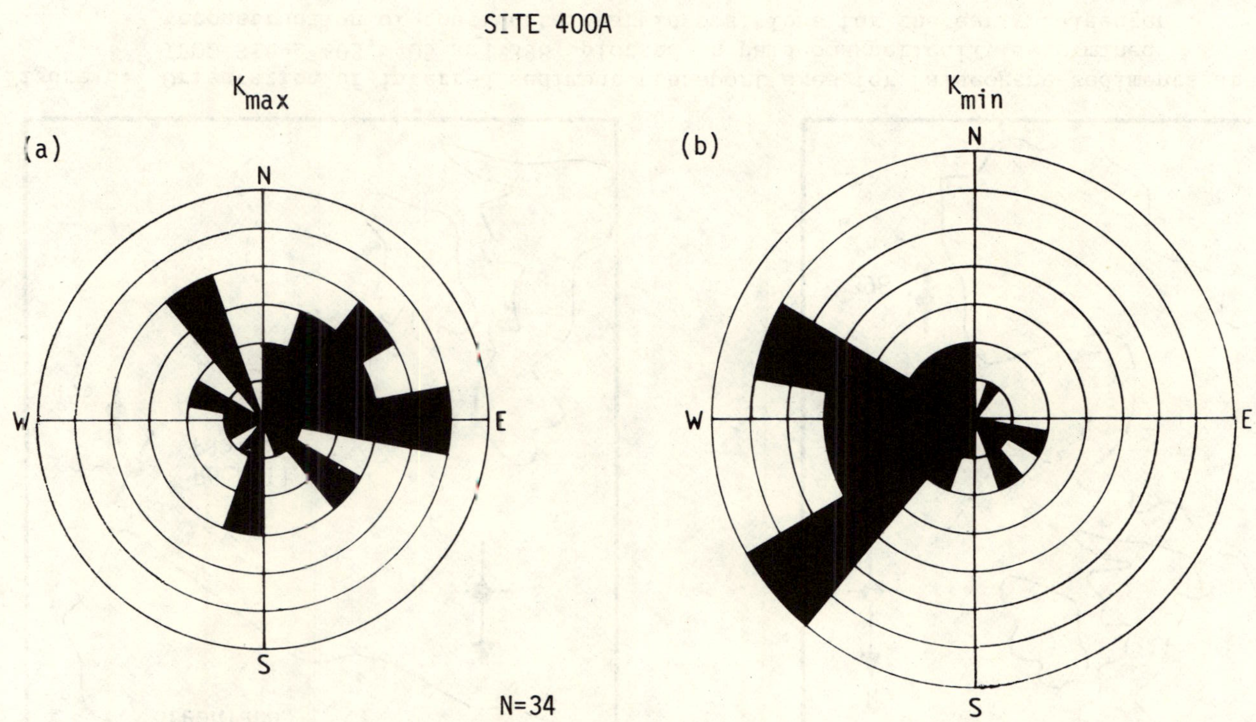


Figure 5. Site 400A Aptian-Albian black shales. Azimuthal distribution of upper ends of $K_{max}$ and $K_{min}$ axes for samples exhibiting "primary style" magnetic fabrics.

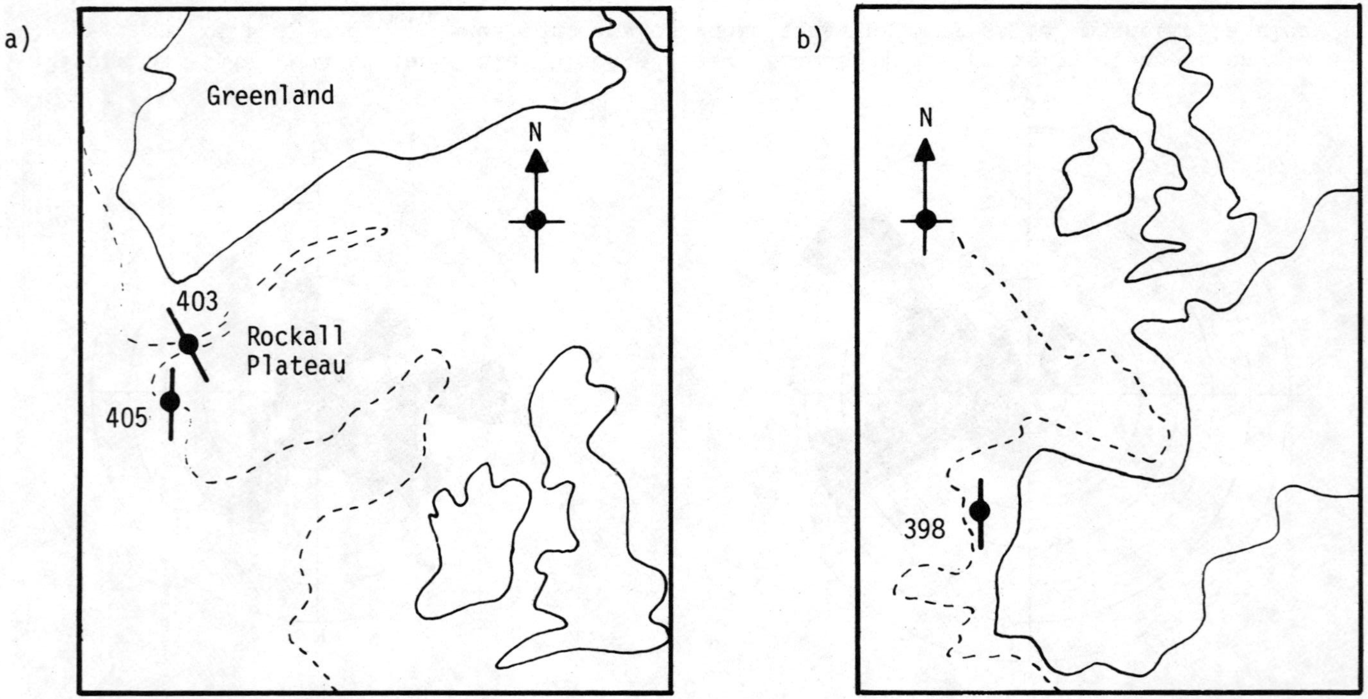


Figure 6. Orientation of inferred sediment transport axes for Palaeogene sediments at IPOD Sites 403, 405 and 398, plotted on palaeomagnetically determined reconstruction of continental margin positions for the early Palaeogene.

## (iv) Results from Site 400A

Site 400A was drilled at the base of the continental slope on the NE margin of the Bay of Biscay. At this site a 120m thick sequence of early Cretaceous (Aptian-Albian) organic-rich black shales was cored, and faunal evidence indicates that deposition of these sediments occurred at water depths of about 2000m. These sediments exhibit a very low degree of bioturbation and 90% of the 38 samples subjected to magnetic fabric investigations possessed "primary-style" fabrics. The maximum susceptibility axes show a slight concentration in the NE quadrant, and define a weak lineation with an average ENE trend (Fig.5). There is a clear tendency for the upward-directed ends of the minimum susceptibility axes to concentrate around a W to WSW azimuth. This situation could be produced as a result of downslope transport in this direction. On an early Cretaceous palaeomagnetic reconstruction of the Biscay margin (Fig.7), this axis lies perpendicular to the general trend of the margin. It may be concluded that the magnetic fabric of the Aptian-Albian organic rich sediments in this area is indicative of deposition at a site of active downslope transport, rather than in a stagnant basin.

## CONCLUSIONS

The application of the magnetic fabric technique to investigating sediment transport mechanism in deep sea sediments is still in its infancy. However, the results presented in this paper have provided some additional insight into the nature of sediment transport processes at certain points on the NE Atlantic margins. Although considerable care has to be exercised in the selection of suitable materials to avoid sediments that have suffered coring disturbance, bioturbation and other forms of mechanical deformation, the potential of the technique in deep sea sedimentological studies is readily apparent.

Table 1. Mean Inclinations of Principal Susceptibility Axes

| Site | $K_{max}$ | $K_{min}$ | Age |
|---|---|---|---|
| 398 | $-3\pm3^{\circ}$ | $-84\pm3^{\circ}$ | Palaeocene-Oligocene |
| 400A | $-6\pm5^{\circ}$ | $-79\pm4^{\circ}$ | Aptian-Albian |
| 403 | $-4\pm3^{\circ}$ | $-82\pm4^{\circ}$ | Late Palaeocene-early Eocene |
| 405 | $-6\pm4^{\circ}$ | $-79\pm3^{\circ}$ | Early-middle Eocene |

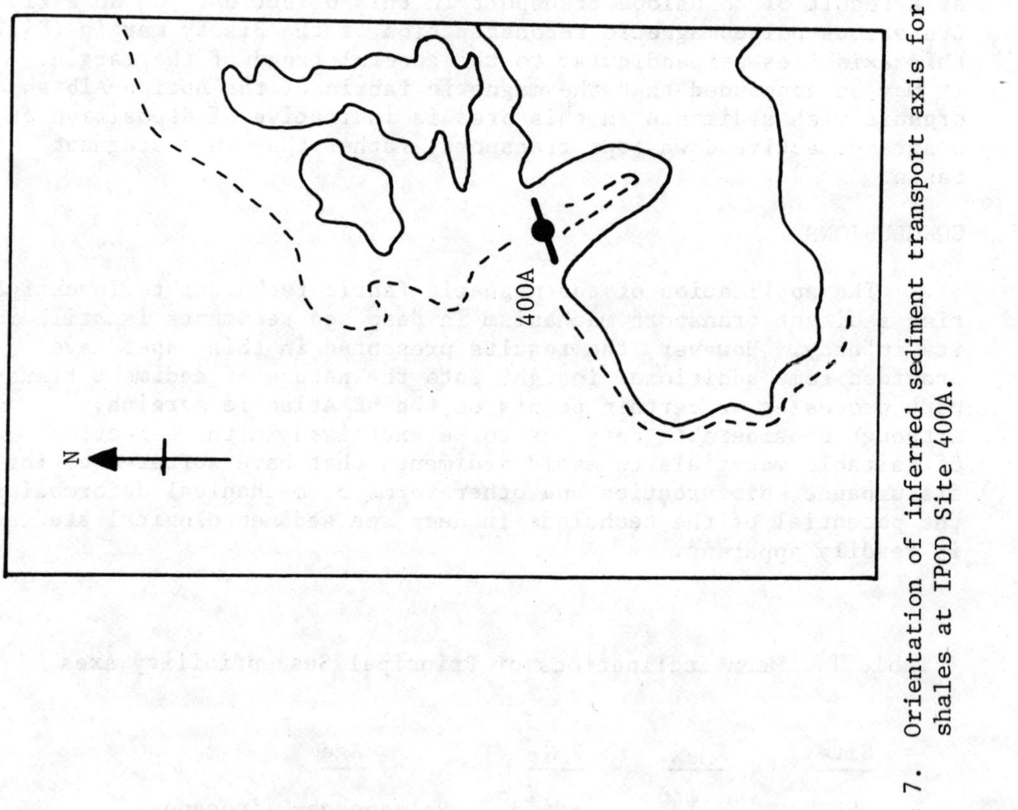


Figure 7. Orientation of inferred sediment transport axis for Aptian-Albian black shales at IPOD Site 400A.

BIBLIOGRAPHY

Hailwood, E. A., 1979. Palaeomagnetism of late Mesozoic to Holocene sediments from the Bay of Biscay and Rockall Plateau, drilled on IPOD Leg 48. In Montadert, L., Roberts, D.G., et al., Initial Reports of the Deep Sea Drilling Project, Vol. XLVIII, Washington (U.S. Government Printing Office), 305-339.

Hailwood, E. A., Bock, W., Costa, L., Dupeuble, P. A., Müller, C. and Schnitker, D., 1979. Chronology and biostratigraphy of Northeast Atlantic sediments, DSDP Leg 48. In Montadert, L., Roberts, D. G., et al., Initial Reports of the Deep Sea Drilling Project, Vol. XLVIII, Washington (U.S. Government Printing Office) 1119-1141.

Hailwood, E. A. and W. O. Sayre, 1979. Magnetic anisotropy and sediment transport directions in North Atlantic early Cretaceous black shales and Eocene mudstones cored on DSDP Leg 48. In Montadert, L., Roberts, D.G., et al., Initial Reports of the Deep Sea Drilling Project, Vol. XLVIII, Washington (U.S. Government Printing Office) 909-918.

Hamilton, N. and A. I. Rees, 1970. The use of magnetic fabric in palaeocurrent estimation. In Runcorn, S. K. (ed.), Palaeogeophysics: New York (Academic Press), 445-464.

King, R. F. and A. I. Rees, 1962. The measurement of the anisotropy of magnetic susceptibility by the torque method. J. Geophys. Res., Vol.67, 1565-1572.

Montadert, L., Roberts, D. G., et al., 1979a. Sites 403 and 404. In Montadert, L., Roberts, D. G., et al., Initial Reports of the Deep Sea Drilling Project, Vol. XLVIII, Washington (U.S. Government Printing Office), 165-210.

Montadert, L., Roberts, D. G., et al., (1979b. Sites 405 and 406. In Montadert, L., Roberts, D. G., et al., Initial Reports of the Deep Sea Drilling Project, Vol. XLVIII, Washington (U.S. Government Printing Office, 211-274.

Potter, P. E. and F. J. Pettijohn, 1977. Palaeocurrents and basin analysis. Springer Verlag, N.Y., 425pp.

Rees, A. I., 1966. The effect of depositional slopes on the anisotropy of magnetic susceptibility of laboratory deposited sands, J. Geol., 74, 856-867.

Rees, A. I. and W. A. Woodall, 1975. The magnetic fabric of some laboratory deposited sediments, Earth Planet. Sci. Lett., 25, 121-130.

Taira, A. and B. R. Lienert, 1979. The comparative reliability of magnetic, photometric, and microscopic methods of determining the orientation of sedimentary grains. J. Sed. Pet., 49, 759-772.

# CENOZOIC MARINE ENVIRONMENTS IN THE NORTH ATLANTIC AND NORWEGIAN-GREENLAND SEA

W. A. Berggren and Detmar Schnitker

Woods Hole Oceanographic Institution, Woods Hole, Mass. 02543 and Brown University, Providence, R.I., 02912
University of Maine, Orono-Walpole, Maine 04573

## ABSTRACT

A warm ocean, with an abundant and diverse flora and fauna invaded the newly forming branches of the North Atlantic, first the Labrador Sea (Late Cretaceous), then, beginning with the Eocene, the northeastern North Atlantic. Land and marine biotas at high latitudes indicate equable climates and low Paleogene latitudinal thermal gradients. The Greenland-Scotland Ridge, forming along with the opening ocean may have acted as an oceanic barrier, but not as an impenetrable hindrance to surface exchange of flora and fauna. Rising rates of biogenic silica deposition during the Eocene suggest high fertility, and thus, upwelling. The global Eocene/Oligocene cooling impoverished North Atlantic planktonic floras and faunas, while the deep water environment was drastically changed, both qualitatively and dynamically. Extinctions, speciations and fundamental depth and abundance readjustments of deep water benthic faunas bear witness to the physicochemical and structural changes of the deep hydrosphere; unconformities and a change in the sedimentation regime suggest a marked intensification of deep currents. It is uncertain whether these events are related to the onset of North Atlantic bottom water formation. Calcareous oozes on Vøring Plateau indicate that substantial influx of North Atlantic surface water into the Norwegian Sea occurred during the middle and late Oligocene, following the opening of a seaway to the Arctic. An early Miocene warming trend was interrupted at the beginning of middle Miocene, expressed in the establishment of cool and temperate microfaunas and floras, as well as in oxygen isotope ratios. A dramatic drop in North Atlantic biological productivity, a change in the style of sedimentation (drift deposits), and the appearance of carbon-isotopically young water coincides with an inferred subsidence of the Iceland-

Faeroe Ridge in the middle Miocene. Initiation of cold water reflux across the ridge was probably the last principal event towards the establishment of the modern global deep water circulation; the benthic microfaunas assumed their present day aspects through a major taxonomic turnover during the middle Miocene.

## INTRODUCTION

The northern North Atlantic presently exerts a strong influence upon oceanic circulation, energy budgets and climate which, when seen in a global perspective, appears to be rather disproportionate to the size of this ocean basin. Whereas the subtropical gyres of other oceans are essentially closed towards high latitudes, the North Atlantic subtropical gyre bleeds off substantial volumes of warm water into the subpolar and polar environment. As a consequence of the accompanying energy flux, most indicators of climate in northwestern Europe are strongly shifted northward of their expected zonal distribution. Positive feedbacks accompany the oceanic cause: storm systems, themselves transfer agents of large amounts of energy, follow the track of warm water into high latitudes (Lamb, 1972); the large areas of high latitude sea and land which have been kept ice and snow free by the warm water masses, contribute to a relative heat gain through their lowered albedo. These North Atlantic waters, which have previously been rendered very saline through evaporative losses in the Caribbean and the Mediterranean, sink after cooling north of the Iceland Ridge and reflux as the principal component of the North Atlantic Deep Water at a rate about equal to that of the Antarctic Bottom Water from the Weddell Sea (Worthington, 1976; Carmack, 1977). The influence of the NADW can be traced through all ocean basins (Reid and Lynn, 1971), and through its derivative, the Antarctic Warm Deep Water, is deeply involved in the formation of Antarctic Bottom Water as well (Foster and Carmack, 1977). The intimate relationship between northern North Atlantic oceanography and northern Hemisphere climatic fluctuations has long been noted; however, Ruddiman and McIntyre (1979) and Rudiman et al. (1980) recently described a phase lag relationship between sea-surface temperatures, sea-ice cover and the buildup and decay of northern Hemisphere glaciers: sea surface temperatures stay high and the sea remains ice-free to supply abundant moisture during periods of rapid ice-buildup, whereas winter ice cover reduces evaporation during periods of deglaciation. While not the cause of the glacial-interglacial contrast, this oceanographic response provides a strong positive feedback, amplifying the ice-buildup and decay processes. Global deep water circulation being so strongly dependent upon the reflux across the Iceland Ridge, can in turn easily be influenced by the climatic and oceanographic surface conditions in the northern North Atlantic. Full glacial conditions and sea-ice cover in the northern North Atlantic can apparently arrest North Atlantic Deep Water formation and force a fundamental rearrangement of the global deep water circulation (Schnitker, 1974,

1980a). In this paper we intend to bring together the essential information on the evolution of the North Atlantic paleoenvironments and paleoceanography from evidently very different conditions during the early Cenozoic to the present, and to show that this evolution occurred largely as a consequence of the paleogeographic development of the North Atlantic through seafloor spreading.

## PALEOGEOGRAPHY

Our understanding of the complex Cenozoic plate tectonic history of the North Atlantic and the North Sea has increased considerably as a result of several studies and summaries that have appeared over the past decade (Laughton, 1971, 1975; Pitman and Talwani, 1972; Ziegler, 1975a, b, 1979; Tucholke and Vogt, 1979; Eldholm and Thiede, 1980; Gradstein and Srivastava, 1980). A brief summary will suffice here for the purpose of this study.

The North Sea was the focus of the taphrogenic rifting during the Mesozoic, which changed to an intracratonic drifting stage in the early Paleocene (latest Danian). Spreading, however, never occurred and during the last 60 my the North Sea has been characterized by regional subsidence and the formation of an intracratonic basin in which terrigenous sediments up to 3.5 km thick have been deposited.

By late Cretaceous time (anomaly 32 = 73 Ma) separation of North America from Greenland began opening the southern Labrador Sea. In the early Eocene (anomaly 24 = 54 Ma) a change from a two to three plate geometry occurred when spreading along the Reykjanes Ridge began, resulting in the separation of Greenland from Rockall and Eurasia, opening the Norwegian Sea. During the Eocene Greenland moved essentially in a NNW direction relative to Svalbard along a regional transform fault. Although there was as yet no deep water connection to the Arctic, the Greenland and Norway Basins were linked by a deep water connection shortly after the initiation of spreading in the early Eocene. The possibility of further connections to the south with the North Atlantic remains enigmatic and is discussed below.

During the late Eocene - early Oligocene (prior to anomaly 13 time) a change in relative motion between Greenland and Eurasia to WNW was associated with cessation of spreading in the Labrador Sea and a return to a two plate geometry with Greenland becoming a part of the North American plate. Surface water connection of the Atlantic with the Arctic probably has existed through the Labrador Sea since late Cretceous time. Seafloor spreading between Greenland and Svalbard did not begin until early Oligocene (anomaly 13 = 36 Ma) but a shallow water connection to the Arctic may have existed prior to that time. The possibility of a surface water connection across the Iceland ridge is discussed in greater detail below.

During the Oligocene and Neogene the North Atlantic and Norwegian-Greenland Sea basins continued to widen. A deep water connection to the Arctic was probably established during the early Miocene.

Basinal filling overcame subsidence rates in the North Sea during the late Paleogene and the gradual withdrawal of the epicontinental seas of Western Europe led to the establishment of the modern day European coastal and North Sea configuration.

## NORTH ATLANTIC CENOZOIC PALEOENVIRONMENTS

### Norwegian-Greenland Sea

Sixteen sites have been drilled in the Norwegian-Greenland Sea by the Deep Sea Drilling Project (Fig. 1). Although discontinuous coring and generally poor recovery characterize the core data in this region, a composite Eocene to Pleistocene record can be reconstructed. The presence of biogenic siliceous deposits from the Eocene to the Pleistocene suggests high primary productivity, and thus high nutrient levels, throughout the Cenozoic. Calcareous nannoplankton oozes are developed briefly during the mid Oligocene on the Iceland Ridge and the Vøring Plateau. Five sites have been considered for this study: 336 and 352 (Iceland Ridge), 348 (Iceland Plateau), 338 and 343 (Vøring Plateau).

A broad spectrum of microfossils has been studied in the Norwegian-Greenland Sea i.al., calcareous nannoplankton, foraminifers, radiolaria, diatoms, silicoflagellates, dinoflagellates (see Table 1).

Paleogene sediments have been recovered at DSDP sites 336, 352, 338 and 343; at DSDP Site 348 terrigenous mudstones with a bathyal agglutinated benthic foraminiferal fauna lie upon basaltic basement dated at 18-19 Ma (Talwani, Udintsev et al., 1976, p. 608).

On the Vøring Plateau early Eocene terrigenous sandy marls are replaced upwards by pelagic siliceous (diatomaceous) ooze in the middle Eocene (Fig. 2). At both sites (338 and 343) a bathyal benthic foraminiferal fauna occurs, having affinities with the lower Eocene faunas of N.W. Europe: *Lenticulina decorata*, *Stilostomella spinulosa*, *Quadrimorphina paleocenica*, *Cancris mauryae*, *Textularia plummerae*, *Spiroplectammina spectabilis*. The latter species ranges through the Eocene at Site 338, comparable to its observed range in the Labrador Sea (Site 112; Miller et al., 1981). Planktonic foraminifers occur sparsely on the Vøring Plateau in early Eocene and belong predominantly to the *Subbotina patagonica* group (referred to as the *Globigerina linaperta* group in the Leg 38 reports) which is the dominant taxon for a brief interval in the early Eocene of N.W. Europe and North Sea, corresponding to planktonic foraminiferal

Figure 1. Map of North Atlantic and Norwegian-Greenland Sea showing location of DSDP Sites discussed in text.

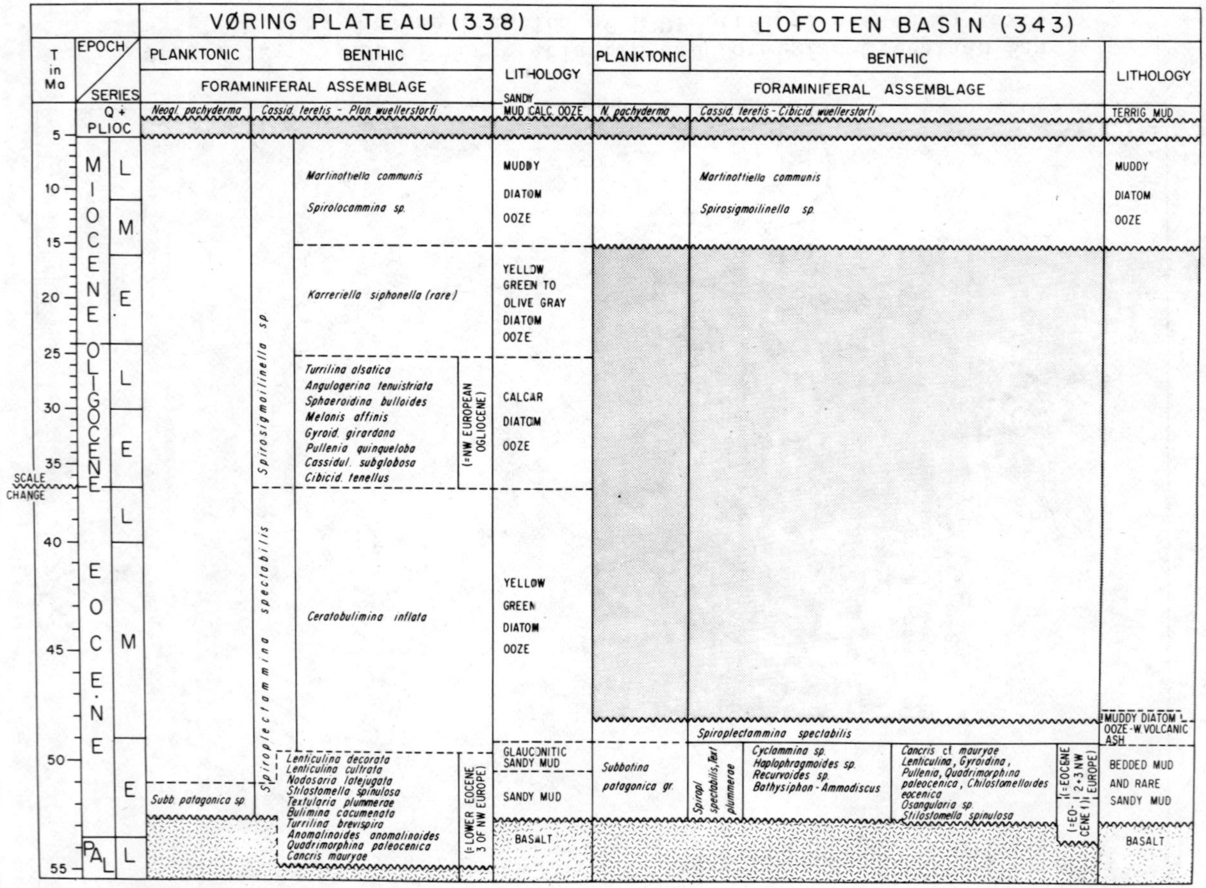


Figure 2. Biostratigraphy and lithostratigraphy of DSDP Sites 338 (VØring Plateau) and 343 (Lofoten Basin).

Zone P7. In the Norway Basin, Core AS-9 (Saito et al., 1967) contains an early Eocene calcareous nannoflora (Müller, 1976), and foraminiferal fauna (*contra* Saito et al., 1967). Reexamination of the planktonic fauna reveals the presence of the *Subbotina patagonica-triangularis* group, various acarininids, and a keeled morozovellid belonging to the *M. aequa-lensiformis* group. Benthic species include *Bathysiphon*, *Rhabdammina*, *Glomospira*, *Ceratobulimina perplexa*, *C. tuberculata*, *Pulsiphonina prima*, *Anomalinoides midwayensis*, *Cibicidoides proprius* (= *alleni*), *Osangularia convexa* and various allomorphinids, chilostomellids, and pleurostomellids - a fauna of mid-bathyal depth, having affinities with other contemporaneous faunas in the Norwegian-Greenland Sea.

A relatively diverse calcareous nannoplankton flora occurs in the lower Eocene sediments of the Vøring Plateau. Characteristic elements include *Imperiaster obscurus*, *Zygolithus dubius*, *Micrantholithus mirabilis*, *Discolithina pulchra*, *Discoasteroides kuepperi*; rare specimens of *Discoaster lodoensis*, *Chiasmolithus solitus*, and *Marthasterites tribrachiatus* have also been recorded enabling correlation with the standard calcareous nannoplankton zonation developed at low latitudes. *Imperiaster obscurus* has been interpreted as having been a "cold water form" and together with *Micrantholithus mirabilis*, *Discolithina fimbriata*, and *D. pulchra*, indicative of neritic environments (Müller; 1976, p. 827). These species occur together in the early Eocene of Rockall Bank and in northern Europe. These faunal and floral data support the existence of a marine communication, either via the Iceland-Faeroe Ridge or via epicontinental seas, between the Norwegian-Greenland Sea and the North Atlantic in the early Eocene, shortly after the initiation of seafloor spreading in this area, although the lack of similarity of radiolarian faunas with contemporaneous faunas in the North Atlantic (Rockall Plateau) and a greater affinity with Siberian faunas suggested an Arctic connection to Bjørklund (1976). Possibly, because many radiolaria are very deep-living plankton (Casey et al., 1979), the shallow water connection of the Norwegian-Greenland Sea may have acted as an effective filter, largely excluding radiolaria, but admitting the siliceous and calcareous phytoplankton and the relatively shallow dwelling planktonic foraminifers.

Low diversity middle and late Eocene calcareous nannoflora in the Norwegian-Greenland Sea led Müller (1976) to suggest separation from the North Atlantic during this time (in fact Müller, *op. cit.*, suggested separation of the two areas until the late Miocene-Pliocene with a short interruption during the mid-late Oligocene, a view which we find difficult to reconcile with other faunal and floral data). However, the presence of *Reticulofenestra umbilica*, *Zygolithus dubius*, *Braarudosphaera bigelowi*, *Isthmolithus recurvus*, *Dictyococcites dictyodus*, *Discolithina pulcheroides*, i. al., among low diversity middle-late Eocene nannofloras would seem to indicate marine connections with the North Atlantic during this interval.

Table 1. Micropaleontologic and biostratigraphic data base from DSDP Leg 38 (Norwegian-Greenland Sea)

| | | |
|---|---|---|
| | Foraminifera | van Hinte (1976): vol. 38: Site Chapters |
| Calcareous | Calcareous Nannoplankton | Müller (1976): vol. 38: 823-841<br>Bukry (1976): vol. 38: 843-855<br>Haq and Lohmann (1978); suppl. to vol. 38: 141-145 |
| | Archaeomonads, Ebridians and endoskeletal dinoflagellates | Perch-Nielsen (1978): suppl. to vol. 38: 147-175 |
| | Radiolaria | Bjørklund (1976): vol. 38: 1101-1167<br>Goll (1978): Suppl. to vol. 38: 177-191 |
| Siliceous | | Dzinoridze, Jouse, Koroleva-Golikova, Kozolova, Nagaeva, Petrushevskaya, and Strelnikova (1978): Suppl. to Leg 38: 289-627 |
| | Diatoms | Schrader and Fenner (1978): vol. 38: 921-1099 |
| | Silicoflagellates | Bukry (1976)*: vol. 38: 843-845<br>Martini and Müller (1976): vol. 38: 857-895 |
| | Dinoflagellates and palynomorphs | Manum (1976): vol. 38: 8971-919 |
| Palynology | Spore and pollen | Koreneva, Zaklinskaya, Bratseva, and Kartoshova (1976): vol. 38: 1169-1193 |
| | Biostratigraphic synthesis | Schrader, Bjørklund, Manum, Martini, van Hinte (1976): vol. 38: 1197-1211 |

*Paper deals with silicoflagellates and calcareous nannoplankton

*paper deals with diatoms and radiolaria

Bathyal benthic foraminiferal faunas having affinities with correlative faunas of the N.W. European Rupelian (early Oligocene) occur at Site 338 (Fig. 2): Turrilina alsatica, Angulogerina tenuistriata, Melonis affine, Gyroidina girardana, Spheroidina bulloides, i. al., have been observed in calcareous-diatomaceous oozes. Calcareous nannoplankton assemblages, associated with these faunas at Site 338 (also Sites 336 and 352A on the Iceland-Faeroe Ridge) are similar to those recorded from the North Atlantic (Leg 12) with Discoaster deflandrei and Chiasmolithus altus (Müller, 1976, p. 827) where they are interpreted as cool water taxa. The general association, however, is considered indicative of a strong influx of North Atlantic water into the Norwegian-Greenland Sea (Müller, 1976). A marked similarity of the high latitude, cool water calcareous nannofloral assemblages (small reticulofenestrids, Coccolithus pelagicus) at Site 338 with contemporaneous mid-Oligocene floras in the North Atlantic was noted by Haq and Lohmann (1978).

These Paleogene assemblages are replaced upward by noncalcareous faunas and floras. A long-ranging, low diversity Neogene agglutinated benthic fauna becomes established that is characterized by Spirosigmoilinella sp., Spirolocammina sp., and Martinottiella communis. This fauna occurs in several of the holes drilled in the Norwegian-Greenland Sea and appears to be characteristic of middle-late Miocene time (e.g., Sites 338, 343, 348, 336). These agglutinated faunas are followed, after an apparent late Miocene-Pliocene hiatus, by late Pleistocene-Holocene calcareous faunas dominated by Neogloboquadrina pachyderma (planktonic) and Cassidulina teretis - Planulina wuellerstorfi (benthic) foraminiferal faunas (see separate discussion below).

On the Icelandic Plateau a relatively complete Neogene section was recovered at Site 348 (Fig. 3), although recovery of various microfossil groups was sporadic and preservation relatively poor, with the result that age determinations are rather tenuous. Above basaltic basement (dated at 18-19 Ma; Kharin et al., 1976, p. 757) is an apparently complete Neogene sequence with the questionable exception of the lower Pliocene. Terrigenous mudstones in the lower part (lower Miocene) contain a bathyal fauna of agglutinated foraminifers (i. al., Ammodiscus, Cyclammina, Cribrostomoides, Bathysiphon). The middle and upper Miocene is characterized by siliceous and calcareous nannofossil oozes and contains a low diversity agglutinated benthic fauna: Martinottiella communis, Spirosigmoilinella and Spirolocammina which occurs commonly during this interval in much of the Norwegian-Greenland Sea. Calcareous benthic and planktonic faunas characterize the so-called "glacial late Pliocene and Pleistocene".

Two holes (Site 336 and 352) were drilled on the northern and southern flank, respectively, of the Iceland-Faeroe Ridge (Fig. 4).

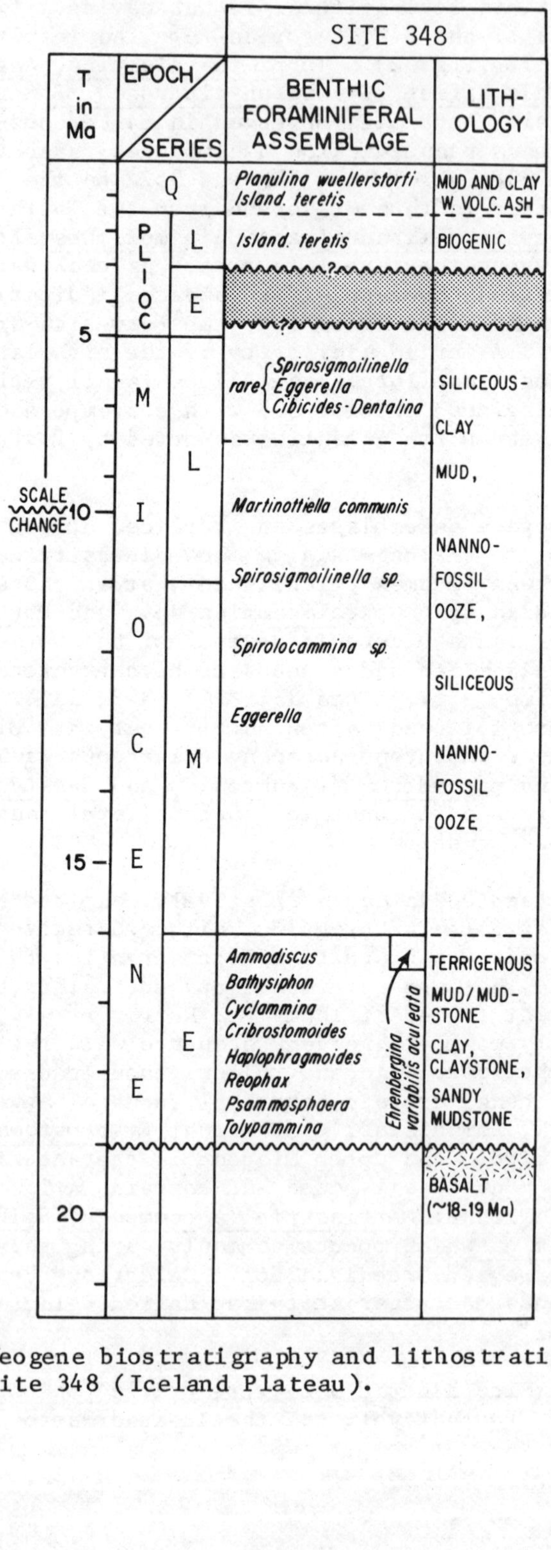


Figure 3. Neogene biostratigraphy and lithostratigraphy of DSDP Site 348 (Iceland Plateau).

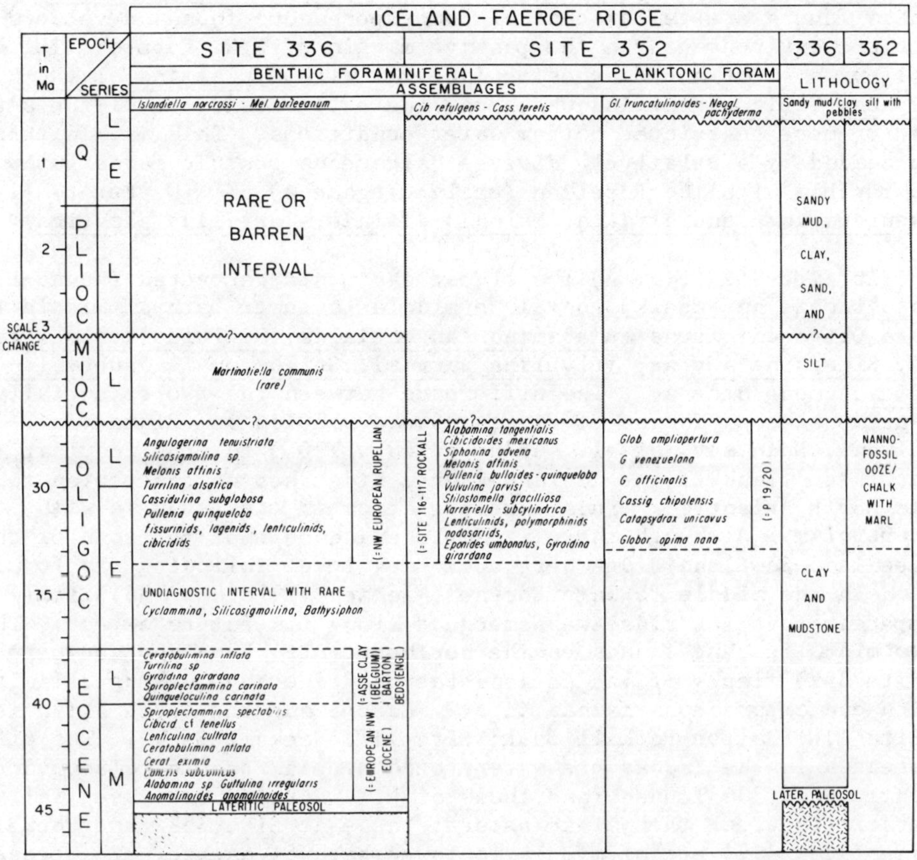

Figure 4. Cenozoic biostratigraphy and lithostratigraphy of DSDP Sites 336 and 352 (Iceland-Faeroe Ridge).

Basement was reached at Site 336 and dated at 40.4 ± 3.2 Ma and 43.4 ± 3.3 Ma (Kharin et al., 1976, p. 757).

At Site 336 (Fig. 4) clays and mudstones with a middle Eocene lower neritic to upper bathyal benthic fauna (with i.al. *Lenticulina cultrata*, *Ceratobulimina inflata*, *C. eximia*, *Cancris subconicus*, *Anomalinoides anomalinoides*, *Spiroplectammina spectablis*) are overlain by a late Eocene fauna (i.al. *Ceratobulimina inflata*, *Turrilina* sp., *Spiroplectammina carinata*, *Gyroidina girardana*). These faunas are reminiscent of contemporaneous faunas developed in neritic environments on the passive margin of N.W. Europe. There follows a short non-diagnostic interval with *Cyclammina*, *Bathysiphon*, *Silicosigmoilina* which may indicate subsidence of the site and/or more restricted bottom water conditions. This is, in turn, succeeded by a relatively diverse calcareous benthic fauna fully comparable with the Rupelian (early Oligocene) of N.W. Europe (i.al. *Angulogerina tenuistriata*, *Melonis affinis*, *Turrilina alsatica*).

At Site 352 (Fig. 4) the oldest sediments recovered (nannofossil ooze and chalk) contain a middle to lower bathyal middle to late Oligocene fauna (*Alabamina tangentialis*, *Cibicidoides mexicanus*, *Siphonina advena*, *Vulvulina jarvisi*, *Stilostomella subcylindrica*, among others). The difference between the two essentially contemporaneous faunas of Sites 336 and 352 led van Hinte (*in* Talwani, Udintsev, et al., 1976, p. 48,. 1205) to conclude that the "northern flank of the Iceland-Faeroe Ridge had no connection to the North Atlantic". This interpretation is treated here with skepticism. Although there is ample evidence that portions of the Greenland-Scotland Ridge were above sea level following its formation in the middle to late Eocene (Nielsen, 1978, p. 786), it is not apparent that the ridge was emergent along its entire extent. The two mid-Oligocene faunas on the northern (Site 336) and southern (Site 352) flanks of the Iceland-Faeroe Ridge have strong affinities with contemporaneous faunas in N.W. Europe and the North Atlantic (Site 116; Hatton-Rockall Bank; Site 117, Rockall Bank). The differences in the faunas are interpreted here as due to paleoenvironment: Site 336 faunas (and those of N.W. Europe) are essentially neritic to upper bathyal in nature; those at Site 352 (and Rockall Sites 116, 117) are of middle to lower bathyal nature. The present depth differential at Sites 352 and 336 is 188 meters. The essential point is, however, that both contain marked affinities with faunal assemblages south of the ridge and some type of marine connection must have existed in this area during this time, and probably as early as the early Eocene (see discussion above and in the section on paleogeography).

All or most of the early Neogene section is missing from both Iceland-Faeroe Ridge sites. At Site 336 rare specimens of *Martinottiella communis* are present in a presumed lower Miocene interval; this is overlain by an essentially barren (or rarely fossiliferous) interval, followed in turn by characteristic late Pleistocene-Holocene calcareous benthic assemblage dominated by Islandiella norcrossi and Melonis barleeanum. At Site 352 a late Pleistocene-Holocene calcareous benthic fauna (Cibicidoides refulgens, Cassidulina teretis) lies directly on the middle Oligocene fauna.

Recent benthic foraminiferal assemblage distribution patterns have been examined by Belanger and Streeter (1980). Present day core-top assemblages were shown to vary significantly with depth despite an apparent lack of vertical gradient changes in hydrographic parameters. A combination of hydrostatic pressure, bottom water temperature and the distribution of labile organic matter (i.e., primary productivity) was suggested as being responsible for the observed distribution patterns. Four overlapping biofacies have been recognized: 1) a Melonis barleeanum-Pullenia bulloides-Islandiella norcrossi fauna (600-1200 m); 2) a Cassidulina teretis fauna (950-1500 m); 3) a Planulina wuellerstorfi-Oridorsalis tener fauna (1250-3200 m); and 4) an Oridorsalis tener fauna (2900-3500 m).

The late Cenozoic history of the Norwegian-Greenland Sea has been studied by Belanger (1981 a,b) using bathyal and abyssal benthic foraminifers, which indicate that the area was essentially as ice-free and productive during the penultimate interglacial (Eemian, isotopic stage 5e) as it is today and at least partially ice-free during an intermediate climate regime (isotopic stages 5a-3). Distribution of calcareous faunas and floras in the Norwegian-Greenland Sea indicates that the CCD was very shallow during the Miocene and early Pliocene. The agglutinated species Martinottiella communis and Spirosigmoilina sp. dominate the generally sparse Miocene benthic foraminiferal assemblages.

Sporadically intercalated are well preserved early Pliocene assemblages, dominated by Cassidulina teretis and Melonis barleeanum. Carbonate preservation was somewwhat improved in the middle Pliocene with the onset of Northern Hemisphere glaciation and markedly improved in the early Pleistocene prior to glacial intensification.

Rockall Plateau

Six sites have been considered from the Rockall Plateau region: Sites 116 (Rockall Basin; Fig. 5), 117 (Hatton-Rockall Bank; Fig. 6), 403-404 (N.W. Rockall Margin; Fig. 7, 8) and 405-406 (S. Rockall Martin).

Early Eocene planktonic foraminiferal faunas of low diversity are dominated by the Subbotina patagonica group; various species of acarininids are also present. Calcareous nannoplankton floras are similar to those in contemporaneous deposits in northern Europe and the Norwegian-Greenland Sea (Müller, 1976).

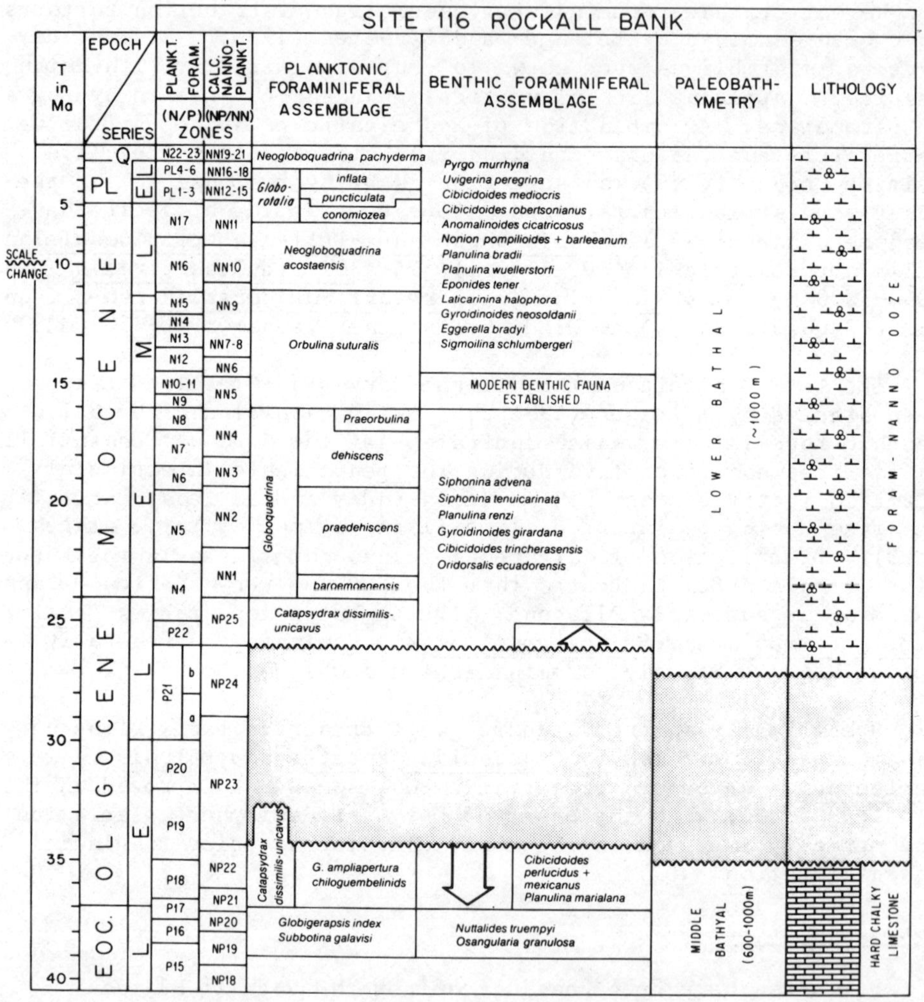


Figure 5. Cenozoic biostratigraphy and lithostratigraphy of DSDP Site 116 (Rockall Basin).

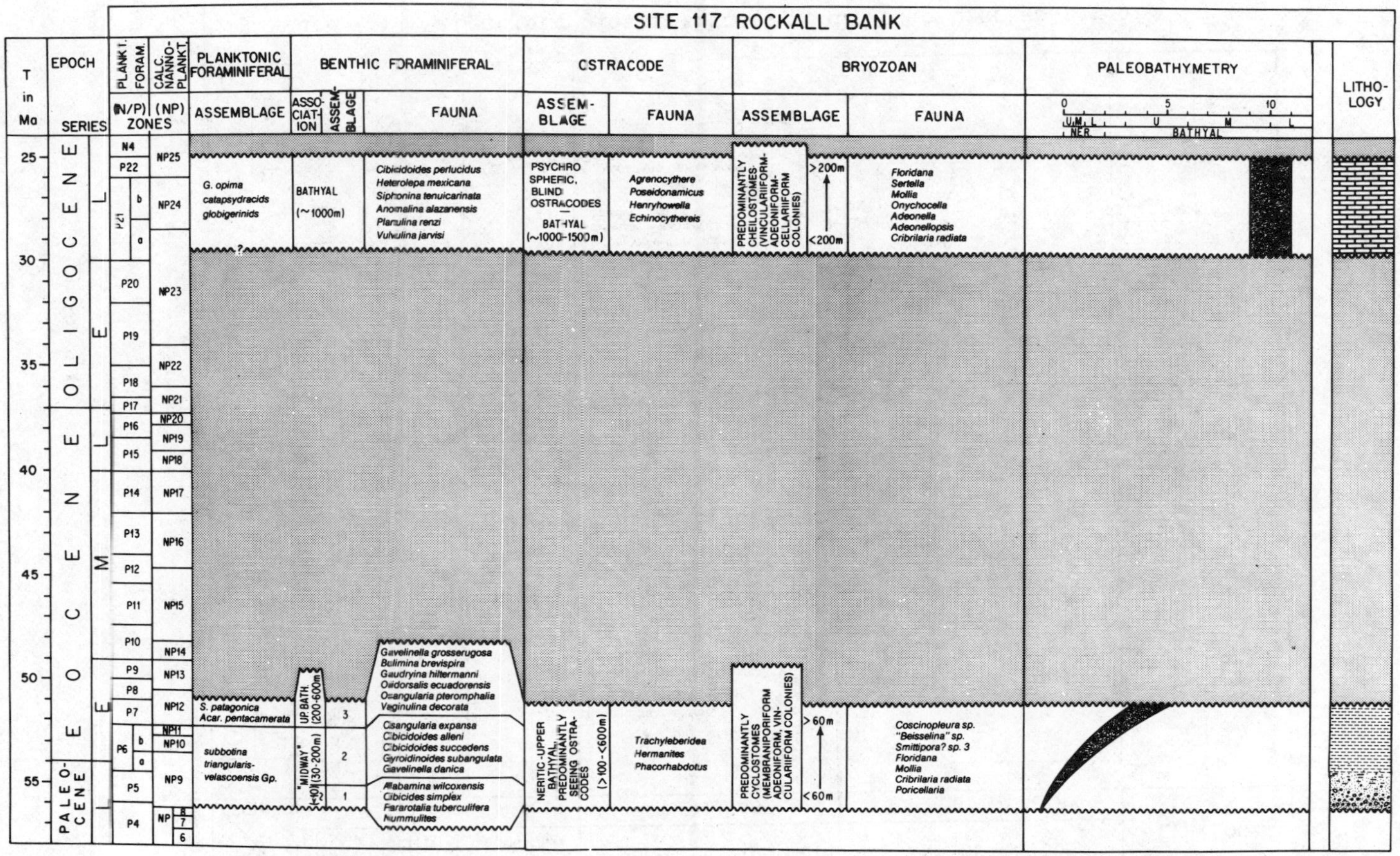


Figure 6. Cenozoic bicstratigraphy and lithostratigraphy of DSDP Site 117 (Rockall Bank).

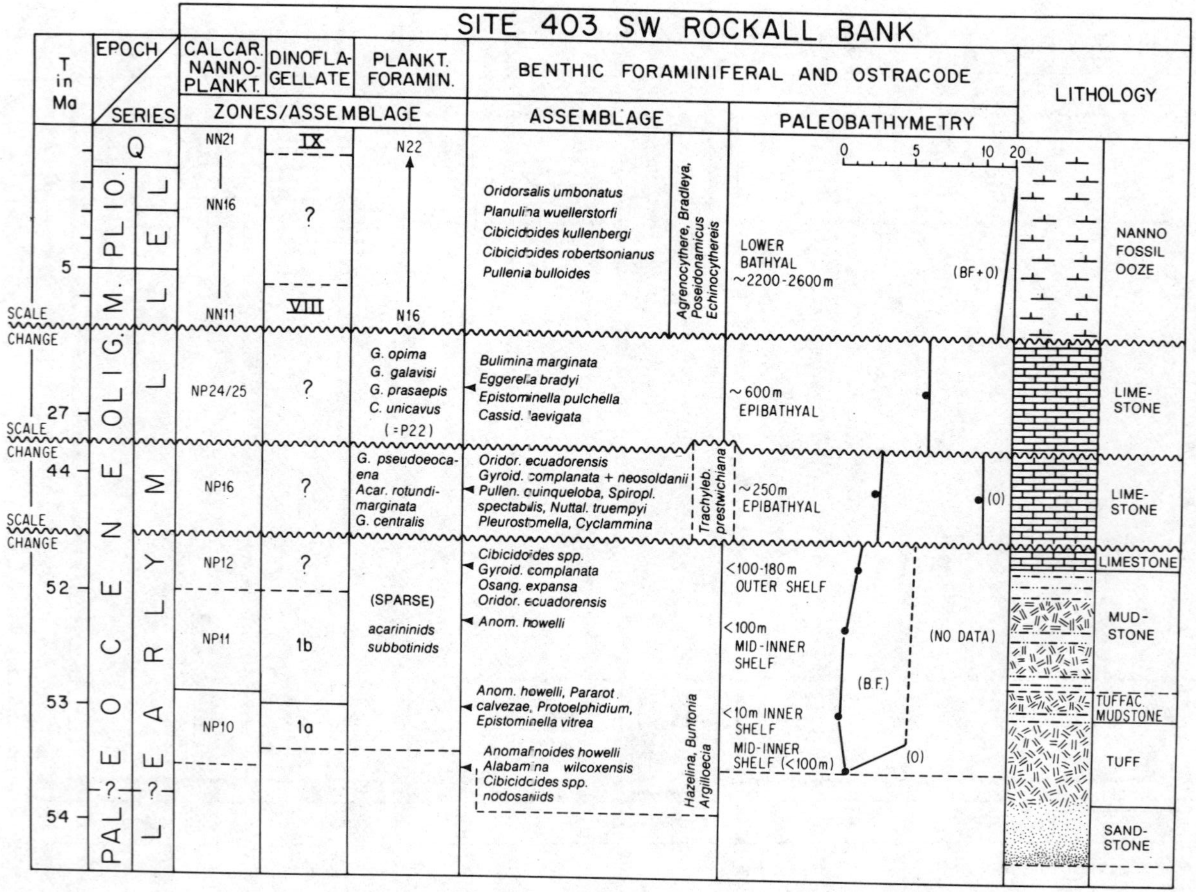


Figure 7. Cenozoic biostratigraphy and lithostratigraphy of DSDP Site 403 (N.W. Rockall Margin).

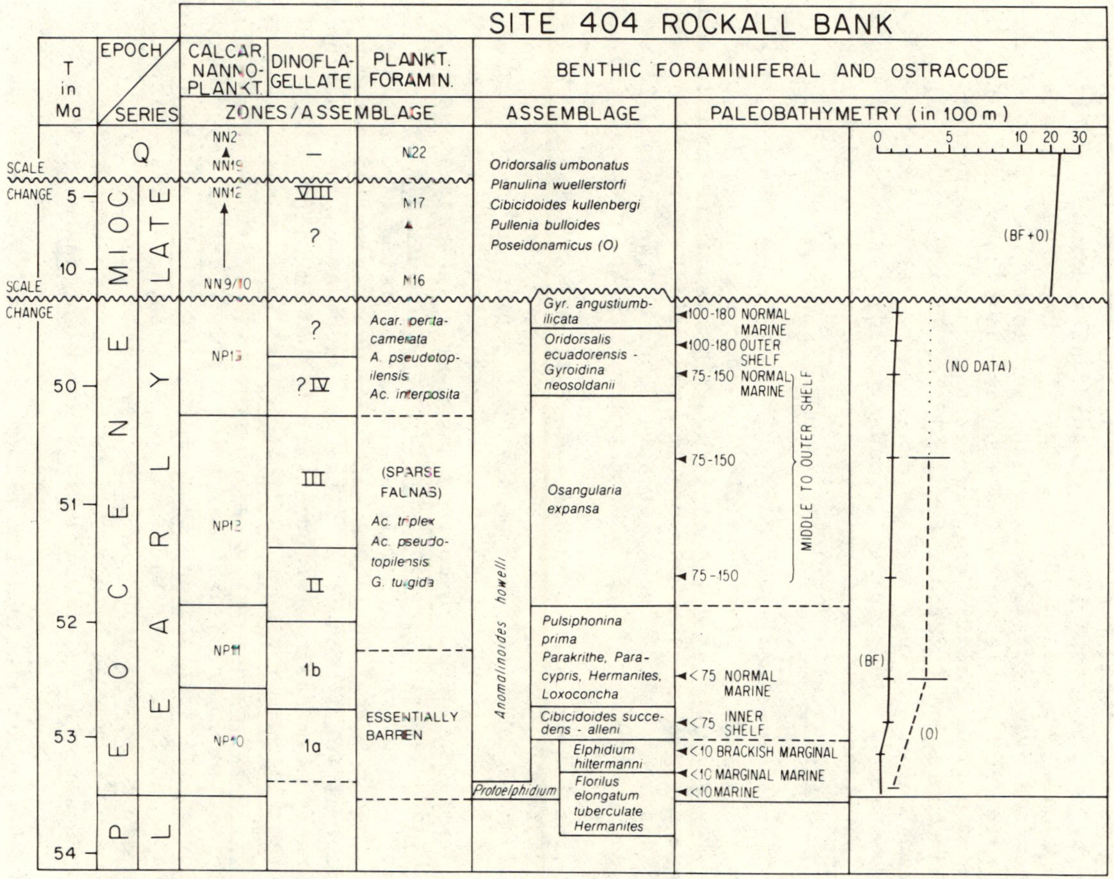


Figure 8. Cenozoic biostratigraphy and lithostratigraphy of DSDP Site 404 (N.W. Rockall Margin).

Oligocene faunas of planktonic foraminifers are characterized by catapsydracids, chiloguembelinids, and several species that are useful in mid-latitude biostratigraphy (*Globigerina ampliapertura, Globorotalia munda*).

Neogene planktonic foraminiferal faunas of Rockall are characterized by temperate species. Globoquadrinids dominate early Miocene faunas, whereas middle and late Miocene assemblages are characterized by various globigerinids and members of the *Globorotalia miozea* and *Neogloboquadrina acostaensis* groups. With the initiation of northern hemisphere glaciation, ca. 3 Ma (Berggren, 1972a) a marked change in planktonic foraminiferal faunas occurs, with the replacement of moderately diverse assemblages with a low diversity fauna dominated by the apparently endemic sinistrally coiled *Neogloboquadrina atlantica*. Accompanying species include *Globigerina bulloides, Globorotalia inflata*, and *Neogloboquadrina pachyderma*.

Relatively similar latest Paleocene-early Eocene benthic foraminiferal faunas with "Midway" affinities occur at Sites 117, 403, and 404 (Fig. 10). The sequential change in faunal composition during the early and, in the case of Site 403, middle Eocene has been used to document the differential subsidence of the Rockall Plateau (Laughton, Berggren et al., 1972; Berggren, 1974; Murray, 1979).

At Site 117 benthic foraminifers and lithofacies show a relatively rapid subsidence from subaerial through littoral (approximately 10 m) and neritic to upper bathyal depths during 4-5 my of the early Eocene (Fig. 6). A suite of predominantly sighted ostracodes (*Hermanites, Phacorhabdotus*) and mostly cyclostomatous bryozoans (*Floridina, Mollia, Poricellaria*, and *Coscinopleura*) give general support to the foraminiferal depth interpretation. Sites 403 and 404 remained somewhat shallower (neritic) over essentially the same time interval (early Eocene). Sites 405 and 406, on the other hand, were already at midbathyal (approximately 1000 m) depth in early Eocene (*Nuttallides truempyi, Stilostomella, Cyclammina, Atlanticythere, Bythocypris*, and *Bairdia*). By the middle Eocene Site 403 had subsided to upper bathyal depth (~500 m) (*Nuttallides truempyi, Spiroplectammina spectabilis, Cyclammina*, and *Trachyleberidea prestwichiana*). At Site 116 late Eocene upper bathyal faunas (*Nuttallides truempyi, Osangularia granulosa*) are succeeded in the early Oligocene by a typical bathyal Oligocene fauna characterized by *Cibicidoides mexicanus, C. perlucidus* and *Planulina marialana*.

Late Oligocene middle-lower bathyal (~1000 m) faunas are present at Sites 116 and 117 (*Cibicidoides mexicanus, Planulina renzi, Siphonina advena, S. tenuicarinata*, and at Site 117 the psychrospheric ostracode fauna *Poseidonamicus, Agrenocythere*, and *Echinocythereis*). Site 403 may have been somewhat shallower (upper bathyal) in the late Oligocene (*Bulimina marginata, Epistominella pulchella*, and *Eggerella bradyi*) (Fig. 9).

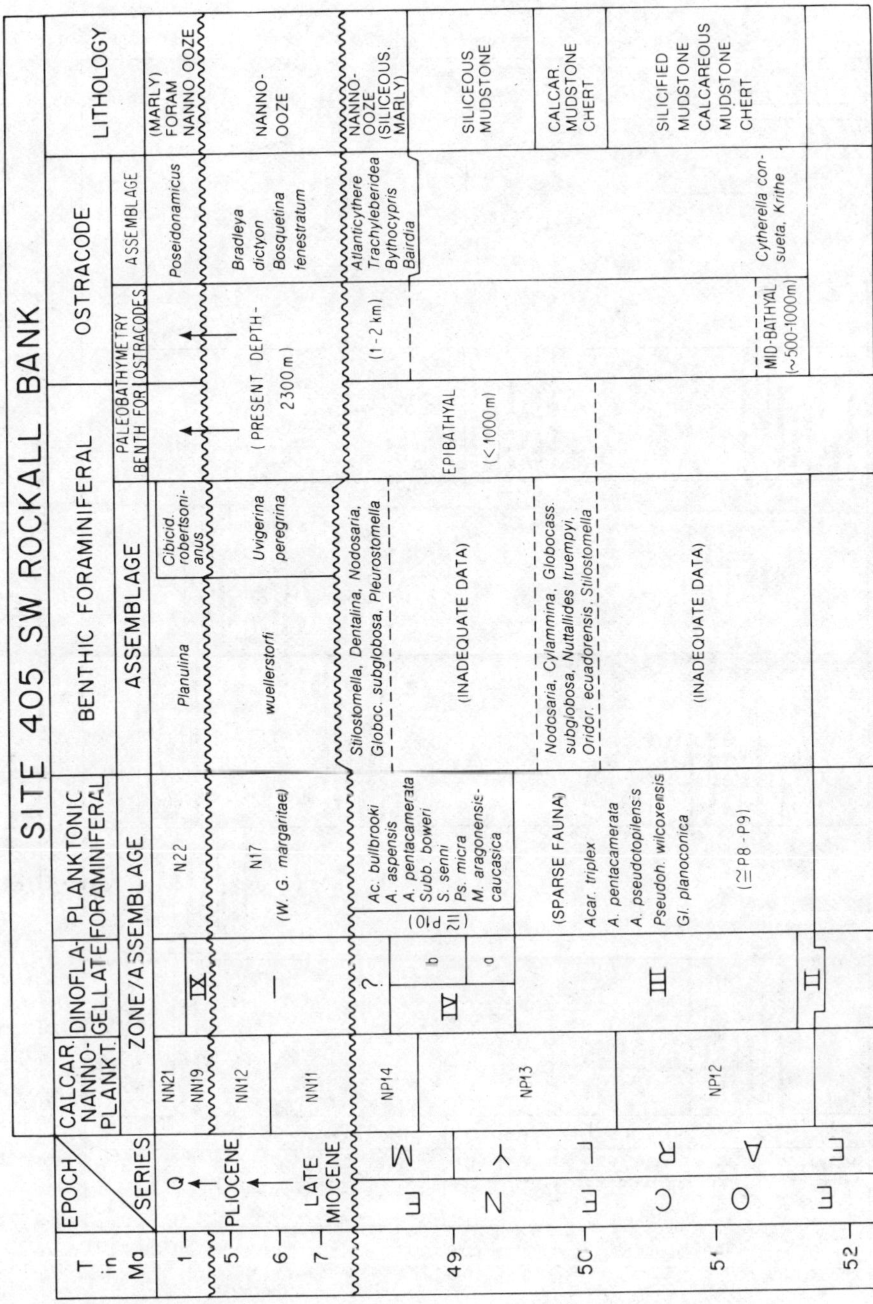


Figure 9. Cenozoic biostratigraphy and lithostratigraphy of DSDP Site 405 (S. Rockall Margin).

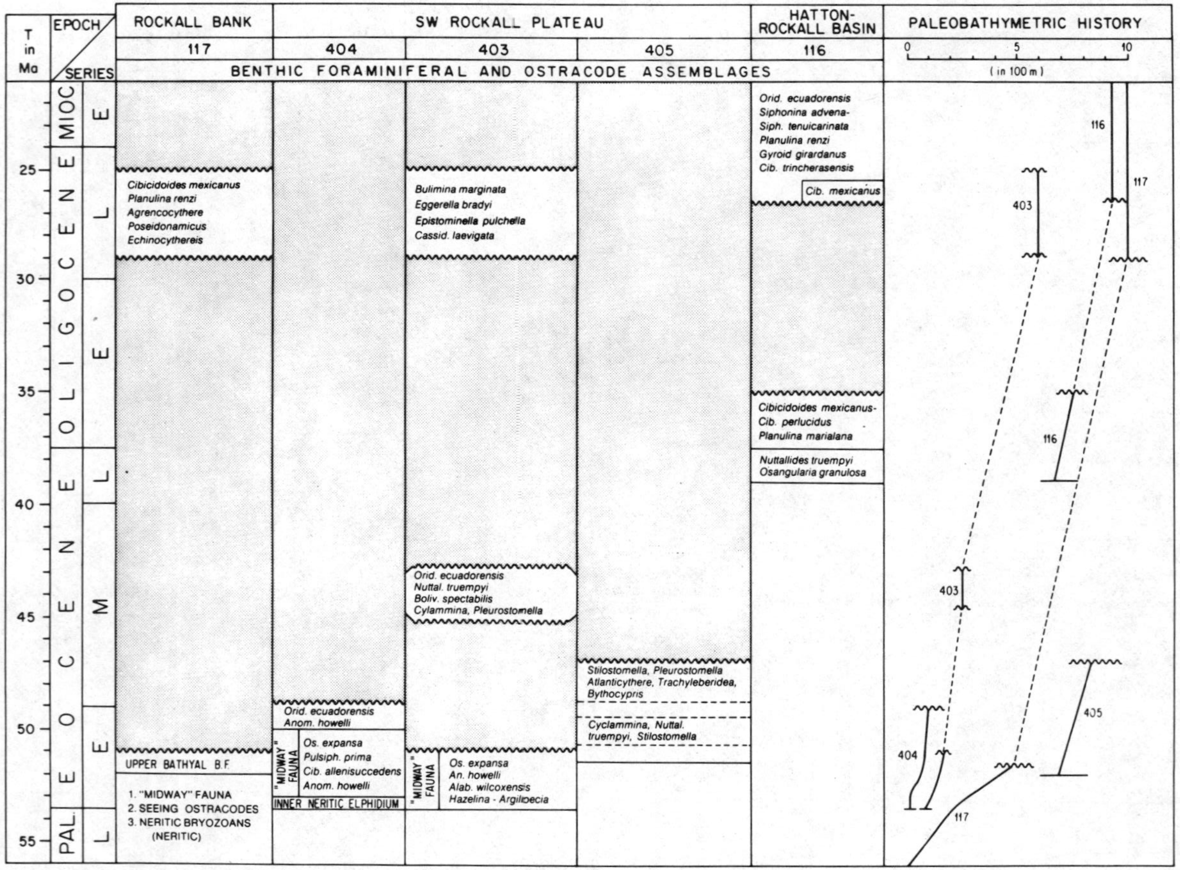


Figure 10. Cenozoic biostratigraphy and subsidence history of Rockall Plateau region.

The essentially complete lower Miocene sequence at Site 116 contains characteristic lower-middle bathyal species (*Cibicidoides trincherasensis*, *Planulina renzi*, *Siphonina advena*, *S. tenuicarinata* and *Oridorsalis ecuadorensis*).

The establishment of the modern day deep water benthic foraminiferal fauna has been recorded at Site 116 (Berggren, 1972a, b) and dated to about 15-16 Ma, near the early-middle Miocene boundary. Typical Neogene foraminiferal species include: *Planulina wuellerstorfi*, *Cibicidoides mediocris* (-?=*C. floridanus*), *C. robertsonianus*, *Melonis pompilioides*, *M. barleeanum*, *Pyrgo murrhina* and the psychrospheric ostracodes *Poseidonamicus*, *Agrenocythere*, *Bradleya*, and *Echinocythereis* (Fig. 5).

## Bay of Biscay

Four sites (118, 119, 400, and 401) have been considered in the context of this study. Sites 118 and 119 (Fig. 11) were drilled in the central part of the Bay of Biscay (118, Biscay abyssal plain; 119 Cantabrian Seamount). Sites 400 and 401 (Fig. 12) were drilled on the Meriadzek terrace near the base of the continental margin of the northwestern Bay of Biscay. All four sites were already near their present day (abyssal) depths in the Paleocene and between them contain a composite Cenozoic abyssal sequence.

During the Paleogene the Bay of Biscay was located near the northern margin of the subtropical belt. High diversity planktonic foraminiferal faunas with keeled morozovellids, hantkeninids, acarininids, and globigerinathecids, were replaced in the Oligocene by low diversity catapsydracid-dominated faunas which appear to be dissolution relict faunas.

Subtropical planktonic foraminiferal faunas continue to characterize the early and mid-Miocene of the Bay of Biscay but temperate elements (the *Globorotalia miozea* group) characterize the late Miocene-early Pliocene and the *Globorotalia puncticulata-G. inflata* group and the characteristic North Atlantic temperate-subpolar form *Neogloboquadrina atlantica* were present in Pliocene time. Dissolution, however, makes an accurate reconstruction of Neogene planktonic foraminiferal paleoceanography of the Bay of Biscay difficult.

A so-called "Velasco type", or deep water, faunal assemblage characterizes the Paleocene through middle Eocene at Sites 400A, 401 (Schnitker, 1979a), the early Eocene at Site 118 and the early and middle Eocene at Site 119 (Berggren, 1972b). Typical forms include the long-ranging *Nuttallides truempyi*, *Bulimina grata*, *B. trinitatensis*, *Alabamina dissonata*, and *Oridorsalis ecuadorensis* (Fig. 11, 12).

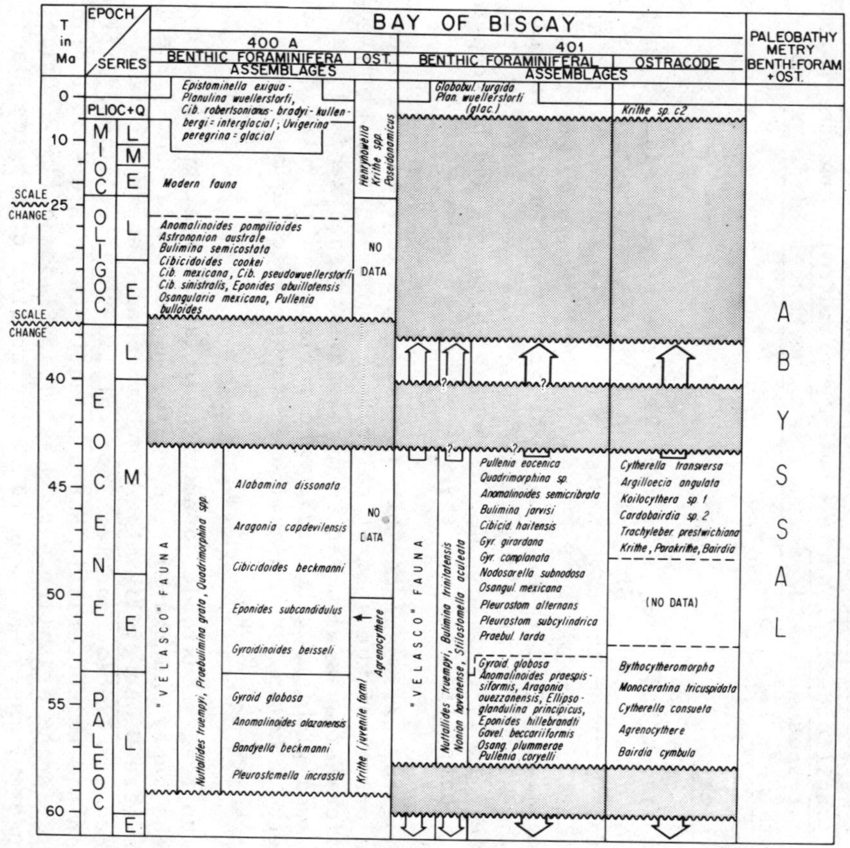


Figure 11. Cenozoic biostratigraphy and lithostratigraphy of DSDP Sites 118 and 119 (Bay of Biscay).

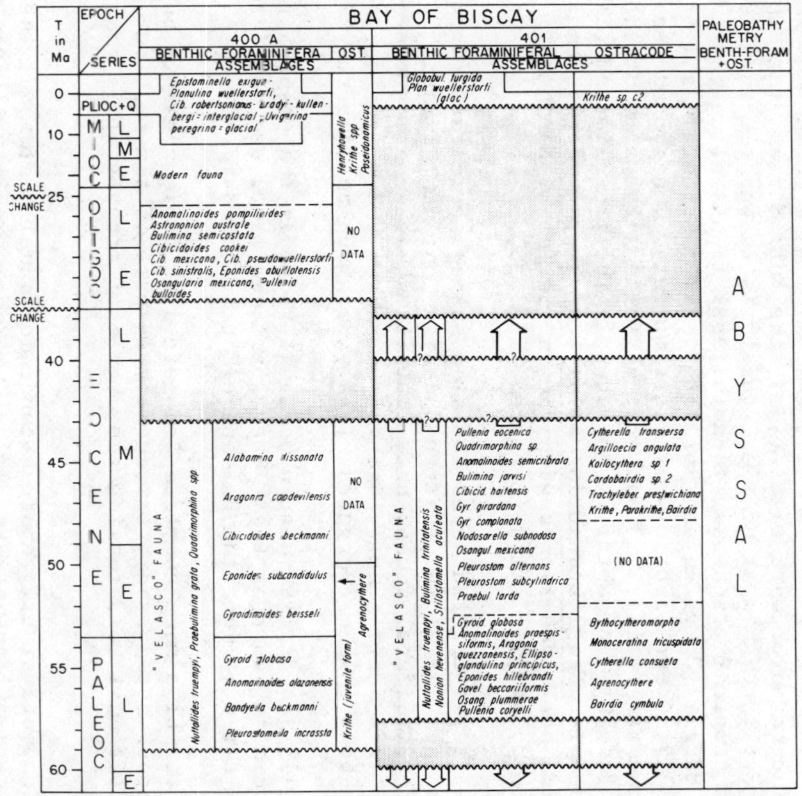


Figure 12. Cenozoic biostratigraphy and lithostratigraphy of DSDP Sites 400 and 401 (Meriadzek Terrace, Bay of Biscay).

Oligocene abyssal faunas occur at Sites 119 and 400A; characteristic species include *Cibicidoides trincherasensis*, *C. grimsdalei*, *C. havanensis*, *C. mexicanus*, *Osangularia mexicana*, and *Anomalinoides pompilioides* (Fig. 11, 12).

The modern deep water benthic foraminiferal faunas, as was the case at the Rockall Site 116, became established near the early/ middle Miocene boundary (ca. 15-16 Ma).

## Labrador Sea

The incursion of warm water (i.e., the Gulf Stream) into the Labrador Sea through most of the Cenozoic is known from the distribution of tropical planktonic foraminiferal faunas along the eastern Canadian Margin (Berggren and Hollister, 1977; Gradstein and Srivastava, 1980). Coeval subtropical and temperate faunas occurred in the southwestern (Site 111) and central (Site 112) parts of the Labrador Sea but were abruptly replaced in the mid Pliocene (ca. 3 Ma) by subpolar *Neogloboquadrina atlantica*, *N. pachyderma*, and *Globigerina bulloides* faunas with the initiation of Northern Hemisphere polar glaciation (Berggren, 1972a,b).

Site 111 was drilled on an isolated submarine feature (Orphan Knoll) at the foot of the eastern Canadian Continental Rise. Benthic foraminiferal faunas show that it subsided during the Paleocene from neritic-upper bathyal depths (~ 200 m) to its present depth of about 1800 m (Berggren, 1972b; Berggren and Aubert, 1976).

Site 112 was drilled near a basement ridge in the central part of the southern Labrador Sea. The oldest sediments recovered there have been dated at near the Paleocene/Eocene boundary.

At Site 112 a more complete Cenozoic section is present, but discontinuous coring and poor recovery in part of the section have made it difficult to determine the degree of stratigraphic continuity.

The Cenozoic stratigraphic record and faunal history at the two sites is very different (Fig. 13). The Eocene at Site 111 contains a diverse fauna of about 80 species which is dominated by *Cibicidoides* (8 species), *Anomalinoides* (6 species), *Pleurostomella* (5 species), *Bulimina* (5 species), *Stilostomella* (4 species each). Numerically dominant taxa include *Nuttallides truempyi*, *Oridorsalis ecuadorensis*, *Gyroidinoides girardanus*, *Bulimina grata*, *Stilostomella curvatura*, *S. aculeata*, *Cibicidoides hercegovinensis*, and *Bulimina orphanensis* (= *B. callahani*). A hiatus of about 45 my separates these deep water Eocene faunas from late Miocene-Pleistocene modern day faunas containing, among others, *Planulina wuellerstorfi*, *Cibicidoides mediocris*, and *Uvigerina peregrina*.

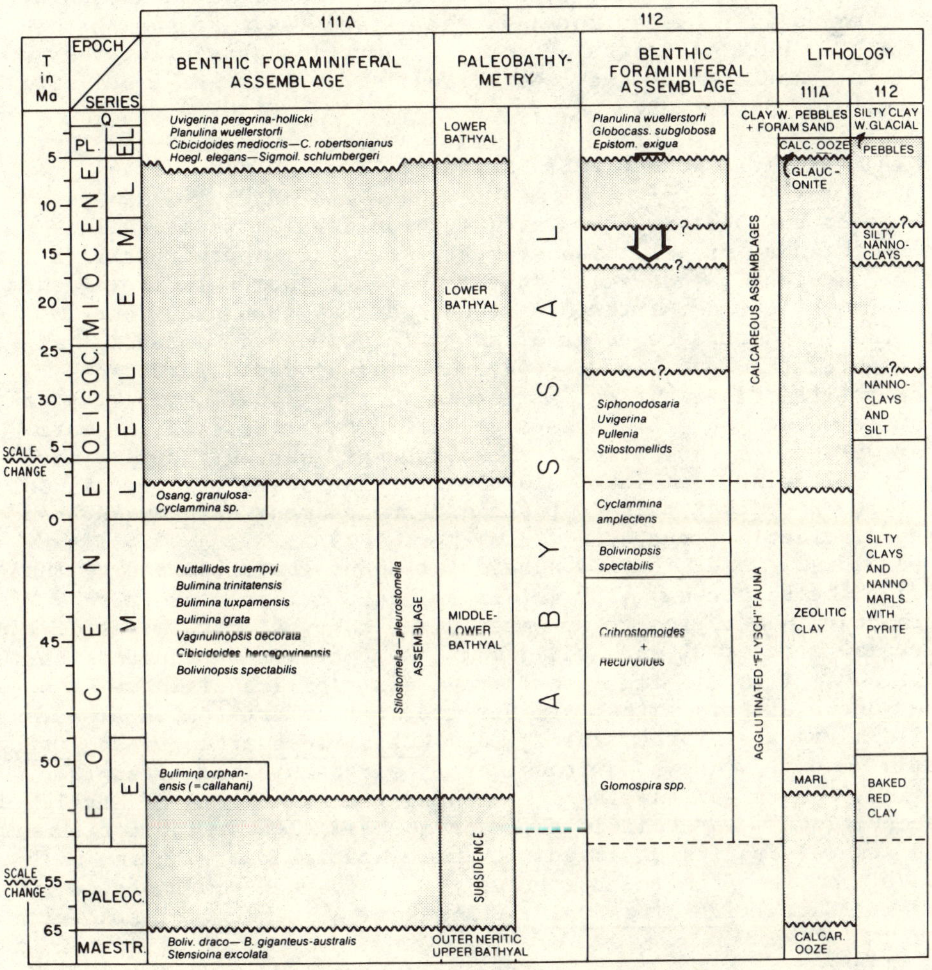


Figure 13. Cenozoic biostratigraphy and lithostratigraphy of DSDP Sites 111 and 112 (Labrador Sea).

At Site 112 a fauna of 37 species of mostly agglutinated foraminifers ranges to the top of the Eocene where it is abruptly replaced by a calcareous benthic assemblage and a calcareous and siliceous (diatom) plankton assemblage (Fig. 13). This benthic fauna, containing *Bolivinopsis spectabilis*, *Cyclammina amplectens*, *Cystammina globigerinaeformis*, *Recurvoides walteri*, *Haplophragmoides walteri*, and *Cribrostomoides subglobosus* resembles the flysch type *Rhabdammina* assemblages known in the flysch basin of the Alpine-Carpathian belt of central Europe, the marginal basins of the eastern Canadian Shelf, the North Sea, and various deep sea sites (Gradstein and Berggren, 1981).

## NORTH ATLANTIC CENOZOIC DEEP ENVIRONMENTS

Paleocene benthic deep water foraminiferal (the so-called Velasco fauna; Fig. 14) faunas were essentially survivors from the late Cretaceous, indicating that the almost wholesale floral and faunal extinctions at the Cretaceous/Tertiary boundary were exclusively a surface event of short duration. It appears that throughout the Cenozoic deep water benthic foraminifers had a cosmopolitan distribution, attesting to the global uniformity of the deep sea biotope. Although oxygen isotope ratios call for warm deep waters and thus weak vertical thermal gradients during the early Paleogene, qualitative and particularly quantitative faunal analyses show that this deep oceanic environment has always been very distinctly structured (Schnitker, 1979a); Biscay Bay Sites 400A and 401 were situated already at about their present depths during the Paleocene, 4399 meters and 2495 meters respectively. Both contain a typical deep water (Velasco-type) fauna, yet of the over 150 species at each site, only 32 species were common to both. Just as striking is the quantitative distribution of those species that occur at both sites: all but two species are common at one site while only rare at the other. Only *Nuttallides truempyi* and *Oridorsalis umbonatus* are ubiquitous, truly eurybathyal species. The shallower site 401 samples are flooded with specimens of *Gavelinella beccariiformis* and contain abundant to common *Osangularia plummerae* and several species of *Aragonia*, while species distribution within coeval samples from the deeper site 400A is much more equable, with perhaps *Nuttallides truempyi* and *Abyssamina* the most common forms.

The end of the Paleocene is marked by a sudden extinction of the most abundant and characteristic species of the "shallow" water fauna at Site 401: *Gavelinella beccariiformis*, all species of *Aragonia*, *Anomalinoides praespissiformis*, *Osangularia plummerae*, *Gyroidinoides globosus* and *Eponides hillebrandti*, while at the same time very few species became extinct at the deep Site 400A. Many new species appeared, but oceanographically interesting is the observation that the occurrence of *Nuttallides truempyi*, *Pleurostomella syubcylindrica*, *Praebulimina tarda* and some others becomes strongly shifted towards the shallower site, whereas especially the

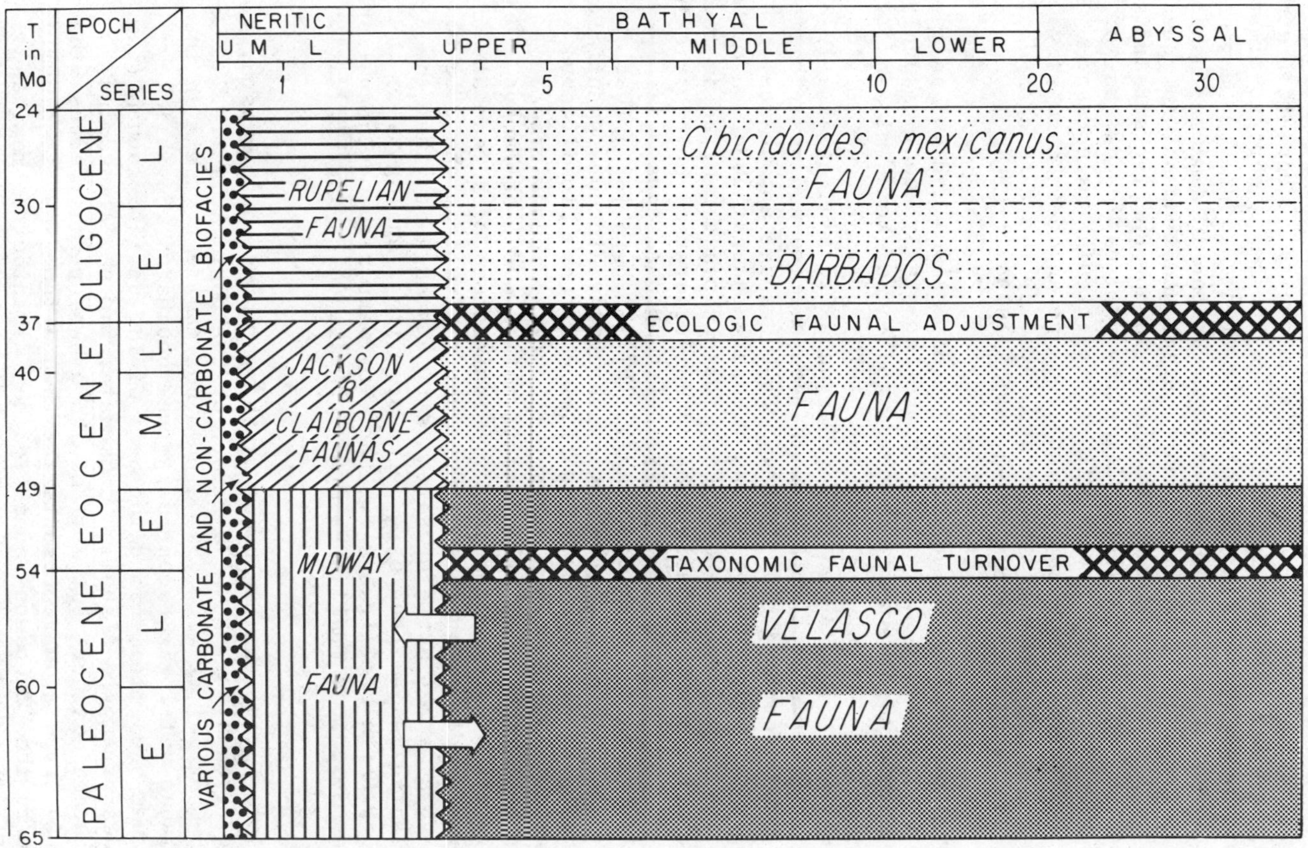


Figure 14. Simplified schematic diagram of Paleogene benthic foraminiferal associations of the North Atlantic and its marginal areas.

species of the Abyssamina plexus (Schnitker and Tjalsma, 1981) become much more restricted to the deep site. If the Paleocene/Eocene extinctions were the result of a "midwater" event, then some surviving species were seeking this new environment, whereas others were avoiding it by restricting their occurrence to greater depths. This faunal event has been observed to occur and in similar fashion throughout the South Atlantic Ocean (Tjalsma and Lohmann, in press) and, although less well documented there, in the Pacific and Indian oceans as well (Fig. 14). No satisfactory explanation has as yet been offered linking this event with any particular paleoceanographic phenomenon or event. A global change in ocean chemistry occurred at about the same time as indicated by the rate change of biogenic sedimentation (Worsley and Davis, 1979).

Throughout the Eocene, the faunas at Sites 400A and 401 in the Bay of Biscay remained qualitatively and especially quantitatively distinct from each other. The general cooling of both, mid and deep water, that commenced at the end of the early Eocene (Savin, 1977; Boersma and Shackleton, 1977; Vergnaud-Grazzini et al., 1978) did not leave a distinct mark upon the composition and distribution of the benthic foraminiferal faunas. However, the appearance of this cool water coincides with, and probably brought about, a marked increase in biosiliceous deposition. These deposits, often in the form of cherts, occur globally in a circum-equatorial belt, tracking the path of the ancient trans-Tethyan circumglobal equatorial current (Berggren and Hollister, 1974, 1977). Notable is that from late early Eocene on until the end of the early Miocene essentially the entire North Atlantic became a site of intense biosiliceous sedimentation as well. Such accumulation of siliceous deposits, according to the basin to basin fractionation model of Berger (1970), together with oxygen isotopic indications of a general temperature gradient and thus of deep water flow from the cold South Atlantic to a warmer North Atlantic, strongly suggests that the North Atlantic was an area of general upwelling, analogous to the present-day North Pacific. Also, the very warm high northern floras and faunas described earlier counter-indicate cold water formation there.

The late Eocene establishment of the psychrosphere (Benson, 1975), so obvious in the $\delta^{18}O$ record of the North Atlantic (Vergnaud-Grazzini et al., 1978), does not manifest itself in the benthic fauna until the Eocene/Oligocene boundary. Several key species that had often formed the bulk of the benthic faunas (i.e., Nuttallides truempyi, Alabamina dissonata, and the Abyssamina plexus) became extinct, but most of the others persist across the line. This transition represents the change from what is denominated here as the "Barbados Fauna" to the "Cibicidoides mexicanus Fauna" (Fig. 14). However, change of depth ranges and relative abundance of many species leaves the impression that the deep abyssal environment was invaded by the mid-depth faunas. Most notable among the arrivals on the abyssal ocean floor are many buliminids, cibicidoidids and Osangularia mexicana. Similarly, a rapid changeover from a predominantly agglutinated Eocene benthic foraminiferal fauna to a calcareous Oligocene assemblage in DSDP Site 112 (Labrador Sea) heralds a drastic change of North Atlantic deep water characteristics.

At about the same time a bottom water event created seismically recognizable erosional unconformities: Horizon $A^u$ in the western North Atlantic (Tucholke and Mountain, 1979) and Reflector R4 in the eastern North Atlantic (Roberts, 1975). Together with the faunal and isotopic evidence of deep water change, these horizons are usually considered clear evidence for the beginning of North Atlantic deep and bottom water formation. It is not clear, however, if, where and how this water may have formed. Considering the near-polar position of the southern Weddell Sea, the probably source of cold bottom water at this time (Schnitker, 1980 b), northern water should have formed at equally high latitudes to achieve competitive densities through cooling (Tethys being still open, and thus not a significant source of highly saline water). If this source area of bottom water had been Baffin Bay (Labrador Sea was probably too far south, and filled with warm water), the occurrence of erosion in the eastern North Atlantic would be difficult to explain. The Norwegian-Greenland Sea, on the other hand, was probably separated from the North Atlantic by the still largely emergent Greenland-Scotland Ridge. A possible passage might have been provided by the Faeroe-Shetland Channel, which might represent a graben structure, predating the Eocene drifting phase of the North Atlantic (Bott, 1971, 1974).

Major changes in the North Atlantic deepwater environment occurred gradually during the late early and middle Miocene:

1) Deposition of biosiliceous sediments nearly ceased, to be replaced by a dominance of carbonate and terrigenous sediments.

2) Deposition and shaping of enormous sediment drifts in both the eastern and western North Atlantic (Jones et al., 1970; Ruddiman, 1972; Roberts, 1975; Shor and Poore, 1979).

3) Establishment of the modern deep water benthic foraminiferal faunas.

4) Enrichment of North Atlantic deep water with $^{13}C$, an indication of "young" water (Blanc et al., 1980).

These observations lead to the conclusion that the North Atlantic has shifted from a mode of general upwelling (estuarine type circulation of Berger, 1970) to a general downwelling (lagoonal type circulation of Berger, 1970) with deep water outflow. The most likely threshold mechanism to bring about this change is, as was suggested by Vogt (1972), the subsidence of the Iceland Ridge to such depth that reflux of cold water from the Norwegian-Greenland Sea became significant. Schnitker (1980 b) noted that North Atlantic deep water could be formed more easily at this time because of the contribution of highly saline Mediterranean outflow that had begun during the early Miocene. With this development of North Atlantic deep water outflow the last fundamental readjustment of the global deep water sphere was accomplished; further changes were due to fluctuations of intensity and responses to climatic cycling.

Although some Paleogene species (e.g., *Osangularia umbonifera*, *Oridorsalis umbonatus*, *Pullenia quinqueloba*) persisted across this environmental change and still comprise an important component of the foraminiferal deep sea benthos, others, among them such distinctive species as *Anomalinoides globulosa*, *Cibicidoides grimsdalei*, *Bulmina jarvisi*, *Gyroidina perampla*, and *Quadratobuliminella pyramidalis*, became extinct. With the rapid appearance of such species as *Cibicidoides robertsonianus*, *Epistominella exigua*, *Melonis pompilioides*, *M. barleeanum*, *Planulina wuellerstorfi*, *Pyrgo murrhina* and *Uvigerina peregrina*, the modern deep water fauna was established. As during the Paleogene, the Neogene deep water benthic faunas are qualitatively and especially quantitatively distinctly structured, but, as Streeter (1973) and Schnitker (1974) have shown, this faunal distinction is in response to the structuring of deep water masses of different physical and chemical character, and in turn allows their use in tracing climatically induced changes of the deep water circulation. It appears that the establishment of an ice cover over the Norwegian-Greenland Sea and the concomitant southward shift of the Gulf Stream (CLIMAP, 1976) prevented the formation of cold, dense water there, and the North Atlantic temporarily ceased to be a source of bottom water and became filled with ageing, low oxygen water, possibly derived from the south.

The record of the benthic environment within the Norwegian-Greenland Sea is very meager. With the exception of the mid and upper Oligocene and some upper Miocene on the relatively shallow Vøring Plateau, no carbonate sediments accumulated until the late Pliocene (Talwani, Udintsev et al., 1976). The occasional Eocene and Oligocene calcareous benthic foraminiferal faunas that do occur in these sequences are qualitatively very similar to coeval faunas from N.W. Europe and Rockall Plateau, but insufficiently known to derive a detailed understanding of bottom water history from them.

The great abundance of biosiliceous sediments in the Norwegian-Greenland Sea attests to its great biological productivity; volcanogenic silica being probably one source of fertility, thorough mixing of the inflowing North Atlantic water as it crosses the Iceland Ridge, another. High organic carbon contents and sulfurization of the mid-Tertiary (pre-middle Miocene) sediments (Emelyanov et al., 1978; Nesterova et al., 1978; Kossouskaya et al., 1978; Rateev et al., 1978; Varentsov, 1978) indicate low oxygen levels of the bottom water, an environment that is similar to that found under present-day upwelling systems, and where, likewise, carbonate sediments are poorly or not at all preserved (Thiede, 1973). The beginning of the mid Miocene reflux of cold water to the North Atlantic is very noticeable in the sediments, in particular the organic carbon and sulfur contents drop, terrigenous input increases, but the accumulation rates of biosiliceous sediments also reach maximum values during the mid and late Miocene and early Pliocene (Nilsen and Kerr, 1978). Failure of water overturn to adequately ventilate the deep basin suggests that the formation of reflux water was (and is) primarily a relatively shallow process, involving mostly the water layers down to sill depth, as described by Peterson and Rooth (1976). The continued presence of a well developed fauna of agglutinated benthic foraminifers, on the other hand, is testimony that bottom water conditions never became completely anoxic.

The late Cenozoic climate deterioration, together with the closure of the Panamanian isthmus and the attendant shunting of Gulf Stream flow to the north, led to the onset of Northern Hemisphere glaciation (Schnitker, 1980 b). In the Norwegian-Greenland Sea this is manifested with a massive occurrence of ice rafted sediment in all DSDP cores, starting as early as latest Miocene/earliest Pliocene at the northernmost Site 344, and all over by middle Pliocene. Along with the appearance of ice rafted sediments a drastic decrease of biosiliceous sediments is noted (Talwani, Udintsev et al, 1976; White, 1978), indicating a sharp decrease in biological productivity. The reduction of productivity was probably caused by the ice cover that must have formed, at least seasonally, over the Norwegian-Greenland Sea. It is also at this time that a differentiation between a western, low productive, and an eastern, more highly productive province becomes apparent. This differentiation marks the onset of the East Greenland Current, with its low salinity Arctic Ocean outflow and almost perennial ice cover.

The reduced input of organic matter into the deep environment lessened the corrosiveness of the bottom waters (or the benthic, interstitial micro-environment) and permitted the accumulation of biogenic calcareous sediments, particularly since the middle Pleistocene.

## NORTH ATLANTIC CENOZOIC GEOGRAPHY, BIOGEOGRAPHY, AND CLIMATOLOGY

Cenozoic biogeography and climatology of the North Atlantic and adjacent land areas is closely related to the continuing fragmentation of Eurasia (which begin in the early Mesozoic) and attendant changes in ocean-continent geometry (Berggren and Hollister, 1974,

1977; Thiede, 1979a, b, 1980). Generally, Mesozoic oceanic circulation was latitudinal (zonal), whereas during the Cenozoic it became predominantly longitudinal (meridional). The development of oceanic gateways and barriers (Figs. 14-17) which gradually decreased the efficiency of poleward heat transfer has played a fundamental role in the progressive climatic change which has taken place over the past 50 my (Schnitker, 1980 b; Berggren, in press).

Syntheses of North Atlantic Cenozoic micropaleontologic biostratigraphy and biogeography include those by Berggren (1972 b, 1977, 1978), Berggren and Hollister (1974, 1977), Haq and Lohmann (1976), Haq et al., (1977), and Haq (1981). Studies of the calcareous and siliceous microplankton and phytoplankton (dinoflagellates) have provided a framework for biostratigraphic zonation schemes suitable for regional correlation. The recent integration of one such scheme to a revised paleomagnetic chronology (Hailwood et al., 1979) for the Cenozoic represents a major step toward an attempt to provide a unified biochronology for the entire Cenozoic.

The present day surface circulation pattern in the North Atlantic, of northward moving warm water along the western boundary and southward moving cool water along the eastern boundary and the resulting asymmetric faunal provincialization is reflected in the Neogene and even enhanced along the western side prior to the middle Pliocene (3 Ma). After the mid Pliocene the Gulf Stream and its tropical flora and fauna were excluded from the Labrador Sea concurrently with the initiation of northern hemisphere continental glaciation. Low latitude standard zonation schemes are generally applicable over much of the North Atlantic during the Paleogene. Whatever difficulties exist stem as much from problems of preservation (as at Site 112, southern Labrador Sea) and paleoenvironment (shallow water sites on Rockall Plateau, 117, 403, 404), as from faunal provincialization. For instance, Eocene planktonic foraminiferal assemblages at Rockall Plateau reflect its subsidence history. Shallow water, near shore early Eocene subbotinid-acarinid assemblages (Sites 117, 403, 404) are replaced in the middle Eocene by assemblages containing sporadic keeled morozovellids (Site 405, 406). The absence of hantkeninids, *Turborotalia cerroazulensis*, *Globigerinatheca semiinvoluta*, i. al., in the late Eocene at Rockall Plateau and their presence over the correlative interval at the deep sites in the Bay of Biscay (118, 119, 400A, 401, and 402) point out the fact that latitudinal biogeographic provincialization did exist in the North Atlantic. However faunal provincialization was a gradual phenomenon and the very strong link between latitudinal distribution and paleoclimate became established only after the Eocene.

In the Norwegian-Greenland Sea, temperate microplankton assemblages were common throughout much of the Cenozoic, with incursions of relatively warmer water assemblages during several episodes of the middle and late Oligocene (see Müller, 1976).

At Rockall, temperate faunas predominated in the Paleogene-early Neogene, whereas subarctic faunas have dominated since the late Pliocene. Faunas at Biscay were tropical in the Paleogene-early Neogene; temperate elements became prominent in the mid Miocene.

Latitudinal shifts in calcareous plankton assemblages (Olsson, 1970a, b); Haq and Lohmann, 1976; Haq et al., 1977) have been interpreted as causally related to changing climate conditions, and recent oxygen isotope investigations would appear to bear out this correlation (Vergnaud-Grazzini et al., 1978). Four marked cooling intervals, denoted by the penetration of high latitude indicator assemblages into lower latitudes, have been recognized in the Paleogene, occurring at the middle Paleocene, middle Eocene, latest Eocene-earliest Oligocene, and middle Oligocene (Figs. 15, 16). On the other hand, notable warming is recorded in the late Paleocene-early Eocene, and a less pronounced warming event during the late Oligocene-early Miocene.

Low pole-to-equator temperature gradients characterize the Paleocene of the North Atlantic, but following an early Eocene temperature maximum, these gradients became steeper as global temperatures dropped.

While during the Paleocene and Eocene the marine climate of the North Atlantic ranged from subtropical at low latitudes to warm temperature at high latitudes, the continental climates of the adjacent land areas were predominantly warm and humid. Evidence for this assessment comes from a variety of sources:

1) In Nugssuaq, West Greenland, early Paleocene foraminiferal faunas of Midwayan affinities, invertebrate faunas [with, among others, large cucullaeids, having affinities with contemporaneous Paris Basin forms (and whose living relatives are found in the in the Indo-Pacific)] and vertebrate faunas (with among others, elasmobranchs, batoids, actinopterygiids) suggest that warm temperature climates extended well into the Labrador Sea (Rosenkrantz, 1970). The occurrence of deciduous conifers (*Metasequoia occidentalis*), whose living representatives are located in central China, in the same beds indicate temperate land climate in western Greenland.

2) At the same time plant assemblages on Spitzbergen, then to the NE of Greenland, were present that were characterized by *Taxodium* swamps, *Sequoia*-gingko-angiosperm mixed forests and, during the Eocene, *Equisetum* meadows, *Osmunda*, *Betula*, *Alnus* (?) bog forests, and *Metasequoia* and deciduous forests rich in plant species (Schweitzer, 1980). The large number of broad-leaved angiosperms in

the Spitzbergen flora, autumn leaf fall, and distinct development of growth rings in conifer woods indicate significant seasonality, with precipitation heaviest during the summer, but still adequate during the winter to allow Sequoia and other evergreen conifers to exist. Mean annual temperatures of 15° to 18°C were estimated for the Paleocene-Eocene of Spitzbergen.

3) The peak of the warming trend during the early Eocene is reflected in the vertebrate faunas from the Eureka Sound Formation on Ellesmere Island in the Canadian Arctic Archipelago, NW of Greenland (Dawson et al., 1976; West et al., 1977; Estes and Hutchinson, 1980; McKenna, 1980). Lower vertebrate species include anguid and varanid lizards, a boid snake, two emydid turtles, the tortoise Geochelone, an anostereine and the durophagous alligator Allognathosuchus. The higher vertebrate fauna is dominated by dermopterids ("flying" "lemurs"), paromomyids (primates), and perissodactyls (Coryphodon, Hyrachyus, and Lambdotherium). Also present are true carnivores (Viverrarus), creodonts, a miacine similar to Uintacyos, and rodents. The faunal data are compatible with the geologic interpretation of a forest environment marginal to, and including coal swamps. The reptiles suggest warm, equable climates.

4) Tropical planktonic foraminiferal faunas and calcareous nannoplankton floras occur in the Eocene of the Labrador Sea (Orphan Knoll, Site 111, Laughton, Berggren, et al., 1972).

5) Occurrence of nearshore calcareous nannoplankton floras in the Norwegian-Greenland Sea (Muller, 1976) and, associated with the same flora, of rare specimens of tropical-subtropical keeled morozovellids (planktonic foraminifers) in piston core AS-9 from the Norway Basin.

6) The presence of lateritic paleosols overlying mid to late Eocene basalts on the Iceland-Faeroe Ridge (Site 336, Nilsen, 1978).

7) The northernmost occurrence (on Rockall Bank) of a true nummulitid species (van Hinte and Wong, 1975) together with shallow water bryozoans (Poricellaria) whose modern representatives are associated with shallow reefs (Håkanson and Cheetham, 1972) in the early Eocene.

8) Presence of nummulitid faunas along western European shores as far north as England and north Germany throughout the Eocene.

9) The presence in the early Eocene London Clay on the SW coast of England, of a tropical flora for which a mean annual temperature of 21°C has been suggested and of a mixed tropical (Tethyan) and temperate (boreal) molluscan fauna in contemporaneous marine sediments.

10) Oxygen isotope studies on material from shallow marginal seas of NW Europe indicate a sharp rise in temperature during the late Paleocene-early Eocene, followed by a gradual decline during the remainder of the Eocene, and a sharp drop across the Eocene/Oligocene boundary (Buchardt, 1978) effectively mirroring similar trends observed in studies of the oceanic environment (Fig. 15). Near shore water temperatures in excess of 25° were suggested for the early-middle Eocene, an estimate which is close to the 27°C estimate of Strauch (1968), based upon shell size of the bivalve *Hiatella* from the middle Eocene of western Europe.

Oxygen isotope studies (Keigwin, 1980; Matthews and Poore, 1980) reveal that a marked decline in surface and bottom water temperatures occurred globally near the Eocene/Oligocene boundary (Fig. 16) which is reflected in a simultaneous world wide major floral and faunal change (Cavelier, 1979). The primary response of North Atlantic paleobiogeography was an enhanced latitudinal provincialization of marine faunas and floras (Fig. 17): southward restriction of nummulitic facies to the Aquitaine Basin, development of a distinct boreal planktonic foraminiferal fauna characterized by catapsydracids and small globigerinids, and southward restriction of the calcareous nannoplankton group of discoasters to the Bay of Biscay, with only a sparse and low diversity flora of cool-water species persisting in the Norwegian-Greenland Sea.

Continued climatic variability during the late Paleogene is shown by successive latitudinal migrations of nannofloral and foraminiferal association boundaries. Haq et al. (1977) inferred an earliest (37-35 my) and a mid (32-28 my) Oligocene cooling, with intervening relative warmth during the early and late Oligocene (Fig. 15). The presence during the mid and late Oligocene of an abundant and diverse nannoflora on the Iceland-Faeroe Ridge, and in particular on the VØring Plateau (Müller, 1976) speaks of a very significant warming, of greater magnitude than can be recognized elsewhere. Müller (1976) noted that the abundance and the geographic restriction of the nannoflora to the eastern margin of the Norwegian Sea was probably the consequence of the dynamic behavior of a current system entering the Norwegian Sea from the south. Such influx of North Atlantic water was only made possible by the creation of an exit to the Arctic by initiation of seafloor spreading between Greenland and Svalbard about 36 Ma. Return of this water to the North Atlantic was possible through Nares Strait, Baffin Bay and Labrador Sea. The generally poor paleontological record for this time in the Baffin Bay-Labrador Sea does not allow recognition of this Arctic outflow, although Gradstein and Srivastava (1980, Fig. 6D) seem to indicate it.

The Neogene started with a warming trend that peaked, in the North Atlantic as well as globally, near the end of the early Miocene (~16 Ma), only to decline to lower temperatures and lesser

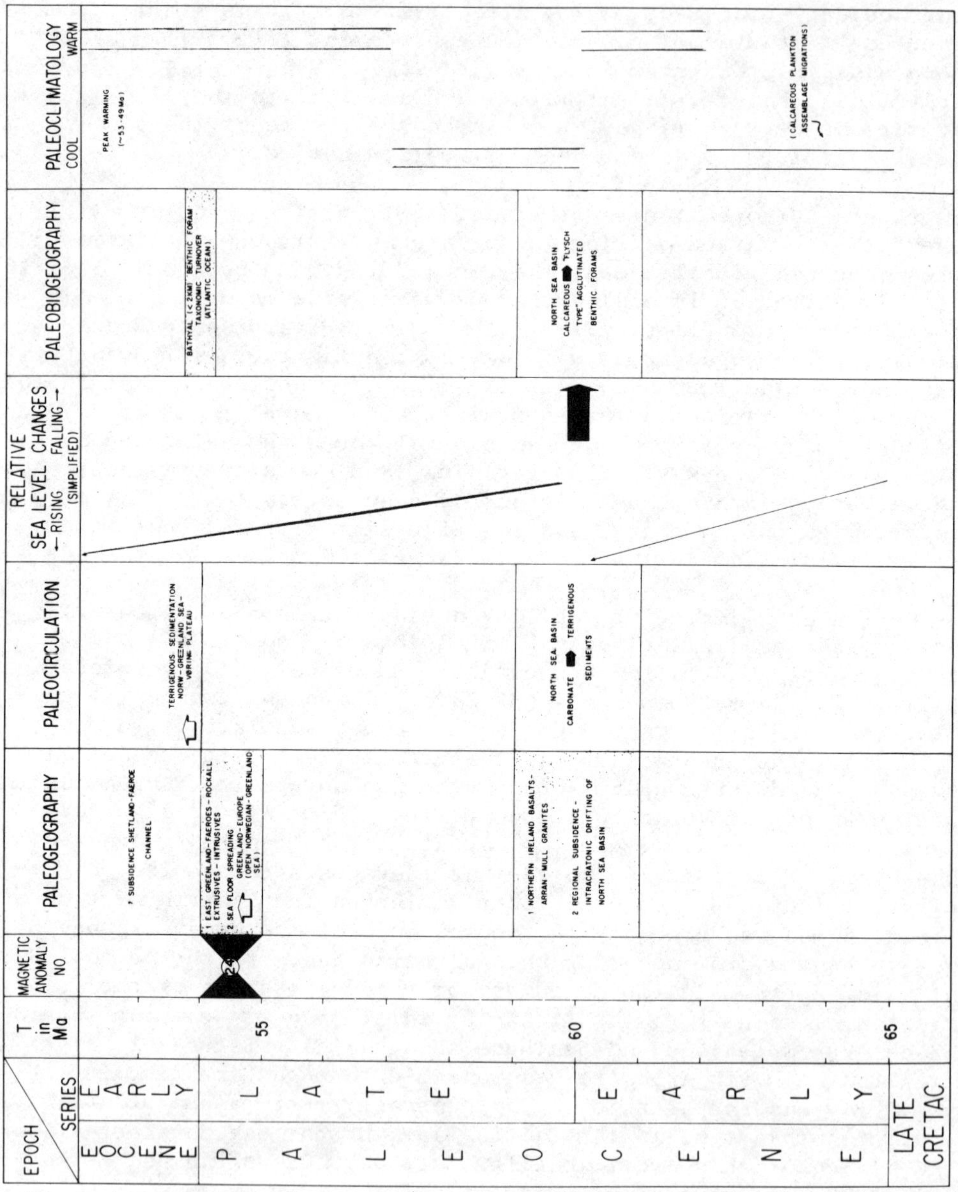
EPOCH
SERIES
T in Ma
MAGNETIC ANOMALY NO
PALEOGEOGRAPHY
PALEOCIRCULATION
RELATIVE SEA LEVEL CHANGES
← RISING FALLING →
(SIMPLIFIED)
PALEOBIOGEOGRAPHY
PALEOCLIMATOLOGY
COOL
WARM
EOCENE
EARLY
PALEOCENE
LATE
EARLY
LATE CRETAC.
55
60
65
24
? SUBSIDENCE SHETLAND-FAEROE CHANNEL
1 EAST GREENLAND-FAEROES-ROCKALL EXTRUSIVES-INTRUSIVES
2 SEA FLOOR SPREADING GREENLAND-EUROPE (OPEN NORWEGIAN-GREENLAND SEA)
1 NORTHERN IRELAND BASALTS-ARRAN-MULL GRANITES
2 REGIONAL SUBSIDENCE-INTRACRATONIC DRIFTING OF NORTH SEA BASIN
TERRIGENOUS SEDIMENTATION NORW-GREENLAND SEA-VØRING PLATEAU
NORTH SEA BASIN
CARBONATE TERRIGENOUS SEDIMENTS
BATHYAL (< 2KM) BENTHIC FORAM TAXONOMIC TURNOVER (ATLANTIC OCEAN)
NORTH SEA BASIN
CALCAREOUS "FLYSCH TYPE" AGGLUTINATED BENTHIC FORAMS
PEAK WARMING (~53-49 Ma)
(CALCAREOUS PLANKTON ASSEMBLAGE MIGRATIONS)

Figure 15. Paleogeographic, paleocirculation, paleobiogeographic, and paleoclimatic events in the North Atlantic: Paleocene.

In this and the following figures upward and downward pointing arrows denote initiation/continuation and cessation/termination of an event(s), respectively. In the column of relative eustatic sea level changes the dark, thick horizontal arrows refer to abrupt relative sea-level lowering, the long thin, inclined arrows to intervals of relative eustatic sea level rising (from Vail et al., 1977). The hachured intervals depicting marine paleoclimatic conditions are based on calcareous plankton data (Haq and Lohmann, 1976; Haq et al., 1977).

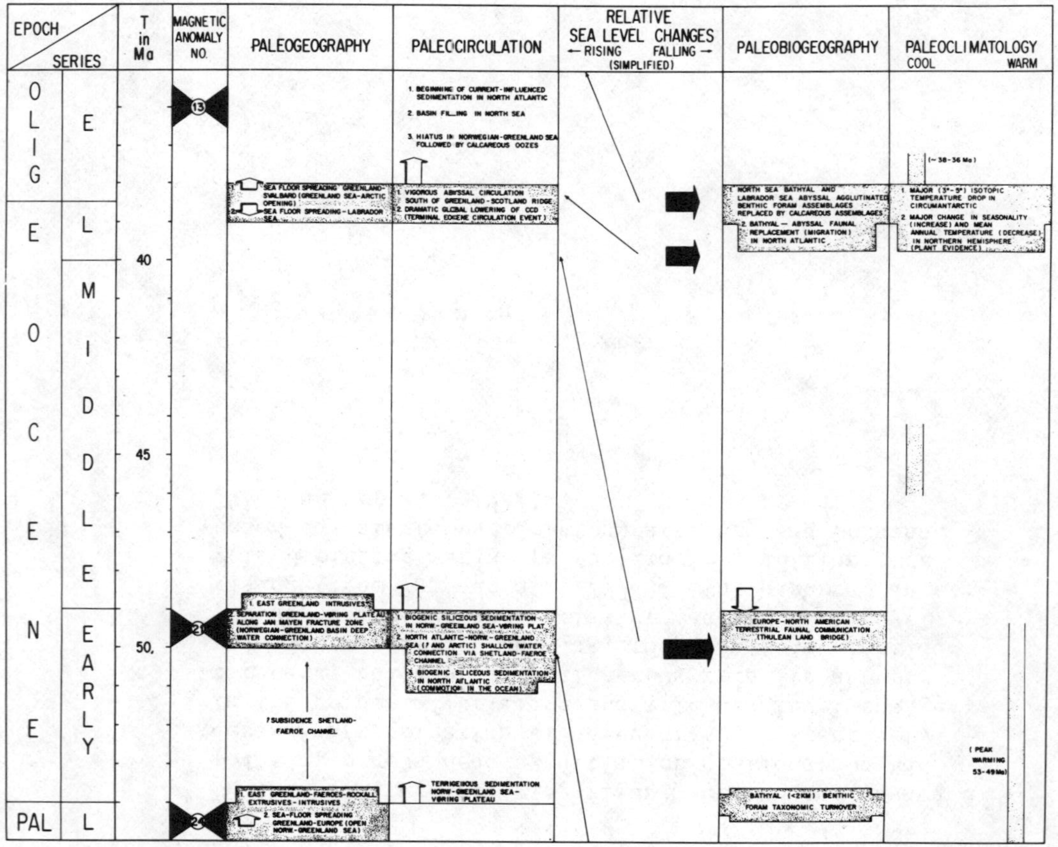


Figure 16. Paleogeographic, paleocirculation, paleobiogeographic, and paleoclimatic events in the North Atlantic: Eocene.

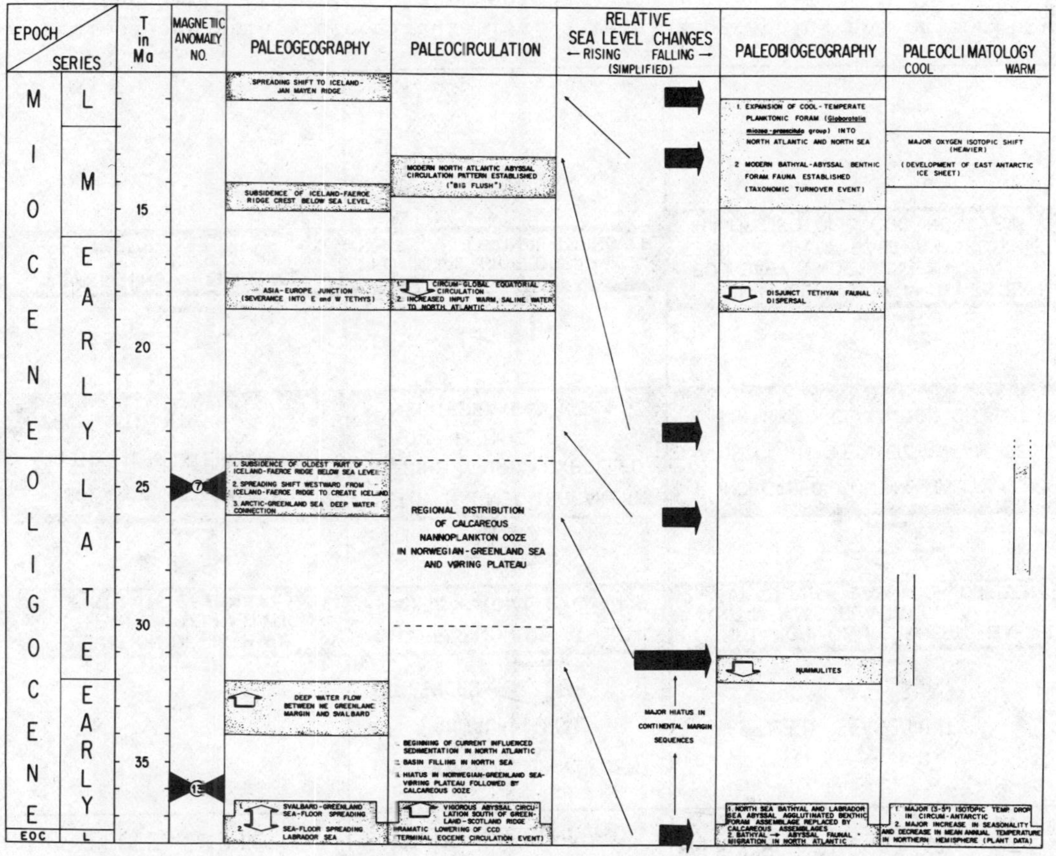


Figure 17. Paleogeographic, paleocirculation, paleobiogeographic, and paleoclimatic events in the North Atlantic: Oligocene to early Miocene.

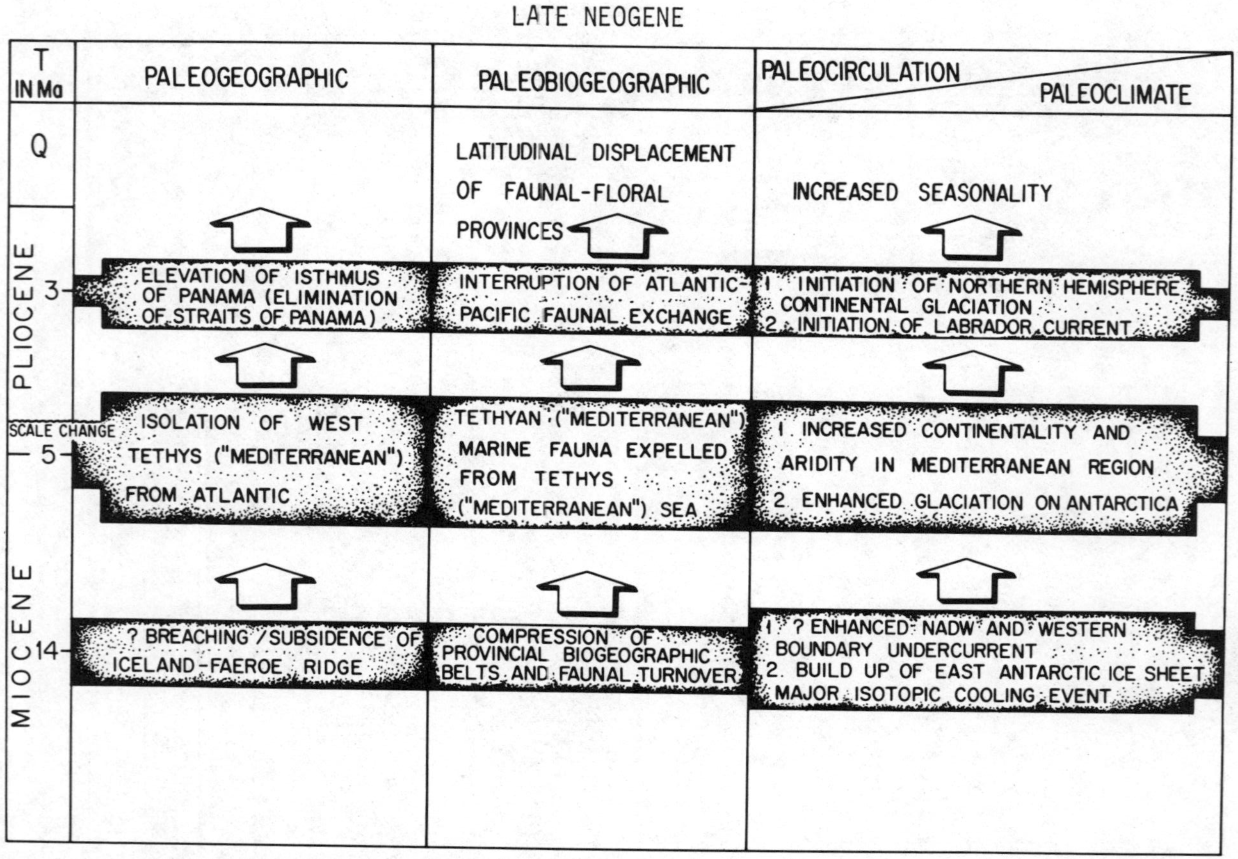


Figure 18. Paleogeographic, paleocirculation, paleobiogeographic, and paleoclimatic events in the North Atlantic: Late Neogene.

equability thereafter (Fig. 17). Superimposed upon this development were, as in the Paleogene, lesser warming and cooling episodes as indicated by the latitudinal migrations of nannofloral assemblages, particularly in the North Atlantic (Haq, 1980). The early Miocene temperature maximum has its reflection in the North Sea in a proliferation of the warm water foraminiferal genus Globigerinoides (particularly of the G. trilobus group). Closure of Tethys between Africa and Europe during the early Miocene (completed by 18 Ma, as indicated by immigration of African mammals into Europe and Asia) rearranged North Atlantic circulation by arresting equatorial circumglobal flow, and, more importantly by adding the now nearly isolated Mediterranean as an evaporation basin to the North Atlantic. Reflux of the warm, salty water rendered the North Atlantic the most saline of the oceans (Schnitker, 1980 b). The isotopic effect, an enrichment in $\delta^{18}O$, can be noted in all oxygen-isotope curves from the North Atlantic (Sites 398, 400A; Vergnaud-Grazzini et al., 1978; Site 116, Rabussier-Lointier, 1980; and in Haq, 1980).

A mid-Miocene cooling is expressed by both isotopic and paleontologic data. The drastic increase in $\delta^{18}O$ at about 15 Ma is thought to represent global cooling as well as the build up of the Antarctic ice sheet (Savin et al., 1975; Shackleton and Kennett, 1975a). In the North Atlantic we see the expression of this cooling trend in the establishment of a mixed-temperate planktonic foraminiferal fauna from the Labrador Sea to Rockall Plateau, characterized by the mid-to high latitude group of Globorotalia miozea. In the North Sea middle Miocene faunas are characterized by members of the closely related Globorotalia zealandica–G. praescitula groups (Fig. 17). The persistence of the "Gulf Stream" off the eastern Canadian margin and the southern Labrador Sea is seen in the continual presence of low latitude tropical foraminifers of the G. fohsi group.

High biological productivity, expressed by high sedimentation rates of diatoms, characterizing the eastern North Atlantic since the Eocene, dropped rather suddenly during the early/middle Miocene, approximately simultaneous with isotopic and faunal expressions of climatic deterioration. This shift from high to low productivity, from siliceous to carbonate sedimentation despite a general rise of the CCD (as indicated by poor preservation of calcareous microfossils in deep sites) speaks of a drastic rearrangement of North Atlantic circulation at that time.

In the late Miocene a pronounced oxygen isotopic shift has been recognized (Shackleton and Kennett, 1975 b) which has been associated with a major expansion of the East Antarctic Ice Shelf. The ensuing lowering of sea level is seen as a major regressive event in peri-Atlantic coastal sections and possibly, is the final cause for closure of the western Mediterranean, resulting in the

Messinian Salinity Crisis (Fig. 18). Perhaps the exclusion of warm Mediterranean influx is responsible for the late Miocene spread of boreal planktonic foraminiferal faunas across the northern part of the North Atlantic, dominated by the Globigerina bulloides and Globorotalia miozea group (Poore and Berggren, 1975, 1976; Poore, 1979). A characteristic element in these faunas is the apparently indigenous species Neogloboquadrina atlantica, a form closely related to N. pachyderma.

Breaching of the isthmus of Gibraltar at the beginning of the Pliocene was followed by a return of warmth to the North Atlantic (Fig. 18). Subtropical and temperate calcareous planktonic floras and faunas coexisted in the early Pliocene of the SW (Orphan Knoll, Site 111) and central (Site 112) parts of the southern Labrador Sea; discoasters return to the Rockall Plateau (Müller, 1979).

Final closure of the isthmus of Panama in the mid-Pliocene intensified the North Atlantic circulation and was followed shortly by the initiation of northern hemisphere glaciation (Schnitker, 1980 b; Fig. 18). Ice rafted debris was deposited by 3 Ma in the Labrador Sea (Berggren, 1972 a), in the Irminger Sea (Shor and Poore, 1979) and on Rockall Plateau (Müller, 1979). The Gulf Stream was expelled from the Labrador Sea by the southward flowing cold Labrador Current, and a distinct subpolar bioprovince established there as well as in the Norwegian-Greenland Sea. Dominant faunal elements were Neogloboquadrina atlantica-N. pachyderma. A boreal faunal province lay somewhat to the south, characterized by members of the Globigerina bulloides and Globorotalia inflata groups. Discoasters disappeared completely from the North Atlantic.

The biogeographic history of the North Atlantic during the past 3 million years has been essentially dominated by rapid and wide fluctuations of faunal province boundaries, in response to glacial interglacial conditions (Fig. 18). These fluctuations were noted to occur as early as the preglacial early Miocene (Savin et al., 1975), became prominent with the mid-Miocene cooling (Woodruff et al., 1981) and were recognized in faunal fluctuations in the late Miocene of Biscay Bay (Schnitker, 1979). This process was amplified about a million years ago as a result of increased severity of climatic cooling (Shackleton and Opdyke, 1977).

## The Greenland-Scotland Ridge: a Paleogeographic and Paleoceanographic Barrier?

The nascent Norwegian Sea was separated from the North Atlantic by a hot-spot generated, fairly flat-topped aseismic ridge. Subsidence and breaching of this barrier must have been a major event in the evolution of North Atlantic oceanography and the exchange of terrestrial vertebrate faunas between Europe and North America. Its significance was recognized by Berggren and Hollister (1974), who noted the temporal coincidence between the development of biogenic siliceous deposition in the North Atlantic (and, indeed the global, equatorial ocean) with the interruption in faunal continuity between European and North American terrestrial vertebrate faunas about 49-50 my ago. They considered this a consequence of the formation of a marine connection between the Arctic and North Atlantic via a breach in the Greenland-Scotland Ridge (the so-called Thulean land bridge of biogeographers) and the initiation of deep circulation ("Commotion in the Ocean") in the North Atlantic (Berggren and Hollister, 1977). Subsequent studies in the Norwegian-Greenland Sea region (primarily as a result of DSDP Leg 38), the North Atlantic (Tucholke and Vogt, 1979; Miller and Tucholke, this volume) and comparison of North Atlantic deep water events with those of other ocean basins (Schnitker, 1980b) have shown that this scenario may be overly simplified.

During the early Eocene, calcareous nannoplankton floras on the Vøring Plateau (DSDP Sites 338 and 343) are similar to contemporaneous floral assemblages on Rockall (DSDP Site 117A) and NW Germany and are characterized by <u>Imperiaster obscurus</u>, a taxon that apparently is indicative of cool and shallow water (Müller, 1976). Saito et al. (1967) report on the similarity of the early Eocene planktonic and benthic foraminiferal assemblages of Norwegian Sea piston core As-9 (66°N, 0° longitude) to contemporaneous assemblages from N.W. Europe. These data suggest that a marine connection may have existed between the Norwegian-Greenland Sea and the North Atlantic as early as the early Eocene. Indeed, a shallow marine (epicontinental) seaway already existed on the margin of Greenland and Scandinavia in the late Cretaceous and Paleocene prior to seafloor spreading. One possible, but perhaps not the only conduit for the necessary faunal exchange may have been the North Sea, even though calcareous microfossils are essentially absent in deposits of its extensive marginal delta system and deep central graben.

Mid-late Oligocene calcareous nannoplankton oozes with similar and relatively warm water floras occur on both the north (DSDP Site 336 and south (DSDP Site 352) flanks of the Iceland Ridge and as far north as the Vøring Plateau (DSDP Site 338) (Müller, 1976). These floras are comparable with contemporaneous floras from the North Atlantic (Leg 12).

Recent studies by Gradstein and Srivastava (1980) suggest that a sea strait linking the Arctic and Atlantic may have existed as early as Campanian-Maastrichtian time, and, judging from the far northward penetration of subtropical planktonic foraminifers, with the surface circulation directed towards the Arctic. McKenna (1980) has drawn attention to the fact that the faunal evidence at Ellesmere Island indicates a warm but cyclic climate, with a weak equator to pole thermal gradient which is difficult to envisage if transfer of large amounts of heat to the polar region during the winter was inhibited or blocked by land barriers or shallow shelves. In a similar manner it must have been possible for relatively warm waters from the North Atlantic to penetrate into the Norwegian-Greenland Sea during the Eocene in order to contribute to the warm, humid climate of that time on Spitzbergen, required by the contemporaneous terrestrial flora.

What role did the Iceland-Faeroe Ridge play as a possible barrier to this exchange? The subsidence of a basaltic aseismic ridge depends upon the age and cooling history of the seafloor upon which it rests (Detrick et al., 1977). Disregarding possible complications introduced by the repeated relocation of the spreading center on the north side of the Iceland Ridge, Nilsen (1978) and Thiede (1980) suggest that the main platform of the Iceland-Faeroe Ridge subsided below sea level during the early Miocene (ca. 24-14 my). Episodes of the subsidence have been recorded from two DSDP holes on the flanks of the ridge. Subaerial lateritic paleosols of Eocene age are overlain by sediments containing late Eocene - late Oligocene outer neritic-upper bathyal benthic foraminiferal assemblages at DSDP Site 336 on the northern flank of the ridge. Middle-lower bathyal benthic assemblages occur in the late Oligocene calcareous nannofossil oozes of DSDP Site 352 which lies on the southern flank of the ridge, below the level of Site 336. These faunas differ primarily in their depth of deposition, but both have close affinities with contemporaneous shallow water faunas in NW Europe and deep sites at Rockall, respectively.

Nilsen (1978) has suggested that the Iceland-Faeroe Ridge (recte Greenland-Scotland Ridge) (= Thulean land bridge) may have been broken or interrupted along its western connection with Greenland. We would suggest, on the other hand, that perhaps one should look in the other direction toward the Faeroe-Shetland channel. If a microcontinent lies beneath the Paleocene lavas of the Faeroe Islands (Bott et al., 1971, 1974; Casten and Nielsen, 1975), supporting the reminder made by Brooks (1980) that the true eastern margin of the North Atlantic is Rockall and the Faeroes, it is possible to envisage the Faeroe-Shetland channel as a graben-like structure formed during the early Tertiary Hebridean phase of volcanism (ca. 55 Ma; Brooks, 1980) associated with pre-drift tensional rifting and subsequent seafloor spreading in the Norwegian-Greenland Sea, beginning about 54 Ma. Subsidence of this area during the early Eocene could have formed a marine connection linking the North Atlantic with the opening Norwegian-Greenland Sea and Svalbard.

A brief interruption in the marine connection between the North Atlantic and Norwegian-Greenland Sea is required to allow the transitory early Eocene trans-Atlantic migration of terrestrial vertebrates (McKenna, 1972, 1975). It may be that this route existed via a subaerial or shallowly submerged "causeway" (?Wyville-Thompson Ridge-Faeroe Bank) linking continental Europe and the Greenland-Scotland Ridge. The major constraint on this "land bridge" is that it was terminated between 49 and 50 Ma (McKenna, 1972, 1975). We may suggest the following scenario: a rapid eustatic drop in sea level (TE1.2 of Vail et al., 1977) about 50 Ma exposed a shallow water or marginally subaerial "causeway" that allowed the transitory trans-Atlantic migration of terrestrial vetebrates. Subsequently the combined effect of subsidence and rising sea level would have quickly terminated the "land bridge" and a marine connection between the North Atlantic and Norwegian-Greenland Sea would have been reestablished.

Marine connections of this new basin with the Arctic are less certain. Prior to the early Oligocene (anomaly 13 = ca. 36 Ma) Svalbard and Greenland were essentially contiguous. Seafloor spreading, accompanied by crustal thinning (Eldholm and Thiede, 1980), began in the early Oligocene with oceanic crust being generated by about 32-34 Ma. It is possible that a shallow marine (epicontinental) connection existed since the Jurassic between the Arctic and Barents Shelf; this connection subsequently linked the Arctic-Barents Sea with the Vøring Plateau at least by early Eocene time (Thiede, 1980). This would vindicate the suggestion made by Berggren and Hollister (1971, 1974) that a deepening surface water connection between the Arctic and North Atlantic was, in fact, initiated in the early Eocene, virtually concomitant with the initiation of seafloor spreading in the Norwegian Sea.

ACKNOWLEDGEMENTS

We would like to thank the organizers and in particular Professor M.H.P. Bott, for their invitation to participate in the NATO Advanced Study Institute on "Structure and Development of the Greenland-Scotland Ridge". We have benefited greatly from discussions with Drs. Bruce Corliss, Bilal Haq, David Johnson, Brian Tucholke, and Mr. Kenneth G. Miller (Woods Hole), and Dr. J. G. Sclater (M.I.T., Cambridge), all of whom have also contributed significantly to the improvement of our text by their critical, constructive reviews.

The research has been supported by grants from the submarine Geology and Geophysics Branch of the National Science Foundation (OCE78-19769 and DES75-13855) (WAB and DS, respectively), and from a consortium of oil companies in support of studies of Cenozoic Benthic Foraminifera (WAB). This is WHOI Contribution No. 4884.

## REFERENCES

Belanger, P.E., (in press), Paleo-oceanography of the Norwegian Sea during the past 130,000 years: Coccolithophorid and foraminiferal data. Boreas.

Belanger, P.E., 1981, Late Cenozoic benthic foraminifera of the Norwegian-Greenland Sea. In: Unpublished Ph.D. thesis, Brown University, Providence, R.I., 294 p.

Belanger, P.E. and Streeter, S.S., 1980, Distribution and ecology of benthic foraminifera in the Norwegian-Greenland Sea. Mar. Micropal., 5 (1980): 401-428.

Benson, R.H., 1975, The Origin of the Psychrosphere as Recorded in Changes of Deep-Sea Ostracode Assemblages. Lethaia, 8: 69-83.

Berger, W.H., 1970, Biogenous Deep-Sea Sediments; Fractionation by Deep-Sea Circulation. Geol. Soc. Amer., 81: 1385-1402.

Berggren, W.A., 1972a, Late Pliocene-Pleistocene glaciation, In "Initial Reports of the Deep Sea Drilling Project", 12, Laughton, A.S., Berggren, W.A., et al., Washington, D.C. (U.S. Government Printing Office): 953-963.

Berggren, W.A., 1972b, Cenozoic biostratigraphy and paleobiogeography of the North Atlantic, In "Initial Reports of the Deep Sea Drilling Project", 12, Laughton, A.S., Berggren, W.A., et al., Washington, D.C. (U.S. Government Printing Office): 965-1001.

Berggren, W.A., 1974, Late Paleocene-Early Eocene benthonic foraminiferal biostratigraphy and paleoecology of Rockall Bank. Micropal. (20(4)· 426-448.

Berggren, W.A., 1977, North Atlantic Cenozoic foraminifera. In "Stratigraphic Micropaleontology of Atlantic Basin Borderlands", Swain, F.M. (ed.), Elsevier Scientific Publishing Co., Amsterdam, 389-410.

Berggren, W.A., 1978, Recent advances in Cenozoic planktonic foraminiferal biostratigraphy, biochronology, and biogeography: Atlantic Ocean. Micropal. (24(4): 337-370.

Berggren, W.A. and Aubert, J., 1976a, Late Paleogene (Late Eocene and Oligocene) benthonic foraminiferal biostratigraphy and paleobathymetry of Rockall Bank and Hatton-Rockall Basin. Micropal. 22: 307-326.

Berggren, W.A. and Aubert, J., 1976b, Eocene benthonic foraminiferal biostratigraphy and paleobathymetry of Orphan Knoll (Labrador Sea). Micropal. 22: 327-346.

Berggren, W.A. and Hollister, C.D., 1971, Biostratigraphy and history of circulation of North Atlantic. Am. Assoc. Petrol. Geol. Bull. 55(2): 331 (abstract).

Berggren, W.A. and Hollister, C.D., 1974, Paleogeography, paleobiogeography and the history of circulation of the Atlantic Ocean, In "Symposium on Geologic History of the Oceans", Hay, W.W. (ed.), Soc. Econ. Pal. Min., Spec. Mem. 20: 126-186.

Berggren, W.A. and Hollister, C.D., 1977, Plate Tectonics and paleocirculation: commotion in the ocean. In "The Present State of

Plate Tectonics", J. Bonin and R.S. Dietz (eds.), Tectonophysics, 38: 11-48.

Bjørklund, K.R., 1976, Radiolaria from the Norwegian Sea, Leg 38 of the Deep Sea Drilling Project. In "Initial Reports of the Deep Sea Drilling Project", vol. 38, Talwani, M., Udintsev (eds.), Washington, D.C., U.S. Government Printing Office: 1101-1168.

Blanc, P.L., Rabussier, D., Vergnaud-Grazzini, C., and Duplessy, J. C., 1980, North Atlantic Deep Water Formed by the Later Middle Miocene. Nature, 283: 553-555.

Boersma, A. and Shackleton, N.J., 1977, Tertiary Oxygen and Carbon Isotope Stratigraphy, Site 357 (Mid Latitude South Atlantic). In "Initial Reports of the Deep Sea Drilling Project", vol. 39: 911-924.

Bott, M.H.P., Browitt, C.W.A. and Stacey, A.,P., 1971, The deep structure of the Iceland-Faeroe Ridge. Mar. Geophys. Res., 1: 328-351.

Bott, M.H., 1974, Structure and evolution of the Atlantic floor between Northern Scotland and Iceland. Norges Geol. Unders., 316: 195-199.

Bott, M.H.P., Sunderland, J., Smith, P.J., Casten, U., and Saxov, S., 1974, Evidence for continental crust beneath the Faeroe Islands. Nature, 248: 202-204.

Brooks, C.K., 1980, Episodic volcanism, epeirogenesis and the formation of the North Atlantic Ocean. Palaeogeogr. Palaeoclimatol., Palaeoecol., 30: 229-242.

Buchardt, B., 1978, Oxygen isotopic Paleotemperatures from the Tertiary period in the North Sea area. Nature, 275: 121-123.

Budyko, M.I., 1974, Climate and Life. Academic Press, New York, 508 pp.

Carmack, E.C., 1977, Water characteristics of the Southern Ocean South of the Polar Front. In "A Voyage of Discovery", M. Angel, (ed.), Pergamon, Oxford and New York. pp. 15-41.

Casey, R., Gust, L., Leavesley, A., Williams, D., Reynolds, R., Duis, T. and Spaw, M.J., 1979, Ecological Niches of Radiolarians, Planktonic Foraminiferans and Pteropods Inferred from Studies on Living Forms in the Gulf of Mexico and Adjacent Waters. Trans. Gulf Coast Assoc. Geol. Soc., 29: 216-223.

Casten, U. and Nielsen, P.H., 1975, Faeroe Islands - a microcontinental fragment? J. Geoph., 41: 357-366.

Caston, V.N.D., 1976, Tertiary sediments of the Vøring Plateau, Norwegian Sea, recovered by Leg 38 of the Deep Sea Drilling Project. In "Initial Reports of the Deep Sea Drilling Project", Talwani, M., Udintsev, G. (eds.), Washington, D.C., U.S. Government Printing Office, 761-782.

Cavelier, C., 1979, La limite Eocene-Oligocene en Europe Occidentale. Science Geol. Mem. 54: 280 p. Strasbourg.

CLIMAP Project Members, 1976, The surface of the Ice-Age Earth. Science, 191: 1131-1144.

Costa, L. and Downie, C., 1979, Cenozoic dinocyst stratigraphy of Sites 403 to 406, Rockall Plateau, IPOD Leg 48. In "Initial Reports Deep Sea Drilling Project", vol. 48, Montadert, L., Roberts, D.G., et al., 1979, Washington, D.C., U.S. Government Printing Office, 513-529.

Dawson, M. (ed.)., 1976, Cenozoic history in and around the northern Atlantic and Arctic Oceans. Palaeogeogr. Palaeoclimatol., Palaeoecol., 30 (3/4): 217-363.

Detrick, R.S., Sclater, J.G. and Thiede, J., 1977, The Subsidence of Aseismic Ridges. Earth Planet. Sci. Lett., 34: 185-196.

Ducasse, O. and Peypouquet, J.P., 1979, Cenozoic ostracodes: their importance for bathymetry, hydrology and biogeography. In "Initial Reports Deep Sea Drilling Project", vol. 48, Montadert, L., Roberts, D.G., et al., 1979. Washington, D.C., U.S. Government Printing Office, 343-363.

Eldholm, O. and Thiede, J., 1980, Cenozoic continental separation between Europe and Greenland. Palaeogeogr. Palaeoclimatol., Palaeoecol., 30: 243-259.

Emelyanov, E.M., Blazchishin, A.I., Kharin, G.S., Lozovaya, N.G. and Zangalis, K.P., 1978, Mineral and chemical composition of sediments of the Voring Plateau, DSDP Leg 38. In "Initial Reports Deep Sea Drilling Project", Supplemental vol. 38, p. 31-54.

Estes, R. and Hutchison, J. M., 1980, Eocene lower vertebrates from Ellesmere Island, Canadian Arctic Archipelago. Palaeogeogr. Palaeoclimatol. Palaeoecol., 30: 325-347.

Foster, T.D. and Carmack, E.C., 1977, Antarctic bottom water formation in the Weddell Sea. In "Polar Oceans, Proc. Polar Oceans Conference", Dunbar, M.J. (ed.), Montreal, 1974, p. 167-177.

Frakes, L.A., 1979, Climates throughout geologic time, Elsevier Scientific Publishing Co., Amsterdam, 310 p.

Gradstein, F.M. and Srivastava, S.P., 1980, Aspects of Cenozoic stratigraphy and paleoceanography of the Labarador Sea and Baffin Bay. Palaeogeogr. Palaeoclimatol., Palaeoecol., 30: 261-295.

Hailwood, E.A., Bock, W., Costa, L., Dupeuble, P.A., Muller, C., and Schnitker, D., 1978, Chronology and biostratigraphy of Northeast Atlantic sediments, DSDP Leg 48. In "Deep Sea Drilling Project", vol. 48, Montadert, L., Roberts, D.G., et al., Washington, D.C., U.S. Government Printing Office, 1119-1141.

Haq, B.U., 1980, Biogeographic history of Miocene calcareous nannoplankton and Paleoceanography of the Atlantic Ocean. Micropal., 26: 414-443.

Haq, B.U., In press, Paleogene paleoceanography: early Cenozoic revisited. Oceanologica Acta.

Haq, B.U. and Lohmann, G.P., 1976, Remarks on the Oligocene calcareous nannoplankton biogeography of the Norwegian-Greenland Sea, DSDP, Leg 38. In "Initial Reports of the Deep Sea Drilling Project", vol. 38 (supplement), Talwani, M., Udintsev, (eds.), Washington, D.C., U.S. Government Printing Office, 141-145.

Haq, B.U., Premoli-Silva, I. and Lohmann, G.P., 1977, Calcareous plankton paleobiogeographic evidence for major climatic fluctuations in the early Cenozoic Atlantic Ocean. Jour. Geophys. Res., 82(27): 3861-3876.

Haq, B.U., Okada, H., and Lohmann, G.P., 1979, Paleobiogeography of the Paleocene/Oligocene calcareous nannoplankton from the North Atlantic Ocean. In "Initial Reports of the Deep Sea Drilling Project", vol. 43, Tucholke, B.E., Vogt, P.R. (eds.), Washington, D.C., U.S. Government Printing Office, 617-629.

Haq, B.U. and Lohmann, G.P., 1976, Remarks on the Oligocene calcareous nannoplankton biogeography of the Norwegian Sea, DSDP, Leg 38. In "Initial Reports of the Deep Sea Drilling Project", vol. 38, Talwani, M., Udintsev, G. (eds.), Washington, D.C., U.S. Government Printing Office, 141-145.

Jones, E.Y.W., Ewing, M., Ewing, J.I. and Eittreim, S.H., 1970, Influences of Norwegian Sea overflow water on sedimentation in the Northern North Atlantic and Labrador Sea. J. Geophys. Res., 75: 1655-1688.

Keigwin, Jr., L.D., 1980, Paleoceanographic change in the Pacific at the Eocene-Oligocene boundary. Nature, 287: 722-725.

Kharin, G.N., Udintsev, G.B., Bogatikov, O.A., Dmitriev, J.I., Raschka, H., Kruezer, H., Mohr, M., Harre, W., and Eckhardt, F.J., 1976, K/Ar ages of the basalts of the Norwegian-Greenland Sea, DSDP Leg 38, In "Initial Reports of the Deep Sea Drilling Project", vol. 38, Talwani, M., Udintsev, G.B. (eds.), Washington, D.C., U.S. Government Printing Office, 755-759.

Kossovskaya, A.G., Timofeev, P.P. and Shutov, V.D., 1978, The lithology and genesis of the sedimentology deposits in the Norwegian Basin and western part of the Lofoten Basin. In "Initial Reports of the Deep Sea Drilling Project", Supplement, vol. 38, p. 67-72.

Krasheninnikov, V.A., 1978, Stratigraphy and planktonic foraminifers of Cenozoic deposits of the Bay of Biscay and Rockall Plateau, DSDP, Leg 48. In "Initial Reports of the Deep Sea Drilling Project", vol. 48, Montadert, L., Roberts, D.G. (eds.), Washington, D.C., U.S. Government Printing Office, 431-450.

Lamb, H.H., 1972, Climate: Present, Past and Future, Vol. 1, Fundamentals and Climate Now. Methuen, London, 596 p.

Laughton, A.S., Berggren, W.A., et al., 1972, In "Initial Reports of the Deep Sea Drilling Project", vol. 12, Washington, D.C., U.S. Government Printing Office, 1243 p.

Matthews, R.K. and Poore, R.Z., 1980, Tertiary $^{18}O$ record and glacio-eustatic sea-level fluctuations. Geology, 8(10): 501-504.

McKenna, M.C., 1972, Eocene final separation of the Eurasian and Greenland-North American land masses. 24th Int. Geol. Congress, 7: 1201-1239.

McKenna, M.C., 1975, Fossil mammals and early Eocene North Atlantic land continuity. Ann. Mo. Bot. gard., 62: 335-353.

McKenna, M.C., 1980, Eocene paleolatitudes, climate and mammals of Ellesmere Island. Palaeogeogr. Palaeoclimatol. Palaeoecol., 30: 349-362.

Miller, K.G. and Tucholke, B.E., In press. Evidence for development of abyssal circulation south of the Greenland-Scotland Ridge.

Miller, K.G., Gradstein, F.M. and Berggren, W.A., In press. Late Cretaceous to early Tertiary agglutinated benthic foraminifera in the Labrador Sea. Micropal.

Müller, C., 1976, Tertiary and Quaternary calcareous nannoplankton in the Norwegian-Greenland Sea, DSDP, Leg 38. In "Initial Reports of the Deep Sea Drilling Project", vol. 38, Talwani, M., Udintsev, G. (eds.), Washington, D.C., U.S. Government Printing Office, 823-828.

Müller, C., 1979, Calcareous nannofossils from the North Atlantic DSDP, Leg 48. In "Initial Reports of the Deep Sea Drilling Project", vol. 48, Montadert, L., Roberts, D.G. (eds.), Washington, D.C., U.S. Government Printing Office, 589-639.

Murray, J.W., 1979, Cenozoic biostratigraphy and paleoecology of sites 403 to 406 based on the foraminifera. In "Initial Reports of the Deep Sea Drilling Project", vol. 48, Montadert, L., Roberts, D.G. (eds.), Washington, D.C., U.S. Government Printing Office, 413-430.

Nesterova, M.P., Scherbokov, F.A., Shevchenko, A.J., Turanskaja, N.W., Kazakova, W.P., Samosudov, A.G., Kuzmina, T.G. and Rudakova, A.N., 1978, Origin of the late Cenozoic sediments of the Icelandic Basin, DSDP, Site 348, Leg 38. In "Initial Reports of the Deep Sea Drilling Project", Supplement, vol. 38: 73-93.

Nilsen, T., 1978, Lower Tertiary laterite on the Iceland-Faeroe Ridge and the Thulean land bridge. Nature, 274: 786-788.

Nilsen, T.H. and Kerr, D.R., 1978, Turbidites, Redbeds, Sedimentary Structures, and Trace Fossils observed in DSDP, Leg 38 cores and the Sedimentary History of the Norwegian-Greenland Sea. In "Initial Reports of the Deep Sea Drilling Project", Supplement, vol. 38: 259-288.

Olsson, R.K., 1970a, Paleocene planktonic foraminifera biostratigraphy and paleozoogeography of New Jersey. Jour. Palaeontol. 46: 589-587.

Olsson, R.K., 1970b, Planktonic foraminifera from the base of the Tertiary, Miller's Ferry and Alabama. Jour. Palaeontol., 44: 598-604.

Peterson, W.H. and Rooth, C.G.H., 1976, Formation and exchange of deep water in the Greenland and Norwegian Seas. Deep Sea Res., 23: 273-283.

Pitman, W.C. III and Talwani, M., 1972, Sea floor spreading in the North Atlantic. Geol. Soc. Am. Bull., 83: 619-646.

Poore, R.Z., 1978, Oligocene through Quaternary planktonic foraminifera biostratigraphy of the North Atlantic, DSDP, Leg 49, In "Initial Reports of the Deep Sea Drilling Project", vol. 49, Luyendyk, B.P., Carr, J.C. (eds.), Washington, D.C., U.S. Government Printing Office, 447-517.

Poore, R.Z. and Berggren, W.A., 1974, Pliocene biostratigraphy of the Labrador Sea: calcareous plankton. Jour. Foram. Res., 4(3): 91-108.

Poore, R.Z. and Berggren, W.A., 1975, Late Cenozoic planktonic foraminifera biostratigraphy and paleoclimatology: Hatton-Rockall Basin. DSDP, Site 116, Jour. Foram. Res., 5(4): 279-293.

Rabussier Lointier, D., 1980, Variations de composition isotopique de l'oxygene et du carbone en milieu marin et coupures stratigraphiques des Cenozoique. These doct. 3e cycle, Univ. Paris, 181 p.

Rateev, M.A., Renngarten, N.V., Shutov, V.D. and Drits, V.A., 1978, Lithology and clay mineralogy of sediments from hole 346. "Initial Reports of the Deep Sea Drilling Project", Supplement, vol. 38, p. 55-65.

Reid, J.L. and Lynn, R.J., 1971, On the influence of the Norwegian-Greenland and Weddell Seas upon the Bottom Waters of the Indian and Pacific Oceans. Deep Sea Res., vol. 18: 1063-1088.

Roberts, D.G., 1975, Marine geology of the Rockall Plateau and Trough. Philos. Trans. R. Soc. London, Ser. A, 278: 447-509.

Roberts, D.G., Montadert, L. and Searle, R.C., 1979, The western Rockall Plateau: stratigraphy and Crustal evolution. In "Initial Reports of the Deep Sea Drilling Project", vol. 48, Washington, D.C., U.S. Government Printing Office, 1061-1088.

Ruddiman, W.F., 1972, Sediment redistribution on the Reykjanes Ridge: Seismic evidence. Bull. Geol. Soc. Amer., vol. 83: 2039-2062.

Ruddiman, W.F., Molfino, B., Esmay, A., Pokras, E., 1980, Evidence bearing on the mechanism of rapid deglaciation. Climatic Change, 3: 65-87.

Ruddiman, W.F. and McIntyre, A., 1979, Warmth of the subpolar North Atlantic Ocean during northern hemisphere ice-sheet growth. Science 204: 173-175.

Saito, T., Burckle, L.K. and Horn, D.R., 1967, Paleocene core from the Norwegian Basin. Nature, 216: 357-359.

Savin, S.M., 1977, The history of the earth's surface temperature during the past one hundred million years. Ann. Rev. Earth Planet. Sci., 5: 319-344.

Savin, S.M., Douglas, R.G. and Stehli, F.G., 1975, Tertiary marine paleotemperature. Geol. Soc. Am. Bull., 86: 1499-1510.

Schnitker, D., 1974, West Atlantic abyssal circulation during the past 120,000 years. Nature, 298: 385-387.

Schnitker, D., 1979, The deep waters of the Western North Atlantic during the past 24,000 years, and the re-initiation of the Western boundary undercurrent. Mar. Micropal., 4: 265-280.

Schnitker, D., 1980a, Quaternary Deep-Sea benthic foraminifers and bottom water masses. Ann. Rev. Earth Planet. Sci. 8: 343-370.

Schnitker, D., 1980b, Global paleoceanography and its deep water linkage to the Antarctic glaciation. Earth Sci. Rev., 16: 1-20.

Schnitker, D. and Tjalsma, L.R.C., 1981, New genera and species of benthic foraminifers from Paleocene and Eocene deep water deposits. J. Foram. Res., 10: 235-241.

Schrader, H.J. and Fenner, J., 1976, Norwegian Sea Cenozoic diatom biostratigraphy and taxonomy: Part 1 - Norwegian Sea Cenozoic diatom biostratigraphy. In "Initial Reports of the Deep Sea Drilling Project", vol. 38, Talwani, M., Udintsev, G. (eds.), Washington, D.C., U.S. Government Printing Office, 921-962.

Schweitzer, H.M., 1980, Environment and climate in the Early Tertiary of Spitsbergen. Palaeogeogr. Palaeoclimatol. Palaeoecol. 30: 297-311.

Sclater, J., Hellinger, S. and Tapscott, C., 1977, The paleobathymetry of the Atlantic Ocean from the Jurassic to present. J. Geol., 85: 509-522.

Shackleton, N.J. and Kennett, J.P., 1975a, Paleotemperature history of the Cenozoic and the Initiation of Antarctic Glaciation: Oxygen and Carbon isotope analyses in DSDP Sites 277, 279 and 281. In "Initial Reports Deep Sea Drilling Project", vol. 29: 743-755.

Shackleton, N.J. and Kennett, J.P., 1975b, Late Cenozoic Oxygen and Carbon Isotopic changes at DSDP Site 284: Implications for Glacial History of the Northern Hemisphere and Antarctica. In "Initial Reports of the Deep Sea Drilling Project", vol. 29: 801-807.

Shackleton, N.J. and Opdyke, N.D., 1977, Oxygen Isotope and Paleomagnetic evidence for early Northern Hemisphere Glaciation. Nature, 270: 216-219.

Shor, A.N. and Poore, R.Z., 1979, Bottom currents and ice rafting in the North Atlantic: Interpretation of Neogene Depositional Environments of Leg 49 Cores. In "Initial Reports of the Deep Sea Drilling Project", vol. 49: 859-872.

Streeter, S.S., 1973, Bottom water and benthonic foraminifera in the North Atlantic-Glacial-Interglacial contrasts. Quat. Res., 3: 131-141.

Streeter, S.S. and Shackleton, N.J., 1979, Paleocirculation of the Deep North Atlantic: 150,000 year record of benthic foramininera and Oxygen-18. Science, 203: 168-170.

Streeter, S.S., Belanger, P.E., Kellogg, T.B. and Duplessy, J.C., In press, Near surface conditions and deep water overflow from the Norwegian Sea: benthic foraminiferal evidence. Jour. Quat. Res.

Strauch, F., 1968, Determination of Cenozoic sea-temperatures using Hiatella arctica (Linne). Palaeogeogr. Palaeoclimatol. Palaeoecol., 5: 213-233.

Talwani, M. and Eldholm, V., 1977, Evolution of the Norwegian-Greenland Sea. Geol. Soc. Am. Bull., 88: 969-999.

Talwani, M. and Udintsev, G., 1976, Tectonic synthesis. In "Initial Reports of the Deep Sea Drilling Project", vol. 38, Talwani, M., Udintsev, G.(eds.), Washington, D.C., U.S. Government Printing Office, 1213-1242.

Talwani, M., Udintsev, G., et al., 1976, Initial Reports of the Deep Sea Drilling Project, vol. 38, Washington, D.C., U.S. Government Printing Office, 1256 p.

Thiede, J., 1973, Planktonic foraminifera in hemipelagic sediments: Shell Preservation off Portugal and Morocco. Geol. Soc. Amer. Bull., 84: 2749-2754.

Thiede, J., 1979a, History of the Atlantic Ocean evolution of an asymmetric zonal paleo-environment in a latitudinal ocean basin. In "Deep Drilling Results in the Atlantic Ocean: Continental Margins and Paleoenvironments, Amer.-Geophys. Union", Talwani, M., Hay, W., Ryan, W.B.F. (eds.), Washington, D.C., Maurice Ewing Series, p. 275-296.

Thiede, J., 1979b, Paleogeography and paleobathymetry of the Mesozoic and Cenozoic North Atlantic Ocean. Geojournal, 3(3): 263-272.

Thiede, J., 1980, Palaeoceanography, margin stratigraphy and paleophysiology of the Tertiary North Atlantic and Norwegian-Greenland Seas. Phil. Trans. R. Soc. Lond., A295: 177-185.

Tjalsma, R.C. and Lohmann, G.P., In Press, Paleocene-Eocene bathyal and abyssal benthic foraminifera from the Atlantic Ocean. Micropal., Spec. Paper, Micropaleontology Press, New York.

Tucholke, B.E. and Mountain, G.S., 979, Seismic stratigraphy, lithostratigraphy and paleosedimentation patterns in the North American Basin. In "Deep Drilling Results in the Atlantic Ocean: Continental Margins and Paleoenvironments", Talwani, M., Hay, W.W., and Ryan, W. (eds.), Washington, D.C., AGU M. Ewing Series 3: 58-86.

Vail, P.R., Mitchum, R.M., and Thompson III, S., 1976, Seismic stratigraphy and global changes of sea level, part 4: global cycles of relative changes of sea level. In "Seismic Stratigraphy-applications to hydrocarbon exploration", Payton, C.E. (ed.), Am. Assoc. Petrol. Geol. Mem. 26: 83-97.

van Hinte, J.E. and Wong, T., 1975, Nummulites rockallensis n. sp. from the Upper Paleocene of Rockall Plateau (North Atlantic). Jour. Foram. Res., 5(2): 90-101.

Varentsov, I.M., 1978, Lithologic-mineralogic studies of the sedimentary deposits from hole 350, DSDP Leg 38. In "Initial Reports of the Deep Sea Drilling Project", Supplement, vol. 38, p. 111-120.

Vergnaud-Grazzini, C., Pierre, C., Letolle, R., 1978, Paleoenvironment of the North-East Atlantic during the Cenozoic: Oxygen and carbon isotope analyses at DSDP Sites 398, 400A and 401. Oceanologica Acta, 1: 381-390.

Vogt, P.R., 1972, The Faeroe-Iceland-Greenland Aseismic Ridge and the Western Boundary Undercurrent. Nature, 239: 79-81.

West, R., Dawson, M. and Hutchison, J., 1977, Fossils from the Paleogene Eureka Sound Formation, N.W.T., Canada: Occurrence, climatic and paleogeographic implications. Milwaukee Bull. Mus. Spec. Bull., Biol. Geol., 2: 77-93.

White, S.M., 1978, Sediments of the Norwegian-Greenland Sea, DSDP, Leg 38. In "Initial Reports of the Deep Sea Drilling Project", Supplement, vol. 38: 293-257.

Woodruff, F., Savin, S.S. and Douglas, R.G., 1981, Miocene stable isotope record: a detailed deep Pacific Ocean study and its paleoclimatic implications. Science, 212: 665-668.

Worsley, T. and Davis, T.A., 1979, Sea level fluctuations and deep sea sedimentation rates, Science, 203: 455-456.

Worthington, L.V., 1976, On the North Atlantic Circulation. Johns Hopkins University Press, Baltimore, 110 p.

Ziegler, P.A., 1975a, North Sea Basin history in the tectonic framework of North-Western Europe. In "Petroleum Geology and the Continental Shelf of North-West Europe", Wood, A.W. (ed.), J. Wiley and Sons, New York, I: 131-151.

Ziegler, P.A., 1975b, Geologic evolution of North Sea and its tectonic framework. Am. Assoc. Petrol. Geol. Bull., 59: 1073-1097.

Ziegler, P.A., 1979, North-Western Europe: Tectonics and basin development. Geol. en Mijnb., 57(4): 589-626.

# DEVELOPMENT OF CENOZOIC ABYSSAL CIRCULATION SOUTH OF THE GREENLAND-SCOTLAND RIDGE

Kenneth G. Miller and Brian E. Tucholke

Woods Hole Oceanographic Institution/Massachusetts Institute of Technology Joint Program in Oceanography Woods Hole, MA USA 02543, and Woods Hole Oceanographic Institution, Woods Hole, MA USA 02543.

## ABSTRACT

Seismic, lithostratigraphic, faunal, and isotopic evidence from the western and northern North Atlantic indicates that formation of northern sources for strongly circulating bottom water began in the late Eocene to early Oligocene. The widely distributed reflector R4 correlates with an unconformity eroded along basin margins at the Eocene/Oligocene boundary. This change in abyssal regime also correlates with a major benthic foraminiferal turnover in the deep southern Labrador Sea (DSDP Site 112) and with a faunal reorganization in the Bay of Biscay. The principal bottom-water source probably was of Arctic origin; it entered the Norwegian Sea following separation of Greenland and Spitsbergen and flowed south across the Greenland-Scotland Ridge through the Faeroe-Shetland Channel and possibly across a sill east of Greenland. This flow may have been supplemented by dense Arctic water entering the basin via Nares Strait and Baffin Bay, and by cooling and sinking of saline surface water south of the Greenland-Scotland Ridge and in the Labrador Sea. Current-controlled sedimentation and erosion, often of a chaotic nature, continued through the Oligocene above reflector R4, but the general intensity of abyssal circulation is thought to have decreased. Above reflector R2 (upper lower Miocene) current-controlled sedimentation became more coherently developed, and a major phase of sedimentary drift development was initiated. We interpret this to be a result of a further general reduction and especially a stabilization of the abyssal circulation, possibly linked with degeneration of numerous fracture-zone conduits that previously funnelled bottom water across the Reykjanes Ridge. The gross nature of the circulation has not changed substantially since the middle Miocene, although it has been punctuated by further climatic and tectonic events.

## INTRODUCTION

Deep and bottom waters formed at high latitudes in the North Atlantic and in the Norwegian-Greenland Seas strongly affect the Holocene sedimentary record throughout the North Atlantic Ocean, especially along the western margins of the basins. It has been known for more than a decade that this current influence also had a significant effect on the sedimentary record well back into the Tertiary (e.g. Jones, et al., 1970). In recent years, an expanding data base of seismic reflection profiles, DSDP boreholes, and piston cores has allowed more extensive study of these current influences, but the definition of even such first-order events as the time of initiation of geologically significant deep circulation in the North Atlantic and the effects of potential tectonic barriers like the Greenland-Scotland Ridge still is in dispute.

Part of the ambiguity is due to a lack of necessary data, but significant dispute also results from varying ways of interpreting existing data. For example, one form of evidence clearly documenting the <u>presence</u> of abyssal circulation is the development of large-scale current-controlled bedforms and sediment drifts. Such drifts were well developed by the early to middle Miocene in both the western and northern North Atlantic (Figures 1, 2). However, it is unclear exactly what change in abyssal circulation the first development of these drifts represents, and there are at least four possibilities:

1) <u>Reduction</u> in current intensity below the critical threshold between erosional/transportational and depositional conditions (especially likely if the drifts are developed above unconformities).
2) <u>Stabilization</u> of abyssal currents compared to earlier erratic flow, with consequent regular development of bedforms and accumulation patterns.
3) <u>Interaction</u> of newly developed flows with existing abyssal currents to form a stable depositional pattern.
4) <u>Intensification</u> of abyssal currents from near-zero flow to trigger sediment transport to areas of stable deposition.

Each of these may be equally plausible in a given situation, and other evidence usually is necessary to resolve the nature of the circulation event.

Perhaps the most convincing evidence for <u>intensification</u> of abyssal circulation is the development of regional deep-sea unconformities, especially along basin margins where currents are geologically most effective. These unconformities may or may not be accompanied by hiatuses in the central parts of ocean basins.

Additional evidence for changes in abyssal circulation also is found in shifts in lithology (e.g. texture and composition) and

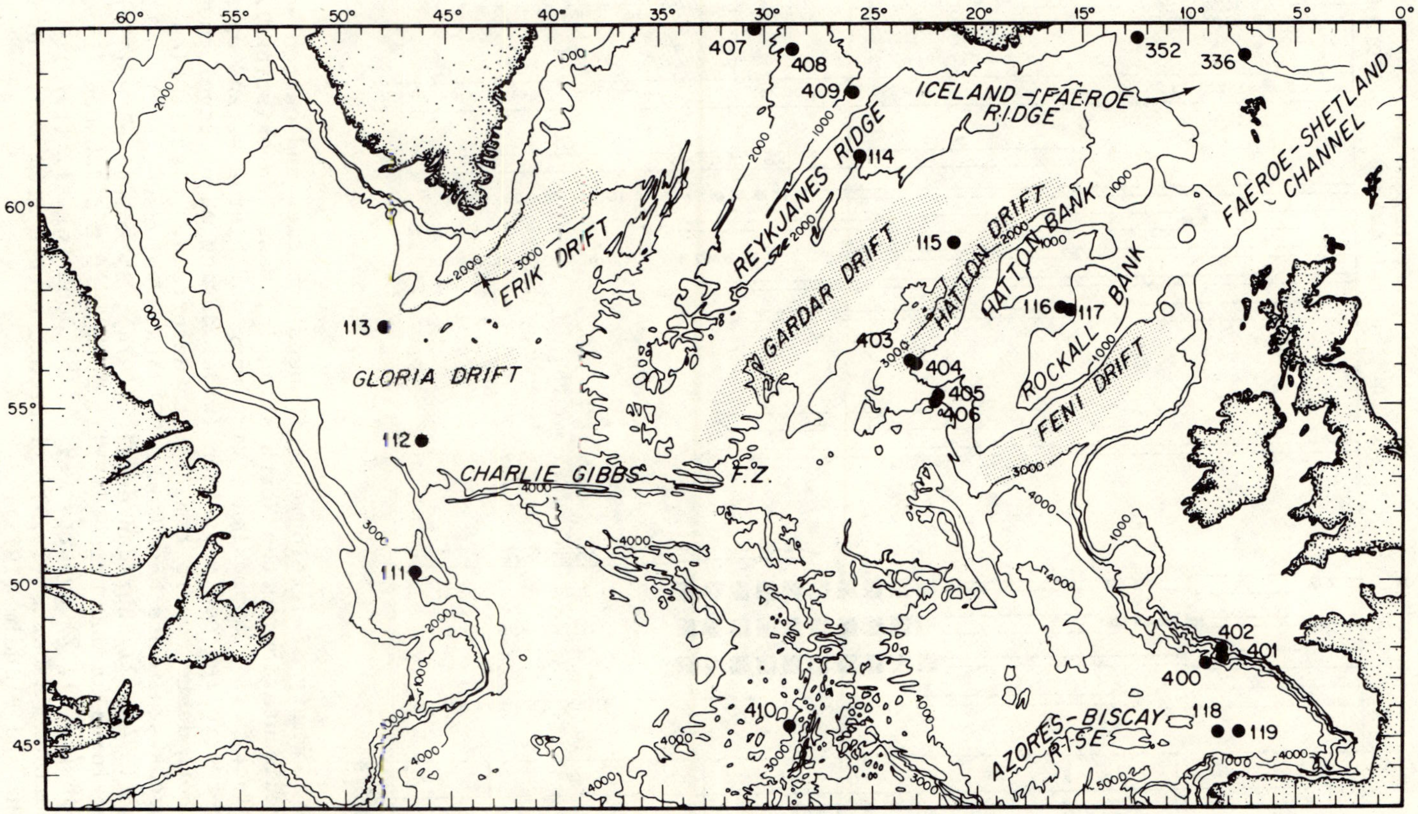


Figure 1. Bathymetric location map of the northern North Atlantic with locations of DSDP sites and major sedimentary drifts. Contour interval 1000 m.

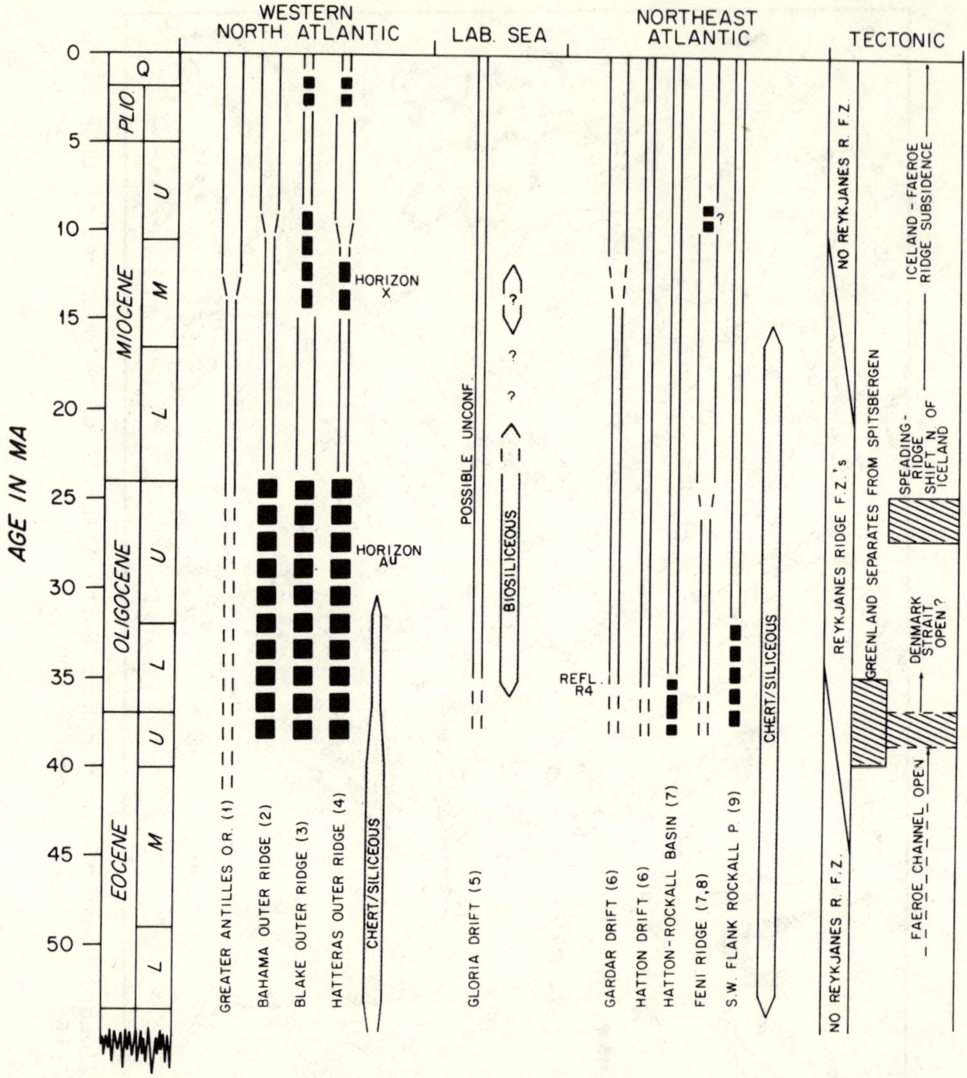


Figure 2. Summary of development of unconformities and current-controlled sedimentation in the North Atlantic, based on existing literature. Black bars - approximate time range of hiatuses, and open bars - current-controlled sedimentation (wider during major development of drifts); both dashed where time limits uncertain. Distribution of bio-siliceous sediments based on DSDP data. References: 1) Tucholke and Ewing (1974); 2) Sheridan et al. (1974); 3) Ewing and Hollister (1972); 4) Tucholke and Laine (1981); 5) Egloff and Johnson (1975); 6) Ruddiman (1972); 7) Roberts (1975); 8) Jones et al. (1970); 9) Montadert and Roberts (1979). Tectonic events from Ruddiman (1972) and Talwani and Eldholm (1977).

benthic fauna (taxonomic and isotopic composition). These effects in themselves rarely are diagnostic of the nature of the circulation change, but they can provide important supporting evidence. For example, Berggren and Hollister (1974) suggested that increased early to middle Eocene deposition of biosiliceous sediments in the North Atlantic (Figure 2) correlates with the first formation, sinking, and upwelling of cold, deep waters in the northern North Atlantic. This correlation is reasonable, but it is possible that the biosiliceous sedimentation was stimulated by other mechanisms. Therefore, the lithologic change by itself is not definitive.

The purpose of this paper is to examine available evidence for the initiation and development of the abyssal circulation in the northern North Atlantic. Based upon evaluation of this evidence, we postulate a preliminary model to explain Eocene to Miocene bottom-water sources, including overflow across the Greenland-Scotland Ridge. Although the model is not definitive, it identifies several major unsolved and testable problems relating to the North Atlantic deep and bottom-water circulation.

## PREVIOUS WORK

Over the past decade, there has been growing debate about when northern sources (North Atlantic, Norwegian-Greenland Sea, or Arctic Ocean) of bottom water began to form. As noted above, Berggren and Hollister (1974) suggested a correlation between the early Eocene deposition of biosiliceous sediments in the western North Atlantic basin and the formation of the first cold deep waters in the North Atlantic. They suggested that the source of bottom water was overflow across the Greenland-Scotland Ridge from the Norwegian-Greenland Sea. Jones et al. (1970) also felt that Norwegian Sea overflow affected sedimentation in Rockall Trough beginning in the late Eocene to early Oligocene, although the overflow was interpreted to influence other parts of the North Atlantic as early as the Paleocene. A widespread reflector in the northern North Atlantic variously termed "R" (Jones et al, 1970; Ruddiman, 1972), or "R4" (Roberts, 1975; Roberts et al., 1979), figures prominently in the interpretation of abyssal circulation history because it appears to divide current-influenced sedimentation above from largely pelagic sedimentation below (Figures 1, 2). This reflector is approximately late Eocene to early Oligocene in age (Ruddiman, 1972; Roberts, 1975; Laughton and Berggren, 1972; Roberts et al., 1979; Miller et al., in press). Vogt (1972) suggested that the apparent increase in abyssal circulation at this time correlated with bottom-water outflow from the Norwegian-Greenland Sea following initial subsidence of the Greenland-Scotland Ridge about 40 Ma.

The first clear evidence for strong bottom-current activity is the erosion of a prominent unconformity (Horizon $A^u$) along the continental rise of eastern North America. Tucholke (1979) and Tucholke and Mountain (1979) demonstrated that this erosion occurred at some point between the late Eocene and earliest Miocene. They suggested that the unconformity was caused by a precursor to the modern Western Boundary Undercurrent (Heezen et al., 1966) which presently contains a combination of Greenland-Scotland Ridge overflow water, Labrador Sea Water, and, in places, Antarctic Bottom Water.

Stratigraphic relations show that by the early Miocene, current-controlled deposition was well established in the Rockall Trough (Feni Drift; Roberts, 1975) and in the western North Atlantic (Blake Ridge; Ewing and Hollister, 1972; Tucholke and Mountain, 1979; Figure 2). By the middle to late Miocene, coherently deposited sediments also accumulated at significantly increased rates on such sedimentary drifts as the Greater Antilles, Bahama, Hatteras, and Gardar Ridges (Figure 2). Based on this observation and on the widespread distribution of a middle Miocene hiatus, Shor and Poore (1979) suggested that the deep circulation of the northeast Atlantic did not attain its present configuration until the early to middle Miocene; they related this to an inferred late Oligocene to early Miocene marine connection across the Iceland-Faeroe Ridge. Vogt (1972) also suggested a significant deepening of this marine connection in the middle to late Miocene (ca. 10-18 Ma) based upon a subsidence model; he related the presumed increase in Norwegian Sea overflow to a middle Miocene change in abyssal circulation noted by Ruddiman (1972) on the Gardar Ridge. Although there is some coherence in the sequence of events outlined above, the existing studies still show a general lack of agreement not only on the timing of initial development of the abyssal circulation, but also on the timing and influence of subsidence of the Greenland-Scotland Ridge.

Similar disagreements appear in interpretations of biostratigraphic, benthic faunal, and isotopic data. Schnitker (1979) inferred from the benthic faunal and isotopic records that a change in abyssal circulation occurred in the middle Miocene. He also (1980a, b) related this change to the first deep connection across the Greenland-Scotland Ridge as suggested by DSDP Leg 38 scientists (Talwani and Udintsev, 1976). Similarly, Blanc et al. (1980) interpreted carbon isotopic data to indicate that the production of oxygenated deep water in the North Atlantic began during the middle Miocene. In contrast, Miller et al. (in press) suggested that significant bottom-water formation began in the North Atlantic in the late Eocene to early Oligocene. Their interpretation was based on a marked shift from agglutinated to calcareous benthic foraminifera at DSDP Site 112 in the Labrador Sea.

Within the varying interpretations outlined above, there are generally well-defined patterns of current-influenced sedimentation (Figure 2). It is apparent that strong abyssal circulation eroded unconformities in the western North Atlantic and portions of the northern North Atlantic as early as the end of the Eocene. Current-influenced sedimentation, perhaps in rather unstable patterns, followed during the Oligocene to early Miocene, and rapid stable accretion of many sediment drifts was established by the middle to late Miocene. In the ensuing discussions, we attempt to refine the interpretation of the significance of these generalized patterns.

## NORTHERN NORTH ATLANTIC OBSERVATIONAL DATA

### Lithofacies/Age Distribution - DSDP Boreholes

In Figures 3 to 6 we have summarized age-versus-lithofacies data for the DSDP boreholes that recovered significant pre-Pliocene sedimentary records in the northern North Atlantic. Age assignments are taken from the appropriate volumes of "Initial Reports of the Deep Sea Drilling Project" with the exceptions noted below.

At Site 117, drilled along the eastern edge of the Hatton-Rockall Basin, Berggren and Aubert (1976) revised the original biostratigraphic zonation by removing core 1 from the early Miocene and placing it in the late Oligocene (Figure 4). This greatly improves correlations between Sites 116 and 117 as suggested by the seismic stratigraphy (Figure 7). We also suggest a revision to the original zonation at Site 116, drilled 40 km west of Site 117. An intra-Oligocene hiatus was inferred between cores 22 and 23 on the basis of the assignment of core 22 to the *Sphenolithus ciperoensis* (NP25) Zone and cores 23 to 25 to the *Erisonia obruta* (NP21) Zone (Perch-Nielson, 1972). Bukry (1972), on the other hand, placed core 25 in the *Helicosphaera reticulata* (NP22) Zone. Roberts (1975) noted this discrepancy and used Bukry's assignment to suggest that sedimentation was slow but continuous in the Oligocene section. Examination of the planktonic foraminiferal data supports the contention of continuous sedimentation. In addition, Berggren (1972) noted the last appearances of *Globigerina ampliapertura* and *Chiloguembelina* in cores 24 and 23, respectively, but he assumed that Perch-Nielson's nannoplankton assignment was correct and therefore thought that the ranges of these taxa were truncated by an intra-Oligocene unconformity. However, if Bukry's assignment is used, then these last appearances may be taken as the *G. ampliapertura* (top of Zone P20) and *Chiloguembelina* (lower Zone P21) datums. Thus, cores 19 to 22 may represent the late Oligocene, cores 23 and 24 the middle Oligocene, and core 25 the early Oligocene. The average Oligocene sedimentation rate, if interpreted in this fashion, is continuous at 9.5 m/Ma (Figure 4).

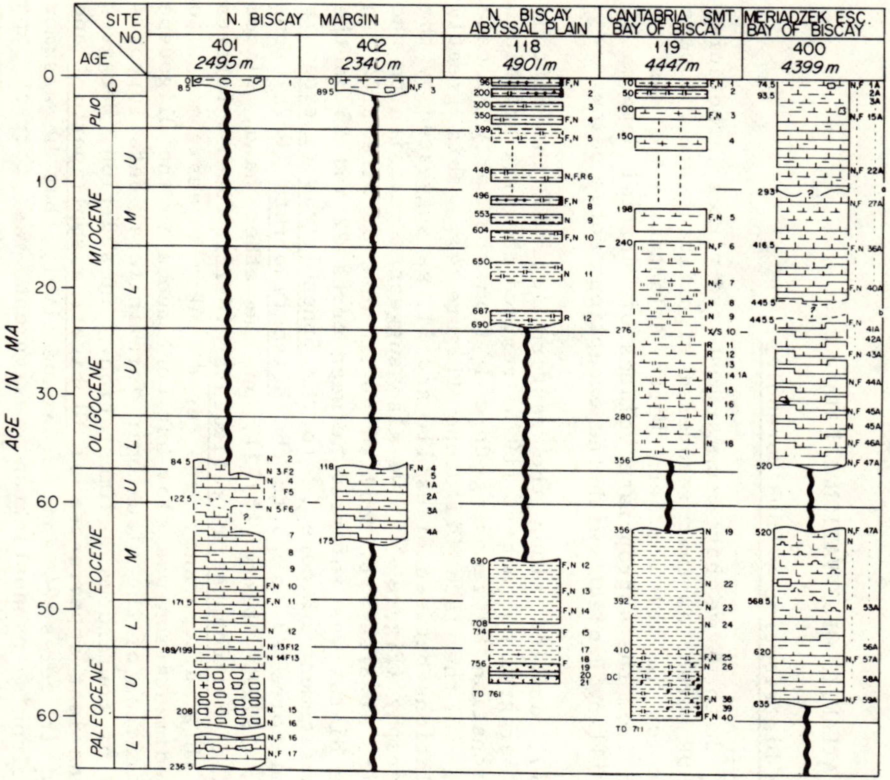


Figure 3. Age versus lithofacies for Tertiary sediments recovered at DSDP sites in the Bay of Biscay. Site locations in Figure 1. Lithologic symbols follow standard "Initial Reports" format. To left of columns are subbottom depths in meters (T.D.=total depth); to right are indicated core number and fossil group on which age is based (N=nannoplankton, F=foraminifers, R=radiolaria, P=palynmorphs, BF=benthic foraminifers, D=diatoms, X/S =extrapolated by sedimentation rate). Heavy wavy lines show hiatuses, dashed lines show inferred continuous sedimentation.

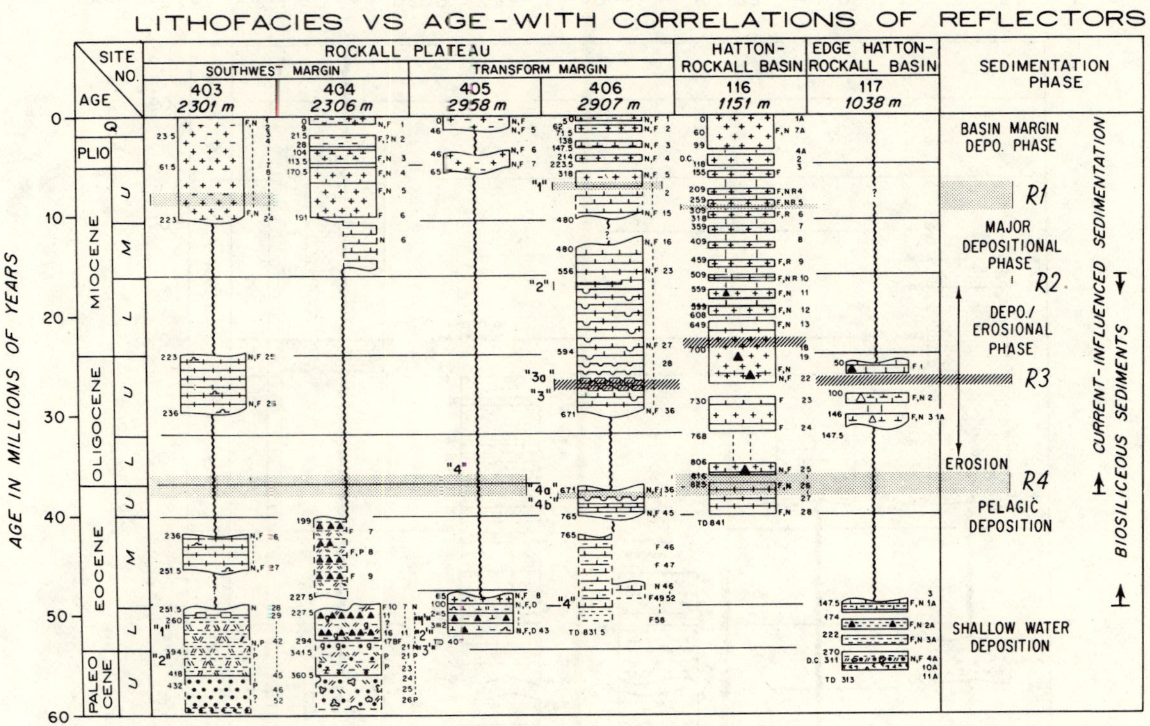


Figure 4. Age versus lithofacies for DSDP sites in Rockall region with correlation of major reflectors. Explanation in Figure 3. Numbers in quotes at left of columns are reflector numbers used in individual site reports by Montadert and Roberts (1979).

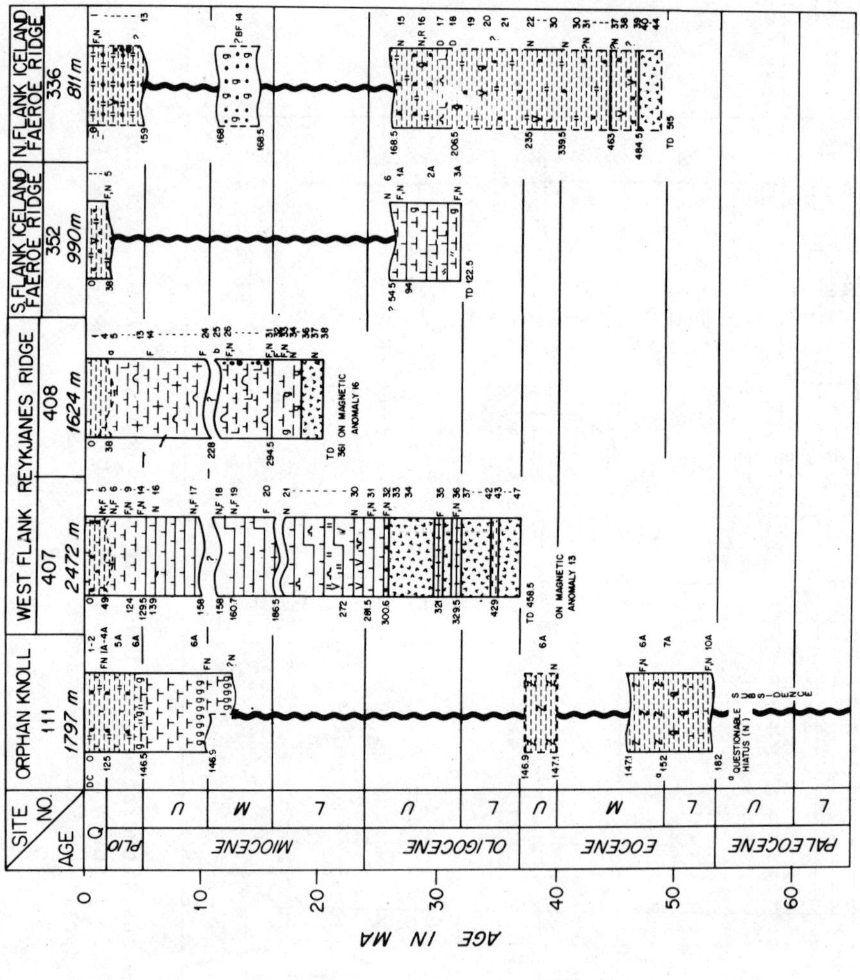


Figure 5. Age versus lithofacies for DSDP sites on Orphan Knoll and Reykjanes Ridge in western North-ern Atlantic and on Iceland-Faeroe Ridge. Explanation in Figure 3.

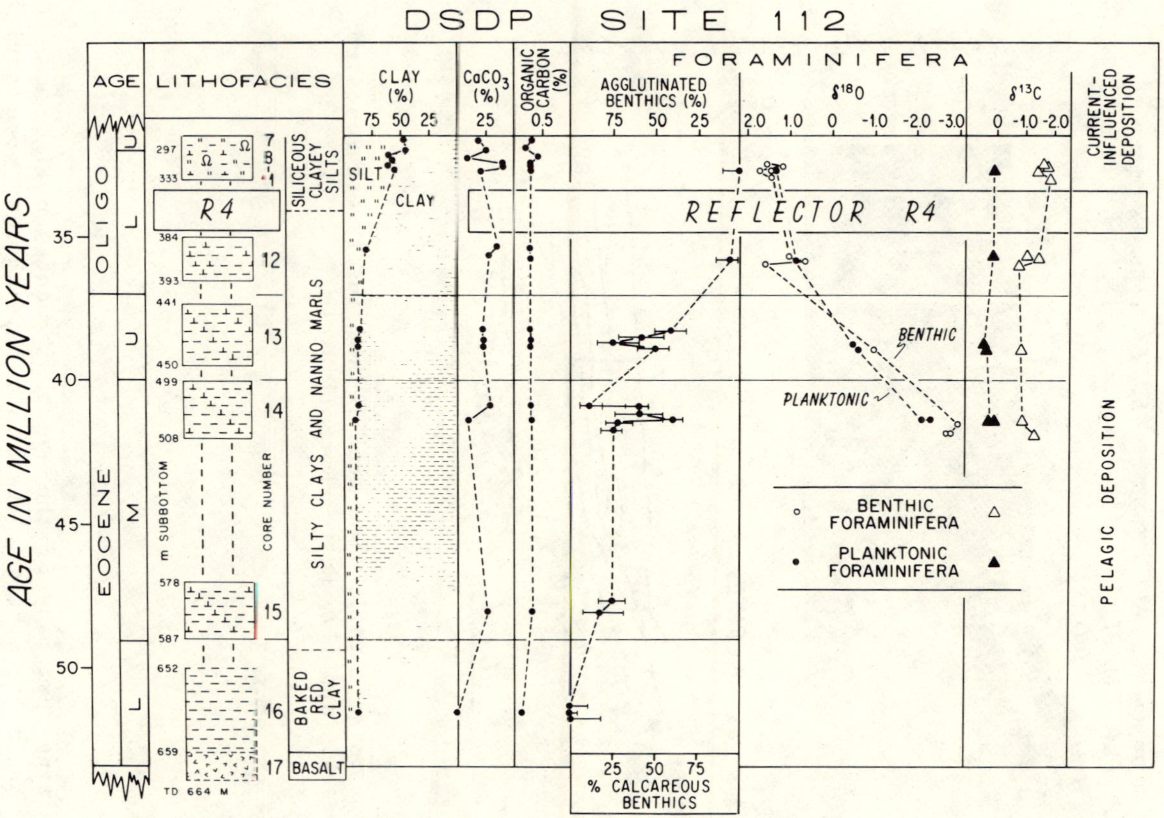


Figure 6. Summary of lithology and foraminiferal data at DSDP Site 112 in Gloria Drift, Labrador Sea. Isotopic values from Miller and Curry (in prep.). Explanation in text.

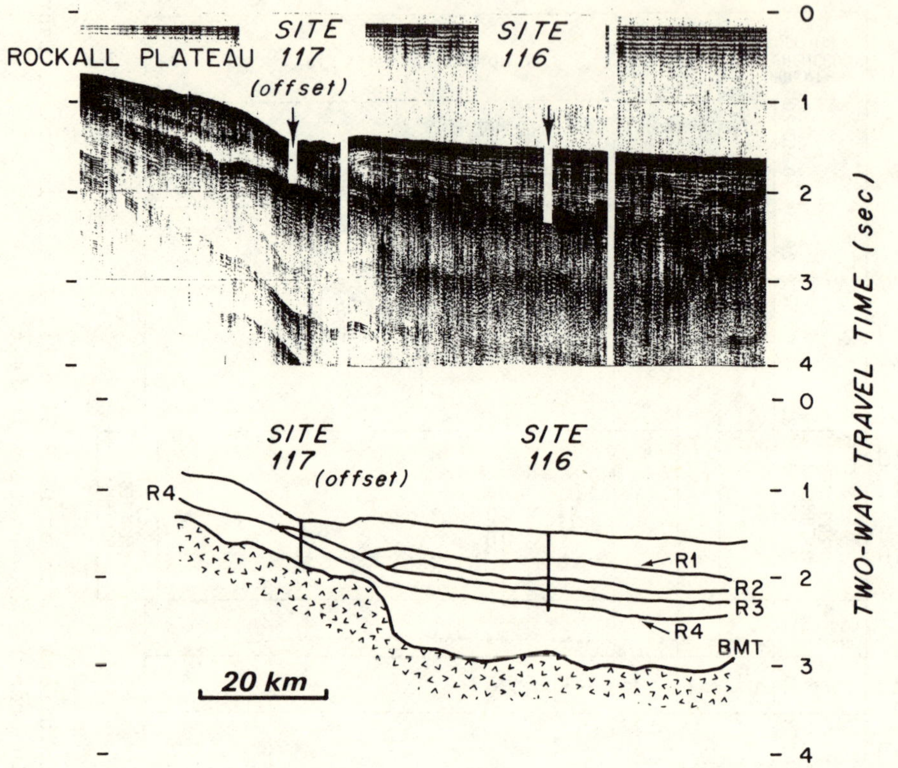


Figure 7. Seismic reflection profile and interpretation across DSDP Sites 116 and 117, Hatton-Rockall Basin.

Recognizing that many of the age assignments in Figures 3 to 6 still include uncertainties, there are nevertheless some striking correlations among the drillsites. The best defined is an unconformity at the Eocene/Oligocene boundary that occurs in every North Atlantic borehole except Sites 112 and 116. The lack of an unconformity in these two sites probably reflects their position away from basin margins. Site 112 is located in the central southern Labrador Sea on Gloria Drift, and Site 116, in contrast to adjacent Site 117, is located away from the margin of the Hatton-Rockall Basin (Figures 1, 7). All other sites are on or near areas where topographic boundary effects should have intensified the abyssal circulation. It is noteworthy that the hiatus appears to be mostly confined to the late Eocene (and earliest Oligocene?) at Sites 118, 119, and 400A; these sites are all in the Bay of Biscay, and they are the only drillsites at present water depths greater than 4000m. The missing time-slice common to all other drillsites (all less than 3000m water depth) is the latest Eocene to early late Oligocene. Thus, pronounced development of unconformities occurred throughout the late Eocene to early Oligocene North Atlantic, but erosion or non-deposition in the deep basin may have slightly preceded erosion at shallower depths.

A second unconformity occurs near the end of the middle Miocene in most drillsites, the notable exceptions being Site 116 in the Hatton-Rockall Basin, and the sites below 4000m water depth in the Bay of Biscay (Sites 118, 119, 400). The hiatus persists back into the middle to early Miocene at most sites, apparently reflecting the degree of sediment removal or the length of a period of non-deposition during this event. In contrast, the upper Miocene to Holocene sedimentary record is reasonably continuous at many sites. We interpret this distribution of hiatuses to indicate that a significant pulse of erosion by bottom currents occurred in the middle Miocene (see also Shor and Poore, 1979). However, less significant and/or local erosional episodes at other times are not precluded.

Lithofacies in the drillsites also show several broad, roughly chronostratigraphic changes (Figures 3-6). The first significant biosiliceous sedimentation occurs in the lower Eocene (e.g., Sites 117, 404, 405; Rockall Plateau region) and cherts also are developed in these largely non-calcareous sediments even though their present depth of burial generally is less than 300m. Biosiliceous detritus occurs patchily throughout the overlying Eocene to early Miocene sedimentary record in most of the northern North Atlantic. The only significant exception occurs in the Labrador Sea (Site 112) where abundant biogenic silica is confined to the Oligocene to Miocene section.

Excepting the siliceous/clayey sections noted above, calcareous sediments occur throughout most of the Tertiary and Quaternary record on the shallow Rockall Plateau. Elsewhere the occurrence of carbonates also tends to correlate with depth of the drillsite. Site 112, for example, has mostly low-carbonate (about 25%) sediments except in the Oligocene where a few values reach 40%. Site 118 in the Bay of Biscay is the deepest in the northern North Atlantic (4901 m) and has low carbonate throughout. Except for the Rockall drillsites, the Quaternary (and often the Pliocene) sedimentary record is largely non-calcareous.

Seismic Stratigraphy

The earliest clear effect of abyssal currents on the sedimentary record of the northern North Atlantic is in the form of sediment drifts (Figures 1, 2) overlying a subhorizontal sedimentary reflector, R4 (Roberts, 1975; = reflector R of Jones et al., 1970, and Ruddiman, 1972). Differential thickening of sediment that is typical of current-controlled deposition occurs above this horizon in the basins east of the mid-ocean ridge, whereas deposition conformable with basement and typical of pelagic accumulation occurs in most of the pre-R4 section (Figures 2, 7; Roberts, 1975; Ruddiman, 1972). West of the mid-ocean ridge and within Gloria Drift in the Labrador Sea, Egloff and Johnson (1975) suggest that "crenulation" of a strong reflector apparently correlative to R4 indicates current-control during or even preceding formation of this surface. However, it is not clear that the crenulation is not a seismic focussing effect created by the overlying intrasediment bedforms.

Reflector R4 is a continuous, interbasinal horizon which can be traced throughout the Rockall region (Roberts, 1975), the Iceland Basin (Ruddiman, 1972), and the Labrador Sea (Egloff and Johnson, 1975). The change in sedimentary regime associated with this reflector therefore must have a regional cause. Like Horizon $A^u$ in the western North Atlantic, reflector R4 may truncate deeper horizons (Roberts, 1975), but the exact relation of reflector R4 to the Horizon-A complex in the western North Atlantic has not been established. The lack of a significant horizon of similar stratigraphic position in the South Atlantic indicates that the paleoceanographic event associated with reflector R4 and Horizon $A^u$ had a North Atlantic genesis. Jones et al. (1970) determined that reflector R4 is pre-late Oligocene in age, and more recent reflector/borehole correlations indicate that it dates to the latest Eocene to early Oligocene (Figures 4, 6; Laughton and Berggren, 1972; Roberts, 1975; Montadert and Roberts, 1979; Miller et al., in press).

Ruddiman (1972) discussed a relatively continuous intermediate reflector (IR) above reflector R4 in the southwestern Iceland Basin. He dated the IR as early Miocene (17 Ma or younger, based on the age of pinchout on oceanic crust) and attributed it to an intensification of abyssal circulation, clearly differentiating this intensification from the initiation of current-controlled deposition at the level of reflector R4. Both Shor and Poore (1979) and Schnitker (1980b) interpreted Ruddiman to mean that current-influenced deposition began in the northern North Atlantic in the early to middle Miocene. However, differential thickening noted in the R4 to IR section (Ruddiman, 1972) clearly documents the initiation of current-controlled sedimentation at reflector R4.

This post-R4 current control also is suggested by the work of Roberts (1975) in the Rockall area. He discussed several intermediate horizons above reflector R4 which he termed 1 through 3 and which he noted were laterally and vertically impersistent. He also noted that internal unconformities present in the reflector 3 to R4 seismic interval suggest that the interval is unconformable with both reflectors 3 and R4. We have identified three horizons in the Iceland Basin/Rockall region above reflector R4 which we term reflectors R1 to R3, corresponding to reflectors 1-3 of Roberts (1975). Reflector R2 also corresponds to the intermediate reflector (IR) of Ruddiman (1972).

We have traced both R4 and the overlying reflectors and tied them to boreholes using seismic reflection data of Lamont-Doherty Geological Observatory and Woods Hole Oceanographic Institution (cruises V28-04, V27-06, C 09-13, and KN51), Deep Sea Drilling Project Legs 12 and 48, and previously published profiles and picks (Roberts, 1975; Montadert and Roberts, 1979). A convenient method for initial determination of subbottom depth of reflectors is to use velocity-regression equations derived from wide-angle sonobuoy measurements. These equations are of the form $V = Vo + KT$, where Vo is the initial velocity, K is the acceleration term, and T is half travel-time in seconds. This can be integrated to determine depth, $H = VoT + KT^2/2$. We used constants derived by Houtz (1980, and unpublished) for the Rockall area, and then modified the depth determination if correlation of the reflector to a nearby lithologic break could be improved, or if auxiliary velocity/density data from the sediments or borehole were available and justified a modification. In either case we restricted the resulting interval velocities to geologically reasonable values. In order to clarify some previously published correlations and to document our new correlations, distribution of reflectors at several drillsites is discussed below.

In the Labrador Sea near Site 112, reflector R4 lies at 0.41 seconds subbottom; Laughton and Berggren (1972) originally correlated the reflector with a hard layer encountered in core 10 (upper lower Oligocene; 315m subbottom). More recently Miller et al. (in press) correlated reflector R4 to the interval between cores 11 and 12 (333-384m) in the lower lower Oligocene section. This yields a more reasonable sediment interval velocity of 1.62 to 1.87 km/s. If a velocity as high as 1.95 km/s is allowed, reflector R4 lies very nearly at the Eocene/Oligocene boundary. The cause of reflector R4 in Site 112 is not clear, although it may represent the upward transition from indurated clays to siltier biosiliceous clays (Figure 6).

At Site 117 on the margin of Hatton-Rockall Basin, reflector R4 lies at 0.17 seconds and reflector R3 at 0.09 seconds subbottom. Reflectors R1 and R2 pinch out between Sites 117 and 116 (Figures

4,7). Using the velocity-regression equation with Vo=1.61 and K=2.28, reflector R4 lies at 147m, which correlates with the unconformity between the early Eocene and late Oligocene. Reflector R3 is close to 75m subbottom, correlating with an uncored interval between cores 1 and 2 (late Oligocene cherty limestones and oozes).

At Site 116, farther from the margin of the basin, four major sedimentary reflectors are present (Figs. 4, 7). Here, reflector R4 was first correlated with increased lithification (oozes to chalks) at 700m subbottom, corresponding to a late Oligocene to early Miocene age (Laughton and Berggren, 1972). Roberts (1975) recorrelated this horizon to the boundary between upper Eocene oozes/chalks and lower Oligocene cherts. At Site 116 reflector R4 lies at 0.79 to 0.81 seconds subbottom, reflector R3 at 0.70 seconds, reflector R2 at 0.60 seconds, and reflector R1 at 0.31 seconds. Although the constants in Houtz' velocity-regression equation yield good correlations of reflectors to lithostratigraphic breaks in the thin sedimentary section at Site 117, this is not the case for the thicker section at Site 116. The high acceleration term (2.28 km/s$^2$) for the Rockall region results in the placement of horizons below 0.3 to 0.4 seconds at anomalously great depths. We adjusted the acceleration term downward to 2.09 km/s$^2$ based on the sonic logs of the very similar section at Site 406 (south flank Rockall Plateau); this is within the standard error of estimate in the measurement of the acceleration term. As a result, depths computed using either acceleration term are very similar above 0.3 seconds (less than 8m difference) but diverge significantly at greater reflection times (71m difference at 0.8 seconds). The computed depth of reflector R4 then is 800-825m (late Eocene to early Oligocene) in agreement with Roberts' (1975) interpretation. Reflector R3 falls at 692m and correlates with the latest Oligocene upward change from cherty chalks to oozes, reflector R2 at 577m matches the top of early Miocene siliceous sediments, and reflector R1 at 284m falls within calcareous oozes in the lower upper Miocene section. This chronostratigraphic placement of horizons at Site 116 correlates well with their placement at Site 406 (Figure 4) where depths were estimated using synthetic seismograms computed from sonic logs (Montadert and Roberts, 1979). It also agrees with the late early Miocene age of reflector R2 (or IR) suggested by Ruddiman (1972).

At Site 406 on the southwest edge of Rockall Plateau, nine significant sedimentary reflectors were noted in multichannel seismic data obtained for DSDP Leg 48 drilling (Montadert and Roberts, 1979). Logging at Site 406 provides good control on the placement of seismic horizons, and our study of single-channel seismic data in this area agrees with these results. Our reflector R4 lies at 0.72 seconds subbottom, reflector R3 at 0.64 seconds, R2 at 0.60 seconds, and reflector R1 at 0.39 seconds. Using velocity-regression constants as for Site 116, the reflectors are at 715m,

622m, 577m, and 354m, respectively (Fig. 3). Reflector R4 is an intra-upper Eocene horizon; it occurs within or near an upward change from marly limestones to diatomites and chalks and is just below the unconformity at the Eocene/Oligocene boundary. This correlation is similar to that of Roberts et al. (1979) who matched R4 with the unconformity. Reflector R3 is an intra-upper Oligocene horizon, reflector R2 is an upper lower Miocene horizon at the upper boundary of silica-rich sediments, and reflector R1 falls within upper Miocene calcareous chalks.

In Site 406, our reflectors R4 to R1 as noted above correspond to reflectors "4a"/"4b", "3a"/3", "2", and "1" identified by Montadert and Roberts (1979) in their higher-resolution multichannel data. Their reflector "4", which corresponds to an unconformity separating upper Eocene from middle Eocene sediments, lies below our reflector R4. Montadert and Roberts (1979) computed synthetic seismograms using sonic logs and they predicted reflectors at 0.72/0.74, 0.64/0.66, 0.61, and 0.39 seconds which corresponded to subbottom depths of 690/710, 600/640, 570, and 350 m. These depths are in good agreement with the depths we obtained using the velocity-regression equation.

Correlation of reflectors is difficult at adjacent Site 405 because poor hole conditions prevented logging above 240m subbottom, and the post-R4 sedimentary record is too thin to rely on the velocity-regression equation. Reflectors R1 to R3 at Site 406 pinch out on R4 where traced to Site 405. A prominent horizon at Site 405 was correlated with the unconformity between upper Miocene and lower Eocene sediments (Montadert and Roberts, 1979) and is thought to be reflector R4. It occurs at less than 0.1 second subbottom and therefore corresponds approximately to the unconformity at 65m (Figure 4).

At Site 403, four reflectors were noted by Montadert and Roberts (1979). The uppermost of these ("1") occurred at 0.3 seconds subbottom and correlated with the top of an early Eocene tuff. We observe two significant reflectors above reflector "1". The first, occurring at 0.17 to 0.19 seconds (144-162m subbottom), falls in the upper Miocene section and it may correlate with reflector R1. The second, at 0.25 to 0.27 seconds (218 to 236 m subbottom), is due to an upward decrease in velocity (shown on the sonic log) associated with the unconformity separating upper Oligocene from middle Eocene strata. This horizon probably correlates with reflector R4. This interpretation agrees with that of Roberts et al. (1979) who correlated reflector R4 with the unconformity at Site 403.

Reflector R4 at Site 403 can be traced to a position 0.23 to 0.25 seconds subbottom (199-218m) at Site 404. This also correlates with the unconformity separating Miocene from Eocene strata as suggested by Montadert and Roberts (1979).

The foregoing correlations suggest that we can characterize each of the major reflectors by rather specific lithologic and age limits. Reflector R4 in the Rockall area shows a strong correlation to the widely distributed unconformity straddling the Eocene-Oligocene boundary (Figures 3-6; see also Roberts et al., 1979), except possibly at Site 406 where it may underlie the unconformity by up to 45m. Away from basin margins the unconformity is not present (e.g. Site 116; Roberts, 1975), and in those regions reflector R4 appears to correlate with the uppermost Eocene to lower Oligocene section. Similarly, the apparently correlative reflector in the Labrador Basin, which we also term R4, dates to the lower part of the lower Oligocene in an apparently continuously deposited hemipelagic clay sequence at Site 112. A reflector similar to R4 in stratigraphic position has been noted in the Bay of Biscay, where it is thought to correlate with the middle Eocene to early Oligocene unconformity (Montadert and Roberts, 1979); however, reflector R4 has not been traced seismically from the Rockall region into the Bay of Biscay.

Reflector R3 is observed only in the Rockall Plateau region where it dates to the latest Oligocene to earliest Miocene. Because the R3-R4 seismic interval commonly has high-amplitude, discontinuous reflectors it often obscures the underlying reflector R4; although reflector R3 has no consistent lithologic correlation, this seismic signature suggests it may cap a "lower" Oligocene sequence of (cherty) chalks variably altered by diagenesis (Roberts, 1975). In addition, the R3-R4 interval commonly wedges out toward basin margins, contorted internal reflectors often appear, and apparent sediment waves occur locally within the Gardar Drift. We interpret this as evidence for continued control of sedimentation by abyssal currents.

Reflector R2 (IR of Ruddiman, 1972, in the Gardar Drift) dates to the latest early Miocene in the reasonably continuous sedimentary sections in the Gardar Drift and at Sites 116 and 406 on Rockall Plateau. Elsewhere, excepting the Bay of Biscay and possibly Site 408 on the west flank of the Reykjanes Ridge, a hiatus is present in this time interval. It is thus possible that R2 is another widespread unconformity, and in many places it may have re-excavated R4. There is some support for this observation in the apparently unconformable attitude of Ruddiman's IR (R2) within the Gardar Drift. Reflector R2 also correlates approximately to the level below which siliceous debris is a significant sedimentary component. The R2-R3 interval lacks the discontinuous, high-amplitude reflectors found in the underlying seismic interval on Rockall, but it retains many of the convoluted reflectors and pinchouts that suggest continued current-controlled sedimentation and erosion.

Reflector R1 is important only in the Rockall Plateau region where it dates to the late Miocene and may represent a lithification boundary (e.g., chalk/ooze) in the calcareous sediments. It is noteworthy that the widespread upper middle Miocene unconformity documented by drilling falls between reflectors R2 and R1 and is not marked by a significant reflector.

The seismic interval between reflector R2 and the seafloor shows the most coherent pattern of current-controlled deposition in the northern North Atlantic, especially in the form of prominent development of sediment drifts and associated sediment waves. Along at least some basin margins (e.g., Sites 403-405) this depositional phase lagged until the late Miocene (about R1 time).

Summary of Evidence for Abyssal Circulation

We can summarize the evidence from DSDP borehole age/lithofacies data and from seismic reflection data in the following manner. Very near the Eocene/Oligocene boundary a widespread unconformity developed, apparently because of erosion by a newly developed system of vigorous abyssal boundary currents. Areas away from basin margins (e.g. Sites 112 and 116) were not eroded but did experience current-controlled deposition. The erosion may have occurred first (latest Eocene) in the deeper parts of the basin (i.e. 4000m, Bay of Biscay), and later (early Oligocene) it affected shallower areas that now are less than 3000m deep. A widespread early Oligocene hiatus also is developed on the continental shelves around the North Atlantic (e.g., northeast U.S. margin - Olsson et al., 1980; Canadian margin - Gradstein and Srivastava, 1980; Europe - Pomerol, 1973; Gulf of Mexico - Murray, 1961). However, there is no known relation between these shallow, sea level-induced hiatuses and the current-induced erosion in the deep basin. It also is not clear that the timing of the two events is exactly the same.

The late Eocene to early Oligocene timing for development of the abyssal circulation is supported indirectly by a major faunal change at the end of the Eocene in the deep Labrador Sea (DSDP Site 112) where predominantely agglutinated Eocene benthic foraminiferal assemblages were replaced by an Oligocene calcareous fauna (Figure 6; Miller et al., in press). It is possible that reflector R4 at Site 112 postdates the faunal exit by up to 2 Ma, but Miller et al. suggest that the changes in fauna and depositional conditions both are expressions of changes in hydrographic regime and development of vigorous abyssal circulation. Miller and Curry (in prep.; Figure 6) also note that a $4^{o}/oo$ shift in oxygen isotopic composition of benthic foraminifera coincides with the faunal change. Some of this isotopic signal may be due to diagenesis and perhaps to ice volume, but part of the shift probably represents a temperature drop which in turn can be attributed to the initiation of northern sources of bottom water. Similar supporting evidence is found in the Bay of Biscay (Site 400A) where a change of benthic assemblages also occurs between the middle Eocene and the early Oligocene (Schnitker, 1979; personal observation). An associated shift in oxygen isotopes also occurs in this region (Site 400A, Vergnaud-Grazzini et al., 1978, 1979; Site 119, Miller and Curry, in prep.); at Site 119 this shift represents a temperature drop of at least 2°C but less than 6°C.

Following the initial erosional phase associated with the developing abyssal circulation, strong abyssal currents continued to control sedimentation patterns throughout the North Atlantic. This is clearly demonstrated between reflectors R4 and R2 by the development of sediment waves and convoluted beds in the Gloria and Gardar Drifts, and possibly in the Rockall Basin, and larger-scale differential sedimentation patterns with pinchouts and possible erosional truncations throughout the basins east of the Mid-Atlantic Ridge (Ruddiman, 1972; Egloff and Johnson, 1975; Roberts, 1975). Except on the Gloria Drift in the Labrador Sea and locally on the Gardar Drift, there was little coherent development of large-scale bedforms, and unconformities may be widely but rather randomly developed. A coherent erosional pulse may have occurred at the level of reflector R2 in the late early Miocene, and another pulse almost certainly occurred in the late middle Miocene above R2; however, neither had much, if any, geologic effect in the Bay of Biscay. Coherently deposited sediment drifts and waves in the northeastern North Atlantic first appear above reflector R2. Because of the well-developed unconformities and associated effects of strong currents below reflector R2, we interpret this observation as a general weakening and stabilization of abyssal flow; the event also appears to mark the onset of near-modern conditions of current-controlled deposition (e.g., Shor and Poore, 1979).

## POTENTIAL BOTTOM-WATER SOURCES

There are three possible bottom-water sources in the northern Atlantic that may have contributed to the late Eocene to early Oligocene development of strong abyssal circulation. These are Greenland-Scotland Ridge overflow, Arctic connection via Baffin Bay/Davis Straits, and bottom-water formation in the northern Atlantic south of the Greenland-Scotland Ridge.

### Greenland-Scotland Ridge Overflow

Considerable debate, based on a variety of differing lines of evidence, exists as to when shallow and deep water connections were established across the Greenland-Scotland Ridge (Figure 8). Muller (1976) noted similarities in early Eocene nannoflora from the Norwegian-Greenland Sea and the North Atlantic. It is not clear whether this was a result of marine connection across the Greenland-Scotland Ridge or of migrations through epicontinental seas such as the North Sea, for Müller also noted a gradual floral differentiation extending from the North Atlantic into the Norwegian-Greenland Sea. This suggests that there was some paleogeographic restriction between the two basins because the early Eocene was a time of very reduced thermal gradients (Haq, in press). The presence of a continuous waterway from the North Atlantic to the Arctic by the middle Eocene also is supported by terrestrial faunal differentiation between Europe and North America/ Greenland (McKenna, 1972; 1975). McKenna (1975; this volume) attributed this differention to the breaking of the "Thulean" land bridge between Greenland and Scotland. The continued presence of a waterway across the Greenland-Scotland Ridge into the early to middle Oligocene is supported by similar nannoflora north and south of the ridge (Haq and Lohmann, 1976; Müller, 1976).

The timing of a sea-level connection across the Greenland-Scotland Ridge (as opposed to around the ridge) has only been estimated by subsidence models. Vogt (1972) used a subsidence model based on a 60 Ma age for initial spreading to suggest that the first surface-water connection between Greenland and Scotland occurred in the late Eocene (37-40 Ma); he also estimated that no "deep" (several hundred meters) connection existed prior to the middle Miocene. Subsequent work by Talwani and Eldholm (1977) has shown that spreading began in this region during anomaly 24 time (53 Ma, time scale of Hailwood et al., 1979; 56.5 Ma, time scale of Hardenbol and Berggren, 1978). Vogt's original estimate of subsidence of the Greenland-Scotland Ridge below sea level by 40 Ma (late Eocene) therefore may be too old by several million years; this revision would place the first surface water connection in the latest Eocene to earliest Oligocene. In contrast, Talwani and Udintsev (1976) estimated from drilling results at Site 336 and from other geologic inferences that subsidence below sea level of even the oldest crust did not occur until 25 Ma; they suggested this as the time that the Thulean land bridge was breached.

The above models both assume that the entire Greenland-Scotland Ridge is oceanic and therefore can be backtracked along a normal age-versus-depth curve for oceanic crust (Sclater et al., 1971; Berger and Winterer, 1974) starting with the initiation of seafloor spreading at anomaly 24 time. However, in the region of the Faeroe Islands and southward, there are complexities that suggest a distinctly different history. Most important is the observation that continental crust probably underlies the "Faeroe Islands Block" (Casten, 1973, Bott et al., 1974, 1976; Casten and Nielson, 1975) and almost certainly underlies Rockall Plateau (Montadert and Roberts, 1979). These continental blocks presumably were separated from the Shetland margin during Mesozoic rifting (e.g. Vogt et al., 1981) thereby forming the Rockall Trough (Roberts, 1975) and possibly the Faeroe-Shetland Channel. This supposition is supported by recent multichannel seismic observations of thick sedimentary sections below "acoustic basement" (Paleocene tuffs and basalts) in the vicinity of the Faeroe-Shetland Channel, Faeroe Bank Channel, and northern Rockall Trough (Figure 9; Ridd, this volume; Roberts, this volume). Subsidence of the Faeroe Channel system therefore could significantly predate the Paleocene spreading phase, and the Faeroe Channel system may have provided a seaway to the Norwegian-Greenland Sea in the Late Mesozoic to Early Cenozoic. The entire region was overprinted by massive Paleocene basalt flows, which for a period may have provided a land bridge for terrestrial faunal migration. By no later than 50 Ma, this land bridge was severed (McKenna, this volume). We speculate that 1) the Faeroe Channel system may have been part of an epicontinental sea in the pre-Anomaly 24 interval, 2) during the initiation of spreading in the Norwegian-Greenland Sea, Paleocene basalts formed the Thulean land bridge (probably the Wyville Thompson Ridge system), and 3) subsidence of this region below sea level broke this terrestial faunal connection and allowed migration of marine flora between the Atlantic and Norwegian-Greenland Sea by early Eocene time (Figure 8). This model presumes that the Paleocene igneous overprinting did not completely "reset" the rate of crustal subsidence in this region. If it had, the sill in Faeroe Bank Channel, now at approximately 800-900 m depth (Figures 10, 11), would initially have been nearly 1500m above sea level; it would not have subsided below sea level until about 30 Ma, which is 20 Ma after McKenna's postulated breach in the Thulean land bridge. Clearly, the age and tectonic history of the Faeroe Channel system and Wyville Thompson Ridge deserve detailed study.

There also is a great deal of uncertainty about the Paleogene configuration and depth of the ridge between Iceland and Greenland (Figures 10, 11). On the edge of the Greenland margin just northeast of the ridge, prograded sediments at least 1 to 2 seconds thick occur. The east Greenland margin is still not well surveyed by seismic data, but recent work by H.C. Larsen, B. Larson (both this volume), and Hinz and Schlüter (1978) indicates that thick prograding sediments have caused both narrowing and shoaling of the Denmark Straits. Accurate age assignments for these sediments and "backstripping" in conjunction with a subsidence model will be necessary to fully evaluate the possibility of a Paleogene marine passage in this area, but it presently seems possible that a significant passage did exist. The area also has an uncertain plate tectonic history. Talwani and Eldholm (1977) suggested that at anomaly 7 time (27Ma) a westward jump of the spreading center occurred on the Greenland-Scotland Ridge, from a central position on the Ridge to a position near the Greenland margin. This particular jump disagrees with magnetic-anomaly identifications on the Ridge by Vogt et al. (1980), but a jump just north of the Ridge still is required. The depth and morphology of any existing passage adjacent to Greenland therefore may have been substantially changed in the late Oligocene by this spreading-center jump and by the subsequent formation of the Iceland Plateau.

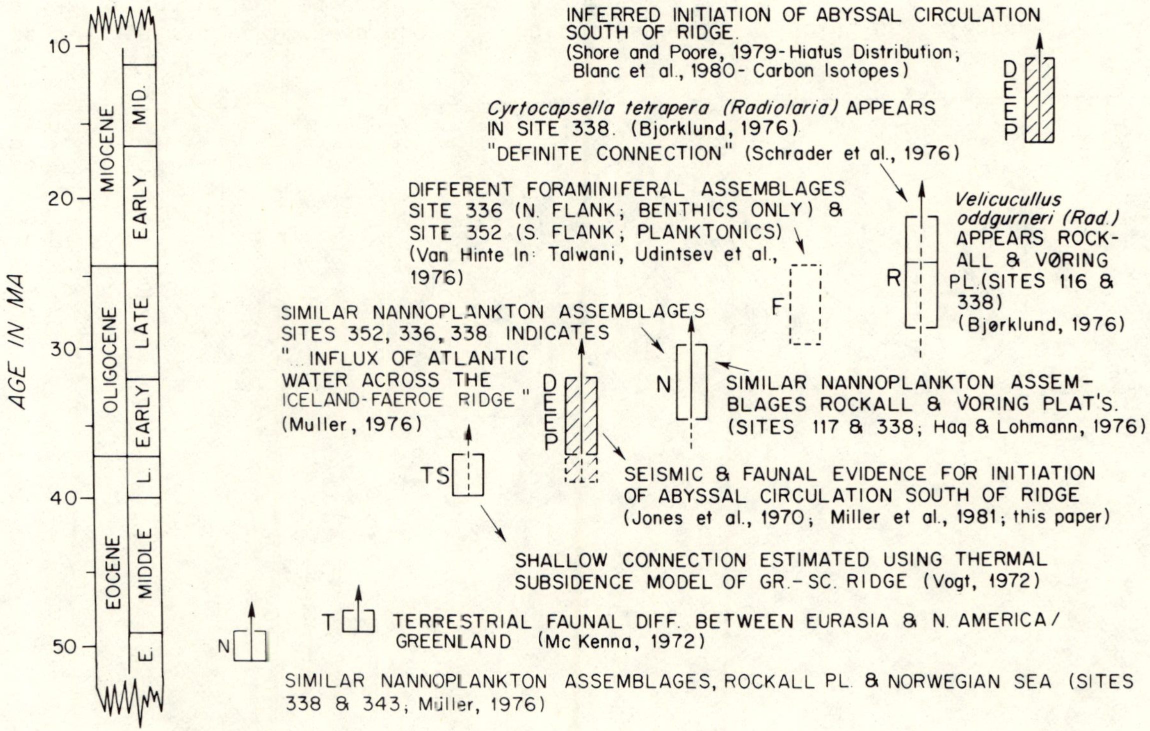


Figure 8. Summary of evidence for marine connection between North Atlantic and Norwegian-Greenland Sea, with reference to Greenland-Scotland Ridge.

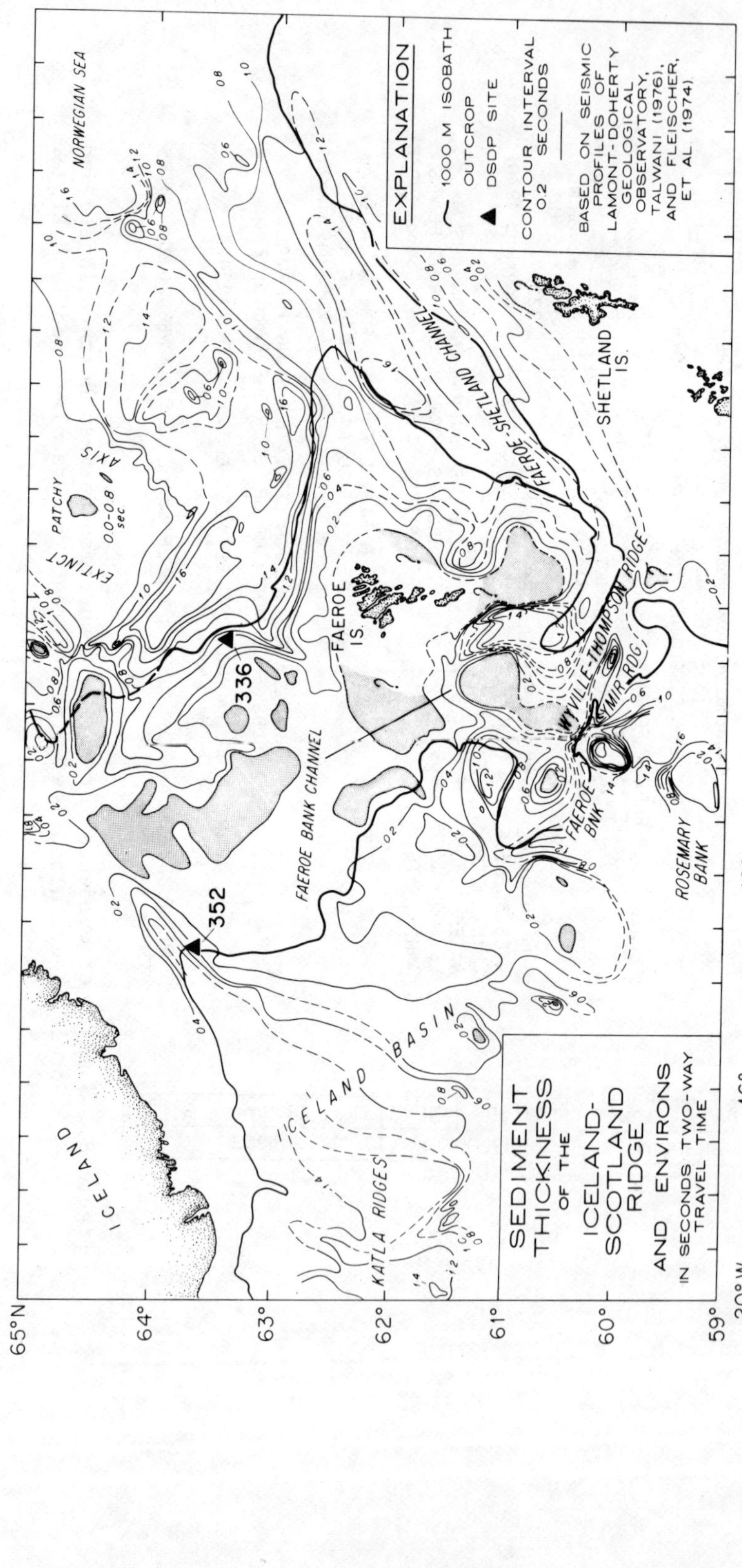


Figure 9. Sediment thickness above "acoustic basement" (including Paleocene basalts) on Iceland-Scotland Ridge, in seconds two-way travel time.

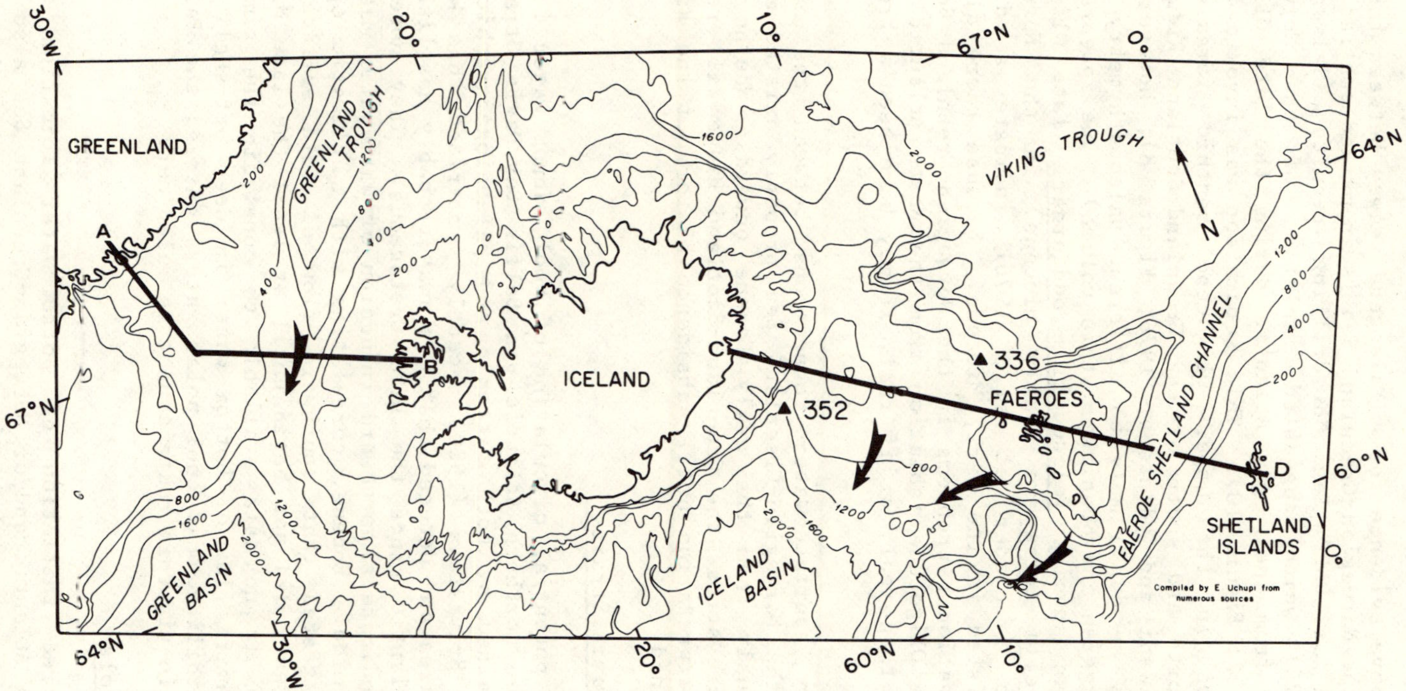


Figure 10. Bathymetry of Greenland-Scotland Ridge, in meters (from Uchupi and Hays, unpublished data). Arrows indicate modern overflow routes. Drillsites shown by triangles. Line A-D shows location of cross-section in Figure 11.

All the above evidence for a Paleogene seaway between the North Atlantic and the Norwegian-Greenland Sea is in marked contrast with interpretation of DSDP Leg 38 drilling results (Talwani and Udintsev, 1976) which suggested isolation of the Norwegian-Greenland Sea from the North Atlantic during the late Oligocene (Sites 336, 352; Figure 10). This separation was invoked because of differences in the upper Oligocene benthic foraminiferal assemblages north and south of the Greenland-Scotland Ridge (van Hinte: in Talwani and Udintsev, 1976; Figure 8); however, this difference may be due simply to different paleobathymetry at the two sites (Berggren and Schnitker, this volume). The appearance of the radiolarian species _Velicucullus_ _oddgurneri_ (late Oligocene) and _Cyrtocapsella_ _tetrapera_ (early Miocene) in the Norwegian-Greenland Sea (Schrader et al., 1976) indicate a definite connection existed in the early Miocene. Because there also are faunal preservational problems in the Norwegian-Greenland Sea, the suggested late Oligocene isolation from the North Atlantic may be more apparent than real, or if real, it may have been a transient event.

In summary, surface-water connections between the northern Atlantic and the Norwegian-Greenland Sea probably existed more or less continuously from the early Eocene onward. Whether these connections were across the Greenland-Scotland Ridge will remain a matter of interpretation until subsidence history of the Ridge is better determined.

## Baffin Bay - Davis Straits

Timing of potential circulation of cool, dense water from the Arctic into the northern Atlantic via Baffin Bay/Davis Straits is poorly constrained by both tectonic and paleobiogeographic data. Gradstein and Srivastava (1980) recently summarized this kind of data for the area and deduced that a seaway probably has linked the Arctic and Atlantic since the Late Cretaceous. They noted that faunal data indicate a northward incursion of warm-water masses in the early to middle Eocene, correlating with a climatic optimum; they also suggested that southward flow of cold surface water (Labrador Current) did not begin until at least the late Miocene. However, there is no data available to constrain the timing of possible southward flow of bottom water in deep off-shelf areas. Thus the possibility that cool polar waters entered the Paleogene abyssal Atlantic via this route cannot be ruled out.

## Climatic Controls

The inferred initiation of bottom-water formation in the northern North Atlantic and/or Norwegian-Greenland Sea also may be related to or accentuated by the climatic history of this region. The Tertiary global climatic record shows that a major cooling

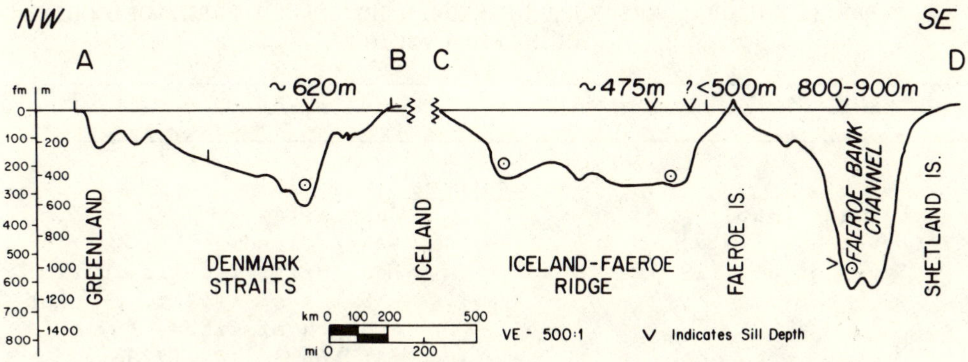


CROSS SECTION: GREENLAND-SCOTLAND RIDGE

Figure 11. Bathymetric cross-section of the Greenland-Scotland Ridge (located in Figure 10) showing maximum present sill depths.

occurred in the middle Eocene to early Oligocene and resulted in an increase in the latitudinal thermal gradient. This cooling has been interpreted from the isotopic, planktonic faunal and floral, and terrestrial floral records (Table 1). A dramatic shift in oxygen isotopic composition of planktonic foraminifera near the Eocene/Oligocene boundary in the southern hemisphere is thought to reflect a major temperature drop at this time (Kennett and Shackleton, 1976). A similar shift in oxygen isotopes in benthic foraminifera has been interpreted as reflecting the initial formation of cold bottom water of southern origin (Kennett and Shackleton, 1976). However, the benthic isotopic shift in the North Atlantic may reflect formation of northern bottom-water sources (this study; Figure 6). It is also possible that both shifts may be attributable in part to ice volume and not to temperature (Matthews and Poore, 1980). In either case, the distribution of planktonic organisms clearly shows a significant cooling of North Atlantic surface waters in the middle Eocene to early Oligocene (Gradstein and Srivastava, 1980; Berggren, 1978; Haq et al., 1977, 1979).

In the modern ocean Norwegian Sea Overflow Water and Labrador Sea Water (which form the bottom and intermediate waters, respectively, in the North Atlantic) form by winter convection of high-salinity subtropical water that has been advected northward and cooled (Worthington, 1976; McCartney and Talley, in press). The source of this low latitude water is thought to be either the Gulf Stream/North Atlantic Current (Worthington, 1976) or Mediterranean Outflow Water (Reid, 1979). A similar situation may have developed in the late Eocene. At that time, the configuration of continents

TABLE 1. GLOBAL EVIDENCE OF MIDDLE EOCENE TO EARLY OLIGOCENE CLIMATIC COOLING

| Author | Location |
|---|---|
| **Isotopes** | |
| Dorman (1966) | Australia. |
| Devereaux (1967) | New Zealand. |
| Savin et al. (1975) | Equatorial Pacific, DSDP Sites 44 (1428m, present depth of site), 167 (3166m), 171 (2283m). |
| Shackleton & Kennett (1975)<br>Kennett & Shackleton (1976) | Southern Ocean, DSDP Site 277 (1214m). |
| Boersma & Shackleton (1977) | Mid-latitude South Atlantic, DSDP Site 357 (2086m). |
| Buchardt (1978) | North Sea. |
| Vergnaud-Grazzini et al. (1978, 1979) | Eastern North Atlantic, DSDP Sites 398 (3910m), 400A (4399m), 401 (2495m). |
| Miller & Curry (in prep.) | Labrador Sea, DSDP Site 112 eastern North Atlantic, DSDP Site 119 (4447m). |
| **Distribution of Calcareous Nannoplankton** | |
| Edwards & Perch-Nielsen (1975) | Southern Ocean. |
| Haq et al. (1977; 1979) | Atlantic. |
| **Distribution of Planktonic Foraminifera** | |
| Jenkins (1968, 1975)<br>Hornibrook (1971) | New Zealand. |
| Kennett (1977, 1978) | Antarctic Ocean. |
| Berggren (1978) | North Atlantic. |
| Gradstein & Srivastava (1980) | Labrador Sea. |
| **Paleobotanical Evidence** | |
| Northern Hemisphere: | |
| Dorf (1964)<br>Wolfe & Hopkins (1967)<br>Addicott, (1969 1970)<br>Wolfe (1978) | Western United States. |
| Tanai & Huzioka (1967) | Japan. |
| Southern Hemisphere | |
| Kemp (1978)<br>Kemp & Barrett (1975) | Antarctic Ocean, Antarctica, and Australia. |

was similar to that of today (Figure 12A), although the basins were slightly smaller and the continental latitudes were 5-10° lower (Sclater et al., 1977). Northward advection of warm, presumably high-salinity water is thought to have occurred in an anticyclonic subtropical gyre (proto-North Atlantic Current). Another possible low-latitude source of warm, high-salinity water was the proto-Mediterranean (Tethys) Sea which lay at latitudes 15° to 30°N in the Eocene/Oligocene (Berggren and Hollister, 1974); such latitudes are the present sites of formation of high-salinity surface and intermediate water in the North Atlantic (Worthington, 1976). Cooling of these high-salinity waters in the northern North Atlantic and/or Norwegian-Greenland Sea presumably occurred during the Eocene/Oligocene phase of climatic cooling and increased latitudinal thermal gradients; based upon the modern analog, the likely sites of most intense cooling (with possible convection and bottom-water formation) would have been in regions where polar air was delivered from the large land masses, i.e. east of North America and Greenland. Thus, bottom-water formation could have occurred not only in the Norwegian-Greenland Sea, but also south of the Greenland-Scotland Ridge and in the Labrador Sea. Unfortunately it is not presently possible to determine whether such northerly advection and cooling generated significant amounts of deep and bottom water in the northern Atlantic. It seems unlikely that the shift in climatic pattern alone was responsible for the dramatic development of abyssal circulation at the Eocene/Oligocene boundary. However, it may have accentuated bottom-water formation, especially in the Labrador-Baffin and Norwegian-Greenland Seas.

## MODEL OF CIRCULATION

Figure 12A summarizes our interpretation of the abyssal circulation at the beginning of the Oligocene. The widespread distribution of reflector R4 and its characteristic correlation with an unconformity indicates that strong abyssal circulation affected the North Atlantic basins both east and west of the mid-ocean ridge. For abyssal circulation to have affected so strongly the sedimentary record in the Rockall region and in the Bay of Biscay, we consider it necessary that the eastern North Atlantic had a major source of bottom water, probably from the Norwegian-Greenland Sea via the Faeroe-Shetland and Faeroe Bank Channels. This timing is coincident with the opening of a deep-water passage between Greenland and Spitsbergen that connected the Arctic Ocean and the Norwegian Sea (Talwani and Eldholm, 1977). The implication therefore is that the bottom water had an Arctic source, possibly supplemented in the Norwegian Sea and south of the Greenland-Scotland Ridge by convective overturn of cooled, saline surface waters. The timing coincidentally agrees with the marine connection through the Faeroe Bank Channel suggested by Vogt (1972). However, as discussed earlier, the passage could have been deeper than Vogt supposed. Thus, once the Greenland-Spitsbergen connection was severed, Arctic water presumably flowed unimpeded into the North Atlantic to form the first strong abyssal circulation in the basin (see Berggren and Hollister, 1974). This "tectonic-threshold" control on introduction of bottom water to the North Atlantic is the most satisfactory explanation of the suddenness and intensity of erosion in the Atlantic basins.

The margin-intensified, anticlockwise circulation in the northeastern Atlantic was contained by the topographic barriers of the continental margin and the mid-ocean ridge (Figure 12A). Bottom water that flowed south along the flank of the mid-ocean ridge probably was turned north along the Azores-Gibraltar Ridge and Azores-Biscay Rise into the Bay of Biscay. Along the first of these, minimum depths presently are about 4300m and maximum depths in narrow passages are 4900m. Backtracking along a North Atlantic empirical age-depth curve (Tucholke and Vogt, 1979) indicates these depths were on the order of 3500m and 4000m, respectively, in the early Oligocene. The same analysis applies to the Azores-Biscay Rise which probably was an even shallower barrier at 3000-3500 m. Thus these ridges must have precluded significant bottom-water exchange with more southerly parts of the eastern North Atlantic at depths below 3000-3500m. By the same token, they blocked any significant input of bottom water to the Bay of Biscay from possible southern sources. Although Schnitker (1980a) preferred such Antarctic sources for bottom water in the northern North Atlantic, it is unlikely that they could have been significant even in the absence of these barriers, given the minor influence of Antarctic Bottom Water in this region under modern, more glaciated conditions.

The circulation probably was most intense along the basin margins, as observed by unconformity development in DSDP boreholes. It is also possible that the first development of bottom currents affected only the deepest part of the basins ( 3500m) as indicated by the possible diachrony in unconformity development below 4000 m and above 3000m as noted earlier. Continued flooding of the basin by dense bottom waters could then have affected progressively shallower seafloor.

Bottom water from the eastern basin probably entered the western basins through major fracture zones in the mid-ocean ridge (e.g. Charlie Gibbs F.Z.) as well as through numerous smaller fracture-zone conduits in the Reykjanes Ridge (Ruddiman, 1972; Figure 12A). Magnetic anomaly patterns suggest that these fracture-zone conduits did not develop until the late Eocene to early Oligocene (Figure 2; Ruddiman, 1972). The abyssal circulation west of the mid-ocean ridge may also have been augmented by overflow across the Greenland-Scotland Ridge adjacent to Greenland and possibly by bottom water derived from the area of Baffin Bay/Nares

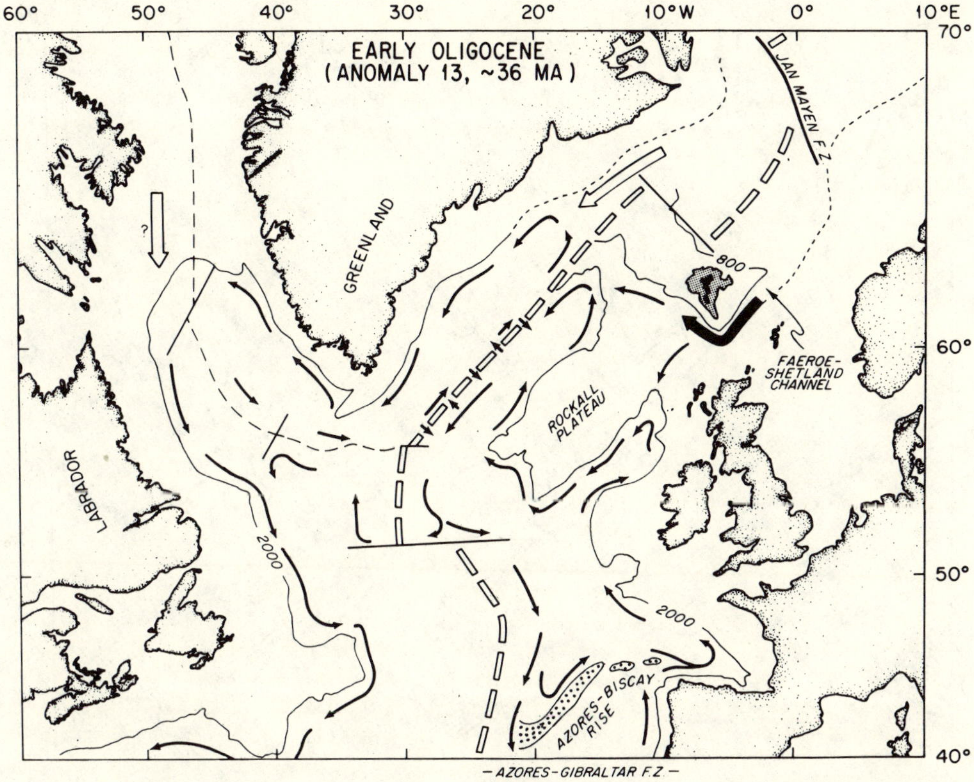


Figure 12A. Model of abyssal circulation in the North Atlantic in the early Oligocene (Anomaly 13 time). Solid and open bold arrows - probable and possible bottom-water sources; solid arrows - inferred abyssal circulation (length and spacing approximately proportional to intensity of circulation). Active spreading centers indicated by open rectangles (i.e with fracture-zone conduits) or continuous parallel lines (no conduits). Major fracture zones are solid lines. Extinct spreading axes are long dashed lines, and approximate ocean-continent boundaries are short dashed lines. Present 2000m bathymetric contour shown for reference. Tectonic data and plate geometries from Ruddiman (1972), Talwani and Eldholm (1977), Srivastava (1978), and Vogt et al. (1980).

Strait (Figure 12A). We suggest that the combined flow from these sources formed the intense abyssal boundary current that eroded the Horizon $A^u$ unconformity in the western North Atlantic, probably in the early to middle Oligocene.

By the late Oligocene (reflector R3), subsidence of the Faeroe Bank Channel permitted increased flow from the Norwegian-Greenland

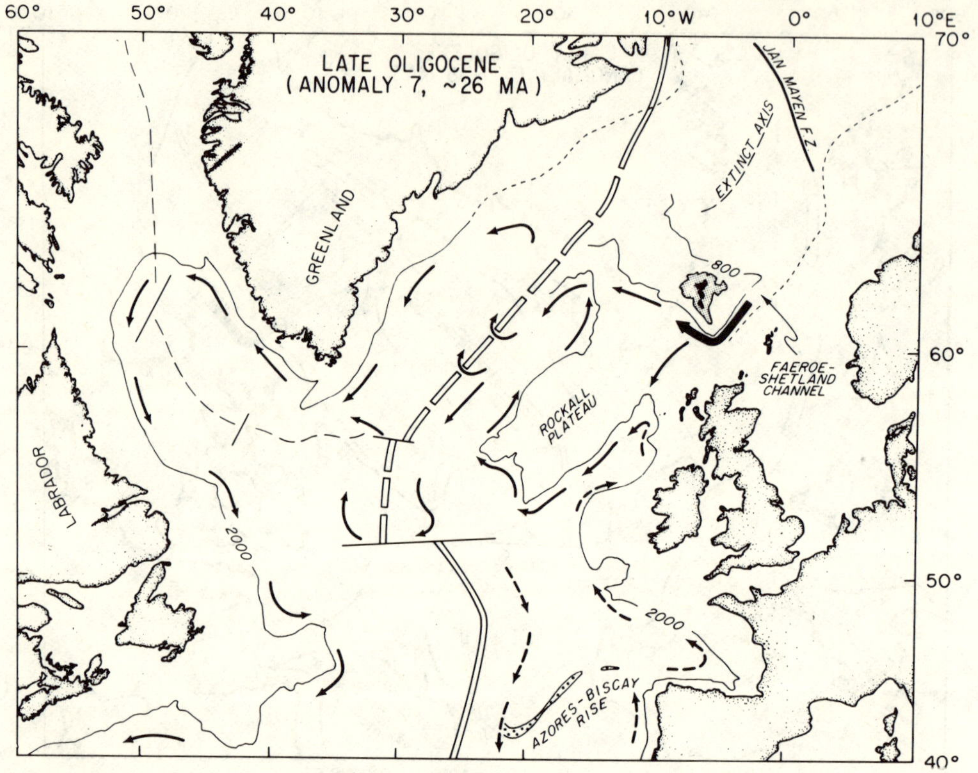


Figure 12B. Model of circulation in the late Oligocene (Anomaly 7 time). Explanation in Figure 12A.

Sea (Figure 12B) into the Rockall region, probably accounting for continued development of unconformities and chaotic current-controlled deposition. However, abyssal-current influence on sedimentation in the Bay of Biscay apparently declined as the Azores-Biscay Rise and Azores-Gibraltar Ridge subsided and failed to divert currents into the Bay of Biscay.

Reykjanes Ridge fracture zones continued to provide conduits for bottom water to enter the western basins, but north of the Greenland-Scotland Ridge a spreading-ridge jump to the Greenland margin may have blocked any significant flow from the Norwegian-Greenland Sea directly into the northwestern basins (Figure 12B). The implied decrease in abyssal boundary flow in the western North Atlantic correlates with current-controlled deposition above the Horizon $A^u$ unconformity in the late Oligocene to early Miocene (Figure 2).

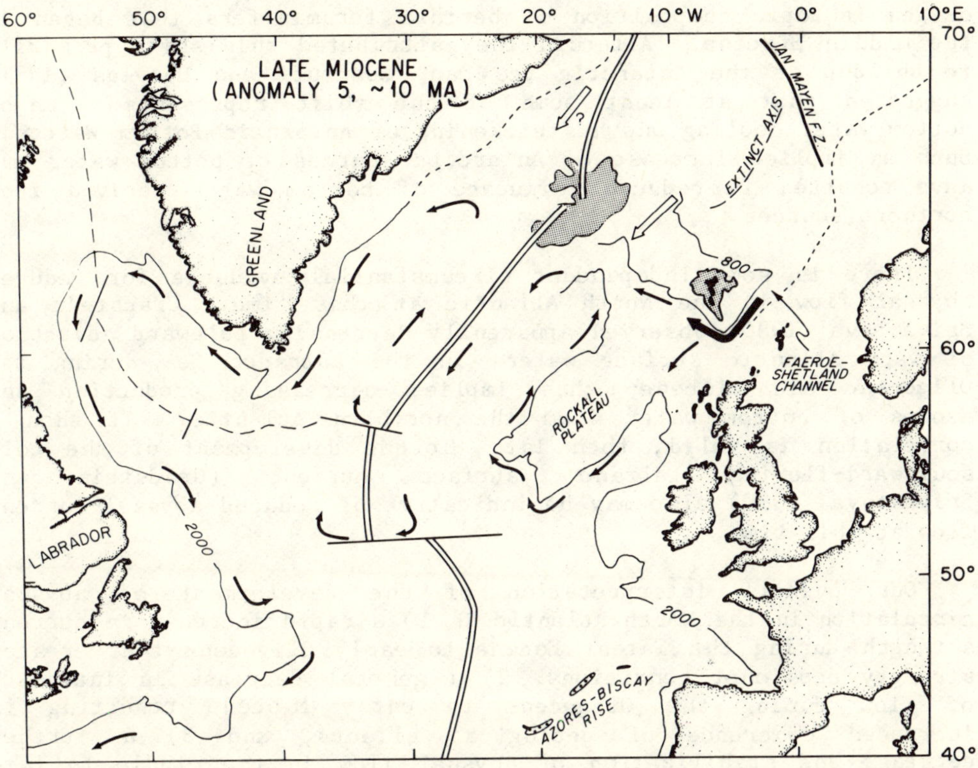


Figure 12C. Model of circulation in the late Miocene (Anomaly 5 time). Explanation in Figure 12A.

During the early to middle Miocene the fracture zone conduits across the Reykjanes Ridge ceased to be developed, and, in agreement with Ruddiman (1972) and Shor and Poore (1979), we correlate this trend toward east-west basin isolation with the coherent development of sediment drifts and waves in the northern and western North Atlantic (Figures 2, 12C). However, we interpret this coherent development to be a result of a general <u>decrease</u> and <u>stabilization</u> of the abyssal circulation. This interpretation is based on the fact that the preceding sedimentary record, characterized by chaotic sedimentation and prominent unconformities, must have required much higher current velocities.

It is not clear how the deterioration of the Reykjanes Ridge fracture zones may have attenuated the abyssal circulation, aside from a general restriction of possible flow conduits, nor is it clear that this was the sole factor responsible for reduced deep circulation. For example, Shackleton and Kennett (1975), Savin et al. (1975), and Woodruff et al. (1981) noted a dramatic shift in oxygen isotopic composition of benthic foraminifera that began in the middle Miocene. Although they attributed this shift primarily to buildup of the Antarctic ice cap, Woodruff and Douglas (1981) suggested that at least some of the shift represents a major bottom-water cooling and a "thickening of Antarctic Bottom Waters." Such an implied increase of Antarctic sources of bottom water may have resulted in reduced influence of bottom water derived from northern sources.

There is some independent circumstantial evidence for reduced abyssal flow in the North Atlantic at this time. Gradstein and Srivastava (1980) observed apparently decreasing poleward advection of warm Atlantic surface waters in the Labrador Sea during the Oligocene and Miocene; this implies decreasing production and exodus of bottom water from the northern Atlantic. If such a correlation is valid, then late Miocene development of the cold southward-flowing Labrador surface current (Gradstein and Srivastava, 1980) also may be indicative of reduced abyssal-current flow at this time.

Our overall interpretation of the development of abyssal circulation in the North Atlantic is 1) a rapid increase in current strength during the latest Eocene to early Oligocene that created strongly erosional conditions, 2) a general decrease in intensity of flow during the Oligocene to early Miocene resulting in increased coherence of geological effects, and 3) a further decrease and stabilization of abyssal flow in the middle to late Miocene with coherent development of sediment drifts that has continued more or less unaltered to the present.

Clearly, any long-term trend in abyssal circulation can be punctuated by tectonic and climatic events, several of which we have discussed. Two additional events are the possible unconformity at reflector R2 and the widespread upper middle Miocene unconformity between reflectors R2 and R1. Either or both of these could be a response to temporarily increased flow caused by such factors as subsidence of the Greenland-Scotland Ridge (e.g. between Iceland and Faeroe Islands, Figure 12C) or to northern hemisphere effects of the middle Miocene climatic cooling. In addition, the Plio-Pleistocene glacial cycles must have had "fine-scale" effects in the uppermost portion of the sedimentary record. At present, however, speculation on anything more detailed than the gross kinds of changes noted above probably is not justified.

## CONCLUSIONS

In the foregoing discussion we presented a model for the development of the abyssal circulation in the northern North Atlantic that is far from being completely developed or correct in detail. However, it does explain the lithostratigraphic, seismic stratigraphic, and floral/faunal data that must be considered in any discussion of the general development of the deep circulation. It also emphasizes the important control that abyssal currents have had on sedimentation patterns since the Eocene.

A better understanding of the abyssal circulation history obviously can be developed by the acquisition of additional, carefully placed geological (borehole) and geophysical data. However, within the province of existing information, we feel that there are three kinds of analyses that will significantly improve our perception of the abyssal circulation history:

1) Basin-wide mapping of the seismic stratigraphic framework in quasi-chronologic intervals, to provide clearer definition of the intra-basin geometries of current effects in relation to tectonic barriers and tectonic threshold events.

2) Improved subsidence models of the Greenland-Scotland Ridge, taking into account such factors as the age of rifting of apparently continental crust beneath the Faeroe Bank Channel, sill depths with appropriate sediment overburden removed, and spreading history and ridge jumps. This will allow more specification of the timing of shallow and deep marine connections across the ridge.

3) Continued benthic faunal, planktonic floral and faunal, and isotopic studies of existing DSDP cores in order to clarify the Tertiary patterns of paleocirculation, the potential northern-hemisphere climatic effects on bottom-water formation, independent of the southern hemisphere climatic and ice-volume record, and the history of surface and deep water connections of the Norwegian-Greenland Sea with the North Atlantic.

We look forward to these and comparable studies as worthwhile tests of our interpretations of the North Atlantic abyssal-circulation history.

## ACKNOWLEDGEMENTS

This research was supported by the United States Navy, Office of Naval Research, under Contract N00014-79-C-0071 and by a graduate fellowship from Phillips Petroleum to K.G. Miller. We are indebted to the conveners of the NATO Advance Research Institute on "Structure and Development of the Greenland-Scotland Ridge" for travel support to attend the meeting and present this paper. We thank W.A. Berggren and D. Schnitker for stimulating discussions, M.L. Bremer, W.B. Curry, M.S. McCartney, and P.E. Belanger for reviewing an early draft of the manuscript, and J.I. Ewing and E. Uchupi for reviewing the final manuscript. Contribution No. 4905 of Woods Hole Oceanographic Institution.

## REFERENCES


Addicott, W.O., 1969, Tertiary climatic change in the marginal northeast Pacific Ocean, Science, 165:583-586.

Addicott, W.O., 1970, Tertiary paleoclimatic trends in the San Joaquin Basin, California, U.S. Geol. Survey Prof. Paper, 644-D:19 p.

Berger, W.H., and Winterer, E.L., 1974, Plate Stratigraphy and the fluctuating carbonate line, in: "Pelagic sediments on land and under the sea," Hsu, K.J., and Jenkyns, H.C., eds., Internat. Assoc. Sedimentologists Spec. Publ. 1:11-48.

Berggren, W.A., 1972, Cenozoic biostratigraphy and paleobiogeography of the North Atlantic, in: "Initial Reports of the Deep Sea Drilling Project," vol. 12, Laughton, A. S., and Berggren, W. A., eds., U. S. Government Printing Office, Washington, D.C.:965-1001.

Berggren, W.A., 1978, Recent advances in Cenozoic planktonic foraminiferal biostratigraphy, biochronology and biogeography: Atlantic Ocean, Micropal., 4:337-370.

Berggren, W.A., and Aubert J., 1976, Late Paleogene (Late Eocene and Oligocene) benthonic foraminiferal biostratigraphy and paleobathymetry of Rockall Bank and Hatton-Rockall Basin, Micropal., 22:307-326.

Bergrren, W.A., and Hollister, C.D., 1974, Paleogeography, paleobiogeography and the history of circulation of the Atlantic Ocean, in: "Studies in Paleo-oceanography," Hay, W.W., ed., Soc. Econ. Pal. Min., Spec. Publ. 20:126-186.

Blanc, P.-L., Rabussier, D., Vergnaud-Grazzini, C., and Duplessy, J.C., 1980, North Atlantic deep water formed by the later middle Miocene, Nature, 283:553-555.

Boersma, A., and Shackleton, N.J., 1977, Tertiary oxygen and carbon isotope stratigraphy, Site 357 (mid-latitude South Atlantic). in: "Initial Reports of the Deep Sea Drilling Project," vol. 39, Supko, P., and Perch-Nielsen, K. eds., U.S. Government Printing Office, Washington, D.C.:911-924.

Bott, M.H.P., Sunderland, J., Smith, P.J., Casten, U., and Saxov, S., 1974, Evidence for continental crust beneath the Faeroe Islands, Nature, 248:202-204.

Buchart, B., 1978, Oxygen isotopic paleotemperatures from the Tertiary period in the North Sea area, Nature, 275:121-123.

Bukry, D., 1972, Further comments on coccolith stratigraphy, Leg XII, Deep Sea Drilling Project, in: "Initial Reports of the Deep Sea Drilling Project," vol. 12, Laughton, A.S., and Berggren, W.A., eds., U.S. Government Printing Office, Washington, D.C.:1071-1084.

Casten, U., 1973, The crust beneath the Faeroe Islands, Nature, Phys. Sci., 241:83-84.

Casten, U. and Nielson, P.H., 1975, Faeroe Islands - a microcontinental fragment?, J. Geophys., 41:357-366.

Devereaux, I., 1967, Oxygen isotope paleotemperature measurements on New Zealand Tertiary fossils, New Zealand J. Sci., 10:988-1011.

Dorf, E., 1964, The use of fossil plants in a paleoclimatic interpretation. in: "Problems in Paleoclimatology," Nairn, A., ed., Interscience, London:16-30 and 46-48.

Dorman, F.H., 1966, Australian Tertiary paleotemperatures, J. Geol., 74:49-61.

Edwards, A.R., and Perch-Nielsen, K., 1975, Calcareous nannofossils from the southern southwest Pacific, Deep Sea Drilling Project, Leg 29, in: "Initial Reports of the Deep Sea Drilling Project," vol. 29, Kennett, J.P., and Houtz, R.E., eds., U.S. Government Printing Office, Washington, D.C.:469-539.

Egloff, J., and Johnson, G.L., 1975, Morphology and structure of the southern Labrador Sea, Can. J. Earth Sci., 12:2111-2133.

Ewing, J.I., and Hollister, C.D., 1972, Regional aspects of deep sea drilling in the western North Atlantic, in: "Initial Reports of the Deep Sea Drilling Project," vol. 11, Hollister, C.D., and Ewing, J.I., eds., U.S. Government Printing Office, Washington, D.C.:951-973.

Gradstein, F.M., and Srivastava, S.P., 1980, Aspects of Cenozoic stratigraphy and paleoceanography of the Labrador Sea and Baffin Bay, Paleogeogr., Paleoclimat., Paleoecol., 30:261-295.

Hailwood, E.A., Bock, W., Costa, L. Dupeuble, P.A., Müller, C., and Schnitker, D., 1979, Chronology and biostratigraphy of northeast Atlantic sediments, DSDP Leg 48, in: "Initial Reports of the Deep Sea Drilling Project," vol. 48, Montadert, L., and Roberts, D.G., eds., U.S. Government Printing Office, Washington, D.C.:1119-1141.

Haq, B.U., in press, Paleogene paleoceanography: Early Cenozoic oceans revisited, Oceanol. Acta.

Haq, B.U., and Lohmann, G.P., 1976, Remarks on the Oligocene calcareous nannoplankton biogeography of the Norwegian-Greenland Sea (DSDP Leg 38), in: "Initial Reports of the Deep Sea Drilling Project," vol. 38, (suppl.), Talwani, M., and Udinstev, G., eds., U.S. Government Printing Office, Washington, D.C.: 141-145.

Haq, B.U., Premoli-Silva, I., and Lohmann, G.P., 1977, Calcareous plankton paleobiogeographic evidence for major climatic fluctuations in the Early Cenozoic Atlantic Ocean. J. Geophys. Res., 82:3861-3876.

Haq, B.U., Okada, H., and Lohmann, G.P., 1979, Paleobiogeography of the Paleocene/Oligocene calcareous nannoplankton from the North Atlantic Ocean, in: "Initial Reports of the Deep Sea Drilling Project," vol. 43, Tucholke, B.E., and Vogt, P.R., eds., U.S. Government Printing Office, Washington, D.C.:617-629.

Hardenbol, J., and Berggren, W.A. 1978, A new Paleogene numerical time scale, Am. Assoc. Petrol. Geol., Studies in Geology 6: 213-234.

Heezen, B.C., Hollister, C.D., and Ruddiman, W.F., 1966, Shaping of the continental rise by deep geostrophic contour currents, Science, 152:502-508.

Hinz, K., and Schlüter, H.-U., 1978, Der Nordatlantik-Ergebnisse geophysikalischer Untersuchungen der Budesanstalt für Geowissenschaften und Rohstoffe and nordatlantischen Kontinentalrändern, Erdoel-Erdgas-Zeitshr., 94:271-280.

Hornibrook, N., 1971, New Zealand Tertiary climate, in "Geology of New Zealand," Suggate, R., ed., N.Z. Geol. Surv. Rept. 47, Wellington.

Houtz, R.E., 1980, Comparison of velocity-depth characteristics in western North Atlantic and Norwegian sediments, J. Acoust. Soc. Am. 68:1409-1414.

Jenkins, D.G., 1968, Planktonic foraminifera as evidence of New Zealand Tertiary paleotemperatures, Tuatara, 16:32-37.

Jenkins, D.G., 1975, Cenozoic planktonic foraminiferal biostratigraphy of the southwestern Pacific and Tasman Sea, DSDP Leg 29, in: "Initial Reports of the Deep Sea Drilling Project," vol. 29, Kennett, J.P., and Houtz, R.E., eds., U.S. Government Printing Office, Washington, D.C.:449-467.

Jones, E.J., Ewing, M., Ewing, J.I., and Eittreim, S.L., 1970, Influences of Norwegian Sea overflow water on sedimentation in the northern North Atlantic and Labrador Sea, J. Geophys. Res., 75:1655-1680.

Kemp, E.M., 1978, Tertiary climatic evolution and vegetation history in the southeast Indian Ocean region, Palaeogeogr., Palaeoclimat., Palaeoecol., 24:169-208.

Kemp, E.M., and Barrett, P.J., 1975, Antarctic glaciation and Early Tertiary vegetation, Nature, 258:507-508.

Kennett, J.P., 1977, Cenozoic evolution of Antarctic glaciation, the circum-Antarctic Ocean, and their impact on global paleoceanography, J. Geophys. Res., 82:3843-3860.

Kennett, J.P., 1978, The development of planktonic biogeography in the Southern Ocean during the Cenozoic, Marine Micropal., 3:301-345.

Kennett, J.P., and Shackleton, N.J., 1976, Oxygen isotope evidence for the development of the psychrosphere 38 Myr ago, Nature, 260:513-515.

Laughton, A.S., and Berggren, W.A., eds., 1972, "Initial Reports of the Deep Sea Drilling Project," vol. 12, U.S. Government Printing Office, Washington, D.C:1243 p.

Matthews, R.K., and Poore, R.Z., 1980, Tertiary $\delta^{18}O$ record and glacio-eustastic sea-level fluctuations, Geology, 8:501-504.

McCartney, M.S., and Talley, L.D., in press, The subpolar mode waters of the North Atlantic Ocean, J. Phys. Oceanogr.

McKenna, M.C., 1972, Eocene final separation of the Eurasian and Greenland-North American land masses, 24th Int. Geol. Congress, 7:1201-1239.

McKenna, M.C., 1975, Fossil mammals and early Eocene North Atlantic land continuity, Ann. Mo. Bot. Gard., 62:335-353.

Miller, K.G., Gradstein, F.M., and Berggren, W.A., in press, Late Cretaceous to Early Tertiary agglutinated benthic foraminifera in the Labrador Sea, Micropal.

Montadert, L., and Roberts, D.G., eds., 1979, "Initial Reports of the Deep Sea Drilling Project," vol. 48, U.S. Government Printing Office, Washington, D.C.:1183 p.

Müller, C., 1976, Tertiary and Quaternary calcareous nannoplankton in the Norwegian-Greenland Sea, DSDP Leg 38, in "Initial Reports of the Deep Sea Drilling Project," vol. 38, Talwani, M., and Udintsev, G., eds., U.S. Government Printing Office, Washington, D.C.:823-828.

Murray, G.E., 1961, "Geology of the Atlantic and Gulf Coastal province of North America," Harper and Brothers, New York:692 p.

Olsson, R.K., Miller, K.G., and Ungrady, T.E., 1980, Late Oligocene transgression of middle Atlantic Coastal Plain, Geology, 8: 549-554.

Perch-Nielsen, K., 1972, Remarks on Late Cretaceous to Pleistocene coccoliths from the North Atlantic, in: "Initial Reports of the Deep Sea Drilling Project," vol. 12, Laughton, A.S., and Berggren, W.A., eds., U.S. Government Printing Office, Washington, D.C.:1003-1070.

Pomerol, C., 1973, "Stratigraphie et paléogéographie - Ere Cenoique (Tertiaire et Quaternaire)," Doin, Paris:269 p.

Roberts, D.G., 1975, Marine geology of the Rockall Plateau and Trough, Phil. Trans. R. Soc. London, Ser. A, 278:447-509.

Roberts, D.G., Montadert, L., and Searle, R.C., 1979, The western Rockall Plateau: stratigraphy and crustal evolution, in: "Initial Reports of the Deep Sea Drilling Project," vol. 48, Montadert, L., and Roberts, D.G., eds., U.S. Government Printing Office Washington, D.C.:1061-1088.

Reid, J.L., 1979, On the contribution of Mediterranean Sea outflow to the Norwegian-Greenland Sea, Deep Sea Res., 26A:1194-1223.

Ruddiman, W.F., 1972, Sediment redistribution on the Reykjanes Ridge: seismic evidence, Geol. Soc. Am. Bull., 83:2039-2062.

Savin, S.M., Douglas, R.G., and Stehli, F.M., 1975, Tertiary marine paleotemperatures, Geol. Soc. Am. Bull., 86:1499-1510.

Schnitker, D., 1979, Cenozoic deep water benthic foraminifers, Bay of Biscay, in: "Initial Reports of the Deep Sea Drilling Project," vol. 48, Montadert, L., and Roberts, D.G., eds., U.S. Government Printing Office, Washington, D.C.:377-413.

Schnitker, D., 1980a, North Atlantic oceanography as a possible cause of Antarctic glaciation and eutrophication, Nature, 284:615-616.

Schnitker, D., 1980b, Global paleoceanography and its deep water linkage to the Antarctic glaciation, Earth-Science Reviews, 16:1-20.

Schrader, H.J., Bjørklund, K., Manum, S., Martini, E., and van Hinte, J., 1976, Cenozoic biostratigraphy, physical stratigraphy and paleoceanography in the Norwegian-Greenland Sea, DSDP Leg 38 paleontological synthesis, in: "Initial Reports of the Deep Sea Drilling Project," vol. 38, Talwani, M., and Udintsev, G., eds., U.S. Government Printing Ofice, Washington, D.C.:1213-1243.

Sclater, J.G., Anderson, R.N., and Bell, N.L., 1971, Elevation of ridges and evolution of the central eastern Pacific, J. Geophys. Res., 76:7888-7915.

Sclater, J., Hellinger, S., and Tapscott, C., 1977, The paleobathymetry of the Atlantic Ocean from the Jurassic to present, J. Geol., 85:509-552.

Shackleton, N.J., and Kennett, J.P., 1975, Paleotemperature history of the Cenozoic and the initiation of Antarctic glaciation: oxygen and carbon isotope analyses in DSDP Sites 277, 279, and 281, in: "Initial Reports of the Deep Sea Drilling Project," vol. 29, Kennett, J.P., and Houtz, R.E., eds., U.S. Government Printing Office, Washington, D.C.:743-755.

Sheridan, R.E., Golovchenko, X., and Ewing, J.I., 1974, Late Miocene turbidite horizon in Blake-Bahama Basin, Am. Assoc. Petrol. Geol. Bull., 58:1797-1805.

Shor, A.N., and Poore, R.Z., 1979, Bottom currents and ice rafting in the North Atlantic: interpretation of Neogene depositional environment of Leg 49 cores, in: "Initial Reports of the Deep Sea Drilling Project," vol. 49, Luyendyk, B.P., and Cann, J.R., eds., U.S. Government Printing Office, Washington, D.C.:859-872.

Srivastava, S.P., 1978, Evolution of the Labrador Sea and its bearing on the early evolution of the North Atlantic, Geophys. J.R. Astr. Soc., 52:313-357.

Talwani, M., and Eldholm, O., 1977, Evolution of the Norwegian-Greenland Sea, Geol. Soc. Am. Bull, 88:969-999.

Talwani, M., and Udintsev, G., eds., 1976, "Initial Reports of the Deep Sea Drilling Project," vol. 38, U.S. Government Printing Office, Washington, D.C.:1256 p.

Tanai, T., and Huzioka, K., 1967, Climatic implications of Tertiary floras in Japan, in: "Tertiary Correlation and Climatic Changes in the Pacific," Hatai, K. (ed.), Sedaski Printing and Publ. Co., Sedaski:89-94.

Tucholke, B.E., 1979, Relationships between acoustic stratigraphy and lithostratigraphy in the western North Atlantic basin, in: "Initial Reports of the Deep Sea Drilling Project," vol. 43, Tucholke, B.E, and Vogt, P.R., eds., U.S. Government Printing Office, Washington, D.C.:827-846.

Tucholke, B.E., and Ewing, J.I., 1974, Bathymetry and sediment geometry of the Greater Antilles Outer Ridge and vicinity, Geol. Soc. Am. Bull., 85:1789-1802.

Tucholke, B.E., and Laine, E.P., in press, Neogene and Quaternary development of the lower continental rise off the central U.S. east coast, Am. Assoc. Petrol. Geol. Mem.

Tucholke, B.E., and Mountain, G.S., 1979, Seismic stratigraphy, lithostratigraphy and paleosedimentation patterns in the North American basin, in: "Deep Drilling Results in the Atlantic Ocean: Continental Margins and Paleoenvironments," Talwani, M.,

Hay, W., and Ryan, W. (eds.), Amer. Geophys. Union, M. Ewing Series 3:58-86.

Tucholke, B.E., and Vogt, P.R., 1979, Western North Atlantic: sedimentary evolution and aspects of tectonic history, in: "Initial Reports of the Deep Sea Drilling Project," vol. 43, Tucholke, B.E., and Vogt, P.R., eds., U.S. Government Printing Office, Washington, D.C.:791-825.

Vergnaud-Grazzini, C., Müller, C., Pierre, C., Letolle, R. and Peypouquet, J.P., 1979, Stable isotopes and Tertiary paleontological paleoceanography in the Northeast Atlantic, in: "Initial Reports of the Deep Sea Drilling Project," vol. 48, Montadert, L., and Roberts, D.G., eds., U.S. Government Printing Office, Washington, D.C.:475-491.

Vergnaud-Grazzini, C., Pierre, C., and Letolle, R., 1978, Paleoenvironment of the northeast Atlantic during the Cenozoic: oxygen and carbon isotope analyses of DSDP Sites 398, 400A, and 401, Oceanol. Acta, 11:381-390.

Vogt, P.R., 1972, The Faeroe-Iceland-Greenland aseismic ridge and the Western Boundary Undercurrent, Nature, 238:79-81.

Vogt, P.R., Johnson, G.L., and Kristjansson, L., 1980, Morphology and magnetic anomalies north of Iceland, J. Geophys., 47:67-80.

Wolfe, J.A., 1978, A paleobotanical interpretation of Tertiary climates in the northern hemisphere, Am. Scientist, 66:694-703.

Wolfe, J.A., and Hopkins, D.M., 1967, Climatic changes recorded by land floras in northwestern North America, in: "Tertiary Correlations and Climatic Changes in the Pacific," 11th Pac. Sci. Congress:67-76.

Woodruff, F., Savin, S.M., and Douglas, R.G., 1981, Miocene stable isotopic record: a detailed deep Pacific ocean study and its paleoclimatic implications, Science, 212:665-668.

Woodruff, F., and Douglas, R., 1981, Miocene deep-sea benthic foraminiferal faunal changes in Pacific, Am. Assoc. Petr. Geol. Bull., 65:1009.

Worthington, L.V., 1976, "On the North Atlantic Circulation," Johns Hopkins Oceanographic Studies No. 6, Johns Hopkins University Press, Baltimore:110 p.

# ON THE LATE PLEISTOCENE EXCHANGE OF WATER ACROSS THE ICELANDIC TRANSVERSE RIDGE

Eric Olausson

Marine Geological Laboratory
University of Göteborg
P.O. Box 33031, S-400 33 Göteborg

## INTRODUCTION

During part of the Tertiary the Faeroe-Icelandic-Greenland transverse was an important obstacle to the exchange of water between the Norwegian-Greenland Seas and the Atlantic proper. The ridge subsided in the Late Tertiary. The sediments from the Pliocene and Early Pleistocene in cores from adjacent areas are fragmentary or missing. Therefore, this discussion is concentrated to the late part of the Pleistocene from which age core data are available. This period can be divided into three main phases, interglacials, exemplified by the Holocene, early stadials (the time up to the maximum of a glaciation) and late stadials (=deglaciations). My discussion is divided into two parts: a) The paleohydrography of the northeastern sector of the Atlantic Ocean, b) evidence from sediment cores. Since a permanent ice cover can only exist on a stratified sea water the paleo-oceanographical budgets deal with fresh and normal sea water components.

## A. THE HOLOCENE WATER BUDGET.

There is a continuous vapor flux across the Isthmus of Panama. The size of it is of the order of 0.1 Sv (1 Sv = $10^6 m^3 s^{-1}$). Steady state conditions require that an equal amount of water is transported back and apparently this takes place via the Arctic Ocean. The fresh water inflow through the Bering Strait is of the order of 0.1 Sv (Stigebrandt, manuscript). The river supply of water to the Arctic Ocean is of the same magnitude. P-E is 0.03 Sv and, thus, negligible in this connection. All fresh water leaves the Arctic Ocean through the Fram Strait (≈45%) and through the Arctic archipelago (≈55%; 1. op.cit.). The total fresh water supply to the Norwegian-Greenland Seas is, thus, around 0.1 Sv or, if all other sources are also taken into account, 0.12 Sv (Mosby 1962).

The inflow of normal sea water to the Norwegian-Greenland Seas is 8-9 Sv from the Atlantic proper and 3 Sv from the Arctic Ocean (Worthington 1970; Mosby 1962 gives lower figures).

A permanent ice cover can only be formed over a stratified sea water body and this stratification must be due to permanent differences in salinity. Due to the aforementioned fresh water supply such a stratification exists in the present Arctic Ocean. An ice sheet, about 4 m thick, covers the present Polar Sea (see Olausson & Jonasson 1969, Olausson 1972).

The Norwegian-Greenland Seas are homohaline, no permanent ice cover can exist there. During winters the thermocline disappears and the very dense Arctic Bottom Water is formed.

## EARLY STADIALS

During a glacial age the "interglacial" hydrographical pattern was disturbed. Cooling, particularly of the North Atlantic, reduced the salinity difference between the North Pacific and the North Atlantic and diminished the surface inflow to the Arctic Ocean as well (see further details in Olausson & Stigebrandt, in prep.). Furthermore, the drop in sea-level lessened and, when the sill of the Bering Strait became dry, finally stopped the inflow from the Pacific. Because of the ice sheet formations on adjacent continents the supply of river water to the Arctic Ocean was reduced down to some 40 per cent. At a somewhat earlier stage (when the river water supply was reduced down to some 60 per cent) the fresh water supply had become too small for keeping the Arctic Ocean stratified and, thus, with an ice cover; the ocean turned into an ice-free stage (details in Olausson 1981, Olausson and Stigebrandt in prep.).

The fresh water discharge to the Norwegian-Greenland Seas was, a consequence, reduced down to some 0.03 Sv, i.e. one fourth of its present size. Apparently the possibility of creating an ice cover over the Norwegian-Greenland Seas was much less during the ice advance phases than at present.

## DEGLACIAL TIMES

This phase differs markedly from the other subages. The temperature was on the increase which may have retarded the vertical mixing and convection. The salinity of the sea water was decreasing due to the meltwater discharge (the average salinity may have been around 36 per mil during the glacial maximum) and this meltwater was, to about 90 per cent, discharged into the North Atlantic.

During the last deglacial period (13,000-9,000 B.P.) the maximum of the meltwater supply occurred during the interstades Bölling (12,300-12,000 B.P.) and Alleröd (11,800-10,900 B.P.) and the Preboreal subage (10,200-9,000 B.P.). The sea-level rose, as appears from many authors, about 1 cm per year (≈0.1 Sv for the North Atlantic). When the sea-level had risen up to the sill of the Bering Strait (during the Alleröd?), that water exchange started again. The fresh water discharged into the Polar Sea by rivers, from melting ice sheets and from the Pacific recreated a stratified water body and, thus, an ice cover over the Arctic Ocean, probably during the Alleröd interstade (see further Olausson 1972). The change in the albedo resulted in the climatic deterioration: Younger Dryas (Olausson & Jonasson 1969).

The fresh water supply to the present Norwegian-Greenland Seas is 0.12 Sv and to this figure we must add the meltwater contribution, something between 0.05 and 0.1 Sv. A slight salinity stratification may have existed for a while during the most intense deglacial time (around and during the Younger Dryas).

## B. EVIDENCE FROM SEDIMENT CORES.

Are there any core data which contradict the paleohydrographical conclusions reached above?

Kellogg (1977) and CLIMAP group (McIntyre et al. 1976) have presented an important amount of core analyses and they reached the opposite conclusion. They suggest that the Norwegian-Greenland Seas were ice-covered 18,000 B.P.. Two of their arguments: 1) a low carbonate stage and 2) certain trends in the grain size can be better interpreted in the opposite way (Olausson 1981).

The third argument, the frequency map of left-coiled Globigerina pachyderma shows only that the conditions could have been similar to the present ice-free (!) Norwegian Sea. The boundary between the Gulf Stream Water Mass and the Subarctic Water Mass 18,000 B.P. (Fig. 1; Olausson & Jonasson 1969, Olausson 1975) agrees well with the CLIMAP data.

The oxygen isotope analyses from Core 288 (Fig. 2) show that the $\delta 18$ of different species with different depth habitats follow each other closely and only during deglacial time Globigerinoides ruber indicates a slight salinity stratification of the water (cfr. above). The isotopic meltwater contamination, which is in evidence from Fig. 2, indicates that the water, in which G. ruber lived, deviated by about 0.5 o/oo from normal salinity at the Weichselian/Holsteinian transition.

The Albatross cores 288-299 (Olausson 1961) have isotopic peak-to-valley amplitudes of 1.4-2.0 per mil (Globorotalia inflata) for the Eemian-Holsteinian stages. Core 288 also shows about 2.4 per mil amplitudes for G. ruber and G. bulloides (Fig. 2) which, at least for G. ruber, is natural since it occupies the shallowest depth habitat where the isotopic contamination should be the highest. The Norwegian Sea cores, discussed by Kellogg et al. (1978), show the same amplitudes. Their analyses, as well as my DDR core 1361, show that the surface water of the Norwegian Sea during the glacial advance phases was isotopically slightly heavier than at present (= the glacial storage effect). The oxygen isotope analyses suggest a more or less homohaline ice-free water of the Weichselian Norwegian Sea.

As appears from Fig. 3 the isotopic meltwater contamination was stronger at the Saale/Eem transition. In core 299 a 1.5 per mil too high isotopic paleotemperature is in evidence; if the $\delta 18$ of the meltwater was -30 o/oo, the lowering of the surface salinity could have been of the order of 1.5 per mil. An isochronous meltwater spike appears in the diagram of Kern 39, from N 63° W 07° (Manze & Strauch 1977), which points to a sudden surface salinity lowering by about 3 per mil north of the Faeroe Islands. Thus, during a short period at the Saale deglaciation the opportunity for a sea ice formation may have existed in certain parts of the Norwegian Sea (-NE Atlantic?). Furthermore, the currents across the Faeroe-Greenland ridge may, for a while, have been reversed. During the same deglacial, as stated before (Olausson 1975), also the currents in the Strait of Gibraltar became reversed.

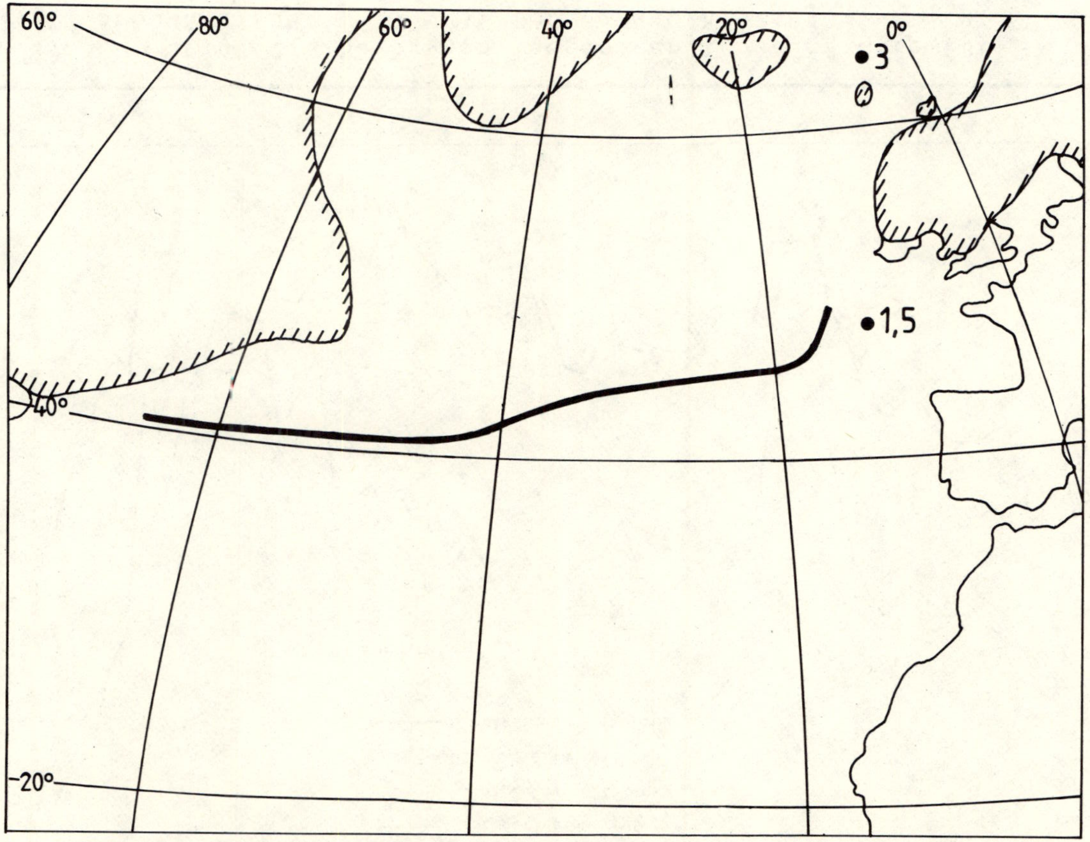


Fig. 1. The boundary between the Gulf Stream Water Mass and the Subarctic Water Mass. The figures are approximate estimations of the sudden drop in surface salinity (in per mil) during the Saale/Eemian transition.

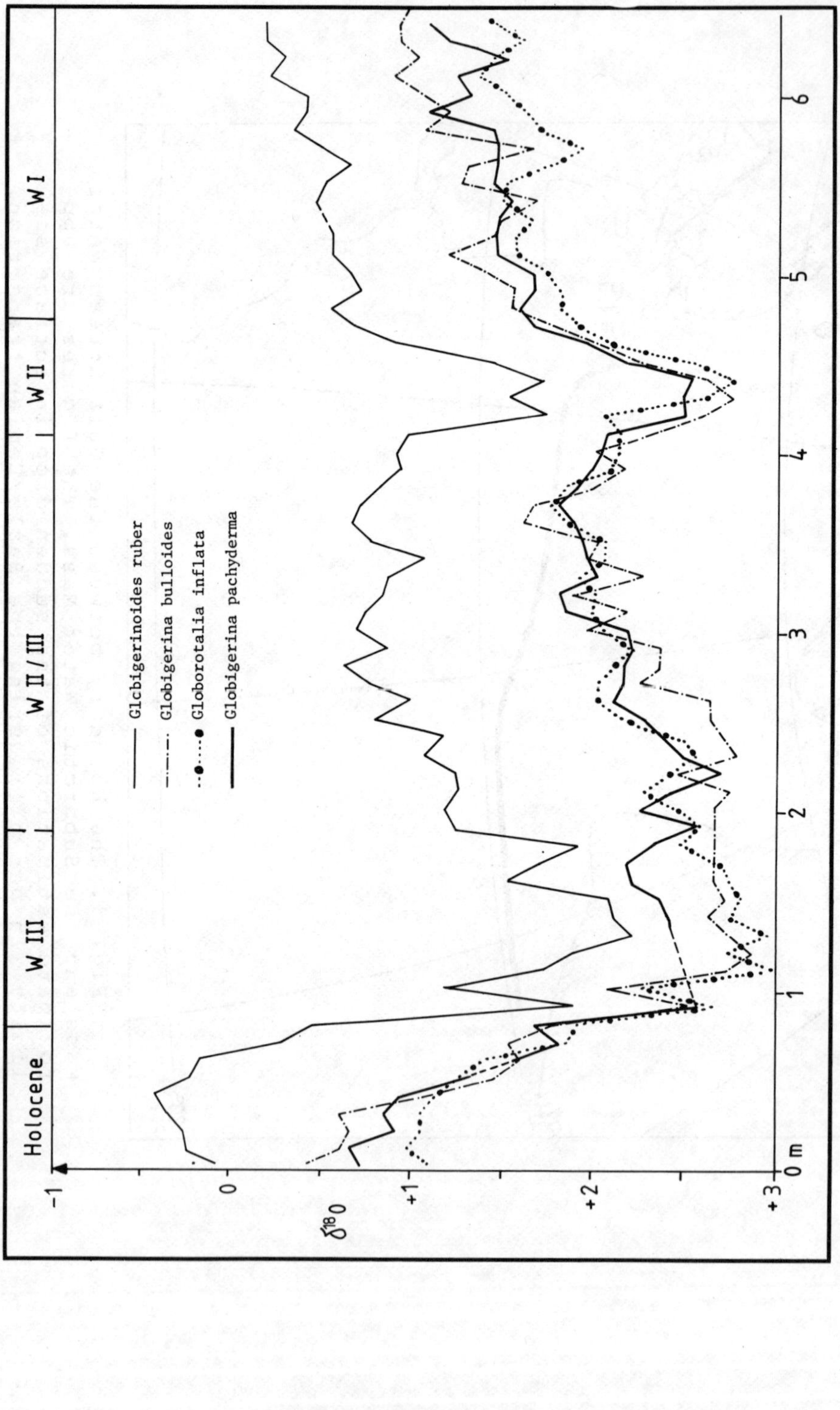


Fig. 2. The oxygen isotope analyses of Core 288 from N 41°08' W 31°50'. Coring depth: 2684 m.

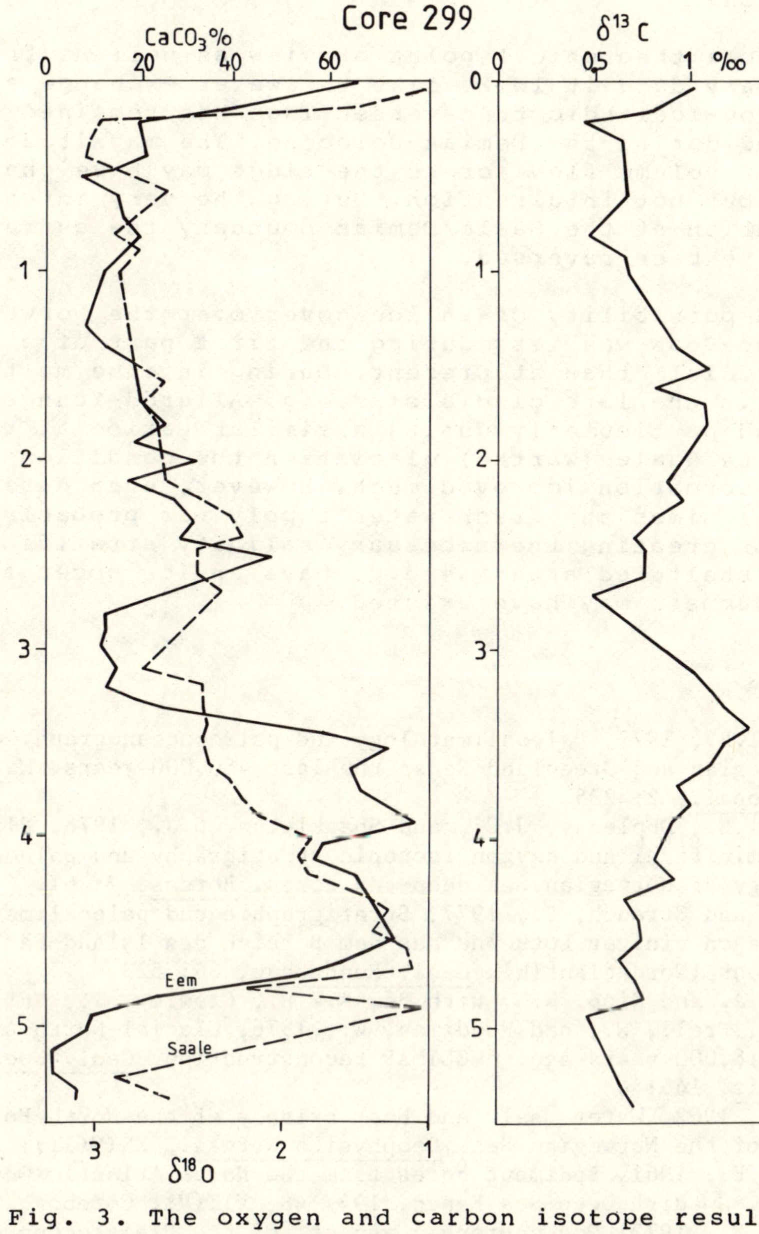


Fig. 3. The oxygen and carbon isotope results and the contents of calcium carbonate in Core 299 (N 47°41' W 12°43'; depth 4288 m). Note the very sudden decrease in the $\delta$18 at the Saale/Eemian transition (core depth of around 5 m). A similar, but less pronounced trend, is also in evidence at the Pleistocene/Holocene Boundary (near the top of the core).

## CONCLUSIONS

From a theoretical point of view as well as from sedimentary data it looks like the water exchange across the Faeroe-Icelandic transverse ridge has remained fairly unchanged during the Eemian-Holocene. The magnitude of the water volume flow across the ridge may have changed in size but not in direction. During the very intense deglaciation at the Saale/Eemian boundary the currents could have been reversed.

The possibility of an ice cover over the Norwegian-Greenland Seas was less during the first part of a glacial (stadial) than at present. During intense melting periods in the late glacials as e.g. Alleröd-Younger Dryas and particularly during a similar period at the end of the Saale (Warthe) glaciation the conditions for sea-ice formation improved much. However, even during deglacial times the fresh water supply was probably too small for creating the necessary salinity stratification; only in sheltered areas as e.g. bays an ice cover also during summers may have existed.

## REFERENCES

Kellogg, T.B., 1977, Paleoclimatology and paleo-oceanography of the Norwegian and Greenland Seas: the last 450,000 years. Marine Micropal., 2: 235.

Kellogg, T.B., Duplessy, J.C., and Shackleton, N.J., 1978, Planktonic foraminiferal and oxygen isotopic stratigraphy and paleoclimatology of Norwegian Sea deep-sea cores. Boreas, 7: 61.

Manze, U. and Strauch, F., 1977, Stratigraphie und paleoklimatische Aussagen einiger Lotkerne aus dem Bereich des Island-Färöer-Rückens (Nordatlantik). Geol. Rundschau, 66: 523.

McIntyre, A. and Kipp, N.G. with Bé, A.W.H., Crowley, T., Gardner, J.V., Prell, W., and Ruddiman, W., 1976, Glacial North Atlantic 18,000 years ago: a CLIMAP reconstruction. Geol. Soc. Amer. Memoir, 145: 43.

Mosby, H., 1962, Water, salt and heat balance of the North Polar Sea and of the Norwegian Sea. Geophysica Norweg., XXIV(11): 289.

Olausson, E., 1961, Sediment cores from the North Atlantic Ocean. Rept. Swedish Deep-Sea Exped. 1947-48, VII(V), Göteborg, 227.

Olausson, E., 1972, Oceanographic aspects of the Pleistocene of the Arctic Ocean. Internord, 12: 151.

Olausson, E., 1975, On the course of a glacial age. Proceed. WMO/IAMP symp. on long-term climatic fluctuations, WMO, 421, Geneva, 47.

Olausson, E., 1981, On the glacial Norwegian-Greenland Seas and the Arctic Ocean. Quaternary studies in Poland (S. Kozarski, ed.). In press.

Olausson, E. and Jonasson, U.C., 1969. The Arctic Ocean during the Würm and early Flandrian. Geol. För. Förh. Stockholm, 91: 185.

Olausson, E. and Stigebrandt, A., Was the Arctic Ocean ice-free during the last glaciation? In prep.

Stigebrandt, A., The flow through the Bering Strait and the control of the salinity difference between the Atlantic and Pacific Oceans. Manuscript.

Worthington, L.V., 1970, The Norwegian Sea as a mediterranean basin. Deep-Sea Res., 17: 77.

# GEOLOGICAL HISTORY OF THE ICELAND-FAEROE-RIDGE

# AND ITS INFLUENCE ON PLEISTOCENE GLACIATIONS

Friedrich Strauch

Geologisch-Paläontologisches Institut der Universität
Corrensstr. 24
D-4400 Münster

## SUMMARY

During the Tertiary the basaltic ridge building a continuous land connection between Greenland and the British Isles separated the Greenland-Norwegian Sea (Scandic Sea) from the Atlantic. This is clearly evidenced by marine and terrestrial biota as well as the geological conditions of the Iceland-Faeroe-Ridge.

Break-down of this barrier at the boundary Plio/Pleistocene led to fundamental changes of oceanic currents and consequently to new meteorologic coincidences and creation of Pleistocene glaciations.

## ZUSAMMENFASSUNG

Der basaltische Rücken zwischen Grönland und den Britischen Inseln trennte als durchgehende Landbrücke die Greenland-Norwegian Sea (Scandic Sea) und den Atlantik im Tertiär, wie marine und terrestrische Biota sowie die geologischen Verhältnisse der Island-Färöer-Schwelle eindeutig belegen.

Der Zerfall dieser Barriere an der Grenze Plio-Pleistozän führte zu grundlegenden Änderungen der ozeanischen Ströme, damit zu neuen meteorologischen Konstellationen und zur Verursachung der pleistozänen Vereisungen.

## GEOLOGICAL HISTORY OF THE ICELAND-FAEROE-RIDGE

The petrographic character of the Iceland-Faeroe-Ridge is -- to a certain extent -- generally known from geophysical examinations as well as analyses of submarine bottom samples. The submarine rise trending NW to SE is part of an anomalous oceanic crust (Bott, 1975; Bott et al., 1971).

Little was known on the history of the rise, and so far only few data were available on its temporal development and on exogene processes forming the rise (Fleischer et al., 1974; Strauch and Wohlfeil, 1978).

Oldest onset of initial development is surely linked with the opening of the northernmost part of the Atlantic (Grønlie, 1979). So very many effusiva were produced over a "hot spot" of Iceland that, due to the expansion of the oceanic basin, the oceanic ridges bilaterally developed. But what part did the basaltic barrier of the Iceland-Faeroe-Ridge play? How far was this barrier able to affect the orographic pattern, the oceanographic and climatic events?

Already in 1970 F. Strauch proved that the basalts of the Iceland-Faeroe-Ridge filled up the developing oceanic scar right into the subaeric region. These, together with other parts of the rise extending from Greenland to the British Isles, built the basaltic Thule land connection, thus separating both oceanic minor basins during the Tertiary (Strauch and Wohlfeil, 1978).

This can be demonstrated by numerous direct and indirect proofs found in the migrational routes of many continental biota that used this basaltic connection. This includes higher vertebrates such as reptiles and mammals as well as invertebrates of the continental environment but also limnic organisms like anodontids. Following temperature drop during the Tertiary, the filtration effect of the continental connection increased. In addition, new migrational routes between Europe and America across the Asian continent blurred the outline of this migrational route, however, it still remained evident in the holarctic region even during warm periods that followed first Pliocene phases of cooling (Strauch, 1972a). This does also count for the floral elements for which the Thule continental connection as a migrational route remains evident until the end of the Tertiary. Thus, an island belt for the dispersal of non-marine organisms can be excluded.

What is considered to be a migrational route for the continental biota, is a firm barrier for marine organisms (Strauch, 1972b). It can be shown that the latter underwent a development, independent of each other, in both secondary basins and their minor seas north and south of the basaltic connection. Up to the late Tertiary only endemic faunas are known from the region of the Greenland/Norwegian Sea -- called the Scandic after Blüthgen (1942) -- and the North Sea. Despite climatic adjustments, these faunas show only little relation to the European or American Atlantic ones, that can be related to an interchange across short span epicontinental routes.

During the Pleistocene, however, the portion of the endemic elements in the North Sea decreased to a negligible amount. American Atlantic elements are completely absent in the Scandic, whereas Pacific ones appear in large numbers during the Pliocene. They invaded right into the North Sea basin. Pleistocene and recent Northwest European marine faunas are even far richer in Pacific elements than American Atlantic waters.

Detailed studies on migrational routes of single faunal elements clearly prove the existence of a continental connection (Strauch, 1972b).

These paleogeographic conditions are also proved by the climatic development in the margins of the Scandic and of Western Europe (Strauch in: Schlickum and Strauch, 1979). In the peripheral areas of the cut-off Scandic basin Pliocene cold phases with glacial invasions become already apparent, while the climate in West Europe at this time remains warm arid due to an increase of the Gulf Stream after the lock up of Central America.

In the uppermost Miocene and Pliocene of Northern Iceland metamorphic heavy minerals are found in shallow water deposits (Strauch, 1963). These minerals could not have been transported over deep oceanic regions but in shallow waters parallel to coastline from continental shores, thus demonstrating a different paleogeographical situation than the present one.

Direct proofs concerning the continental character of the Iceland-Faeroe-Ridge during the Tertiary are found in waters between both island complexes.

Besides magnetic abnormalities that do not point to subaquatic deposition (Fleischer et al., 1974) and besides bottom formations (Nilsen, 1978), the basalts in the ridge area show intercalated coal-bearing and limnic deposits with single and poorly preserved fresh water gastropods and Tertiary sporomorpha (Strauch and Wohlfeil, 1978; Wohlfeil, 1981).

When did this continental connection start to break down? First penetration of both parts of the ocean basin through this continental basaltic barrier can be fairly exactly determined by migrational movements of the organisms and by the earliest incidence of faunal mixing. Faunal mixing is found in deposits that correspond to the middle Red Crag of the British Isles close to the Plio/Pleistocene boundary. Here, the migrational routes of molluscs are also precisely exhibited.

## THE SIGNIFICANCE OF THE RIDGE IN CONNECTION WITH PLEISTOCENE GLACIATIONS

This coincidence of destruction and initial break of the continental connection is not linked casually and causally with the Plio/Pleistocene boundary.

With the break-down of the Thule basaltic connection, due to crustal movements, weathering and marine abrasional processes, an exchange of Scandian with Atlantic waters could, for the first time, take place from the Plio/Pleistocene boundary onwards. The North Atlantic Stream invaded into the Scandic basin and the East Greenland Stream influenced western parts of the North Atlantic. Warm and cold waters passed along one another over an area of some 1000 kms, thus changing the meteorological conditions, and were newly adjusted.

The position of the eastward drifting cyclones -- as they are known today -- established. The climate of West and Central Europe, still markedly arid during the Pliocene (Strauch in: Schlickum and Strauch, 1979), changed to humid. A distinct cooling, especially during the Upper Tertiary, took place resulting in the availability of appropriate precipitations. These could be snow falls and could lead to glaciation due to back-coupling effects.

The morphology as well as the sedimentary side covers of the Iceland-Faeroe-Ridge reflect the Pleistocene glacial event of this region (Manze and Strauch, 1977). Glacial littoral formations are found at the slopes above deep sea deposits of the interglacial and, finally, even non-marine influenced morainal materials in the topmost crest (Strauch et al., 1971). Even during last glaciation large, if not all parts of the ridge were covered by ice and, according to own investigations, the western region certainly represented a characteristic glaciation centre. North and South of the ridge, glaciers drifting from SE' direction concentrated and built a close ice pack belt (Strauch and Wohlfeil, 1978; Wohlfeil, 1981).

Subsidence of oceanic level, glaciation of this ridge and probably also of the Greenland-Iceland-Ridge, again led to complete separation of both parts of the oceanic basin, thus resuming the initial orographic and oceanographic Tertiary conditions. The initial cool climate had to reverse again from humid to arid. Precipitations ceased and glaciation maximum had passed. In connection with the North Atlantic Stream, farther thrust to SE, respectively the branch of the Irminger Stream which approached from SE during the glacial washing around the Iceland-Faeroe-Ridge, the ice melted and the oceanic level rose again. A late glacial to early interglacial overflow to the North is clearly evidenced by this rise of level (Strauch and Wohlfeil, 1978). At the complete overflooding of the ridge, same conditions, as did happen after the break-down of the Thule continental connection at the Plio/Pleistocene boundary, could recur. Re-humidisation led back to an unstable interglacial climate and consequently to the starting point of further glaciation.

During the Pleistocene the water routes between both parts of the ocean basin enlarged. Oceanic areas between Greenland and Iceland as well as between Iceland and the British Isles widened. Increasing drift, crustal movements, glacial erosion in the region of the ridge, and marine abrasions are accountable for this development.

Widening of oceanic regions between Scandic (Greenland-Norwegian-Sea) and Atlantic, in addition to a relatively cool global climate, inevitably led to stronger, respectively longer glaciations and more instable interglacial conditions during the Pleistocene; consequently warm periods became shorter unto the present time. Pleistocene sediments of the Iceland-Faeroe-Ridge point to an essential participation of this ridge, where control of Pleistocene glaciation is concerned.

## REFERENCES

Bott, M. H. P., 1975, Structure and evolution of the Atlantic floor between northern Scotland and Iceland, Norsk geol. Unders., 316:195-199.

Bott, M. H. P., Browitt, C. W. A., and Stacey, A. P., 1971, The deep structure of the Iceland-Faeroe-Ridge, Mar. geophys. Res., 1:328.

Blüthgen, J., 1942, Tatsachen und Deutungen zur Geschichte des Skandik (= Europäischen Nordmeeres), Geol. Meere u. Binnengewässer, 5:83-117.

Fleischer, U., Holzkamm, F., Vollbrecht, K., and Voppel, O., 1974, Die Struktur des Island-Färöer-Rückens aus geophysikalischen Messungen, Dt. hydrogr. Z., 27:97-113.

Grønlie, G., 1979, Tertiary paleogeography of the Norwegian -Greenland Sea, Norsk Polarinst. Skr., 170:3-24.

Manze, U., and Strauch, F., 1977, Stratigraphie und klimatische Aussagen einiger Lotkerne aus dem Bereich des Island-Färöer-Rückens (Nordatlantik), Geol. Rdsch., 66:523-551.

Nilsen, T. H., 1978, Lower Tertiary Laterite on the Iceland -Faeroe-Ridge and the Thulean land bridge, Nature, London, 274:786-788.

Schlickum, W. R., and Strauch, F., 1979, Die Land- und Süßwassermollusken der pliozänen Deckschichten der rheinischen Braunkohle, Abh. senckenb. naturforsch. Ges., 536:1-144.

Strauch, F., 1963, Zur Geologie von Tjörnes (Nordisland), Sonderveröff. geol. Inst. Univ. Köln, 8:1-129, (Geology of Tjörnes, Transl., U.S. geol. Surv., Calif., 1964).

Strauch, F., 1970, Die Thule-Landbrücke als Wanderweg und Faunenscheide zwischen Atlantik und Skandik im Tertiär, Geol. Rdsch., 60:381-417.

Strauch, F., 1972a, Zum Klima des nordatlantisch-skandischen Raumes im jüngeren Känozoikum, Z. dt. geol. Ges., 123: 163-177.

Strauch, F., 1972b, Phylogenese, Adaptation und Migration einiger nordischer mariner Molluskengenera (*Neptunea*, *Panomya*, *Cyrtodaria* und *Mya*), Abh. senckenb. naturforsch. Ges., 531:1-211.

Strauch, F., Vollbrecht, K., and Zobel, B., 1971, Meeresgeologie, in: Forschungsschiff "Meteor", Fahrt 20, Ber. üb. wiss. Arb., 44-52, Hamburg.

Strauch, F., and Wohlfeil, K., 1978, Zur geologischen Geschichte und Bedeutung der Island-Färöer-Schwelle (N-Atlantik), Sonderveröff. geol. Inst. Univ. Köln, (Schwarzbach Festschr.), 33:241-258.

Wohlfeil, K., 1981, Die Sedimente der Island-Färöer-Schwelle und ihre geologische und ozeanographische Bedeutung, Inaug. Diss., Köln, 1-222.

# TERRIGENOUS LATE QUATERNARY SEDIMENT COMPONENTS NORTH AND SOUTH OF THE SCOTLAND-GREENLAND RIDGE AND IN THE NORWEGIAN SEA

D. Eisma and S.J. van der Gaast

Netherlands Institute for Sea Research

P.O. Box 59, Texel, Netherlands

## INTRODUCTION

The Quaternary sediments around the Scotland-Greenland Ridge are only known from two DSDP sites on the north and south flank (sites nrs. 336 and 352 of leg 38), from nearby site 337 in the Norway Basin (Talwani et al. 1976), and from core- and grab-samples. On sites 336 and 352 Plio-Pleistocene deposits lie on top of pelagic Oligocene deposits with a possible hiatus during the Miocene and Pliocene. The Plio Pleistocene consists of sandy muds and muds with an admixture of coarse pebbels, indicating pelagic deposition of terrigenous material with contributions from ice-rafting. On site 337 47 m of Plio-Pleistocene clays, sandy muds and muds with ice-rafted pebbles cover an almost complete Tertiary section. Other sampling (cores and surface samples) around the Scotland-Greenland Ridge has reached into Saale glacial deposits at best, except a few cores that penetrated more than 450.000 years of deposition (Kellogg 1977). Our knowledge on the Quaternary deposits in this area is therefore chiefly limited to the Late Quaternary. This paper reviews the existing data for this period and present the results of clay mineral analysis and Coccolith counts in a number of cores collected around the Iceland-Faeroe Ridge and in the Norwegian Sea.

### Surface Deposits

Sediments on the crest of the Scotland-Greenland Ridge including the Wyville-Thomson Ridge, the shallower parts of the Faeroe-Shetland Channel and Denmark Strait, and the adjacent shelves of Greenland and Scotland-Shetland, are sandy, containing

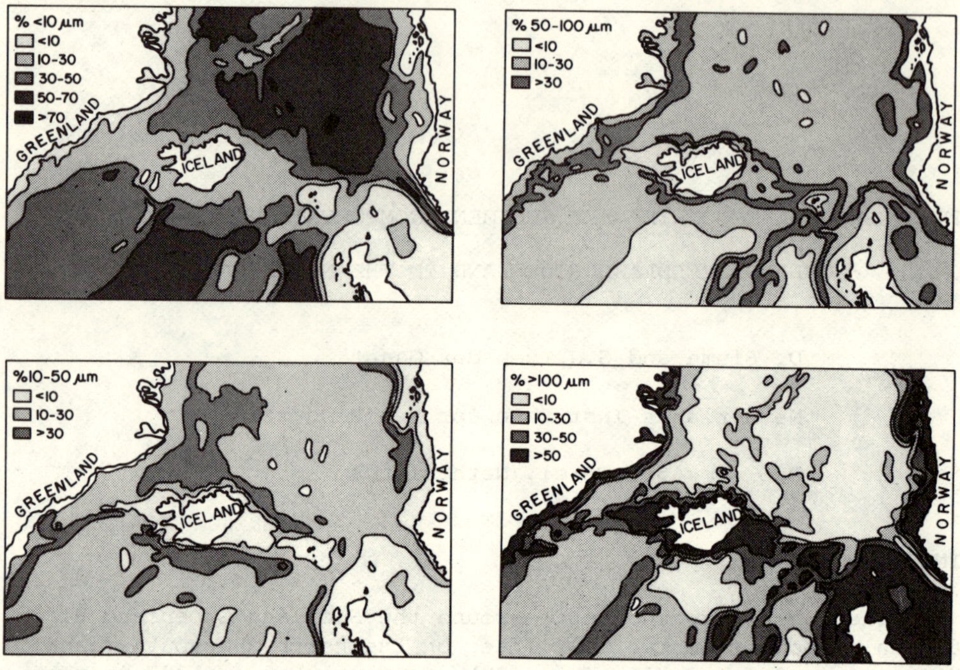


Fig. 1. Distribution of particle size in surface sediments (after Lisitzin 1969)

gravel as well as coarse silts and grading along the flanks into finer silts and muds with greater waterdepth (Bøggild 1900, Lisitzin 1969 ; fig. 1). Size distributions are bimodal or trimodal indicating mixtures of material from different sources and/or different modes of transport. On the ridge, from Greenland to the Faeroes, a major part of the sediment is of volcanic origin ; ash, glass, scoria (Böggild 1900, Shurko 1966, Horowitz 1974 ; fig. 2). A second important component is ice-rafted material containing gravel, rock fragments and minerals that cannot have come from Iceland or the Faeroes (Horowitz 1974). Calciumcarbonate content is less than 10 % but increases with waterdepth to more than 50 % in the Norway Basin and the Northern Atlantic Ocean, where a calcareous ooze is present (Gorshkova 1960, Kellogg 1976, 1980 ; fig. 2), the volcanic and ice-rafted components correspondingly decreasing. The calciumcarbonate is chiefly of organic origin : Foraminifera, Coccoliths, in the shallower areas also Mollusc shells, shell fragments and calcareous fossils. The southern

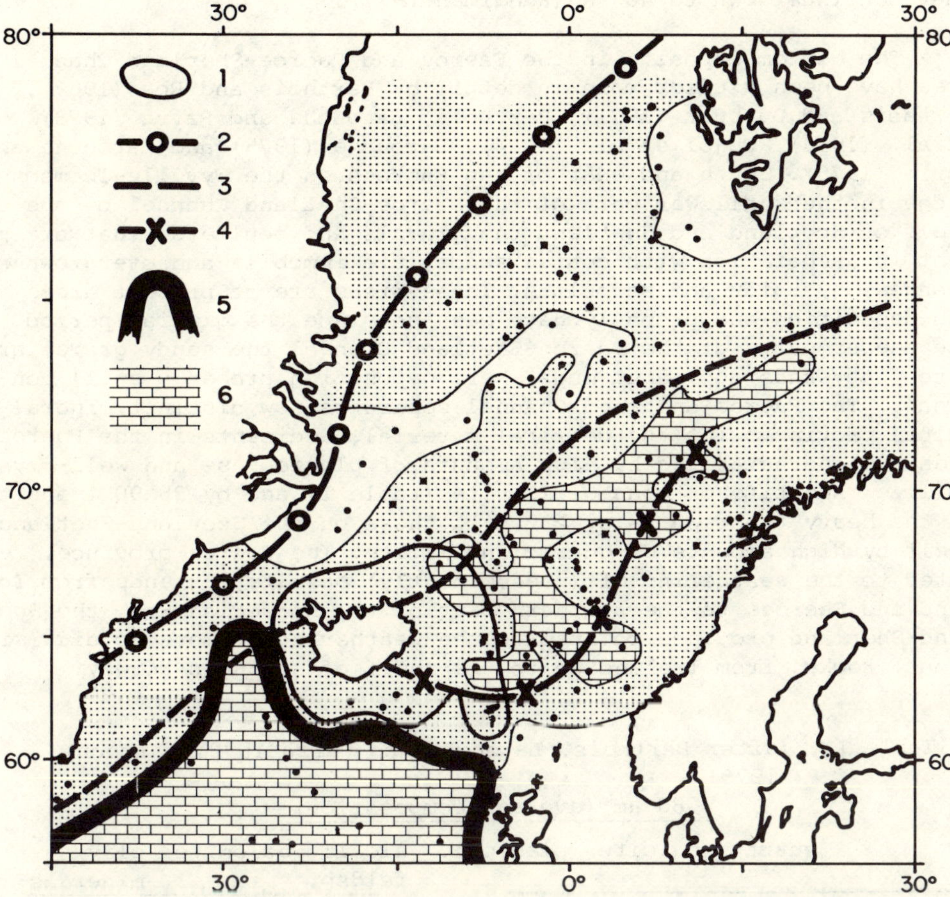


Fig. 2. Distribution of sediment types in surface sediments (after Bøggild 1900, Kellogg 1976). 1 = distribution of strong volcanic admixture, 2 = sea ice limit in august ( ⩾ 0.5 cover) 3 = sea ice limit in february ⩾ 0.1 cover) 4 = maximum sea ice extension in 18$^{th}$-19$^{th}$ centuries, 5 = southern limit of glacial marine sediment, 6 = Foraminiferal clay, marl or ooze; the remainder is sandy clay, marl or ooze (dotted area) or clay and Foraminiferal clay (blank).

limit up to where ice-rafted sediments are present at the surface, lies just south of the Scotland-Greenland Ridge (fig. 4). Further south, and in the central part of the Norwegian Sea, they are covered by a thin layer of calcareous mud but below this layer they continue down to 40° N (Ruddiman 1977).

The bottom deposits in the Faeroe and Faeroe-Shetland Channel area have been studies more in detail by Berthois and Roa (1967), Berthois and Du buit (1971), Berthois, Latouche and Parra (1973), Moyes et al. (1974), Pujol et al. (1974), Latouche (1975) and Latouche and Parra (1976). South and east of the Faeroes on the Wyville-Thomson Ridge in the shallower part of the Faeroe-Shetland Channel on the the shelf of Scotland and Shetland the gravels and boulders, that are present in association with coarse sands, are immobile and overgrown with sponges, hydroids and serpulids. These areas are at present areas of non-deposition and probably have been so since the glacial period. In the deeper parts of the Faeroe-Shetland Channel the sandy gravel grades into fine sands and sandt muds; here deposition probably still continues. The Faeroe-Shetland Channel separates tow distinct mineralogical provinces int the detrital material. Sediments in the Faeroe area are characterized by high amounts of plagioclase and well-crystallized smectite or montmorillonite (table 1) and by 35-90 % augite in the heavy fraction (s.g. > 2.9.), those on the Scotland-Shetland shelf by high amounts of quartz and illite. The Faeroe-province, related to the weathering of (predominantly) basalts, extends from Iceland and Faeroes to the southern part of the Norwegian Sea, the Scotland-Shetland province related to the weathering of more acidic (silicious) rocks, from Ireland up to the shelf of southern Norway.

Table 1 . (after Berthois, Latouche and Parra 1973, Latouche 1975)

< 80 µm (average composition)

| | quartz | calcite | plagiocl. | alcali feldsp. | dolomite | clay minerals |
|---|---|---|---|---|---|---|
| Faeroe area | 23 | 25 | 18 | 3 | 2 | 29 |
| Scotland-Shetland area | 60 | 21 | 8 | 1 | tr. | 10 |

< 2 µm (average composition)

| | mica/illite | chlorite | smectites (montm.) | kaolinite | amphib. |
|---|---|---|---|---|---|
| Faeroe area | 26 | 17 | 41 | 10 | 5 |
| Scotland-Shetland area | 40 | 25 | 15 | 10 | 10 |

Latouche (1975) indicatesthat around Faeroe montmorillonite contents may reach 50-100 % (illite 0-30 %).

The Holocene surface deposits vary in thickness from several cm to nearly 1 meter. From data given by Kellogg (1980) a minimum input of terrigenous material during the Holocene can be estimated for the Norwegian Sea. Taking the onset of Holocene hemipelagic carbonate deposition as 11.000 BP and assuming the bulk density to be 0.8 gr/cm$^3$ (Ruddiman 1977), the input rate of sand (> 62 μm) for 15 cores scattered through the Norwegian Sea, is 62-204 mg/cm$^2$/1000 yrs, or 440-1448 mg of total terrigenous sediment per cm$^2$ per 1000 yrs, using a conversion factor of 7.1 (Ruddiman 1977). Applying a timetransgressive scale extrapolated from the data of Ruddiman and McIntyre (1973), with hemipelagic deposition starting in the southern part of the Norwegian Sea around 13.000 BP and in the north around 6.000 BP, the input rate of sand becomes 76-300 mg/cm$^2$/1000 yrs, or 540-2130 mg/cm$^2$/1000 yrs of total terrigenous sediment. The latter rates are probably more realistic but the differences are not large and both types of calculations give a minimum input rate that is of the same order of magnitude as the input rate during the latest interglacial period (96-215 mg sand per cm$^2$ per 1000 yrs ; Ruddiman 1977). Since during the latest interglacial conditions were very similar to those at present (Kellogg 1980), the rates calculated here for the Holocene may closely appromixate the actual ones, suggesting that loss of sediment from the top of the cores during sampling has not been very large. Relatively high input rates were found at four sites (out of nineteen) : just south of Denmark Strait where input rates of sand have been exceptionally high all through the late Quaternary (Ruddiman 1977 ; cores V 28-14 and V 29-210 with Holocene rates of 587 and 900 mg sand/cm$^2$/1000 yrs), at two sites near Jan Mayen Island (V 23-74 and V 28-38 with 430 and 386 mg/cm$^2$ 1000 yrs), and at a site near to the slope towards the Barentz Sea (V 27-60 ; 880 mg sand/cm$^2$/1000 yrs). Since within the Norwegian Basin only at a few sites such high rates were found, they can be regarded as only of local significance. On this basis the average yearly input rate into the Norwegian Sea can be estimated as in the order of 16-20 million ton (dryweight ; c.f. Ruddiman 1977).

The present (Holocene) input of terrigenous material comes a) from glaciers on Greenland and Spitzbergen and ice-rafting (chiefly in the W and N of the Norwegian Sea), b) from the continental shelf, and the crest of Scotland Greenland Ridge by currents and large waves or eroded from the coast, c) from volcanism (Iceland), d) from the continents and large islands (Iceland, Spitzbergen) through the air (dust), e) from the North Sea, and possibly, the Barentz Sea, and f) from the North Atlantic and Arctic Oceans.

The present sea-ice limits in august and february are indicated in fig. 2. Lamb (1963) has shown that in the 18th and 19th centuries these limits were situated more to the SE, whereas during the "little ice-age" in the 17th century they may have extended even further E. A large part of the Norwegian Sea has therefore been temporarily covered by sea-ice during the Holocene and recent ice-berg deposits can be expected in that area. Further S and E the glacial-marine deposits are relict, as can be clearly seen around Faeroe (Berthois and Du Buit 1971). Besides by icerafting terrigenous material has been dispersed by downslope mass flows, atmospheric transport and currents. North of the Scotland-Greenland Ridge, in contrast to the North Atlantic Ocean south of the Ridge, there are no indications for significant deposition or reworking by contour currents (Damuth 1978).

Not enough data are available for the Norwegian Sea to make a balanced sediment budget, as has been done for the North Sea (Eisma 1981), but some supply can be quantitatively estimated. The total amount of volcanic material outside the direct vicinity of Iceland and the Faeroes (where the volcanic admixture in the 50-500 µm fraction is at least 50 % ; fig. 2) is in order of 10 %, indicating an average yearly supply of several million tons. Dust supply, in analogy of dust supply to the Arctic (Mullen et al. 1972) is also in the order of several million tons per year. From the North Sea supply is 2-3 million tons per year at most (Eisma 1981), whereas supply by inflow from the North Atlantic through the Faeroe-Shetland Channel, using the figure of $2 \times 10^6$ $m^3$/sec given by Meincke (this volume) rather than Worthington's (1970) figure of $8 \times 10^6$ $m^3$/sec, is in the order of 3 million tons/yr, taking 0.05 mg/l as the average concentration of inorganic suspended matter in the North Atlantic surface waters (Jacobs and Ewing 1969). This indicates that at least half of the total supply comes from other sources (ice-rafting, erosion of coasts, shelf and Ridge, the Barentz Sea, inflow from the Arctic and through Denmark Strait), taking into account that not all supply is necessarily deposited.

Besides by ice-rafting and downslope mass flow, terrigenous material is dispersed by wind and by currents. Since volcanic ash and glass, primarily injected into the atmosphere, are a widespread admixture on the Scotland-Greenland Ridge as well as in and along the North Atlantic Ocean and the Norwegian Sea, this points to a significant eolian contribution. Subsequently ash and glass have been incorporated in turbidites (Horowitz 1974), in glacial marine sediments (Ericson et al. 1964) and in hemipelagic deposits (Bjørklund et al. 1979).

The present surface current pattern includes northgoing currents through the Faeroe-Shetland Channel (the Norwegian Current) and Denmark Strait, the southgoing East Greenland Current, and a more complicated current pattern north of Iceland (Worthington 1970). Through the Faeroe-Freenland Channel, across Iceland and Faeroe Ridge and through Denmark Strait ther is southward overflow along the bottom into the North Atlantic. Between Iceland and Faeroes this bottomflow is weak and irregular and there is also some northward flow (Tait 1967, Müller et al. 1974). The southward moving bottom water, because of friction and the Coriolis force, tends to follow the submarine relief, flowing around Faroe Rockall Bank and the Reykjanes Ridge as far south as 52°N. These contour currents are strong enough for transporting, reworking and depositing bottom sediment in the North Atlantic, which is reflected in the modelling of the large drift deposits and in the turbidity of the near-bottom water (Jones et al. 1970, Londsdale and Hollister 1979). On the Iceland-Faeroe Ridge bottom velocities reach 20-100 cm/sec (Dietrich 1956, 1967, Joseph 1967, Müller et al. 1974, Meincke, this volume), which is sufficient to erode sandy deposits and to bring fine material in suspension. Winnowing of fine material from the top of the Ridge and dispersal along the flanks results in the grading of the bottom sediments on both sides of the Ridge (fig. 1) in a similar way as along the continental shelf, although the Ridge lies approximately 200 meter lower. On the top and the southern flank of the Iceland-Faeroe Ridge, in the Faeroe-Shetland Channel and in Denmark Strait bottom photographs and echgrams show scour, ripples and lineations that indicate sediment movement by strong currents(Damuth 1978). Erosion is also indicated by gullies and submarine valleys on the SW flank of the Iceland-Faeroe Ridge, that are absent on the relatively smooth bottom of the NE flank. The surface deposits on the SW flank are also somewhat coarser than those on the NE flank (Dietrich 1956, Jarke 1958). Moyes et al. (1974) and Latouche and Parra (1976) assume that smectite (montmorillonite)-rich sediment is transported southward from the ridge by the overflow water along the bottom and is deposited as far south as 52°-54°N. This would explain the generally high montmorillonite concentrations in North Atlantic surface deposits also found by Biscaye (1965). Low montmorillonite contents in the glacial sediments of the same area would be the result of a reduction or absence of the overflow during the glacial period. That this is not the result of a more arctic weathering regime during the glacial period resulting in low smectite concentrations, can be seen near to the Faeroe where the glacial deposits contain high percentages of smectite (Pujol et al. 1974, Latouche and Parra 1976). This indicates that also during the glacial period the Faeroe area was a source of smectite but that during that period the southward transporting mechanism was absent or reduced.

The influence of the Norwegian current on carbonate deposition in the Norwegian Sea and its virtual absence during the glacial period has been studied extensively by Kellogg (1975,

1976, 1977, 1980, Kellogg et al. 1978). Moyes et al. (1974) suggest that a high montmorillonite (smectite) content in Holocene sediments of the southern Norwegian Sea is caused by supply of montmorillonite from the Faeroes by the Norwegian Current. In the glacial sediments the montmorillonite concentrations are low. Latouche and Parra (1976), however, arrive for this area at a southward dispersal of bottom sediment during the Holocene by bottom flow through the Faeroe-Shetland Channel During the glacial period dispersal of material from the Faroe would have been more radial without prominent directions of transport.

## Pleistocene Deposits

During the Weichsel glacial period (roughly 115.000 - 13.000 BP) carbonate deposition was low, the calciumcarbonate content being 10-15 % in most of the glacial deposits in the Norwegian Sea and the North Atlantic, with higher values around 82.000 BP during the early glacial period. Part of the calciumcarbonate in the glacial sediments is an admixture of Cretaceous calcareous fossils and chalk, which have been ice-rafted to the Scotland-Greenland Ridge area and the Norwegian Sea probably from Denmark and/or England (Ericson et al. 1964).

During the glacial, ice-rafting was the most important process supplying coarse as well as fine terrigenous material and transporting it over large distances, Cretaceous shale fragments from the Spitzbergen area being present in glacial marine deposits in the southern Norwegian Sea (Ericson et al. 1964). In individual cores the content of non-carbonate coarse material ( > 62 μm) is highly variable, owing to the variable nature of ice-rafting, but these irregularities are smoothed out by the fine non-carbonate fraction, which is probably redistributed by currents (Kellogg 1975). Calculations by Ruddiman (1977) indicate that on a yearly basis roughly the same amounts of glacial marine sediments were deposited in the Norwegian Sea through all substages of the glacial period as well as during the latest (Eemian) interglacial (125.000 - 115.000 BP) with only a small (± 10 %) increase around 80.000 BP. This is surprising because just south of Denmark Strait and in the North Atlantic, where terrigenous deposition during the Eemian and the early glacial period (in mg/cm$^2$/1000 yr) was of approximately the same magnitude as in the Norwegian Sea, there was a three- to fivefold increase during the second half of the glacial period (57.000-13.000 BP), in the North Atlantic around 45$^o$-50$^o$N and near Denmark Strait starting already somewhat earlier (around 75.000 BP). This corresponds to a drastic change in ice conditions in the North Atlantic and to a general increase in the size of the northern hemisphere glaciers (Ruddiman 1977). Since in the Norwegian Sea, the deposition rate remained very much the same, with a small change to more glacial-marine deposition around 80.000 BP, the state of the Norwegian Sea during the glacial period may have been much the same as today. An ice cover like in the present Arctic, would not maintain a deposition rate which is ca. 6 times larger than the deposition rate below the Arctic pack ice (Bjørklund and Goll, 1979).

Not all glacial-marine deposits in the Norwegian Sea are the result of ice-rafting. Holtedahl and Sellevoll (1972) found on the Norwegian shelf and slope glacial deposits of presumably Weichselian age associated with morainic ridges down to ca. 500 m waterdepth. Here deposition from floating ice, as indicated by pebbles from the Oslo-region, has been comparatively small, most of the material, containing crystalline rocks from the mainland, having been deposited by inland ice with a maximum thickness of the deposit of probably several hundred meters. Downslope sliding of this morainic material, of which the scars have been found along the shelf edge (Løvlie and Holtedahl 1980, Rokoengen 1980), has contributed to the glacial marine deposits on the basin floor. Similar conditions may have existed off Iceland and Greenland and the small increase in glacial marine deposition around 80.000 BP may have been caused by the growing glaciers reaching the outer continental shelf and the upper slope by that time. Damuth (1978) from echosounding records and pistoncores concluded that throughout the Norwegian Sea gravity-controlled mass-flows (turbidite deposition, slumping, sliding) and ice-rafing were most important, downslope processes having resulted in a considerable redistribution of bottom sediment along the slopes and the adjacent basin floor.

In the Faeroe-Shetland Channel area the boundary between the Faeroe-mineralogical province and the Scotland-Shetland province shifted during the glacial period towards the Faeroes. Illite-rich material was deposited on the Wyville-Thomson ridge, while further south as well as in the southern Norwegian Sea the montmorillonite (smectite) percentage decreased so that the extension of the Faeroe province became smaller. During the latest (Eemian) interglacial, however, the Faeroe province had again a similar extension as in the surface deposits (Moyes et al. 1974, Latouche and Parra 1976). Also with regard to the calcium-carbonate content the Eemian sediments are very similar to the Holocene deposits.

The late-glacial sediments on the southern Norwegian shelf and slope show the change from Weichselian glacial to Holocene warm conditions (Holtedahl 1959, Holtedahl et al. 1972, 1974, Mangerud et al. 1979, Bjørklund et al. 1979, summary in Bjørklund and Goll 1979, and Rokoengen 1980). The retreat of the inland ice started on the shelf shortly after 13.500 BP but the ice readvanced two, probably three times untill the final retreat from the shelf around 12.200 BP. Laminated sands and sands containing pebbles were deposited on top of the glacial deposits,

Fig. 3. Position of cores used for clay mineral analysis.

probably by sediment-laden water flowing downslope from the icefront. Then there was an abrupt change to deposition of hemipelagic clays with increasing amounts of microfossils (Foraminifera, Coccoliths, Radiolaria, Diatoms). This change occured around 11.000-12.000 BP and agrees well with the inferred northward shift along the Norwegian shelf of the position of the polar front between 13.000 and 11.000 BP (Ruddiman and McIntyre 1973).

## Clay Minerals and Coccolith content

Since the Iceland-Faeroe Ridge is a source of fine-grained material characterized by high concentrations of smectite (montmorillonite), there may have been a detectable northward dispersal,as well as southward,during the Holocene (by the Norwegian Current) and the latest interglacial. The data of Moyes et al. (1974) and Latouche and Parra (1976) are inconclusive. Clay mineral analysis of the fraction < 2 μm at 25 cm intervals in core V 27-86, situated below the axis of the Norwegian Current and with a maximum warm-water influence during the latest interglacial (Kellogg 1980), indicated a strong relation between the concentration of swelling minerals and calciumcarbonate content. This led us to analysis of a number of cores collected around the Iceland-Faeroe ridge and in the Norwegian Sea (fig. 3) and of surface samples from the same area (fig. 4). Since the results did not fit the model of Kellogg (1980) with two short periods of warm-water influx (during the Holocene and the latest interglacial), the Coccolith content was determined in the same samples, providing an indication of surface water conditions. The core samples were made available by the Lamont-Doherty Geological Observatory. Most of these cores had previously been studied by Kellogg (1975, 1978, 1980 : Foraminifera, $CaCO_3$ and oxygen isotopes). In the cores that had not been studied before total calciumcarbonate content was also determined.

Clay mineral composition was determined at 50 % relative humidity with a Philips diffractometer and Co Kα radiation in the size fraction < 2 μ (separated off by centrifuge and Ca-exchanged). The percentage composition was calculated from the peakheights of swelling minerals (± 14,6 Å), mica/illite (9.95 Å) and kaolinite (7.04 Å). Chlorite was determined in about half of the samples that were analysed later at 0 % relative humidity. This was done to separate the swelling illite reflections from those of smectite ; the chlorite 001 peak then appears separately at 14.20 Å. The chlorite content was found to be 10 % at most, showing small variations that did not influence the systematic differences found in the peakheights of swelling minerals and kaolinite, which at 50 % relative humidity include chlorite. The chlorite content was found to be much lower than estimated by Berthois et al. (1973) and Latouche

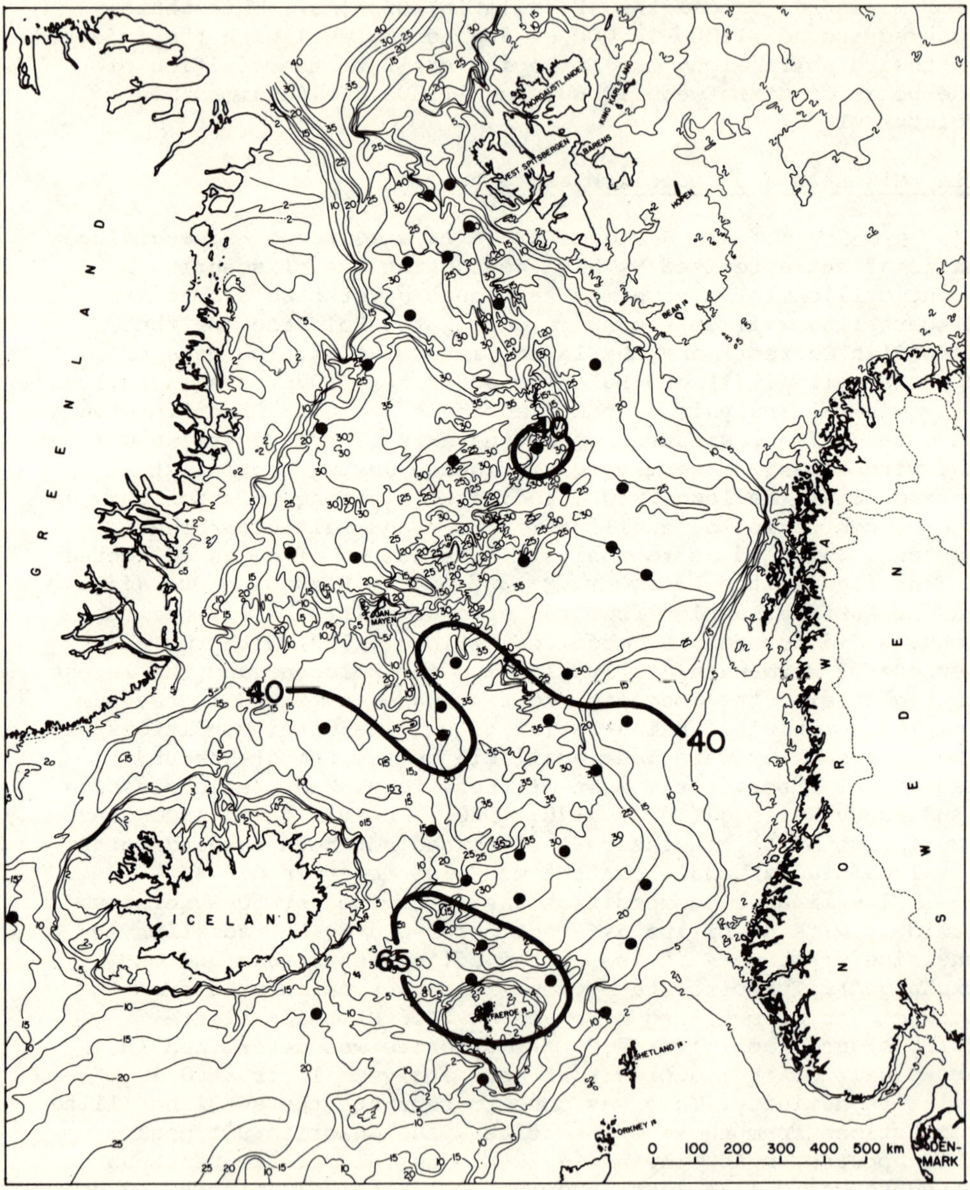


Fig. 4. Percent of swelling minerals (smectite + swelling illite) in surface sediments.

(1975), and the kaolinite content correspondingly higher.

Coccolith content was determined by suspending a small amount of sediment in distilled water, desintegrating it by a weak ultrasonic treatment. After homogenization, a portion was filtered through a 20 µm nylon sieve and then over a 0.4 µm Nuclepore filter that was subsequently dried at 60°C. A piece cut from the middle of the filter was mounted on an aluminium sample holder, coated with gold and the Coccolith content was determined with a Cambridge 150 Scanning Electron Microscope by counting the number of Coccoliths in 100-200 fields of vision. For the first series of samples the amount of sediment on the filter was approximately the same, which involved exact weighing and quantitative dilution. In later series this was not continued because of the time involved but a thin film of sediment < 20 µm was brought on the Nuclepore filter and a correction was made for differences in particle density on the filter. In that way comparable, semiquantitative data were obtained that were sufficient for the comparisons made below. *Gephyrocapsa caribbeanica* and *Coccolithus pelagicus* were predominantly present (McIntyre et al. 1970, 1972), with some occasional specimens of *C. leptoporus* and *E. huxleyi*. Calciumcarbonate content was determined by dissolving a small amount of sediment in 12 % HCL and determining Ca by titration with EGTA in a Corning Calcium Analyser.

In the surface deposits the highest contents of swelling minerals, up to 80 %, occur along the north flank of the Iceland-Faeroe Ridge, decreasing norhward to less than 40 % (minimum found 29 % ; fig. 4). In the two cores collected south of the Iceland-Faeroe Ridge (27-102 and V 27-98; Fig. 5) the content of swelling minerals is generally high but variable. The low carbonate content of 10-15 % and the virtual absence of Coccoliths indicate that the sediments are largely glacial marine. They contain volcanic glass at all levels and V 27-102 has ash layers and scoria in the sandy layers. The layering, the only rare occurrence of ice-rafted pebbles, and an occasional ca 1 cm thin layer with planktonic Foraminifera in the upper 10-15 cm indicates that the top of the cores contain largely reworked older sediments, the reworking being probably Holocene. Cores 27-42, V 28-33 and V 23-76, NE of the Faeroes, are very similar to V 27-102 and V 27-98, but the content of swelling minerals gradually decreases from the top downward (fig. 6). In core V 23-75, N of the Faeroes, there is a toplayer of ca. 30 cm with high contents of calciumcarbonate and swelling minerals and a sharp transition to a layer with low contents of both (fig. 6). In all these cores the ratio between the content of mica/illite and kaolinite (I/K ratio) remains approximately the same throughout the core whereas the content of swelling minerals varies independently.

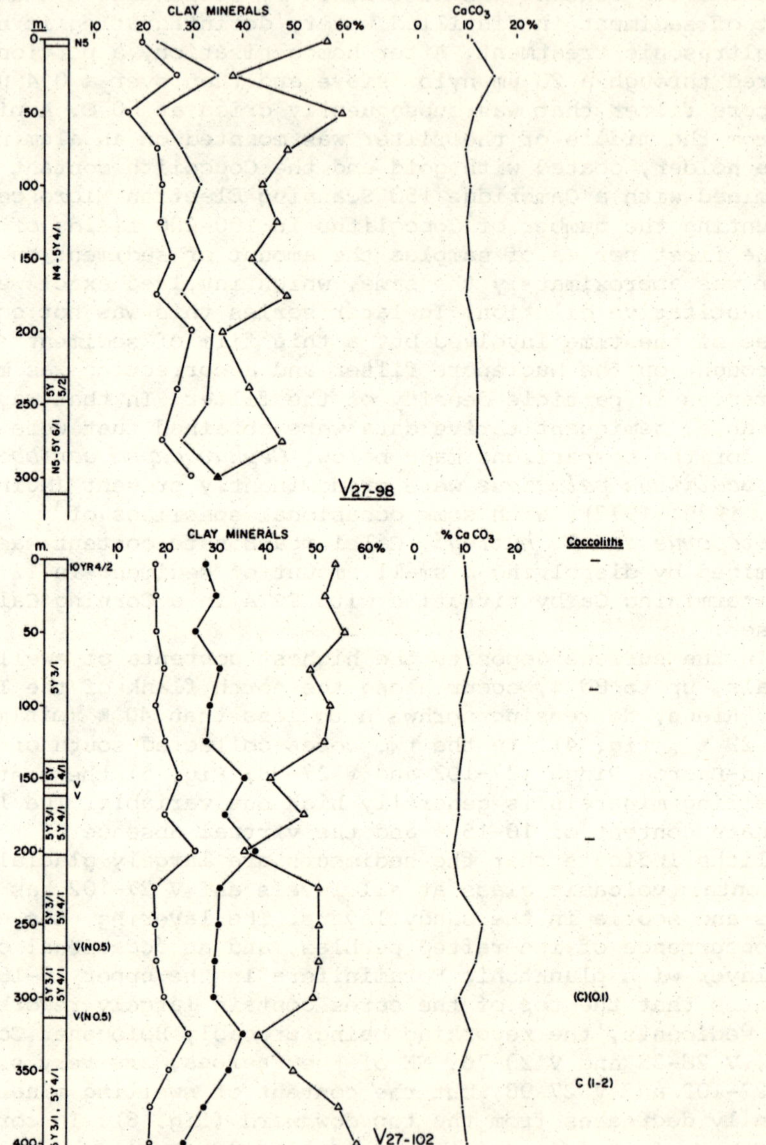


Fig. 5. Profiles of two cores located SW of the Iceland-Faeroe Ridge. V = volcanic glass or ash; C,(C),-, and clay minerals indicated as in Fig. 7.

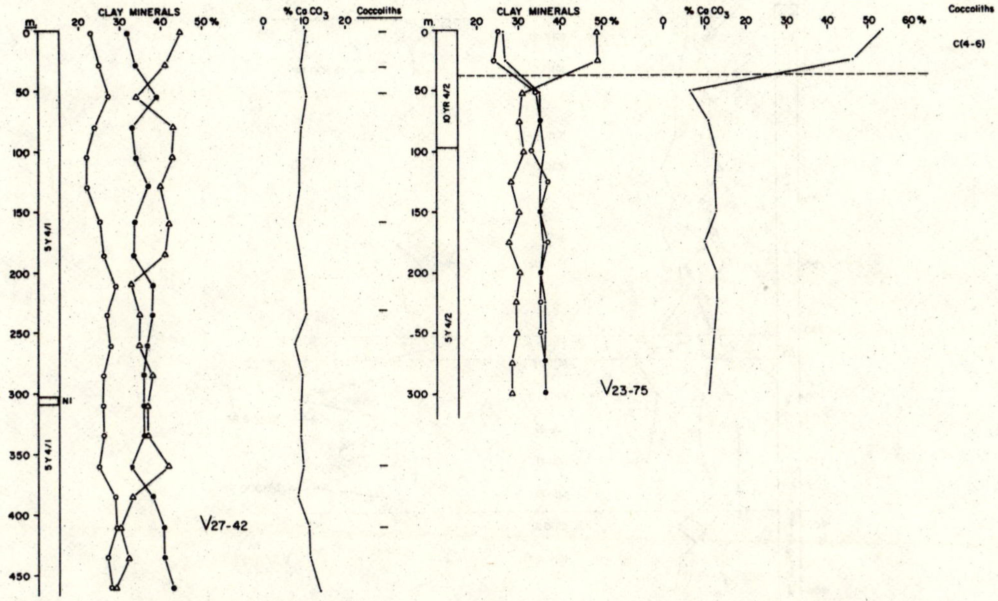

Fig. 6. Profiles of cores located E and NE of the Iceland-Faeroe Ridge. Clay minerals indicated as in Fig. 7. Profiles of V 28-33 and V 23-76 are very similar to the profile of V 27-42.

The I/K ratio varies between the cores from 0.84-1.23 in V 23-76 to 1.48-1.59 in V 27-102, indicating that although in each core mica/illite and kaolinite come from the same source (and the swelling minerals largely from another) the source areas have different I/K ratio's.

N of the Faeroes and V 23-75 there are four cores that follow the main direction of the Norwegian Current and penetrate the glacial and as well as the latest interglacial deposits (fig. 7,8). V 27-86 and V 27-85 from the area with > 40 % swelling minerals in the surface deposits, show a similar general pattern : high concentrations of swelling minerals (above 40 %) where the carbonate content is high, and low where the carbonate is low (fig. 7). An exception is a dark grey layer (colour N4), present in both cores, with high concentrations of kaolinite as well as swelling minerals and low concentrations of mica/illite and carbonate. Coccolith concentrations are high where carbonate contents are high, but also in the glacial deposits, where carbonate contents are low, some Coccoliths can be found, although in very low concentrations. In the glacial sediments, and

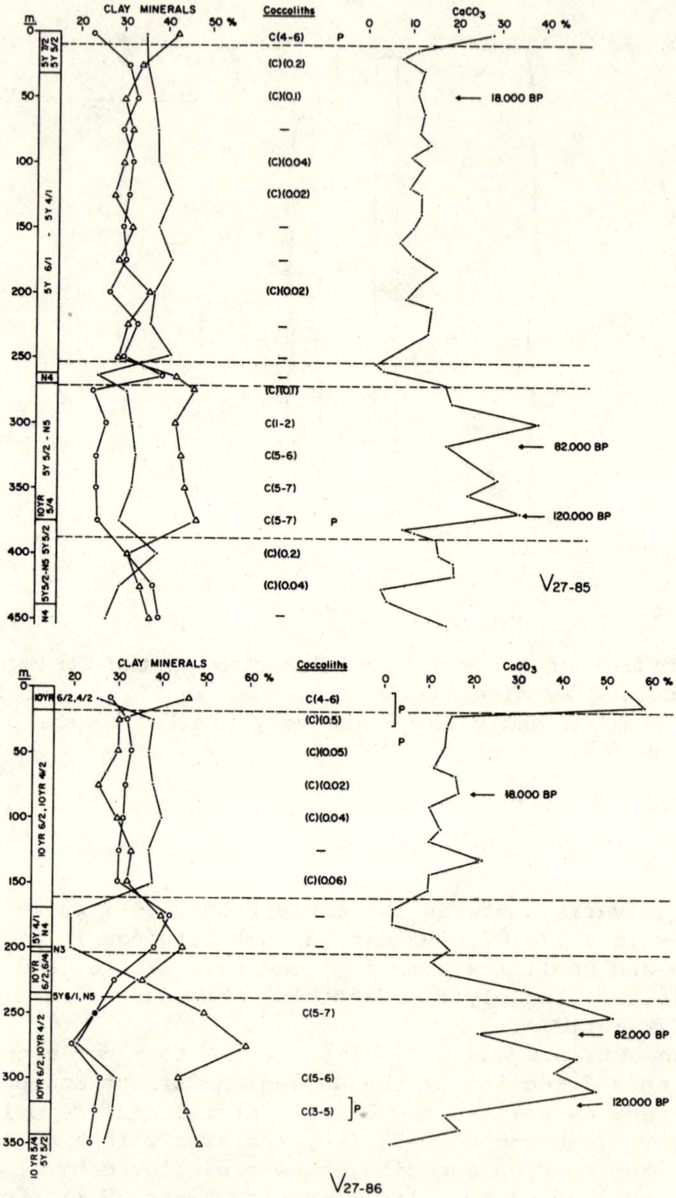


Fig. 7. Profiles of cores V 27-85 and V 27-86 from the Norwegian Sea. Data on calciumcarbonate content and dating from Kellogg (1980). C=Coccoliths present in high numbers, (C) = only few Coccoliths present, - = no Coccoliths found. Numbers in brackets indicate average range of number of Coccoliths per field of vision. P = high percentage of dextral G.pachyderma (after Kellogg 1980). Δ - Δ Swelling minerals, .-. mica/illite, o - o kaolinite.

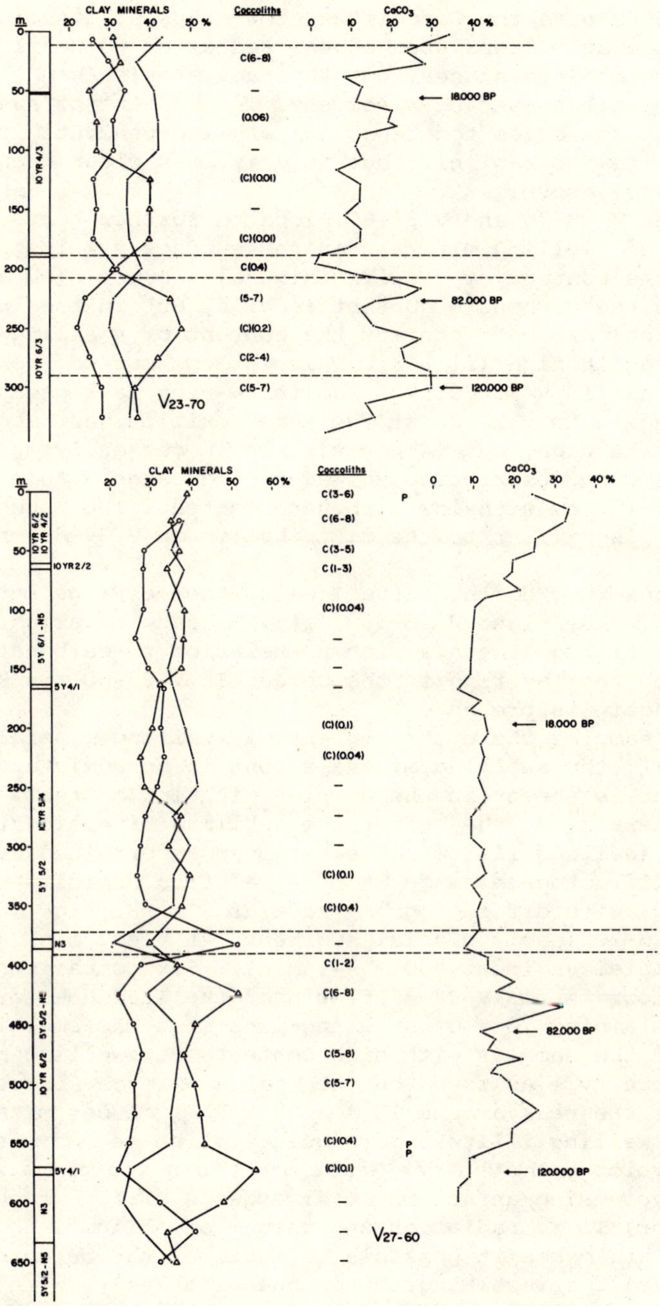


Fig. 8. Profiles of cores V 23-70 and V 27-60 from the Norwegian Sea. C,(c),-,P and clayminerals as in fig.7.

especially at the transitions from carbonate-highs to carbonate-lows and the reverse, the Coccoliths show signs of dissolution such as irregularly crenulated edges, reduction to nearly a ring and intermediate stages. But the same samples usually also contain some well preserved specimens. The content of swelling minerals in these cores too tends to vary independently from the mica/illite and kaolinite but this is less clear than in the cores discussed above.

In cores V 23-70 and V 27-60, located further north where the content of swelling minerals in the surface samples is less than 40 %, the content of swelling minerals is not high at the top although the carbonate content is high, but in the lower zone with high carbonate content the content of swelling minerals is again high (fig. 8). This is less marked, however, in V 27-60 than in V 23-70. The kaolinite-rich dark greay layer is present again in V 27-60 at the same position just above the lower carbonate high, but is not visible in our analyses of V 23-70. The core description, however, indicates a thin sandy layer at 189-191 cm with low carbonate content. The Coccolith distribution is similar to the distribution in V 27-86 and V 27-85.

Two cores from north of the Iceland-Faeroe Ridge but more to the west (V 23-58 and V 28-56 ; fig. 9) show a variable content of swelling minerals with no relation to carbonate content or to the layering in the cores. In V 23-58 the kaolinite-rich layer again is present.

In the samples where the content of swelling minerals is high (> 40 %), the swelling minerals contain an admixture of smectite that is absent in the samples with low contents of swelling minerals. In the latter the swelling minerals are almost exclusively swelling illite. This has been ascertained by comparing diffractograms made at 50 % relative humidity (50 % RH ; fig. 10) with diffractograms made at 0 % RH. Fig. 10 shows this for V 23-70 (early glacial sediment, 45 % sw. min., and for V 27-86 (glacial sediment, 30 % sw. min.). For comparison the the diffractograms for V 27-39 from near the Iceland-Faeroe Ridge. are given containing 70 % of swelling minerals. The smectite admixture in the samples with high contents of swelling minerals is of the same type as found on the Iceland-Faeroe Ridge.

Because the peak around 10 Å at 0 % RH includes mica/illite as well as swelling illite, a more quantitative method was applied recently developed by Van der Gaast and Vaars (in preparation), which involves using an automatic divergence slit, a graphite monochromator, Ti Kα radiation and vacuum or heliumflushed apparatus. This makes it possible to subtract the background and to compare the resulting curves quantitatively. With this method the mica/illite peak was subtracted. Fig. 11 represents the resulting curves for the swelling minerals when background-corrected x-ray diagrams for 50 % RH are subtracted from those obtained at 0 % RH. V 27-86/75 cm and V 27-70/50 cm, both

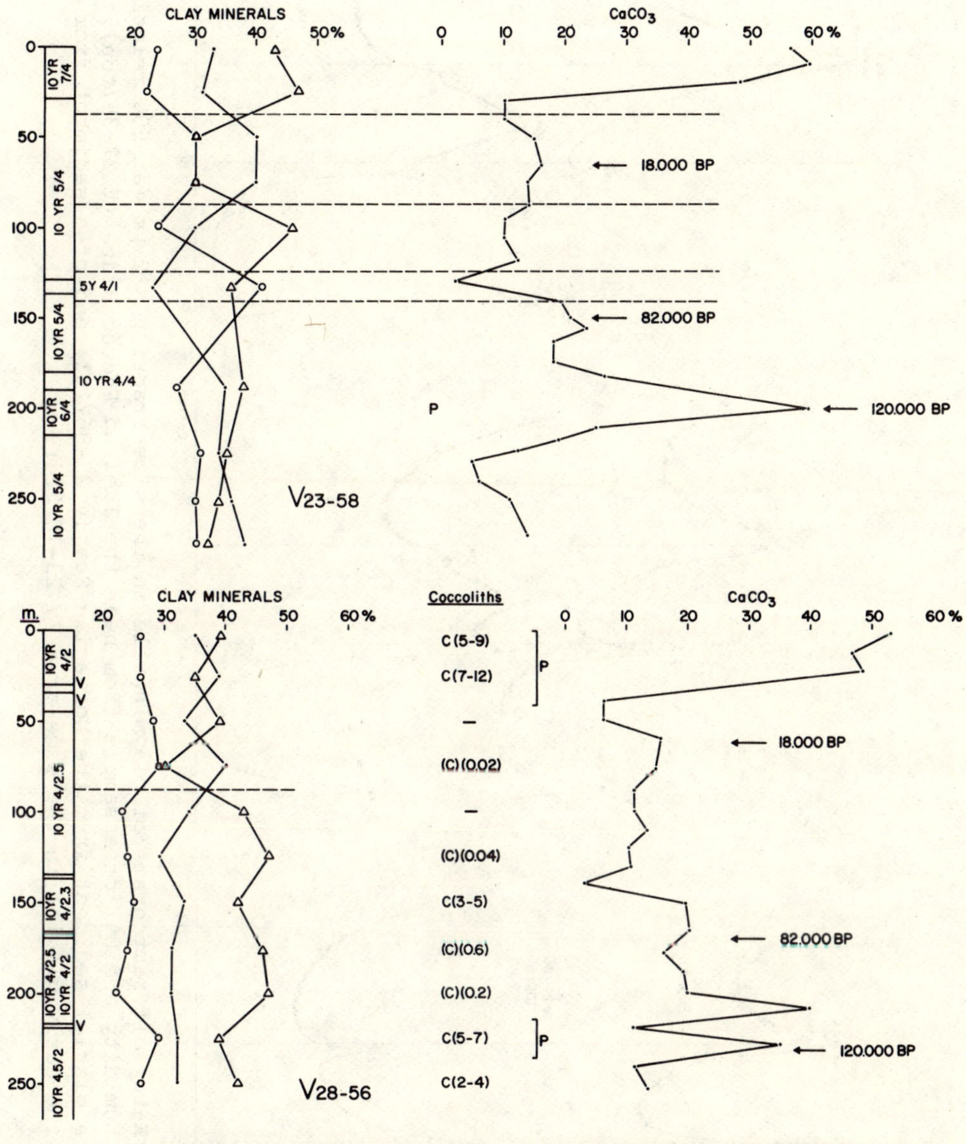


Fig. 9. Profiles of two cores located north of the Iceland-Faeroe Ridge. V = volcanic glass or ash; data on calciumcarbonate content and dating from Kellogg (1980); C, (c), –, P and clay minerals indicated as in Fig. 7.

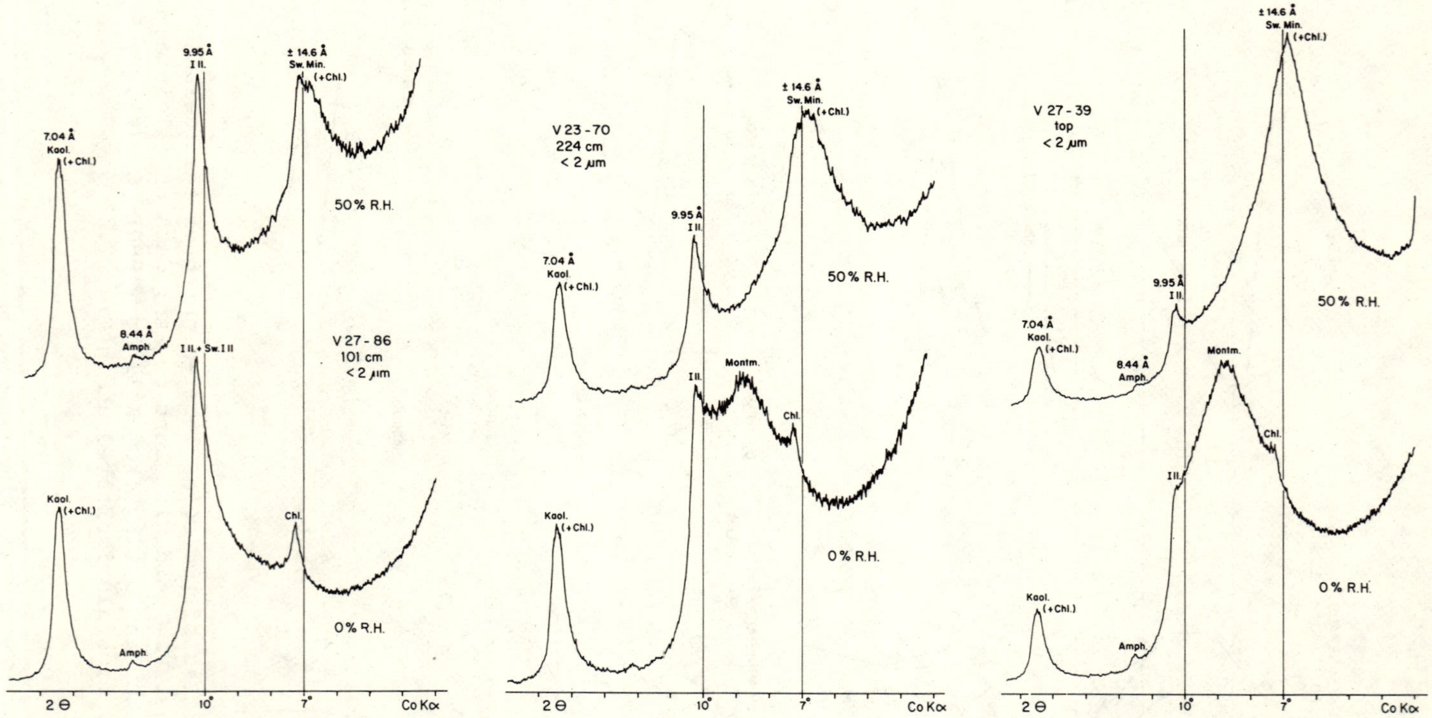


Fig. 10. X-ray diffractograms of clayminerals in the size fraction < 2 μm (at 50% relative humidity and 0%) of a sample from near the Iceland-Faeroe Ridge (V 27-39/top), of a glacial sediment from the Norwegian Sea (V 27-86/101) and of a sediment from the early glacial in the same area (V 23-70/224).

glacial sediments, show high amounts of swelling illites and no or only a very low amount of smectite, whereas early glacial sediments from V 27-86/249 cm and V 23-70/250 cm show swelling illite with a strong admixture of smectite.

Assuming the Iceland-Faeroe Ridge to be the source of the admixture of smectite, the following picture emerges. During the warm periods (Holocene and, in analogy, the latest interglacial) there is supply of smectite from the ridge to the north by a northgoing current, that during the latest interglacial was somewhat more vigorous and possibly more directed towards NE than during the Holocene. Kellogg (1980) on the basis of data from Foraminifera came to the same conclusion viz. the latest interglacial. This northward transport persisted the early glacial (approximately isotopic stages 5d-5a), but during the glacial period up to the Holocene the northward transport was absent or strongly reduced, Around the Iceland-Faroe Ridge the dispersal of smectite is not only dependent on the presence of absence of northward flow but also on more regional flow conditions and, most probably, downslope massflow. Cores V 23-58 and V 28-56, situated more to the west and with an irregular admixture of smectite, contain high concentrations of volcanic material (ash, glass), which occurs further east in much lower concentrations. The irregular supply of smectite may have come from the same volcanic source or sources. Further east, and aspecially in the surface samples, the smectite distribution indicates that the Iceland-Faeroe Ridge is the only surce of more than local importance.

The origin of the sediment in the kaolinite-rich layer is not known. Holtedahl et al. (1974) found clays of similar composition on the Norwegian shelf at 64°06' N in glacial till of presumably Late Weichsel age, together with rock fragments from locally eroded Middle Jurassic to Early Tertiary bed rock containing the same clay minerals and fossils as found in the matrix of the till. This till lies near to deep scars where slumping has occurred: the central Norwegian shelf therefore seems a most likely source area for the kaolinite-rich sediment in the Norwegian Sea basins.

## Consequences for Currents and Ice-Conditions during the Late Quaternary

The above picture implies that during the latest interglacial as well as during the early glacial period a current pattern existed, with northward surface flow as well as southward overflow, very much in the same way as it exists now. This contradicts with Kellogg's (1980) conclusion that during the early glacial period there was no strong Norwegian current, only some weak gyres. It agrees better with the conclusion of (Ruddiman et al. (1980) that at the end of the early glacial,

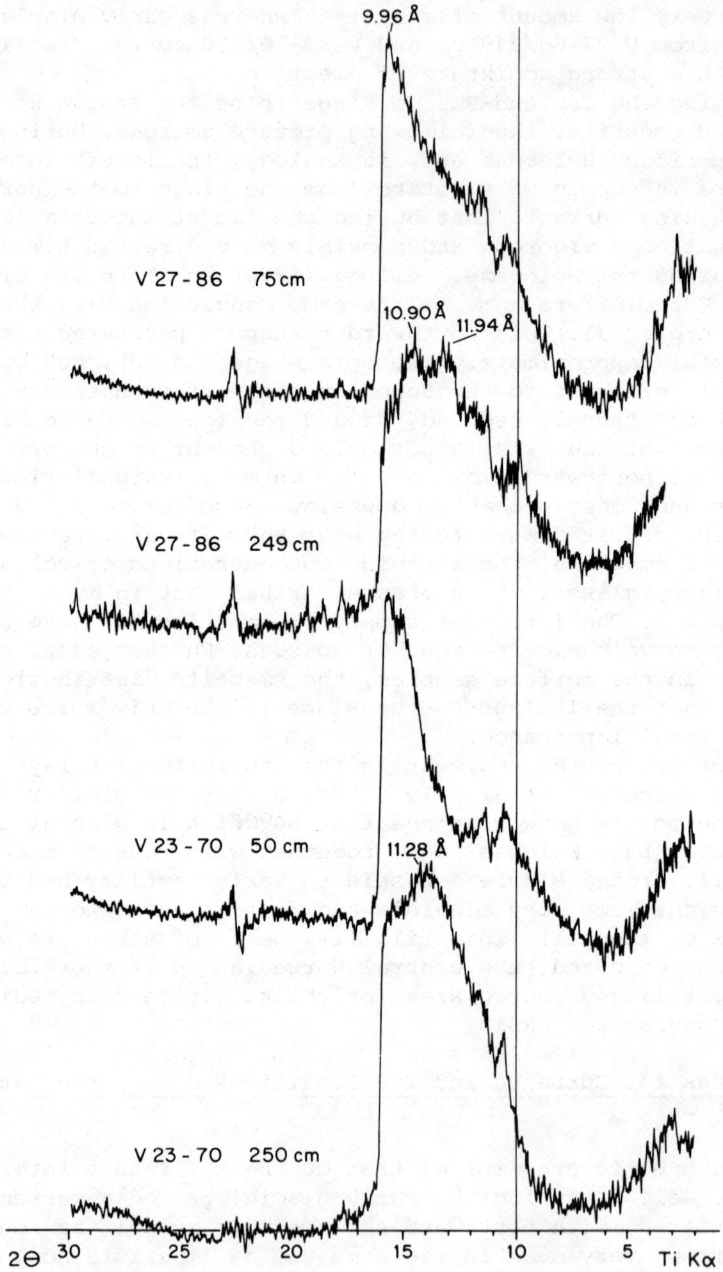


Fig. 11. Subtracted background-corrected X-ray curves (at 50% RH and 0% RH) for the swelling minerals in two glacial and two early glacial samples.

during stage 5a around 75.000 BP, at least the southeastern Norwegian Sea was seasonally open water with warm water conditions just to the south at 60°N. Streeter and Shackleton (1979) found indications in the North Atlantic that deep water formation and overflow from the Norwegian Sea occurred at that time. The clay mineral distributions support the latter conclusion, whereas the Coccolith content indicates that at least as far north as the area around core V 27-60 off northern Norway, open water conditions existed up to the end of the early glacial period. The fact that Coccoliths are abundantly present where only very few Foraminifera are found, points to low salinity conditions in the surface water, the Coccolithophorids having a much broader salinity tolerance than Foraminifera (Berggren pers. comm.). More in detail, however, Coccolith concentration (or carbonate content) and clay mineral concentrations do not entirely match, which is not surprising. Whether clay mineral dispersal is visible in cores depends on the impact of the transporting agent as well as on the degree of dilution by clay minerals from other sources, whereas growth of Coccolithophorids also depends on the availability of nutrients, temperature, salinity, light conditions, the presence of open water. The data presented here are inconclusive as to whether a reduction or absence of the overflow may have occurred within the early glacial period (in approximately stages 5c and 5d) as suggested by the data of Streeter and Shackleton (1979).

To explain the presence of small amounts of Coccoliths in the glacial sediments, (from the same species as in the interglacial and early glacial sediments) there are several possibilities :

- influx from the south by slowly northward flowing water without significant transport of smectite,
- reworking of interglacial or early glacial sediments,
- occasional growth of Coccolithophorides in the Norwegian Sea during the main glacial period,
- regular growth of Coccolithophorids in the Norwegian Sea during the main glacial period but reduction of the Coccolith content in the bottom sediments through dissolution.

Because of the nearly constant influx of terrigenous material during the interglacial and glacial periods dilution cannot explain the low Coccolith concentrations. Continuing reworking of interglacial and early glacial deposits during the glacial period seems unlikely because up to now these have only been found buried below glacial marine deposits. Depending on the importance that is attached to carbonate dissolution, two situations seem likely :

a) an open Norwegian Sea, approximately like the present one, but with reduced or absent northward flow and regular growth of Coccolithophorids, the Coccoliths being largely dissolved

b) a Norwegian Sea largely covered by ice-bergs, preventing growth of Coccolithophorids, and also with reduced or absent northward flow, but being occasionally more open, allowing growth of Coccolithophorids for very short periods.

## Conclusions

Bottom deposits on the Scotland-Greenland Ridge are sandy, containing gravel and grade into sandy muds and muds with greater waterdepth. Most of the Norwegian Sea has been temporarily covered by floating ice during the Holocene, except in the S and E where the glacial-marine deposits are relict, dating from the glacial period. Around Iceland and the Faeroes there is a strong admixture of volcanic ash and glass (> 50 %). In the deeper parts N and S of the Scotland-Greenland Ridge a carbonate ooze is being formed (> 30 % $CaCO_3$). The minimum Holocene input rate of terrigenous material (540-2130 mg/cm$^2$/1000 yrs) is of the same magnitude as the input rate during the latest interglacial and the glacial period. The average yearly input rate of terrigenous sediment in the Norwegian Sea is estimated as 16-20 million ton. Less than half of this comes from volcanism, from the adjacent continents and from Iceland in the form of eolian dust, from the North Sea, and from the North Atlantic through the Faroe-Shetland Channel. The major part comes from ice rafting, erosion of coasts, continental shelves and the Scotland-Greenland Ridge, with probably some contribution from the Barentz Sea, from the Arctic Ocean and from the North Atlantic through Denmark Strait. Pleistocene deposits, as far as known, are largely glacial marine but locally redistributed by downslope mass flow and by currents.

Two mineral provinces can be distinguished in the detrital material : a basaltic province on the Iceland-Faeroe Ridge with high concentrations of smectite, plagioclase and augite, and an acidic province along the Scotland-Shetland shelf with high concentrations of illite, swelling illite and quartz. The distribution of these mineral provinces, especially the distribution of smectite, and the concentration of Coccoliths in the Late Quaternary sediments, reflect the changing current patterns during that period. During the latest interglacial there was approximately the present pattern, with northward flow of warmer water, probably more vigorously than now, along the eastern half of the Norwegian Sea, and southward overflow along the bottom through the Faeroe-Shetland Channel and across the Iceland-Faeroe Ridge. There are strong indications that after the latest interglacial this system persisted with the Norwegian Sea remaining open, perhaps temporarily reduced or interrupted, until the end of the early glacial period (± 75.000 BP). During the main glacial period the northward flowing current was absent or very weak with either an open Norwegian Sea, approximately like the present one, with regular growth of Coccolithophorides and subsequent dissolution of most of the Coccoliths, or a largely ice-covered Norwegian Sea but occasionally more open, allowing growth of Coccolithophorida for very short periods.Ice rafting must have remained important throughout the glacial period to allow for the high sedimentation rates as compared to the present Arctic which is covered by pack ice. During the Holocene the northward flow again is attested in the bottom sediments of the Norwegian Sea by an increased deposition of smectite north of the Scotland-Greenland Ridge.

Acknowledgements

We are very grateful to the Lamont-Doherty Geological Observatory and in particular to Dr. Floyd W. McCoy for making core samples available and to Dr. W.F. Ruddiman and Dr. A. McIntyre for a fruitful discussion. To Dr. R.R. Schmidt of Utrecht University we are much indebted for his assistance in the determination of Coccoliths. Mrs. R. Gieles-Witte, Miss J. van Ieperen and Mr. J. Kalf did most of the preparatory work for the X-ray and SEM-analysis, Mr. F. Eijgenraam made the computer-program used in clay mineral analysis, Mr. H. Hobbelink, Mr. R.P.D. Aggenbach and Mr. B. Verschuur made the drawings, Mrs. J. Hart did the typing, and Mr. J. Pieters (Electronmicroscope Laboratory Utrecht University) assisted with the SEM, which we gratefully acknowkledge.

REFERENCES

Berthois, L., and Roa, P., 1967, Interprétation des analyses de mineraux lourds par pourcentage pondérés dans la région des Faroër-Ecosse-Hébrides, C.R. Acad. Sc. Paris, 265: Série D, 195-197.

Berthois, L., and Du Buit, M.H., 1971, Contribution à l'étude de la sédimentation en Atlantique nord, Bull. Inst. Géol. Bassin Aquitaine, 11: 365-374.

Berthois, L., Latouche, Cl., and Parra, M., 1973, Etude minéralogique et géochimique de quelques sédiments de la zone comprise entre les Archipels Faeroe, Shetland, Orcades, et Hébrides. Contribution à la connaissance de l'origine des sédiments dans le domaine profond de l'Atlantique Nord-oriental. Bull. Inst. Géol. Bassin Aquitaine, 14: 3-17.

Biscaye, P.E., 1965, Mineralogy and sedimentation of Recent deepsea clay in the Atlantic Ocean and adjacent seas and oceans, Geol. Soc. Amer. Bull., 76: 803-832.

Bjørklund, K.R., and Goll, R.M., 1979, Ice Age climates of the Norwegian-Greenland Sea, GeoJournal 3.3: 273-286.

Bjørklund, K.R., Thiede, J., and Holtedahl, H., 1979, Deglaciation of the southeastern Norwegian Sea towards the end of the last glacial age, Boreas, 8: 133-135.

Boeggild, O.B., 1900, The deposits of the sea-bottom, The Danish Ingolf-Expedition, 1(3), 89 p. Copenhagen.

Damuth, J.E., 1978, Echo character of the Norwegian-Greenland Sea: relationship to Quaternary sedimentation, Mar. Geol., 28: 1-36.

Dietrich, G., 1956, Überströmung des Island-Färöer-Rückens in Bodennähe nach Beobachtungen mit dem Forschungsschiff "Anton Dohrn" 1955/56, D. Hydr. Zeitschr. 9: 78-89.

Dietrich, G., 1967, The International "Overflow" Expedition (ICES) of the Iceland-Faeroe Ridge, May-June 1960, a review, Rapp. Pr.-verb. Réun. Cons. Perm. Int. l'Explor. Mer, 157: 268-274.

Eisma, D., 1981, Supply and deposition of suspended matter in the North Sea, in: Spec. Publ. int. Assoc. Sed. 5, ed. S.D. Nio, R.T.E. Schüttenhelm and T.C.E. van Weering, in press.

Eldholm, O., and Windisch, Ch.C., 1974, Sediment distribution in the Norwegian-Greenland Sea, Geol. Soc. Amer. Bull., 85: 1661-1676.

Ericson, D.B., Ewing, M., and Wollin, G., 1964, Sediment cores from the Arctic and Subarctic Seas, Science 144 (3623) : 1183-1192.

Fleischer, U., Holzkamm, F., Vollbrecht, K., and Voppel, D., 1974, Die Struktur des Island-Färöer-Rückens aus geophysikalischen Messungen, D. Hydr. Zeitschr., 27: 97-113.

Gorshkova, T.I., 1960, Sediments of the Norwegian Sea, Proc. 21 Int. Geol. Congr., 10 (Submarine Geology): 16-22.

Holtedahl, H., 1959, Geology and paleontology of Norwegian Sea bottom cores, Jour. Sed. Petrol. 29: 16-29.

Holtedahl, H., and Sellevoll, M., 1972, Notes on the influence of glaciation on the Norwegian continental shelf bordering on Norwegian Sea, Ambio spec. vol., 31-38.

Holtedahl, H., Haldorsen, S. and Vigran, J.O., 1974, Two sediment cores from the Norwegian continental shelf between Haltenbanken and Frøyabanken (64°06'N, 7°39'E), Norges geol. unders. 304: 1-20.

Horowitz, A., 1974, The geochemistry of sediments from the northern Reykjanes Ridge and the Iceland-Faroes Ridge, Mar. Geol., 17: 103-122.

Jacobs, M.B., and Ewing, M., 1969, Suspended particulate matter : concentration in the major oceans. Science, 163, 380-383.

Jarke, J., 1959, Sedimente und Mikrofaunen im Bereich der Grenzschwelle zweier ozeanische Räume, dargestellt an einem Schnitt über den Island-Faröer-Rücken (Nordatlantische Ozean - Rosengarten - Europäisches Nordmeer), Geol. Rundschau 47: 234-249.

Jones, E.J.W., Ewing, M., Ewing, J.I., and Eittreim, S.L., 1970, Influences of Norwegian Sea overflow water on sedimentation in the northern North Atlantic and Labrador Sea, Jour. Geoph. Res., 75: 1655-1680.

Joseph, J ., 1967, Current measurements during the International Iceland-Faroe Ridge Expedition, 30 May to 18 June, Rapp. Pr.-verb. Réun. Cons. Perm. Int. l'Explor. Mer, 157: 157-172.

Kellogg, Th.B., 1975, Late Quaternary climatic changes in the Norwegian and Greenland Seas, in: Climate of the Arctic, S.A. Bowling and G. Weller ed., Univ. Alaska Press, Fairbanks, 3-26.

Kellogg, Th.B., 1976, Late Quaternary climatic changes: evidence from deep-sea cores of Norwegian and Greenland Seas, Geol. Soc. Amer. Mem., 145: 77-110.

Kellogg, Th.B., 1977, Palaeoclimatology and paleo-oceanography of the Norwegian and Greenland Seas: the last 450.000 years, Mar. Micropal., 2: 235-249.

Kellogg, Th.B., 1980, Paleoclimatology and paleo-oceanography of the Norwegian and Greenland seas: glacial-interglacial contrasts, Boreas, 9: 115-137.

Kellogg, Th.N., Duplessy, J.C., and Shackleton, N.J., 1978, Planktonic foraminiferal and oxygen isotopic stratigraphy and paleoclimatology of Norwegian Sea deep-sea cores, Boreas, 7: 61-73.

Lamb, H.H., 1963, Mapping methods applied to the study of climatic variations and vicissitudes, in: The Changing Climate, selected papers by H.H. Lamb, ch. 4, Methuen & Co, London.

Latouche, C., 1975, Les mineraux argileux des sédiments actuels de l'Atlantique nord-oriental et du S de la Mer de Norvège, Proc. Int. Clay Conf., 45-54.

Latouche, C., and Parra, M., 1976, Mineralogie et geochimie des sédiments quaternaires de l'océan Atlantique nord-oriental (Mer de Norvège-Golfe de Gascogne) - essais d'interprétations sédimentologiques, Mar. Geol. 22: 33-69.

Lisitzin, A.P., 1969, ed., Bottom sediment maps of the Atlantic Ocean (in Russian), State Geodetic and Cartographic Service of the USSR. Moskou.

Løvlie, R., and Holtedahl, H., 1980, Apparent paleomagnetic low-inclunation excursion in a pre-consolidated continental shelf sediment, Physics Earth Planet. Int., 22: 137-143.

Lonsdale, P., and Hollister, Ch.D., 1979, A near-bottom traverse of Rockall Trough: hydrographic and geologic interferences, Oceanologica Acta, 2: 91-105.

Mangerud, J., Larsen, E., Longva, O., and Sønstegaard, E., 1979, Glacial history of western Norway 15,000-10,000 B.P., Boreas, 8: 179-187.

McIntyre, A., Bé, A.W.H., and Roche, M.B., 1970, Modern Pacific Coccolithophorida: a paleontological thermometer, Trans. N.Y. Acad. Sci., 32: 720-731.

McIntyre, A., Ruddiman, W.F., and Jantzen, R., 1972, Southward penetration of the North Atlantic Polar Front: faunal and floral evidence of large-scale surface water mass movements over the last 225.000 years, Deep-Sea Res., 19: 61-77.

Moyes, J., Duplessy, J.C., Gonthier, E., Latouche, C., Maillet, N., Parra, M., and Pujol, C., 1974, Les sédiments profonds actuels et pleistocene recent de l'Atlantique nord-oriental et du sud de la Mer de Norvège, 2e Coll. Int. Expl. Océans, Bordeaux, 4 (Bx201) : 1-14.

Muller, R.E., Darby, D.A., and Clark, D.L., 1972, Significance of atmospheric dust and ice rafting for Arctic Ocean sediment, Geol. Soc. Amer. Bull., 83, 205-212.

Müller, T.J., Schott, F.A., Siedler, G., and Koltermann, K.P., 1974, Observations of overflow on the Iceland Faeroe Ridge, "Meteor" Forsch. Ergebn. A 15: 49-55.

Pujol, Cl., Duprat, J., Gonthier, E., Peypouquet, J.P., and Pujos-Lamy, 1974. Résultats préliminaires de l'étude effectuée per l'Institut de Géologie du Bassin d'Aquitaine concernant la mission FAEGAS (25 juin-17 juillet 1973) dans l'Atlantique nord-est. Bull. Inst. Géol. Bassin Aquitaine, 16: 65-94.

Rokoengen, K., 1980, De øvre lags geologi på kontinentalsokkelen utenfor Møre og Romsdal, Cont. Shelf Inst. Publ. 105, 49 p.

Ruddiman, W.F., 1977, Late Quaternary deposition of ice-rafted sand in the subpolar North Atlantic (lat 40° to 65°N), Geol. Soc. Am. Bull., 88: 1813-1827.

Ruddiman, W.F., and McIntyre, A., 1973, Time-transgressive deglacial retreat of polar waters from the North Atlantic, Quat. Res., 3: 117-130.

Ruddiman, W.F., McIntyre, A., Niebler-Hunt, V., and Durazzi, J.T., 1980, Oceanic evidence for the mechanism of rapid northern hemisphere glaciation, Quat. Res., 13: 33-64.

Shurko, I.I., 1966, Terrigenous-volcanic minerals in modern sediments of the North Atlantic, Dokl. Akad. Nauk SSSR 171: 461-464.

Streeter, S.S., and Shackleton, N.J., 1979, Paleocirculation of the Deep North Atlantic: 150,000 year record of benthic foraminifera and oxygen-18, Science, 203: 168-171.

Tait, J.B., 1967, ed., The Iceland-Faroe Ridge International (ICES) "Overflow" Expedition, May-June, 1960, Rapp. Pr.-verb. Réun. Cons. Perm. Int. l'Explor. Mer, 157, 274 p.

Talwani, M., Udintsev, G., et al., 1976, Initial Reports of the Deep Sea Drilling Project, XXXVIII, Washington.

Worthington, L.V., 1970, The Norwegian Sea as a mediterranean basin, Deep-Sea Res. 17: 77-84.

# THE MODERN CURRENT REGIME ACROSS THE GREENLAND-SCOTLAND RIDGE

Jens Meincke

Institut für Meereskunde an der Universität Kiel

Düsternbrooker Weg 20, 2300 Kiel 1, F.R.G.

## INTRODUCTION

The Greenland-Scotland ridge system is not only a topographical boundary between the basins of the Norwegian-Greenland seas and the basins of the mid-latitude North Atlantic, but it also plays a decisive role in the location of the zone separating water masses of different climatological origin. This is readily demonstrated in Figure 1 showing the ridge system to be an area of strong temperature contrasts. The most important water masses merging there are characterized in Figure 2 by means of their temperature and salinity characteristics. The two extremes are Atlantic waters with high salinities and high temperatures originating from south of the ridge and Polar waters with low temperatures and low salinities relating their origin to surface waters influenced by melting and run-off into the Polar Sea. The two Arctic waters (intermediate and deep) received their relatively high salinities from Atlantic waters and have obtained their low temperatures from cooling in the Greenland-Norwegian seas.

Figure 1 demonstrates, that the water mass boundaries do not correspond to a line simply following the crest of the Greenland-Scotland ridge. This is because from a kinematic point of view the ridge system consists of three different parts: Denmark-Strait is basically a straight channel cutting across the Greenland-Iceland sill. The Iceland-Faeroe ridge is a wide and fairly uniform ridge with only a few notches cutting across its sill. The Faeroe-Scotland ridge consists of the relatively deep Faeroese channels (Faeroe-Shetland Channel and Faeroe-Bank Channel) meandering around the southeastern tip of the Iceland-Faeroe ridge with the Faeroe slope on one side and the Shetland slope and the Wyville-Thomson ridge on the

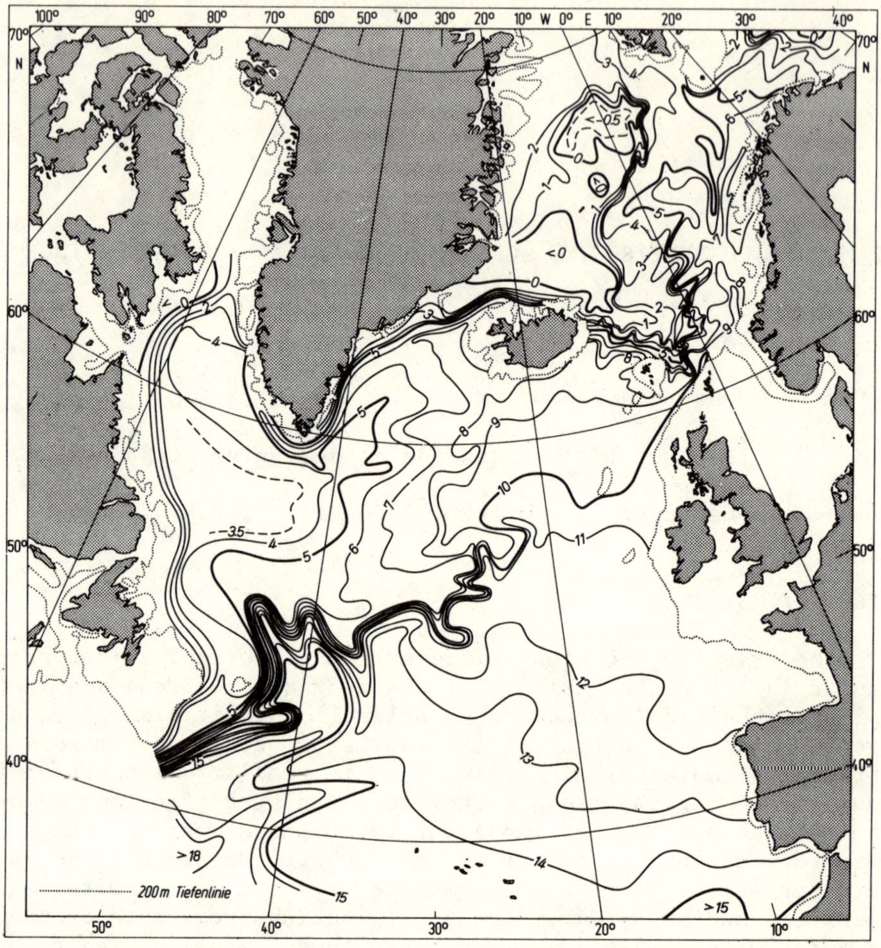


Fig. 1. Temperature distribution at 200 m depth for the northern North Atlantic in summer 1958 (from Dietrich, 1969).

other side. The kinematic role of the topography consists of its control of the mean flow direction. Currents tend to flow basically parallel to the isobaths with the shallower water to the right when looking downstream (on the northern hemisphere). In Figure 1 the zones of strong temperature gradients which are a result of currents advecting waters of different temperature are in fact parallel to the major topographic features as described above.

Thus the subsequent discussion of the flow regime in the Greenland-Scotland ridge area will be separated for Denmark-Strait, the Iceland-Faeroe ridge and the Faeroese channels. A profitable tool for this discussion has proven to be the tracing of characteristic water

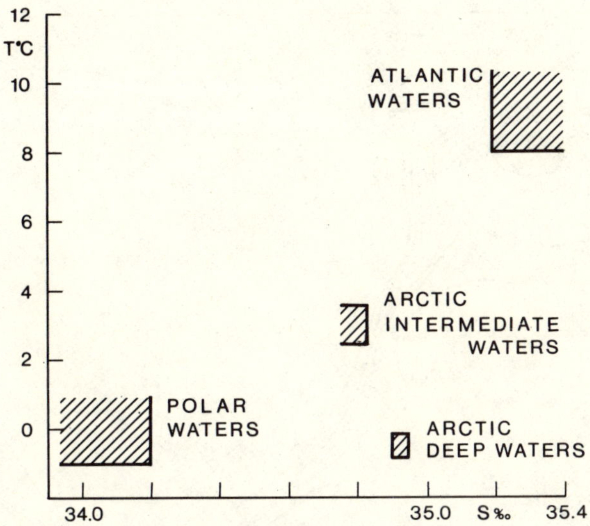


Fig. 2. Temperature-salinity characteristics of the principle water masses observed in the Greenland-Scotland ridge area.

masses especially in the near bottom layers. By means of the results from direct current measurements it will be possible to quantify the results of the water mass analysis. In addition the current records will be used to discuss the kinematics of the vigorous fluctuations within the current regime of the Greenland-Scotland ridge. The data base for this investigation has become available from several international field programmes, especially the Overflow '73 and the MONA-project conducted under the auspices of the International Council for the Exploration of the Sea (ICES, 1976, Willebrand et al., 1980).

## DENMARK-STRAIT

Because of its relatively simple topography Denmark-Strait shows the clearest mean flow pattern. The cold southward flows are leaning against the Greenland slope, i.e. the Polar waters in the near surface layers and the Arctic intermediate and deep waters in the deeper layers. Figure 3 represents a temperature section across Denmark-Strait showing the sharp gradients separating the cold overflow water from the warm northward flowing water of Atlantic origin which leans on to the Icelandic slope. The horizontal distribution of the maximum portion of Arctic deep water is shown in Figure 4. It demonstrates the reservoir of this water in the Iceland Sea basin north of the sill and it shows the narrow band of overflow water following the isobaths of the Greenland slope. The currents in Denmark-Strait are strongest in the near bottom layer over the Greenland-slope. Current measurements have shown most speed values in the interval between 40 to 95 cm $sec^{-1}$ (see Figure 5), the maximum ob-

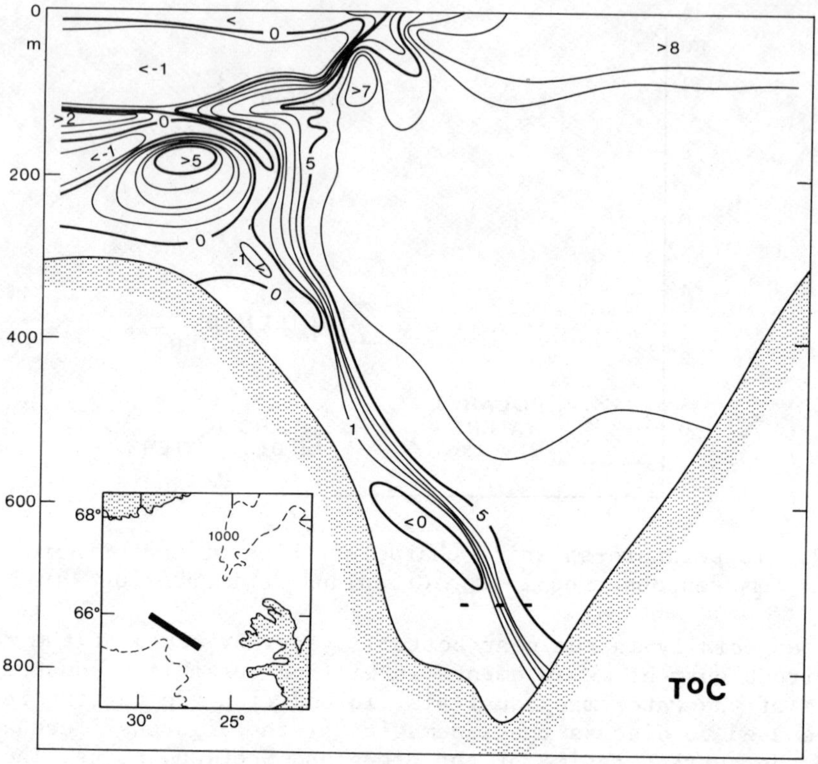


Fig. 3. Vertical section of temperature across Denmark-Strait (after Ross, 1978).

served value was 167 cm $sec^{-1}$ at 100 m off the bottom (Ross, 1978). Over the Iceland slope the northward flow has speeds of order 10 cm $sec^{-1}$.

On the Greenland slope the near bottom currents show strong fluctuations superimposed on the mean flow. From records of up to one year duration the most energetic variations in local current measurements have a period of 1.8 days with an amplitude around 25 cm $sec^{-1}$. It is caused by baroclinic instabilities, a feature inherent to flows in stratified waters above sloping bottom (Smith, 1976). These local fluctuations are different from fluctuations of the integrated transport of the overflow waters, that was derived from averaging over all current observations within waters of temperatures less than 2° C. The transports vary between 1 and 7 x $10^6$ $m^3$ $sec^{-1}$ with a mean of 2.5 x $10^6$ $m^3$ $sec^{-1}$. These variations occur in bursts of typically 1 day duration at intervals of several days. The cause is not yet fully understood (Ross, 1978).

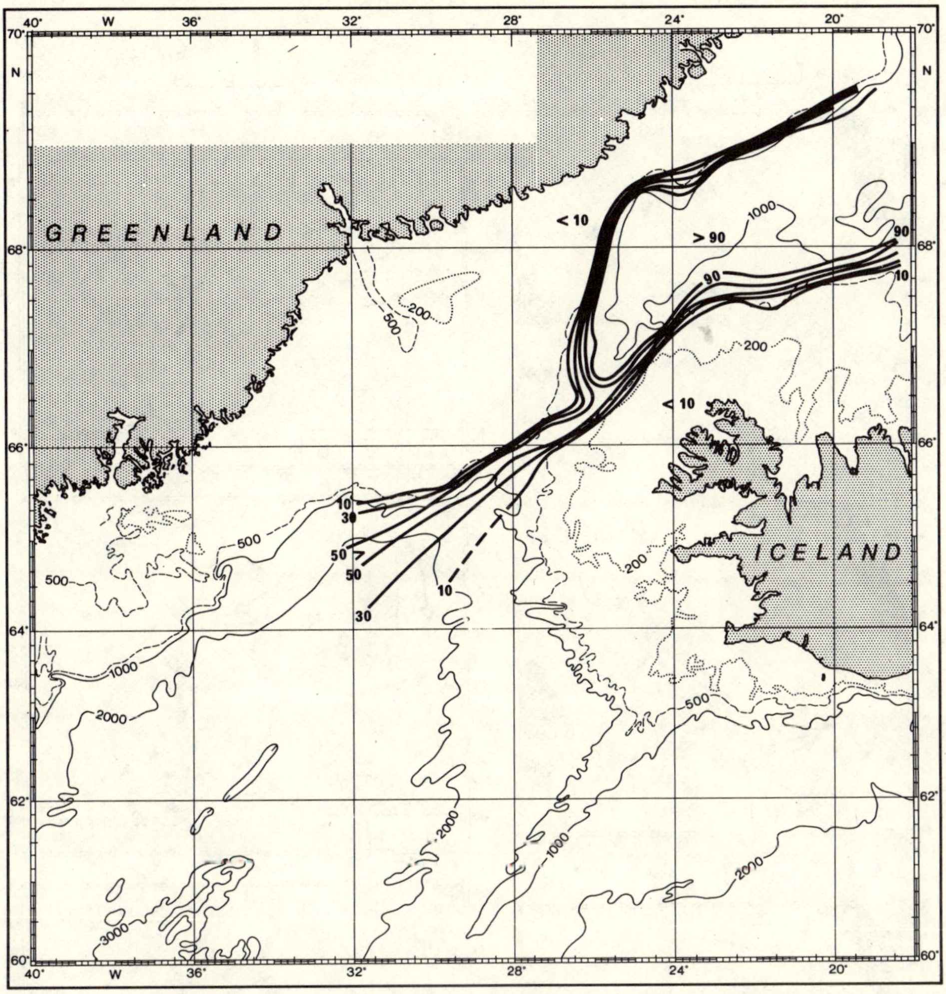


Fig. 4. Horizontal distribution of the layer with maximum portion of Arctic deep water in Denmark-Strait. Isolines represent percentages (after Müller, 1978).

## THE ICELAND-FAEROE RIDGE

The Iceland-Faeroe ridge with its sill depth of less than 500 m is the shallowest portion of the Greenland-Scotland ridge system. Whereas the topography of the Arctic and the Atlantic flank is generally smooth, the sill topography is characterized by elevations of 50 to 100 m height with a spatial scale of order 5 to 20 nm which leave inbetween them a series of notches cutting across the sill. From the large scale distribution of Arctic deep water (for the region of the Iceland-Scotland ridge it is traditionally named

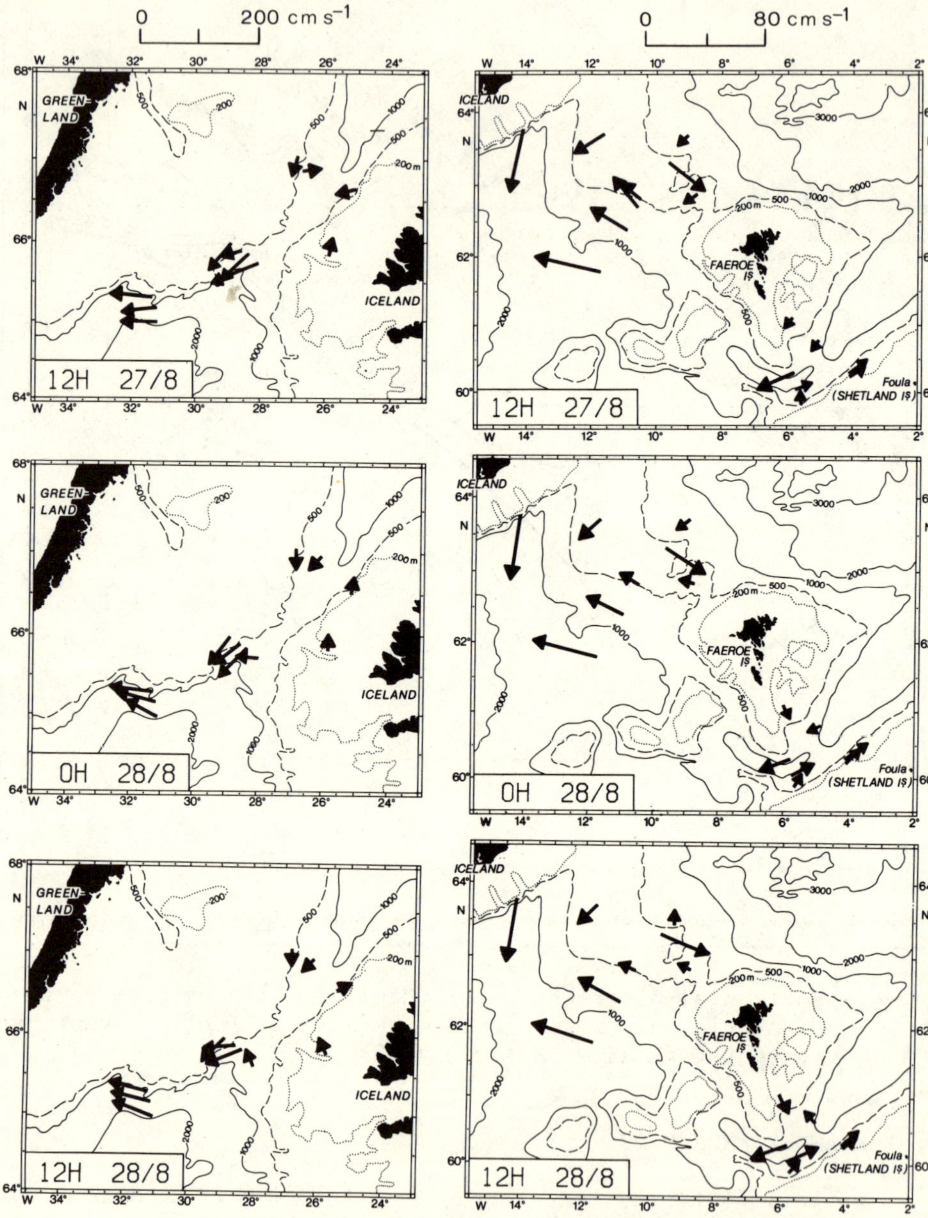


Fig. 5. Near bottom currents for three consecutive periods of 12 hours each on August 27/28, 1973. The tidal signal has been removed by applying a 25-hour running mean of equal weights. Left: Denmark-Strait. Right: Iceland-Scotland ridge (from Ross and Meincke, 1979).

Norwegian Sea deep water) represented in Figure 6 it is evident that the ridge is an effective obstacle. The spreading of deep water does follow the bathymetric contours in a clockwise sense with the Faeroese channels providing the link between the flows over the Arctic and the Atlantic flank. Smaller scale hydrographic surveys have revealed

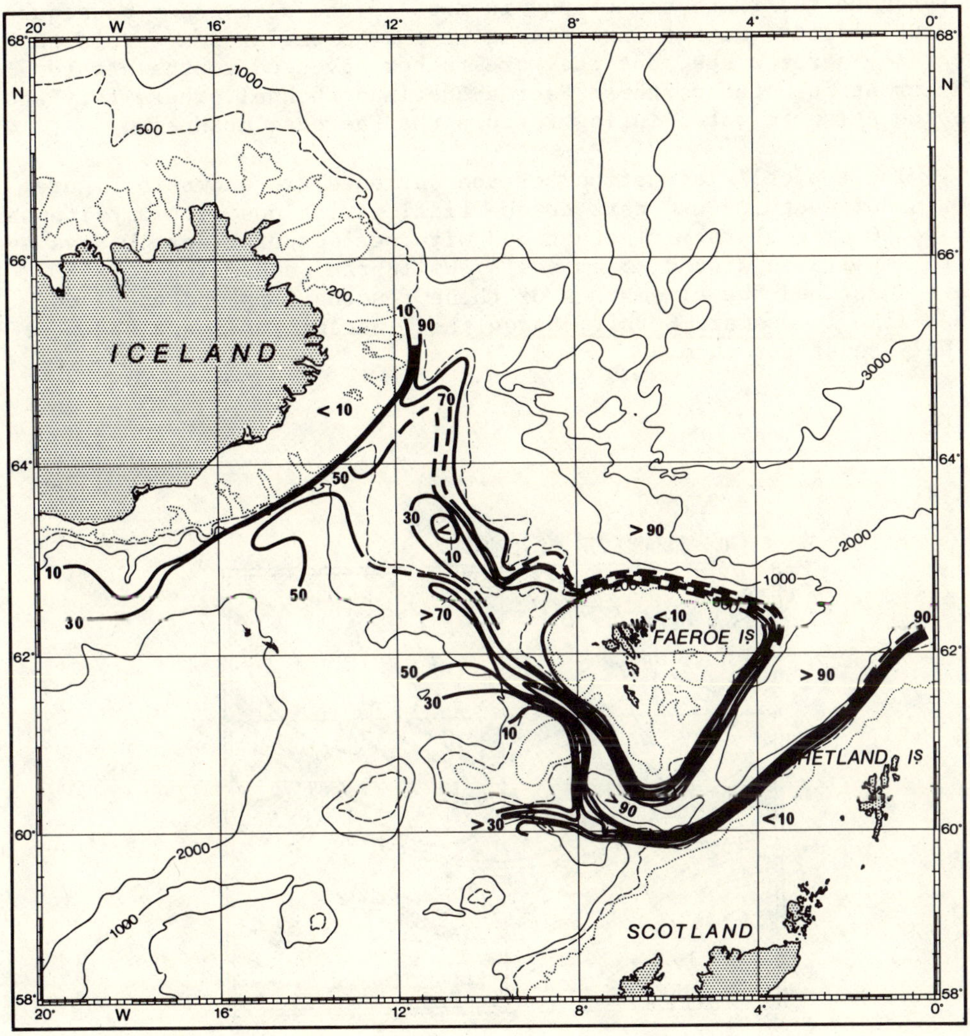


Fig. 6. Horizontal distribution of the layer with maximum portion of Arctic deep water in the Iceland-Scotland ridge area. Isolines represent percentages (after Müller, 1978).

trickles of overflows with varying intensity following the notches across the sill. An example for this is presented in Figure 7. The results of an analysis of water mass distribution and local current measurements have been combined into Figure 8 which shows the magnitude of the near bottom currents and the dominant role of the small scale topography in determining the flow direction. A similar analysis was performed for the near surface layers. Figure 9 shows the frontal zone between Atlantic and Arctic surface waters to be aligned to the ridge topography in most of the area. As a consequence Atlantic waters do not directly enter the Norwegian Sea across the Iceland-Faeroe ridge, but they are rather advected southeastward along the front into the northern Faeroe-Shetland Channel, where they add to the Atlantic water inflow through the Faeroese channels.

The typical fluctuations of the currents are shown in Figure 10. A dominant feature next to the tidal variations with amplitudes up to 60 cm $sec^{-1}$ are fluctuations with periods of order 10 days and with amplitudes around 25 cm $sec^{-1}$. Willebrand and Meincke (1980) have described the kinematics of these fluctuations and concluded that like in Denmark-Strait, baroclinic instabilities are the most likely cause for them.

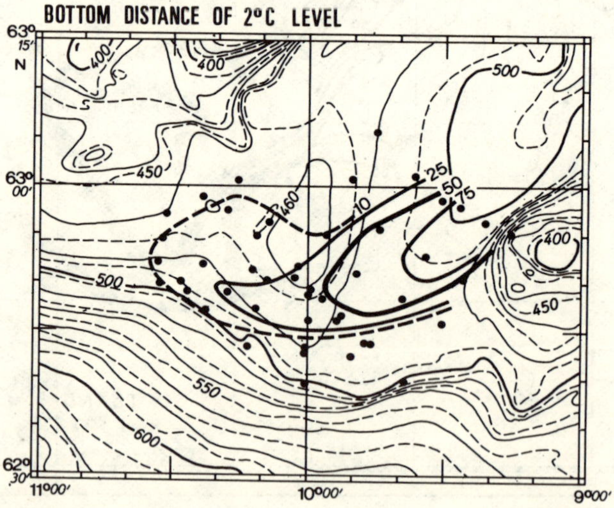


Fig. 7. Horizontal distribution of water with temperatures less than 2° C (mainly Arctic deep water) through a notch of the sill of the Iceland-Faeroe ridge. Isolines represent bottom distance of 2° C-layer (from Müller et al., 1974).

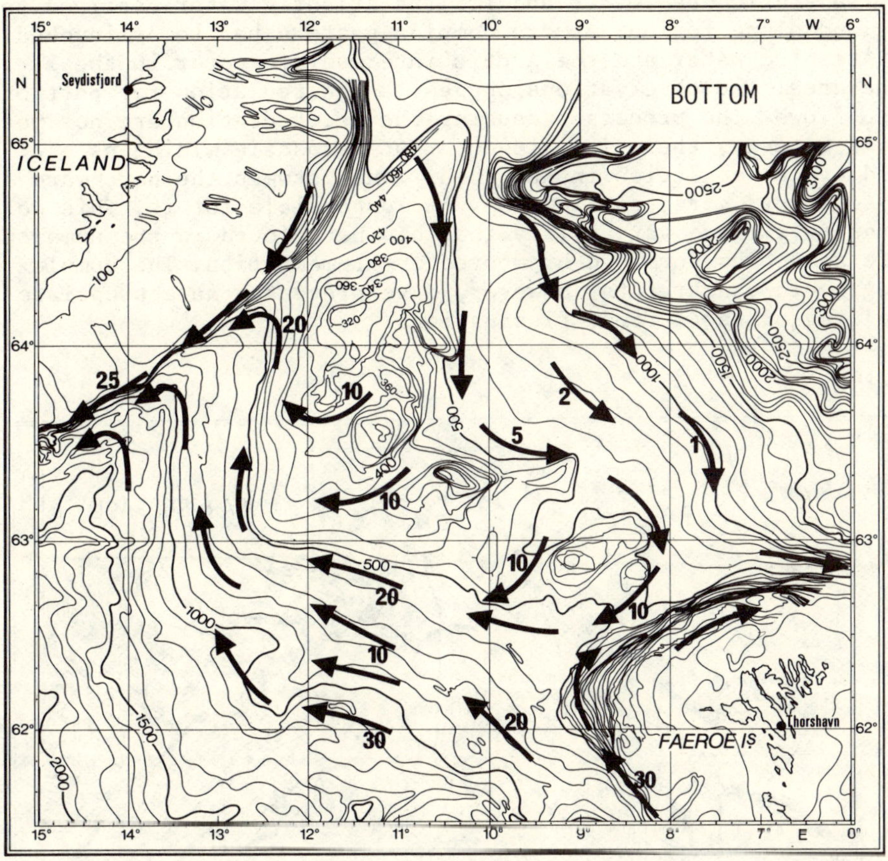


Fig. 8. Schematic representation of near bottom currents in the Iceland-Faeroe ridge area. The construction of the arrows is based on a combined analysis of water masses and direct current observations. Numbers give mean speed values at locations where current meter records were available (from Hansen and Meincke, 1979).

## THE FAEROESE CHANNELS

The Faeroese channels are the deepest connection across the Greenland-Scotland ridge system with a rather complex topography. Figure 6 has already revealed the channels to be the major drain of Arctic deep water out of the Norwegian Sea basins. Its coupling to the flow pattern in the Iceland-Faeroe ridge area has also been already demonstrated by Figure 5. Dooley and Meincke (1981) have combined current measurements and water mass analyses into a schematic representation of the circulation and the transports in the Faeroese channels. It is shown in Figure 11. Simple flow pattern was found for the Arctic deep water and for the Atlantic water confined to the Shetland slope region. A more complex pattern had to be invoked for the Atlantic water and the Arctic intermediate water in the Faeroe slope area. The observations suggest a recirculation for part of these flows. The processes causing the recirculation are not yet known, however, there is evidence that mesoscale mixing as e.g. induced by instabilities in the shear zone between the northward and the southward directed flows may be responsible for it. This corresponds to a large variability that has been found in the numerous current records obtained by moored instrumentation. The complex topographic constraints, however, have prevented an appropriate modelling of the current fluctuations.

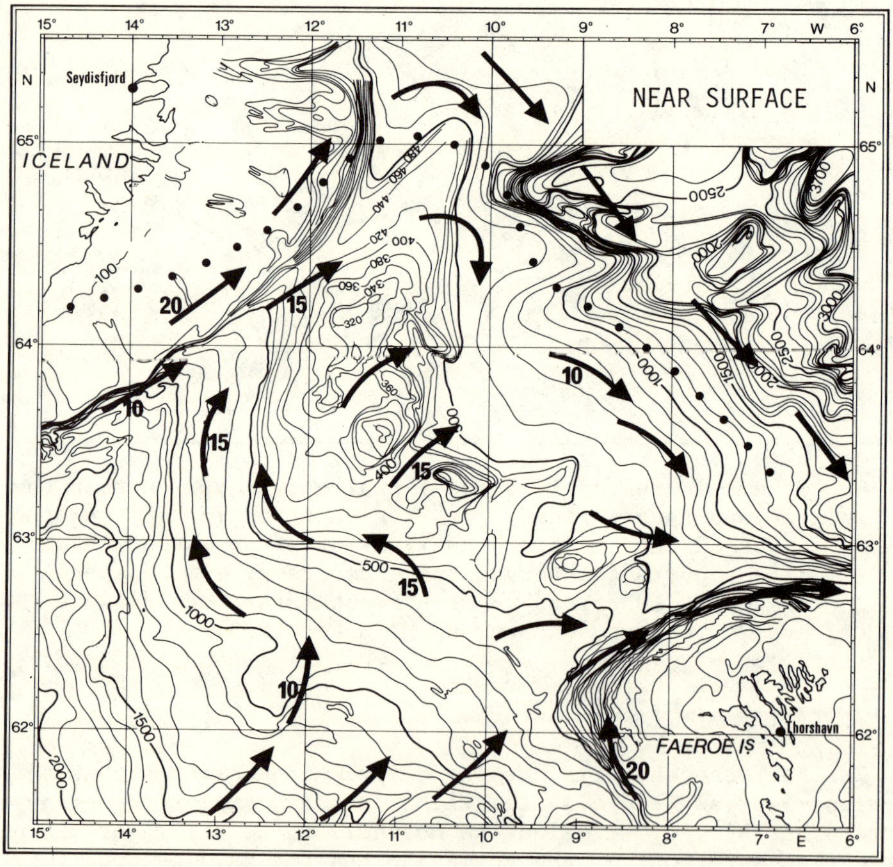


Fig. 9. Schematic representations of near surface currents in the Iceland-Faeroe ridge area. Construction like mentioned for Figure 8. Dotted line represents schematic location of front between Atlantic and Arctic waters.

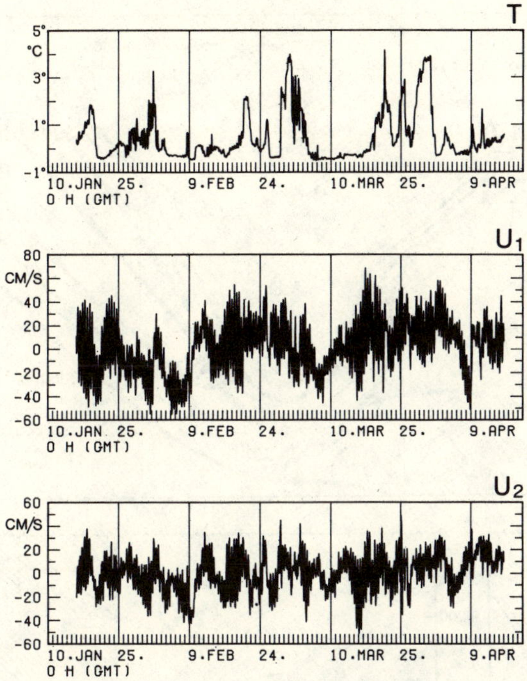


Fig. 10. Portions from a one-year record of current components ($u_1$ and $u_2$) and temperature (T) at position 63°11' N, 09°02' W showing typical fluctuations at a fixed level 17 m above bottom (from Willebrand and Meincke, 1980).

## DISCUSSION

A reasonably detailed description of the distribution of water masses and the corresponding flow pattern in the Greenland-Scotland ridge area could be given on the basis of a data set which has become available since 1973. Special emphasis has been given to the overflows of Arctic deep waters, since they are mostly affected by the ridge topography and they are most relevant for the North Atlantic deep circulation. Figure 12 summarizes these cold overflows. A comparison of Figure 12 with Worthington's (1970) "classical" presentation does not show significant differences. However, this is not true for the inflow of Atlantic water through the Faeroese channels. Worthington (1970) used an inflow of $8 \times 10^6$ m$^3$ sec$^{-1}$ for his heat budget of the Norwegian Sea, whereas Tait's (1957) estimate on data from the period 1927 to 1952 and Dooley and Meincke's (1981) estimate yielded 2.3 and $2.0 \times 10^6$ m$^3$ sec$^{-1}$ respectively. Dooley and Meincke (1981) discussed this difference and emphasized a revision of Worthington's estimate.

The present paper also discussed the fluctuations of the flow

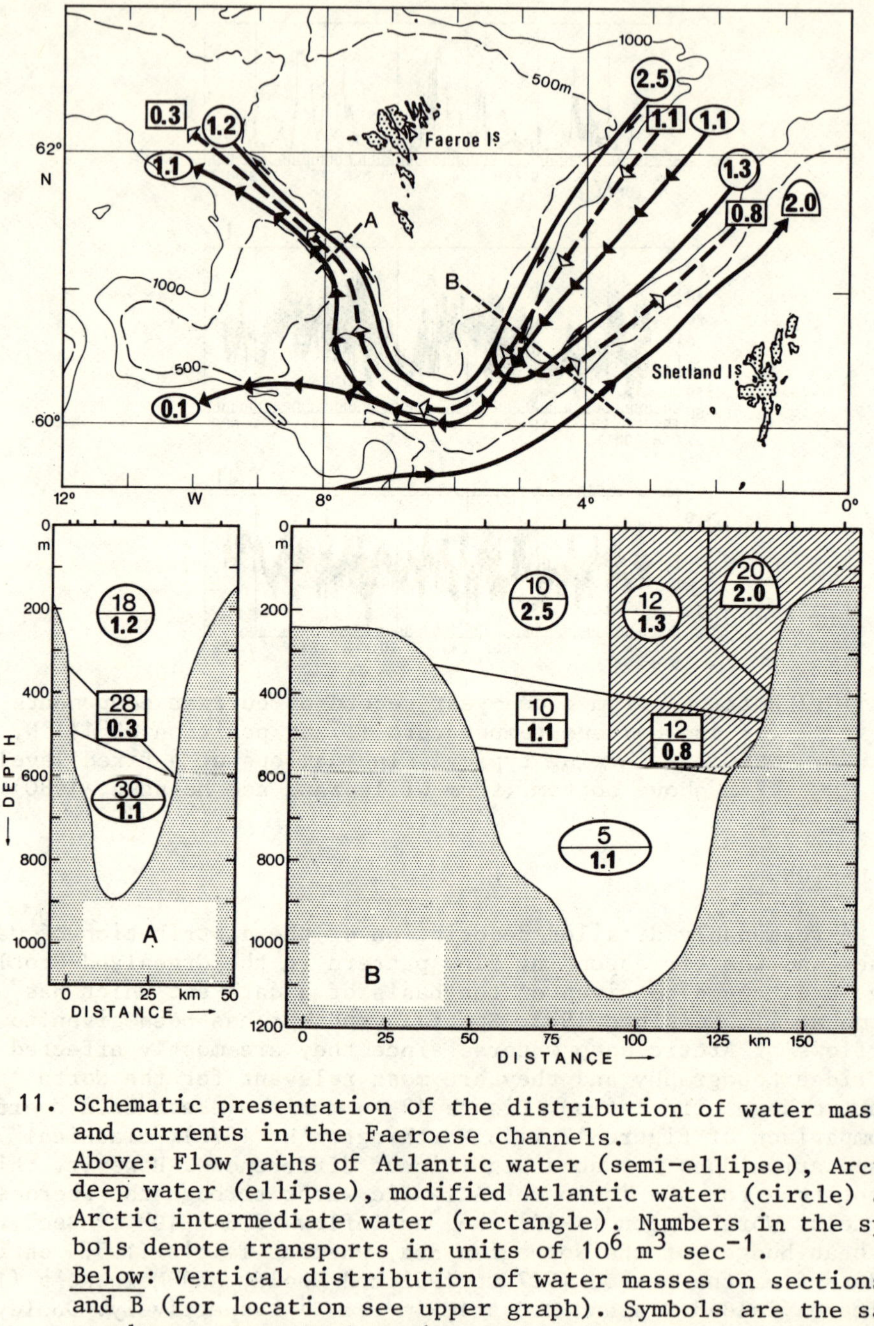


Fig. 11. Schematic presentation of the distribution of water masses and currents in the Faeroese channels.
Above: Flow paths of Atlantic water (semi-ellipse), Arctic deep water (ellipse), modified Atlantic water (circle) and Arctic intermediate water (rectangle). Numbers in the symbols denote transports in units of $10^6$ $m^3$ $sec^{-1}$.
Below: Vertical distribution of water masses on sections A and B (for location see upper graph). Symbols are the same as above, upper numbers in symbols denote mean flow speeds, lower numbers denote transports in units of $10^6$ $m^3$ $sec^{-1}$. Shaded area: flow into the Norwegian Sea. Non-shaded area: flow into the North Atlantic (from Dooley and Meincke, 1981).

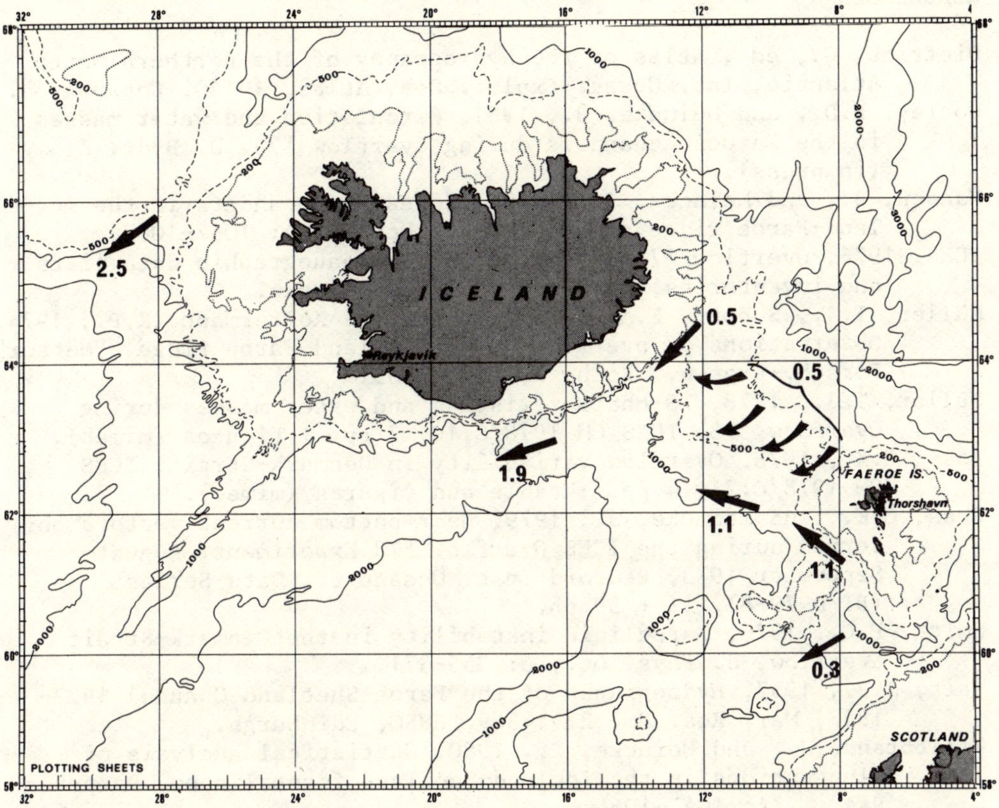


Fig. 12. Charts showing the major overflows of Arctic deep water and their transports in units of $10^6\ m^3\ sec^{-1}$.

across the Greenland-Scotland ridge. The investigation of recent current meter records attributed most of the local variations to instability mechanisms of shearing flows interacting with sloping bottom. Since the scales involved are of orders kilometers and days it is evident that analyses of standard hydrographic surveys of this region have usually lead to strongly biased distributions of hydrographic variables.

A concluding remark is concerned with one important aspect of future research in this region. So far there is no information on seasonal changes of the flow pattern. In particular no signals in the cold overflows have been revealed that could be linked to the formation of intermediate and deep waters in the Greenland-Norwegian seas. This requires monitoring-type studies. The present knowledge should be a sufficient basis for launching them.

REFERENCES


Dietrich, G., ed., Atlas of the hydrography of the northern North Atlantic, Int. Counc. Explor. Sea, Atlas, 1-140, Copenhagen.

Dooley, H.D., and Meincke, J., 1981, Circulation and water masses in the Faroese channels during Overflow '73, D. Hydr. Z. (in press).

Hansen, B., and Meincke, J., 1979, Eddies and meanders in the Iceland-Faroe ridge area, Deep-Sea Res., 26A: 1067-1082.

ICES, 1976, Overflow '73 Inventory, ICES Oceanographic data lists and inventories, No. 29: 1-203, Copenhagen.

Müller, T.J., Schott, F.A., Siedler, G., and Koltermann, K.P., 1974, Observations of overflow on the Iceland-Faroe ridge, "Meteor" Forsch.-Ergebn., Reihe A, 15: 49-55.

Müller, T.J., 1978, TS-characteristics and water masses during Overflow '73, ICES CM 1978/C:10, 7 pp. + figures (mimeo).

Ross, C.K., 1978, Overflow variability in Denmark-Strait, ICES CM 1978/C:21, 4 pp. + table and figures (mimeo).

Ross, C.K., and Meincke, J., 1979, Near-bottom current vectors observed during the ICES Overflow '73 Experiment, August-September 1973, Bedford Inst. Oceanogr., Data Series (BI-D-79-8), ii + 39 pp.

Smith, P.C., 1976, Baroclinic instability in the Denmark-Strait overflow, J. Phys. Oc., 6: 355-371.

Tait, J.B., 1957, Hydrography of the Faroe-Shetland Channel 1927-1952, Mar. Res. No. 2: 1-309, HMSO, Edinburgh.

Willebrand, J., and Meincke, J., 1980, Statistical analysis of fluctuations in the Iceland-Scotland frontal zone, Deep-Sea Res., 27A: 1047-1066.

Worthington, V., 1970, The Norwegian Sea as a mediterranean basin, Deep-Sea Res., 17: 77-84.

# CHAPTER 4

# CONCLUSIONS

# DISCUSSION, CONCLUSIONS, AND RECOMMENDATIONS

Svend Saxov

Laboratory of Geophysics
Aarhus University
Finlandsgade 6-8
DK-8200 Aarhus N, Denmark

One point of major interest to the participants in this NATO Advanced Research Institute was along which lines future research should develop. It was generally agreed that the evolution of the Greenland-Iceland-Faeroe-Scotland Ridge can be approached satisfactorily only by interdisciplinary cooperation. It was also agreed upon that the Ridge is a key area for understanding and future prediction of climate and oceanic water circulation.

In order to obtain a better knowledge of the Ridge region and to improve understanding of the history and development of the area, a summary of some suggested future studies is given below.

## PREFACE

The evolution of the Greenland-Iceland-Faeroe-Scotland Ridge which may be considered to be a key area in marine geoscience, has caused many problems, most of which are unsolved or only partly solved.

Geological problems like the origin of the volcanic North Atlantic province, the history of water exchange across the Ridge between the Norwegian-Greenland Sea and the main North Atlantic Basin, modern sediment dispersal around the Ridge, and investigation of terrestrial sediments for new fossil vertebrate evidence, are only partly understood and need more research.

Additional seismic surveys are needed to determine the change of crustal thickness and structure with age, to investigate the deep crustal and mantle transitions, and to supply information on the geology (continental or oceanic).

Geochemical and petrological sampling along the Ridge to study the early stages of the evolution of the Icelandic hot spot and to date the formation of the Ridge with better precision are desirable.

The age, character and thickness of the sediments need to be determined from seismic profiling and drilling.

## Morphological and Structural Investigation

The Greenland-Iceland-Faeroe-Scotland Ridge is a major structural feature crossing the entire northern Atlantic Ocean (Fig. 1). Its origin is unknown although many geophysical and geochemical investigations have been centred on or near this feature. The possible origin ranges from hot spot volcanics superimposed on a normally accreting ridge to transform related tectonics which could have structurally controlled the injection, filling, and growth of the Ridge through space and time. It is clear that a large gap in knowledge remains. This gap is most pronounced in the detailed morphological and structural mapping of the region. The orientation of structural trends (faults, folds, their offsets, and orientations), the general morphologic grain or fabric of the ocean floor, and whether or not the structures are rift parallel, transform parallel or oblique to the above should unambiguously indicate the mode of accretion of the Ridge. A panoply of tools exist that can be utilized to map the detailed morphology and structure of the Ridge. These range from a seabeam system to Gloria and smaller-scale side scan sonar system.

The combined use of these tools will allow the textural and morphological features of the seabed to be superimposed on the bathymetric maps generated by multibeam and should help to reveal (with other evidence) the true origin and development of this present-day aseismic Ridge through space and time.

## Volcanic Rock Sampling

There is increasing geological, petrological, geochemical, and geophysical evidence from Iceland, East Greenland, and the Faeroe Islands suggesting that the evolution of the Ridge has been influenced by cyclic or repetitive changes in spreading conditions, ranging from active (hot spot type) to passive (midocean ridge type) of spreading. Complex tectonic and volcanic activities similar to those observed currently on Iceland appear to have influenced the formation of the whole Ridge.

Progress in further understanding these processes and the evolution of the Ridge is presently hampered by lack of samples. Only two DSDP holes have been drilled. Both are on the flank of the Ridge,

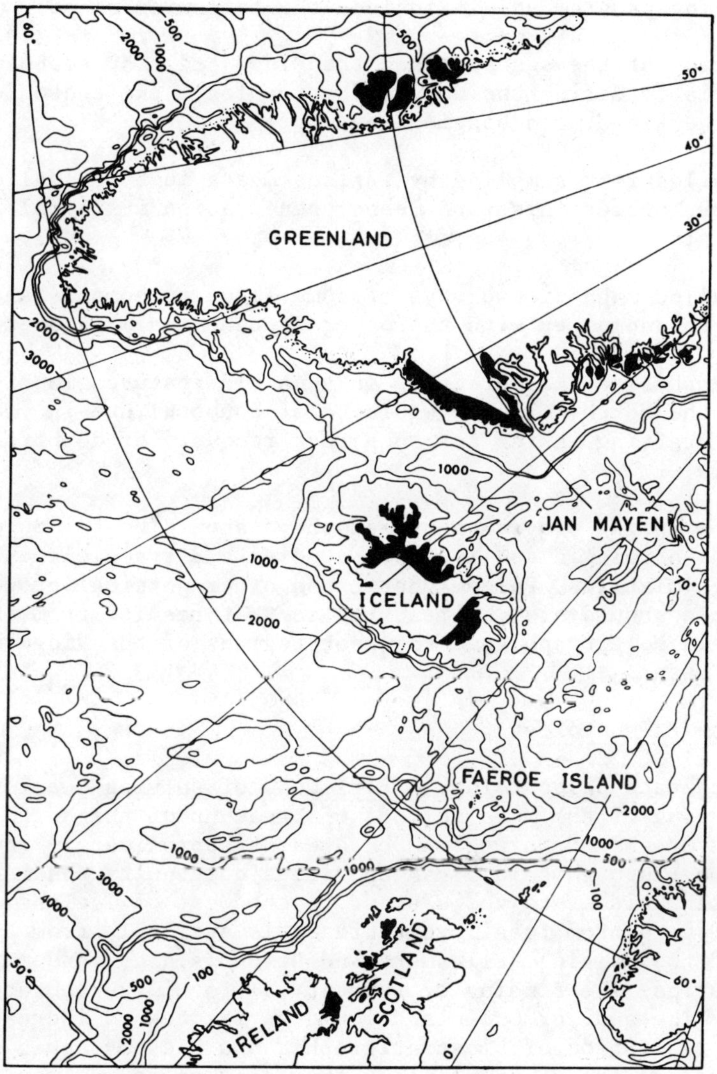


Fig. 1. The North Atlantic basalt province.
Depths in m.

and only one hole (336) managed to penetrate the basement, and only for a few tenths of a metre. Additional sampling can best be done along the sediment bare Iceland-Faeroe Ridge, where the age appears to decrease from SE to NW from 55 to 20 (or more) Ma. We urge that every effort should be made in the near future to sample the Ridge. The sampling program should include the following steps:

1. Mapping of the seabed along the Ridge crest by seabeam and Gloria to define the detailed morphology, the tectonic grain of the Ridge, and basement exposures.

2. Detailed rock sampling by various means such as shallow drill holes, submersibles and deeper penetration drill holes, and dredging.

3. Detailed magnetic surveys at some sites to compare short wave length anomalies with the other (geochemical) parameters.

We emphasize that dredging on such old features in high latitudes in the North Atlantic would be of doubtful use in view of intense ice rafting making any such rock recovery of doubtful provenance.

The sampling program proposed above should be followed by coordinated petrologic and isotopic studies and trace element analyses including volatiles. Paleomagnetic and other physical properties of these rocks should also be measured so that results from the samples can be extrapolated to wider regions of the Ridge by geophysical means.

## Vertebrate Paleontology

Vertebrate paleontology offers hope of solving the location of the beginning of the main episode of break-up of the North Atlantic Ocean and Greenland/Norwegian Sea area at approximately anomaly 24/25 time. Through this break-up marine continuity resumed in the early Eocene either to the east or to the west of Iceland. At present, evidence from fossil vertebrates is available from Europe (south of the Baltic), Ellesmere and Devon Islands, and mainland North America. The finding of terrestrial fossil vertebrates on Svalbard, Iceland, or Greenland would go far toward increasing the power of resolution of biostratigraphic and biogeographic analysis and offers promise of pinpointing the location as well as the timing of the first break in the former continuous land between North America and Europe.

Moreover, there is an excellent possibility of sampling an extinct and potentially useful terrestrial fauna on Iceland. The following internationally supported projects are therefore recommended:

1. Air reconnaissance of northwestern Iceland, followed by surface prospecting of sediments occurring interbedded in dated and paleomagnetically well-known volcanic rocks.

2. Air reconnaissance and surface prospecting of early Tertiary sediments beneath, within, and above the Blosseville volcanics of East Greenland. The Scoresby Sund and Kangerdlugssuaq areas are at present the most promising areas to begin prospecting for fossil vertebrates, but other appropriate deposits exist as far north along the coast as Sabine Island.

3. Further exploration of the Judge Daly Peninsula area of Ellesmere Island is appropriate in order to calibrate the motion of the Nares Strait Lineament.

4. Exploration of the upper part of the Eureka Sound Formation of Central Banks Island, Arctic Canada, is appropriate.

## Sedimentological Investigations

The proposed coordinated geological studies of the Ridge environs recognize the potential of sedimentological investigations as a means for elucidating the history of subsidence of the Ridge and concomitant changes in paleocirculation patterns. Although the general pattern of present-day bottom water circulation and sediment distribution in this area is now reasonably well-established, it will be important to extend this information backwards in time, through the Neogene, to investigate the influence of glacial episodes on the sedimentary history, and into the Paleocene to explore the way in which oceanic circulation between the Norwegian Sea and Atlantic Ocean was affected by the construction and subsidence of the aseismic ridge. In addition to standard mineralogical and grain size studies, grain orientation investigations, for example using magnetic fabric techniques and paleomagnetic orientation methods, should be undertaken in order to fully investigate the changes patterns of bottom transport in the area throughout the Tertiary and Quaternary.

The above programs will require sediment cores from key localities on the crest and flanks of the Ridge and adjacent areas, and would be best achieved by a program of IPOD hydraulic piston coring. A carefully planned programme of short piston coring in this region should complement these investigations and be directed towards a study of the nature and timing of the sedimentary processes that have led to the development of the thick sediment drifts, and the progressive development of the major bottom water overflow routes and sediments transport paths across the Ridge throughout the Tertiary.

## Geological Programme

A geological programme involving three parts: (a) long term suspended sediment studies and bottom photography (b) piston coring of quaternary sediments, and dredging of oceanic surface rocks primarily along the Iceland-Faeroe Ridge crest and (c) shallow and deep drilling to sample unaltered volcanic rocks and the sedimentary sequences along the northern and southern flanks of the Ridge. (a) and (b) should be concluded in conjunction with the geophysical programme given below. For (c) we propose 15 holes (say 5 shallow ~100 m holes along crest and 5 holes on each flank penetrating the sediments into ~50 m of basement) see Fig. 2. The drilling programme should take place in the third year.

Contemporaneous with this programme studies concerning continued faunal, biostratigraphic, isotopic and lithostratigraphic studies in order to clarify the Tertiary pattern of paleocirculation in the North Atlantic should be carried out also to document more clearly the history of surface and deep water connections of the Norwegian-Greenland Sea with the North Atlantic. Improved subsidence models of the Ridge system and paleogeographic reconstructions of the region are also required. Models for regional paleo-circulation and paleo-geography need to be carefully examined and improved with new studies.

## Geophysical Programme

The normal bathymetric pattern of oceanic ridges and basins is interrupted by shallow bathymetric ridges which cross the oceans and which lack seismic activity. While the normal oceanic structure and development is now reasonably well understood in terms of sea-floor spreading and plate tectonics, the origin and structure of these transverse features, which are an important part of the ocean, are little understood. The Greenland-Iceland-Faeroe-Scotland Ridge is the best known and best studies of these features, even though its origin, structure and development are still poorly understood. These aseismic ridges are important, not only because of the problem of their origin, but also because of their shallow depth, which profoundly influences the shallow and deep water circulation and consequently the sedimentation, floral and faunal distribution, and climate. There is evidence that some of these ridges stood above sea-level when formed. It is for these reasons that an intensive study of the Greenland-Iceland-Faeroe-Scotland Ridge is important

To contribute to a solution of some of these problems a series of seismic investigations during two successive years is proposed. The seismic survey should coincide in time with the proposed geological programme points (a) and (b). These seismic investigations are summarized as follows:

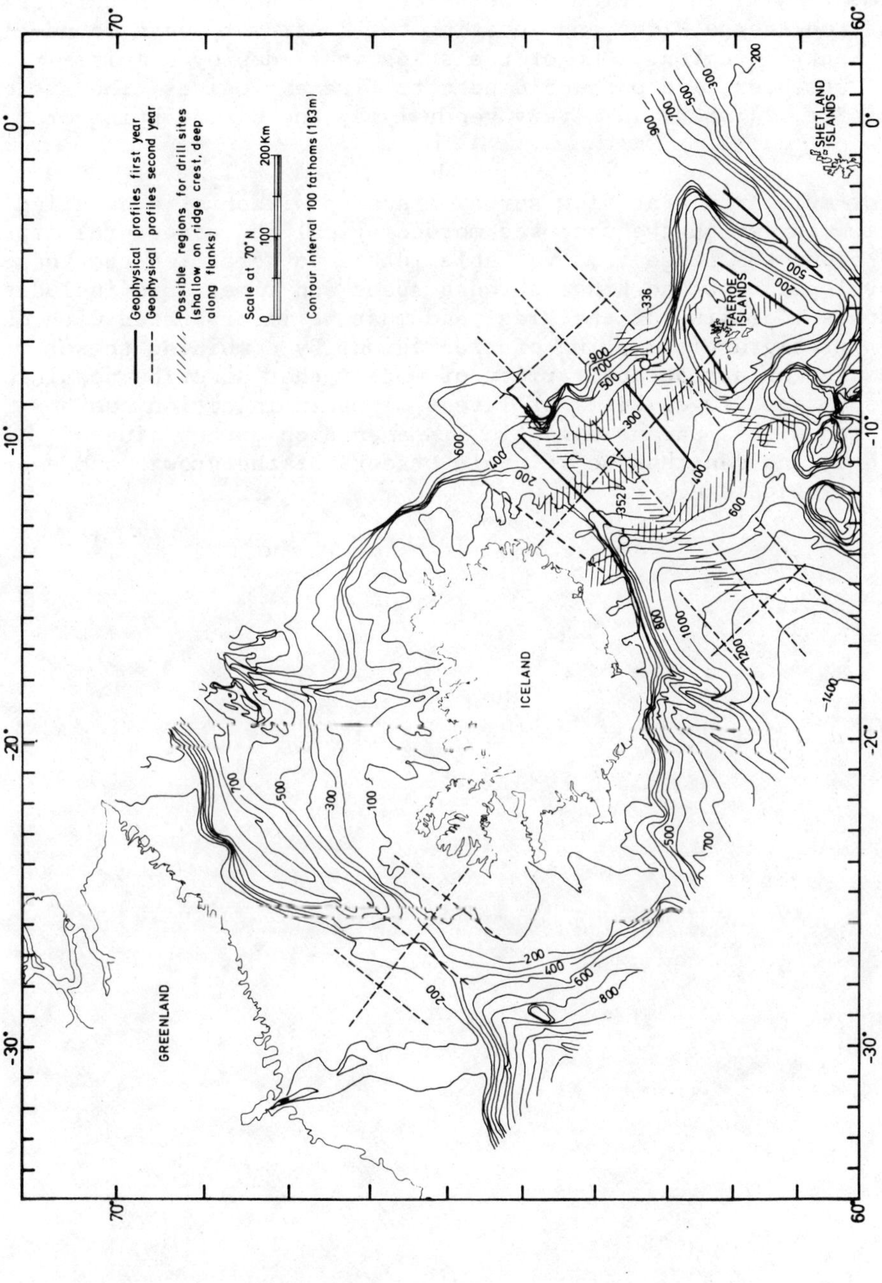


Fig. 2. Geophysical and geological outline map.

A two-ship seismic experiment involving the three techniques (a) vertical incidences multichannel reflection (b) expanding spread experiments, and (c) constant offset experiments. These experiments will in particular be concerned with the structure of the Iceland-Faeroe Ridge and locating the continent-ocean boundary in central regions. One of the ships would deploy a multichannel streamer, and one would need to fire explosives. The attached map (Fig. 2) shows in a very general way the distribution of lines.

For multigrad heat flow survey a superposition of a detailed heat flow survey on the proposed morphological and structural maps at and along the Ridge is a valuable method to determine the thermal development of the Ridge through space and time. This includes subsidence modelling of the Ridge and must be incorporated with detailed stratigraphic mapping of the region. By combining these methods we can narrow the variety of models that describe possible origins of the Ridge. Dates of latest magmatic injection can be correlated with apparent magnetic lineaments on either side of the Ridge, and can thus be valuable tracers of the growth and evolution of the Ridge.

MAGNETIC ANOMALIES OVER ICELAND AND SURROUNDING SEAS

(Text to Accompany Map)

A.G. Nunns[1], M. Talwani[1], G.R. Lorentzen[2], P.R. Vogt[3], T. Sigurgeirsson[4], L. Kristjánsson[4], H.C. Larsen[5], D. Voppel[6]

[1]Lamont-Doherty Geological Observatory of Columbia University, Palisades, New York 10964, USA

[2]Code 8200, U.S. Naval Oceanographic Office, NSTL. Station Bay Saint Louis, Mississippi 39522, USA

[3]Code 5110, Naval Research Laboratory Washington D.C. 20375, USA

[4]Science Institute, University of Iceland Dunhaga 3, Reykjavik, Iceland

[5]The Geological Survey of Greenland Øster Volgade 10, DK-1350 Copenhagen K, Denmark

[6]Deutsches Hydrographisches Institut Postfach 220, 2000 Hamburg, Federal Republic of Germany

## INTRODUCTION

During the past two decades many shipborne and airborne magnetic surveys have been carried out in the vicinity of Iceland, but no comprehensive compilation has been available. As an accompaniment to the present volume we present a polar stereographic chart at a scale of 1 : 2 000 000 on which most of the available data are displayed in the form of residual anomaly profiles along track. The chart has been colored to indicate the continuity of positive and negative anomalies. Bathymetric contours in fathoms (Grønlie and Talwani, 1979) are also shown.

The preparation of this map has been an international cooperative effort. An initial version, based mainly on published data, was presented at the NATO Advanced Research Institute meeting. Attendees from Iceland, Denmark, Germany and the USA arranged for the provision of further data from their national institutes. Addition of the extra information, much of it previously unpublished, more than doubled the data coverage and made possible the delineation of anomaly trends over nearly the entire area of the map. The final compilation was carried out at Lamont-Doherty Geological Observatory but this effort was subordinate to the collective task of acquiring and reducing the data, which was carried out at the respective institutes, mainly under the supervision of the joint contributors.

The present paper is restricted to a brief description of the map but it is hoped that a more complete analysis will be prepared in the future, including a detailed consideration of geological and geophysical data with the aim of elucidating some of the outstanding problems of the structure and development of Iceland and the surrounding seas.

## MAP PROJECTION

A conformal polar stereographic projection was chosen in order to minimize distortion and also to facilitate the incorporation of the Icelandic aeromagnetic survey data which were available in chart form. In this projection, lines of longitude radiate at equal angles from the geographic pole and a line of latitude, $\phi$, is concentric with the pole at a radial distance of 605.1 cotan ($\phi/2$) cm. The scale of 1 : 2 000 000 is with respect to an earth radius of 6361 km, appropriate to the mid-latitude of 64.5°N.

## DATA SOURCES

Fig. 1 is a trackline chart coded according to institution. Detailed survey areas are also outlined but the tracklines are not shown for the detailed surveys. Profiles which appear on the anomaly map within the outlined areas, but not on the trackline chart, belong to the detailed surveys.

The shipborne investigations carried out by Lamont-Doherty Geological Observatory in the vicinity of Iceland have been described by Talwani et al. (1971), Talwani and Eldholm (1977) and others. Of the U.S. Naval marine profiles included in the map, the Lynch profiles over the East Greenland margin have been described by Johnson et al. (1975b). The Bartlett 75 profiles over the Reykjanes Ridge and possibly some of the other Navy profiles have not been previously published. From the Deutsches Hydrographisches Institut we have incorporated profiles from the Komet 28 survey south of the Iceland-Faeroe Ridge (Voppel et al., 1979) and previously unpublished profiles from the Meteor 48 and 54 (1978 and 1979; south Norway Basin) and Meteor 45 (1977; Reykjanes Ridge) surveys. The Jean Charcot profiles, originally collected by the Bureau National des Données

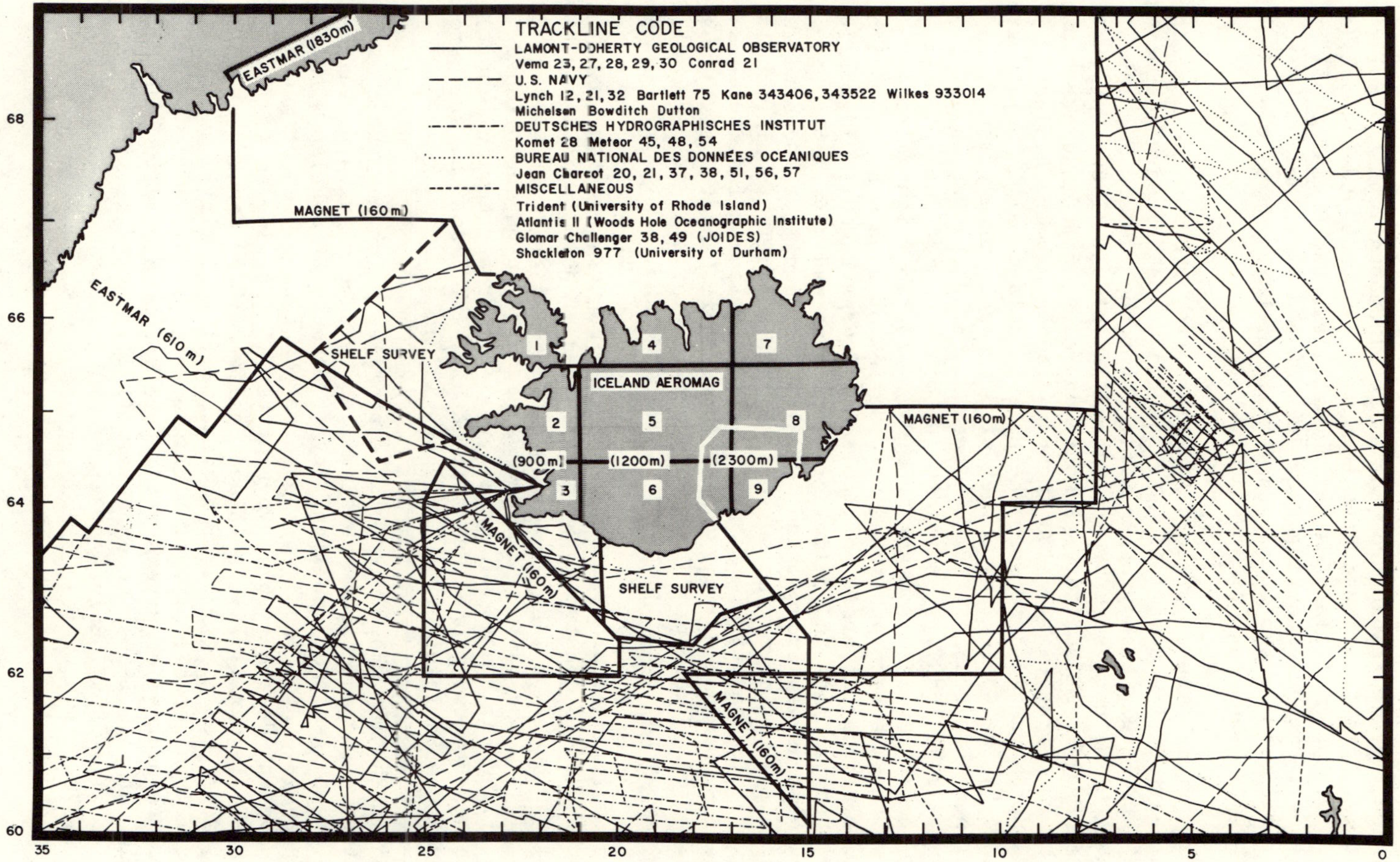


Fig. 1. Data sources. Detailed survey areas are outlined in bold and end at coastlines unless otherwise indicated. Flight heights for aeromagnetic surveys are shown in brackets.

Océaniques, were obtained from the National Geophysical and Solar Terrestrial Data Center and have been used by Nunns and Peacock (in press). Of the miscellaneous marine profiles, the Glomar Challenger profiles were measured during DSDP legs 38 and 49. The Shackleton 9/77 profiles collected by the University of Durham have been described by Nunns and Peacock (in press).

Data from three major aeromagnetic survey programs form an important part of the compilation. The Project MAGNET data collected by the U.S. Naval Oceanographic Office over the Iceland Plateau (north of $65^{o}$N) have been presented by Vogt et al. (1980) but the detailed Project MAGNET surveys south and east of Iceland are shown here for the first time. Also included in the map are Project EASTMAR aeromagnetic data over the East Greenland margin which were collected by the Geological Survey of Greenland. The planning and execution of Project EASTMAR have been described by Larsen and Thorning (1979; 1980). Central to the map are profiles acquired during the Icelandic aeromagnetic survey program (Sigurgeirsson, 1970) which have been published as a series of sheets by the Science Institute of the University of Iceland (Sigurgeirsson, 1970-1982). Unpublished profiles in preliminary form were made available for aeromagnetic survey areas 1 and 4. Also included in the chart are data acquired during marine surveys of the Icelandic insular shelf which were a cooperative effort of various Icelandic institutions and the U.S. Defense Mapping Agency Topographic Center (Kristjánsson, 1976 a, b, c). These profiles are shown in Fig. 2 at the same relative scale as the aeromagnetic data over Iceland.

In addition to profile data, existing magnetic contour maps have been used to constrain the delineation of positive and negative anomalies; including maps of the Iceland-Faeroe Ridge (Fleischer et al., 1974), the eastern Iceland Plateau (based on unpublished Meteor 1972 data; see Voppel et al., (1979) for a trackline chart), the Tjörnes Fracture Zone (McMaster et al., 1977) and the area around the Reykjanes Peninsula (Kristjánsson, 1978).

While the map shows most of the profiles which have been collected in the area, some were omitted because of data unavailability or navigational unreliability or because of overlap with later surveys. Additional marine survey data have been presented by Johnson and Heezen (1967), Vogt et al. (1970), Ingles (1971), Johnson and Tanner (1971), Bott and Ingles (1972), Johnson et al. (1972), Meyer et al. (1972), Featherstone (1976), Mirlin and Melikhov (1976), Talwani et al. (1979), Mirlin et al. (1980) and Armstrong (1981). Aeromagnetic surveys which have not been used include those of Heirtzler et al. (1966), Godby et al. (1968), Avery et al. (1968), Serson et al. (1968), Haines et al. (1970), Johnson et al. (1975a), Larsen (1978) and Navrestad and Jørgensen (1979). Ground magnetic measurements in Iceland have been collected and described by

Gudmundsson (1967), Angenheister et al. (1977) and Becker (1980).

## DEPICTION OF ANOMALIES

The residual anomalies are presented at two scales: 1 cm : 8000 nT over Iceland and 1 cm : 2130 nT elsewhere. In all cases the trackline is the base for the residual anomalies which are projected either at right angles to track or at a constant azimuth, depending on the data source, as detailed below.

The Icelandic aeromagnetic residual anomalies were redrawn from sheets at a scale of 1 : 250 000 on which both residual and regional field profiles were shown. The profiles are projected either due north or due east except in two areas: (1) in southwest Iceland (sheet 3) the anomalies are projected at right angles to the tracks, (2) in northwest Iceland (sheet 1) the anomalies are projected north-east in the area to the east of the trackline discontinuity at about 23°W.

Most of the profiles outside of Iceland were drawn from computer plots produced specifically for the compilation. The Project EASTMAR data were plotted by the Geological Survey of Greenland as profiles projected at right angles to track. The remaining computer drawn profiles were produced at Lamont-Doherty Geological Observatory. To simplify the appearance of the chart the anomalies were projected parallel to regional trends; due north for the Project MAGNET data over the Iceland Plateau and N 40°E elsewhere.

A relatively small number of the profiles outside of Iceland were available in chart form only and are depicted with less accuracy than the computer drawn profiles. The profiles which have been used from the Icelandic insular shelf surveys (Figs. 1 and 2) were re-scaled to 1 cm : 2130 nT. The anomalies for areas 7 and 9 were digitized and projected N 40°E. The profiles from other areas (2, 3 and 10) were simply rescaled at right angles to track. It should be noted that the shelf survey data are smoothed, in contrast to the other data on the map. The data from the *Meteor 48* and *54* were also available in chart form, fortuitously at the correct relative scale and are projected at right angles to track.

Navigational accuracy is estimated at ± 0.3 km for the Iceland aeromagnetic survey (Sigurgeirsson, 1970), better than ± 0.5 km for Project EASTMAR (Larsen and Thorning, 1980) and better than ± 2 km for Project MAGNET (Vogt et al., 1980). Most of the marine surveys employed either satellite navigation or superior systems and should have achieved accuracies of ± 0.5 km or better. Rather large navigational errors were inferred for an old survey line running from 60°N, 26.4°W to 65.7°N, 0°W and were allowed for in contouring.

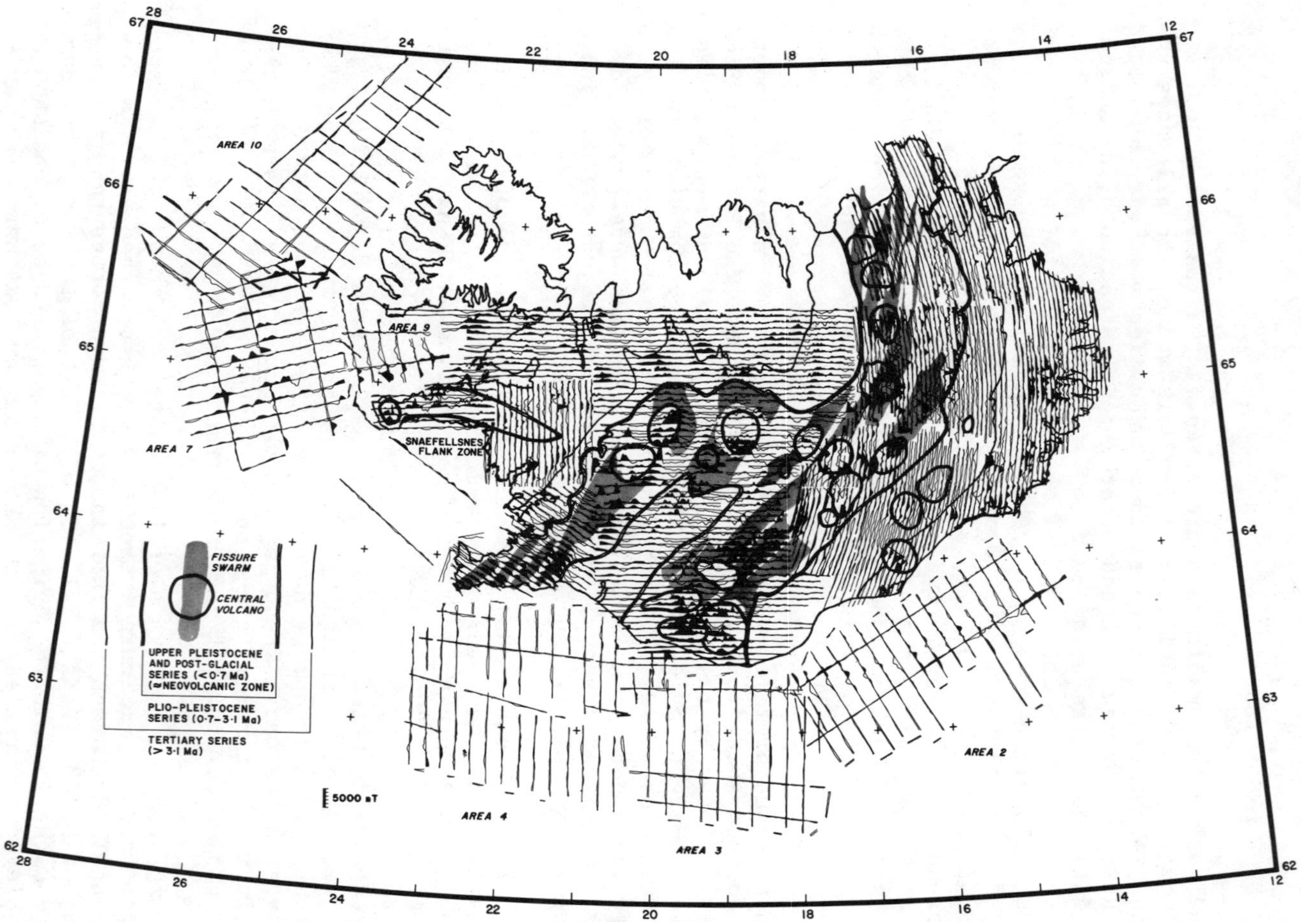


Fig. 2. Icelandic aeromagnetic and shelf survey magnetic anomalies in relation to simplified geological boundaries after Saemundsson (1979).

## TEMPORAL CORRECTIONS AND REGIONAL FIELD

Temporal corrections were applied to data from the Icelandic aeromagnetic and shelf surveys and to data from Projects EASTMAR and MAGNET (north of 65°N only). A high data quality was maintained for Project EASTMAR by reflying lines for which temporal variations exceeded 5 nT in 5 minutes (Larsen and Thorning, 1980). The remainder of the profiles on the map were not corrected although a few obviously disturbed profiles were omitted.

The Icelandic aeromagnetic and shelf survey residual anomalies were computed with respect to a regional field described by Sigurgeirsson (1970), based on the aeromagnetic survey of Haines et al. (1970). The Project MAGNET profiles over the Iceland Plateau were computed with respect to the American World Chart regional field, 1975 (Peddie and Fabiano, 1976). For Project EASTMAR a modified IGRF was used (Thorning, 1977). All the remaining profiles were computed with respect to the IGRF 1975; (IAGA Division I, Study Group, 1976) although in some instances *ad hoc* adjustments were made to remove persistent biases. The marine profiles were plotted without offset except in the west Reykjanes Basin (west of anomaly 6) where a positive bias of about 300 nT was removed. Positive biases of 100-200 nT were also removed from the Project MAGNET profiles farther east.

Some of the remaining biases were allowed for in the depiction of positive and negative anomaly areas. The delineation of stripes over the Iceland Plateau was based on anomalies which were plotted at a greater relative scale with offsets varying from 120 nT in the west to zero in the southeast; in this way significant magnetic stripes were recognized in the area of subdued anomalies in the western Iceland Plateau. In areas of identifiable lineated anomalies some license was employed in the coloring of anomalies, in order to indicate the inferred polarity of the oceanic crust. Thus, for example, anomalies 13 and 19 have been delineated even where the residual anomalies remain negative, as long as a well defined peak is present. In areas where recognizable anomalies are not present the colors are based simply on residual anomaly polarity. In complex areas such as Iceland, one tends to see anomaly trends preferentially at right angles to track, but the depicted trends are very close to those determined by Rutten (1975) in his analysis of three component aeromagnetic profiles (Serson et al., 1968). Over the Iceland-Faeroe Ridge, a preliminary delineation of trends, based only on the profile data was in good agreement with the contour chart of Fleischer et al. (1974) which was then used to refine the map.

Because of the variety of regional fields and adjustments which have applied and the absence of consistent temporal corrections, the map should not be used for analysis of long wavelength variations in the magnetic field.

## DISCUSSION

In this section a discussion of the anomaly patterns is presented. Fig. 3 shows the main tectonic and physiographic elements of the area, together with key magnetic lineations, which are numbered according to the system of Pitman et al. (1968), as on the main map. Reference may be made to Vogt et al. (1981) for a recent review of the geology and geophysics of the Iceland environment.

### Reykjanes Basins

The magnetic anomalies about the Reykjanes Ridge are extremely linear (Heirtzler et al., 1966; Godby et al., 1968) and can be identified out to anomaly 22 in the map area (Talwani et al., 1971; Herron and Talwani, 1972; Johnson et al., 1975b; Featherstone et al., 1977; Voppel et al., 1979). The anomaly sequence 18-7 does not appear to be as intensely disturbed by closely spaced fractures as it is farther south (Vogt and Avery, 1974; Voppel and Rudloff, 1980) but fractures with small offsets may be present.

South and southwest of the map area anomaly 22 is flanked on both sides of the axis by anomalies 23 and 24 and these anomalies may continue northward in a disturbed fashion. An alternative suggestion (Featherstone et al., 1977; Voppel et al., 1979) is that anomaly 24 was duplicated on the eastern side of the Reykjanes Ridge by an early westward shift of axis. If this was not the case, and there is only one anomaly 24 lineation in the East Reykjanes Basin (as shown by the tentative numbering on the map) then it is difficult to explain the strong positive anomaly farther landward, as available evidence suggests that spreading between Greenland and Eurasia commenced during the reversed period preceding anomaly 24 (Vogt et al., 1981).

### Norway Basin

The identification of anomalies in the Norway Basin follows Nunns and Peacock (in press). According to Nunns (this volume), fairly regular spreading in the Norway Basin was followed after anomaly 20 time by a period of simultaneous fan-shaped spreading about the Aegir and Kolbeinsey axes, during which the Jan Mayen block rotated anticlockwise away from Greenland. Fan-shaped spreading ended when the Aegir axis became extinct at about anomaly 7 time (Talwani and Eldholm, 1977) and was followed by regular spreading on the Iceland Plateau. The fan-shaped pattern of anomalies 18 and younger in the Norway Basin (Talwani and Eldholm, 1977) is particularly evident on the present map.

### Iceland Plateau

Over the Iceland Plateau, the identification of anomalies younger than 7 follows Vogt et al. (1980), who have discussed many aspects of

Fig. 3. Physiographic and tectonic elements of the map area. Fracture zones are shown dashed.

the anomaly pattern in detail. Over the southwestern part of the Iceland Plateau, there are positive lineations which may correspond to the 7-12, 13 and 15-18 anomaly bands, formed during the early fan-shaped spreading phase. These are complementary in position to the southernmost part of the Norway Basin, where spreading was slowest, and it is therefore appropriate that the anomaly widths are only slightly less than those in the Reykjanes Basins. No anomalies older than 7 have been identified over the eastern Iceland Plateau. Over the Jan Mayen block, which is thought to be underlain mainly by pre-anomaly 24 crust, there is a pattern of weak, sinuous lineations, some of which correlate with the basement ridges mapped by Grønlie et al. (1979) and may therefore be of topographic origin.

## Greenland-Scotland Ridge

The magnetic anomaly patterns over the Greenland-Scotland Ridge are more enigmatic than those to the north and south. In the following discussion we start with Iceland and then consider the older parts of the ridge. A full understanding of the complex magnetic anomaly pattern over Iceland will depend on an overview of the geology, geochronology and magnetostratigraphy of the island, and although a considerable amount of the necessary information has been collected (see Saemundsson (1979) and Kristjánsson and McDougall (1982), for example), the task is not yet complete. The following preliminary account relies mainly on published material and on fairly obvious observations of the map itself.

Fig. 2 shows magnetic anomalies over Iceland in relation to geological boundaries which have been simplified after Saemundsson (1979). In particular, the 0.7 and 3.1 Ma boundaries have been simplified and extrapolated across areas of ice and outwash, which are not shown. As on the original map, some of the central volcanoes have been inferred beneath icecaps, but the fissure swarms have not been continued beneath ice. The volcanoes associated with the fissure swarms over the Reykjanes Peninsula have been omitted as they are less well developed than the other volcanoes. Numerous extinct central volcanoes occur in the Tertiary lava pile but are not shown.

## Icelandic Neovolcanic Zone

Large positive anomalies over Iceland are generally associated with the central volcanoes and fissure swarms of the neovolcanic zone.

The central volcanoes are complex features which may have, for example, cores of acidic, basic and/or hydrothermally altered rock surrounded by a lenticular mass of basic or intermediate lava, cut across by minor intrusions (Saemundsson, 1979). Their magnetic signatures apparently reflect this sort of complexity; e.g. central high, central low, low ringed partly or wholly by high etc.

The fissure swarms are closely related to the central volcanoes

(Björnsson, et al., 1979) and are elongate zones within which rifting and intrusion are taking place episodically, with or without extrusion of lava. Generally, but not always, magnetic highs are associated with the fissure swarms and magnetic lows with the areas in between. A magnetized layer at least 2 km thick is necessary to explain the highs (Kristjánsson and Watkins, 1977). The lows may be due to the presence at depth of either weakly magnetized material (tuffs, clastics or alteration products) or pre-Brunhes epoch reversely magnetized rock (Kristjánsson, 1972; Becker, 1980). The latter possibility is in accord with Walker's (1975) suggestion that post-glacial spreading has taken place mainly within the active or recently active fissure zones rather than diffusely across the whole neovolcanic zone. In this case, the oceanic axial anomaly would have an analogy in the anomaly over a single fissure swarm, with the important differences being that on Iceland several distinct active zones can exist along a line parallel to the spreading direction and active zones tend to die out and be replaced elsewhere (Walker, 1975).

## Oceanic Axial Anomalies in Relation to Iceland

At the southwestern tip of Iceland, the continuation of the Reykjanes Ridge axial anomaly into the en echelon magnetic anomaly pattern of the Reykjanes Peninsula is obvious (Sigurgeirsson, 1967; Talwani et al., 1971) but to the north the situation is less clear in the area of the Tjörnes Fracture Zone (Saemundsson, 1974).

McMaster et al. (1977) have surveyed the Tjörnes Fracture Zone in detail and have concluded that the activity is relayed eastward from the Kolbeinsey Ridge to the northern neovolcanic zone by way of a number of overlapping active zones (Fig. 3). The Kolbeinsey axial anomaly weakens at 66.7°N and continues southward as a very weak lineation above a sediment filled trough (McMaster et al., 1977). To the east, a positive lineation runs in a southeasterly direction from the Kolbeinsey axis to near the Icelandic coast at about 17.5°W. This lineation is associated with submarine indications of rifting and volcanism and is thought to represent a currently active zone, whereas the more westerly lineation probably became dormant during the Brunhes epoch (McMaster et al., 1977). From the south end of the oblique lineation, activity probably steps farther east to the area of the intense positve anomaly in Axafjördur and thence south into the neovolcanic zone.

It can be argued that the Axafjördur anomaly is anomaly 5 (Vogt et al., 1980) as it is similar in appearance to anomaly 5 west of the Kolbeinsey axis and lies at an appropriate distance to the east. However there is no direct continuity between the Axafjördur anomaly and the anomaly 5 lineation farther north, which becomes irregular and appears to veer east at about 67°N. An axial origin for the Axafjördur anomaly is supported by the observation that earthquake swarms during the recent Krafla rifting episode extended northward from the Krafla fissure to beneath the anomaly (Björnsson et al.,

1977), but it could be that the rifting is taking place within old anomaly 5 age crust.

## Icelandic Volcanic Pile

Outside of the neovolcanic zone the interpretation of magnetic anomalies over Iceland is uncertain at present. In view of the many unusual features of the Icelandic spreading environment (such as multiple and shifting axes, laterally extensive lava flows, prevalent central volcanoes and modifications due to erosion and alteration) the absence of persistent, identifiable linear anomalies is not surprising. Magnetic anomalies in the older parts of Iceland may arise from (1) tilted series of lava flows, (ii) central volcanoes and (iii) dyke swarms (Kristjánsson, 1976a). There are a number of places where magnetic anomalies can be correlated with crustal polarity as determined in magnetostratigraphic studies.

The magnetic low flanking the Reykjanes Peninsula is generally attributed to crust formed during the Matuyama epoch which preceded the Brunhes normal polarity epoch (Piper, 1973; Kristjánsson, 1978). Within the negative zone to the north, a prominent circular positive anomaly (64.2°N, 22.5°W) is associated with the Stardalur central volcano which is of Olduvai event age, correlating with marine magnetic anomaly 2 (Fridleifsson and Kristjánsson, 1972).

There are also prominent linear positive anomalies associated with the two major fjords north of the Reykjanes Peninsula in an area where detailed magnetostratigraphy has been carried out (Saemundsson and Noll, 1974; McDougall et al., 1977; Kristjánsson et al., 1980). The more southerly of these lineations (Hvalfjördur) is associated with outcropping tilted lavas of Gauss age (Piper, 1973; Kristjánsson, 1978) and thus may be correlated with marine anomaly 2A, although there does not appear to be direct continuity with anomaly 2A about the Reykjanes Ridge. The positive anomalies in the vicinity of the more northerly fjord (Borgafjördur) may be correlatives of marine anomalies 3 and 3A, although this is less certain.

Farther north, the lava pile dips toward an extinct axis on the Snaefellnes Peninsula Saemundsson, 1979). North of the peninsula, the prominent anomaly crossing Breidafjördur Bay can be reasonably correlated with anomaly 5 (Kristjánsson, 1978). Anomaly 5 does not continue in any obvious way across the northwest peninsula of Iceland to join the lineation about the Kolbeinsey axis; one possible continuation would involve two eastward jumps.

Anomalies over the central northern part of Iceland should be interpreted with caution because of axis shifts which have occurred in this region (Saemundsson 1974; 1979), but it is possible that correlatives of anomalies 2A and 3 are present just west of the northern neovolcanic zone.

To the east of the neovolcanic zone, Piper (1973) has suggested that a correlative of anomaly 5 may be present on the profiles of Serson et al. (1968), in an area of eastern Iceland where intensive magnetostratigraphic research has been performed (Dagley et al., 1967; McDougall et al., 1976; Watkins and Walker, 1977; Piper et al., 1977). A recognizable anomaly 5 is not present on the low level aeromagnetic profiles shown on the present map. This is possibly because the flight lines are perpendicular to the topographic trends rather than to the crustal isochrons. However, a careful analysis of topographic anomalies should allow the upper crustal polarity to be estimated. One of the main positive anomalies in eastern Iceland appears to be associated with the Breidalur central volcano at about 65°N, 14.5°W.

## Icelandic Insular Shelf

Magnetic anomalies over the Icelandic insular shelf have been discussed by Kristjánsson (1976 a, b, c; 1978). A number of large positive anomalies off the west coast (mainly area 7, Fig. 2) have associated gravity anomalies and are probably related to central volcanoes. Fleischer et al., (1974) delineated a large number of similar features over the Iceland-Faeroe Ridge.

A remarkable feature which has emerged from the present compilation is the sweep of anomaly trends around eastern Iceland; broadly parallel to the shelf edge. It is not certain whether the offshore lineations are due mainly to isochronous polarity zones or to basement features which are cut across by crustal isochrons.

To the south of Iceland, the arcuate anomaly labelled 'E' in Fig. 3 is probably the magnetic edge effect of a buried basement step (Kristjánsson, 1976c) which defines the southern limit of the high-standing Icelandic platform. Farther east and north, however there are parallel anomalies which cannot be explained by a single basement step and it is possible that some of these may be correlatives of regular marine anomalies. This interpretation is suggested by the broad parallelism that exists between the offshore anomalies and the onshore anomaly trends and isochrons (including the eastern neovolcanic zone).

A possible southward continuation of anomaly 7 from the Iceland Plateau is shown on Fig. 3 and the main map; this lies an appropriate distance east of the -10 Ma isochron in eastern Iceland (Fig. 10, after Saemundsson (1979)). The anomaly pattern is suggestive of an active, rather than passive spreading origin for Iceland (Walker, 1975).

## Greenland-Iceland and Iceland-Faeroe Ridges

Considering now the older parts of the Greenland-Scotland Ridge, it is noted that in a general way, the character of the anomaly pattern over the Greenland-Iceland Ridge is similar to that over the

Iceland-Faeroe Ridge, although the Project EASTMAR profiles, collected at a height of 610 m, are smoother than the marine and low level aeromagnetic profiles over the Iceland-Faeroe Ridge.

Some of the anomalies in the Reykjanes Basins may continue onto the Greenland-Iceland and Iceland-Faeroe Ridges without major offsets, as shown on Fig. 3 but not on the main map (Voppel et al., 1979; Vogt et al., 1980; Nunns, this volume). However the example of Iceland should caution one against accepting the continuity of anomaly bands as evidence for the continuity of crustal isochrons.

The line separating disturbed magnetic anomalies over the Greenland-Scotland Ridge from coherent anomalies in the Reykjanes Basins lies in deep water (∿ 1100 fathoms) at the outer margins of the basins but gets progressively shallower towards the Reykjanes Ridge, where it meets the coast of Iceland. Not taking into account erosion and sedimentation, the normal subsidence for anomaly 23 age crust is about 1400 fathoms (Cochran and Talwani, 1977), so this relationship is broadly consistent with a subaerial or shallow water origin for the irregular anomalies over the Greenland-Scotland Ridge.

ACKNOWLEDGEMENTS

We would like to thank collectively, the many people who were involved in collecting and processing the magnetic anomaly data used in the map, and the various governmental agencies which funded the surveys. The compilation was supported by the NATO Science Advisory Panel for Marine Sciences and Lamont-Doherty Geological Observatory. We express special thanks to Professor S. Saxov for his cooperation. Critical reviews by Drs. J. L. LaBrecque and G. K. Westbrook are gratefully acknowledged. We also thank Mr. M. Braun for his meticulous photographic work and Mrs. A. Alvarez for typing the manuscript.

Contribution No. 3303, Lamont-Doherty Geological Observatory.

REFERENCES

Angenheister, G., Petersen, N., Schönharting, G., 1977, Zur Interpretation der Anomalien des Erdmagnetfeldes in Island, Fortschr. Mineral., 54:54.

Armstrong, T. L., 1981, "A marine geophysical investigation of the continental margin of East Greenland (63°N to 69°N)", Ph. D. thesis, University of Durham.

Avery, O. E., Burton, G. D., and Heirtzler, J. R., 1968, An aeromagnetic survey of the Norwegian Sea, J. geophys. Res., 73:4583.

Becker, H., 1980, Magnetic anomalies (ΔZ) in NE Iceland and their interpretation based on rock magnetic investigations, J. geophys., 47:43.

Björnsson, A., Johnsen, G., Sigurdsson, S., and Thorbergsson, G., 1979, Rifting of the plate boundary in north Iceland 1975-1978, J. geophys. Res., 84:3029.

Björnsson, A., Saemundsson, K., Einarsson, P., Tryggvason, E., 1977, Current rifting episode in north Iceland, Nature, 266:318.

Bott, M. H. P., and Ingles, A., 1972, Matrix methods for joint interpretation of two-dimensional gravity and magnetic anomalies with application to the Iceland-Faeroe Ridge, Geophys. J. R. astr. Soc., 30:55.

Cochran, J. R., and Talwani, M., 1977, Free-air gravity anomalies in the world's oceans and their relation to residual elevation, Geophys. J. R. astr. Soc., 50:495.

Dagley, P., Wilson, R. L., Ade-Hall, J. M., Walker, G. P. L., Haggerty, S. E., Sigurgeirsson, T., Watkins, N. D., Smith, P. J., Edwards, J., and Grasty, R. L., 1967, Geomagnetic polarity zones for Icelandic lavas, Nature, 216:25.

Featherstone, P. S., 1976, "A geophysical investigation of the southeast Greenland margin," Ph. D. thesis University of Durham.

Featherstone, P. S., Bott, M. H. P., and Peacock, J. H., 1977, Structure of the continental margin of southeastern Greenland, Geophys. J. R. astr. Soc., 48:15.

Fleischer, U., Holzkamm, F., Vollbrecht, K., and Voppel, D., 1974, Die Struktur des Island-Färöer-Rückens aus geophysikalischen Messungen, Dt. hydrogr. Z., 27:97.

Fridleifsson, I. B., and Kristjánsson, L., 1972, The Stardalur magnetic anomaly, SW-Iceland, Jökull, 22:69.

Godby, E. A., Hood, P. J., and Bower, M. E., 1968, Aeromagnetic profiles across the Reykjanes Ridge southwest of Iceland, J. geophys. Res., 73:7637.

Grønlie, G., Chapman, M., and Talwani, M., 1979, Jan Mayen Ridge and Iceland Plateau: origin and evolution, Norsk Polarinst. Skr., 170:145.

Grønlie, G., and Talwani, M., 1979, Bathymetry of the Norwegian-Greenland Sea, Norsk Polarinst. Skr., 170:3.

Gudmundsson, G., 1967, Magnetic anomalies, in: "Iceland and mid-ocean ridges," S. Björnsson, ed., Soc. Sci. Isl. Rit., 38:7.

Haines, G. V., Hannaford, W., and Serson, P. H., 1970, Magnetic anomaly maps of the Nordic countries and the Greenland and Norwegian Seas, Publ. Dom. Obs. Ottawa, 39:121.

Heirtzler, J. R., Le Pichon, X., and Baron, J. G., 1966, Magnetic anomalies over the Reykjanes Ridge, Deep Sea Res., 13:427.

Herron, E. M., and Talwani, M., 1972, Magnetic anomalies on the Reykjanes Ridge, Nature, 238:390.

IAGA Division I Study Group, 1976, International Geomagnetic Reference Field, 1975, Geophys. J. R. astr. Soc., 44:733.

Ingles, A. D., 1971, "The interpretation of magnetic anomalies between Iceland and Scotland," Ph. D. thesis, University of Durham.

Johnson, G. L., and Heezen, B. C., 1967, Morphology and evolution of the Norwegian-Greenland Sea, Deep-Sea Res., 14:755.

Johnson, G. L., McMillan, N. J., and Egloff, J., 1975 a, The continental margin of East Greenland, in: "Canada's Continental Margins and Offshore Petroleum Exploration," Yorath, C. J., Parker, E. R., and Glass, D. J., eds., Can. Soc. Petrol. Geol. Mem., 4:391.

Johnson, G. L., Sommerhoff, G., and Egloff, J., 1975 b, Structure and morphology of the West Reykjanes Basin, Mar. Geol., 18:175.

Johnson, G. L., Southall, J. R., Young, P. W., and Vogt, P. R., 1972, The origin and structure of the Iceland Plateau and Kolbeinsey Ridge, J. geophys. Res., 77:5688.

Johnson, G. L., and Tanner, B., 1971, Geophysical Investigations on the Iceland-Faeroe Ridge, Jökull, 21:45.

Kristjánsson, L., 1976 a, Marine magnetic surveys off the west coast of Iceland, Soc. Sci. Isl. Greinar, 5:23.

Kristjánsson, L., 1976 b, Central volcanoes on the western Icelandic shelf, Mar. geophys. Res., 2:285.

Kristjánsson, L., 1976 c, A marine magnetic survey off southern Iceland, Mar. geophys. Res., 2:315.

Kristjánsson, L., 1978, Magnetic trends around the Reykjanes Peninsula, SW-Iceland, Report RH-78-14, University of Iceland Science Institute, Reykjavik.

Kristjánsson, L., Fridleifsson, I. B., and Watkins, N. D., 1980, Stratigraphy and paleomagnetism of the Esja, Eyrarfjall and Akrafjall Mountains, SW-Iceland, J. Geophys., 47:31.

Kristjánsson, L., and McDougall, I., 1982, Some aspects of the late Tertiary geomagnetic field in Iceland, Geophys. J. R. astr. Soc., 68:273.

Kristjánsson, L., and Watkins, N. D., 1977, Magnetic studies of basalt fragments recovered by deep drilling in Iceland, and the "magnetic layer concept". Earth Planet. Sci. Lett., 34:365.

Larsen, H. C., 1978, Offshore continuation of East Greenland dyke swarm and North Atlantic Ocean formation, Nature, 274:270.

Larsen, H. C., and Thorning, L., 1979, Project EASTMAR: planning of an aeromagnetic survey off East Greenland, Rapp. Grønlands geol. Unders., 95:93.

Larsen, H. C., and Thorning, L., 1980, Project EASTMAR: acquisition of high sensitivity aeromagnetic data off East Greenland, Rapp. Grønlands geol. Unders., 100:91.

McDougall, I., Saemundsson, K., Johannesson, H., Watkins, N. D., and Kristjánsson, L., 1977, Extension of the geomagnetic polarity time scale to 6.5 m.y.: K-Ar dating, geological and paleomagnetic study of a 3,500-m lava succession in western Iceland, Geol. Soc. Am. Bull., 88:1.

McDougall, I., Watkins, N. D., Walker, G. P. L., and Kristjánsson, L., 1976, Potassium-Argon and paleomagnetic analysis of Icelandic lava flows: limits on the age of anomaly 5, J. geophys. Res., 81:105

McMaster, R. L., Schilling, J.-G., E., and Pinet, P. R., 1977, Plate boundary with Tjörnes Fracture Zone on northern Iceland's insular margin, Nature, 269:663.

Meyer, O., Voppel, D., Fleischer, U., Closs, H., and Gerke, K., 1972, Results of bathymetric, magnetic and gravimetric measurements between Iceland and 70°N, Dt. hydrogr. Z., 25:193.

Mirlin, Ye. G., and Melikhov, V. R., 1976, New data on the nature of the magnetic anomalies in the North Atlantic, Oceanology

(English Edition), 16:52.

Mirlin, Ye, G., Popov, K. V., and Finger, D. L., 1980, Age of the ocean floor in the Icelandic region, Oceanology (English Edition), 19:696.

Navrestad, T., and Jørgensen, F., 1979, Aeromagnetic investigations on the Jan Mayen Ridge, in: "Norwegian Sea Symposium," Norwegian Petroleum Society, Oslo.

Nunns, A.G., and Peacock, J. H., in press, Correlation, identification and inversion of magnetic anomalies in the Norway Basin, Earth Evol. Sci.

Peddie, N. W., and Fabiano, E. B., 1976, A model of the geomagnetic field for 1975, J. geophys. Res., 81:2539.

Piper, J. D. A., 1973, Interpretation of some magnetic anomalies over Iceland, Tectonophysics, 16:163.

Piper, J. D. A., Fowler, M. G., and Gibson, I. L., 1977, Dyke magnetization, magnetostratigraphy and upper-crustal structure in the Reydarfjordur area of Eastern Iceland, Tectonophysics, 40:227.

Pitman, W. C. III, Herron, E. M., and Heirtzler, J. R., 1968, Magnetic anomalies in the Pacific and sea-floor spreading, J. geophys. Res., 73:2069.

Rutten, K., 1975, Two dimensionality of magnetic anomalies over Iceland and Reykjanes Ridge, Mar. geophys. Res., 2:243.

Saemundsson, K., 1974, Evolution of the axial rifting zone in northern Iceland and the Tjörnes Fracture Zone, Geol. Soc. Am. Bull., 85:495.

Saemundsson, K., 1979, Outline of the geology of Iceland, Jökull, 29:7.

Saemundsson, K., and Noll, H., 1974, K/Ar ages of rocks from Húsafell, western Iceland, and the development of the Húsafell central volcano, Jökull, 24:40.

Serson, P. H., Hannaford, W., and Haines, G. V., 1968, Magnetic anomalies over Iceland, Science, 162:355.

Sigurgiersson, T., 1967, Aeromagnetic surveys of Iceland and its neighborhood, in: "Iceland and mid-ocean ridges," S. Björnsson, Ed., Soc. Sci. Isl. Rit, 38:91.

Sigurgeirsson, T., 1970, Aeromagnetic survey of SW Iceland, Science in Iceland, 2:13.

Sigurgeirsson, T., 1970-1982, "Iceland aeromagnetic profiles," University of Iceland Science Institute, Reykjavik. 9 sheets at 1:250 000 scale as follows:
1, North West Iceland (in prep.),
2, West Central Iceland (1979),
3, South West Iceland (1970),
4, North Central Iceland (in prep.),
5, Central Iceland (1980),
6, South Central Iceland (1980),
7, North East Iceland (1982),
8, East Central Iceland (1981),
9, South East Iceland (1980),

Talwani, M., and Eldholm, O., 1977, Evolution of the Norwegian-Greenland Sea, Geol. Soc. Am. Bull., 88:969.

Talwani, M., Udinstev, G., Mirlin, Ye. G., Beresnev, E., Kanayev, V. F., Chapman, M., Grønlie, G., and Eldholm, O., 1979, Survey of sites 346, 347, 348, and 350: The area of the Jan Mayen Ridge and Icelandic Plateau, in: Talwani, M., Undinstev, G., et al., "Initial Reports of the Deep Sea Drilling Project," supplement to vols. 38, 39, 40 and 41:465.

Talwani, M., Windisch, C. C., and Langseth, M. G., 1971, Reykjanes Ridge crest: a detailed geophysical study, J. geophys. Res., 76:463.

Thorning, L., 1977, A brief description of the computer programs used by GGU in the treatment of aeromagnetic data, Rapp. Grønlands geol. Unders., 75:88.

Vogt, P. R., and Avery, O. E., 1974, Detailed magnetic surveys in the northeast Atlantic and Labrador Sea, J. geophys. Res., 79:363.

Vogt, P. R., Johnson, G. L., and Kristjánsson, L., 1980, Morphology and magnetic anomalies north of Iceland, J. Geophys., 47:67.

Vogt, P. R., Ostenso, N. A., and Johnson, G. L., 1970, Magnetic and bathymetric data bearing on sea-floor spreading north of Iceland, J. geophys. Res., 75:903.

Vogt, P. R., Perry, R. K., Feden, R. H., Fleming, H. S., and Cherkis, N. Z., 1981, The Greenland-Norwegian Sea and Iceland environment, in: "The Ocean Basins and Margins, vol. 5," A. E. M. Nairn, M. Churkin, Jr., and F. G. Stehli, eds., Plenum Publishing Corporation, New York.

Voppel, D., and Rudloff, R., 1980, On the evolution of the south of 60°N between 40 and 12 million years before present, J. Geophys., 47:61.

Voppel, D., Srivastava, S. P., and Fleischer, U., 1979, Detailed magnetic measurements south of the Iceland-Faeroe Ridge, Dt. hydrogr. Z., 32:154.

Walker, G. P. L., 1975, Excess spreading axes and spreading rate in Iceland, Nature, 255:468.

Watkins, N. D., and Walker, G. P. L., 1977, Magnetostratigraphy of Eastern Iceland, Am. J. Sci., 277:513.

INDEX